Methods in Plant Biochemistry

Volume 2
Carbohydrates

METHODS IN PLANT BIOCHEMISTRY

Series Editors

P. M. DEY
Department of Biochemistry, Royal Holloway and Bedford New College, UK

J. B. HARBORNE
Plant Science Laboratories, University of Reading, UK

1 Plant Phenolics: J. B. HARBORNE

2 Carbohydrates: P. M. DEY

3 Enzymes of Primary Metabolism: P. J. LEA

4 Lipids, Membranes and Aspects of Photobiology: J. L. HARWOOD and J. R. BOWYER

Methods in Plant Biochemistry

Series editors
P. M. DEY and J. B. HARBORNE

Volume 2
Carbohydrates

Edited by

P. M. DEY

Department of Biochemistry, Royal Holloway and Bedford New College, UK

ACADEMIC PRESS
Harcourt Brace Jovanovich, Publishers
London   San Diego   New York
Boston   Sydney   Tokyo   Toronto

ACADEMIC PRESS LIMITED
24–28 Oval Road
London NW1 7DX

US edition published by
ACADEMIC PRESS INC
San Diego, CA 92101

British Library Cataloguing in Publication Data is available

ISBN 0-12-461012-9

Filmset by Bath Typesetting Limited, Bath, Avon
Printed by Galliard (Printers) Ltd, Great Yarmouth, Norfolk

Contents

Contributors

P. Albersheim, Complex Carbohydrate Research Center and Department of Bio-chemistry, University of Georgia, 220 Riverbend Road, Athens, GA 30602, USA

G. Avigad, Department of Biochemistry, UMDNJ-Robert Wood Johnson Medical School, Piscataway, NJ 08854, USA

G. A. Barber, Department of Biochemistry, Ohio State University, Columbus, OH 43210, USA

E. Beck, Lehrstuhl Pflanzenphysiologie, Universität Bayreuth, Universitätsstrasse 30, D 8580 Bayreuth, FRG

W. Blaschek, Institute of Pharmacy, University of Regensburg, D 8400 Regensburg, FRG

S. C. Churms, Department of Organic Chemistry, University of Cape Town, Ronde-bosch 7700, South Africa

A. Darvill, Complex Carbohydrate Research Center and Department of Biochemistry, University of Georgia, 220 Riverbend Road, Athens, GA 30602, USA

P. M. Dey, Department of Biochemistry, Royal Holloway and Bedford New College (University of London), Egham, Surrey TW20 0EX, UK

A. D. Elbein, Department of Biochemistry, The University of Texas Health Science Center, San Antonio, TX 78284, USA

D. S. Feingold, Department of Microbiology, Biochemistry and Molecular Biology, University of Pittsburgh, School of Medicine, Pittsburgh, PA 15261, USA

G. Franz, Institute of Pharmacy, University of Regensburg, D 8400 Regensburg, FRG

M. J. Gidley, Unilever Research Laboratory, Colworth House, Sharnbrook, Bedford, MK44 1LQ, UK

M. C. Hawley, Department of Chemical Engineering, Michigan State University, East Lansing, MI 48824, USA

H. Hopf, Botanisches Institut, Universität München, Menzinger Strasse 67, D 8000 München 19, FRG

J. Karkalas, Department of Bioscience and Biotechnology, University of Strathclyde, James P. Todd Building, 131 Albion Street, Glasgow G1 1SD, UK

G. P. Kaushal, Department of Biochemistry, The University of Texas Health Science Center, San Antonio, TX 78284, USA

D. T. A. Lamport, MSU-DOE Plant Research Laboratory, Michigan State University, East Lansing, MI 48824, USA

F. A. Loewus, Institute of Biological Chemistry, Washington State University, Pullman, WA 99164-6340, USA

N. K. Matheson, Department of Agricultural Chemistry, The University of Sydney, Sydney, N.S.W. 2006, Australia

R. H. McDowell, Chemistry Department, Royal Holloway and Bedford New College (University of London), Egham, Surrey TW20 0EX, UK

W. R. Morrison, Department of Bioscience and Biotechnology, University of Strathclyde, James P. Todd Building, 131 Albion Street, Glasgow G1 1SD, UK

M. O'Neill, Complex Carbohydrate Research Center and Department of Biochemistry, University of Georgia, 220 Riverbend Road, Athens, GA 30602, USA

E. Percival, Chemistry Department, Royal Holloway and Bedford New College (University of London), Egham, Surrey TW20 0EX, UK

H. G. Pontis, Centro de Investigaciones Biológicas, Fundación para Investigaciones Biológicas Aplicadas (FIBA), Casilla de Correo 1348, 7600 Mar del Plata, Argentina

R. Pont Lezica, Department of Biology, Washington University, St. Louis, MO 63130, USA

L. Quesada-Allué, Instituto de Investigaciones Bioquímicas, Fundacíon Campomar, Buenos Aires, Argentina

G. Robinson, Department of Food Research and Technology, Silsoe College, Silsoe, Bedford MK45 4DT, UK

G. L. Rorrer, Department of Chemical Engineering, Oregon State University, Corvallis, OR 97331, USA

P. Ryden, AFRC Institute of Food Research, Colney Lane, Norwich NR4 7UH, UK

S. M. Selke, Department of Packaging, Michigan State University, East Lansing, MI 48824, USA

R. R. Selvendran, AFRC Institute of Food Research, Colney Lane, Norwich NR4 7UH, UK

A. M. Stephen, Department of Organic Chemistry, University of Cape Town, Rondebosch 7700, South Africa

R. J. Sturgeon, Department of Biological Sciences, Heriot-Watt University, Riccarton, Edinburgh EH14 4AS, UK

D. C. Vogt, Department of Organic Chemistry, University of Cape Town, Rondebosch 7700, South Africa

Preface to the Series

Scientific progress hinges on the continual discovery and extension of new laboratory methods and nowhere is this more evident than in the subject of biochemistry. The application in recent decades of novel techniques for fractionating cellular constituents, for isolating enzymes, for electrophoretically separating nucleic acids and proteins and for chromatographically identifying the intermediates and products of cellular metabolism has revolutionised our knowledge of the biochemical processes of life.

While there are many books and series of books on biochemical methods, volumes specifically catering for the plant biochemist have been few and far between. This is particularly unfortunate in that the isolation of DNA, enzymes or metabolites from plant tissues can often pose special problems not encountered by the animal biochemist. For a long time, the Springer series *Modern Methods in Plant Analysis*, which first appeared in the 1950s, provided the only comprehensive guide to experimental techniques for the investigation of plant metabolism and plant enzymology. This series, however, has never been completely updated; a second series has recently appeared but this is organised on a techniques basis and thus does not provide the comprehensive coverage of the first series. One of us (JBH) wrote a short guide to modern techniques of plant analysis *Phytochemical Methods* in 1976 (second edition, 1984) which showed the need for an expanded comprehensive treatment, but which by its very nature could only provide an outline of available methodology.

The time therefore seemed ripe to us to produce an entirely new multi-volume series on methods of plant biochemical analysis, which would be both thoroughly up-to-date and comprehensive. The success of *The Biochemistry of Plants*, edited by P. K. Stumpf and E. E. Conn and published by Academic Press, was an added stimulus to produce a complementary series on the methodology of the subject. With these thoughts in mind, we planned individual volumes covering: phenolics, carbohydrates, amino acids, proteins and nucleic acids, terpenoids, nitrogen and sulphur compounds, lipids, membranes and light receptors, enzymes of primary and secondary metabolism, plant molecular biology and biological techniques in plant biochemistry. Thus we have tried to cover all the major areas of current endeavour in phytochemistry and plant biochemistry.

The main aim of the series is to introduce to the scientist current knowledge of techniques in various fields of biochemically-related topics in plant research. It is also intended to present the historical background to each topic, to give experimental details of methods and analyses and appraisal of them, pointing out those methods that are

ix

most suitable for immediate application. Wherever possible illustrations and structures have been used and one or more case treatments presented. The compilation of known data and properties, where appropriate, is included in many chapters. In addition, the reader is directed to relevant references for further details. However, for the sake of clarity and completeness of individual reviews, some overlap between chapters of volumes has been allowed.

Finally, we extend our warmest thanks to our volume editors for undertaking the important task of organising each volume and cooperating in preparing the contents lists. Our special thanks go to the staff of Academic Press and to the many colleagues who have made this project a success.

P. M. DEY
J. B. HARBORNE

Preface to Volume 2

This volume on plant carbohydrates contains 18 chapters. The subject matter can be roughly divided into three parts: first, the simple sugars, their derivatives and related compounds; second, the polysaccharides; and third, polysaccharide organisation and analysis.

The first section covers monosaccharides, cyclitols, branched sugars and sugar alcohols, nucleotide sugars, lipid-linked saccharides, disaccharides and oligosaccharides. In Chapter 1, Sturgeon reviews those methodological aspects of monosaccharides required by the laboratory researcher working on the chemistry or biochemistry of carbohydrates. This includes isolation, generation from complex carbohydrates, methods of separation and visualisation and quantitation by chemical and enzymic methods. The second chapter on nucleotide sugars, by Feingold and Barber, describes both chemical and enzymic methods for obtaining unlabelled or isotopically labelled compounds. The authors have also included methods for preparing relevant enzymes that can be used for the synthesis of nucleotide sugars. Similarly, Elbein and Kaushal have reviewed the topic of polyprenyl sugar derivatives in Chapter 3, giving a theoretical background of the topic as well as presenting a comprehensive discussion of the methodology. Chapter 4 on disaccharides by Avigad deals mainly with sucrose, trehalose, maltose, melibiose and some miscellaneous sugars and is a critical and exhaustive account of the methodologies pertaining to these sugars. The next chapter on oligosaccharides by Dey concentrates on α-galactosyl sucrose derivatives. A limited number of methods are discussed here in order to avoid duplication; the reader should turn to Chapters 1, 2 and 4 for general techniques in this area. The metabolic role of cyclitols, which include inositols, is being increasingly recognised; thus the methodologies relating to this small group of polyhydroxylated compounds are becoming more and more important. Loewus has presented a concise review of these compounds in Chapter 6. The branched-chain sugars and sugar alcohols are a small group of compounds found in the plant kingdom; some have been recognised as part of plant cell-wall architecture, but others (including their derivatives) play vital metabolic roles. This topic has been presented here by Beck and Hopf in probably the first detailed account on the subject.

The section on polysaccharides begins with Chapter 8 on cellulose. Cellulose and starch have been well investigated over many years and a vast literature is available. Comprehensive reviews on these two glucose-based polysaccharides have been presented in Chapters 8 and 9, respectively. Chapter 10 on the fructans by Pontis is unique in that it is probably the first up-to-date review of experimental techniques on this topic. The following chapter on mannose-based polysaccharides, presented by Matheson, includes methodologies related to mannans, glucomannans, galactomannans and galactoglucomannans. Some of these polymers, for example galactomannans, are commercially important as gel inducers. However, they play a significant role *in vivo* as reserve carbohydrates. The primary cell wall pectic polysaccharides have been reviewed in Chapter 12 by Albersheim and his coworkers. His group is at the forefront of this field of research, and therefore very recent developments can be found in this chapter. These polysaccharides constitute not only a major part of cell wall architecture, but also fragments obtained from them, termed 'oligosaccharins', show potent biological activities. Pont Lezica and Quesada-Allué have presented a contribution on the difficult topic of chitin. Although this polymer is found almost exclusively in lower plants and insects, its inclusion in this volume is justified because of its potential importance in plant–pathogen interactions. In Chapter 14 Stephen *et al.* have presented a comprehensive account of methods related to plant exudate gums. These gums have much use in the food industry. Likewise, algal polysaccharides, reviewed by Percival and McDowell in Chapter 15, have considerable commercial importance.

The third section has three chapters. Chapter 16 by Selvendran and Ryden is a critical account of methodologies pertaining to plant cell wall analysis. The thorough treatment of the subject includes chemical, physical and enzymic methods. In Chapter 17 Rorrer *et al.* have described how hydrogen fluoride can be used for cleaving glycosidic bonds of cell wall polysaccharide components. This approach is potentially important in biomass conversion and in analysing glycoproteins. Finally, in Chapter 18, Gidley and Robinson have presented a detailed review on *in vitro* interactions between polysaccharides. The commercial importance of this topic stems from the exploitation of the gel formation that occurs during such interactions.

I gratefully acknowledge the efforts and hard work of the authors, renowned in their fields, in preparing the reviews for this volume. I am sure that all these efforts have been worthwhile and the contributions will be of much value to the reader. I also thank the personnel at Academic Press for their co-operation, patience and assistance. My special thanks go to Professors J. B. Harborne and J. B. Pridham for their valuable advice.

PRAKASH M. DEY

1 Monosaccharides

R. J. STURGEON

Department of Biological Sciences, Heriot-Watt University, Riccarton, Edinburgh EH14 4AS, UK

I. OCCURRENCE

Monosaccharides are not normally found free in large amounts in extracts of plant material. Glucose and fructose are usually the most abundant. As they have a central role in metabolism they may also be found as their phosphorylated derivatives. Other

METHODS IN PLANT BIOCHEMISTRY Vol. 2
ISBN 0-12-461012-9

monosaccharides such as mannose, galactose, xylose, arabinose, ribose, apiose, rhamnose, fucose, glucuronic acid, galacturonic acid and 2-acetamido-2-deoxyglucose may also be detected but they are usually found glycosidically linked in the form of oligo- or polysaccharides or glycoconjugates (glycolipids and glycoproteins). All of the monosaccharides mentioned above occur in the D-configuration with the exception of arabinose, rhamnose and fucose which are normally found in the L-configuration. The sugars normally exist in the six-membered ring pyranose configuration with the exception of arabinose, ribose and fructose which are usually found in the five-membered ring furanose configuration. In addition, since the biosynthesis of the glycan chains of carbohydrates involves the transfer of monosaccharides from the corresponding nucleotide sugar, these latter derivatives may also be detected. Both glucose and fructose form phosphorylated sugars, glucose 6-phosphate and fructose 6-phosphate, mediated by the action of the enzyme hexokinase. These phosphorylated sugars are usually metabolised, either via glycolysis, the oxidative pentose phosphate pathway or tricarboxylic acid cycle to produce energy, or are involved in the biosynthesis of a range of oligo- and polysaccharides. Sucrose, a non-reducing disaccharide, is the principal sugar of vascular plants, and is found throughout plants often in high concentration. Sucrose acts as a source of glucosyl residues for the synthesis of the nucleotide sugar UDP-glucose which is the key sugar nucleotide intermediate involved in the formation of all other sugar nucleotides and hence in the formation of other oligo- and polysaccharides. The fructosyl residues derived from sucrose are involved in the production of fructans. Fructans, starch, galactomannans, xyloglucans and galactans comprise the main reserve polysaccharides in plant materials, and are extractable from plant tissue using mild conditions. The remaining major polysaccharide fractions of the plant are associated with the cell walls and include diverse polysaccharides, such as cellulose, 'hemicelluloses' and 'pectins'. The hemicelluloses, originally named because of their presumed chemical relationship to cellulose, are now known to include polysaccharides such as glucans, arabinoxylans and arabinogalactans. Many of these polysaccharides, together with pectins, are part of the amorphous matrix of the plant cell wall. Extraction of these polysaccharides normally requires hot alkali (0.2–4.0 M) to effect efficient dissociation and solubilisation from the cell wall. The pectins, a class of polysaccharides rich in arabans, rhamnogalactans and pectic acid, are extracted using mild treatment, such as hot or cold water, often in the presence of chelating agents.

A. Composition

Colorimetric analysis can be used to identify broad classes of sugars, and is frequently used in following a fractionation procedure during purification of crude extracts of plant material. In some of the methods, strong acid is included in the reagent, thus rendering unnecessary prior hydrolysis of the carbohydrate (polysaccharide or glycoconjugate). Hydrolysis occurs followed by decomposition to produce a chromogen (pyrrole or furan type) which condenses with the reagents to generate a colour. The phenol–sulphuric acid (see Section III.A.2) and L-cysteine–sulphuric acid (see Section III.A.3) assays for hexoses and the orcinol–hydrochloric acid assays (see Section III.A.5) for pentoses are developed in this manner.

   These are empirical methods, requiring calibration with known monosaccharides. All members of the same class will produce the chromogens. In addition, the molar response for individual sugars will be different.

II. DEPOLYMERISATION

A. Acid Hydrolysis

Acid hydrolysis is usually a compromise between total depolymerisation of the material and eventual destruction of some of the monosaccharides released due to the differing stabilities of individual monosaccharides and the different strengths of the bonds linking them. Mild acid hydrolysis by 0.1 M oxalic acid at 70°C for 1 h (Aspinall *et al.*, 1953), or 0.05 M sulphuric acid at 80°C for 1 h (Codington *et al.*, 1976) is usually sufficient to release furanosyl-linked monosaccharides. Stronger conditions (1 M sulphuric acid at 100°C for 4 h) are necessary for the release of aldoses from neutral polysaccharides such as glucans, galactomannans, arabinoxylans. Alternatively 2 M trifluoroacetic acid at 100°C for 3 h has been widely advocated for the hydrolysis of plant material (Albersheim *et al.*, 1967). Carbohydrates containing uronic acid residues pose different problems, due to the greater resistance to hydrolysis of the uronosyl glycosidic bond. Frequently poor yields of the monosaccharide components are obtained and side products are observed. The introduction of a prior hydrolysis step using formic acid gives better yields (Radharkrishnamurthy and Berenson, 1963). Amino sugars are frequently components of glycoconjugates, occurring as 2-acetamido-2-deoxyhexoses. During acid hydrolysis, *N*-deacetylation occurs and the resulting 2-amino-2-deoxyglycosides are resistant to hydrolysis. The complete hydrolysis of amino sugar-containing glycoconjugates usually requires treatment with 4 M hydrochloric acid at 100°C for periods of up to 6 h (Spiro, 1973). These strong conditions are required because on de-*N*-acetylation of amino sugars the amino group becomes protonated and this reduces the susceptibility of the glycosidic linkage to acid hydrolysis. Unfortunately, at the same time considerable degradation of neutral carbohydrates and uronic acids takes place. When amino sugars are present in the sample it is recommended that hydrolysis conditions are optimised with respect to time and molarity of acid. Minimal degradation occurs if the monosaccharides, including amino sugars, are released as methyl glycosides from polymeric material in the presence of methanolic hydrogen chloride (Clamp *et al.*, 1972).

B. Enzymic Depolymerisation

Frequently hydrolysis of plant material by enzymes may be an attractive alternative to acid hydrolysis. Advantages of the use of enzymes include the use of mild conditions of pH and temperature (25–80°C) with no destruction of the liberated monosaccharide units. In the case of enzymic hydrolysis of polysaccharide material, usually no single enzyme is capable of quantitatively converting the material to the monosaccharide level. Enzymes hydrolysing polysaccharide material are classified into two types, the endo-glycanases which bring about a rapid decrease in the molecular size (by hydrolysing internal linkage positions) of the polysaccharide, and the exo-glycanases which normally have an action pattern of release of mono- or disaccharide residues from the non-reducing end of the polysaccharide. α-Amylase is a well characterised endo-α-(1→4)-D-glucanase which hydrolyses starch in a random manner releasing oligosaccharides still bearing α-(1→6)-glucosyl residues. Glucoamylase is an exo-α-(1→4)-D-glucosidase which hydrolyses terminal 1,4-linked-D-glucose residues successively from the non-reducing ends of these oligosaccharides with the release of glucose. The enzyme is also capable of rapidly hydrolysing the α-(1→6)-D-glucosidic bonds. Thus, starch can be determined after a complete hydrolysis to glucose with α-amylase and glucoamylase (Dintznis and Harris, 1981). In a similar way inulin can yield fructose and glucose only, on treatment with a fungal inulinase (Nakamura *et al.*, 1978). Cereal mixed-linkage glucans β-(1→3)(1→4)-D-glucan, yields glucose only on treatment with a commercial cellulase preparation (Bamforth, 1983).

The hydrolysis of oligosaccharide material is normally easier to achieve, with exo-glycosidases only being required. Sucrose can be quantitatively converted to glucose and fructose by the action of invertase (a β-D-fructofuranosidase). Maltose is hydro-lysed to glucose by α-glucosidase, and raffinose releases galactose, glucose and fructose on treatment with α-galactosidase and invertase. Several factors have to be considered when an enzymic approach is to be adopted. Many commercially available enzymes are marketed as concentrated culture filtrates produced from microorganisms. They may contain a wide variety of enzymes. For instance, a 'cellulase' from *Trichoderma reesei* has been used to hydrolyse cereal β-glucans to glucose. A contaminating amylogluco-sidase which will hydrolyse α-(1→6)-D-glucosidic linkages in amylopectin has to be removed from the crude enzyme before it can be used in an analytical system measuring β-glucan in the presence of starch (Martin and Bamforth, 1981). In addition, it is always wise to check (by thin layer chromatography or high performance liquid chromato-graphy that the enzymic reaction has gone to completion, producing only monosacchar-ide. A number of the commonly available enzymes for the enzymic degradation of plant carbohydrates are listed in Table 1.1.

III. COLORIMETRIC ASSAYS

These assays measure the amounts of general classes of carbohydrates in samples, e.g. aldoses (hexoses and pentoses), ketoses, and uronic acids. In each case the amount of sugar being determined is compared to a standard sugar of that class. In the measure-ment of hexose, for example, using the phenol–sulphuric acid procedure (see Section III.A.1) glucose is frequently used as the standard hexose for comparative purposes. All other neutral hexoses will give a positive reaction in the assay but the amount of chromogen produced will vary from hexose to hexose.

A. Aldose Determinations

1. Phenol–sulphuric acid assay

The method is based on that of Dubois *et al.* (1956), but on a microscale.

Reagents:
  1. Phenol, 5% (w/v), in water. This solution is stable indefinitely.
  2. Concentrated sulphuric acid.
  3. Standard hexose solution (200 µg ml^{-1})

Procedure: The sample (0.5 ml) containing 10–100 µg carbohydrate is mixed with Reagent 1 (1.0 ml). Reagent 2 (2.5 ml) is added directly onto the surface of the solution

TABLE 1.1. Enzymes used for the hydrolysis of oligo- and polysaccharides.

Systematic name	Trivial name	EC No.[a]
Agarase 3-glycanohydrolase	Agarase	3.2.1.81
1,4-α-D-Glucan glucanohydrolase	α-Amylase	3.2.1.1
1,4,-α-D-Glucan maltohydrolase	β-Amylase	3.2.1.2
α-L-Arabinofuranoside arabinofuranohydrolase	α-L-Arabinofuranosidase	3.2.1.55
1,4-(1,3:1,4)-β-D-Glucan 4glucanohydrolase	Cellulase	3.2.1.4
β-D-Fructofuranoside fructohydrolase	β-D-Fructofuranosidase (Invertase)	3.2.1.26
α-D-Galactoside galactohydrolase	α-D-Galactosidase	3.2.1.22
1,3(1,3:1,4)-β-D-Glucan 3(4)-glucanohydrolase	Endo-1,3(4)-β-D-Glucanase	3.2.1.6
1,4-α-D-Glucan glucohydrolase	Exo-1,4-α-D-Glucosidase (Glucoamylase)	3.2.1.3
α-D-Glucoside glucohydrolase	α-D-Glucosidase (Maltase)	3.2.1.20
β-D-Glucoside glucohydrolase	β-D-Glucosidase	3.2.1.21
2,1-β-D-Fructan fructanohydrolase	Inulinase	3.2.1.7
1,4-β-D-Mannan mannanohydrolase	Endo-1,4-β-Mannanase	3.2.1.78
β-D-Mannoside mannohydrolase	β-D-Mannosidase	3.1.1.25
1,4-β-D-Xylan xylanohydrolase	Endo-1,4-β-D-Xylanase	3.2.1.8
1,4-β-D-Xylan xylohydrolase	Exo-1,4-β-D-Xylanase	3.2.1.37

[a] Enzyme nomenclature (1984). Recommendations of the Nomenclature Committee of the International Union of Biochemistry, Academic Press, New York

vigorously, using a syringe. Heat is evolved causing hydrolysis of any oligo- or polysaccharide material. The solutions are allowed to stand for 5 min, and cooled and the absorbance is determined at 490 nm for hexoses. Aldoses, ketoses and uronic acids produce chromophores in this reaction (pentoses, deoxyhexoses and uronic acids give colours with the maximum absorbance at 450 nm).

Interference in the assay is observed with protein, non-carbohydrate reducing agents, heavy metal ions and azide.

2.   Orcinol–sulphuric acid assay

The methods described by Winzler (1955) and Svennerholm (1956) have been modified.

Reagents:
1. Orcinol (1.6%; v/v) in water.
2. Sulphuric acid (60%; v/v).
3. Standard hexose solution (200 µg ml^{-1}).
4. Prior to use, 1 vol of Reagent 1 and 7.5 vols of cold Reagent 2 are mixed. This solution is stable for several days if kept in a dark bottle.

Procedure: The sample (0.5 ml) containing 10–100 µg hexose is mixed with 4.25 ml orcinol–sulphuric acid Reagent 4. After mixing, the tubes are heated at 80°C for 15 min, cooled rapidly and the absorbance is determined at 420 nm.

Increased sensitivity can be obtained by measuring absorbance at 505 nm provided deoxy sugars and uronic acids are absent from the samples.

3.   L-Cysteine–sulphuric acid assay for hexoses

This method is based on that of Dische *et al.* (1949).

Reagents:
1. Ice-cold sulphuric acid (86%; v/v).
2. L-Cysteine hydrochloride (700 mg l^{-1}) in Reagent 1.
3. Standard hexose solution (100 µg ml^{-1}).

Procedure: Both samples and reagents should be pre-cooled to 4°C. Samples (0.2 ml, containing 0.2–20 µg hexose) are mixed with Reagent 2, 1.0 ml. After mixing, the reactants are heated at 100°C for 3 min in glass-stoppered test tubes. After rapid cooling of the mixture to room temperature the absorbance is read at 415 nm.

Chromogens, with absorbance maxima in the range 380–430 nm, produced by pentoses, heptoses and uronic acids interfere in this reaction.

4.   L-Cysteine–sulphuric acid assay for pentoses

Pentoses may be determined in the presence of hexoses using a modification of Section III.A.3 (Dische, 1949).

Reagents:
1. Concentrated sulphuric acid.

2. Aqueous L-cysteine hydrochloride (30 mg ml^{-1}).
3. Standard pentose solution (80 µg ml^{-1}).

Procedure: Reagent 1 (1 ml) is added to sample (0.25 ml) containing 0.2–20 µg pentose and mixed whilst cooling with cold water. The reaction mixture is kept at room temperature for 1 h, prior to addition of Reagent 2 (0.025 ml), followed by further mixing. After 30 min at room temperature, the absorbance is determined at 415 nm (if hexoses are present) and at 390 nm. The difference in absorbances at 390 and 415 nm is used to calculate the pentose content of the sample.

5. Orcinol–hydrochloric acid assay for pentoses

A modification of the original Bial reaction was reported by Brown (1946).

Reagents:
1. Aqueous trichloracetic acid (10%; w/v).
2. A freshly prepared aqueous solution of orcinol (0.2%; w/v) and ferric ammonium sulphate (1.15%; w/v) in 9.6 M hydrochloric acid.
3. Standard pentose solution (200 µg ml^{-1}).

Procedure: Samples (0.2 ml) containing 4–40 µg pentose are mixed with Reagent 1 (0.2 ml) and heated at 100°C for 15 min. On cooling to room temperature Reagent 2 (1.20 ml) is added and after mixing, the reactants are heated at 100°C for a further 20 min. The solutions are then cooled to room temperature, and the absorbance determined at 660 nm. Hexoses absorb strongly at 520 nm, at which wavelength the chromogen produced by pentoses is minimal.

B. Ketose Determinations

1. L-Cysteine–carbazole–sulphuric acid assay

The method was first described by Dische and Borenfreund (1951).

Reagents:
1. L-Cysteine hydrochloride (1.5%; w/v) in water).
2. Sulphuric acid (45 ml diluted with 19 ml water).
3. Carbazole (0.12%; w/v in 95% ethanol).
4. Standard ketose solution (10 µg ml^{-1}).

Procedure: Samples (0.1 ml) containing 0–1 µg keto sugar are mixed with Reagent 1 (0.02 ml). Reagent 2 (0.60 ml) is added with cooling and shaking, followed by Reagent 3 (0.02 ml). After mixing, the reactants are kept at room temperature for 24 h, and the absorbance is determined at 560 nm.

2. Anthrone–sulphuric acid assay

In the original method (Van Handel, 1967), anthrone–sulphuric acid was used to determine fructose and fructose-containing polymers. The sensitivity of the method may be increased using the following modified procedure (Somani *et al.*, 1987).

Reagent:
1. Anthrone (10 mg) and L-tryptophan (10 mg) are dissolved in sulphuric acid (75%; v/v, 100 ml).

Procedure: Samples (0.9 ml containing 5–100 nmol D-fructose) are mixed with Reagent 1 (2.0 ml). The reactants are incubated at 55°C for 90 min and after cooling, the absorbances are measured at 520 nm.

The absorbance is approximately three times higher than that obtained using the standard anthrone procedure, in which tryptophan is omitted. Glucose and some other aldoses interfere to give colours about 1% of the D-fructose values.

C. Amino Sugars

Amino sugars, such as 2-amino-2-deoxy-D-glucose and 2-amino-2-deoxy-D-galactose (glucosamine and galactosamine, respectively) usually occur in nature as the *N*-acetylated derivatives, frequently as part of the glycan chains of glycoproteins. When these sugars are released as monosaccharides, following acid hydrolysis, they undergo *N*-deacetylation, thus giving 2-amino-2-deoxy-hexoses. They can be determined by using a colorimetric procedure such as the Elson–Morgan assay (see Section III.C.1), whilst 2-acetamido-2-deoxy-hexoses, which are released from polymers by enzymic action, are determined by the Morgan–Elson assay (see Section III.C.2).

1.   Elson–Morgan assay

Whilst many modifications exist of the original Elson and Morgan (1933) procedure, the most commonly used one is that of Rondle and Morgan (1955).

Reagents:
1. Redistilled pentane-2,4-dione (acetylacetone) (2 ml) is dissolved in sodium carbonate (1 M, 100 ml). This reagent should be freshly made daily.
2. 4-(*N*,*N*-Dimethylamino) benzaldehyde (0.8 g) is dissolved in ethanol (30 ml) and concentrated hydrochloric acid (30 ml).

Procedure: Samples originating from hydrochloric acid hydrolysis of material (see Section II.A) should be taken to dryness several times by rotary evaporation. Final traces of acid may be removed by storing the samples in a vacuum desiccator over sodium hydroxide pellets for 18 h. Samples (0.25 ml, containing 0–80 μg 2-amino-2-deoxyhexose) are mixed with Reagent 1 (0.25 ml) in stoppered tubes which are then heated at 100°C for 20 min. After cooling to room temperature, ethanol (1.75 ml) and Reagent 2 (1 ml) are added, followed by thorough mixing. The tubes are then incubated at 65°C for 10 min, cooled to room temperature and the absorbances determined at 530 nm. Interference with the assay is observed from mixtures of hexoses and amino acids.

2. Morgan–Elson assay

Increased sensitivity of the original method of Morgan and Elson (1934) is achieved by using the modification of Reissig *et al.* (1955).

Reagents:
1. Potassium tetraborate (0.2 M).
2. 4-(*N*,*N*-Dimethylamino) benzaldehyde (10 g) is dissolved in glacial acetic acid–10 M hydrochloric acid (7 : 1; v/v; 100 ml). This solution is stable for several weeks if stored in a dark bottle. Immediately before use, the reagent is diluted (1 : 9) with glacial acetic acid.

Procedure: Samples (0.25 ml, containing 0–5 µg 2-acetamido-2-deoxyhexose) are mixed with Reagent 1 (0.05 ml) and heated at 100°C for 3 min. After cooling, Reagent 2 (1.5 ml) is added. After further mixing the tubes are incubated at 37°C for 20 min, cooled to room temperature and the absorbances are determined at 585 nm.

D. Uronic Acid

1. Carbazole assay

The carbazole method modified by Bitter and Muir (1962) can be used for the analysis of hexuronic acids.

Reagents:
1. Concentrated sulphuric acid (98 ml) is added to sodium tetraborate (0.95 g in 2.0 ml water). This reagent is stable if refrigerated.
2. Carbazole (recrystallised from ethanol; 125 mg dissolved in ethanol (100 ml)).

Procedure: Samples and reagents should be pre-cooled to 4°C prior to mixing. Reagent 1 (1.50 ml) is added to samples (0.25 ml, containing 0–20 µg hexuronic acid) with mixing and cooling. The reactants are heated at 100°C for 10 min, and cooled rapidly to 4°C before addition of Reagent 2 (0.05 ml). After mixing, the solutions are heated at 100°C for 15 min, cooled and the absorbance determined at 525 nm. D-Glucurono-6,3-lactone is used as the standard. Both hexoses and, to a lesser extent, pentoses interfere in this assay.

E. Reducing Group Determinations

The amount of reducing groups in a sample of carbohydrates is frequently measured in aqueous extracts of plant material or after acid or enzymic hydrolysis of oligo- or polysaccharide samples. Reducing oligo- and some low molecular weight polysaccharides will react in these assays.

1. Alkaline copper sulphate

The modification of the Nelson (1944) and Somogyi (1952) procedure as reported by Robyt and Whelan (1968) is recommended.

Reagents:

1. Anhydrous sodium carbonate (25 g), sodium potassium tartrate (25 g), sodium bicarbonate (20 g) and anhydrous sodium sulphate (200 g) are dissolved in water and made up to 1 litre.
2. Copper sulphate pentahydrate (30.0 g) is dissolved in water (200 ml) containing four drops of concentrated sulphuric acid.
3. Ammonium molybdate (25 g) is dissolved in water (450 ml) to which concentrated sulphuric acid (21 ml) is added. Sodium arsenate (3 g) is dissolved separately in water (25 ml) and added slowly with stirring to the ammonium molybdate solution. The entire solution is diluted to 500 ml with water and maintained at 37°C over night and stored in a dark bottle.
4. Reagent 2, 1.0 ml is added to Reagent 1 (25 ml) prior to use.
   Reagents 1–3 are stable at room temperature for several months.

Procedure: Samples (1.0 ml, containing 0–80 µg reducing sugar) are mixed with an equal volume of Reagent 4, and heated at 100°C for 20 min. After rapid cooling to room temperature, Reagent 3 (1.0 ml) is added and the mixture is shaken until carbon dioxide is no longer evolved. The solution is allowed to stand for 10 min before addition of water (10 ml) and the absorbances are determined at 600 nm.

2.   3,5-Dinitrosalicylic acid assay

This assay is described by Bernfeld (1955).

Reagent:

   3,5-Dinitrosalicylic acid (0.25 g) and sodium potassium tartrate (Rochelle salt) (75 g) dissolved in sodium hydroxide (2 M, 50 ml), and diluted to 250 ml with water.

Procedure: Samples (0.1 ml containing 0–500 µg hexose) are mixed with the reagent (1.0 ml), and heated at 100°C for 10 min. After cooling the absorbance is measured at 570 nm.

3.   2,2'-Bicinchoninate assay

The method of Sinner and Puls (1978) has been modified by Waffenschmidt and Jaenicke (1987).

Reagents:

1. Sodium carbonate (31.75 g) and sodium hydrogen carbonate (12.1 g) are dissolved in water (approx. 450 ml). Disodium 4,4'-dicarboxy-2,2'-biquinoline [2,2'-bicinchoninate] (0.97 g) is added and the mixture is made up with water to 500 ml. This reagent is stable for approximately 4 weeks.
2. Cupric sulphate pentahydrate (0.625 g) and L-serine (0.63 g) are dissolved in water and made up to 500 ml. The reagent is stable on storage in a brown bottle at 4°C for approximately 4 weeks.
3. The working reagent is made daily by mixing equal volumes of Reagents 1 and 2.

TABLE 1.2. Enzymic estimation of monosaccharides.

Monosaccharide	Enzyme	EC No.	Cofactor requirement	Reference
L-Arabinose	β-D-Galactose dehydrogenase	1.1.1.48	NAD⁺	Melrose and Sturgeon (1983), Sturgeon (1984)
D-Fructose	Hexokinase, glucose 6-phosphate dehydrogenase, phosphoglucose isomerase	2.7.1.1., 1.1.1.49, 5.3.1.9.	ATP, NADP⁺	Sturgeon (1980a)
	Fructose dehydrogenase	1.1.99.11	—	Ameyama et al. (1981)
L-Fucose	Fucose dehydrogenase	1.1.1.122	NAD⁺	Schachter (1975)
D-Galactose	Galactose oxidase	1.1.3.9	—	
	β-D-Galactose dehydrogenase	1.1.1.48	NAD⁺	Sturgeon (1980b)
L-Galactose	L-Fucose dehydrogenase	1.1.1.122	NAD⁺	Sturgeon (1990)
D-Glucose	Hexokinase, glucose 6-phosphate dehydrogenase	2.7.1.1., 1.1.1.49	ATP, NADP⁺	Sturgeon (1980a)
	D-Glucose dehydrogenase	1.1.1.47	NAD⁺	Price and Spencer (1979)
	D-Glucose oxidase	1.1.3.4	—	Trinder (1969)
D-Mannose	Hexokinase, glucose 6-phosphate dehydrogenase phosphoglucose isomerase phosphomannose isomerase	2.7.1.1., 1.1.1.49 5.3.1.9. 5.3.1.8	ATP, NADP⁺	Sturgeon (1980a)
L-Rhamnose	L-Rhamnose dehydrogenase		NAD⁺	Pittner and Turecek (1987)

Procedure: Samples (0.5 ml, containing 0–25 nmol sugar) and Reagent 3 (0.5 ml are added to screw-capped vials. The lids are tightly screwed on, and the mixtures are heated at 100°C for 15 min in an aluminium heating block. After cooling to room temperature for 20 min the absorbances are read at 520 nm. The method is approximately 10 times more sensitive than the alkaline copper method (see Section III.E.1). Neither borate nor phosphate buffers interfere in the assay.

F. Enzyme Assays

Many monosaccharides can be estimated by enzymic procedures (Table 1.2). Most of the enzymes are available commercially as highly purified proteins and some are produced in kit form. Two classes of enzymes, the oxidases and dehydrogenases (oxidoreductases), are used in all of the procedures. The oxidases, which require the presence of molecular oxygen, act by oxidising the monosaccharide with concomitant production of hydrogen peroxide. A second enzyme, peroxidase, hydrolyses the hydrogen peroxide and oxidises a leuco dye to produce a chromogen, the amount of which is proportional to the amount of monosaccharide oxidised (Fig. 1.1).

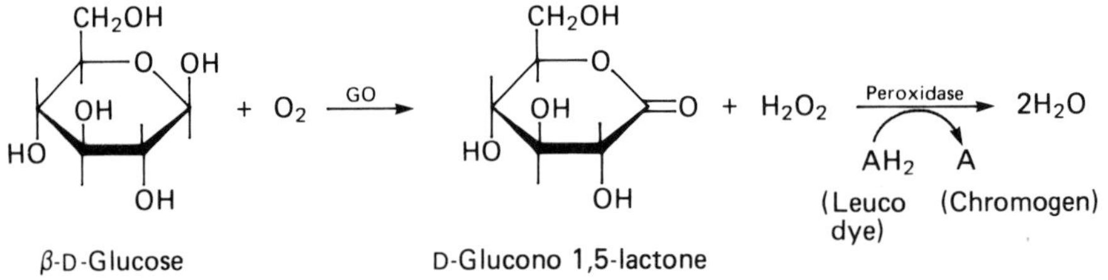

FIG. 1.1. The oxidation of D-glucose by glucose oxidase.
GO = glucose oxidase.

The dehydrogenases, on the other hand, have a cofactor requirement, either nicotinamide adenine dinucleotide (NAD^+) or nicotinamide adenine dinucleotide phosphate ($NADP^+$). On oxidation of the substrate, for example, D-galactose by D-galactose dehydrogenase, an equimolar amount of the cofactor is reduced and from its absorbance at 340 nm the amount of sugar can be directly obtained (Fig. 1.2).

The specificity of the enzyme may be absolute, as is the case with β-D-glucose oxidase, when β-D-glucose is the only known monosaccharide to be oxidised by the enzyme. In other cases, several monosaccharides may act as substrates for an enzyme, and in these cases there will be a very close structural similarity in the monosaccharides recognised by the enzyme. Thus D-galactopyranose, D-fucose and L-arabinopyranose and oxidised by D-galactose dehydrogenase, D-glucose and D-xylose by D-glucose dehydrogenase and L-fucose and L-galactose by L-fucose dehydrogenase. When both monosaccharides are present in a sample, e.g. D-galactose and L-arabinose, each monosaccharide can be determined by using two enzymes, one with absolute specificity for one of the monosaccharides, and the other capable of acting on both sugars. Thus in a mixture containing D-galactose and L-arabinose, the D-galactose, but not L-arabinose, is oxidised completely by D-galactose oxidase. After oxidation of the hexose, L-arabinose is measured in the same sample using D-galactose dehydrogenase.

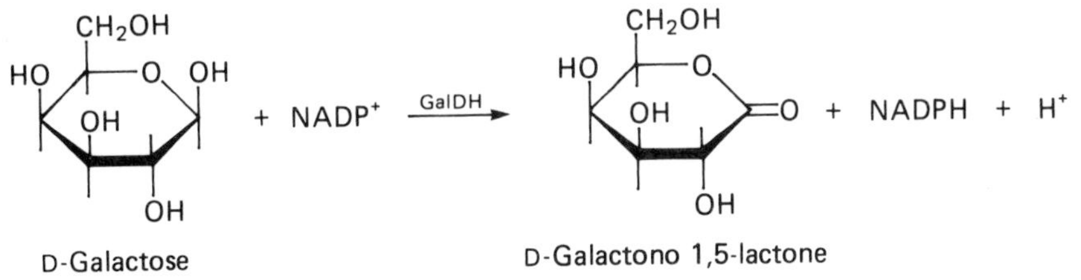

FIG. 1.2.   The oxidation of D-galactose by galactose dehydrogenase.
GalDH = D-galactose dehydrogenase.

Another monosaccharide mixture which can be analysed in this way is glucose in the presence of xylose: glucose is first oxidised by glucose oxidase, and xylose is then oxidised by glucose dehydrogenase.

One of the most widely used enzymes for monosaccharide analysis is glucose 6-phosphate dehydrogenase. When used in conjunction with hexokinase, phosphogluco isomerase and phosphomannose isomerase, it is possible to determine the levels of glucose, fructose and mannose in the same mixture (see Section III.F.2). Many of the enzymic analytical reactions go to completion over short incubation times (2–15 min) and the end-point of the reaction is measured. In these cases, enzymes of high specific activity, and low K_m values are obtainable. In other systems, where the end-point may only be achieved after 1–2 h, it is easier to measure the initial velocity of the reaction over 1–2 min. A linear relationship exists between the initial velocity and monosaccharide concentration.

The analysis time of enzymic spectrophotometric methods is usually fairly short but limitations may include the relatively high cost of enzymes and cofactors, and the unsuitability for handling large numbers of samples. The direct measurement of generated reduced cofactor can be replaced by indirect colorimetric procedures. Cairns (1987), in measuring glucose using the hexokinase method, utilised the intermediary electron carrier phenazine methosulphate to couple to $NADP^+$ reaction to the production of a formazan dye from a tetrazolium salt. The reaction can be carried out in plastic microtitre plates. The advantage in using such a system is that the total volume per well of the plate is 0.2 ml compared with 1.0 ml for a spectrophotometric cell, thus making a considerable saving in the amount of enzymes and cofactors required. In addition, 96 samples can be handled at one time, and the absorbances of formazan formed can be measured automatically using a microtitre plate reader.

1.   D-Glucose oxidase assay

D-Glucose oxidase (β-D-glucose: oxygen 1-oxidoreductase, EC 1.1.3.4) oxidises β-D-glucose to D-glucono-1,5-lactone and produces hydrogen peroxide. Dahlqvist (1961) developed a coupled enzymic method using peroxidase (donor: hydrogen peroxidase oxidoreductase, EC 1.11.1.7) to catalyse the oxidation of the peroxide with the concomitant production of a chromogen which is determined colorimetrically. The donor used was 2-dianisidine, a toxic molecule which can be replaced by a number of less hazardous substances, such as 4-aminoantipyrine (Sugimura and Hirano, 1977) and 2,2′-azino-di-(3-ethyl benzythiazolin-6-sulphonic acid) (ABTS) (Okuda *et al.*, 1977) which have the added advantage of increased sensitivity.

Reagent:

1. Tris-hydrochloric acid buffer (0.1 M) is prepared by dissolving Tris (2-hydroxy-methyl-2-amino-propane 1,3-diol, 1.21 g) in water (50 ml), and adjusting to pH 7.0 with concentrated hydrochloric acid. D-Glucose oxidase (10 mg, 40 units mg^{-1}), peroxidase (5 mg, 250 units mg^{-1}) and ABTS (5 mg) are added and the reagent is diluted to 100 ml. The reagent is stable at 4°C for several weeks.

Procedure: Samples (0.1 ml, containing 0.5–50 μg glucose) are mixed with Reagent 1 (2.9 ml) and incubated at 37°C for 30 min. Hydrochloric acid (5 M, 1 ml) is added to stop the reaction and the absorbances are measured at 440 nm, and compared with absorbances produced from standard amounts of glucose.

2. *Hexokinase–glucose 6-phosphate dehydrogenase assay*

This method will measure mixtures of glucose, fructose and mannose (Sturgeon, 1980a). Addition of hexokinase, HK (ATP: hexose 6-phosphotransferase, EC 2.7.1.1) to solutions of the three hexoses and adenosine triphosphate (ATP) results in the production of the corresponding hexose 6-phosphate and adenosine diphosphate (ADP) (Fig. 1.3, Reaction 1). On addition of glucose 6-phosphate dehydrogenase (D-glucose 6-phosphate: NADP-1-oxidoreductase EC 1.1.1.49) and NADP^{+}, the glucose 6-phosphate alone is converted to 6-phosphogluconate and NADPH (Reaction 2). The increase in absorbance measured after addition of the second enzyme corresponds to the glucose content of the sample.

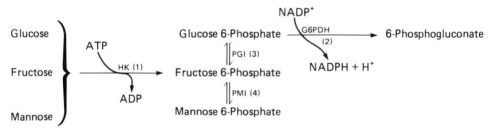

FIG. 1.3. The oxidation of D-glucose, D-fructose and D-mannose by the hexokinase–glucose 6-phosphate dehydrogenase method. 1, Hexokinase; 2, glucose 6-phosphate dehydrogenase; 3, phosphogluco isomerase; 4, phosphomanno isomerase.

On addition of phosphogluco isomerase PGI (D-glucose 6-phosphate ketol isomerase, EC 5.3.1.9) the fructose 6-phosphate is converted to glucose 6-phosphate. This results in a further change in absorbance arising from the additional production of NADPH (Reaction 3) which is a measure of the fructose content of the sample. On addition of phosphomanno isomerase, PMI (D-mannose 6-phosphate:ketol isomerase, EC 5.3.1.8), the mannose 6-phosphate is converted, first to fructose 6-phosphate (Reaction 4), then to glucose 6-phosphate (Reaction 3) with the eventual production of 6-phosphogluconate and NADPH. This final change in absorbance is due to the mannose.

Reagents:

1. Triethanolamine buffer (0.05 M) is prepared by dissolving triethanolamine hydrochloride (0.9 g) and magnesium chloride hexahydrate (0.14 g) in water (70 ml). The

pH is adjusted to 7.6 with sodium hydroxide (5 M) before diluting to 100 ml with water.

2. Adenosine triphosphate (0.017 M) is prepared by dissolving ATP (disodium salt, 10 mg) in water (1 ml).
3. Nicotinamide adenine dinucleotide phosphate (0.011 M) is prepared by dissolving NADP$^+$ (disodium salt (10 mg)) in water (1 ml).
4. Hexokinase, HK (> 140 units mg^{-1}; 10 mg protein ml^{-1}).
5. Glucose 6-phosphate dehydrogenase, G6PDH (> 140 units mg^{-1}: 1 mg protein ml^{-1}).
6. Phosphogluco isomerase, PGI (> 350 units mg^{-1}; 1 mg protein ml^{-1}).
7. Phosphomanno isomerase, PMI (> 60 units mg^{-1}; 10 mg protein ml^{-1}).

Procedure: Reagent 1 (2.7 ml), Reagent 2 (0.1 ml), Reagent 3 (0.1 ml) and sample (0.1 ml, containing up to a total of 60 μg glucose, fructose and mannose) are mixed in a 1 cm light-path silica cell. The initial absorbance (A_1) is measured at 340 nm. Reagent 5 (0.02 ml) is added and after further mixing, the absorbance is measured until no further change occurs (usually 2–3 min). The final absorbance A_2 is measured. The absorbance change $A_2 - A_1 = A$ is used in the calculation of the amount of glucose in the sample. Reagent 6 (0.02 ml) is added and after mixing, the absorbance is again measured until no further change is observed. The final absorbance $A_3 - A_2 = A'$ is used to calculate the amount of fructose in the sample. Reagent 7 (0.02 ml) is added and after mixing the absorbance is measured until no further change is observed. The final absorbance $A_4 - A_3 = A''$ is used to calculate the amount of mannose in the sample.

Calculation: The amount of sugar (μmol) in the cell is given by $(A \times v) / (E \times d)$, where v is the total volume of liquid in the cell, $E = 6.22 \text{ cm}^2 \text{ μmol}^{-1}$, d = light-path (cm) and $A = A$, A' or A''.

Comments: In addition to the glucose, fructose and mannose, hexokinase will phosphorylate 2-amino-2-deoxy-D-glucose and 2-deoxy-D-glucose. However, glucose 6-phosphate dehydrogenase is specific for glucose 6-phosphate.

3. Galactose oxidase assay

The oxidation of galactose by D-galactose oxidase (D-galactose: oxygen 6-oxidoreductase, EC 1.1.3.9) was originally described by Avigad et al., 1962) and an analytical system based on the measurement of hydrogen peroxide was reported by Roth et al. (1965).

As in the glucose oxidase assay, earlier toxic donor molecules for the peroxidase step have been replaced by ABTS.

Reagent:
1. Sodium phosphate buffer (0.1 M, pH 6.0) is prepared by dissolving sodium dihydrogen phosphate (1.21 g) and disodium hydrogen phosphate heptahydrate (0.33 g) in water (80 ml). Galactose oxidase (10 mg, 20 units mg^{-1}) peroxidase (5 mg, 250 units mg^{-1}) and ABTS (5 mg) are added and the reagent is diluted to 100 ml with water.

Procedure: The reagent (2.9 ml) is mixed with sample (0.1 ml, containing 0–50 μg D-galactose) and incubated at 37°C for 1 h. The absorbance is measured at 440 nm, and compared with the absorbances produced from standard amounts of galactose.

Other galactose-containing carbohydrates and related sugars are oxidised, often at higher rates (Amaral *et al.*, 1966; Avigad *et al.*, 1961). These include 2-acetamido-2-deoxy-D-galactose, and oligo- and polysaccharides bearing non-reducing terminal D-galactosyl or 2-acetamido-2-deoxy-D-galactosyl residues.

IV. CHROMATOGRAPHY OF MONOSACCHARIDES

The chromatographic separation and identification of monosaccharides is one of the most important aspects of carbohydrate analysis (Ghebregzabher *et al.*, 1976; Verhaar and Kuster, 1981). Historically, chromatography of sugars has progressed from paper chromatography (PC) to thin layer chromatography (TLC) to gas–liquid chromatography (GLC) and finally to high performance liquid chromatography (HPLC). A preliminary qualitative examination of the composition of monosaccharides is frequently required. Factors such as low cost, simple equipment and ease of detection of the products are the main reasons for the continued popularity of the PC and TLC techniques. Although PC was one of the earliest chromatographic procedures developed for the separation of sugars (Partridge, 1947, 1948) it has been largely superseded by TLC. This is because the relatively short development times required for separations and the use of a variety of inorganic supports on TLC plates allows use of a wider range of reagents containing corrosive sprays for the detection of sugars (Dutton *et al.*, 1965; Ertel and Horner, 1962). The development of GLC for sugars had to await the preparation of volatile derivatives in quantitative yield. In early reports, separations of fully methylated methyl-glycopyranosides derived from pentoses and hexoses were achieved (Kircher, 1960). Gunner *et al.* (1961) demonstrated that anomeric glycose acetates gave separate peaks and derivatives of epimers had different retention times. Several products can frequently be detected in the GLC analysis of a single glycose, arising from the formation of anomeric derivatives of possible furanose and pyranose ring forms (Morrison and Perry, 1966). Complex patterns are therefore detected when several monosaccharides are present. Simplified monosaccharide separations are achieved by using cyclic sugar derivatives (Murphy and Pennock, 1972).

Many of the improvements in the use of GLC as a separational tool relate to the availability of a wide variety of methods for the production of volatile derivatives, improvements in column design and stationary phases, and improved detection systems. The GLC columns commonly used are of two main types. Conventional glass-packed columns are packed with uniformly sized particles of diatomaceous earth which support a thin film of the stationary liquid phase. In the second type a column (Capillary or open tubular) is constructed of a fused silica. Two types of open tubular column exist. The wall-coated open tubular column (WCOT) has the internal wall coated with a thin film of stationary phase. With the porous-layer open tubular column (PLOT), the internal wall of a glass or steel column is coated with a thick layer of porous material which is coated with a stationary phase producing a support-coated open tubular column (SCOT).

The original detectors were based on thermal conductivity systems and they have generally been replaced by flame ionisation, electron capture and mass spectrometric systems giving sensitivity increased by several orders of magnitude. The signals from the detector are amplified and recorded on a chart recorder. The amounts of separated sugars are obtained by inclusion of an internal standard, prior to the derivitisation procedure, and measuring, either the recorded peak heights or peak areas relative to the standard. Pentaerythritol is commonly used as the internal standard when dealing with plant extracts.

Liquid chromatography systems in which a liquid mobile phase moves under gravity have been used in sugar separations for many years. Anion-exchange resins in the borate form were used to separate carbohydrates as their borate complexes. Limitations on the development of the technique include the use of liquids of relatively high viscosity, and poor packing materials which compact and result in slow flow rates. Diffusion of the separating species ensues, resulting in poor resolution and low sensitivity. HPLC allows increased speed of elution and improved resolution, achieved by use of specially designed pumps for eluant delivery, and column fittings and silica packings capable of withstanding the high pressures used. Analysis times are short and solvent consumption is low. HPLC has been reported to be superior to GLC as a technique for carbohydrate analysis (Folkes and Taylor, 1982). Some of the advantages of the technique include direct sample injection, no preparation of volatile derivatives and no need to operate at high temperature during the chromatographic procedure.

Much valuable information on experimental conditions and separational parameters involved in PC, GLC and HPLC has been recorded by Churms (1982).

A. Paper Chromatography

1. General procedure

To ensure reproducible results, the solution to be chromatographed must be free from non-carbohydrate materials. A preliminary purification may be necessary. If sulphuric acid has been used in the hydrolysis of oligo- or polysaccharide material, neutralisation of the acid is usually achieved by addition of barium carbonate, but excess barium ions must be removed on small ion-exchange columns. On the other hand, if trifluoroacetic acid has been used for hydrolysis, this volatile acid can be removed by evaporation under reduced pressure. In early reports on PC of carbohydrates the organic phases of biphasic solvents were used to irrigate paper strips. These have now all been superseded by monophasic solvent systems which consist of at least three components, water, a water-miscible—and a water-immiscible—organic solvent. The mobility of the sugars, from the origin of the chromatogram is generally pentose > hexose > disaccharide > trisaccharide, etc. Separations of sugars are most frequently carried out by descending PC on Whatman No. 1 paper when 2–5 µg loadings of individual sugars are optimal for detection. Improved separations are sometimes achieved by multiple development of chromatograms; the papers are dried between chromatographic runs, and the solvent is allowed to travel over the paper a second or third time. A number of solvent systems, together with the retention data for sugars, are given in Table 1.3.

2. Visualisation of sugars on paper chromatograms

After chromatographic separation of the sugars, it is customary to dip paper chromatograms in reagents, rather than to spray the reagents for visualisation of sugars.

TABLE 1.3. Relative mobilities[a] of some monosaccharides and alditols by descending paper chromatography.

Sugar	Solvent 1	Solvent 2	Solvent 3	Solvent 4
D-Galactose	0.8	0.8	0.8	1.5
D-Glucose	1.0	1.0	1.0	1.0
D-Mannose	1.1	1.4	1.2	1.2
D-Fructose	1.2	1.4	1.3	1.7
D-Arabinose	1.1	1.4	1.8	2.3
D-Xylose	1.3	1.8	2.0	2.5
D-Ribose	1.5	2.2	—	4.4
L-Rhamnose	1.6	—	3.3	3.4
L-Fucose	—	2.0	2.3	—
D-Glucitol	0.9	—	—	2.4
D-Mannitol	1.0	—	—	1.9
Xylitol	—	—	—	3.2
D-Glucuronic acid	0.14	—	0.4	—
D-Galactouronic acid	0.11	—	0.3	—

Solvents: 1, ethyl acetate–pyridine–water (10:4:3; v/v) (Hough and Jones, 1962); 2, butan-1-ol–pyridine–water (6:4:3; v/v) (Hough and Jones, 1962); 3, ethyl acetate–acetic acid–pyridine–water (10:3:3:2; v/v) (Jarvis and Duncan, 1974); 4, nitromethane–acetic acid–ethanol–water, saturated with boric acid (8:1:1:1; v/v) (Robyt, 1975).
[a] Values given are R_{glc} (mobilities relative to glucose).

(a) *Silver nitrate reagent (sensitivity c. 1 µg sugar).* A modification of the method of Trevelyan *et al.* (1950) is used.

Reagents:
1. Saturated aqueous silver nitrate (0.1 ml) is added to acetone (20 ml).
2. Sodium hydroxide (2 g) is dissolved in water (< 1 ml) and made up to 100 ml with methanol.
3. Sodium hydroxide (2 g) and pentaerythritol (4 g) is dissolved in water (20 ml). Ethanol (80 ml) is added with mixing.
4. Sodium thiosulphate (240 g), sodium sulphite (10 g) and sodium bisulphite (25 g) are added to water (1 litre).

Procedure: The dried chromatogram is dipped in Reagent 1 and allowed to dry (2–5 min), before being dipped in Reagent 2. When boric acid is one of the components of a solvent, Reagent 2 is replaced by Reagent 3. Reducing sugars produce brown spots almost instantaneously. Non-reducing sugars react slowly; their visualisation may be enhanced by holding the papers over a steam bath for 1–2 min. Within 10 min, the paper should be dipped in Reagent 4 to remove background colour, washed for 15 min in running water and air dried.

(b) p-*Anisidine-phthalic acid (sensitivity c. 0.5 µg hexoses and 0.2 µg pentoses and uronic acids).*

Reagent:
    p-Anisidine (1.2 g) and phthalic acid (1.6 g) are dissolved in ethanol (100 ml).

Procedure: The paper is dipped in the reagent, and heated at 100°C for 10 min. Pentoses give red spots, hexoses and deoxyhexoses give yellow to green spots and uronic acids give brown spots.

(c) *Periodate-pentane-2,4-dione (acetylacetone) (sensitivity about 1 μg for hexoses).*

Reagents:
   1. Periodic acid (40%; w/v, 2 ml) and pyridine (2 ml) are added to acetone (100 ml).
   2. Ammonium acetate (15 g), glacial acetic acid (0.3 ml) and pentane-2,4-dione (acetylacetone) (1 ml) are added to methanol (100 ml).

Procedure: The chromatogram is dipped through Reagent 1 and allowed to dry before dipping in Reagent 2. After 10–15 min most sugars including alditols give yellow-green spots when examined under ultraviolet light. Any substance capable of liberating formaldehyde, e.g. alditols after oxidation with periodate, will react with this reagent.

B. Thin Layer Chromatography

1. General procedure

A wide range of different supports have been used for the TLC of carbohydrates. Microcrystalline cellulose has the resolution of PC but with faster separation times. The same solvent systems and visualising reagents which are used for PC can be used for TLC. An alternative chromatographic system depends on inorganic supports such as silica gel, where resolution depends on both partition effects and adsorption effects. Carbohydrate separations can be improved by the incorporation of inorganic salts, e.g. weak acid, into the silica plate. Ovodov *et al.* (1967) have systematically investigated the effects of a number of inorganic salts on the chromatographic behaviour of sugars on TLC. Phosphates are commonly incorporated into silica plates, as are borates (Ghebregzabher *et al.*, 1974), molybdates and tungstates (Mezzetti *et al.*, 1971).

The thin layer plates may be prepared within the laboratory, but in general pre-coated plates may be purchased and give excellent reproducibility. A very wide range of solvent systems have been reported for TLC separations of sugars, especially those used for separations on inorganic supports. A selection of these solvent systems, together with retention data for sugars, is given in Table 1.4.

2. Visualisation of sugars on TLC plates

Reagents used for PC are normally also suitable for TLC. When inorganic supports are used for the chromatographic separations, more corrosive spray reagents, incorporating sulphuric acid into anisaldehyde (Stahl and Kaltenbach, 1961), naphthol-resorcinol (Pastuska, 1961) or phenol (Adachi, 1965) can be used.

(a) *Diphenylamine–aniline–phosphoric acid (sensitivity c. 1 μg sugar).*

Reagents:
   1. Diphenylamine (2 g) and aniline (2 ml) are dissolved in acetone (100 ml).
   2. Orthophosphoric acid (85%; w/v) (10 ml).

TABLE 1.4. Relative mobilities[a] of some monosaccharides and alditols by TLC.

Sugar	Solvent 1	Solvent 2	Solvent 3	Solvent 4
D-Galactose	0.9	0.8	0.9	0.7
D-Glucose	1.0	1.0	1.0	1.0
D-Mannose	1.2	1.2	1.1	1.2
D-Fructose	1.2	1.2	1.2	1.1
D-Arabinose	1.3	1.3	1.3	—
D-Xylose	1.3	1.6	1.5	—
D-Ribose	1.5	—	1.3	—
L-Rhamnose	1.6	2.1	1.8	—
L-Fucose	1.4	—	1.6	—
D-Glucitol	—	—	0.8	—
D-Mannitol	—	—	1.0	—
D-Galactitol	—	—	0.9	—
D-Xylitol	—	—	1.3	—
D-Glucuronic acid	1.0	—	0.8	—
D-Galacturonic acid	0.9	—	0.5	—

Solvents: 1, formic acid–2-butanone–*tert*-butanol–water (3:6:8:3; v/v) on cellulose plates (Petre *et al.*, 1972); 2, ethyl acetate–pyridine–water (20:7:5; v/v) on cellulose plates (Raadsveld and Klomp, 1971); 3, benzene–acetic acid–ethanol (2:2:1; v/v) on silica gel-Kieselguhr plates (Haldorsen, 1977); 4, 2-propanol–acetone–lactic acid (0.1 M) (2:2:1; v/v) on pre-coated silica plates, impregnated sodium dihydrogen phosphate (0.5 M) (Hansen, 1975).
[a] Values given are R_{glc} (mobilities relative to glucose).

Procedure: Prior to use, Reagents 1 and 2 are mixed. The dried TLC plate is sprayed, air dried and then heated at 100°C for 10 min. Aldoses produce blue-grey spots, and ketoses give red spots.

(b) *Vanadium pentoxide–sulphuric acid (sensitivity c. 0.25–1 µg sugar).* This method was described by Haldorsen (1977).

Reagent:

Vanadium pentoxide (1.8 g) is dissolved in sodium carbonate (1 M, 30 ml) with heating. After cooling, sulphuric acid (2.5 M, 40 ml) is added and the mixture is diluted with water to 100 ml. The plates are sprayed with this reagent and examined at room temperature. All reducing sugars give blue spots.

(c) *Periodate–pentane 2,4-dione (sensitivity c. 1 µg for hexoses and alditols).* This method was developed by Weiss and Smith (1967) The reagents are the same as described in Section IV.A.2.C except that they are sprayed onto plates.

The reagent will react with most sugars, giving yellow-green spots when examined under UV light. Any substance liberating formaldehyde after oxidation with periodate will react with this reagent.

C. Gas–Liquid Chromatography

Gas–liquid chromatography (GLC) has been widely used for the separation and quantitative determination of complex mixtures of carbohydrates. The separation of carbohydrates is based on differences in their distribution (due to adsorption/partition) between a gas phase and a liquid phase, which is coated on a solid matrix. Since the carbohydrates are polar non-volatile compounds, they have first to be derivatised to produce a product which is volatile under the operating conditions of the GLC. Sub-nanomolar amounts of sugars can be analysed by this technique.

1. Column types

The effectiveness of solute interaction with the stationary phase of a column is determined by the design and dimensions of the column. Two main types of GLC columns are in use. Glass columns (1–3 m × 2–4 mm i.d.) are normally packed with diatomaceous earth which has been coated with a thin film of stationary phase (Table 1.5). The second type of column, capillary or open tubular, may be made of glass or fused silica. Two main types of open tubular columns are usually available. Wall-coated open tubular columns (WCOT) have the internal walls coated with a thin film of stationary phase. In the second type of open tubular column, the internal wall is coated with a layer of porous material, and is then coated with a stationary phase to produce a support-coated open tubular column (SCOT) (Table 1.6). The dimensions of these columns vary but are usually 25–50 m × 0.25–0.50 mm i.d. The main advantages of WCOT columns compared with packed columns include greater resolving power and speed of separation.

TABLE 1.5.   Packed columns and stationary phases used in carbohydrate analysis.

Derivative	Stationary phase	Column support	Reference
Alditol acetate	5% XE-60	Chromosorb W	Sawardeker *et al.* (1965)
	10% Carbowax 20M	Chromosorb W	Sawardeker *et al.* (1965)
	3% ECNSS-M	Chromosorb W	Sawardeker *et al.* (1965)
	5% OV-275	Chromosorb W	Ochiai (1980)
TMS-alditol	3% SE-52	Diataport S	Cayle *et al.* (1968)
TMS-oxime	0.5% OV-17	Chromosorb G	Petersson (1974)
Acetyl-oxime	5% QF-1	Supelcoport	Manius *et al.* (1979)
TMS	3% SE-30	Diatoport S	Bhatti *et al.* (1970)
Trifluoroacetate	20% SE-30	Chromosorb W AW	Zanetta *et al.* (1972)

2. Stationary phases

A wide variety of stationary (liquid) phases is available. It is customary to chromatograph polar derivatives (acetates, etc.) on polar phases and non-polar derivatives (TMS-ethers, etc.) on non-polar phases.

3. Derivatives

A single glycose may produce several products on derivatisation, for example as the acetyl derivatives (Gunner *et al.*, 1961). This is due to the formation of anomeric

TABLE 1.6. Open tubular columns used for carbohydrate analysis.

Derivative	Stationary phase	Column	Reference
Alditol acetate	OV-275	WCOT vitreous silica	Blakeney et al. (1982)
	SP-1000	WCOT glass capillary	Lomax and Conchie
		WCOT fused silica	(1982)
	Polysiloxane	SCOT glass capillary	Green et al. (1981)
	PEG 20M	WCOT fused silica	Oshima et al. (1981)
	Carbowax 20M	WCOT fused silica	Oshima et al. (1982)
TMS-alditol	OV-101	WCOT fused silica	Bradbury et al. (1981)
TMS-oxime	Methyl silicone	Fused silica capillary	Li and Andrews (1986)
	SP-2250	WCOT fused silica	Schaffler and Morel du Boil (1981)
TMS-methioxime	SP-2250	Glass capillary	Pelletier and Cadieux (1982)
	SP-2100	Fused silica	Pelletier and Cadieux (1982)
Acetyl oxime	SE-30	WCOT	Pfaffenberger et al. (1975)
TMS	SE-30	WCOT fused silica	Cowie and Hedges (1984)
	SE-54	WCOT glass capillary	Comparini et al. (1983)
Trifluoroacetyl oxime	OV-225	Glass capillary	Schwer (1982)
Trifluoroacetate	SE-54	WCOT glass capillary	Eklund et al. (1977)

WCOT, wall-coated open tubular column; SCOT, support-coated open tubular column.

derivatives of possible furanose and pyranose ring forms (Morrison and Perry, 1966). This has generally been regarded as a drawback for the analysis of carbohydrates, especially when packed columns are used. With the superior resolving power of WCOT columns, the requirement for the production of single derivatives of sugars is less important. If the carbonyl group of the monosaccharide is first reduced with sodium borohydride to the corresponding sugar alcohol before acetylation to the alditol acetate, this eliminates the formation of multiple derivatives from a single sugar. Although this method is widely used, the operator should be aware of certain limitations. For example, although the preliminary reduction step is quantitative, borate ions derived from the sodium borohydride interfere in the acetylation step. A ketose will normally produce two alditols (fructose on reduction yields glucitol and mannitol), whilst glucitol is derived from both glucose and sorbose. Another approach to the production of single derivatives is the conversion of aldoses to the corresponding methyloxime at C-1 followed by acetylation of free hydroxyl groups (Murphy and Pennock, 1972). Although glucose, fructose, mannose and xylose give single peaks, some carbohydrates give two peaks due to the presence of both *syn-* and *anti-*forms (Schwer, 1982). Trimethylsilyl (TMS) derivatives are the most popular volatile derivatives of sugars used for GLC analyses. Sweeley *et al.* (1963) formed these derivatives by reacting monosaccharides with a mixture of hexamethyldisilazane (HMDS) and trimethyl-chlorosilane (TMCS) in anhydrous pyridine. In early reports on the production of TMS ethers of monosaccharides, the necessity of maintaining anhydrous conditions was stressed. However, it is now known that the presence of small amounts of water is permissible, provided an excess of silylating agent is used (Brobst and Lott, 1966).

It is necessary to re-*N*-acetylate amino sugars which are present in samples which have been prepared by acid hydrolysis irrespective of the type of derivative being prepared.

(a) *TMS-ethers*. The preparation of the derivatives is a variation of the method of Sweeley *et al.* (1963).

Reagents:
1. Anhydrous pyridine (dried over potassium hydroxide pellets).
2. Hexamethyldisilazane (HMDS).
3. Trimethylchlorosilane (TMCS).

Procedure: Carbohydrate samples (1–5 mg) are added to Teflon-capped vials. If the samples are in aqueous solutions, they should be taken to dryness over a stream of dry nitrogen. To the dried samples are successively added Reagent 1, (1.0 ml), Reagent 2, (0.2 ml) and Reagent 3 (0.1 ml). The mixtures are shaken vigorously for about 30 s and allowed to stand for 5 min. The ammonium chloride which is produced in the reaction settles out at the bottom of the vials. Solvent tailing due to the presence of pyridine is frequently observed, making interpretations of the more volatile sugars difficult. For this reason it is advisable after derivitisation to replace the pyridine with another solvent, such as *n*-hexane. The samples should be re-evaporated using a stream of dry nitrogen before dissolving the TMS-ethers in *n*-hexane (0.2–0.5 ml).

For the preparation of derivatives from samples, such as syrups, where water is present, an alternative silylating reagent composed of anhydrous pyridine–hexamethyldisilazane–trifluoracetic acid (10:9:1; v/v) is used. It is necessary to maintain the carbohydrate content at least 1 mg ml^{-1} reagent and up to 2 mg water ml^{-1} of reagent.

TMS-derivatives of uronic acids give complicated chromatographic patterns. The problem of multiple peaks can be eliminated by reduction of the uronic acids to the corresponding aldonic acids, prior to lactonisation and trimethyl silylation, using the following procedure reported by Perry and Hulyalkar (1965).

Hexuronic acids (1–5 mg) are dissolved in water (2 ml) and converted to the barium salts by addition of barium carbonate (150 mg). After stirring the solution at 65°C for 10 min, insoluble material is removed by filtration and the soluble hexuronic acid salts are reduced to the corresponding aldonic acids by the addition of sodium borohydride (50 mg). After 1 h at room temperature the excess borohydride is destroyed by addition of a cation-exchange resin (H^{+} form) (2 × 1 cm). The resin and solution are then added to a small (6 × 1 cm) column of the same resin, and aldonic acid is eluted with water. The combined eluate and washings are evaporated to dryness under reduced pressure. Boric acid is removed from the residue as volatile methyl borate, by redissolving the products in dry methanol (5 ml) and evaporating under reduced pressure. This procedure is repeated four times. The aldonic acids are converted to the corresponding aldono-1,4-lactones by adding hydrochloric acid (11 M, 0.5 ml) and evaporating the solution to dryness *in vacuo*. The lactones are then converted to the TMS-derivatives using either of the two silylating reagents already described.

(b) *Alditol acetates*. Two procedures, those of Sloneker (1972) (Method A) and of Blakeney *et al.* (1983) (Method B), are described.

Method A

Reagents:
1. Sodium borohydride (5 mg ml^{-1}).
2. Anhydrous pyridine–acetic anhydride (1:1; v/v).

Procedure: Samples (1–2 mg) in water (1 ml) are mixed with Reagent 1 (1 ml). The reduction is allowed to proceed at room temperature for 1 h. Excess reagent is then destroyed by addition of acetic acid (2 M, 0.1 ml) and the products are passed through a column of cation-exchange resin in the H$^+$ form (2 × 1 cm). The column is washed with water (5 × 1 ml) and the combined eluate and washings are evaporated to dryness under reduced pressure. Borate is removed from the product as volatile methyl borate by addition of dry methanol (2 ml) followed by re-evaporation. This step is repeated four times. Reagent 2 (0.2 ml) is added and the solution is heated at 100°C for 3 h. On cooling the solution may be injected directly into the gas chromatograph.

Method B

Reagents:
1. Sodium borohydride (1 g) is dissolved in anhydrous dimethylsulphoxide (100 ml) at 100°C and is stable if kept dry at 4°C. The anhydrous dimethylsulphoxide is prepared by storage of the material over molecular sieve type 4A.
2. Acetic anhydride.
3. 1-Methylimidazole.

Procedure: Samples (200–500 µg) are dissolved in aqueous ammonia (1 M, 0.1 ml) and reduced to the alditols by addition of Reagent 1 (1 ml). The solution is incubated at 40°C for 1.5 h when excess sodium borohydride is destroyed by the addition of glacial acetic acid (0.1 ml). Reagent 2 (2 ml) and Reagent 3 (0.2 ml) are then added with thorough mixing. The reaction is allowed to proceed for 10 min when excess acetic anhydride is decomposed by the addition of water (5 ml). The solution is cooled (4°C) and mixed with dichloromethane (1 ml). The peracetylated products are removed from the settled lower layer and stored in a septum-capped vial at −20°C, prior to analysis.

(c)   *TMS-oximes.* The method is adapted from that reported by Sweeley *et al.* (1963).

Reagents:
1. Anhydrous pyridine.
2. Hydroxylamine hydrochloride.

Procedure: Samples (1–5 mg) are added to Teflon-capped vials. If the samples are in aqueous solution, they should be taken to dryness in a stream of dry nitrogen. To the dried samples are added Reagent 1 (1.0 ml) and Reagent 2 (10 mg). The capped vials are then heated at 70°C for 0.5 h after which the solution is cooled. Silylation is achieved by addition of HMDS (0.2 ml) and TMCS (0.1 ml) (see Section IV.C.3.a). The solvent can be removed by evaporation in a stream of dry nitrogen and the samples are then dissolved in *n*-hexane (0.2–0.5 ml).

Laine and Sweeley (1973) have prepared O-methyl oximes in an analogous manner, in which samples (1 mg) in Reagent 1 (0.05 ml) are mixed with O-methylhydroxylamine hydrochloride (1 mg) in Teflon-capped vials and heated at 80°C for 2 h, after which the TMS ether was prepared by addition of N,O-bis(trimethylsilyl)-trifluoroacetamide (BSTFA) (0.1 ml) followed by heating at 80°C for a further 0.25 h.

(d) *Trifluoracetyl oximes.* The method is adapted from that reported by Schwer (1982).

Reagents:
 1. O-Methylhydroxylamine hydrochloride.
 2. Sodium acetate.
 3. Trifluoroacetic anhydride.

Procedure: Samples (1 mg) are added to Teflon-capped vials. If the samples are in aqueous solution, they should first be taken to dryness in a stream of dry nitrogen. To the dried samples are added Reagent 1 (3 mg), Reagent 2 (6 mg), and water (0.1 ml). The capped vials are heated at 60°C for 1 h, after which the water is evaporated in a stream of nitrogen. Methanol (0.1 ml) is added, and samples are again taken to dryness. Traces of water are removed by addition of benzene, evaporated as the azeotrope. Reagent 3 (0.03 ml) and ethyl acetate (0.015 ml) are added. The capped vials are kept at room temperature for 2 h or at 4°C for 12 h. The derivatives are stable at 0°C for several months.

D. Liquid Chromatography

Several types of systems can be covered by the term 'liquid chromatography'. They include ion-exchange chromatography, high performance liquid chromatography (HPLC) and gel permeation chromatography. Only the ion-exchange and high performance liquid chromatographies have proved of real value in the separation of monosaccharides.

1. Low pressure ion-exchange chromatography

Anion-exchange chromatography was first applied to the separation of monosaccharides as their borate complexes by Khym and Zyll (1952). Extremely long separation times limited the widespread use of that procedure. However, modifications of the techniques, such as low pumping pressures maintained by hydraulic pressure or peristaltic pumps, and higher column temperatures (55–70°C) are capable of reducing the analysis time to 2–18 h.

2. Anion-exchange chromatography

Borate buffers form complexes with carbohydrates producing negatively charged complexes. The differing affinities of these complexes on ion-exchange resins can be exploited for the separation of monosaccharides. By use of a column (13 × 0.8 cm) of quaternary ammonium ion-exchange resin Kennedy and Fox (1980) were able to separate seven monosaccharides and four disaccharides in under 6 h, using a series of borate buffers of increasing pH and ionic strength to elute the carbohydrate borate complexes. The carbohydrate is detected either by collection of fractions followed by a manual assay, or by mixing the eluant with a stream of orcinol–sulphuric acid reagent followed by colorimetric detection (see Section III.A.2). Concern is frequently expressed as to whether alkaline conditions may cause interconversions of reducing sugars, especially when extended analysis times are involved. Hough *et al.* (1972) used a borate–chloride gradient at pH 7.0 to separate nine monosaccharides. Simple polyols including glucitol, galactitol, mannitol, xylitol and arabinitol, may also be separated on this column (Hough *et al.*, 1975). Since uronic acids already bear a charge, they can be separated on anion-exchange resins in the absence of borate ions. Mopper (1978a) separated a variety of uronic acids at pH 8.5 in acetate buffer. Post-column detection of the alditols was achieved by release of formaldehyde by periodate oxidation followed by production of the chromogen formed in the Hantzch reaction with pentane 2,4-dione.

A wide variety of aldoses, ketoses, alditols, cyclitols and oligosaccharides have been measured in the effluents from borate anion-exchange columns after periodate oxidation (Nordin, 1983). Peaks may be detected by an absorbance decrease at 260 nm caused by consumption of the periodate ion. The sensitivity of the reagent allows less than 1 nmol of carbohydrate to be measured.

3. *High performance liquid chromatography (HPLC)*

(a) *General principles.* Modern HPLC has evolved from the development of specially designed column packings which allow the controlled flow of eluant at pressures up to 6000 psi (41 MPa) in narrow bore columns. Concurrent development has required the production of equipment and fittings for the reproducible handling of small volumes, together with detection systems capable of measuring the presence of nanograms of product in microlitres of eluant.

The analytical column, on which the separation takes place, contains packing of 5 μm particle size and usually has an efficiency of about 4×10^4 plates per metre. The eluate flows immediately into the detector which measures one of a number of physical properties of either the separated sugars, or of their derivatives.

(b) *Column packings.* Most of the HPLC column packings are based on microporous silica. Fully porous particles of closely defined shape and of 5 or 10 μm diameter are used. Bonded phases are produced by covalent linkage of a stationary phase to the silica. Silyl ester bonds were originally used for these support materials, but their instability when used under aqueous acid or alkaline conditions has reduced their popularity. Columns packed with supports bearing silyl ether groups or silicon–carbon bonds, which are more resistant to hydrolysis, exhibit greater stability and are more reliable. Alkylated, cyano- and amino-bonded stationary phases in combination with aqueous methanol or aqueous acetonitrile as mobile phase have been used for many carbohydrate separations. Chemically bonded particles readily aggregate on contact with hydrophilic molecules, thus altering the packing conditions. This problem is usually overcome by inserting a guard column between the injection point and the analytical column. *In situ* preparation of amino columns using eluants containing low concentrations of aliphatic amines was used by Uobe *et al.* (1980) to ensure the stability of such columns by constantly renewing the surface of the stationary phase. Reverse phase partition chromatography can be carried out on an ion-exchange porous polymer. Separations using cation-exchange resins are comparable to that of chemically bonded silica. Alteration of the counter ion results in changes in elution profiles. Cation-exchangers in lithium, sodium, calcium, lead and silver forms have been used for carbohydrate separations of materials derived from fermentation processes and food manufacturing. A ligand-exchange mechanism is considered to be involved in these separations (Macrae and Dick, 1981).

Honda *et al.* (1981a) have demonstrated that anion-exchange of carbohydrate–borate complexes can separate all nine naturally occurring aldoses. One drawback of this system is the widening of peaks due to continuous dissociation of complexes during elution.

Hydrophobic properties can be introduced into monosaccharides by the formation of derivatives such as benzoates (Oshima and Kumanotani, 1983), per-4-nitrobenzoates (Nachtmann and Budna, 1977) peracetates (Thiem *et al.*, 1978) and pernaphthoates (Golik *et al.*, 1983). These derivatives can then be separated by adsorption or by partition chromatography on silica. Phenylisocyanates (Bjorkqvist, 1981) *N*-methyl-benzyl glycosamines (Batley *et al.*, 1982), *O*-benzyl and *O*-methyl-oximes (Chen and McGinnis, 1983) and dansylhydrazones (Mopper and Johnson, 1983) have also been used for the separation of mono- and oligosaccharides mainly by reverse phase chromatography. Examples of some of these systems are listed in Table 1.7.

(c) *Detection.* Since the majority of simple carbohydrates possess neither chromophores nor fluorophores, they cannot be detected in the normal ultraviolet or visible light regions. Some of the more commonly used detector systems are cited in Table 1.8, and are described below.

(i) *Direct methods.* A widely used, but relatively insensitive, detector is based on measuring changes in refractive index (McGinnis and Fang, 1980). The detector, which covers a wide linear range and has a lower limit of sensitivity at about 10 nmol of monosaccharide, may be adequate when reasonable amounts of material are available for chromatography. In addition the detector is sensitive to changes in column temperature and solvent composition, and its use is thus normally limited to the qualitative monitoring of carbohydrates which are eluted isocratically from columns. Monosaccharides do absorb at wavelengths in the far UV, with absorbance maxima at about 188 nm. With the appropriate instrumentation, monosaccharides can normally be detected at wavelengths between 192 and 205 nm. The sensitivity of detection is of the same order as for refractive index measurements, but in the application of this technique, only a limited number of very pure solvents can be used as eluants.

(ii) *Indirect methods.* Post-column labelling of carbohydrates is the method of choice when a greater sensitivity than can be obtained by direct detection is required. Although in some cases a high degree of instrumentation is required, it is the system of choice for the detection of separated carbohydrates. Most of the methods used are based on existing detection systems used in the manual analysis of carbohydrates. When aqueous solvents are used, as in borate complex anion-exchange chromatography, it is common to form chromogens in the presence of strong acids to produce furfural, or derivatives

TABLE 1.7. HPLC systems used for monosaccharide separations.

Mode	Support material	Mobile phase	Detector	Reference
Anion-exchange (borate)	DA-X4	0.5 M Borate Borate	Post-column derivatisation Post-column derivatisation to 2-Cyanoacetamide	Honda et al. (1981a) Mopper (1978b)
Cation-exchange				
Ca^{2+}	AG50W X4	Water	Refractometer	Ladisch et al. (1978)
Ca^{2+}	HPX-87C			Baker and Himmel (1986)
Ag^{+}	HPX-65A	Pb^{2+} ion		Van Riel and Olieman (1986)
Ca^{2+}	Sugar-Pak-1	Water	UV 190 nm	Owens and Robinson (1985)
Microparticulate silica	LiChrosorb Si60	Water–acetonitrile	Refractometer	Van Olst and Joosten (1979)
	Partisil 5	Ethyl acetate	UV 260 nm, pre-column derivatisation to dimethyl phenylsilyl groups	White et al. (1983a,b)
	LiChrosorb Si100	Methanol–water	UV 352 nm, pre-column derivatisation to dinitrophenylhydrazones	Karamanos et al. (1987)
Bonded phase silica	Amino-cyano-phase	Acetonitrile–water	Refractometer	Rabel et al. (1976)
	Supelcosil LC-18	Acetonitrile–water	UV 254 nm, pre-column derivatisation of perbenzoates	Gisch and Pearson (1988)
	Micropak NH$_2$	Acetonitrile–water	Refractometer; fluorimeter	Slabach and Robinson (1983)
	LiChrosorb NH$_2$	Acetone–water–acetic acid	Refractometer	Moriyasu et al. (1984)

TABLE 1.8. Post-column detection of carbohydrates.

Detector	Carbohydrate detected	Reagent	Wavelength (nm)	Sensitivity (nmol)	Reference
Refractive index	Monosaccharides, alditols	—	—	<10	Honda (1984)
U.V.	Monosaccharides	—	192–200	10	Honda (1984)
Photometric	Monosaccharides	Cuprammonium	280–310	2.5	Grimble et al. (1983)
	Monosaccharides	Orcinol	420	1.0	Simatupang (1979)
	Monosaccharides	Phenol	480	1.0	Smith et al. (1978)
	Monosaccharides	Anthrone	640	1.0	Krammer et al. (1978)
		2-Cyanoacetamide	280	1.0	Honda et al. (1980)
	Alditols	Periodate-2,4-pentanedione	412	2.0	Honda et al. (1983a)
Fluorimetric	Alditols	Periodate-2,4-pentanedione	a	0.5	Honda et al. (1983a)
	Monosaccharides	2-Cyanoacetamide	b	0.01–0.1	Honda et al. (1980)

[a] Excitation at 410 nm, emission at 503 nm.
[b] Excitation at 331 nm, emission at 383 nm.

thereof, which then condense with either orcinol (Simatupang, 1979), phenol (Smith et al., 1978) or anthrone (Krammer et al., 1978). Aldoses and ketoses, but not alditols or 2-amino-2-deoxy hexoses, are detected using these procedures. Whilst the linear range of the colour reactions is wide, the use of systems which do not depend on the use of strong acids is frequently preferred. One method which measures reducing power of aldoses depends on the complexation of the cuprous ion, arising from the reduction of cupric ions by the aldoses with 2,2'-bicinchoninate (Mopper, 1978b). The method has been adapted to reverse-phase partition chromatography in aqueous ethanol (Mopper, 1978a). The reaction between monosaccharides and cuprammonium, which is instantaneous at room temperature, has been studied by Grimble et al. (1983). Using a suitable post-column pump, the technique is not sensitive to changes in solvent composition, as with gradient elution, and is applicable to the analysis of mono- and oligosaccharides and also other 1,2-diols. Post-column labelling of alditols and aldonic acids is best carried out by reaction with the periodate ion to liberate formaldehyde, which then undergoes the Hantzsch reaction in the presence of pentane 2,4-dione and ammonium salts, to form lutidine. The reaction product can be measured either photometrically or fluorimetrically (Honda et al., 1983a). The development of fluorimetric methods has produced the most sensitive methods for detecting separated carbohydrates. Although methods involving reactions with aliphatic amines such as 1,2-diaminoethane (Mopper et al., 1980) and ethanolamine (Kato and Kinoshita, 1980) have been developed, elevated temperatures (150°C) are necessary to produce the derivatives. Probably the most promising fluorogenic reagent is 2-cyanoacetamide which reacts with reducing sugars to produce cyano-pyridine and/or cyano-pyrrolidine derivatives (Honda et al., 1981a). The products of the reaction are formed under relatively mild conditions, namely in borate buffer pH 8 at 100°C. The method will successfully monitor aldoses (Honda et al., 1981a,b), amino sugars (Honda et al., 1983c) and uronic acids (Honda et al., 1983b) in eluates of borate complex anion-exchange chromatography. High concentrations of acetonitrile, as encountered in reverse phase partition chromatography, quench the fluorescence of the derivatives, thus preventing fluorescent detection. However, no interference in photometric analysis occurs (Honda et al., 1984). Four sensitive post-column derivatisation methods which operate in the presence of borate buffer are fully described below.

2,2'-Bicinchoninate method
This method is based on the work of Mopper (1978b) and Sinner and Puls (1978), and is used for the determination of reducing sugars.

Reagents:
   1. Disodium 2,2'-bicinchoninate (1.3 g) is dissolved in water (500 ml). Anhydrous sodium carbonate (62.3 g) is added, and when dissolved, the solution is made up with water to 1 litre.
   2. Aspartic acid (3.7 g) and anhydrous sodium carbonate (5 g) are dissolved in water (100 ml). Cupric sulphate (1 g) in water (40 ml) is added to the aspartic acid solution, and after mixing, the volume is adjusted with water to 150 ml.

The reagents are stable for several months. The working reagent, prepared several hours before use by mixing Reagent 1 (230 ml) and Reagent 2 (10 ml), is kept in a dark bottle and is stable for 1–2 weeks before the background absorbance increases above acceptable levels. The reagent must be degassed prior to use.

Procedure: When monosaccharides are separated on quaternary ammonium resins as the borate complexes they will be eluted in borate buffer (0.5 M, pH 8.6; Mopper, 1978b). At the column outlet the eluate, at a flow rate of 24 ml h^{-1}, is mixed with the reagent at a flow rate of 36 ml h^{-1} through a Y-shaped connector which in turn is connected to a Teflon reaction coil (6 m × 0.3 mm i.d.). The reaction coil is immersed in a bath of ethylene glycol monoethyl ether heated to 124°C. This reaction takes about 2 min. The product is then determined spectrophotometrically at 562 nm. The sensitivity of the reagent is about 100 pmol at 0.1 absorbance units.

Cuprammonium method
The method is based on the work of Grimble *et al.* (1983) and measures reducing sugars and 1,2-diols.

Reagent:
A stock solution is prepared by dissolving cupric sulphate (5 g) in water (100 ml) and mixing it with concentrated ammonia (35%; w/v, NH$_3$, sp.gr. 0.88). The working solution is prepared from the stock solution (100 ml) by addition of concentrated ammonia solution (250 ml) and water (650 ml). Before use, the reagent is filtered through a 0.2 μm pore size glass fibre filter.

Procedure: By means of a suitable post-column pump, delivering reagent at the same rate as the column eluate, the two reactants are mixed via a Y-shaped joint at the column exit. The reaction is immediate. The absorbance is measured spectrophotometrically at 280–310 nm. Although measurements at the lower wavelengths give the highest sensitivity, higher background absorbances are encountered. At 310 nm the sensitivity of the method is 2.5 nmol of glucose. The technique, which does not require the use of corrosive reagents or elevated reaction temperatures, is not sensitive to changes in solvent composition.

Periodate-pentane 2,4-dione method
The method is based on the work of Honda *et al.* (1983a) and measures alditols.

Reagents:
1. Sodium metaperiodate (2.68 g) is dissolved in water (250 ml).
2. Ammonium acetate (37.5 g) and sodium thiosulphate pentahydrate (14.4 g) are dissolved in water (245 ml) and mixed with pentane 2,4-dione (5 ml).

Procedure: Column eluate (1 ml min^{-1}) and Reagent 1 (0.5 ml min^{-1}) are mixed via a Y-shaped joint and passed through a Teflon coil (10 m × 0.15 mm i.d.) at room temperature. This allows oxidation to proceed for about 80 s. Reagent 2 (0.5 ml min^{-1}) is introduced via another Y-shaped joint and the mixture is passed through a second Teflon coil (10 m × 0.5 mm i.d.) which is immersed in a glycerol bath thermostatted at 100°C. The reaction time for this stage is about 1 min. The product is measured spectrophotometrically at 412 nm where the sensitivity is about 2 nmol alditol. Alternatively fluorimetric measurements can be made at 410 nm (excitation) and 503 nm (emission).

Provided periodate oxidation takes place at room temperature, only minor interference arises from aldoses.

2-Cyanoacetamide method
The method is based on the work of Honda et al. (1981a) and measures aldoses, amino sugars (Honda et al., 1983c) and uronic acids (Honda et al., 1983b).

Reagents:
1. 2-Cyanoacetamide (10 g) previously decolorised with activated charcoal and re-crystallised from methanol, is dissolved in water (100 ml). The solution is stable for at least four weeks if stored in a dark bottle at 4°C.
2. Boric acid (15.5 g) dissolved in water (300 ml) is adjusted to pH 8.5 by addition of potassium hydroxide (2 M) and diluted with water to a final volume of 500 ml (0.5 M borate).

Procedure: Neutral monosaccharides
Column eluate (0.6 ml min^{-1}) is mixed successively with Reagent 1 (0.5 ml min^{-1}) and Reagent 2 (0.5 ml min^{-1}) via Teflon Y-shaped joints. The mixture is then passed through a Teflon reaction coil (10 m × 0.5 mm i.d.) contained in a thermostatted glycerol bath at 100 ± 1°C followed by passage through an air-cooled Teflon coil (1 m × 0.5 mm i.d.).

The absorbance is then measured at 280 nm, where the detection limit is 0.2 nmol of reducing sugar. Alternatively, fluorimetric measurements can be made at 331 nm (excitation) and 383 nm (emission) where the limit of sensitivity is 0.05 nmol. If acetonitrile is used in the elution system from the analytical column, the fluorescence cannot be measured due to quenching by the solvent.

Procedure: Uronic acids
If the uronic acids have been separated on a quaternary ammonium-type resin as the borate complexes with 1.3 M borate, Reagent 2 is omitted and Reagent 1 is 1% 2-cyanoacetamide. Column eluate (1.0 ml min^{-1} is mixed with Reagent 1 through a Y-shaped connector at 0.5 ml min^{-1} before reacting in an identical manner to that described for neutral monosaccharides.

Procedure: Amino sugars
If the amino sugars have been separated on polystyrene sulphonate columns in the presence of borate buffer at pH 7.5, Reagent 2 is made 0.6 M with respect to borate and the procedure outlined for neutral monosaccharides is used.

(iii)   *Pre-column derivatisation.* Whilst pre-column derivatisation is used to increase the sensitivity of detection of carbohydrates, it does not always give improved separation of the individual components in a mixture compared with that for underivatised components. Some of the methods reported for pre-column derivatisation do not produce the desired chromophoric derivatives in high yields. In selecting a method for pre-column derivatisation, it is important to ensure that the chosen method will produce only one derivative. For example, some earlier methods used to perbenzoylate glucose produced a series of partially benzoylated esters, each one of which will give a separate peak on HPLC. Pre-column derivatisation has been used to prepare a wide range of derivatives including benzoates (Oshima and Kumanoti, 1983), peracetates (Thiem *et al.*, 1978), phenylisocyanates (Bjorkqvist, 1981) and pyridylamino compounds (Takemoto *et al.*, 1985). Separations may be achieved on either derivatised or underivatised silica columns. Two of these methods are described below.

Perbenzoates
The method is based on the work of Jentoft (1985) and is used for the determination of reducing sugars. Monosaccharides may be directly perbenzoylated, or first converted to the methylglycosides prior to perbenzoylation.

Reagents:
1. Methanolic hydrogen chloride (1 M) is prepared by adding acetyl chloride (2.3 ml) to anhydrous methanol (47.3 ml).
2. Benzoic anhydride (1 g) and dimethylaminopyridine (0.5 g) are dissolved in pyridine.

Procedure: Preparation of methylglycosides. Samples, containing up to 1 µmol total sugar in a 1 ml vial are dried in a stream of dry nitrogen. Reagent 1 (0.2 ml) is added to the vials which are then closed and heated at 80°C for 4 h. On cooling, methanolic *t*-butanol (80%; v/v, 0.1 ml) is added and the products are again taken to dryness. If amino sugars are present in the samples, they are re-*N*-acetylated by addition of dry methanol (0.1 ml), pyridine (0.04 ml) and acetic anhydride (0.4 ml). The sealed vials are kept at room temperature for 1 h before taking the samples to dryness over a stream of nitrogen. Toluene (0.04 ml) is added and the samples are again taken to dryness.

Procedure: Perbenzoylation. Samples of monosaccharides, prepared as above, or methylglycosides are added to silanised vials. Reagent 2 (0.1 ml) is added. The vials are capped and incubated at 37°C for 2 h. Excess benzoic anhydride is destroyed by the addition of water (0.9 ml) and incubation at room temperature for 30 min. Benzoylated sugars are separated from reaction by-products by applying the reaction mixture to a small reverse phase extraction column inserted into a vacuum manifold. The columns are washed with water (92 ml) and the derivatives are eluted into silanised tubes by addition of acetonitrile (2 ml). Samples are dried in a stream of nitrogen and redissolved in a suitable volume of acetonitrile, prior to HPLC analysis, monitoring the eluate with a UV detector (230–240 nm). The method is suitable for detecting 1–10 nmol monosaccharide. It is not used for detecting aldonitriles containing five carbons or more.

Pyridylamino derivatives
The method is based on the work of Takemoto *et al.* (1985) and is used for the determination of neutral and amino sugars.

Reagents:
1. 2-Aminopyridine (0.5 g) and concentrated hydrochloric acid (0.4 ml) are added to water (11 ml).

2. Sodium cyanoborohydride (0.02 g) is dissolved in water (1 ml).

Procedure: If amino sugars are present in the samples they are re-*N*-acetylated. Samples are added to tapered, screw-capped vials, and taken to dryness over a stream of nitrogen. Reagent 1 (0.005 ml) is added and the sealed vials are heated at 100°C for 15 min. The vials are opened and Reagent 2 (0.002 ml) is added before the vials are re-sealed and heated at 90°C for 8 h. The reactants are diluted with water (0.02 ml), and the entire contents of vials are transferred to columns of TSK-GEL G2000, equilibrated in ammonium acetate, 0.02 M, pH 7.5. The excess reagents are eluted from the column with this buffer at a flow rate of 0.5 ml min^{-1}. The sugar derivatives are eluted with the same buffer, and taken to dryness before being redissolved in a suitable volume of water prior to chromatography.

REFERENCES

Adachi, S. (1965). *J. Chromatogr.* **17**, 295–299.
Albersheim, P., Nevins, D. J., English, P. D. and Karr, A. (1967). *Carbohydr. Res.* **5**, 340–345.
Amaral, D., Kelly-Falcoz, F. and Horecker, B. L. (1966). *Methods Enzymol.* **9**, 87–92.
Ameyama, M., Shinagawa, E., Matsushita, K. and Adachi, O. (1981). *J. Bacteriol.* **145**, 814–823.
Aspinall, G. O., Hirst, E. L., Percival, E. G. V. and Telfer, R. G. J. (1953). *J. Chem. Soc.* 337–342.
Avigad, G., Asensio, C., Amaral, D. and Horecker, B. L. (1961). *Biochem. Biophys. Res. Commun.* **4**, 474–477.
Avigad, G., Amaral, D., Asensio, C. and Horecker, B. L. (1962). *J. Biol. Chem.* **237**, 2736–2743.
Baker, J. O. and Himmel, M. E. (1986). *J. Chromatogr.* **357**, 161–182.
Bamforth, C. W. (1983). *J. Inst. Brew.* **89**, 391–392.
Batley, M., Redmond, J. W. and Tseng, A. (1982). *J. Chromatogr.* **253**, 124–128.
Bernfeld, P. (1955). *Methods Enzymol.* **1**, 149–158.
Bhatti, R., Chambers, R. E. and Clamp, J. R. (1970). *Biochim. Biophys. Acta* **222**, 339–347.
Bitter, T. and Muir, H. M. (1962). *Anal. Chem.* **4**, 330–334.
Bjorkqvist, B. (1981). *J. Chromatogr.* **218**, 65–71.
Blakeney, A. B., Harris, P. J., Henry, R. J., Stone, B. A. and Norris, T. (1982). *J. Chromatogr.* **249**, 180–182.
Blakeney, A. B., Harris, P. J., Henry, R. J. and Stone, B. A. (1983). *Carbohydr. Res.* **113**, 291–299.
Blaschek, W. (1983). *J. Chromatogr.* **256**, 157–163.
Bradbury, A. G., Halliday, D. J. and Medcalf, D. G. (1981). *J. Chromatogr.* **213**, 146–150.
Brobst, K. M. and Lott, C. E. (1966). *Cereal Chem.* **43**, 35–43.
Brown, A. H. (1946). *Arch. Biochem.* **11**, 269–278.
Cairns, A. J. (1987). *Anal. Biochem.* **167**, 270–278.
Cayle, T., Viebrock, F. and Schiaffino, J. (1968). *Cereal Chem.* **45**, 154–161.
Chen, C. C. and McGinnis, G. D. (1983). *Carbohydr. Res.* **122**, 322–326.
Churms, S. C. (1982). In "CRC Handbook of Chromatography: Vol 1, Carbohydrates" (G. Zweig and J. Sherma, eds), pp. 1–272. CRC Press, Boca Raton, Florida.
Clamp, J. R., Bhatti, T. and Chambers, R. E. (1972). *In* "Glycoproteins, Part A", 2nd edn (A. Gottschalk, ed.), pp. 300–321. Elsevier, Amsterdam.
Codington, J. F., Linsley, K. B. and Silber, C. (1976). *Meth. Carbohydr. Chem.* **7**, 226–232.
Comparini, I. B., Centini, F. and Pariali, A. (1983). *J. Chromatogr.* **279**, 609–613.
Cowie, G. L. and Hedges, J. I. (1984). *Anal. Chem.* **56**, 497.
Dalqvist, A. (1961). *Biochem. J.* **80**, 547–551.
Dintzis, F. R. and Harris, C. C. (1981). *Cereal Chem.* **58**, 467–470.
Dische, Z. (1949). *J. Biol. Chem.* **181**, 379–392.
Dische, Z. and Borenfreund, E. (1951). *J. Biol. Chem.* **192**, 583–587.
Dische, Z. Shettles, L. B. and Osnos, M. (1949). *Arch. Biochem.* **22**, 169–184.

Dubois, M., Gilles, K. A., Hamilton, J. K., Rebers, P. A. and Smith, F. (1956). *Anal. Chem.* **28**, 350–356.
Dutton, G. G. S., Gibney, K. B., Reid, P. E. and Slessor, K. N. (1985). *J. Chromatogr.* **20**, 163–165.
Eklund, G., Josefsson, B. and Roos, C. (1977). *J. Chromatogr.* **142**, 575–585.
Elson, L. A. and Morgan, W. T. J. (1933). *Biochem. J.* **27**, 1824–1828.
Ertel, H. and Horner, L. (1962). *J. Chromatogr.* **7**, 268–269.
Folkes, D. J. and Taylor, P. W. (1982). *In* "HPLC in Food Analysis" (R. Macrae, ed.), pp. 149–166. Academic Press, London.
Ghebregzabher, M., Rufini, S., Ciuffini, G. and Lato, M. (1974). *J. Chromatogr.* **95**, 51–58.
Ghebregzabher, M., Rufini, S., Monaldi, B. and Lato, M. (1976). *J. Chromatogr.* **127**, 133–162.
Gisch, D. J. and Pearson, J. D. (1988). *J. Chromatogr.* **443**, 299–308.
Golik, J., Liu, H. W., Dinovi, M., Furukawa, J. and Nakanishi, K. (1983). *Carbohydr. Res.* **118**, 135–146.
Green, C., Doctor, V. M., Holzer, G. and Oro, J. (1981). *J. Chromatogr.* **207**, 268–272.
Grimble, G. K., Barker, H. M. and Taylor, R. H. (1983). *Anal. Biochem.* **128**, 422–428.
Gunner, S. W., Jones, J. K. N. and Perry, M. B. (1961). *Can. J. Chem.* **39**, 1892–1899.
Haldorsen, K. M. (1977). *J. Chromatogr.* **134**, 467–476.
Hansen, S. A. (1975). *J. Chromatogr.* **107**, 224–226.
Hobbs, J. S. and Lawrence, J. G. (1972). *J. Sci. Food Agric.* **23**, 45–51.
Honda, S. (1984). *Anal. Biochem.* **146**, 47.
Honda, S., Matsuda, Y., Takahashi, M., Kakehi, K. and Ganno, S. (1980). *Anal. Chem.* **52**, 1079–1082.
Honda, S., Takahashi, M., Kakehi, K. and Ganno, S. (1981a). *Anal. Biochem.* **113**, 130–138.
Honda, S., Takahashi, M., Nishimura, Y., Kakehi, K. and Ganno, S. (1981b). *Anal. Biochem.* **118**, 162–167.
Honda, S., Takahashi, M., Shimada, S., Kakehi, K. and Ganno, S. (1983a). *Anal. Biochem.* **128**, 429–437.
Honda, S., Suzuki, S., Takahashi, M., Kakehi, K. and Ganno, S. (1983b). *Anal. Biochem.* **134**, 34–39.
Honda, S., Konishi, T., Suzuki, S., Takahashi, M., Kakehi, K. and Ganno, S. (1983c). *Anal. Biochem.* **134**, 483–488.
Honda, S., Suzuki, S. and Kakehi, K. (1984). *J. Chromatogr.* **291**, 317–325.
Hough, L. and Jones, J. K. N. (1962). *Methods Carbohydr. Chem.* **1**, 21–31.
Hough, L., Jones, J. V. S. and Wusteman, P. (1972). *Carbohydr. Res.* **21**, 9–17.
Hough, L., Ko, A. M. Y. and Wusteman, P. (1975). *Carbohydr. Res.* **44**, 97–100.
Jarvis, M. C. and Duncan, H. J. (1974). *J. Chromatogr.* **92**, 454–456.
Jentoft, N. (1985). *Anal. Biochem.* **148**, 424–433.
Karamanos, N. K., Tegenidis, T. and Antonopoulos, C. A. (1987). *J. Chromatogr.* **405**, 221–228.
Kato, T. and Kinoshita, T. (1980). *Anal. Biochem.* **106**, 238–243.
Kennedy, J. F. and Fox, J. E. (1980). *Methods Carbohydr. Chem.* **8**, 3–12.
Khym, J. X. and Zill, L. P. (1951). *J. Am. Chem. Soc.* **73**, 2399–2400.
Khym, J. X. and Zill, L. P. (1952). *J. Am. Chem. Soc.* **74**, 2090–2094.
Kircher, H. W. (1960). *Anal. Chem.* **32**, 1103–1106.
Krammer, K. J., Spiers, R. D. and Childs, C. N. (1978). *Anal. Biochem.* **86**, 692–696.
Ladisch, M. R., Huebner, A. L. and Tsao, G. T. (1978). *J. Chromatogr.* **147**, 185–193.
Laine, R. A. and Sweeley, C. C. (1973). *Carbohydr. Res.* **27**, 199–213.
Li B. W. and Andrews, K. N. (1986). *Chromatographia* **21**, 596–598.
Lomax, J. A. and Conchie, J. (1982). *J. Chromatogr.* **236**, 385–394.
Macrae, R. and Dick, J. (1981). *J. Chromatogr.* **210**, 138–145.
McGinnis, G. D. and Fang, P. (1980). *Methods Carbohydr. Chem.* **8**, 33–43.
Manius, G. J., Lui, T. M. Y. and Wen, L. F. L. (1979). *Anal. Biochem.* **99**, 365–371.
Martin, H. L. and Bamforth, C. W. (1981). *J. Inst. Brew.* **87**, 88–91.
Melrose, J. and Sturgeon, R. J. (1983). *Carbohydr. Res.* **118**, 247–253.
Mezzetti, T., Lato, M., Rufini, S. and Ciaffini, G. (1971). *J. Chromatogr.* **63**, 329–342.
Mopper, K. (1978a). *Anal. Biochem.* **86**, 597–601.

Mopper, K. (1978b,. *Anal. Biochem.* **87**, 162–168.

Mopper, K. and Johnson, L. (1983). *J. Chromatogr.* **256**, 27–38.

Mopper, K., Dawson, R., Liebezeit, G. and Hansen, H. F. (1980). *Anal. Chem.* **52**, 2018–2022.

Moriyasu, M., Kato, A., Okada, H. and Hashimoto, Y. (1984). *Anal. Lett.* **17**, 689–699.

Morgan, W. T. J. and Elson, L. A. (1934). *Biochem. J.* **28**, 988–995.

Morrison, I. M. and Perry, M. B. (1966). *Can. J. Biochem.* **44**, 1115–1126.

Murphy, D. and Pennock, C. A. (1972). *Clin. Chim. Acta* **42**, 67–75.

Nachtmann, F. and Budna, K. W. (1977). *J. Chromatogr.* **136**, 279–287.

Nakamura, T., Maruki, S., Nakatsu, S. and Ueda, S. (1978). *J. Agric. Chem. Soc. Japan* **52**, 581–587.

Nelson, N. (1944). *J. Biol. Chem.* **153**, 375–380.

Nordin, P. (1983). *Anal. Biochem.* **131**, 492–498.

Ochiai, M. (1980). *J. Chromatogr.* **194**, 224–227.

Okuda, J., Miwa, I., Maeda, K. and Tokui, K. (1977). *Carbohydr. Res.* **58**, 267–270.

Oshima, R. and Kumanotani, J. (1983). *J. Chromatogr.* **265**, 335–341.

Oshima, R., Yoshikawa, A., Kumanotani, J. (1981). *J. Chromatogr.* **213**, 142–145.

Oshima, R., Kumanotani, J. and Watanabe, C. (1982). *J. Chromatogr.* **250**, 90–95.

Ovodov, Y. S., Evtushenko, E. V., Vaskovsky, V. E., Ovodova, R. G. and Soloveva, T. F. (1967). *J. Chromatogr.* **26**, 111–115.

Owens, J. A. and Robinson, J. S. (1985). *J. Chromatogr.* **338**, 303–314.

Partridge, S. M. (1947). *Nature (London)* **158**, 270–271.

Partridge, S. M. (1948). *Biochem. J.* **42**, 238–250.

Pastuska, G. (1961). *Z. Anal. Chem.* **179**, 427–429.

Pelletier, O. and Cadieux, S. (1982). *J. Chromatogr.* **231**, 225–235.

Perry, M. B. and Hulyalkar, R. K. (1965). *Can. J. Biochem.* **43**, 573–584.

Petersson, G. (1974). *Carbohydr. Res.* **33**, 47–61.

Petre, R., Dennis, R., Jackson, B. P. and Jethwa, K. R. (1972). *Planta Med.* **21**, 81–88.

Pfaffenberger, C. D., Szafranek, J. and Horning, E. C. (1975). *Anal. Biochem.* **63**, 501–512.

Pittner, F. and Turecek, P. L. (1987). *Appl. Biochem. Biotechnol.* **116**, 15–24.

Price, C. P. and Spencer, K. (1979). *Ann. Clin. Biochem.* **16**, 100–105.

Raadsveld, C. W. and Klomp, H. (1971). *J. Chromatogr.* **57**, 99–106.

Rabel, F. M., Caputo, A. G. and Butts, E. T. (1976). *J. Chromatogr.* **126**, 731–740.

Radhakrishnamurthy, B. and Berenson, G. S. (1963). *Arch. Biochem. Biophys.* **101**, 360–362.

Reissig, J. L., Strominger, J. L. and Leloir, L. F. (1955). *J. Biol. Chem.* **217**, 959–966.

Robyt, J. F. (1975). *Carbohydr. Res.* **40**, 373–374.

Robyt, J. F. and Whelan, W. J. (1968). In "Starch and its Derivatives", 4th edn, pp. 432–470. Chapman and Hall, London.

Rondle, C. J. M. and Morgan, W. T. J. (1955). *Biochem. J.* **61**, 586–589.

Roth, H., Segal, S. and Bertoli, D. (1965). *Anal. Biochem.* **10**, 32–52.

Sawardeker, J. S., Sloneker, J. H. and Jeanes, A. (1965). *Anal. Chem.* **37**, 1602–1604.

Schachter, H. (1975). *Methods Enzymol.* **41**, 3–10.

Schaffler, K. J. and Morel du Boil, P. G. (1981). *J. Chromatogr.* **207**, 221–229.

Schwer, H. (1982). *J. Chromatogr.* **236**, 355–360.

Simatupang, M. H. (1979). *J. Chromatogr.* **180**, 177–183.

Sinner, M. and Puls, J. (1978). *J. Chromatogr.* **156**, 197–204.

Slabach, T. D. and Robinson, J. (1983). *J. Chromatogr.* **282**, 169–177.

Sloneker, J. H. (1972). *Methods Carbohydr. Chem.* **6**, 20–24.

Smith, D. F., Zopf, D. A. and Ginsburg, V. (1978). *Anal. Biochem.* **85**, 602–608.

Somani, B. L., Khanade, J. and Sinha, R. (1987). *Anal. Biochem.* **167**, 327–330.

Somogyi, M. (1952). *J. Biol. Chem.* **153**, 375–380.

Spiro, R. G. (1973). *Methods Enzymol.* **28B**, 3–43.

Stahl, E. and Kaltenbach, U. (1961). *J. Chromatogr.* **5**, 351–355.

Sturgeon, R. J. (1980a). *Methods Carbohydr. Chem.* **8**, 135–137.

Sturgeon, R. J. (1980b). *Methods Carbohydr. Chem.* **8**, 131–133.

Sturgeon, R. J. (1984). In "Methods of Enzymatic Analysis", 3rd edn (H. U. Bergmeyer, ed.), Vol. VI, pp. 427–431. Verlag Chemie, Weinheim.

Sturgeon, R. J. (1990). *Carbohydr. Res.* **200**, (in press).

Sugimura, M. and Hirano, K. (1977). *Clin. Chim. Acta* **75**, 387–391.

Svennerholm, L. (1956). *J. Neurochem.* **1**, 42–53.

Sweeley, C. C., Bentley, R., Makita, M. and Wells, W. W. (1963). *J. Am. Chem. Soc.* **85**, 2497–2507.

Takemoto, H., Nase, S. and Ikenata, T. (1985). *Anal. Biochem.* **145**, 245–250.

Thiem, J., Schwentner, J., Karl, H., Siervers, A. and Reimer, J. (1978). *J. Chromatogr.* **155**, 107–118.

Tomita, K., Kamei, S., Nagata, K., Okuno, H., Shiraishi, T., Motoyama, A., Ohkubo, A. and Yamanaka, M. (1987). *J. Clin. Biochem. Nutr.* **3**, 11–16.

Trevelyan, W. E., Procter, D. P. and Harrison, J. S. (1950). *Nature (London)* **166**, 444–445.

Trinder, P. (1969). *Ann. Clin. Biochem.* **6**, 24–27.

Uobe, K., Nishida, K., Inoue, H. and Tsutsui, M. (1980). *J. Chromatogr.* **193**, 83–88.

Van Handel, E. (1967). *Anal. Biochem.* **19**, 193–194.

Van Olst, H. and Joosten, G. E. (1979). *J. Liquid Chromatogr.* **2**, 111–115.

Van Riel, J. A. M. and Olieman, C. (1986). *J. Chromatogr.* **362**, 235–242.

Verhaar, L. A. T. and Kuster, B. F. M. (1981). *J. Chromatogr.* **220**, 313–328.

Waffenschmidt, S. and Jaenicke, L. (1987). *Anal. Biochem.* **165**, 337–340.

Weiss, J. B. and Smith, I. (1967). *Nature* **215**, 638.

White, C. A., Vass, S. W., Kennedy, J. K. and Large, D. G. (1983a). *J. Chromatogr.* **264**, 99–109.

White, C. A., Vass, S. W., Kennedy, J. K. and Large, D. G. (1983b). *Carbohydr. Res.* **119**, 241–247.

Winzler, R. J. (1955). *Methods Biochem. Analysis*, **2**, 279–311.

Zanetta, J. P., Breckenridge, W. C. and Vincendon, G. (1972). *J. Chromatogr.* **69**, 291–304.

2 Nucleotide Sugars

DAVID S. FEINGOLD[1] and GEORGE A. BARBER[2]

[1]*Department of Microbiology, Biochemistry and Molecular Biology, University of Pittsburgh, School of Medicine, Pittsburgh, PA 15261, USA*

[2]*Department of Biochemistry, Ohio State University, Columbus, OH 43210, USA*

METHODS IN PLANT BIOCHEMISTRY Vol. 2
ISBN 0-12-461012-9

I. INTRODUCTION

It is now apparent that nucleotide sugars are involved in biosynthesis of the majority of complex saccharides, including those found in plants. In this chapter we shall describe methods of isolation, characterisation, and preparation of typical classes of nucleotide sugars. Although compounds found in plants will be emphasised, the treatment will not be restricted to the Plantae, and examples from work done with other organisms will also be presented as appropriate.

II. HISTORICAL

The most abundant constituents of plants are carbohydrates. They can represent up to 90% of plant dry weight and fulfil a multitude of functions. Polysaccharides (e.g. cellulose, hemicellulose, mannan) act as structural elements, while others are involved in intracellular binding (pectin), wound sealing (callose, certain plant gums and mucilages), or energy storage (starch and the disaccharide sucrose). Carbohydrates also contribute to the solubilisation of otherwise hydrophobic compounds like those present in such glycosides as digitoxin (e.g. cardiac glycosides).

Plant carbohydrates are composed of a large variety of individual monosaccharide units, comprising among others aldopentoses, aldo- and ketohexoses, 6-deoxyaldohexoses, dideoxyaldohexoses, amino sugars, uronic acids, and branched chain sugars. By far the most abundant monosaccharide moiety is D-glucose, since it represents the sole building block of cellulose, callose, starch, and one half of the sucrose molecule.

The first *in vitro* synthesis of a polysaccharide (glycogen) was accomplished by Cori *et al.* (1939), with the glycogen phosphorylase-catalysed transfer of D-glucosyl moieties from α-D-glucosyl phosphate (Glc-1-P) to a glycogen acceptor. Shortly thereafter Hanes (1940) was able to demonstrate *in vitro* starch synthesis using Glc-1-P and extracts from potato. Although these were epoch making discoveries, subsequent work revealed that *in vivo* glycogen phosphorylase and starch phosphorylase effected the degradation rather than the biosynthesis of glycogen or starch. Nonetheless, these observations established the principle of polymer synthesis by glycosyl transfer from precursors with high group transfer potential and provided a theoretical basis for continued investigation. The group transfer concept was strengthened by the demonstration of the ability of sucrose to serve as a D-glucosyl donor in dextran synthesis catalysed by an enzyme from the Gram-positive coccus *Leuconostoc mesenteroides* (Hehre and Sugg, 1942) and also as a D-fructosyl donor in the biosynthesis of levan catalysed by an enzyme from a Gram-negative rod (Hestrin *et al.*, 1943). The synthesis of sucrose by an enzyme preparation (sucrose phosphorylase) from the Gram-negative bacillus *Pseudomonas saccharophila* (Wolochow *et al.*, 1949), was shown to occur by transfer of a D-glucosyl moiety from α-D-glucosyl phosphate to the acceptor D-fructose. However, no evidence for the presence of similar activity could be found in plants (W. Z. Hassid, pers. commun.). Thus, despite the ability of biochemists to achieve the *in vitro* synthesis of certain oligo- and polysaccharides, until the 1950s mechanisms for the biosynthesis of complex saccharides remained a mystery. In addition, little was known concerning the origin of the various types of sugars which form the building blocks of the complex saccharides of plants.

On the basis of early work which showed the presence of structurally related hexosyl, hexuronosyl, and pentosyl moieties in the same or closely associated polysaccharides, e.g. cellulose and xylan, it was proposed that pentosans such as xylan were formed from hexosan precursors via oxidation of the hexosyl moieties to uronosyl residues followed by decarboxylation of the latter to yield pentose polymers. Hirst (1942) pointed out that since differences of ring forms and anomeric linkages between the putative precursors and products were inconsistent with oxidation and decarboxylation of monosaccharide residues linked in a polymer, C-6 oxidation of hexoses and subsequent decarboxylation of the hexuronic acids might occur at the monosaccharide level. The resulting monomers would then be assembled into polysaccharides by some unspecified mechanism. However, this proposal provided no insight into *how* the carbohydrate interconversions might occur or what processes were involved in polymer formation.

Such insight came mainly from the work of a small group of researchers in Buenos Aires under the leadership of the late Luis Leloir who were investigating the conversion of a α-D-galactopyranosyl phosphate (Gal-1-P) to D-glucose 6-phosphate (Glc-6-P) by extracts of the lactose-utilising yeast *Saccharomyces fragilis*. After showing that the cofactor for the Glc-6-P → Glc-1-P interconversion is D-glucose 1,6-bisphosphate, these workers turned their attention to the formation of Glc-1-P from Gal-1-P. In a note published in February 1950 in *Nature* it was proposed that a cofactor in the reaction was uridine 5′-(α-D-glucopyranosyl pyrophosphate) (UDP-Glc) (Cardini *et al.*, 1950). Later that same year a full paper describing the isolation and complete characterisation of UDP-Glc appeared (Caputto *et al.*, 1950). [Interestingly, Park and Johnson (1949) had reported that addition of penicillin to a culture of *Staphylococcus aureus* caused formation of a number of acid-labile uridine-containing compounds later identified as uridine 5′-(α-2-acetamido, 2-deoxy-D-muramyl pyrophosphate) derivatives (Park, 1952a,b,c).] The role of these compounds in the biosynthesis of bacterial peptidoglycan is now well understood (Strominger, 1959).

When UDP-Glc was incubated with extracts from *S. fragilis*, it was reversibly converted to uridine 5′-(α-D-galactopyranosyl pyrophosphate) (UDP-Gal) (Leloir, 1951). The conversion of Gal-1-P to Glc-6-P was proposed to occur as follows:

$$Gal\text{-}1\text{-}P + UDP\text{-}Glc \rightleftharpoons Glc\text{-}1\text{-}P + UDP\text{-}Gal$$

$$UDP\text{-}Gal \rightleftharpoons UDP\text{-}Glc$$

$$Glc\text{-}1\text{-}P \rightleftharpoons Glc\text{-}6\text{-}P$$

$$\text{Sum} \qquad Gal\text{-}1\text{-}P \rightleftharpoons Glc\text{-}6\text{-}P$$

The first reaction, surmised but not specifically demonstrated by Leloir, was described later by Munch-Petersen *et al.* (1953), who also were the first to show the *de novo* formation of UDP-Glc from UTP and Glc-1-P. This reaction: UTP + Glc-1-P → UDP-Glc + PPi, catalysed by an enzyme (UDP-glucose pyrophosphorylase, EC 2.7.7.9) present in *S. fragilis*, is the prototypical reaction for the *de novo* formation of sugar nucleotides. The second reaction in the sequence, catalysed by UDP-glucose 4-epimerase (EC 5.1.3.2), involves the interconversion of UDP-Glc and UDP-Gal. The D-glucosyl and D-galactosyl moieties of UDP-Glc and UDP-Gal differ only in the configuration at C-4; the hydroxyl group is equatorial in the D-*gluco* and axial in the D-*galacto* configuration. An important clue to the reaction mechanism was provided by

Maxwell (1957) who showed a requirement for NAD^+. It is now known that the epimerisation occurs *via* reversible NAD^+-linked oxido reduction at C-4 of the glycosyl moiety (Feingold, 1982). Demonstration that NAD^+ is necessary for 4-epimerisation had far-reaching implications since, as subsequent work has shown, most modifications of the structure of the glycosyl moieties of sugar nucleotides involve the participation of pyridine nucleotide-linked oxidoreductions.

Buchanan *et al.* (1952) and Buchanan (1953) noted that during the photosynthetic conversion of $^{14}CO_2$ to carbohydrate, UDP-Glc was labelled more rapidly than sucrose phosphate, another of the compounds isolated from the photosynthesis reaction mixture. These workers suggested that 'compounds of the UDPG (UDP-Glc) type could be concerned in the formation of sugars and their subsequent incorporation into polysaccharides'. This view of the role of sugar nucleotides as glycosyl donors was confirmed by Dutton and Storey (1953) who demonstrated that uridine 5'-(α-D-glycopyranosyluronic acid pyrophosphate) (UDP-GlcA) is required for the formation of glucuronides catalysed by extracts of liver. Hard on the heels of this announcement Leloir and Cabib (1953) described the synthesis of trehalose 6-phosphate by transfer of the D-glucosyl moiety of UDP-Glc to D-glucose 6-phosphate catalysed by an extract from yeast. The synthesis of sucrose itself was conclusively demonstrated by Cardini *et al.* (1955) with the report that wheat germ extracts catalyse the synthesis of sucrose (or sucrose 6-phosphate) from UDP-Glc and D-fructose (or D-fructose 6-P). The isolation of guanosine 5'-(α-D-mannopyranosyl pyrophosphate) (GDP-Man) from yeast (Cabib and Leloir, 1954) and of UDP derivatives of amino sugars, e.g. uridine 5'-(2-acetamido-2-deoxy-α-D-glucopyranosyl pyrophosphate) (UDP-GlcNAc) (Cabib *et al.*, 1952), further established the central position of sugar nucleotides in the formation and transfer of carbohydrate moieties. Glaser and Brown (1955) showed that UDP-GlcNAc could function as a glycosyl donor in the synthesis of hyaluronate, catalysed by extracts of Rous sarcoma. The same sugar nucleotide was the sole substrate in chitin synthesis catalysed by extracts from *Neurospora crassa* (Glaser and Brown, 1957). Niemyer (1955) suggested that UDP-Glc rather than Glc-1-P was the true glucosyl donor in the synthesis of glycogen; this was conclusively shown by Leloir *et al.* (1959). The first demonstration of the participation of UDP-Glc in polymer synthesis in plants was the synthesis of callose (β-1,3-D-glucan) catalysed by preparations from mung bean seedlings and other plants (Feingold *et al.*, 1958b).

In the meantime Hirst's (1942) hypothesis concerning the origin of the monosaccharide residues of plant polysaccharides was being investigated. Neish (1955) and Altermatt and Neish (1956) found that D-[^{14}C]glucose was converted by wheat plants to the D-xylosyl moieties of xylan with loss of C-6 and little randomisation of C-1 through C-5. Also, D-glucuronolactone was converted to D-xylosyl moieties by corn coleoptile with loss of C-6 (Slater and Beevers, 1958). Ginsburg *et al.* (1956) isolated UDP-Glc and uridine 5'-(α-D-xylopyranosyl pyrophosphate) (UDP-Xyl) from mung bean seedlings. Neufeld *et al.* (1957) showed that, in addition, the seedlings contained UDP-Gal and uridine 5'-(β-L-arabinopyranosyl pyrophosphate) (UDP-Ara) as well as enzymes capable of catalysing the 4-epimerisation of the glycosyl moieties of these sugar nucleotides. Further support for the role of sugar nucleotides in the formation of sugar moieties was provided by Solms and Hassid, (1957) and Solms *et al.* (1957), who demonstrated that the seedlings also contained UDP-GlcA. This compound is formed in plants either by the NAD^+-linked oxidation of UDP-Glc (Strominger and Mapson,

1957) or via pyrophosphorolysis from α-D-glucopyranosyl uronic acid phosphate (GlcA-1-P) and UTP (Solms *et al.*, 1957; Feingold *et al.*, 1958a). UDP-GlcA was shown to be converted to a mixture of uridine 5′-(α-D-galactopyranosyluronic acid pyrophosphate) (UDP-GalA), UDP-Xyl, and UDP-Ara by extracts frum mung bean seedlings (Neufeld *et al.*, 1958). These findings supplied actual compounds and reactions to substantiate, at least partially, Hirst's (1942) revision of the older oxidation–decarboxylation hypothesis for the biosynthesis of plant polysaccharides.

The early findings were followed by an ever increasing flood of reasearch. Nucleotide sugars were isolated from a large variety of organisms, both prokaryotic and eukaryotic, and their biosynthesis was explored in detail. Their function as donors of carbohydrate moieties in the biosynthesis of oligosaccharides, polysaccharides, proteoglycans, glycolipids, and a large variety of simple and complex glycosides of plants, has now been amply demonstrated. The high water-mark of nucleotide sugar research was reached in the late 1970s, and the number of new publications concerned with these compounds has been constantly decreasing. Since they are connected with such a large variety of metabolic processes, nucleotide sugars have now become common laboratory reagents and are used almost daily in many investigations. Nonetheless, in the realm of plant biochemistry some intriguing unsolved problems involving nucleotide sugars remain. Perhaps the most interesting and enigmatic of these is the biosynthesis of the L-arabinofuranosyl moiety. This structure is widely distributed among plant polysaccharides, yet its mode of formation remains a complete mystery.

Another problem of considerable significance, at least quantitatively, concerns the synthesis of cellulose and other polysaccharides of the plant cell wall. These are all presumably derived from nucleotide sugar precursors, but the details of the overall processes are still surprisingly obscure. Attempts to solve these and similar problems will doubtless occupy plant biochemists and physiologists for some time to come.

III. SUMMARY OF THE BIOSYNTHESIS OF NUCLEOTIDE SUGARS IN HIGHER PLANTS

Current knowledge of the biosynthesis of sugar nucleotides in higher plants is summarised in Fig. 2.1. Many of the compounds and reactions depicted have been demonstrated in other organisms, both prokaryotic and eukaryotic. Some reactions, like the sucrose biosynthesis and utilisation sequences, are found uniquely in plants. Yet another series of reactions confined to the plant world is the so-called inositol oxidation pathway. This pathway (*myo*-inositol → D-glucuronate → GlcA-1-P → UDP-GlcA), which permits the plant to by-pass UDP-Glc in the synthesis of UDP-GlcA, contributes significantly to cell wall uronide and pentose in many plant tissues (Loewus and Dickinson, 1982). Certain other reactions outlined in Fig. 2.1, such as the formation of UDP-D-apiose from UDP-GlcA, are also found only in plants; the branched chain aldopentose D-apiose is solely a plant produce.

Reactions of nucleotide sugars analogous to those which occur in plants are found throughout the living world; in many instances, work with plant systems has pointed the way to similar or identical biosynthetic pathways in mammalian tissues. For example, UDP-Xyl, the donor of the protein–polysaccharide linking residue in the biosynthesis of a variety of proteoglycans of animals, was first isolated from plants (Ginsburg *et al.*,

1956). UDP-Xyl synthesis by decarboxylation of UDP-GlcA likewise was demonstrated for the first time in plant tissues (Neufeld *et al.*, 1958); subsequently the enzyme involved, UDP-glucuronate decarboxylase (EC 4.1.1.35), was detected in the basidiomycete *Cryptococcus laurentii* (Ankel and Feingold, 1966), bacteria (Fan and Feingold, 1972) and various animal tissues (Bdolah and Feingold, 1965; Castellani *et al.*, 1967; Silbert and Deluca, 1967; John *et al.*, 1977).

Plants have proven to be a rich source of a nucleotide sugars. These compounds have been shown in every viable plant tissue hitherto investigated. An extensive, but not comprehensive, list of nucleotide sugars isolated from or identified in (many) higher plants is presented in Table 2.1. Consistent with their central position in plant carbohydrate metabolism (Fig. 2.1), uridine nucleotides were the most abundant nucleotide sugars in almost every tissue investigated, with UDP-Glc usually present in the highest concentration.

As shown in Fig. 2.1, many nucleotide sugars can be formed by the action of specific pyrophosphorylases. These enzymes catalyse the following general reaction type: (N is a nucleoside) NTP + glycopyranosyl phosphate $\rightleftharpoons$ NDP-glycopyranose + inorganic pyrophosphate (PPi). Actually pyrophosphorylases catalyse transfer of a nucleotidyl moiety between specific phosphate acceptors; during the reaction, pyrophosphate linkages in the nucleotide substrates are cleaved. Thus pyrophosphorylases are properly named NDP:α- (or β)-D-(or L)-glycopyranosyl phosphate nucleotidyl transferases.

The 'primary' nucleotide sugars (Kochetkov and Shibaev, 1973) adenosine 5'-(α-D-glucopyranosyl pyrophosphate) (ADP-Glc), thymidine 5'-(α-D-glucopyranosyl pyrophosphate) (TDP-Glc), UDP-Glc, UDP-GlcNAc and GDP-Man are synthesised by pyrophosphorolysis from the corresponding glycosyl phosphates. The latter are formed directly from D-fructose 6-phosphate produced by photosynthetic CO_2 fixation. 'Secondary' nucleotide sugars are formed by enzymatic modification of the glycosyl moieties of the primary compounds. They can also result from the action of specific pyrophosphorylases which participate in the so-called salvage pathways in plants. Monosaccharides released (e.g. during seed germination) by degradation of complex saccharides are salvaged by conversion to nucleotide sugars (e.g. UDP-GlcA, UDP-Gal) by the action of specific kinases and phosphorylases. The nucleotide sugars are then recycled into biosynthetic pathways. Nucleotide sugars in addition to those mentioned above which are substrates for plant pyrophosphorylases include: UDP-Gal, UDP-GlcA, UDP-GalA, UDP-Xyl, UDP-Ara, TDP-Glc, GDP-Man, adenosine 5'-(2-acetamido-2-deoxy-α-D-glucopyranosyl pyrophosphate) (ADP-GlcNAc), guanosine 5'-(β-L-fucopyranosyl pyrophosphate) (GDP-Fuc), cytidine 5'-(α-D-glucopyranosyl pyrophosphate) (CDP-Glc), and guanosine 5'-(α-D-glucopyranosyl pyrophosphate) (GDP-Glc).

FIG. 2.1 (Opposite) Reactions in plants which form nucleotide sugars. Reactions for which evidence is fragmentary or preliminary are shown by dashed lines. The direction in which the reaction is most likely to proceed under normal physiological conditions is indicated by the arrows. Free L-arabinose, D-galactose, D-glucose, D-glucuronate, D-galacturonate, and D-xylose result from the hydrolytic degradation of complex glycosides or sugar phosphates. D-glucuronate also is formed by oxidation of *myo*-inositol in the so-called 'inositol oxidation pathway'.

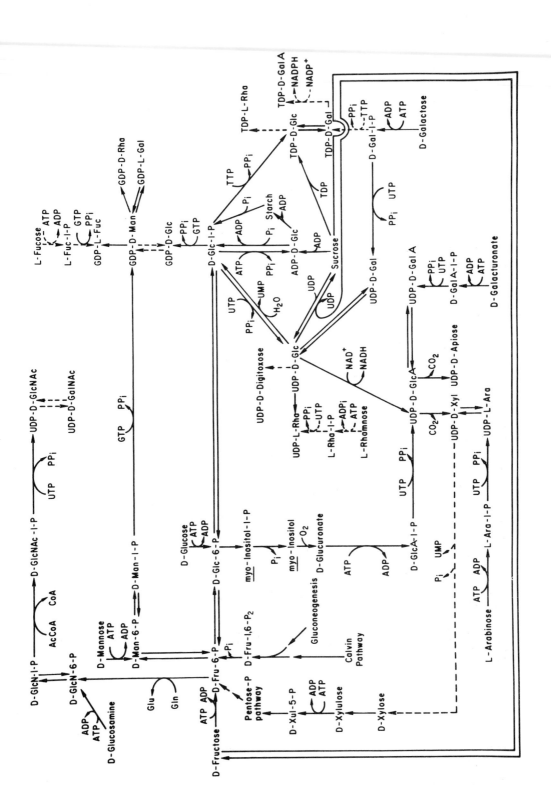

TABLE 2.1. Nucleotide sugars demonstrated in plants (Feingold and Avigad, 1980).

Sugar	Nucleotide	Plant
L-arabinose	UDP	Mung bean seedlings (Ginsburg *et al.*, 1956); brown algae (Lin and Hassid, 1966); pea seeds (Hampe and Gonzalez, 1967); parsley (Sanderman and Grisebach, 1968a,b); *Digitalis* leaves (Franz and Meier, 1969); cotton hair (Franz, 1969); *Chlorella* (Sanwal and Preiss, 1969); larch cambium and xylem (Cumming, 1970).
	ADP	Brown algae (Lin and Hassid, 1966); *Chlorella* (Sanwal and Preiss, 1969).
	GDP	Brown algae (Lin and Hassid, 1966).
D-ribose	ADP	Pea seeds (Brown, 1962); potato starch grains (Rees and Duncan, 1965; Cassells and Harmey, 1968); larch cambium (Cumming, 1970).
D-xylose	UDP	Mung bean seedlings (Ginsburg *et al.*, 1956; Grégoire *et al.*, 1965); sugar beet root (Katan and Avigad, 1965); pea seeds (Hampe and Gonzalez, 1967); parsley (Sanderman and Grisebach, 1968a,b); *Chlorella* (Sanwal and Preiss, 1969); *Digitalis* leaves (Franz and Meier, 1969); larch cambium and xylem (Cumming, 1970); corn roots (Roberts and Butt, 1970); strawberry leaves (Isherwood and Selvendran, 1970); bean leaves (Wallis and Bradbeer, 1970); wheat aleurone (Collins *et al.*, 1972); *Lemna minor* (Kindel and Watson, 1973).
	GDP	Strawberry leaves (Selvendran and Isherwood 1967; Isherwood and Selvendran, 1970).
D-galactose	UDP	Sugar beet leaves (Buchanan *et al.*, 1952); mung bean seedlings (Ginsburg *et al.*, 1956; Grégoire *et al.*, 1965); *Chlorella* (Buchanan *et al.*, 1952; Pakhomova *et al.*, 1965; Sanwal and Preiss, 1969); red algae (Su and Hassid, 1962a,b); pea seeds (Brown, 1962, 1965; Hampe and Gonzalez, 1967); corn grain (Dankert *et al.*, 1964b, 1972); sugar beet roots (Katan and Avigad, 1965); Jerusalem artichoke tubers (Taniguchi *et al.*, 1967); parsley (Sanderman and Grisebach, 1968a,b); *Digitalis* leaves (Franz and Meier, 1969); cotton hair (Franz, 1969); larch cambium and xylem (Cumming, 1970); strawberry leaves (Isherwood and Selvendran, 1970); fenugreek seeds (Sioufi *et al.*, 1970); barley coleoptiles and corn roots (Roberts *et al.*, 1971a); wheat aleurone (Collins *et al.*, 1972); sugar cane cells (Maretzki and Thom, 1978); hybrid *Triticale* grain (Sharma and Bhatia, 1978).
	ADP	Corn grain (Passeron *et al.*, 1964; Dankert *et al.*, 1964b); *Chlorella* (Pakhomova *et al.*, 1965); larch cambium and xylem (Cumming, 1970).
D-galactose	GDP	Strawberry leaves (Selvendran and Isherwood, 1967; Isherwood and Selvendran, 1970); *Chlorella* (Sanwal and Preiss, 1969); larch cambium (Cumming, 1970).

Continued

TABLE 2.1—continued.

Sugar	Nucleotide	Plant
D-glucose	UDP	Sugar beet leaves (Buchanan *et al.*, 1952); *Chlorella* (Buchanan *et al.*, 1952; Kauss and Kandler, 1962; Pakhomova *et al.*, 1965; Sanwal and Preiss, 1969); common beans (Sebesta and Sorm, 1959); mung bean seedlings (Ginsburg *et al.*, 1956; Grégoire *et al.*, 1965); wheat, barley and oat plants (Bergkvist, 1956, 1957); red algae (Su and Hassid, 1962a,b); pea seeds (Brown, 1962, 1965; Hampe and Gonzalez, 1967); wheat seedlings (Keys, 1963); rice grain (Murata *et al.*, 1964, 1966); sugar beet roots (Katan and Avigad, 1965); brown algae (Lin and Hassid, 1966); strawberry leaves (Selvendran and Isherwood, 1967; Isherwood and Selvendran, 1970); Jerusalem artichoke tubers (Taniguchi *et al.*, 1967); wheat leaves (Wang, 1967); parsley (Sanderman and Grisebach, 1968a,b); wheat grain (Jenner, 1968); *Digitalis* leaves (Franz and Meier, 1969); sycamore cells (Brown and Short, 1969); cotton hair (Franz, 1969); cucumbers (Matsumoto *et al.*, 1969); bean leaves (Wallis and Bradbeer, 1970); fenugreek seeds (Sioufi *et al.*, 1970); larch tissues (Cumming, 1970); sweet potato (Murata, 1977); Yucca exudate (Becker *et al.*, 1971); corn grain (Dankert *et al.*, 1972); wheat aleurone (Collins *et al.*, 1972); Konjac corm (Murata, 1975); *Nicotiana tabacum* callus (Palmer, 1976); *Ricinus* phloem sap (Mengel and Haeder, 1977); black gram seeds (Ashihara, 1977); *Triticale* grain (Sharma and Bhatia, 1978); sorghum seedlings (Bhatia and Uppal, 1979).
D-glucose	GDP	Sugar beet root (Katan and Avigad, 1965); brown algae (Lin and Hassid, 1966); strawberry leaves (Selvendran and Isherwood, 1967; Isherwood and Selvendran, 1970); larch cambium (Cumming, 1970); *Triticale* grain (Sharma and Bhatia, 1978).
	TDP	Sugar beet root (Katana and Avigad, 1965).
	ADP	*Chlorella* (Kauss and Kandler, 1962; Pakhomova *et al.*, 1965; Sanwal and Preiss, 1969); corn grain (Recondo *et al.*, 1963; Dankert *et al.*, 1972); rice grain (Murata *et al.*, 1963, 1964, 1966); potato starch grain (Rees and Duncan, 1965; Cassells and Harmey, 1968); brown algae (Lin and Hassid, 1966); wheat grain (Jenner, 1968); larch cambium (Cumming, 1970); sorghum seedlings (Bhatia and Uppal, 1979); sweet potatoes (Murata, 1977); konjac corm (Murata, 1975); *Triticale* grain (Sharma and Bhatia, 1978).
D-mannose	GDP	Red algae (Su and Hassid, 1962a,b); mung bean seedlings (Grégoire *et al.*, 1963, 1965); *Chlorella* (Pakhomova *et al.*, 1965; Sanwal and Preiss, 1969); brown algae (Lin and Hassid, 1966); strawberry leaves (Selvendran and Isherwood,

Continued

TABLE 2.1—continued.

Sugar	Nucleotide	Plant
		1967; Isherwood and Selvendran, 1970); Jerusalem artichoke tubers (Taniguchi *et al.*, 1967); larch cambium (Cumming, 1970); fenugreek seeds (Sioufi *et al.*, 1970); konjac corm (Murata, 1975).
D-mannose	ADP	Corn grain (Dankert *et al.*, 1964b).
	UDP	*Triticale* grain (Sharma and Bhatia, 1978).
L-galactose	GDP	Red algae (Su and Hassid, 1962a,b).
D-fructose	UDP	Dahlia tubers (Gonzalez and Pontis, 1963); germinating pea seeds: *Impatiens balsamina* leaf nodes (Brown and Mangat, 1967); Jerusalem artichoke tubers (Taniguchi *et al.*, 1967; Uemura *et al.*, 1967); parsley (Sanderman and Grisebach, 1968a); larch cambial and xylem cells (Cumming, 1970); strawberry leaves (Isherwood and Selvendran, 1970); wheat aleurone (Collins *et al.*, 1972).
	ADP	Larch (Cumming, 1970).
L-fucose	GDP	Brown algae (Lin and Hassid, 1966); corn roots (Kirby and Roberts, 1971).
L-rhamnose	UDP	Brown algae (Kauss, 1965); pea seeds (Hampe and Gonzalez, 1967); parsley (Sanderman and Grisebach, 1968a,b).
2-deoxy-D-glucose	UDP	*Picea* (spruce) cells treated with 2-deoxy-D-glucose (Zemek *et al.*, 1976).
D-digitoxose (2,6 dideoxy-D-ribohexose)	UDP	*Digitalis purpurea* leaves (Franz and Meier, 1969).
D-apiose	UDP	Parsley (Sanderman and Grisebach, 1968a,b); *Lemna minor* (Kindel and Watson, 1973).
D-glucuronic acid	UDP	Mung bean seedlings (Solms and Hassid, 1957; Solms *et al.*, 1957; Grégoire *et al.*, 1965); *Chlorella* (Pakhomova *et al.*, 1965; Sanwal and Preiss, 1969); strawberry leaves (Selvendran and Isherwood, 1967; Isherwood and Selvendran, 1970).
D-galacturonic acid	UDP	Mung bean seedlings (Neufeld and Feingold, 1961; Grégoire *et al.*, 1965); cotton hair (Franz, 1969); sycamore cells (Brown and Short, 1969); larch cambium (Cumming, 1970).
	TDP	Sugar beet roots (Katan and Avigad, 1965).
D-mannuronic acid	GDP	Brown algae (Lin and Hassid, 1966).
L-guluronic acid	GDP	Red algae (Su and Hassid, 1962a,b); brown algae (Lin and Hassid, 1966).
2-deoxy-2-acetamido-D-glucose	UDP	Mung bean seedlings (Solms and Hassid, 1957; Solms *et al.*, 1957; Grégoire *et al.*, 1963; Roberts and Pollard, 1975); barley and oat plants (Bergkvist, 1957); brown algae (Kauss, 1965); *Chlorella* (Pakhomova *et al.*, 1965; Sanwal and Preiss, 1969); Jerusalem artichoke tubers (Taniguchi *et al.*, 1967); corn grain (Dankert *et al.*, 1964a, 1972);

Continued

TABLE 2.1—continued.

Sugar	Nucleotide	Plant
		bean roots (Roberts, 1970); barley coleoptiles and corn roots (Roberts et al., 1971a,b)
	ADP	Corn grain (Passeron et al., 1964; Dankert et al., 1964a, 1972)
2-deoxy-2-acetamido-D-galactose	UDP	Dahlia tubers (Gonzalez and Pontis, 1963); Chlorella (Pakhomova et al., 1965); mung bean seedlings (Grégoire et al., 1963); corn grain (Dankert et al., 1972)
Oligosaccharides cellobiose	UDP	Larch cambium (Cumming, 1970)
D-glucose + L-arabinose	UDP	Larch cambium (Cumming, 1970)
D-glucose + L-mannose	GDP	Larch cambium (Cumming, 1970)

This list may not be complete, particularly with regard to the listing of the most commonly found nucleosides (such as, UDP-Glc). Nucleotide sugars are included in this table if they have been demonstrated chromatographically, chemically, or enzymatically in extracts of higher plants and some related photosynthetic organisms. Cases where the presence of a particular nucleotide sugar is suggested because a plant extract contains an enzyme which can act on it as a substrate (e.g. in pyrophosphorylation, oxido-reduction and transglycosylation reactions) are not mentioned in this table. For a comprehensive list of nucleotide sugars, see Kochetkov and Shibaev (1973), Nikaido (1975).

IV. GENERAL METHODS OF NUCLEOTIDE SUGAR RESEARCH IN PLANTS

A. Extraction of Nucleotide Sugars from Plant Tissues

During isolation from plant tissues, nucleotide sugars must be extracted under conditions which afford minimal degradation and avoid contamination by interfering substances. Early methods involved initial extraction with hot 50% (final concentration) ethanol, followed by a lengthy and complex series of operations to eliminate undesired contaminants (Ginsburg et al., 1956) prior to actual separation of the nucleotides by column chromatography and paper chromatography and electrophoresis.

B. Concentration and Preliminary Purification of Nucleotide Sugars (and Nucleotides)

Specific adsorption and desorption of nucleotides from charcoal has been a common feature of most isolation methods. Cumming (1970) obtained good recovery of nucleoside monophosphates and uridine diphosphate sugars by this method, but significant proportions of other adenosine or guanosine derivatives were irreversibly bound to the charcoal. Further, adenosine and guanosine nucleotide sugars were more strongly adsorbed than the corresponding unglycosylated nucleotides. This may account for the less frequent identification of those compounds in plant extracts or their apparent presence in such small quantities. A material that may prove superior to charcoal for the adsorption of nucleotides is the non-ionic resin Amberlite XAD, which is a styrene–divinyl benzene copolymer. This resin has been used for the removal of salts from purines, pyrimidines and nucleosides (Mills, 1982), and recently it was also found to retard the movement of sugar nucleotides significantly when it was employed in a column (M. Appleton, pers. commun.).

Nucleotide solutions are brought to pH 4–4.5 with 1 N HCl and loaded on a column of activated carbon, nucleotide desalting grade (Aldrich Chemical Company, Inc., Milwaukee, WI, USA). One gram of charcoal will adsorb at least 30 μmol of UDP-GlcA (Lin and Hassid, 1966). The column is washed exhaustively with ice-cold water and the nucleotides are then eluted with ice-cold 50% ethanol–0.5% ammonium hydroxide (sp. gr. 0.88). The eluate is brought to pH 6.5 with cold concentrated formic acid and reduced in volume at 37°C *in vacuo*. The resulting aqueous solution is lyophilised to eliminate ammonium formate.

The initial extraction of nucleotides from plant tissues is often effected with hot aqueous ethanol. Cumming (1970) found that the hot solution activated various phosphatases in the preparation that degraded labile phosphorylated compounds. Consequently, she used the method of Cole and Ross (1966) in which the extraction is carried out in cold ethanol acidified with formic acid. Storage of that extract for 20 h at −30°C inactivated the phosphatases. Formic acid was later removed as the ammonium salt by vacuum sublimation.

The plant tissue is frozen in liquid nitrogen and macerated in a blender with cold 80% formic acid and poured directly into a large quantity (e.g. 50 vols) of ethanol at −20°C. The extract is stored at −30°C or below for at least 24 h to inactivate phosphatases; subsequently, the pH is brought to 6.5 with ammonium hydroxide (sp. gr. 0.88), freed of precipitate by centrifugation, and reduced in volume *in vacuo* until an aqueous solution is obtained. The latter is freed from ammonium formate by lyophilisation. The residue is dissolved in water and subjected to further purification.

Separation of mixtures of nucleotide sugars is facilitated and simplified by prior elimination of extraneous compounds containing terminal phosphate groups. This can be achieved by treating mixtures of nucleotides and nucleotide sugars eluted from charcoal with phosphomonoesterase, thereby converting nucleoside mono-, di- and triphosphates to nucleosides and inorganic orthophosphate but leaving nucleotide sugars intact. These low molecular weight compounds can then be separated from the nucleotide sugars by gel chromatography. An example of this procedure is given in Section VI.D.1.

C. Separation of Nucleotide Sugars

1. Low pressure liquid chromatography

Initial separation of nucleotide sugars is usually done by anion-exchange column chromatography. The most frequently used chromatographic media are the strongly basic resins Dowex-1 or Dowex-2 in the chloride or formate form. DEAE-Sephadex has also been used (e.g. Cumming, 1970). The sample is adsorbed on the resin from a dilute neutral solution and subsequently eluted with increasing concentrations of a solution containing the anion corresponding to the counter-ion of the resin, e.g. sodium or lithium chloride for chloride form resins, formic acid or sodium formate for formate form resins. The separated compounds, or mixtures of compounds, detected by their absorbance at 260 nm (or 275 nm), are concentrated by charcoal adsorption–desorption and subjected to further separation. Many variations of the ion-exchange separation have been described by different investigators. A typical separation is given in the next paragraph, using Dowex-1 (formate). A similar separation on Dowex-2 (chloride) is presented in Section V.B. For more detail the reader should consult individual publications.

Dowex-1 (chloride) resin (Dow Chemical Co., Midland, MI, USA) $\times 4$ or $\times 8$, 200 to 400 mesh, is freed of fine particles by repeated suspension in water and decantation. The resin is converted to the formate form by washing with 3 M ammonium formate until free of chloride ion, followed by washing with water. A slurry of the resin in water is used to form a column [length $= (10–20) \times$ diameter]. Coloured impurities are removed from the resin by washing the packed column with 2–4 column vols of 88% formic acid until the effluent is colourless. Water is then run through the column until the effluent pH is between 5 and 6.

The sample is applied to the column as a dilute (0.01–0.05 M) neutral (pH 6–7) solution at a flow rate of 0.5 to 0.8 ml min^{-1} cm^{-2} of column cross-section. [The optimum amount of material for chromatography on Dowex columns has not been established; however, as an example, a 200 ml bed volume column of Dowex-formate resin (8 $\times$, 200–400 mesh) was used to separate the nucleotides from 5 kg of 4 day-old etiolated mung bean seedlings (Neufeld and Feingold, 1961)]. If the sample applied contains non-ionic and/or coloured material, these can be removed by washing the column with 50% ethanol followed by at least 4 bed volumes of water. Elution of the column is usually with a linear concentration gradient of formic acid/ammonium formate, most simply obtained by using two identical bottles connected to each other via bottom outlets. The outlet from the mixer bottle flows onto the column; mixing is achieved with a magnetic stirrer. The two bottles are filled with the same volume of liquid at the outset; an identical descending liquid level is maintained as the liquid flows onto the column. The volume of liquid in the mixer (and in the reservoir also) is usually 20–25 times the column volume. In a separation of nucleotides from 2.5 kg of cambial tissue of the European larch, Cumming (1970) used a 4×40 cm column of Dowex-1 ($\times 4$, formate form, 200–400 mesh), which was eluted with 1 litre of 1 M formic acid in the mixer and 1 litre of 0.8 M ammonium formate in 1 M formic acid in the reservoir. The flow rate is usually adjusted to 0.5–1.0 ml^{-1} min^{-1} cm^{-2} of resin bed cross-section, and fractions corresponding to 5–20% of the column bed volume are collected.

An automatic fraction collector is used in conjunction with a light-absorption monitor set at 260 nm (although adenosine has only 50% of its maximum absorbance at 260 nm, the absorption maxima of cytidine, guanosine, and uridine are all at or near 260 nm). The fractions emerging under peaks are pooled and treated as appropriate, i.e. concentrated by use of charcoal and further purified if necessary.

2. Paper electrophoresis

Electrophoresis of nucleotide sugars on paper is a simple, rapid and inexpensive method for the isolation, separation and characterisation of such compounds. It also offers the considerable advantage that the movement of most nucleotides and their derivatives can be predicted with some accuracy.

An apparatus developed by Crestfield and Allen (1955) has been widely copied and used for this purpose. Its exact counterpart was never produced commercially, but it has been reproduced in many university machine shops around the world. It consists of a frame of plastic and plywood supporting a plate made of a pane of single ply window glass about 24×40 cm. Tap water flows through a perforated copper tube beneath the glass and serves to cool the plate. A rectangular plexiglass vessel at each end of the plate holds the appropriate buffer and a platinum electrode assembly. (Detailed plans can be obtained from Dr G. A. Barber.) A filter paper strip is moistened with the buffer and laid on the glass surface so that no bubbles form between it and the glass. Each end of the paper is submerged in buffer in one of the tanks. The underside of the glass must be cleaned carefully with a solvent such as dichloroethane so that the cooling water flows in a sheet across its surface. The sample (about $20\ \mu l\ cm^{-1}$) is applied at one end of the moistened paper, the paper is covered with a plexiglass sheet to prevent drying out and a DC voltage (1200–1500 V) is applied to the platinum electrodes. The higher the voltage applied, the more quickly the compounds move and the less diffusion occurs. However, the greater the voltage, the greater the heating of the paper and evaporation of the buffer. Hence, the voltage applied is limited by the efficacy of the cooling system. A number of commercially available electrophoresis apparatuses are also available. Most of these use a metal plate which is cooled by circulating a refrigerated liquid. The electrophoresis paper is prevented from contact with the metal by sandwiching it between polyethylene sheets.

When subjecting phosphate compounds to electrophoresis on paper, it is important that heavy metals first be removed from the paper. They will complex with phosphate and thus retard the movement of phosphorylated compounds. Schleicher and Schuell No. 589 Orange Ribbon, a commercially washed paper, was quite satisfactory for the purpose, but it is currently not manufactured. Whatman No. 1 or 3MM can be used, but it should first be washed with EDTA or with oxalic acid and subsequently several times with water.

TABLE 2.2.   The net charges on several representative nucleotides and hexose monophosphate at various pH values.

Compound	pH	pK of ionisable groups other than phosphate (Dunn and Hall, 1968)	Total charge including phosphates	Approximate net charge
UDP	3.7	9.4	−2	−2
UDP	2.7	9.4	−2	−2
CDP	5.0	4.6, c. 9	−2, + 0.28	−1.72
CDP	3.7	4.6, c. 9	−2, + 0.89	−1.11
CDP	2.7	4.6, c. 9	−2, + 1	−1.0
GDP	3.7	2.9, 9.6	−2, + 0.13	−1.87
GDP	2.7	2.9, 9.6	−2, + 0.63	−1.37
ADP	3.7	3.9, ?	−2, + 0.63	−1.37
ADP	2.7	3.9, ?	−2, + 0.86	−1.14
Hexose monophosphate[a]	3.7	c. 2, 7	−2	−2
Hexose monophosphate[a]	2.7	c. 2, 7	c. 2, 7	−1.6

[a] No ionisable groups other than phosphate are relevant under these conditions.

TABLE 2.3. Electrophoretic mobility of some phosphorylated compounds and uronic acids.[a,b]

	Buffer 1 (pH 3.6)[c]	Buffer 2 (pH 5.8)[c]
Uridine nucleotides		
UTP	1.6	1.6
UDP	1.4	1.5
UMP	0.8	0.9
UDP-GlcA	1.4	1.6
UDP-GalA	1.3	1.5
UDP-Xyl; UDP-Ara	1.2	1.1
UDP-Glc; UDP-Gal	1.1	1.0
UDP-GlcNAc	1.0	1.0
Adenosine nucleotides		
ATP	1.2	1.4
ADP	0.9	1.1
AMP	0.3	0.7
NAD^+	0.5	0.5
NADH	0.9	0.9
$NADP^+$	0.9	0.9
NADPH	1.1	1.2
Guanosine nucleotides		
GTP	1.4	1.5
GDP	1.1	1.1
GMP	0.6	0.7
GDP-Man	0.9	1.0
Phosphates		
Pyrophosphate	2.1	2.1
Orthophosphate	1.6	1.6
GlcA-1-P	1.5	1.7
GalA-1-P	1.4	1.6
Xyl-1-P; Ara-1-P	1.1	1.0
Glc-1-P; Gal-1-P	1.1	1.0
GlcNAc-1-P	0.9	1.0
Uronic acids		
GlcA	0.9	1.0
GalA	0.7	1.0

[a] Electrophoresis performed on Whatman No. 1 paper, previously washed as described.

[b] The mobility is recorded as a function of the mobility of picrate, caffeine being used as an indicator for electro-osmosis. The mobility is calculated as:

$$\frac{\text{distance between compound and caffeine}}{\text{distance between picrate and caffeine}}$$

[c] The buffers are prepared as follows: Buffer 1 is prepared by dissolving 378 g of ammonium formate and 200 ml of 98% formic acid in 3 litres of solution. This stock solution is diluted 20-fold for usage. Buffer 2 is prepared by dissolving 308.4 g of ammonium acetate and 30 ml of glacial acetic acid in 2 litres of solution. This solution is diluted 10-fold for use in electrophoresis.

Ammonium formate and ammonium acetate have been employed to a great extent as buffers for the separation of nucleotide sugars and sugar phosphates. The pKs of formic and acetic acids (3.7 and 5) are in suitable regions, and the ammonium salts of those acids can readily be removed from papers by virtue of the volatility of their components.

For molecules of essentially equal mass, the greater the number of charges on any molecule, the farther it will move on the paper. (In the case of the nucleotides and sugar phosphates, that movement is toward the positive pole since under the conditions of the electrophoresis, net charges on the molecules are negative.) The number of charges can be calculated at any given buffer pH from the pKs of the various ionisable groups on the molecule. Because of their greater mass, the purine nucleotides often can be easily separated from pyridine nucleotides by paper electrophoresis.

The net charges of each of several nucleoside diphosphates and monophosphohexose molecules at different pH values are given in Table 2.2. Movement of all these compounds is generally expressed relative to the movement of picric acid, a bright yellow compound and a strong acid that is completely ionised at these buffer pHs. Caffeine is employed as an indicator of electro-osmosis, since it is immobile under the conditions of electrophoresis and absorbs ultraviolet light. Relative mobilities of a number of nucleotide sugars are given in Table 2.3.

GDP-Glc presents a particular separation problem because at pH 3.7 it cannot be separated from Glc-1-P (or any other hexosyl phosphate). However, at pH 2.7 the nitrogen at position 7 of the guanine ring (pK 2.9) acquires a partial positive charge; the net charge on the GDP-Glc is then about -1.5 and it moves more slowly than the hexopyranosyl phosphates. Mobilities relative to picric acid are: hexose phosphate, 0.9; GDP-Glc, 0.8; UDP-Glc, 1.12; ADP-Glc, 0.7; CDP-Glc, 0.7.

After electrophoresis, the paper is air-dried and compounds of interest are located by their UV absorption, or radioactivity if they are labelled with a radioisotope. To recover the compound, the appropriate strip of paper is cut out and eluted with water onto a planchet or in a small beaker. The material is evaporated to dryness *in vacuo* in a vacuum desiccator over concentrated sulphuric acid and pellets of potassium hydroxide. The desiccator is then evacuated further with an oil pump and allowed to stand overnight. That removes most of the ammonium acid formate buffer and leaves the phosphorylated compound on the planchet as its ammonium salt.

3.   Paper chromatography

Chromatography of nucleotide sugars on paper is usually carried out in a descending direction on washed paper. Schleicher and Schuell No. 589 White or Blue Ribbon are washed papers that have proved satisfactory. Nucleotides are conveniently located on the chromatograms by their ultraviolet (UV) absorption under a small, short-wave UV lamp. The solvent system most frequently used was developed by Paladini and Leloir (1952) and consists of 95% ethanol–1 M ammonium acetate, pH 7.0 (7:3; v/v). The ratio of the constituents and the pH can be varied to effect different separations. In those solvent systems nucleotides generally move in the order of increasing R_f as follows: NTP, NDP, NMP and NDP-sugar where N is a pyrimidine or purine nucleoside. Most of the residue of ammonium acetate on the papers can be removed by irrigating the papers with 95% ethanol for 24 h so that the solvent runs off the end of the paper. The process is accelerated by serrating the bottom edge of the chromatogram with scissors or pinking shears.

Some other solvent systems have been used for special cases. For example, GDP-Man and GDP-L-galactose could only be separated by chromatography with *iso*butyric acid–1 M ammonium hydroxide (10:6; v/v) after multiple development for several days (Barber, 1975).

Another system is based upon the formation of borate complexes with the sugar moiety of the nucleotide using borate salts of organic bases in the solvent. It has been employed to separate glucose, mannose and galactose nucleotides of the same purine or pyrimidine base (Carminatti *et al.*, 1965). As is the case with the *iso*butyric acid solvent, development times as long as 7 days are sometimes required. The separation of UDP-Glc, UDP-Gal, UDP-Xyl and UDP-Ara can be achieved by paper chromatography in ethylene glycol dimethyl ether–methylethyl ketone–0.5 M morpholinium tetraborate, pH 8.6, in 0.01 M EDTA (7:2:3; v/v/v). After separation, each nucleotide sugar is eluted and freed of salts by paper chromatography in methanol–1 M ammonium acetate, pH 3.8 (2:1; v/v). Specific examples of the chromatographic systems mentioned here are presented later in this chapter.

4. Thin layer chromatography

For analytical purposes, thin layer chromatography offers a number of advantages over paper chromatography or paper electrophoresis. It is more sensitive than the other two methods by a factor of 10; quantities of unlabelled nucleotide as small as 0.1 μg can be detected. Furthermore, complex mixtures of nucleotides that yield to no other method can be separated by anion-exchange thin layer chromatography. The time usually required for these separations is also much shorter, often taking only 30–60 min.

For the resolution of mixtures of nucleotides, the procedure of Randerath and Randerath (1964) has proved especially effective. The method involves the anion-exchange chromatography of nucleotides on thin layers of poly(ethyleneimine)cellulose layers (0.5 mm thick) on 20 × 20 cm sheets of glass; such layers backed with plastic are now available commercially. The nucleotide mixture (about 1 μl) of a 2 mM solution of the sodium or lithium salt is applied in a small spot 3 cm from the corner of the plate. The chromatogram is developed in a closed jar with 0.2 M lithium chloride for 2 min, with 1.0 M lithium chloride for 6 min and with 1.6 M lithium chloride for about 65 min to a line about 13 cm from the origin. The portions of the layer not needed for further development in the second direction are scraped off the plate. To remove residual lithium chloride before the next step, the plate is washed by immersing it in a dish of anhydrous methanol and agitating it gently for 15 min. After drying in the air, the chromatogram is developed in the second direction with increasing concentrations of sodium formate buffer, pH 3.5: 0.5 M for 30 s, 2 M for 2 min, and 4 M until the solvent reaches 15 cm from the origin. The plate is dried in a stream of hot air and examined under an ultraviolet (254 nm) lamp for nucleotides. Those appear as dark blue spots under these conditions, but when the plate is exposed to HCl fumes for several minutes, only guanine derivatives fluoresce a brilliant light blue. By this procedure a clear separation of the following 22 nucleotides can be achieved: NAD, CMP, AMP, CDP-Glc, UMP, UDP-GlcNAc, ADP, NADP, UDP-Glc, GMP, GDP-Man, CDP, ADP, UDP-GlcA, UDP, IDP, GDP, CTP, ATP, UTP, ITP, GTP. About 0.5–2 mol of each compound is detectable.

A modified version of this method was used in the analysis of the nucleotides in a culture of *Salmonella typhimurium* (Bochner and Ames, 1982). Their method produces a chromatographic fingerprint of the nucleotide pool in the cell that should be readily adaptable to other cell (e.g. plant) cultures.

The procedure is carried out as follows: a culture of the bacteria is grown overnight (or for a suitable time, depending upon growth rate) on a medium low in carbon sources. It is used to inoculate heavily (15%) fresh medium containing glucose, salts and ^{32}P-labelled KH_2PO_4 (150–500 µCi µmol^{-1}). The culture is incubated well into the logarithmic phase and then extracted with formic acid. Small quantities of the extract (5 µl) are applied to plastic-backed poly(ethyleneimine) cellulose sheets (obtained commercially) which are formed into cylinders held by rubber bands at top and bottom. They are developed in two directions. In the first direction the nucleotides are separated by their electrical charge with an ionic solvent buffered at pH 8 (e.g. 0.75 M Tris base, 0.45 M HCl). Salts in the developing solvent are removed by washing the plate in anhydrous methanol. In the second dimension nucleotides are separated by the solubility of the purine or pyrimidine base. For that separation the most effective solvent is a saturated aqueous solution of ammonium sulphate adjusted to pH 3.5 with ammonium bisulphate and with EDTA added to bind traces of metal ions. After development, radioactive compounds are located by exposure of the sheets to X-ray film at $-80°C$ in a cassette between intensifying screens. Radioactive areas on the thin-layer plates are cut out, eluted with water, and the eluted material is used as desired. Characterisation of the ^{32}P-labelled nucleotide sugars is described in Section IV.D.

5.   High performance liquid chromatography

(a)   *Extraction of nucleotide sugars from plant tissues for separation by high performance liquid chromatography.* With the advent of high performance liquid chromatography (HPLC) for the separation and quantitation of nucleotides, many of the extraction methods available were found to be incompatible with the use of microparticle analytical columns, and special extraction techniques had to be devised.

Meyer and Wagner (1985) have described a procedure for the determination by HPLC of nucleotide pools in plant tissue in which extraction is effected with 0.4 M perchloric acid. Purification (i.e., elimination of interfering contaminants) of the extract is achieved on a phenyl bonded column, and perchloric acid is removed from the purified extract by careful neutralisation with potassium hydroxide. Although effective, this technique may be less convenient than that developed by Olempska-Beer and Freese (1984) for yeast cells in which extraction was accomplished with formic acid in n-butanol. The latter method should prove valuable for most cell types, since yeast cells are known to be particularly resistant to extraction. Nonetheless, the extraction procedure presented here is that of Meyer and Wagner (1985), as it is the only published method which has been specifically designed for HPLC separation and determination of nucleotides in plant tissues.

The extraction procedure is described for cells of *Nicotiana tabacum* W 38 grown in suspension culture. Cells are filtered on a G-3 sintered glass filter, washed with water, and stored at $-70°C$ after freezing with liquid nitrogen. Other plant tissues (e.g. leaves or roots) can be similarly frozen and stored. The extraction procedure involves four separate operations. All subsequent procedures are done at 4°C or in ice.

After grinding in a mortar cooled with liquid nitrogen, the tissue is weighed and transferred to a Potter–Elvejhem glass homogeniser. Chilled 0.4 M perchloric acid is added (10 ml g^{-1} frozen tissue); after homogenisation, the mixture is held at 0–4°C for 10 min and then centrifuged in a Sorvall SS34 rotor at 18 000 rpm for 10 min.

A column (*c.* 15 ml g^{-1} tissue) of phenylsilane bonded silica gel (1.25 g g^{-1} tissue) is prepared and washed with 1 column volume of methanol, at a rate of *c.* 3 ml min^{-1}, using a peristaltic pump. The column is equilibrated with 5 ml of 0.4 M perchloric acid, and 1.25 ml of the clarified extract is applied and eluted with 0.4 M perchloric acid. Ten millilitres of eluate is collected for neutralisation. The eluate is very carefully brought to pH 5.5–6.5 with cold potassium hydroxide; less concentrated potassium hydroxide is used as titrant as the pH approaches 5.5. The precipitated potassium perchlorate is removed by centrifugation, and the supernatant fluid is lyophilised to dryness. The dry material is dissolved in 0.63 ml 10 mM hydrochloric acid, chilled, and the precipitated traces of potassium perchlorate are removed by centrifugation. The supernatant solution is stored frozen at $-70°C$.

(b) *HPLC*. High performance liquid chromatography has been successfully employed in the separation and quantitation of nucleotides, and improvements in the technology are constantly developed. A system which has been specifically designed for examination of nucleotides in plant tissues is that of Meyer and Wagner (1985).

Equipment required is an HPLC apparatus with gradient-forming ability and an automatic recording spectrophotometer capable of measuring at 260 nm. The column (25 cm × 4 mm i.d.) used is Hypersil APS (Applied Science, State College, PA, USA) with a guard column (3 cm × 4 mm i.d.) filled with Vydac-201RP, 30–40 μm (Machery and Nagel, Duren, FRG). The buffers used for ion-exchange chromatography on the Hypersil column are: (A) 5 mM KH_2PO_4, pH 3.0; and (B) 0.5 M KH_2PO_4, pH 4.0. New columns are washed with the latter buffer for 6 h before use, and gradient elution profiles are varied according to the age of the column. Thus the gradient elution profile for a new column starts with an isocratic phase of Buffer A (5 min) followed by a linear gradient from A to B (20 min for a new column and up to 60 min for a very old column). After every three runs, the column is washed for 2 min with 5% phosphoric acid and then for 10 min with Buffer A.

The columns are calibrated using mixtures of nucleotide sugars (0.2–0.4 nmol). The column is calibrated with mixtures of known compounds—the following are retention times (in min) for some typical nucleotide sugars: CDP-Glc, 19.4; UDP-GlcNAc, 20.5; UDP-Glc, 21.3; ADP-Glc, 22.4; GDP-Glc, 23.9; UDP-GlcA, 30.4; UDP-GalA, 31.2. Other HPLC schemes have been proposed for separation of nucleotide sugars. Payne and Ames (1982) have described a method for extraction and separation of nucleotides from Gram-negative bacteria which uses reverse-phase and ion-paired HPLC and which might be applicable to investigation of plant extracts.

D. Characterisation of Nucleotide Sugars

Until quite recently small quantities of sugar nucleotides obtained from natural sources or synthesised chemically were isolated and purified by electrophoresis and partition chromatography on paper. A combination of electrophoresis in ammonium formate buffer at pHs ranging from 2.7 to 5.8, and chromatography with 95% ethanol–1 M ammonium acetate, pH 7 (7:3; v/v), permitted the purification of most sugar nucleotides. This was particularly effective if the authentic compounds were available for comparison.

After isolation and tentative identification by these methods, nucleotide sugars can be further characterised by the identification and quantification of the component parts of the molecule released by acid or enzymic hydrolysis. For example, a nucleoside diphosphate hexose upon mild acid hydrolysis (pH 2, 100°C for 10 min) would be expected to yield equal amounts of hexose, nucleoside monophosphate and ortho-phosphoric acid. Similarly, the intact nucleotide can be treated with a purified nucleotide pyrophosphatase (phosphodiesterase) from snake venom and will yield a glycosyl phosphate and the nucleoside monophosphate. The nucleoside can be characterised by its chromatographic mobility and its characteristic UV absorption spectrum. If the extinction coefficient for the nucleoside is known, its quantity can also be determined from its optical absorbancy at the appropriate wavelength. The nature and quantity of the associated sugar or sugar phosphate is also ascertained by paper chromatography, paper electrophoresis and colorimetric analysis. Typical of such an analysis are the data reported by Kornfeld and Glaser (1961) for the characterisation of TDP-Glc: (μmol μmol^{-1} thymidine); thymidine (the reference compound), 1.0; acid labile phosphate, 0.98; total phosphate, 2.04; α-D-glucopyranosyl 1-phosphate, 1.0; D-glucose, 1.0.

Nucleotides labelled solely with ^{32}P can be identified by application of a number of enzymatic and chemical methods (Bochner and Ames, 1982). Radioactive areas on the thin layer plates corresponding to nucleotide sugars are scraped from the plate, and the nucleotides are eluted with water and subjected to various analyses.

^{32}P-Labelled materials removed from the thin layer sheets are characterised by the following criteria:

1. The labelled compound is shown to co-chromatograph with an authentic standard.
2. It is absorbed by charcoal, a property of nucleotides.
3. It gives the expected sensitivity or resistance to oxidation by periodate.
4. It is or is not susceptible to hydrolysis catalysed by one or more of the following five enzymes:
   (a) bacterial alkaline phosphatase (hydrolyses terminal phosphate esters non-specifically);
   (b) 5′-nucleotidase (hydrolyses 5′ monophosphates);
   (c) snake venom phosphodiesterase (hydrolyses phosphodiesterase bonds and also, α-, β-pyrophosphate linkages of nucleoside 5′ polyphosphates);
   (d) nuclease P-1 (this enzyme has 3′-monophosphatase activity);
   (e) ribonuclease T$_2$ (this enzyme catalyses the hydrolysis of the α-, β-pyro-phosphate linkages of nucleoside 3′-polyphosphates and also has the usual phosphodiesterase activity).

An application of the most up-to-date instrumental techniques to the character-isations of a sugar nucleotide has recently been described (Okuda et al., 1986). The compound, identified as uridine 5′-(2,3-diacetamido-2,3-dideoxy-α-D-glucopyranosyl uronic acid pyrophosphate), was obtained from the bacterium Pseudomonas aeruginosa where it is presumably involved in the biosynthesis of O-specific lipopolysaccharides. It was isolated from 600 g of wet cells suspended in buffer and broken in a French pressure cell. Nucleotides were adsorbed on charcoal, eluted and adsorbed to a column of Dowex-1(Cl$^-$) anion-exchange resin, eluted with NaCl and desalted by adsorption again on charcoal and re-elution. The compound was isolated and further purified by chromatography in three solvent systems, electrophoresis on paper at pH 4.8, and paper chromatography one more time. Finally, the aqueous solution of the sugar nucleotide was passed through a small column of Dowex-50 (H$^+$) to remove cations, made pH 8.5

with aqueous ammonia and evaporated to dryness *in vacuo*. The yield was 11.3 mg.

The identification of the compound was based upon the following evidence:

1. The positive fast-atom-bombardment mass spectrum showed a peak at $m/2$ 663 corresponding to the $(M + 1)^+$ ion of the calculated values for free acid $(C_{19}H_{28}N_4O_{18}P_2)$ of the proposed structure.

2. The carbon-13 NMR spectrum indicated the presence of 19 C atoms (four of uracil, five of ribose phosphate and 10 of 2,3-diacetamido-2,3-dideoxy D-glucuronic acid). Assignments were made by comparison with the spectra of authentic UMP and 2,3-diacetamido-2,3-dideoxy-D-glucuronic acid.

3. The sugar nucleotide was treated with dinucleotide nucleotidohydrolase, and a nucleoside monophosphate was crystallised as the sodium salt. The ultraviolet and infrared absorption spectra and the molar rotation were all identical with those of UMP.

4. The glycosyl phosphate released in Step 3 was isolated as a solid. Its proton NMR spectrum showed 11 protons, including six C-methyl protons of two acetamido groups and five ring protons. Signals were assigned by double-resonance experiments. Coupling constants for the ring protons indicated an α-glucopyranosyl configuration for the sugar. The couplings of the phosphorous nucleus with H-1 and H-2 further supported the α-anomeric configuration.

5. The reducing sugar released from the hexosyl phosphate upon treatment with orthophosphoric monoester phosphohydrolase (phosphomonoesterase) was formed with an equal quantity of orthophosphoric acid. The fast atom bombardment spectrum of the sugar and its infrared, proton magnetic resonance, and circular dichroic spectra were all identical to those of the authentic monosaccharide. The conclusion then was that the compound was uridine 5'-(α-D-2,3-diacetamido-2,3-dideoxy)glucopyranosyluronic acid pyrophosphate.

The nuclear magnetic resonance spectra of a variety of nucleoside diphosphohexoses and their components have been investigated (Lee and Sarma, 1976) using Fourier transform techniques. The chemical shifts and coupling constants of: UDP-Glc, UDP-GlcNAc, UDP-Gal, UDP-GlcA, UDP-Man, UDP-GalA, ADP-Glc, ADP-Man, GDP-Glc, GDP-Man, and CDP-Glc and of the nucleotide diphosphate and hexosyl phosphate components of those compounds are presented. No significant interactions between the hexose and nucleoside parts of the molecule were detected. The tables should serve as primary sources of reference data for the analysis of sugar nucleotides by NMR.

V. CHEMICAL SYNTHESIS OF NUCLEOTIDE SUGARS

A. Chemical Synthesis of Glycopyranosyl Phosphates

Aldopyranosyl phosphates can be prepared by the method of MacDonald, which involves fusion of the fully acetylated aldose with crystalline orthophosphoric acid (MacDonald, 1966). The preparation of α-D-glucopyranosyl phosphate is described here; similar procedures are used to prepare other glycosyl phosphates. More information is available in the article by MacDonald (1966).

Crystalline phosphoric acid (Matheson, Coleman and Bell, East Rutherford, NJ, USA) is dried overnight *in vacuo* over magnesium perchlorate and melted at 50°C in a 250 ml round bottom flask connected to a vacuum pump through a cold trap (dry ice–acetone). β-D-Glucopyranose pentaacetate (5.0 g) is added, and the contents of the flask are stirred *in vacuo* (oil pump) at 50°C for 2 h. To the cooled syrupy product is added 205 ml of ice-cold 2 M lithium hydroxide; the syrup is dispersed by vigorous mixing and then held at 25°C for 12 h.

Lithium phosphate is filtered off through Celite, the precipitate is washed with 50 ml of cold 0.01 M lithium hydroxide and the filtrates are combined. The cold solution is passed through (at 4°C) a column (1.9 × 30 cm) of Dowex 50W-H$^+$, and the column is eluted with 200 ml of water. The column effluent is run into a stirred solution of cyclohexylamine (10 ml in an equal volume of water). The solution is concentrated *in vacuo* and the resultant syrup is dissolved in 3 ml of water; 100 ml of absolute ethanol is added to the solution with stirring. The solution is allowed to reach room temperature and then held at 5°C (refrigerator) until crystallisation is complete (*c.* 24 h). The crystals of the cyclohexylammonium salt are filtered at the pump, washed with cold absolute ethanol, and dried over calcium chloride.

B. Nucleotide Sugar Synthesis by the Phosphoromorpholidate Procedure

Nucleotide sugars can be readily synthesised by the reaction of nucleoside 5′-phosphoro-morpholidates with the desired glycosyl phosphate in an anhydrous organic solvent. Morpholidates are available from commercial sources (e.g. Sigma), which means that the investigator need only supply the glycosyl phosphate as the other reaction partner. The coupling procedure is very simple and the only critical consideration is maintenance of strictly anhydrous conditions. The following example of the synthesis of UDP-Glc (Moffatt, 1966) is typical. To ensure that rigorously anhydrous conditions are maintained, it is helpful to carry out the operations in such a way that the reactants are never exposed to other than dry air.

An aqueous solution of glycosyl phosphate (e.g. 1 mmol of dipotassium α-D-gluco-pyranosyl phosphate.H$_2$O) is converted to the pyridinium salt by passage through a column (1 × 5 cm) of Dowex 50 (pyridinium form). The column is washed with water and the eluate and washings are concentrated to a volume of 5 ml to which 15 ml of pyridine and 1 mmol of tri-*n*-octylamine are added. Water and pyridine are removed *in vacuo* and the syrupy residue is made anhydrous by thrice-repeated addition and (vacuum) evaporation of dry pyridine. Finally, the residue is dissolved in 5 ml of dry pyridine. This solution is added to 0.33 mmol of the 4-morpholine *N'N*-dicyclohexyl carboxamidine salt of uridine 5′-phosphomorpholidate, rendered anhydrous by the same procedure. The pyridine is evaporated once more and the mixture is dissolved in 5 ml of dry pyridine and allowed to stand at 25°C for 2 days. Pyridine is then removed by vacuum distillation and the oily reaction mixture is dispersed in 25 ml of water, and 150 mg of lithium acetate is added with continuous stirring. After ether extraction to remove the tri-*n*-octylamine, the aqueous phase is loaded onto a column (2 × 10 cm) of Dowex-2 resin (chloride). After washing the column with 250 ml of water, the column is eluted with a linear gradient of lithium chloride. The mixing vessel initially contains 2 l of 0.01 M lithium chloride in 3 mM HCl and the reservoir contains 2 l of 0.1 M lithium chloride in 3 mM HCl. Fractions of 20 ml are collected, and the effluent is monitored at 260 nm. UDP-Glc is eluted at a salt concentration of approximately 0.06 M. The eluate is brought to pH 4.0 with 1.0 M lithium hydroxide and evaporated to dryness at 30°C, followed by evacuation with an oil pump, whereupon a dry white solid, still contaminated with lithium chloride, is obtained. It can be freed from this contamination by chromatography in water on a column of BioGel P-2 or by repeated solution in methanol (5 ml) and precipitation with acetone (30 ml) and ether (5 ml). Dilithium UDP-Glc.H$_2$O is finally isolated by vacuum evaporation.

The morpholidate coupling procedure has been reduced to micropreparative scale by a number of workers. Such procedures are often convenient for preparation of nucleotide sugars labelled in the glycosyl moiety if the labelled glycosyl phosphate is available. A general procedure has been described by Elbein (1966). As in the macro method, maintenance of anhydrous conditions is essential for success. The process is identical to that just described, except that the reaction is run on the 1–2 μmol scale and the reaction temperature is maintained at 60°C for 6 h. At the end of the reaction the pyridine is removed by evaporation *in vacuo* and to the reaction mixture is added 10 ml of water containing sodium acetate (about 20% more than the total amount of tri-*n*-octylamine and nucleoside monophosphate morpholidate used (e.g. 4.5 μmol for a reaction involving 1 μmol of glycosyl phosphate)). After thrice-repeated ether extraction of the aqueous layer, the ether extract is extracted with water. The combined aqueous layers are lyophilised, and the residue is dissolved in water and subjected to paper electrophoresis in 0.2 M ammonium formate, pH 3.6. The band of interest is eluted with water and chromatographed in 95% ethanol–1 M ammonium acetate, pH 7.5 (7:3; v/v). Further purification, if necessary, may be accomplished as described earlier (Section IV.C).

VI. PREPARATION OF INDIVIDUAL NUCLEOTIDE SUGARS

A. Nucleoside Diphosphohexoses

1. GDP-D-glucose (GDP-Glc)

This nucleotide sugar is available by chemical synthesis, albeit in low yield (Moffat, 1966). GDP-D-Glucose labelled with ^{14}C in the glucosyl moiety is prepared more readily by enzymic methods.

An enzyme extract from developing seeds of maize brings about the formation of GDP-Glc from GTP and α-D-glucopyranosyl phosphate, but in the initial studies of the system, its activity was found to vary considerably from batch to batch (Barber, 1985). It was discovered that much of the enzyme that precipitated in an ammonium sulphate step did not dissolve upon dialysis of the suspension of the ammonium sulphate precipitate. The enzyme remained active but appeared to be adsorbed to some extraneous material. This difficulty was overcome by introducing a fractionation with polyethylene glycol before the ammonium sulphate precipitation step. The corn kernels most recently used in this preparation had been frozen and stored at −20°C for 3.5 years and showed no great loss of activity. In a typical preparation 70 g kernels are ground in three 5-s bursts of a food homogeniser (Osterizer) on the 'hi' setting. The homogenate is squeezed through several layers of cheesecloth and centrifuged at

17 000 × g for 20 min. The supernatant fluid is treated with 1 M manganese chloride to bring the final concentration to 0.05 M, and the considerable precipitate is removed by centrifugation. The solution is made 4% in monomethylpolyethyleneglycol (average mol. wt. 5000) by additions from a 50% solution of the polymer. The voluminous precipitate which results after the mixture stands for a few minutes on ice is removed by centrifugation. More polymer solution is added to 8%, the precipitate is collected by centrifugation and is dissolved in 25 ml 0.1 M Tris–HCl–0.001 M EDTA buffer, pH 7.6. Solid ammonium sulphate is added with stirring to 40%, and after 1 h, the mixture is centrifuged for 10 min at 17 000 × g. Much of the desired protein floats or adheres to the wall of the centrifuge tube so that the supernatant solution must be decanted with great care to avoid losing the protein. The latter is suspended in a small volume of buffer (0.7 ml) and dialysed against two 500 ml volumes of 0.025 M Tris–HCl–0.001 M EDTA buffer, pH 7.5, for 5 h. The dialysed solution is frozen in aliquots without further treatment.

GDP-Glc is prepared as follows: a reaction mixture for the enzymic synthesis of GDP-Glc labelled uniformly with ^{14}C in the glucosyl moiety contains: 0.036 μmol α-D-[U-^{14}C]glucopyranosyl phosphate (c. 9 μCi), 0.45 μmol GTP, 0.45 μmol CoCl$_2$ and 60 μl enzyme preparation from 5 g maize kernels in 0.025 M Tris–HCl–0.001 M EDTA buffer, pH 7.5, in a total volume of 90 μl. It is incubated on a planchet in a moist atmosphere for 30 min at 37°C. The mixture is applied in a band of about 1 × 5 cm to a sheet of filter paper on the electrophoresis apparatus, and the product is separated and isolated as described elsewhere in this chapter. Under these conditions about one-third of the ^{14}C is recovered in GDP-Glc.

2.   GDP-D-mannose (GDP-Man)

GDP-Man is a donor of the D-mannosyl moiety in the synthesis of plant glycoproteins, glycolipids and polysaccharides. It is also the precursor of GDP-Fuc and GDP-L-Gal (Feingold, 1982). GDP-Man can be prepared by chemical synthesis.

A relatively simple method has been developed for the enzymatic synthesis of this sugar nucleotide labelled with ^{14}C or ^{3}H in the D-mannosyl moiety. Although the labelled compound is available commercially, it is quite expensive. The procedure involves the use of a crude enzyme mixture from yeast supplemented with purified hexokinase to carry out the following series of reactions:

$$\text{Glc} \xrightarrow[\text{ADP}]{\text{ATP}} \text{Glc-6-P} \longrightarrow \text{Fru-6-P} \longrightarrow \text{Man-6-P} \xrightarrow[\text{PPi}]{\text{GTP}} \text{Man-1-P} \longrightarrow \text{GDP-Man}$$

The enzyme extract from yeast is prepared by the method of Braell et al. (1976). Fresh baker's or brewer's yeast (Anheuser Busch Corp., St. Louis, MO, USA) or reconstituted dried brewer's yeast (225 g) is crumbled into liquid nitrogen, and the nitrogen is driven off under a stream of air. The frozen cells are suspended in 250 ml 0.05 M K$_2$HPO$_4$ and stirred for 22 h in the cold to bring about their autolysis. The suspension is centrifuged at 16 000 × g for 30 min, and the residue is discarded. The supernatant fluid is adjusted to pH 6.0 with 2 M acetic acid, brought to 50% saturation with solid ammonium sulphate and stirred for 1 h. The precipitate is removed by centrifugation at $16\,000 \times g$ for 1 h. The supernatant fluid is made 70% saturated with solid ammonium sulphate, stirred for 1 h and the precipitate is recovered by centrifugation as before. It is suspended in a small amount of 0.01 M Tris–HCl buffer, pH 7.5, and dialysed for 6 h against three 2 litre volumes of that buffer. The dialysed enzyme mixture is frozen in aliquots and stored at $-20°C$. It remains active for many months under these conditions.

Reactions are carried out after Grier and Rasmussen (1982) as modified by Chang *et al.* (1985). A typical reaction mixture contains 40 µl M Tris–HCl buffer, pH 7.5, 13 µl M MgCl$_2$, 4 mg GTP (Na salt), 3 mg 3-phosphoglyceric acid (cyclohexylammonium salt), 3 µl 0.014 M α-D-glucose 1,6-*bis*phosphate, 20 units yeast hexokinase, 0.1 µmol D-glucose (^{14}C- or ^{3}H-labelled) and water to a total volume of 125 µl. The mixture is incubated at about 23°C for 30 min. After complete phosphorylation of the glucose, 125 µl of the yeast enzyme extract is added, and the mixture is incubated for 3 h at 37°C. The reaction is stopped by adding the mixture to an equal volume of 95% ethanol. A precipitate is removed by centrifugation, ethanol is evaporated under a stream of air, and the aqueous solution is reduced to a small volume *in vacuo*. Labelled GDP-Man is isolated by electrophoresis of the solution on paper at pH 3.6 by the techniques described earlier.

3. GDP-L-galactose (GDP-L-Gal)

GDP-L-Gal has been isolated from the edible snail *Helix pomatia* (Goudsmit and Neufeld, 1966); it previously had been shown in the red alga *Porphyra perforata* by Su and Hassid (1962b). However, it has not yet been found in a higher plant. Nonetheless, the monosaccharide L-galactose is widespread in plants in glycosidic linkage and an enzyme that catalyses the synthesis of GDP-L-Gal has been isolated from the green alga *Chlorella pyrenoidosa* and also is present in extracts of flax (Barber, 1971). It seems likely that GDL-L-Gal is a common constituent of plants, although not present in very large amounts.

The enzyme catalysing the synthesis of GDP-L-Gal can be prepared from *Chlorella* as follows: cells of *Chlorella pyrenoidosa* 395 are cultured autotrophically in a fermenter in a sterile salts medium (Barber, 1975). Eleven litres of medium are inoculated with 1 litre of a subculture of the alga (A at 652 nm $= 1.0$; about 3×10^7 cells ml^{-1}). The culture is agitated at 180 rpm, gassed with 1.5% CO$_2$ in air, and illuminated by a semicircular bank of twelve 15 W 'Gro-lux' bulbs (Sylvania) at 4 cm from the surface of the cylindrical vessel. In about 48 h at 23°C, the culture should reach the density of the inoculum, which represents the late log phase of growth. The cells are harvested by centrifugation and washed with cold water. The packed cells can be stored at $-20°C$. To extract protein, the cells are thawed and suspended with a Teflon pestle in a buffer of 0.05 M Na/K phosphate: 0.02 M-2-mercaptoethanol, pH 6.9 (20 ml per 8 g fresh weight of cells). They are broken by two passages through a French pressure cell at about 15 000 psi (~ 103 MPa). The homogenate is centrifuged for 20 min at $20\,000 \times g$, and the precipitate is discarded. The supernatant solution is treated with solid ammonium sulphate and the fraction precipitating between 40 and 60% saturation is retained and dialysed overnight against two 1 litre volumes of 0.01 M potassium phosphate buffer, pH 7. The enzyme preparation can be lyophilised at this stage and stored for an indefinite period at $-20°C$ without significant loss of activity. This ammonium sulphate fraction usually contains another enzyme or enzymes that degrade GDP-L-galactose to β-L-galactopyranosyl-1-phosphate and, presumably, GMP (Hebda *et al.*, 1979). That activity can be eliminated by chromatography of the fraction on a column of hydroxylapatite. A solution of 130 mg of the ammonium sulphate fraction in 3 ml of 0.01 M potassium phosphate buffer is applied to a 1.5 × 16 cm column of Bio-Gel HTP. The column is eluted with a step gradient of potassium phosphate buffer, pH 7, of increasing concentration, and 12 ml fractions are collected. After elution of the column with 360 ml of 0.01 M buffer and 120 ml of 0.02 M buffer, the epimerase appears in the first 50 ml of the 0.03 M buffer eluate. The degradative enzyme is not removed from the hydroxylapatite until it is eluted with 0.15 M potassium phosphate buffer. Activity of the GDP-D-mannose/GDP-L-galactose epimerase is estimated by the extent of conversion of GDP-D-[^{14}C]Man to the GDP-L-[^{14}C]Gal. A typical reaction mixture contains 0.2–0.4 mg lyophilised enzyme extract containing potassium phosphate buffer, 0.01 μmol GDP[^{14}C]D-mannose (*c*. 0.035 μCi) in 20 μl 0.025 M mercaptoethanol–0.025 M EDTA (Na salt), pH 8.3, in a sealed capillary tube. It is incubated for 20 min at 37°C and the contents of the tube are added to 200 μl of 0.1 M trifluoroacetic acid with 0.2 μmol each of carrier D-mannose and D-galactose and heated at 100°C for 15 min. The hydrolysate is chromatographed on paper with *n*-propanol–ethyl acetate–water (7:1:2; v/v); the monosaccharides are located by their reaction with the *p*-anisidine.HCl reagent and the appropriate areas are cut from the paper and their radioactivities measured in a scintillation spectrometer.

GDP-Man and GDP-L-Gal cannot be readily separated; furthermore, the equilibrium of the reaction catalysed by the epimerase is such that less than 35% of the mixture of nucleotide sugars is GDP-L-Gal. Therefore, the enzymic method is not a good one for preparing GDP-L-Gal.

A more practical synthesis has been carried out chemically by the condensation of GMP-morpholidate with peracetylated β-L-galactopyranosyl phosphate (Hebda *et al.*, 1979). The most difficult part of the procedure is in the maintenance of strictly anhydrous conditions throughout. In this particular case, however, there is another difficulty in that all of the common methods for the synthesis of glycosyl phosphates of galactose result in mixtures of the α- and β-anomers. The β-anomer is kinetically favoured, but when the reaction time is made long enough to give a suitable yield, the proportion of the α-anomer increases considerably. A marked improvement in the process was effected by synthesis of β-L-galactopyranosyl phosphate via an *ortho*-ester intermediate. *Ortho*-esters are only formed at the 1,2-position when the substituents are *trans*. Upon phosphorylation, an inversion of the anomeric configuration results and the product is almost entirely the desired anomer. Reaction of that *ortho*-ester directly with dry phosphoric acid gives the β-L-phosphate with only a trace of the α-anomer.

L-Galactose (available commercially) is converted to the peracetylated derivative (a mixture of anomers) in essentially theoretical yield (Hudson and Dale, 1915). The addition of ice to the reaction mixture gives a sticky precipitate that crystalises upon stirring, and the crystals are collected by filtration. The acetylated sugar is all in the pyranose form, 20% α-, 80% β-. That mixture is used to form 2,3,4,6-tetra-*O*-acetyl-α-L-galactopyranosyl bromide (Hudson and Dale, 1915). The bromacetyl sugar (8.22 g) is dissolved in nitromethane (20 ml) and a mixture of *sym*-collidine (4.65 ml) and dry ethanol (4.65 ml) with oven-dried molecular sieves (0.5 nm) and stirred at 25°C for about 24 h. Diethyl ether is added, the collidinium hydrobromide is filtered off and the excess of collidine is removed by vacuum evaporation at 45°C. The product 3,4-di-*O*-acetyl-1,2-*O*-(ethyl orthoacetyl)-α-L-galactopyranose (about 5 mmol), is dissolved in 11 ml tetrahydrofuran and added with stirring to a solution of dry phosphoric acid (60 mmol) and 0.6 g phosphorus pentoxide (4.2 mmol) in 15 ml tetrahydrofuran at 5°C. After reaction for 5 min, the mixture is made alkaline with cold aqueous 2 N lithium hydroxide, and lithium phosphate is removed by filtration. L-Galactopyranosyl phosphates in the mixture are in the proportion 95% β- and 5% α-. The anomers are separated (0.5–0.8 mmol) by anion-exchange chromatography on a column (3.6 × 54 cm) of AG1-X8 100–200 mesh (BioRad Laboratories, Richmond, CA, USA) in the bicarbonate form eluted with a linear gradient of triethylammonium bicarbonate buffer, pH 6.8, starting with 1 litre of 0.175 M buffer in the mixing vessel and 1 litre of 0.3 M buffer in the reservoir. Twelve millilitre samples are collected at a flow rate of 2 ml min^{-1}. Every fifth tube is assayed for acid-labile phosphate (MacDonald, 1968) and appropriate fractions are pooled. Triethyl ammonium bicarbonate buffer is removed by lyophilisation of the mixtures.

The orthoester method has not been adapted to the synthesis of micro quantities of sugar phosphates, however, so that ^{3}H- or ^{14}C-labelled β-L-galactopyranosyl phosphate must be made in some other way. An enzymic synthesis suggests itself since β-L-galactopyranosyl phosphate is formed from GDP-Man and an enzyme extract from *Chlorella pyrenoidosa* prepared in the absence of ethylenediaminetetra-acetic acid (EDTA). Presumably the extract contains a nucleotide pyrophosphatase which is inhibited by EDTA. The action of that enzyme also apparently pulls the epimerase reaction (GDP-Man ⇌ GDp-L-Gal) to near completion by removing GDP-L-galactose as it is formed. Under the conditions described below, about 80% of GDP-[^{14}C]Man is converted to β-L-[^{14}C] galactopyranosyl phosphate. The reaction mixture contains: 0.001 μmol GDP-[^{14}C]Man (0.015 μCi), 2 μmol Tris–HCl buffer, pH 7.6, and 0.4 mg (25 μl) of lyophilised *Chlorella* (40–60% $(NH_4)_2SO_4$ fraction). It is incubated for 60 min at 23°C. β-L-[^{14}C]Galactopyranosyl phosphate is isolated from the reaction mixture by electrophoresis on paper at pH 2.7. It can be eluted and converted to GDP-L-[^{14}C]Gal by the morpholidate procedure described earlier.

4. UDP-D-glucose (UDP-Glc)

UDP-Glc is the most abundant nucleotide sugar found in plants. As one of the primary nucleotide sugars (Kochetkov and Shibaev, 1973), it is the precursor of UDP-gal, UDP-GlcA, UDP-Rha (and probably UDP-D-digitoxose (Feingold, 1982)). It is also the D-glycosyl donor in the biosynthesis of a variety of glycosides (e.g. sucrose) and polysaccharides (e.g. callose) in plants. UDP-Glc is commercially available either unlabelled or labelled with ^{3}H or ^{14}C in the D-glucosyl moiety. It can be prepared chemically from UMP morpholidate and Glc-1-P (see Section V.A). It can also be prepared from UTP and Glc-1-P, using UDP-glucose pyrophosphorylase and inorganic pyrophosphatase. This synthesis is useful for the preparation of specifically labelled UDP-Glc and can easily be accomplished with purchased enzymes.

A typical reaction mixture in 1.0 M Tris–acetate buffer, pH 7.8, contains 10 μmol UTP, 5 μmol ^{3}H- or ^{14}C-labelled Glc-1-P, and 1 μmol magnesium acetate in a total volume of 0.5 ml. Five units of UDP-glucose pyrophosphorylase (Type X, Sigma Chemical Company, St. Louis, MO, USA) in 25 μl of 1.0 M Tris–acetate buffer, pH 7.8, and 25 units of inorganic pyrophosphatase (Sigma) in 25 μl of 1.0 M Tris–acetate are added, and the reaction mixture is incubated at 37°C for an appropriate time (usually not more than 30 min). The mixture is then subjected to paper electrophoresis in 0.1 M ammonium acetate, pH 5.8; the radioactive band corresponding in mobility to UDP-Glc is eluted and purified by paper chromatography in 95% ethanol–1 M ammonium acetate, pH 7.0 (7:3; v/v). This method can also be used to prepare UDP-Glc labelled in the uridyl or phosphate moiety, using suitably labelled UTP.

UDP-Glc can also be prepared *directly* from D-glucose in a single reaction mixture, using highly purified commercially obtainable enzymes. The procedure is a modification of that for the preparation of UDP-GlcA, described in Section VI.D.1. The test mixture is identical to that described in the synthesis of UDP-GlcA. However, the mixture used in the actual preparation has the following differences. NAD$^+$ and UDP-glucose dehydrogenase are omitted, whereas the amount of inorganic pyrophosphatase added is increased four-fold. The method of isolation of UDP-Glc from the reaction mixture depends on the reaction scale. Either combined paper electrophoresis and chromatography or the more lengthy purification procedure described for the preparation of UDP-GlcA can be used as appropriate.

5. UDP-D-galactose (UDP-Gal)

Like UDP-Glc, UDP-Gal can be purchased either unlabelled or labelled with ^{3}H or ^{14}C in the D-galactosyl moiety. It is also available via the usual organic synthetic route (see Section V).

An uneconomical but convenient method of preparing UDP-Gal from UDP-Glc involves the use of UDP-glucose-4-epimerase (EC 5.1.3.2). Since the equilibrium constant of the reaction UDP-Glc ⇌ UDP-Gal is 0.28, the yield of UDP-Gal is unsatisfactory. However, if UDP-Gal specifically labelled in the D-galactosyl moiety is required (e.g. UDP-[5-^{3}H]Gal), this method is convenient.

To prepare UDP-Gal using UDP-glucose-4-epimerase, the substrate in 0.1 M sodium glycinate buffer, pH 8.5, is incubated in the presence of the enzyme (from galactose adapted yeast, Sigma) for time sufficient to attain equilibrium. After separation of nucleotide sugars from the reaction mixture using charcoal, UDP-Gal is isolated by paper chromatography with a morpholinium borate solvent as described in Section IV.C.3.

B. Nucleoside Diphosphohexosamines

1. UDP-N-acetyl-D-glucosamine (UDP-GlcNAc)

UDP-GlcNAc has been detected in many plant tissues; it serves as donor of the GlcNAc moiety of glycoproteins and plant glycolipids (Feingold, 1982). The compound is available commercially either unlabelled or labelled with ^{3}H or ^{14}C in the glycosyl moiety. UDP-GlcNAc can be prepared chemically by the methods outlined in this chapter (see also Roseman et al., 1961). It is also available via a number of enzymatic syntheses. Of these, the most useful is that described by Lang and Kornfeld (1984). This method has the additional advantage of enabling the convenient and economical preparation of UDP-GlcNAc labelled with ^{32}P in the β-position as well as the more usual glycosyl-labelled material. The following description is basically the method of Lang and Kornfeld (1984); some minor modifications have been included to facilitate purification of the product.

Prior to beginning the reaction, 28 units of hexokinase (yeast, 127-809, Boehringer-Mannheim Biochemicals, Indianapolis, IN, USA) and 160 units of phosphogluco-mutase (rabbit muscle, 108-383, Boehringer-Mannheim, FRG) are mixed in a 6 mm dialysis sac and dialysed for 2 h at 4°C against 500 ml of 10 mM Tris-HCl, pH 8.0; 1 mM 2-mercaptoethanol:1 mM $MgCl_2$:0.1 mM EDTA. To minimise loss of material and to avoid dilution, the dialysis sac volume should be kept as small as possible and the sac should be kept taut. Twenty-five units of UDP-glucose pyrophosphorylase (yeast, U-8501, Sigma) and 100 units of inorganic pyrophosphatase (yeast, I-4503, Sigma) are resuspended in 80 μl of the same buffer and kept at 4°C.

The following example is for preparation of β-^{32}P-labelled UDP-GlcNAc. Obviously, the same method can be used to prepare UDP-GlcNAc labelled either in the uridyl or glucosaminyl moieties, depending upon the precursor used. Five mCi of [γ^{32}P]ATP is dried *in vacuo* at 25°C in a 1.5 ml polypropylene tube. To the dried ATP are added: 1 M Tris-HCl, pH 8.0 (10 μl); 0.25 M $MgCl_2$ (2 μl); 0.2 M 2-mercaptoethanol (2.5 μl); bovine serum albumin (20 μg); 0.1 M glucosamine (10 μl); 0.1 M ATP, neutralised to pH 6.5 with NaOH (2.5 μl); 0.3 mM glucose 1,6-bisphosphate (G5750, Sigma) (1.3 μl); 0.01 M EDTA (5 μl), and 46.7 μl of the dialysed hexokinase/phosphoglucomutase mix to yield a final volume of *c.* 100 μl. After 60 min at 37°C, 16 μl of 0.1 M UTP (neutralised to pH 6.5 with NaOH) and 44 μl of the pyrophosphorylase mixture are added to yield a final volume of *c.* 160 μl. After an additional 2 h at 37°C, the remaining pyrophosphorylase/pyrophos-phatase mixture is added and incubation at 37°C is continued for a further 2 h, then 53 μl of saturated sodium bicarbonate is added and the mixture is mixed by vortexing. Acetic anhydride (A6404, Sigma) which has been stored in an evacuated desiccator at room temperature (20 μl) is mixed with 500 μl of water by vortexing and 70 μl of the mixture is immediately transferred to the reaction mixture; the latter is vigorously vortexed and held at 25°C for 5 min and finally immersed in a boiling water bath for 3 min, cooled, and lyophilised. One hundred μl of 1 M ammonium bicarbonate (pH 7.5) and 2 μl of 1 M $MgCl_2$ are added, followed by 3 μl of alkaline phosphatase (Sigma) (0.2 units ml^{-1}, dialysed at 4°C for 12 h against 0.01 M Tris–HCl, pH 7.5). After 1 h at 34°C, the mixture is subjected to electrophoresis for 3 h at 20 V cm^{-1} on Whatman 3MM paper in 0.2 M ammonium formate, pH 3.6. The band corresponding to UDP-GlcNAc is eluted and chromatographed on a 1 mm layer of cellulose on a thin layer plate in 95% ethanol–1 M ammonium acetate, pH 7.5 (3:2; v/v). The area corresponding to UDP-GlcNAc is scraped from the plate, eluted in water, and chromatographed on a PEI thin layer plate (Machery Nagel) in a solution containing sodium tetraborate.10H_2O (6 g), boric acid (3 g) ethyleneglycol (25 ml), and water (70 ml). After location and elution from the PEI plate with 0.5 M ammonium bicarbonate, pH 8.5, the labelled UDP-GlcNAc is isolated by adsorption–desorption from charcoal and lyophilisation.

C. Nucleoside Diphosphodeoxyhexoses

1. GDP-L-fucose (GDP-Fuc)

L-Fucose has been found widely distributed in the polysaccharides and glycoproteins of higher plants (Miller, 1973), and enzyme extracts from several species bring about the synthesis of its nucleotide derivative, GDP-Fuc (Barber, 1968, 1969). Although small amounts of that compound labelled with ^{14}C or ^{3}H can be obtained commercially, larger quantities are not readily available. The most successful chemical synthesis has been carried out by Nuñez and Barber (1976) who produced gram quantities of GDP-Fuc in 40% yield from L-fucose. They followed the procedure of Prihar and Behrman (1973) and phosphorylated 2,3,4-Tri-acetyl β-L-fucopyranose directly with o-phenylene phosphorochloridate, removed the phenylene group by oxidation with lead tetra-acetate instead of bromine and deacetylated with lithium hydroxide. The lead tetra-acetate modification increases the yield of the required β-L-fucopyranosyl phosphate five-fold. The compound is purified by chromatography on a column of Dowex-1 bicarbonate eluted with triethylammonium bicarbonate buffer. α-L-Fucopyranosyl phosphate is eluted at 0.22 M buffer concentration and the β-anomer at 0.24 M. Excess buffer is removed from the eluate by repeated evaporation from water. The triethylammonium salt of the β-anomer can then be used directly in the condensation with GMP-morpholidate to form GDP-Fuc.

2.   UDP-L-rhamnose (UDP-Rha)

The chemical synthesis of those L-rhamnosyl nucleotides that predominate in nature, TDP-L-rhamnose (TDP-Rha) and UDP-L-rhamnose (UDP-Rha), was impeded by the lack of a method for phosphorylating L-rhamnose, or any hexose with the manno-configuration at C-2, in the β-configuration. That difficulty was overcome by Prihar and Behrman (1973) who were able to isolate the β-hemiacetal anomers of various hexoses and phosphorylate the hemiacetal group directly. (Details of their method are given above in the section on GDP-L-Fuc). In light of that development, it is surprising that the complete synthesis of UDP-Rha or TDP-Rha has yet to be carried out.

In any case, to obtain small amounts of ^{14}C-labelled UDP-[^{14}C]Rha for experiments with plant systems, synthesis is effected by the epimerisation and reduction of the D-glucosyl moiety of UDP-[^{14}C]Glc with an enzyme preparation from tobacco leaves. Fifty grams of blades from leaves of the upper three or four nodes of 6 to 8-week-old tobacco plants are ground in a mortar with sea sand and 50 ml 0.15 M Tris–HCl–0.05 M 2-mercaptoethanol buffer, pH 7.5. The homogenate is squeezed through cheesecloth and centrifuged at 15 000 × g for 20 min, and the precipitate is discarded. A further precipitate formed by treatment of the supernatant solution with 0.05 vols of manganese chloride is removed by centrifugation. The supernatant solution is made 50% saturated with ammonium sulphate by addition of the solid salt, and the precipitate is suspended in a small volume of buffer and dialysed for 2 h against 1 litre of 0.025 M Tris–HCl–0.01 M mercaptoethanol buffer, pH 7.5. The dialysed protein is passed over a 1 × 10 cm column of Sephadex G-25; a green pigment remains associated with the protein and is used to indicate its effluence from the column. The eluate is made 50% saturated with ammonium sulphate and the precipitate is dialysed overnight against 1 litre of the dilute buffer. The final volume of the preparation is about 0.5 ml and contains about 30 mg protein. Twenty-five microlitres of that extract is incubated for 60 min at 37°C with 3 μl 0.1 M NADPH and 2 μl 0.01 M UDP-[U-^{14}C]Glc (c. 1 μCi). UDP-[^{14}C]Rha is isolated from the reaction mixture by electrophoresis on paper at pH 3.7 and purified by paper chromatography with ethanol–1 M ammonium acetate, pH 7.5 (7:3; v/v). About 15% of the labelled substrate is recovered in UDP-[^{14}C]Rha by this procedure.

D. Nucleoside diphosphohexuronic acids

1.   *UDP-D-glucuronic acid (UDP-GlcA)*

UDP-GlcA has been isolated from a number of plant tissues (Table 2.1). In plants UDP-GlcA is synthesised by the NAD-linked oxidation of UDP-Glc, catalysed by UDP-glucose dehydrogenase, or by the action of UDP-glucuronate pyrophosphorylase on UTP and GlcA-1-P (Feingold, 1982). It is the direct precursor in plants of UDP-GalA, UDP-Xyl, and UDP-Apiose. Additionally, UDP-GlcA serves as the donor of the D-glucuronosyl moiety in the biosynthesis of plant gums and hemicellulose.

UDP-GlcA can be purchased either unlabelled or labelled with ^{14}C in the D-glucuronosyl moiety. It can be prepared via a number of different routes. Chemically, UDP-GlcA is available from the reaction of UMP morpholidate and GlcA-1-P. The latter is prepared by oxidation of Glc-1-P in a stream of oxygen in the presence of platinum oxide (Marsh, 1952). Adam's platinum oxide catalyst is activated by hydrogenation at atmospheric pressure; the activated catalyst can be stored under water for up to one week without loss of activity. A solution of dipotassium α-D-glucopyranosyl phosphate (0.34 g) in 20 ml H_2O is oxidised at 40–45°C in a stream of pure O_2 gas in the presence of 0.1 g of catalyst. Vigorous stirring is maintained during the oxidation to keep the catalyst in suspension, and the O_2 is admitted to the inverted, cone-shaped reaction vessel through a coarse sintered glass disk at the apex of the cone. The pH of the mixture is maintained at 7.5–9.0 by the intermittent addition of 1% aqueous potassium carbonate. At the completion of the reaction (after *c.* 5 h), the filtered solution is evaporated to yield a gum. The latter is dissolved in 50% aqueous methanol and 1.5 ml of absolute ethanol is added. An oil separates which crystallises rapidly at 0°C (0.27 g). Recrystallisation from aqueous methanol as above yields pure tripotassium α-D-glucopyranosyluronate phosphate. This compound is converted to the tri-*n*-octylammonium salt as follows. An aqueous solution containing 1 mmol of the tripotassium salt is passed through a 1 × 5 cm column of Dowex 50 (pyridinium) resin. The column is eluted with water and the eluate and washings are concentrated to 5 ml. Pyridine (15 ml) and 1 mmol of tri-*n*-octylamine are added and the clear solution is evaporated to dryness *in vacuo*; this procedure is repeated three more times. The dried tri-*n*-octylammonium salt of GlcA-1-P in anhydrous pyridine is reacted with UMP morpholidate, and UDP-GlcA is isolated from the reaction mixture as described in Section V.A. Yields of 66% have been reported (Roseman *et al.*, 1961).

UDP-GlcA may be obtained by oxidation of UDP-Glc, using UDP-glucose dehydrogenase (EC 1.1.1.22) (Zalitis *et al.*, 1972). UDP-GlcA can also be prepared directly from D-glucose in a single reaction mixture, using commercially available, highly purified enzymes (e.g. those marketed by Sigma). The procedure is especially useful for making UDP-GlcA from D-glucose labelled with ^{3}H or ^{14}C. The following solutions (in 0.1 M Tris buffer, pH 7.5) are used in the procedure. Reagents: 0.01 M ATP, 0.01 M UTP, 1.5 mM D-glucose 1,6-bisphosphate, 0.08 M $MgCl_2$, 0.03 M phosphoenolpyruvate, and 0.28 M cysteine (freshly prepared). Enzymes (in 0.5 M Tris buffer, pH 7.5): hexokinase, 400 units ml^{-1}; pyruvate kinase, 800 units ml^{-1}; inorganic pyrophosphatase, 1000 units ml^{-1}; UDP-glucose pyrophosphorylase (Type X from Baker's yeast), 100 units ml^{-1}; phosphoglucomutase from rabbit muscle, 100 units ml^{-1} UDP-glucose dehydrogenase, 20 units ml^{-1} in 0.5 M Tris buffer, pH 7.5, 0.01 mM in 2-mercaptoethanol.

Before beginning the synthesis proper, the reaction mixture is tested to determine whether all the enzymes are active. To 60 µl of freshly prepared 0.18 M cysteine (33 mg ml^{-1}) in 0.1 mM Tris–HCl, pH 7.5, is added 7.5 µl of phosphoglucomutase solution and the mixture is held at 25°C for 1 h. The following are added, in the order given, to 0.2 ml of 0.1 M Tris–HCl, pH 7.5: 0.01 M ATP, 30 µl; 0.03 M phosphoenol pyruvate, 50 µl; 0.01M UTP, 50 µl; 0.08 M MgCl$_2$, 2 µl; 1.6 mM D-glucose 1,6-bisphosphate, 1 µl; 0.01 M NAD, 60 µl; cysteine–phosphoglucomutase mixture, 65 µl; hexokinase, 2 µl; pyruvate kinase, 2 µl; inorganic pyrophosphatase, 2.5 µl; and UDP-glucose dehydrogenase, 25 µl. A 0.4 ml portion of the mixture is added to a 0.5 ml capacity spectrophotometer cuvette, 0.5 µl of 0.01 M D-glucose is added and the reaction mixture is mixed thoroughly. To the mixture in the cuvette is added 5 µl of UDP-glucose pyrophosphorylase, the contents of the cuvette are mixed and the absorbance is monitored at 340 nm. If there is an appreciable absorbance increase within 10–12 min, the enzymes are active and the main reaction may be started.

To carry out the large-scale synthesis, the labelled (^{3}H or ^{14}C) D-glucose (from 2–5 mmole) is dried in a small (10–25 ml) flask and dissolved in 2 ml of 0.1 M Tris–HCl, pH 7.5. Phosphoglucomutase (0.075 ml) is activated for 1 h at 25°C by the addition of 0.6 ml of 0.28 M cysteine. In the meantime, 10-fold volumes of the ingredients listed for the test reaction are added to the flask, followed by the phosphoglucomutase–cysteine mixture. The reaction is started by addition of 30 µl of the UDP-glucose pyrophosphorylase solution, and incubation at 37°C is continued for 1 h. A drop of toluene is added to the reaction mixture and the reaction is allowed to proceed at 25°C for an additional 12–18 h. At this time, 5 µl of the reaction mixture is mixed with an equal volume of 0.02 mM UDP-GlcA and subjected to paper electrophoresis at pH 3.6 for 1.5 h at a potential of 30 V cm^{-1}. Complete conversion of the radioactive substrate to UDP-[^{14}C]GlcA is indicated by the presence of a single radioactive compound with the electrophoretic mobility of UDP-GlcA. If unreacted substrate or radioactive intermediates are detected, completion of the conversion may occur with further incubation. However, the reaction described usually proceeds to completion during the initial incubation period.

The reaction mixture is heated at 100°C for 2 min and precipitated protein is removed by centrifugation after cooling to 25°C. The supernatant solution is brought to pH 8.0 with Tris base, 100 µl of alkaline phosphatase (Type 111-S, Sigma) (250 units ml^{-1}) and a drop of toluene are added, and the mixture is held at 25°C for 12 additional hours. It is finally brought to a small volume ($\sim$0.5 ml) at 37°C *in vacuo* and loaded onto a 5 × 60 cm Biogel P-2, 200–400 mesh (Bio-Rad Laboratories, Richmond, CA, USA) column, packed in and eluted with distilled water. Fractions of 9–10 ml are collected and monitored for adsorption at 260 nm, as well as for radioactivity. The radioactive fractions are pooled, concentrated *in vacuo* at 37°C and finally subjected to electrophoresis on Whatman 3MM paper at pH 5.8 in 0.1 M ammonium acetate. The single radioactive band is eluted with water and UDP-GlcA is recovered by lyophilisation.

2.   *UDP-D-galacturonic acid (UDP-GalA)*

UDP-GalA has been demonstrated in a number of plant tissues (Table 2.1). Since the major role of the compound in plants is in the biosynthesis of pectic substances (Ericson and Elbein, 1980), its distribution is ubiquitous. In plants UDP-GalA is synthesised *de novo* by the 4-epimerisation of UDP-GlcA; it also is formed from UTP and GalA-1-P in a reaction catalysed by UDP-galacturonate pyrophosphorylase (Feingold et al., 1958a). UDP-GalA is not commercially available. It can be prepared chemically, as described for UDP-GlcA, by coupling UMP morpholidate and GalA-1-P. The latter is prepared exactly as described for GlcA-1-P by the catalytic oxidation of Gal-1-P.

UDP-GalA can also be made by the 4-epimerisation of UDP-GlcA, using an enzyme preparation from radish roots (Neufeld, 1966). This method is useful for the preparation of small quantities of isotopically labelled UDP-GalA. Although a more active enzyme can be prepared from cells of *Anabaena floss-aquae* (Gaunt et al., 1974), radish root preparations are much more convenient to use.

Assay: Enzyme (15 µl) and 15 µl of UDP-[^{14}C]GlcA (0.3 mM, at least 30 µCi mmol^{-1}) are mixed and incubated at 37°C. At appropriate times (over a 2-h period), the mixture is subjected to paper electrophoresis in 0.1 M ammonium formate, pH 3.6, at 30 V cm^{-1} for 2 h. The compound which migrates with the mobility of UDP-GalA is located by autoradiography and quantitated by strip counting, densitometry of the radioautogram, or by elution from the paper and scintillation counting.

Enzyme preparation: Radish roots (available commercially) are cut up and homogenised in a Waring blender in 0.01 M phosphate buffer, pH 7.0 (1 weight of roots per 0.7 vols of buffer). The slurry is filtered through multiple layers of cheesecloth and freed of particulates by centrifugation at 2000 × g for 5 minutes. The supernatant fluid is then centrifuged at 18 000 × g for 30 min, and the precipitate is suspended in 0.1 M Tris buffer, pH 7.4 (5 µl g^{-1} of radish root). This preparation maintains its activity for at least a year when stored frozen at −20°C.

The assay conditions described are suitable for the larger scale preparation of UDP-[^{14}C]GalA. After separation of the product by paper electrophoresis at pH 3.6 and elution, further purification by paper electrophoresis at pH 5.8 in 0.1 M ammonium acetate buffer aids in eliminating trace contaminants.

E. Nucleoside diphosphopentoses

1. UDP-D-xylose and UDP-L-arabinose (UDP-Xyl and UDP-Ara)

UDP-D-Xylose and its 4-epimer, UDP-L-arabinose, have been isolated from a number of plant sources (Table 2.1). UDP-Xyl is formed in plants by decarboxylation of UDP-GlcA, and UDP-Ara arises from the 4-epimerisation of UDP-Xyl. These are the only reactions known to form UDP-Xyl and UDP-Ara *de novo* in plants; UDP-Ara is also available by pyrophosphorolysis from UTP and Ara-1-P (Feingold, 1982). The nucleotide sugars are donors of the D-xylopyranosyl and L-arabinopyranosyl moieties in the biosynthesis of plant biopolymers. The origin of the most commonly present form of L-arabinose, the L-arabinofuranosyl moiety, has not yet been elucidated (Aspinall *et al.*, 1972).

UDP-Xyl is commercially available either unlabelled or labelled with ^{14}C or ^{3}H in the D-xylosyl moiety, but UDP-Ara is not currently (1988) listed as a stock item by biochemical supply companies. UDP-Xyl and UDP-Ara may be prepared by chemical synthesis from Xyl-1-P or β-L-arabinopyranosyl phosphate (Ara-1-P). 1,2,3,4-Tetra-*O*-acetyl D-xylopyranose or 1,2,3,4-tetra-acetyl L-arabinopyranose (Wright and Khorana,

1958) is converted to Xyl-1-P or Ara-1-P by heating with crystalline phosphoric acid according to the procedure of MacDonald (1966). Following O-deacetylation with LiOH, the product is isolated as the bis cyclohexylammonium salt; the latter compound is converted to the desired UDP-pentose by reacting with UMP-morpholidate as described previously (see Section V.B).

Enzymatic decarboxylation of UDP-GlcA is an alternative method of making UDP-Xyl. The enzyme, UDP-glucuronate decarboxylase (EC 4.1.1.35), is conveniently prepared from the basidiomycete *Cryptococcus laurentii*. The preparation described here yields a partially purified protein which, however, is perfectly adequate for preparation of UDP-Xyl, especially UDP-Xyl labelled with ^{3}H or ^{14}C in the D-xylosyl residue.

Assay: The most convenient assay is that described by John *et al.* (1977) in which $^{14}CO_2$ released from UDP-GlcA uniformly labelled with ^{14}C in the GlcA moiety is trapped and counted. The assay mixture, in a total volume of 0.3 ml of 0.1 M sodium phosphate buffer (pH 7.0) containing $0.5 \, g \, l^{-1}$ of EDTA and $0.5 \, g \, l^{-1}$ of freshly prepared glutathione, is 2 mM in NAD^+ and 2 mM in UDP-[^{14}C]GlcA (0.2 µCi). A small tube with 0.2 ml of a mixture of ethanolamine and 2-methoxyethanol (1:2; v/v) is placed inside the larger tube which contains the assay mixture; the larger tube is then sealed with a serum stopper with a needle septum. The reaction (at 25°C) is started by injecting enzyme solution through the septum into the reaction mixture; after an appropriate time, 0.2 ml of 2 M HCl is injected to stop the reaction and the reaction mixture is held at 37°C for 2 h. Scintillation counting in a scintillation fluid containing $5.5 \, g \, l^{-1}$ of 2,5-diphenyloxazole in 2-methoxyethanol–toluene (1:2; v/v) is used to quantitate the trapped $^{14}CO_2$ in the smaller tube.

Cryptococcus laurentii (Northern Regional Research Laboratories, Peoria, IL USA; strain Y1401) is grown at 28°C with aeration in the following medium: 2% (w/v) glucose, 0.1% urea, 0.1% KH_2PO_4, 0.05% $MgSO_4.5H_2O$, and $0.2 \, mg \, l^{-1}$ thiamine–HCl. When the culture has reached an absorbency of 0.9 (450 nm), the cells are harvested by centrifugation and washed by suspension in 0.1 M Na/K phosphate buffer, pH 7.0, 1 mM in 2-mercaptoethanol and EDTA followed by centrifugation. All subsequent operations are performed at 0–4°C.

Packed, washed cells are suspended in an equal volume of 'Buffer' (0.1 M Na/K phosphate buffer, pH 7.0, containing 0.5 ml 2-mercaptoethanol and 0.5 g EDTA per litre) and disrupted at 0–4°C by sonication [e.g. with a Heat Systems Ultrasonics, Inc. (Farmingdale, NY) Sonifier, Model W-375] at full power for sufficient time to disrupt the cells (usually 4–8 min). The turbid supernatant liquid obtained after centrifugation at $28\,000 \times g$ for 20 min is centrifuged for 1 h at $105\,000 \, g$. To the clear supernatant fluid is added, with stirring, sufficient 0.5 M manganese chloride to make the final concentration 2.5 mM. After an additional 5 min of stirring, the precipitate is removed by centrifugation and discarded. Solid ammonium sulphate ($231 \, g \, l^{-1}$) is added to the clear supernatant fluid and the precipitate is discarded. To the supernatant liquid is added solid ammonium sulphate ($125 \, g \, l^{-1}$), the precipitate is dissolved in a minimal volume of buffer and used in the preparation of UDP-Xyl.

An example of the use of the partially purified enzyme to prepare UDP-[^{14}C]Xyl from UDP-[^{14}C]GlcA follows (Ankel and Feingold, 1966). An alternative technique for the introduction of ^{3}H label into UDP-Xyl is to carry out the decarboxylation in ^3H_2O, which introduces one ^{3}H atom at C-5 of the D-xylosyl moiety (Schutzbach and Feingold, 1970). To 0.4 ml of 0.03 M UDP-[^{14}C]GlcA is added 2.6 ml of Buffer followed by freshly prepared 0.01 M NAD^{+} (0.1 ml). If necessary, the pH is adjusted to 7.0, and the solution is brought to 37°C. Enzyme (3.0 ml) is added, and after sufficient time at 37°C (e.g. 1–2 h) to convert the substrate completely to UDP-[^{14}C]Xyl, the reaction mixture is heated at 100°C for 1 min and the precipitated protein is removed by centrifugation. The nucleotides present in the reaction mixture are isolated by adsorption–desorption from charcoal, and UDP-[^{14}C]Xyl is purified by an appropriate method, e.g. paper electrophoresis at pH 5.8, followed by paper chromatography in ethanol–1 M ammonium acetate, pH 7.0 (7:3; v/v) (Fan and Feingold, 1970).

UDP-Ara may be obtained by the 4-epimerisation of UDP-Xyl followed by separation of the two UDP-pentoses by paper chromatography (Fan and Feingold, 1970). This method is particularly suitable for the preparation of small quantities of UDP-Ara labelled with ^{14}C or ^{3}H in the L-arabinosyl moiety. Raw wheat germ is used as the enzyme source. (All operations are done at 0–4°C.) Two hundred grams of wheatgerm is stirred for 1.5 h with 1500 ml of 0.1 M sodium phosphate, pH 7.0, containing 0.5 g l^{-1} of EDTA and 0.5 ml l^{-1} of 2-mercaptoethanol (buffer). The slurry is filtered through multiple layers of cheesecloth and to the turbid solution is added, with stirring, sufficient 0.5 M manganese chloride to yield a final concentration of 0.015 M. After a further 5 min of stirring, the mixture is centrifuged (10 000 × g for 30 min) and the supernatant liquid is treated with solid ammonium sulphate (298 g l^{-1}). The precipitate is removed by centrifugation and solid ammonium sulphate (96 g l^{-1}) is added to the supernatant solution. The precipitate is collected and dissolved in a minimal volume of buffer (enzyme).

In a typical preparation of UDP-Ara, UDP-[^{14}C]Xyl (1 mg) is incubated with enzyme in a total volume of 5 ml of buffer for 24 h at 25°C. A small drop of toluene is added to prevent bacterial growth. Ethanol (15 ml) is added to the mixture, and the precipitate is discarded. The supernatant liquid is concentrated to a small volume at 37°C *in vacuo* and chromatographed on Whatman 3MM filter paper in 95% ethanol–1 M ammonium acetate, pH 7.5 (7:3; v/v). The radioactive ultraviolet-quenching band is eluted from the paper with water, the eluate concentrated by freeze-drying, and the residue is rechromatographed on Whatman 3MM paper in 1-propanol–0.2 M morpholinium borate at pH 8.6 in 0.01% EDTA (13:7; v/v) (Aspinall *et al.*, 1972). The slower radioactive UV-quenching band corresponds to UDP-Ara; it is eluted with water and freed of morpholinium borate by electrophoresis on Whatman 3MM paper in 0.1M ammonium formate, pH 3.6. The single radioactive UV-quenching band is eluted from the electrophoretogram with water, and the UDP-[^{14}C]Ara is recovered by lyophilisation.

F. UDP-D-apiose (UDP-Api)

UDP-Api is the source of the branched chain pentose D-apiose, found glycosidically bound in many higher plant types (Duff, 1965). UDP-Api is formed upon decarboxylation of UDP-GlcA. The enzyme responsible, UDP-apiose/UDP-xylose synthase, also converts some of the substrate to UDP-Xyl. These activities are inseparable, therefore the reaction product always contains UDP-Api and UDP-Xyl (Matern and Grisebach, 1977), which must be separated to obtain UDP-Api. The enzyme has been purified over 1400-fold from cell suspensions of parsley (Matern and Grisebach, 1977). Less pure enzyme is adequate for UDP-Api preparation; the method given here is essentially that of Baron *et al.* (1973).

The rather cumbersome assay is based on determination of the content of radioactive apiose in the mixture of sugars obtained by acid hydrolysis of reaction mixtures with UDP-[^{14}C]GlcA as substrate. Enzyme is added to a solution composed of 10 µl UDP-[^{14}C]GlcA (300 µCi µmol^{-1}), 10 µl NAD$^+$ (0.02 M), 10 µl dithioerithritol (0.01 M), and 20 µl Tris–HCl (0.5 M, pH 7.8). In crude extracts the assay mixture should also be 10 mM in KCN. After 10 min at 30°C, 20 µl of glacial acetic acid is added and the mixture is held at 100°C for 15 min. The hydrolysate is chromatographed as a 7 cm band for 12 h in ethyl acetate–pyridine–water (8:2:1; v/v/v), using apiose, xylose, and arabinose as reference sugars. After completed chromatography, the reference sugars are detected using analine phthalate spray reagent (1.66 g of phthalic acid and 0.93 g of redistilled aniline in 100 ml water-saturated butanol; the sprayed paper is heated at 110°C until colours develop). The radioactivity present in the apiose and xylose/arabinose zones is quantitated by strip counting in a suitable apparatus or by cutting the paper into strips 1 cm wide perpendicular to the direction of chromatography and counting them in toluene-2,5-diphenyloxazole (5 g l^{-1}). Background is estimated by the counts in a zone between the apiose and xylose zones which is equal to the pentose zone in width.

Although the enzyme has been purified from both *Lemna minor* and cell cultures of parsley, young parsley leaves can also be used as an enzyme source. All procedures are done at 4°C. About 250 g of fresh young parsley leaves and stems is homogenised in a Waring blender with 500 ml of 0.25 M Tris–HCl buffer, pH 8.0; 10 mM 2-mercaptoethanol; 1 mM KCN. The slurry is filtered through four layers of cheesecloth and centrifuged at 40 000 × g for 45 min. The supernatant solution is stirred for 20 min with 25 g of Dowex 1X-2 (Cl$^-$ form equilibrated with 0.25 M Tris–HCl, pH 8.0). After decantation through glass wool, the solution is centrifuged at 10 000 × g for 20 min. To the supernatant liquid is added a saturated solution of ammonium sulphate (pH 8.0) to 40% saturation. The precipitate is discarded and the supernatant solution is brought to 50% saturation with saturated ammonium sulphate solution, and the resulting precipitate is collected by centrifugation at 35 000 × g for 30 min. The precipitate is dissolved in a minimal amount of buffer, clarified by centrifugation at 10 000 × g for 15 min and loaded onto a column (10 ml of gel per ml protein solution) of Sephadex G-25, packed in and eluted with 20 mM Tris–HCl, pH 8.0, 1 mM in 2-mercaptoethanol. The desalted protein solution (enzyme) contains the desired activity.

A typical preparation of high specific activity UDP-Api (Kindel and Watson, 1973) is made as follows. Reaction mixtures contain 2.5 nmol of UDP-[^{14}C]GlcA (0.5 µCi), 0.6 µmol of NAD$^+$, and 0.5 ml of enzyme in a total volume of 0.6 ml. After a sufficient time of incubation at 25°C to convert the substrate to products completely, the reaction mixture is brought to pH 6–6.5 with acetic acid. The reaction mixture is streaked onto Whatman 3MM paper which has been washed with 0.01 M EDTA at pH 7.0, followed by demineralised water. The paper is subjected to descending paper chromatography for 5 days in ethanol: 1 M ammonium acetate, pH 7.5 (5:2; v/v). Five radioactive zones are present: ($R_{\text{UDP-Glc}}$ values are given in parentheses): UDP-GlcA (0.45), unknown compound (0.64), UDP-Ara (0.9), UDP-Xyl (1.0), UDP-Api (1.2). The UDP-Api zone, which always contains some UDP-Xyl, is eluted in 1 mM potassium phosphate, pH 6.0, to minimise degradation. No more than 60–80% of the apiose formed in the reaction can be recovered as UDP-Api. The remainder is converted to UMP and cyclic apiose phosphate during the chromatography procedure. UDP-Api is extremely unstable, especially in the presence of alkali, and to date it has not been isolated as a stable compound and free from UDP-Xyl.

REFERENCES

Altermatt, H. S. and Neish, A. C. (1956). *Can. J. Biochem. Physiol.* **34**, 405–413.
Ankel, H. and Feingold, D. S. (1966). *Biochemistry* **5**, 182–189.
Ashihara, H. (1977). *Z. Pflanzenphysiol.* **81**, 199–211.
Aspinall, G. O., Cottrell, I. W. and Matheson, N. K. (1972). *Can. J. Biochem.* **50**, 574–580.
Barber, G. A. (1968). *Biochim. Biophys. Acta* **165**, 68–75.
Barber, G. A. (1969). *Biochemistry* **8**, 3692–3695.
Barber, G. A. (1971). *Arch. Biochem. Biophys.* **147**, 619–623.
Barber, G. A. (1975). *Arch. Biochem. Biophys.* **167**, 718–722.
Barber, G. A. (1985). *FEBS Lett.* **183**, 129–132.
Baron, D., Streitberger, U. and Grisebach, H. (1973). *Biochim. Biophys. Acta* **293**, 526–533.
Bdolah, A. and Feingold, D. S. (1965). *Biochem. Biophys. Res. Commun.* **21**, 543–546.
Becker, D., Kluge, M. and Ziegler, H. (1971). *Planta* **99**, 154–162.
Bergkvist, R. (1956). *Acta Chem. Scand.* **10**, 1303–1316.
Bergkvist, R. (1957). *Acta Chem. Scand.* **11**, 1457–1464.
Bhatia, I. S. and Uppal, D. (1979). *Plant Sci. Lett.* **16**, 59–66.
Bochner, B. R. and Ames, B. N. (1982). *J. Biol. Chem.* **257**, 9759–9769.
Braell, W. A., Tyo, M. A., Krag, S. S. and Robbins, P. W. (1976). *Anal. Biochem.* **74**, 484–487.
Brown, E. G. (1962). *Biochem. J.* **85**, 633–640.
Brown, E. G. (1965). *Biochem. J.* **95**, 509–514.
Brown, E. G. and Mangat, B. S. (1967). *Biochim. Biophys. Acta* **148**, 350–355.
Brown, E. G. and Short, K. C. (1969). *Phytochemistry* **8**, 1365–1372.
Buchanan, J. G. (1953). *Arch. Biochem. Biophys.* **44**, 140–149.
Buchanan, J. G., Bassham, J. A., Benson, A. A., Bradley, D. F., Calvin, M., Dans, L. C., Goodman, M., Hayes, P. M., Lynch, V. H., Norris, L. T. and Wilson, A. T. (1952) *In* "Phosphorus Metabolism", Vol. II (W. D. McElroy and B. Glass, eds), pp. 440–466. Johns Hopkins Press, Baltimore.
Cabib, E. and Leloir, L. F. (1954). *J. Biol. Chem.* **206**, 779–790.
Cabib, E., Leloir, L. F. and Cardini, C. E. (1952). *Cienc. Invest.* **8**, 469–471.
Caputto, R., Leloir, L. F., Cardini, E. C. and Paladini, A. C. (1950). *J. Biol. Chem.* **184**, 333–350.
Cardini, C. E., Caputto, R., Paladini, A. C. and Leloir, L. F. (1950). *Nature (London)* **165**, 191–192.
Cardini, C. E., Leloir, L. F. and Chiriboga, J. (1955). *J. Biol. Chem.* **214**, 149–155.
Carminatti, H., Passeron, S., Dankert, M. and Recondo, E. (1965). *J. Chromatogr.* **16**, 126–129.
Cassells, A. C. and Harmey, M. A. (1968). *Arch. Biochem. Biophys.* **126**, 486–491.
Castellani, A. A., Calatroni, A. and Righetti, P. G. (1967). *Ital. J. Biochem.* **16**, 5–11.
Chang, S., Broschat, K. O. and Serif, G. S. (1985). *Anal. Biochem.* **144**, 253–257.
Cole, C. V. and Ross, C. (1966). *Anal. Biochem.* **17**, 526–539.
Collins, G. G., Jenner, C. F. and Paleg, L. G. (1972). *Plant Physiol.* **49**, 398–403.
Cori, C. F., Schmidt, G. and Cori, G. T. (1939). *Science* **89**, 464–465.
Crestfield, A. M. and Allen, F. W. (1955). *Anal. Chem.* **27**, 422–423.
Cumming, D. E. (1970). *Biochem. J.* **116**, 189–198.
Dankert, M., Passeron, S., Recondo, R. and Leloir, L. F. (1964a). *Biochem. Biophys. Res. Commun.* **14**, 358–362.
Dankert, M., Goncales, I. R. J. and Recondo, E. (1964b). *Biochim. Biophys. Acta* **81**, 78–85.
Dankert, M., Passeron, S. and Recondo, E. (1972). *An Assoc. Quim. Argent.* **60**, 257–271.
Duff, R. B. (1965). *Biochem. J.* **94**, 768–771.
Dunn, D. B. and Hall, R. H. (1968). *In* "Handbook of Biochemistry" (H. A. Sober, ed.). Chem. Rubber Co., Cleveland.
Dutton, G. J. and Storey, I. D. E. (1953). *Nature (London)* **4188**, 191–192.

Elbein, A. L. (1966). *Methods Enzymol.* **8**, 142–145.
Ericson, M. C. and Elbein, A. D. (1980). *In* "The Biochemistry of Plants", Vol. 3 (J. Preiss, ed.), p. 592. Academic Press, New York.
Fan, D.-F. and Feingold, D. S. (1970). *Plant Physiol.* **46**, 592–595.
Fan, D.-F. and Feingold, D. S. (1972). *Arch. Biochem. Biophys.* **148**, 576–580.
Feingold, D. S. (1982). *Encyclopedia of Plant Physiol.* **13A**, 1–76.
Feingold, D. S. and Avigad, G. (1980). *In* "The Biochemistry of Plants", Vol. 3 (J. Preiss, ed.), pp. 101–169. Academic Press, New York.
Feingold, D. S., Neufeld, E. F. and Hassid, W. Z. (1958a). *Arch. Biochem. Biophys.* **78**, 401–406.
Feingold, D. S., Neufeld, E. F. and Hassid, W. Z. (1958b). *J. Biol. Chem.* **233**, 783–788.
Feingold, D. S., Neufeld, E. F. and Hassid, W. Z. (1960). *J. Biol. Chem.* **235**, 910–913.
Franz, G. (1969) *Phytochemistry* **8**, 737–741.
Franz, G. and Meier, H. (1969). *Biochim. Biophys. Acta* **184**, 658–659.
Gaunt, M. A., Maitra, U. S. and Ankel, H. (1974). *J. Biol. Chem.* **249**, 2366–2372.
Ginsburg, V., Stumpf, P. K. and Hassid, W. Z. (1956). *J. Biol. Chem.* **223**, 977–983.
Glaser, L. and Brown, D. H. (1955). *Proc. Natl. Acad. Sci. USA* **41**, 253–260.
Glaser, L. and Brown, D. H. (1957). *Biochim. Biophys. Acta* **23**, 449–450.
Gonzalez, H. S. and Pontis, H. G. (1963). *Plant Physiol.* **54**, 186–191.
Goudsmit, E. M. and Neufeld, E. F. (1966). *Biochim. Biophys. Acta* **121**, 192–195.
Grégoire, J., Grégoire, J., Limozin, H. and Van, L. V. (1963). *C. R. Acad. Sci.* **Ser. D257** 3508–3511.
Grégoire, J., Grégoire, J., Limozin, H. and Van, L. V. (1965). *Bull. Soc. Chim. Biol.* **47**, 195–212.
Grier, T. J. and Rasmussen, J. R. (1982). *Anal. Biochem.* **127**, 100–104.
Hampe, M. M. V. and Gonzalez, N. S. (1967). *Biochim. Biophys. Acta* **148**, 566–568.
Hanes, C. S. (1940). *Proc. Roy. Soc. Ser. B* **128**, 421–450.
Hebda, P. A., Behrman, E. J. and Barber, G. A. (1979). *Arch. Biochem. Biophys.* **194**, 496–502.
Hehre, E. J. and Sugg, J. Y. (1942). *J. Exp. Med.* **75**, 339–353.
Hestrin, S., Avineri-Shapiro, S. and Aschner, M. (1943). *Biochem. J.* **37**, 450–456.
Hirst, E. L. (1942). *J. Chem. Soc.* **208**, 70–78.
Hudson, C. S. and Dale, J. K. (1915). *J. Am. Chem. Soc.* **64**, 1851–1856.
Isherwood, F. A. and Selvendran, R. R. (1970). *Phytochemistry* **9**, 2265–2269.
Jenner, C. F. (1968). *Plant Physiol.* **43**, 41–49.
John, K. V., Schwartz, N. B. and Ankel, H. (1977). *J. Biol. Chem.* **252**, 6707–6710.
Katan, R. and Avigad, G. (1965). *Isr. J. Chem.* **3**, 110.
Kauss, H. (1965). *Biochem. Biophys. Res. Commun.* **18**, 170–173.
Kauss, H. and Kandler, O. (1962). *Z. Naturforsch.* **17b**, 858–860.
Keys, A. J. (1963). *J. Exp. Bot.* **14**, 14–28.
Kindel, P. K. and Watson, R. R. (1973). *Biochem. J.* **133**, 227–241.
Kirby, E. C. and Roberts, R. M. (1971). *Planta* **99**, 211–221.
Kochetkov, N. K. and Shibaev, V. N. (1973). *Adv. Carbohydr. Chem. Biochem.* **28**, 307–399.
Kornfeld, S. and Glaser, L. (1961). *J. Biol. Chem.* **236**, 1795–1799.
Lang, L. and Kornfeld, D. (1984). *Anal. Biochem.* **140**, 264–269.
Lee, C. H. and Sarma, R. H. (1976). *Biochemistry* **15**, 697–704.
Leloir, L. F. (1951). *Arch. Biochem. Biophys.* **33**, 186–190.
Leloir, L. F. and Cabib, E. (1953). *Nature (London)* **4188**, 37–38.
Leloir, L. D., Olarvarria, J. M., Goldenberg, S. H. and Carminatti, H. (1959). *Arch. Biochem. Biophys.* **81**, 508–520.
Lin, T. Y. and Hassid, W. Z. (1966). *J. Biol. Chem.* **241**, 3283–3293.
Loewus, F. A. and Dickinson, D. B. (1982). *Encyclopedia of Plant Physiol.* **13A**, 193–216.
MacDonald, D. L. (1966). *Methods Enzymol.* **8**, 121–125.
MacDonald, D. L. (1968). *Carbohydrate Res.* **6**, 376–381.
Maretzki, A. and Thom, M. (1978). *Plant Physiol.* **61**, 544–548.
Marsh, C. A. (1952). *J. Chem. Soc.*, 1578–1582.
Matsumoto, H., Wakiuchi, N. and Takahashi, E. (1969). *Physiol. Plant.* **22**, 537–545.
Matern, U. and Grisebach, H. (1977). *Eur. J. Biochem.* **74**, 300–312.
Maxwell, E. S. (1957). *J. Biol. Chem.* **229**, 139–151.

Mengel, K. and Haeder, H. E. (1977). *Plant Physiol.* **59**, 282–284.

Meyer, R. and Wagner, K. G. (1985). *Anal. Biochem.* **148**, 269–276.

Miller, L. P. (1973). *In* "Phytochemistry", Vol. I (L. P. Miller, ed.), pp. 297–375. Van Nostrand Reinhold, New York.

Mills, G. C. (1982). *J. Chromatogr.* **242**, 103–110.

Moffat, J. G. (1966). *Methods Enzymol.* **8**, 136–142.

Munch-Petersen, A., Kalckar, H. M., Cutolo, E. and Smith, E. E. B. (1953). *Nature (London)* **172**, 1036–1037.

Murata, T. (1975). *Agric. Biol. Chem.* **39**, 1401–1406.

Murata, T. (1977). *Agric. Biol. Chem.* **41**, 1995–2002.

Murata, R., Minamikawa, T. and Akazawa, T. (1963). *Biochem. Biophys. Res. Commun.* **13**, 439–443.

Murata, T., Minamikawa, T., Akazawa, T. and Sugiyama, T. (1964). *Arch. Biochem. Biophys.* **106**, 371–378.

Murata, T., Sugiyama, T., Minamikawa, T. and Akazawa, T. (1966). *Arch. Biochem. Biophys.* **113**, 34–44.

Neish, A. C. (1955). *Can. J. Biochem. Physiol.* **33**, 658–666.

Neufeld, E. F. (1966). *Methods Enzymol.* **8**, 276–277.

Neufeld, E. F. and Feingold, D. S. (1961). *Biochim. Bioiphys. Acta* **53**, 589–590.

Neufeld, E. F., Ginsburg, V., Putman, E. W., Fanshier, D. and Hassid, W. Z. (1957). *Arch. Biochem. Biophys.* **69**, 602–616.

Neufeld, E. F., Feingold, D. S. and Hassid, W. Z. (1958). *J. Am. Chem. Soc.* **80**, 4430.

Niemyer, H. (1955). "Metabolismo de los Hydrates de Carbono en el Higado", p. 148. Universidad de Chile, Santiago, Chile.

Nikaido, H. (1975). *In* "Handbook of Biochemistry and Molecular Biology: Lipids, Carbohydrates, Steroids", 3rd edn (G. D. Fasman, ed.), pp. 446–458. Chemical Rubber Press, Cleveland, OH.

Nuñez, H. A. and Barker, R. (1976). *Biochemistry* **15**, 3843–3847.

Okuda, S., Murata, S. and Suzuki, N. (1986). *Biochem. J.* **239**, 733–738.

Olempska-Beer, Z. and Freese, E. B. (1984). *Anal. Biochem.* **140**, 236–245.

Pakhomova, M. V., Zaitseva, G. N. and Al'bitskaya, O. N. (1965). *Biokhimiya* **30**, 1204–1221.

Paladini, A. and Leloir, L. F. (1952). *Biochem. J.* **51**, 426–430.

Palmer, C. E. (1976). *Z. Pflanzenphysiol.* **77**, 345–349.

Park, J. T. (1952a). *J. Biol. Chem.* **194**, 877–884.

Park, J. T. (1952b). *J. Biol. Chem.* **194**, 885–895.

Park, J. T. (1952c). *J. Biol. Chem.* **194**, 897–904.

Park, J. T. and Johnson, M. J. (1949). *J. Biol. Chem.* **179**, 585–592.

Passeron, S., Recondo, E. and Dankert, M. (1964). *Biochim. Biophys. Acta* **89**, 372–374.

Payne, S. M. and Ames, B. N. (1982). *Anal. Biochem.* **123**, 151–161.

Prihar, H. S. and Behrman, E. J. (1973). *Biochemistry* **12**, 997–1002.

Randerath, E. and Randerath, K. (1964). *J. Chromatogr.* **16**, 126–129.

Recondo, E., Dankert, M. and Leloir, L. F. (1963). *Biochem. Biophys. Res. Commun.* **12**, 204–207.

Rees, W. R. and Duncan, H. J. (1965). *Biochem. J.* **94**, 19.

Roberts, R. M. (1970). *Plant Physiol.* **45**, 263–267.

Roberts, R. M. (1971a). *Arch. Biochem. Biophys.* **145**, 685–692.

Roberts, R. M. and Butt, V. S. (1970). *Planta* **84**, 250–262.

Roberts, R. M. and Pollard, W. E. (1975). *Plant Physiol.* **55**, 431–436.

Roberts, R. M., Heiseman, A. and Wicklin, C. (1971a). *Plant Physiol.* **48**, 36–42.

Roberts, R. M., Connor, A. B., and Cetorelli, J. J. (1971b). *Biochem. J.* **125**, 999–1008.

Roseman, S., Distler, J. J., Moffatt, J. G. and Khorana, H. G. (1961). *J. Am. Chem. Soc.* **83**, 659–663.

Sanderman, H. and Grisebach, H. (1968a). *Eur. J. Biochem.* **6**, 404–410.

Sanderman, H. and Grisebach, H. (1968b). *Biochim. Biophys. Acta* **156**, 435–436.

Sanwal, G. G. and Preiss, J. (1969). *Phytochemistry* **8**, 707–723.

Schutzbach, J. S. and Feingold, D. S. (1970). *J. Biol. Chem.* **245**, 2476–2482.

Sebesta, K. and Sorm, F. (1959). *Chem. Commun.* **24**, 2781–2789.

Selvendran, R. R. and Isherwood, F. A. (1967). *Biochem. J.* **105**, 723–738.

Sharma, K. P. and Bhatia, I. S. (1978). *Ind. J. Biochem. Biophys.* **15**, 133–135.

Silbert, J. E. and DeLuca, S. (1967). *Biochim. Biophys. Acta* **141**, 193–196.

Sioufi, A., Percheron, F. and Courtois, J. E. (1970). *Phytochemistry* **9**, 991–999.

Slater, W. G. and Beevers, H. (1958). *Plant Physiol.* **33**, 146–151.

Solms, J. and Hassid, W. Z. (1957). *J. Biol. Chem.* **228**, 357–364.

Solms, J., Feingold, D. S. and Hassid, W. Z. (1957). *J. Am. Chem. Soc.* **79**, 2342–2343.

Strominger, J. L. (1959). *C. R. Trav. Lab. Carlsberg* **31**, 181–192.

Strominger, J. S. and Mapson, L. W. (1957). *Biochem. J.* **66**, 567–572.

Su, J. C. and Hassid, W. Z. (1962a). *Biochemistry* **1**, 468–474.

Su, J. C. and Hassid, W. Z. (1962b). *Biochemistry* **1**, 474–480.

Taniguchi, H., Uemura, Y. and Nakamura, M. (1967). *Agric. Biol. Chem.* **31**, 231–239.

Uemura, Y., Nakamura, M. and Funahashi, S. (1967). *Arch. Biochem. Biophys.* **199**, 240–252.

Wallis, M. E. and Bradbeer, J. W. (1970). *J. Exp. Biol.* **21**, 1039–1047.

Wang, D. (1967). *Can. J. Biochem.* **45**, 721–728.

Wolochow, H. E., Putman, E. W., Doudoroff, M., Hassid, W. Z. and Barker, H. A. (1949). *J. Biol. Chem.* **180**, 1237–1242.

Wright, R. S. and Khorana, H. G. (1958). *J. Am. Chem. Soc.* **78**, 811–816.

Zalitis, J., Uram, M., Bowser, A. M. and Feingold, D. S. (1972). *Methods Enzymol.* **28**, 430–437.

Zemek, J., Stremen, J. and Hricova, D. (1976). *Z. Pflanzenphysiol.* **77**, 95–98.

3 Lipid-linked Saccharides in Plants: Intermediates in the Synthesis of N-linked Glycoproteins

ALAN D. ELBEIN and GUR P. KAUSHAL

Department of Biochemistry, The University of Texas Health Science Center, San Antonio, TX 78284, USA

I. INTRODUCTION

Lipid-linked saccharides have now been found in a large number of diverse organisms, including bacteria, fungi, yeasts, protozoa, insects, birds, mammals and higher plants

METHODS IN PLANT BIOCHEMISTRY Vol. 2
ISBN 0-12-461012-9

(Hemming, 1974a; Waechter and Lennarz, 1976; Parodi and Leloir, 1979; Pont Lezica et al., 1986). All of these lipid-linked saccharides have the same general structure [i.e. polyisoprenyl-phosphoryl-sugar or polyisoprenyl-pyrophosphoryl-sugar(s)], but they differ in the specific sugars that are attached to these lipids. There appear to be two general roles for these types of intermediates in biosynthetic reactions: one is as an intermediate glycosyl donor for the formation or polymerisation of extracellular polysaccharides such as peptidoglycans, lipopolysaccharides, teichoic acids, capsular polysaccharides, etc. (Nakaido and Hassid, 1971; Troy, 1979), while the other role is to serve as a precursor in the glycosylation of the N-linked glycoproteins (Hubbard and Ivatt, 1981). In either case, it is likely that the lipid-linked saccharides function to transport the hydrophilic sugars from the cytoplasm into the hydrophobic environment of the endoplasmic reticulum membranes where polymerisation reactions are believed to occur.

Plant cells produce cell wall polysaccharides, extracellular polysaccharides (i.e. slime polysaccharides) and N-linked glycoproteins, and lipid-linked saccharides have been implicated in the synthesis of both the N-linked glycoproteins and the cell wall polysaccharides (Elbein, 1979; Lehle and Tanner, 1983). However, the vast majority of studies on the lipid-linked saccharide intermediates in plants has involved their critical role in the biosynthesis of the oligosaccharide chains of the N-linked glycoproteins. For that reason, and since the role of these intermediates in cell wall polysaccharide synthesis has been covered in a recent review (James et al., 1985), this chapter will concentrate on their involvement in glycoprotein biosynthesis. Much of the initial work in this field was done in animal systems and much more information is available concerning the 'dolichol pathway' in animal tissues. Therefore some of those studies will be reviewed here. However, since that literature is so extensive, it cannot be covered in detail. In addition to the reviews already cited above, the reader is urged to examine a number of other excellent reviews on this subject in order to obtain a better perspective of the whole field (Struck and Lennarz, 1980; Olden et al., 1982; Snider, 1983; Vliegenthart et al., 1983; Kornfeld and Kornfeld, 1985; Tanner and Lehle, 1987).

II. GENERAL METHODS USED IN THESE STUDIES

Most of the studies on the lipid-linked saccharides have been of a biosynthetic nature and have involved incubating microsomal preparations of a given tissue with a radioactive nucleoside diphosphosugar, such as GDP-[^{14}C]mannose, UDP-[^{14}C]glucose or UDP-[^{3}H]GlcNAc. In fact, much of our understanding on the pathway of synthesis and the structures of the intermediates have come from these biosynthetic studies (Elbein, 1979; Struck and Lennarz, 1980; Lehle and Tanner, 1983; Kornfeld and Kornfeld, 1985). Of course, a great variety of procedures have been utilised for preparing microsomal fractions and for doing the incubations, but a general protocol used in our laboratory for mung bean sprouts is as follows: About 400 g of sprouts are blended for 20 s in 200 ml of 50 mM Tris-HCl buffer, pH 7.5, containing 8% sucrose, 0.5% polyvinylpyrrolidone, 1 mM EDTA, 1 mM $MgCl_2$ and 0.5 mM dithiothreitol. The homogenate is strained through 8 layers of cheesecloth and centrifuged at $3000 \times g$ to remove large particles and whole cells. The resulting supernatant liquid is then centrifuged at $105\,000 \times g$ for 45 min to obtain the membrane pellet (i.e. microsomal fraction). This membrane fraction is generally suspended in 50 mM Tris buffer, pH 7.5, containing 10% glycerol and 0.5 mM dithiothreitol. To assay for the formation of lipid-linked saccharides, 0.1 ml of this microsomal enzyme is incubated in 50 mM Tris buffer, pH 7.5, containing 2 mM $MgCl_2$, 0.1 ml of 0.5% bovine serum albumin and 0.1 to 0.5 μCi of nucleoside diphosphosugar (GDP-[^{14}C]mannose, UDP-[^{14}C]glucose or UDP-[^{3}H]GlcNAc), all in a final volume of 0.4 ml. In some cases, especially when lipid acceptors are added to the incubation mixtures, detergents such as Triton X-100 or NP-40 are included in the incubation mixtures, usually at 0.1 to 0.5%. These lipid acceptors, such as dolichyl-pyrophosphoryl-GlcNAc, GlcNAc-GlcNAc-pyrophosphoryl-dolichol and Man-β-GlcNAc-GlcNAc-pyrophosphoryl-dolichol, are usually added to those incubations when solubilised enzyme preparations are used in the incubations, but in some cases, they may also be added to particulate enzyme preparations. The purpose of adding these lipids to the incubations is, of course, to provide acceptors for the sugars. When the lipid-acceptors are included in the incubations, they are added to the tubes first, and the solvent (usually $CHCl_3$) is removed under a stream of nitrogen. Then, 0.1 ml of 0.5% Triton X-100 is added and the lipid is suspended by vigorous mixing. Other reaction components are then added and the incubations are done at 37°C for various times. The reactions are terminated by the addition of 0.6 ml of H_2O and 2 ml of 1:1 mixture of $CHCl_3$–CH_3OH (Forsee and Elbein, 1975; Chambers and Elbein, 1975).

In general, the lipid-linked saccharides can be separated into groups, based on the number of sugar residues that they contain (i.e. on their solubility in organic solvents), by extraction of the incubation mixtures with $CHCl_3$–CH_3OH–H_2O solvents of increasing polarity. Thus, the lipid-linked monosaccharides (i.e. polyprenyl-phosphoryl-mannose or glucose, or polyprenyl-pyrophosphoryl-GlcNAc) are usually extracted by adding sufficient amounts of $CHCl_3$, CH_3OH and H_2O to the incubation mixtures to give a 1:1:1 (v/v) mixture of these solvents. After thorough mixing, the layers are separated by centrifugation, and the lower or $CHCl_3$ layer is removed and saved. The upper layer and the interface are re-extracted with $CHCl_3$ by the addition of another 1.5 ml portion of $CHCl_3$ and, after centrifugation, the lower layer is combined with the first lower layer. This combined $CHCl_3$ layer contains mostly the lipid-linked monosaccharides, but some of the smaller oligosaccharide-lipids may also be partially extracted into this fraction. Thus, small amounts of GlcNAc-GlcNAc-pyrophosphoryl-polyisoprenol and Man-β-GlcNAc-GlcNAc-pyrophosphoryl-polyisoprenol may also be found in the $CHCl_3$ layer (Parodi et al., 1972; Waechter et al., 1973; Hsu et al., 1974; Behrens and Tabora, 1978).

After removal of the $CHCl_3$ layer as described above, several ml of absolute methanol are added to the upper phase and interface in order to solubilise any remaining $CHCl_3$ in this fraction. The particulate material, which contains most of the lipid-linked oligosaccharides and the glycoproteins, is then isolated by centrifugation and the supernatant liquid is removed and discarded. The pellet is then washed three times with 50% CH_3OH to remove the radioactive nucleoside diphosphosugars and their degradation products. Finally, the pellet is washed with absolute methanol and suspended in $CHCl_3$–CH_3OH–H_2O (10:10:3; v/v) to extract the lipid-linked oligosaccharides. After allowing this mixture to stand at room temperature for several hours, the pellet is again removed by centrifugation, and the $CHCl_3$–CH_3OH–H_2O is removed and saved. The pellet may then be washed with methanol, 10% trichloroacetic acid, and 50% methanol to remove all traces of contamination before the glycoproteins are isolated (Spiro et al., 1976; Sharma et al., 1976; Herscovics et al., 1977a).

Extraction of radioactivity by one of these solvents is not definitive proof for the formation of polyprenol-linked saccharides, since sugars can be incorporated into a number of different types of glycolipids by microsomal enzyme preparations. For example, many plant extracts catalyse the transfer of glucose from UDP-[^{14}C]glucose into steryl glucosides and acylated steryl glucosides, and these radioactive lipids would be extracted into the first chloroform extract (Hou, et al., 1968; Laine and Elbein, 1971). Thus, in order to prove that the radioactivity is really in polyisoprenyl-sugar intermediates, the radioactive glycolipids must be further purified, analysed and characterised.

One property of the polyisoprenyl-sugar intermediates that is quite distinctive is the fact that they are all negatively charged due to the presence of phosphoryl or pyrophosphoryl groups. As a result, these compounds bind to columns of DEAE-cellulose (in the acetate form). For the lipid-linked monosaccharides, one applies the sample in $CHCl_3$–CH_3OH (1:1; v/v) (after pre-washing the column in $CHCl_3$–CH_3OH), washes the column well with this solvent to remove any neutral lipids, and then elutes the charged lipids with a gradient of sodium acetate (usually 0–0.5 M) in $CHCl_3$–CH_3OH (1:1; v/v). Dolichyl-phosphoryl-mannose and dolichyl-phosphoryl-glucose emerge from this column in the same position (usually at about 0.1 M sodium acetate), but dolichyl-pyrophosphoryl-GlcNAc binds more tightly to the column and requires higher concentrations of salt for elution. The polyisoprenyl-pyrophosphoryl oligosaccharides can also be purified and partially characterised by chromatography on columns of DEAE-cellulose, but in this case the column is washed with $CHCl_3$–CH_3OH–H_2O (10:10:3; v/v), and the sample is applied in this solvent. Again, after thorough washing of the column with solvent to remove the neutral lipids, the charged lipids are eluted with a gradient of sodium acetate in the same solvent. The lipid-linked oligosaccharides that have the larger-sized oligosaccharides emerge from the column earliest, i.e. at the lower salt concentrations, since they have a lower charge per mass. Thus, one can obtain some separation of these intermediates into different groups on the basis of the size of the oligosaccharide (Caccam et al., 1980; Schutzbach et al., 1980; Rearick et al., 1981b; Sasak et al., 1984). Minor contaminants such as phospholipids that may be present in the lipid-acceptor fraction can be removed by saponification of the fraction with 0.1 N NaOH in $CHCl_3$–CH_3OH–H_2O (10:10:3; v/v) at 37°C for 20 min. Further purification of the lipid acceptors can be achieved by gel filtration on columns of LH-20 or by HPLC.

Another property of the polyisoprenyl-sugar intermediates that is useful for their characterisation is the fact that the sugars are attached to the lipid in acid-labile bonds. Thus, the glucose or mannose can be released from the polyisoprenyl-phosphoryl-sugar by heating at 100°C for 15 min in quite dilute acid (0.02–0.05 N HCl). Likewise, the oligosaccharides or the GlcNAc (or GlcNAc-GlcNAc) are released from the polyisoprenyl-pyrophosphoryl-oligosaccharides under the same hydrolytic conditions. On the other hand, the sugars found in most other lipids (i.e. steryl-glucosides, galactosyl-diglycerides, cerebrosides, etc.) are present in acid-stable bonds and usually require 1–3 N HCl at 100°C for 1 or 2 h to liberate the sugars. After hydrolysis of the glycolipid in 50% propanol containing 0.02–0.05 N HCl at 100°C for 15–20 min, the mixture is cooled, neutralised with NaOH, and extracted with $CHCl_3$ by the addition of 1 vol of this solvent to give a $CHCl_3$–propanol–H_2O mixture of 1:1:1 (v/v). Release of radioactivity into the aqueous phase (upper phase) is an indication that the sugars were attached in acid-labile bonds and that the glycolipid was probably a polyisoprenyl-phosphoryl- or pyrophosphoryl-linked saccharide. The radioactive sugars or oligosaccharides liberated by acid can be identified by paper chromatography, by HPLC, or by gel filtration on columns of Biogel P-4 (Caccam *et al.*, 1980; Schutzbach *et al.*, 1980; Rearick *et al.*, 1981a,b; Sasak *et al.*, 1984). The polyisoprenyl-linked saccharides can also be identified directly by thin layer chromatography in a variety of solvent systems (Forsee and Elbein, 1973; Lehle and Tanner, 1978). For example, one can chromatograph these lipids on silica gel plates in a neutral solvent [CHCl$_3$–CH$_3$OH–H$_2$O (65:25 4; v/v)], an acidic solvent [CHCl$_3$–CH$_3$OH–acetic acid–H$_2$O (25:15:4:2; v/v)], and a basic solvent [CHCl$_3$–CH$_3$OH–NH$_4$OH (75:25:4; v/v)] in order to obtain information about their structure. The polyprenyl-phosphoryl-sugars usually show an intermediate migration in the neutral solvent (R_f = 0.3), a rapid migration in the acidic solvent (R_f = 0.75) and a slow migration in the basic solvent (R_f = 0.1). On the other hand, the uncharged glycolipids such as the steryl glucosides usually show similar migration rates in all three solvents since there is no charged group involved in the migration properties of these neutral lipids (Elbein *et al.*, 1975; Elbein, 1980). Thus, by some combination of the above methods, one can obtain sufficient data to determine the nature of the biosynthetic products. Other methods such as mass spectral identification of lipids and sugars, or NMR analysis, are of great help and significance if one can obtain sufficient amounts of material. However, with these types of intermediates, the paucity of material is frequently a handicap. Of course, the ultimate proof of the polyisoprenyl-nature of the glycolipids is to isolate the glycolipid and show that it will serve as a sugar donor when added back to *in vitro* glycoprotein biosynthesising systems.

III. OCCURRENCE AND STRUCTURE OF THE LIPID PORTION OF THE LIPID-LINKED SACCHARIDES

Plants contain a great variety of naturally occurring polyisoprenols. All of these compounds have a carbon skeleton made up of isoprene units that are linked in a head-to-tail fashion and contain a primary alcohol group on the α-isoprene unit. Figure 3.1 shows the general structure of a polyisoprenol, where *n*, that is the number of isoprene units, can vary from 4 to 20 or 21. Table 3.1 lists some of the polyprenols that have been identified in plants and indicates the source as well as something about the structure. Most of the polyprenols, with the exception of solanesol and spadicol, contain both Z and E double bonds. Usually the first three isoprene units at the end of the molecule have E double bonds and the remaining isoprene units have Z double bonds. However, in the case of solanesol and spadicol, all of the double bonds are E. The α-isoprene unit also varies in the polyprenols in that it may be saturated or unsaturated (allylic). Most of the polyprenols, with the exception of the dolichols, have allylic α-isoprene units, and these types of lipids serve as sugar carriers and sugar donors in the biosynthesis of various bacterial cell wall polymers, such as lipopolysaccharide, pepetidoglycan, teichoic acid and so on. On the other had, the dolichols usually contain 13–21 isoprene units with the α-isoprene being saturated. These latter polyprenols are the ones that are usually found in animal cells and in animal tissues, but they may also occur in fungi, algae and higher plants (Hemming, 1974b).

Any of the polyprenols, i.e. those that have allylic α-isoprene units or those that have saturated α-isoprene units, can act as acceptors of sugars in the *in vitro* (i.e. cell-free

TABLE 3.1. Characteristics of some plant polyprenols.

Polyprenol	Source	No. of isoprene units	No. of double bonds		α-Residue
			E	Z	
Betaprenol	*Beta vulgaris*	10–13	3	6–9	Allylic
Betulaprenol	*Betula verrucosa*	6–9	2	3–6	Allylic
Castaprenol	*Aesculus hippocastanum*	10–13	3	7–9	Allylic
Dolichol	*Phytophora cactorum*	13–17	3	8–12	Saturated
Ficaprenol	*Ficus elastica*	10–13	3	6–9	Allylic
Hexahydropolyprenol	*Aspergillus fumigatus*	19–23	2	14–18	Saturated
Solanesol	*Nicotiana tabacum*	9	9	0	Allylic

systems) studies. Apparently as long as the polyprenol is in the phosphorylated form, it can serve reasonably well as a glycosyl acceptor with many microsomal preparations. For example, ficaprenyl-phosphate works as well as a mannosyl acceptor from GDP-mannose with a mung bean particulate enzyme as does dolichyl-phosphate (Butterworth and Hemming, 1968; Dallner *et al.*, 1972; Forsee and Elbein, 1973; Brett and Leloir, 1977). Other phosphorylated polyprenols would probably be equally effective with these cell-free extracts.

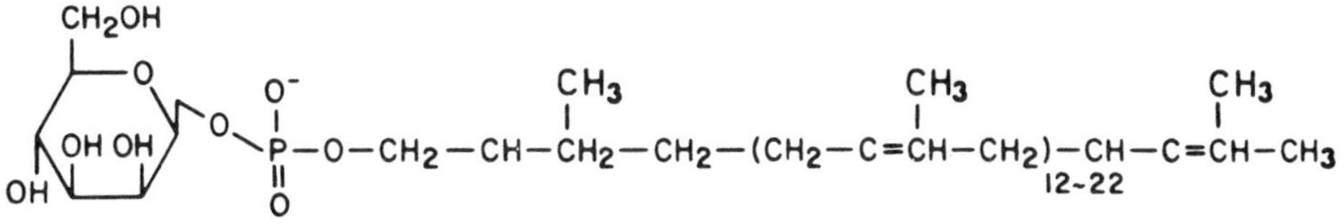

FIG 3.1. Structure of mannosyl-phosphoryl-dolichol, a common polyisoprenol precursor of N-linked glycoprotein biosynthesis.

On the other hand, it seems likely that the polyprenols that are actually involved in N-linked glycosylation in eukaryotic cells are the dolichols. This is easily established in animal cells where allylic polyprenols are not found and the dolichol family represents virtually the only polyisoprenyl-phosphates present in these cells. Interestingly enough in spite of this, retinyl-phosphate can serve as a mannosyl acceptor in animal cells, and retinyl-phosphoryl-mannose has been isolated from animal tissues and also synthesised in animal cell extracts (DeLuca *et al.*, 1973). But there is no evidence that the retinyl-phosphoryl-mannose serves any physiological function (DeLuca *et al.*, 1973), and the same enzyme appears to synthesise both dolichyl-phosphoryl-mannose and retinyl-phosphoryl-mannose (Stoll *et al.*, 1985). In several cases, the polyisoprenyl-linked sugars have been isolated from animal tissues and the polyisoprenyl-phosphate has been obtained after degradation of the glycolipid by mild acid hydrolysis. In these studies, the lipid was identified as a dolichol by mass spectrometry, or other chemical methods (Richards and Hemming, 1972; Evans and Hemming, 1972).

In plant systems, the available evidence also suggests that the lipid carrier of the sugars is a dolichol derivative. Initially, this evidence was based on various chemical treatments that can distinguish allylic polyprenyl-phosphates from those with an α-saturated isoprene unit. For example, the allylic polyprenyl-phosphates are labile to mild acid hydrolysis which releases the phosphate group, while those α-saturated isoprene units are not affected by this treatment. Although plants contain both types of polyprenyl-phosphates, after this treatment the lipid fraction was still able to act as an acceptor of sugars, indicating that phosphate had not been cleaved from all of the polyprenols. In addition, the allylic polyprenyl-glycosyl-phosphates (for example, ficaprenyl-phosphoryl-mannose) are degraded by treatment with 50% phenol, or by catalytic hydrogenation, both of which release the sugar moiety. However, the non-allylic derivatives are not affected by this treatment (Garcia *et al.*, 1974; Hopp *et al.*, 1977). The earlier studies in plants (Pont Lezica *et al.*, 1975; Lehle *et al.*, 1976) did indicate that the carrier lipids were of the saturated types and further work involving biosynthesis from mevalonate (Daleo and Pont Lezica, 1977) and mass spectrometry (Delmer *et al.*, 1978) have supported these data.

There is some evidence in plant systems to suggest that the polyprenyl-phosphate that carries mannose is different from that for GlcNAc (Ericson *et al.*, 1978). Thus, the polyprenyl-phosphate isolated after mild acid hydrolysis of the mung bean polyprenyl-phosphoryl-mannose stimulated the *in vitro* incorporation of mannose from GDP-mannose into lipid-linked monosaccharide, but had no effect on GlcNAc incorporation from UDP-GlcNAc into lipid-linked saccharides. On the other hand, the lipid obtained from polyprenyl-pyrophosphoryl-GlcNAc stimulated the *in vitro* incorporation of GlcNAc from UDP-GlcNAc into lipid, but not that of mannose from GDP-mannose. However, it is not known at this time what differences exist between these lipids. It may be that certain chain length dolichols function as mannose carriers whereas other chain length dolichols carry GlcNAc.

IV.　BIOSYNTHESIS AND METABOLISM OF DOLICHOLS

Again, much of the information on the biosynthesis and metabolism of the dolichols has come from studies done in animal systems. As indicated in Fig. 3.2, the biosynthesis of dolichol-phosphate involves the conversion of acetate (i.e. acetyl-CoA) to farnesyl-pyrophosphate, using the same series of reactions that participate in the biosynthesis of cholesterol. These reactions involve the conversion of acetate to hydroxymethylglutarylCoA, and mevalonic acid to farnesyl-pyrophosphate. At this point, the pathway branches to form either squalene and then cholesterol, or to form 2,3-dehydrodolichyl-pyrophosphate and then dolichyl-phosphate (Grange and Adair, 1977). The synthesis of cholesterol, and presumably also of dolichol, is regulated at the level of hydroxymethylglutaryl-CoA reductase, and probably also at other enzymes specific to either branch of the specific pathway.

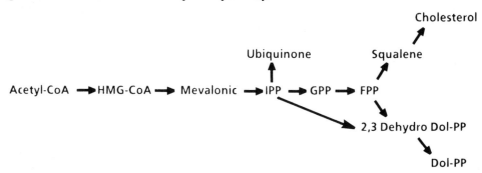

FIG 3.2.　Outline of the pathway for biosynthesis of dolichyl-phosphate and other polyprenyl-phosphates.

Much less is known about the biosynthesis of polyprenols in plants. Early information on the biosynthesis and the stereochemistry of the E,Z-polyprenols was obtained (Gough and Hemming, 1970a) using a plant system. These workers showed that the 4-S hydrogen atom of mevalonate is retained when a Z-double bond is formed, whereas the hydrogen is lost in the biosynthesis of an E-double bond. Similar results were found for dolichol formation in rat liver (Gough and Hemming, 1970b). The *in vitro* synthesis of dolichyl-phosphate from isopentenyl diphosphate was reported using extracts from garden peas (Pont Lezica *et al.*, 1975), whereas in algae the synthesis could be achieved using mevalonate as a precursor (Hopp et al., 1978). These early studies indicated that membrane fractions were required for synthesis and that the mitochondrial-rich fractions showed the best activity (Burgos and Morton, 1962). Other studies have shown that the levels of free dolichol are higher in mitochondria than in other organelles (Butterworth and Hemming, 1968). On the other hand, the Z-prenyltransferase of rat liver that adds isopentenyl diphosphate units to farnesyl diphosphate is localised mostly in an endoplasmic reticulum-rich fraction, and only a minor part of this activity is in the mitochondrial fraction (Wong and Lennarz, 1982). Apparently the biosynthesised dolichol is rapidly transported to a mitochondrial-lysosomal fraction and accumulates there. In fact, careful separation of various organelles by Ficoll gradient centrifugation demonstrated that most of the dolichol was in the lysosomes (Wong et al., 1982).

In vitro studies on dolichol synthesis in hen oviduct resulted in the isolation of 2,3-dehydrodolichyl-phosphate, a likely intermediate in the biosynthesis of dolichyl-phosphate (Grange and Adair, 1977). However, it is not known whether this compound is the precursor for dolichyl-phosphate, nor whether the precursor for the α-isoprene reductase is the diphosphate or the monophosphate derivative of 2,3-dehydrodolichol. It is also not known whether this compound is a precursor in plants or whether some of the unsaturated polyprenols might serve as precursors to the dolichols. At this time, we also do not know what role the various polyprenols play in plants, i.e. if the dolichols serve as sugar carriers for glycoproteins and polysaccharides, what role do the allylic polyprenols play? Finally, one must wonder how the synthesis of the various polyprenols, i.e. allylic and saturated, is regulated?

In animal tissues, there is a phosphatase (dolichyl-phosphate phosphatase) that can convert dolichyl-phosphate to dolichol (Adrian and Keenan, 1979), and this enzyme may account for the free dolichol that is found in various tissues. Free dolichol can be phosphorylated by a CTP-dependent dolichol kinase (Allen et al., 1978), or it can be acylated by a dolichol acyltransferase (Keenan and Kruczek, 1976).

V. THE LIPID-LINKED SACCHARIDE PATHWAY

A. Lipid-linked Monosaccharides

The reactions involved in the formation of various lipid-linked saccharides are presented in Fig. 3.3. This series of reactions leads to the formation of the final lipid-linked oligosaccharide, $Glc_3Man_9(GlcNAc)_2$-pyrophosphoryl-dolichol, and transfer of this oligosaccharide to protein. Most of the reactions shown in Fig. 3.3 were derived from studies with animal tissues or animal cells grown in culture media. However, many of these reactions have now been demonstrated in plant tissues as well.

In terms of the lipid-linked monosaccharides, there are three sugars that are universally found as monomers attached to the lipid carriers, i.e. dolichyl-phosphoryl-mannose (also referred to as mannosyl-phosphoryl-dolichol), dolichyl-phosphoryl-glucose and dolichyl-pyrophosphoryl-GlcNAc. The dolichyl-PP-GlcNAc is the first lipid intermediate in the pathway shown in Fig. 3.3 and ultimately is converted to $Glc_3Man_9(GlcNAc)_2$-PP-dolichol. On the other hand, the dolichyl-P-mannose and dolichyl-P-glucose function as donors of mannose and glucose for the assembly of the lipid-linked oligosaccharides (see below) (Nakaido and Hassid, 1971; Hemming, 1974a; Waechter and Lennarz, 1976; Parodi and Leloir, 1979; Troy, 1979; Elbein, 1979; Hubbard and Ivatt, 1981; Lehle and Tanner, 1983; Pont Lezica *et al.*, 1986).

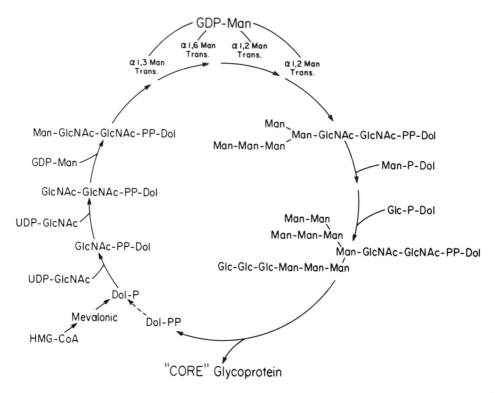

FIG 3.3. Pathway for the biosynthesis of the lipid-linked saccharides leading to the formation of $Glc_3Man_9(GlcNAc)_2$-PP-dolichol.

The first studies that indicated that sugar-containing lipids might be transient intermediates in the synthesis of complex carbohydrates came almost simultaneously in yeast (Tanner, 1969), animals (Caccam *et al.*, 1969; Zatz and Barondes, 1969), and plants (Villemez and Clark, 1969; Kauss, 1969). These early reports indicated that membrane preparations from the various tissues catalysed the transfer of the mannose portion of GDP-[^{14}C]mannose into lipid-soluble material. The product of these reactions (i.e. the 'mannolipid') was labile to mild acid hydrolysis but stable to mild alkaline treatment (i.e. saponification), and the formation of this mannolipid was reversed by addition of GDP. These data suggested that the mannose was present in an 'activated' or 'high-energy' type of linkage (Richards, *et al.*, 1971; Waechter *et al.*, 1973; Baynes *et al.*, 1973). In addition, the kinetics of synthesis and turnover of this lipid showed that the mannose incorporation reached a maximum very rapidly, and then declined, suggesting that the mannolipid was functioning as an intermediate in the synthesis of other mannose-containing compounds. As indicated above, more detailed studies on the structure of the mannolipid from animal and plant systems indicated that the lipid carrier was a dolichol and the compound was dolichyl-phosphoryl-mannose. The reaction responsible for the formation of this compound is:

$$\text{GDP-mannose} + \text{dolichyl-phosphate} \rightleftharpoons \text{dolichyl-P-mannose} + \text{GDP}$$

In some cell-free systems, the mannosyl transferase that catalyses the above reaction is fairly specific for dolichyl-P as the mannosyl acceptor, whereas in other systems the specificity is much less rigid. For example, with crude extracts from chick retina, the addition of exogenous dolichyl-P stimulated the formation of dolichyl-P-mannose, but the addition of exogenous ficaprenyl-P, citronellyl-P, farnesyl-P, etc., did not stimulate mannose incorporation into lipid (Kean, 1977). On the other hand, with extracts of thyroid glands, exogenous dolichyl-P, at $50 \, \mu M$, gave a 16-fold increase in mannose incorporation into lipid whereas exogenous ficaprenyl-P, also at $50 \, \mu M$, stimulated 11-fold (Adamany and Spiro, 1975). In hen oviduct preparations, $300 \, \mu M$ dolichyl-P stimulated the formation of dolichyl-P-mannose 6 to 7-fold, while ficaprenyl-P gave a 2 to 3-fold stimulation of mannolipid formation (Waechter et al., 1973). Such variability in the effects of added lipid is also seen in various other systems, including plants, indicating that in many cases the mannosyl transferase is not that discriminating and can use other polyprenols as sugar acceptors. As indicated above, these other polyprenols are probably not as likely to be physiological substrates since they are probably not found in many eukaryotic cells. They are, however, found in various plant tissues. The mannosyl transferase that catalyses the above reaction has a strong requirement for a divalent cation. Usually Mg^{2+} is best, but Mn^{2+} is also effective. However, in some cell-free extracts, Mn^{2+} may be equal to or better than Mg^{2+} at stimulating this reaction. The metal ion requirement has been of some value in helping to understand the biosynthetic pathway since it has been possible to inhibit the formation of dolichyl-P-mannose by adding EDTA, or some other chelating agent, to cell-free extracts. Thus, when microsomes are incubated with GDP-[^{14}C]mannose in the presence of EDTA, mannose incorporation into lipid is severely inhibited but some mannose is still incorporated into $CHCl_3$–CH_3OH–H_2O (10:10:3; v/v/v) phase. When the oligosaccharide portion of the lipid-linked oligosaccharides is examined under these conditions, the major species found is a $Man_5(GlcNAc)_2$ (Chambers et al., 1977). This kind of study gave the first suggestion that some of the mannose residues in the lipid-linked oligosaccharides might come directly from GDP-mannose, rather than from dolichyl-P-mannose (see below). The synthesis of dolichyl-P-mannose in cell-free extracts is also inhibited by the antibiotic, amphomycin (Kang et al., 1978a). This inhibition may be due to the fact that amphomycin ties up the available dolichyl-P (Banerjee et al., 1981), or it could be due to a chelation effect by amphomycin. At any rate, when cell-free extracts are incubated with GDP-[^{14}C]mannose in the presence of amphomycin, mannose incorporation into dolichyl-P-mannose is almost completely inhibited, but mannose is still incorporated into lipid-linked oligosaccharides. The major oligosaccharide found associated with the lipids is a $Man_5(GlcNAc)_2$, again suggesting that the first five mannoses come directly from GDP-mannose (Kang et al., 1978a,b).

Diumycin is another antibiotic that has been reported to inhibit the formation of dolichyl-P-mannose. This antibiotic has only been used in a few studies to inhibit lipid-linked saccharides, in vitro. Using a membrane preparation or a solubilised enzyme from yeast, diumycin was found to inhibit the synthesis of dolichyl-P-mannose. The inhibition of this reaction was somewhat better with the solubilised enzyme preparation. In the presence of diumycin, transfer of mannose from GDP-mannose to preformed dolichyl-PP-(GlcNAc)$_2$ still occurred, but mannose transfer from dolichyl-P-mannose to serine or threonine residues on the protein was also inhibited (Babczinski, 1972). On the other hand, with a solubilised enzyme preparation from *Acanthamoeba*, diumycin inhibited mannose and GlcNAc transfer from their sugar nucleotides into lipid-linked monosaccharides, but the formation of dolichyl-P-glucose was only slightly inhibited. The antibiotic also apparently blocked the transfer of the second GlcNAc to dolichyl-PP-GlcNAc, and this reaction was even more sensitive than the GlcNAc-1-P transfer (Villemez and Carlo, 1980).

The dolichyl-P-mannose synthase has been solubilised from a number of different tissues (Heifetz and Elbein, 1977 and Haselbeck and Tanner, 1982), and has been partially purified (about 880-fold purification) from rat liver (Jensen and Schutzbach, 1985). The partially purified enzyme has a K_m for GDP-mannose of about 0.69 μM, and an apparent K_m for dolichyl-P of 0.3 μM. GMP, GDP and GTP inhibited mannosyl-transfer 50% at concentrations of 16 μM, 1.3 μM and 3 μM, respectively. Attempts to further purify this enzyme, as well as other enzymes in the lipid-linked oligosaccharide pathway, have been fraught with difficulty apparently due to the lability of these enzymes after solubilisation (see below for other enzymes). As a result, very little is known about the absolute requirements of these proteins and little is known about how these enzymes, or the whole pathway, are controlled.

In terms of possible mechanisms of regulation and the subcellular localisation or orientation of the GDP-mannose : dolichyl-P mannosyltransferase, the partially purified enzyme was incorporated into liposomes, and the reaction that it catalyses (i.e. the formation of dolichyl-P-mannose) was studied in these 'intact' vesicles (Haselbeck and Tanner, 1982). It was possible to demonstrate a mannose transfer from external GDP-mannose, via membrane-bound dolichyl-P to internal GDP. Using double labelled [^{3}H]GDP-[^{14}C]mannose in this experiment, it was found that only the [^{14}C]mannose entered the lumen of the liposomes. Since this reaction required membrane-bound dolichyl-P, and non-phosphorylated dolichol was inactive, the assumption was made that the transferase catalysed both the transfer of mannose to dolichyl-P to form dolichyl-P-mannose, and then the 'flip-flop' of this dolichyl-P-mannose to the luminal side of the vesicle. The non-catalysed translocation of polyprenols has been shown to occur extremely slowly in model membrane systems (McCloskey and Troy, 1980), suggesting that such a rearrangement must be catalysed by a protein translocator. The enzymatically-formed dolichyl-P-mannose, produced by calf pancreas microsomes, was compared both chemically and enzymatically with the dolichyl-P-mannose synthesised chemically. These studies indicated that the enzymatic product was dolichyl-α-D-mannopyranosyl-phosphate (Herscovics *et al.*, 1975).

In 1970, it was shown that rat liver microsomes incorporated glucose from UDP-[^{14}C]glucose into material that was soluble in $CHCl_3–CH_3OH$. This compound was subsequently shown to be dolichyl-P-glucose (Behrens and Leloir, 1970). Thus, the following reaction is responsible for the synthesis of this compound:

$$\text{UDP-glucose} + \text{dolichyl-P} \rightleftharpoons \text{dolichyl-P-glucose} + \text{UDP}$$

This glycolipid was characterised by chemical methods and compared to the chemically synthesised compound. The biosynthetic material was determined to be dolichyl-β-D-glucopyranosyl-phosphate (Herscovics et al., 1977a). The synthesis of this compound has also been demonstrated in various other animal tissues (Nakaido and Hassid, 1971; Hemming, 1974a; Waechter and Lennarz, 1976; Elbein, 1979; Parodi and Leloir, 1979; Troy, 1979; Hubbard and Ivatt, 1981; Lehle and Tanner, 1983; Pont Lezica et al., 1986), and this glucolipid is also biosynthesised in various plant tissues. However, in these latter cases, its formation is more difficult to demonstrate, due to the synthesis of other types of glucolipids. Thus, many plant tissues also incorporate glucose from UDP-glucose into steryl glucosides, and these glucolipids are also extractable into the $CHCl_3$–CH_3OH–H_2O phase (Kaushal et al., 1988). As indicated earlier in this chapter, the steryl glucosides can be easily distinguished from the dolichyl-linked saccharides on the basis of acid stability as well as charge, but to obtain this additional information requires isolation and further characterisation of the biosynthetic or isolated lipid. Interestingly enough, although the synthesis of dolichyl-P-glucose has been known for some time, and although the presence of glucose on the lipid-linked oligosaccharides is apparently critical for their transfer to protein, very few studies have been reported on the solubilisation or purification of this enzyme. In fact, although there is some evidence from previous studies (Behrens et al., 1975; Waechter and Scher, 1976; Herscovics et al., 1977a,b) to indicate that the glucoses are donated from dolichyl-P-glucose, the system has not been carefully dissected in such a way that it is clear that all three glucoses come from this glucosyl donor. In order to show this conclusively, it will be necessary to purify the individual glucosyl transferases and determine their substrate specificities and requirements.

The enzyme that catalyses the synthesis of dolichyl-P-glucose was purified about 1700-fold from crude extracts of MOPC-315 plastocytoma tissue, or about 200-fold starting from the microsomes of this tissue (Gold and Green, 1983). Nevertheless, in spite of this purification, the glucosyltransferase was still not homogeneous, although it showed an absolute requirement for exogenous dolichyl-P and also was activated by choline-containing phospholipids (phosphatidylcholine, lysophosphatidylcholine and sphingomyelin). The enzyme exhibited a K_m of 0.79 μM for the substrate, UDP-glucose, and of 0.65 μM for the other substrate, dolichyl-P, both in the presence of 4 mg ml^{-1} phosphatidylcholine. The dye, Remazol Blue, acted as a competitive inhibitor of this enzyme with respect to UDP-glucose. Showdomycin is another compound that was found to inhibit the synthesis of dolichyl-P-glucose and dolichyl-P-mannose in membrane preparations of pig aorta, but the synthesis of dolichyl-PP-GlcNAc was not very sensitive to this inhibitor (Kang et al., 1979). The inhibition caused by showdomycin is probably due to the reactivity of its maleimide group towards sulphydryl groups on the sensitive enzymes. In extracts of Volvox, dolichyl-P-glucose formation was also sensitive to inhibition by showdomycin, but in this case, the synthesis of dolichyl-PP-GlcNAc was equally sensitive (Muller et al., 1981).

The UDP-GlcNAc : dolichyl-P GlcNAc-1-P transferase that synthesises dolichyl-pyrophosphoryl-GlcNAc represents the first enzyme in the biosynthesis of $Glc_3Man_9(GlcNAc)_2$-PP-dolichol (see Fig. 3.3), and therefore would seem to be a likely enzyme to be involved in the regulation of this pathway. As a result, a number of studies have been done on this enzyme, and a number of attempts have been made to purify it from various tissues. The enzyme was initially studied in microsomal fractions from a variety of tissues including liver (Behrens et al., 1971b; Molner et al., 1971; Palamarczyk and Hemming, 1975; Zatta et al., 1976;) brain (Waechter and Harford, 1977), oviduct (Chen and Lennarz, 1977), and mung bean seedlings (Forsee and Elbein, 1975). The formation of the GlcNAc-PP-dolichol occurs according to the following reaction:

$$\text{UDP-GlcNAc} + \text{dolichyl-P} \rightleftharpoons \text{dolichyl-PP-GlcNAc} + \text{UMP}$$

It should be noted that this reaction is different from the other two reactions described above in that a sugar-phosphate rather than a sugar is transferred to the dolichyl-P. However, as in the case of dolichyl-P-mannose and dolichyl-P-glucose formation, the addition of dolichyl-P to cell-free extracts stimulates the synthesis of dolichyl-PP-GlcNAc in many cases. As noted above (see Section III), there is some evidence in plants to suggest that the lipid carrier for GlcNAc may be different from the lipid carrier for mannose. At this stage, the nature of this difference has not been determined and could relate to the chain length of the polyprenols. Finally, it should be pointed out that the formation of GlcNAc-PP-dolichol can be reversed by the addition of UMP to the incubation mixtures, whereas the reversal of dolichyl-P-mannose or dolichyl-P-glucose formation requires the addition of the nucleoside diphosphate (GDP or UDP). The GlcNAc-1-P transferase has also been solubilised from several different sources (Heifetz and Elbein, 1977; Carlo and Villemez, 1979; Keller et al., 1979; Palamarczyk et al., 1980; Sharma et al., 1982; Arakawa and Mourkerjea, 1984), and has been minimally purified in a few cases. The problem in terms of purification is related to the poor stability of these enzymes once they have been solubilised from the membranes. Only in the case of the GlcNAc transferase from Acanthamoeba was the stability such that the enzyme could be stored for months at 4°C with little or no loss of activity. However, to date there have been no reports that indicate that the enzyme has been purified from that source. Recently, it has been possible to stabilise the aorta GlcNAc-1-P transferase to a significant extent by adding 20% glycerol containing 20 µg of phosphatidylglycerol per mg of protein to the buffer solutions used in the purification and storage of the enzyme (Kaushal and Elbein, 1985). Under those conditions, the enzyme could be maintained at 4°C for at least 6 days with negligible losses in activity. On the other hand, in the absence of stabilisers, the transferase lost most of its activity within 24 h when stored at 4°C, or in the frozen state. The aorta GlcNAc-1-P transferase was purified about 65-fold using the above-described conditions for stabilising the protein. During the purification, a heat-stable factor was separated from the enzyme upon chromatography on columns of DEAE-cellulose, Although this 'factor' has not yet been identified, adding it back to the enzyme preparation resulted in a five-fold stimulation in activity.

The GlcNAc-1-P transferase, after solubilisation or purification, is also stimulated by, or requires, a phospholipid for activity. The earlier studies on the lung enzyme showed that treatment of the solubilised transferase with phospholipase A_2 resulted in a time-dependent loss of activity (Plouhar and Bretthauer, 1982). Much of this activity could be restored by the addition of phosphatidylglycerol. With the partially purified aorta GlcNAc-1-P transferase, there was an almost absolute requirement for a phospholipid for activity (Kaushal and Elbein, 1985). Phosphatidylglycerol was the best activator of the transferase, whereas phosphatidylinositol also stimulated, but much less so. The phospholipids with charged head groups (phosphatidylserine, phosphatidylethanolamine and phosphatidylcholine) were ineffective in stimulating the enzyme.

The GlcNAc-1-P transferase would seem like the most likely enzyme to serve as a control point for the lipid-linked saccharide pathway since it is the first enzyme in the reaction series. However, since this enzyme has been so difficult to purify, little is currently known about its regulation. Nevertheless, studies by Kean (1980, 1982) have shown that the incorporation of GlcNAc-1-P, from UDP-GlcNAc, into GlcNAc-PP-dolichol by microsomal preparations of chick retina is stimulated 7- to 15-fold by adding GDP-mannose plus dolichyl-P, or dolichyl-P-mannose alone, to these incubations. Presumably these microsomes contained the mannosyltransferase activity that could convert the GDP-mannose to mannosyl-P-dolichol which was the real effector. A considerable stimulation in the activity of the partially purified aorta GlcNAc-1-P transferase was also observed upon addition of dolichyl-P-mannose to the enzyme incubations (Kaushal and Elbein, 1985). This stimulation was not observed when dolichyl-P-mannose was replaced by dolichyl-P, and, since the enzyme preparation did not contain any detectable mannosyltransferase activity, nor any dolichyl-P-mannose synthase activity, it seems unlikely that the stimulation could be due to the formation or effect of GDP-mannose. However, the meaning of the stimulation and how it could fit into the control of this pathway still remains a mystery. The GlcNAc-1-P transferase is specifically and potently inhibited by the nucleoside antibiotic, tunicamycin (Tamura, 1982). This antibiotic is composed of uracil, a fatty acid, and two glycosidically linked sugars. The sugars are GlcNAc and an unusual 11 carbon aminodeoxydialdose called tunicamine. The tunicamine is attached to uracil in an N-glycosidic bond, and is itself substituted at two positions. At the anomeric carbon, the GlcNAc is linked in an O-glycosidic bond, while a long chain fatty acid is bound in amide linkage to the amino group (Ito et al., 1979).

The site of action of tunicamycin was initially demonstrated by Tkacz and Lampen using a microsomal enzyme preparation of calf liver (Tkacz and Lampen, 1975). This antibiotic inhibited the first step in the lipid-linked saccharide pathway, i.e., the formation of GlcNAc-PP-dolichol. Tunicamycin had the same site of action in membrane preparations of plants (Ericson et al., 1977), and in chick oviduct membranes (Struck and Lennarz, 1977), but the enzyme that adds the second GlcNAc to form GlcNAc-GlcNAc-PP-dolichol was not affected by tunicamycin, nor were the GlcNAc transferases that add terminal GlcNAc residues to the mannose chains (Lehle and Tanner, 1976; Struck and Lennarz, 1977). In addition, this inhibitor did not affect the phospho-N-acetylglucosamine transferase that adds GlcNAc-1-P to terminal mannose residues on high-mannose chains of lysosomal hydrolases (Reitman and Kornfeld, 1981).

The mechanism of action of tunicamycin was examined with the solubilised GlcNAc-1-P transferase from pig aorta (Heifetz et al., 1979) or hen oviduct (Keller et al., 1979), and also with the membrane-bound enzyme from chick embryo (Takatsuki and Tamura, 1982). With the aorta enzyme, competitive inhibition could not be demonstrated, possibly because of the very strong affinity of the tunicamycin for the GlcNAc-1-P transferase (Heifetz et al., 1979). However, with the antibiotic streptoviridin, which is an anologue of tunicamycin with shorter chain length fatty acids and with a lowered affinity for the enzyme, it was possible to show competitive inhibition, and to reverse the inhibition by the addition of UDP-GlcNAc (Elbein et al., 1979). The addition of UDP-GlcNAc, but not dolichyl-P, protected the solubilised oviduct transferase from inactivation by tunicamycin (Keller et al., 1979). The suggestion was made that tunicamycin acted as a tight binding, reversible inhibitor, and might be a substrate–product transition state analogue (Takatsuki and Tamura, 1982). Kinetic studies with the GlcNAc-1-P transferase gave a K_m for UDP-GlcNAc of 3×10^{-6} M, and an apparent K_i for tunicamycin of about 5×10^{-8} M (Heifetz et al., 1979).

These enzymes involved in the formation of the lipid-linked monosaccharides are probably localised in the endoplasmic reticulum of the cell, although this has not been conclusively established in most cases. However, rate zonal sedimentation and isopycnic banding in linear sucrose density gradients in the presence of 1 mM EDTA indicated that the dolichyl-P-mannose and dolichyl-PP-GlcNAc synthases were localised in the endoplasmic reticulum in maize endosperm cells (Riedell and Miernyk, 1988). It will be important to purify these enzymes to homogeneity and to prepare antibodies against them so that more conclusive localisation studies on these enzymes can be done. Such antibodies could also be used to determine the orientation or 'sidedness' of these proteins in the membranes.

Particulate preparations from peas have been reported to catalyse the transfer of glucose-1-P from UDP-[^{14}C]glucose to an endogenous acceptor to produce glucosyl-pyrophosphoryl-dolichol (Romero and Pont Lezica, 1976). When the UDP-glucose was labelled with ^{32}P in the β-phosphate and ^{14}C in the glucose, both labels were found in the lipid fraction. Interestingly enough, this same enzyme preparation also transferred glucose alone to form glucosyl-phosphoryl-dolichol (Hopp et al., 1978). The endogenous lipids were extracted from peas and purified on columns of DEAE-cellulose. The purified lipid fraction acted as a glucose acceptor using enzyme preparations from either plants or animals (Pont Lezica et al., 1975). By various chemical criteria, the acceptor lipid was characterised as a dolichyl-P. The suggestion was made that the glucosyl-PP-dolichol was a precursor to cellulose.

In addition to dolichyl-P-mannose, dolichyl-P-glucose and dolichyl-PP-GlcNAc, other lipid-linked monosaccharides have occasionally been reported in eukaryotic cells. For example, oviduct enzyme fractions can transfer the xylose portion of UDP-xylose to dolichyl-P to form dolichyl-P-xylose (Waechter et al., 1974). However, to date, no function for this xylolipid has been found and it seems likely that the dolichyl-P-glucose synthase mistakes the xylose portion of UDP-xylose for the glucose of UDP-glucose, since these two sugars differ only by the hydroxymethyl group. An earlier report described the formation of a number of lipid-linked saccharides when plant extracts were incubated with any of the following radioactive sugar nucleotides: UDP-xylose, UDP-arabinose, UDP-galactose, UDP-glucuronic acid or GDP-glucose (Villemez and Clark, 1969). However, the products of these various reactions have not been characterised, and thus far, these reactions have not been confirmed in any other systems. Other lipid-linked monosaccharides that have been reported in plants or other eukaryotes are dolichyl-P-galactose (McEvoy et al., 1977), dolichyl-PP-N-acetylmannosamine (Palamarczyk and Hemming, 1975) and polyprenol-linked fucose (Green and Northcote, 1979). In these cases also, the function of the lipid-linked saccharide is not known, nor have these compounds been found in other systems.

B. Lipid-linked Oligosaccharides

As indicated above and as shown in Fig. 3.3, the lipid-linked monosaccharides, dolichyl-PP-GlcNAc, dolichyl-P-mannose and dolichyl-P-glucose, are precursors for the formation of the lipid-linked oligosaccharide, $Glc_3Man_9(GlcNAc)_2$-PP-dolichol

(Nakaido and Hassid, 1971; Hemming, 1974a; Waechter and Lennarz, 1976; Elbein, 1979; Parodi and Leloir, 1979; Troy, 1979; Hubbard and Ivatt, 1981; Lehle and Tanner, 1983; Pont Lezica *et al.*, 1986). This oligosaccharide-lipid is the product of the lipid-linked saccharide or 'dolichol' pathway and serves as the donor of oligosaccharide to the appropriate asparagine residues on the protein.

The lipid-linked oligosaccharides were first detected by virtue of the fact that their solubility in $CHCl_3$–CH_3OH–H_2O is different from that of the lipid-linked monosaccharides (Behrens and Leloir, 1970). Thus, as indicated earlier these oligosaccharide-lipids are not readily extracted by the $CHCl_3$–CH_3OH–H_2O (1:1:1; v/v) used to isolate the lipid-linked monosaccharides, but instead are extractable with this solvent in the proportions of 10:10:3 (v/v). The early reports on this pathway showed that mannose from GDP-mannose, or glucose, from UDP-glucose, was transferred into 10:10:3 soluble products (for example, see Parodi *et al.*, 1972; Waechter *et al.*, 1973; Forsee and Elbein, 1973; Hsu *et al.*, 1974; Sharma *et al.*, 1976; Spiro *et al.*, 1976; Lehle and Tanner, 1978; Schutzbach *et al.*, 1980). The radioactivity was released from the lipid-soluble material by mild acid hydrolysis and was shown to be present in various sized oligosaccharides by paper chromatography and by gel filtration.

A large number of studies have since been done on the lipid-linked saccharide pathway in animals and plants that have led to a reasonable understanding of this series of reactions. The formation of dolichyl-PP-GlcNAc represents the first step in this pathway, and this lipid intermediate functions as the acceptor for the other sugars of the $Glc_3Man_9(GlcNAc)_2$-PP-dolichol. Thus, a second GlcNAc is added from UDP-GlcNAc, to give GlcNAc-β1,4-GlcNAc-PP-dolichol (Leloir *et al.*, 1973; Herscovics *et al.*, 1978). In plants, the incorporation of GlcNAc from UDP-GlcNAc into both GlcNAc-PP-dolichol and $(GlcNAc)_2$-PP-dolichol has been reported in various tissues, including soybean cotyledons (Bailey *et al.*, 1980), *Phaseolus vulgaris* cotyledons (Ericson and Delmer, 1977), pea cotyledons (Beevers and Mense, 1977) suspension-cultured soybean cells (Kaushal and Elbein, 1986b) and mung bean seedlings (Lehle *et al.*, 1976; Forsee *et al.*, 1976; Brett and Leloir, 1977).

The GlcNAc transferase that catalyses the addition of this second GlcNAc was solubilised from the microsomal fraction of mung bean seedlings with 0.1% Triton X-100, and was purified about 140-fold on columns of DEAE-cellulose and hydroxylapatite. This partially-purified enzyme preparation was stable when stored in 20% glycerol containing 0.5 mM dithiothreitol, and was free of both GlcNAc-1-P transferase and the mannosyl transferase that adds the β-linked mannose. The pH optimum for the enzyme was 7.4 to 7.6, while the K_m for the dolichyl-PP-GlcNAc was about 2 μM, and that for UDP-GlcNAc about 0.25 μM. The GlcNAc transferase also showed a strong requirement for detergent and was stimulated to some extent by divalent cations, especially Mg^{2+}. In contrast to the GlcNAc-1-P transferase, this GlcNAc transferase was not inhibited by tunicamycin, nor was it affected by amphomycin, bacitracin or showdomycin. It was, however, inhibited by diumycin with 50% inhibition occurring at about 15 μg ml^{-1} of this antibiotic (Kaushal and Elbein, 1986a). The first mannose that is added to the lipid-linked oligosaccharides is added in a β-linkage to the GlcNAc to form the trisaccharide-lipid, Man-β1,4-GlcNAc-β1,4-GlcNAc-PP-dolichol. The synthesis of this trisaccharide-lipid from GDP-mannose was previously demonstrated in microsomes from liver (Levy *et al.*, 1974), and from oviduct (Chen and Lennarz, 1976), and this enzyme was solubilised and partially purified from yeast (Sharma *et al.*, 1982) and from pig aorta (Kaushal and Elbein, 1986c). The 116-fold purified enzyme from aorta was inactive when GDP-mannose was replaced by dolichyl-P-mannose, indicating that the sugar nucleotide was the real mannosyl donor. The K_m for this substrate was about 1×10^{-7} M while that for the lipid acceptor, GlcNAc-GlcNAc-PP-dolichol, was about 1×10^{-6} M. This mannosyl transferase showed an almost absolute requirement for the divalent cation, Mg^{2+}, whereas Ca^{2+} and Mn^{2+} were only slightly active. The synthesis of trisaccharide-lipid has also been demonstrated with particulate preparations of cotton fibres (Forsee and Elbein, 1975), mung bean seedlings (Lehle et al., 1976; Hori and Elbein, 1982; Kaushal and Elbein, 1987) and developing bean cotyledons (Beevers and Mense, 1977).

The β-mannosyltransferase was also solubilised from the microsomal fraction of suspension-cultured soybean cells and purified about 700-fold by a combination of conventional methods as well as affinity chromatography on columns of GDP-adipic acid-hydrazide-Sepharose (Kaushal and Elbein, 1987). The purified enzyme was still not homogeneous, but was fairly stable to storage in the presence of 20% glycerol and 0.5 mM dithiothreitol. The transferase required either a detergent (Triton X-100) or a phospholipid (phosphatidylcholine) for maximum activity, but the effects of these two were not additive. In terms of the phospholipid, both the head group and the acyl chain appeared to be important since phosphatidylcholines with 18 carbon unsaturated fatty acids were the most effective. Like the animal enzyme described above, this plant mannosyltransferase showed an absolute requirement for a divalent cation, with Mg^{2+} being the best metal ion. The K_m for GDP-mannose with this purified enzyme was about 1.7×10^{-6} M and for GlcNAc-GlcNAc-PP-dolichol about 9×10^{-6} M. No activity was seen when GDP-[^{14}C] mannose was replaced with dolichyl-P-[^{14}C]mannose. The β-mannosyltransferase was inhibited by guanosine nucleotides, such as GDP-glucose, GDP, GMP and GTP, but various uridine and adenosine nucleotides were without effect. It is not clear that these inhibitions have any physiological or regulatory function. The purified dolichyl-P-mannose synthase from liver described above was also inhibited by various guanosine nucleotides (Jensen and Schutzbach, 1985).

The next four α-linked mannoses to form the Man$_5$(GlcNAc)$_2$-PP-dolichol (i.e. Man α-1,2 Man α-1,2 Man α1,3[Man α-1,6]Man β-1,4GlcNAc β-1,4GlcNAc-PP-dolichol) are donated from GDP-mannose rather than dolichyl-P-mannose (Elbein, 1979; Kornfeld and Kornfeld, 1985). As indicated above, this was initially implied by the fact that microsomal fractions incubated with GDP-[^{14}C]mannose in the presence of EDTA synthesise the above Man$_5$(GlcNAc)$_2$-PP-dolichol but cannot elongate it unless provided with exogenous dolichyl-P-mannose (Chambers et al., 1977). Similar kinds of results were obtained with amphomycin, an antibiotic that inhibits the synthesis of dolichyl-P-mannose from GDP-mannose in these microsomal preparations (Kang et al., 1978). In addition, a Thy-1$^-$ mutant mouse lymphoma cell was isolated that was not able to synthesise dolichyl-P-mannose due to a deficiency in the dolichyl-P-mannose synthase. This mutant accumulated the Man$_5$(GlcNAc)$_2$-PP-dolichol (or the corresponding glucosylated lipid-linked saccharide). However, when microsomal fractions of this mutant were provided with dolichyl-P-[^{14}C]mannose, they were able to transfer the mannose to lipid-linked saccharides to give the Man$_{6-9}$(GlcNAc)$_2$-PP-dolichol intermediates (Chapman et al., 1980). These studies indicated that the mutant cell line contained all of the mannosyltransferases involved in elongation of the Man-β-GlcNAc-GlcNAc-PP-dolichol, but was missing the mannosyl donor necessary for elongation of Man$_5$(GlcNAc)$_2$-PP-dolichol to higher homologues. In addition to this the mannosyl-transferases involved in the assembly of the $Man_5(GlcNAc)_2$-PP-dolichol have been solubilised from several different tissues and the reactions have been studied with a number of different substrates. GDP-mannose was shown to be the necessary mannosyl donor, and no mannose transfer to form the Man_5-lipid was observed with dolichyl-P-mannose (Spencer and Elbein, 1980; Prakash *et al.*, 1984). Finally, two mannosyltransferases—the one that adds the α-1,2-linked mannoses (Schutzbach *et al.*, 1980) and the one that adds the first α-1,3-linked mannose (Jensen and Schutzbach, 1981)—were purified from rat liver and shown to utilise GDP-mannose as the mannosyl donor.

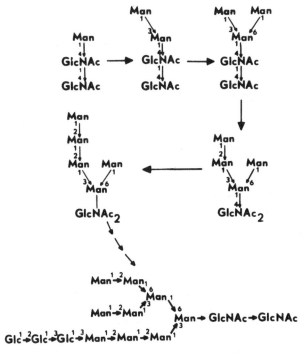

FIG 3.4. Proposed sequence of addition of mannose residues to the Man-GlcNAc-GlcNAc-PP-dolichol to give the $Man_5(GlcNAc)_2$-PP-dolichol.

Although most of the enzymes involved in assembly of the $Man_9(GlcNAc)_2$-PP-dolichol have not been purified, the sequence of addition of the mannose residues to the Man-β-GlcNAc-GlcNAc-PP-dolichol to give the $Glc_3Man_9(GlcNAc)_2$-PP-dolichol has been proposed to proceed as shown in Fig. 3.4. This series involves an ordered addition of sugar units to build up the $Man_9(GlcNAc)_2$ structure. These data were obtained by incubating CHO cells with [2-^{3}H]mannose and then isolating the lipid-linked oligosaccharides. The oligosaccharides were then released by mild acid hydrolysis and each of the oligosaccharide species was separated from the mixture and characterised individually by a number of different chemical methods. Thus, the tetrasaccharide was shown to be a Man α-1,3Man β-1,4-GlcNAc-GlcNAc, the pentasaccharide to be a Man α-1,3(Man α-1,6)Man β-1,4-GlcNAc-GlcNAc, and so on (see Fig. 3.4) (Chapman *et al.*, 1979). Assuming that all of these intermediates arise via the biosynthetic pathway and none result from degradative reactions, then the ordered addition of mannose would be as shown in the figure.

While the first five mannoses are derived from GDP-mannose, the next four mannoses appear to be donated by dolichyl-P-mannose. Thus, extracts of the Thy-1 lymphoma cells were incubated with GDP-[^{14}C]mannose and dolichyl-P-[^{3}H]mannose and the labelled lipid-linked oligosaccharides were isolated. Each oligosaccharide was then characterised by digestion with α-mannosidase and by acetolysis. These studies indicated that the major pathway for conversion of Man$_5$(GlcNAc)$_2$-PP-dolichol to Man$_9$(GlcNAc)$_2$-PP-dolichol involved the addition of an α-1,3-linked mannose to the 1,6-mannose branch. An α-1,2-linked mannose is then added to the 3-linked mannose, followed by another α-1,6-linked mannose to the 1,6-mannose branch. Finally, the mannose additions are completed by the addition of the 2-linked mannose to the newly added 1,6-mannose linkage. Of course, all of these mannoses are added in α-linkage, and all of them appear to come from dolichyl-P-mannose (Rearick et al., 1981b). During the characterisation of these oligosaccharides, these workers found some heterogeneity in the oligosaccharide structures. That is, in this in vitro synthesising system, the addition of mannose residues to the growing lipid-linked oligosaccharide was not as highly ordered as it appeared to be in the intact cell. As a result, evidence was obtained for the presence of more than one isomer in several of the purified oligosaccharide species. The authors speculate that the mannosyltransferases were highly specific for the mannosyl donor but less specific for the oligosaccharide that acted as the mannosyl acceptor. Similar results in terms of heterogeneity of the oligosaccharide produced in the in vitro systems has been reported by Vijay and coworkers (Vijay et al., 1980; Vijay and Perdew, 1982) using membrane preparations of lactating mammary gland incubated with radioactive UDP-[^{3}H]GlcNAc and GDP-[^{14}C]mannose. It seems possible that these minor oligosaccharide isomers really do represent an alternative pathway(s) of lipid-linked oligosaccharide assembly and that this pathway might be under a different type of control in the cell. However, it is not known whether such isomeric oligosaccharides also exist in vivo.

In plants, much less is known about the steps in the assembly of the Man$_9$(GlcNAc)$_2$-PP-dolichol. Lipid-linked oligosaccharides were synthesised with particulate enzyme preparations from mung bean seedlings in the presence of GDP-[^{14}C]mannose. The oligosaccharides were released from the lipids by mild acid hydrolysis and purified by several passages through columns of Biogel P-4. Five different oligosaccharides were purified in this way and were sized (compared to standards) as Man$_7$(GlcNAc)$_2$, Man$_5$(GlcNAc)$_2$, Man$_3$(GlcNAc)$_2$, Man$_2$(GlcNAc)$_2$ and Man(GlcNAc)$_2$. These oligosaccharides were characterised on the basis of various enzymatic treatments with endoglucosaminidase H and α- and β-mannosidases and by various chemical treatments. From these data, the structures of the oligosaccharides were deduced to be: Man β-GlcNAc-GlcNAc, Man α-1,3-Man β-GlcNAc-GlcNAc, Man α-1,2-Man α-1,3-Man β-GlcNAc-GlcNAc, Man α-1,2-Man α-1,2-Man α-1,3(Man α-1,6)Man β-GlcNAc-GlcNAc, Man α-1,2Man α-1,2Man α-1,3(Man α-1,3[Man α-1,6]Man α-1,6)Man β-GlcNAc-GlcNAc (Hori et al., 1982). These data suggest slight differences in the pathway of biosynthesis of the lipid-linked oligosaccharides in plants as compared to animals.

The final reactions in the assembly of the Glc$_3$Man$_9$(GlcNAc)$_2$-PP-dolichol are the addition of the three glucose residues. A number of studies have shown that the presence of glucose on the lipid-linked oligosaccharide facilitates the transfer of oligosaccharide to protein (Turco et al., 1977; Spiro et al., 1979; Staneloni et al., 1980b,

1981; Trimble et al., 1980; Murphy and Spiro, 1981; Sharma et al., 1981; Lehle and Bause, 1984; Ballou et al., 1986). However, the presence of glucose on the oligosaccharide is not an absolute requirement for its transfer to protein. For example, shorter oligosaccharides including the N,N'-diacetylchitobiosyl portion of the lipid-linked saccharide could be transferred to protein acceptors in cell-free preparations of oviduct (Chen and Lennarz, 1977) and yeast (Trimble et al., 1980). In fact, in yeast membranes, both the disaccharide-lipid, GlcNAc-GlcNAc-PP-dolichol, and the oligosaccharide-lipid, $Glc_3Man_9(Glc-NAc)_2$-PP-dolichol, exhibited the same K_m (Sharma et al., 1981). In addition, Robbins and coworkers (Huffaker and Robbins, 1983; Runge et al., 1984; Runge and Robbins, 1986) have shown that yeast mutants that are defective in various steps in the synthesis of the lipid-linked saccharides can transfer non-glucosylated oligosaccharides ranging in size from $Man_{1-2}(GlcNAc)_2$ to $Man_9(GlcNAc)_2$ to the protein. Non-glucosylated lipid-linked oligosaccharides are also transferred to protein in various protozoa such as Trypanosoma cruzi, Crithidia fasciculata and Leishmania mexicana (Parodi et al., 1981, 1983, 1984). In some of these organisms, glucosylation appears to occur after the oligosaccharide has been transferred to the protein. Since this addition of glucose is of a transient nature and the glucose residues are removed shortly after their transfer to the protein, it is not clear what role glucosylation of the protein plays, or whether it might somehow be involved in targeting or processing of the indicated protein. In terms of glucosylation of the lipid-linked oligosaccharides, the role of glucose is apparently to expedite transfer of oligosaccharide to protein. Several studies have presented evidence to indicate that the glucoses are donated from dolichyl-P-glucose rather than from UDP-glucose. In liver microsomes incubated with labelled dolichyl-P-glucose, all three glucose-containing lipid-linked oligosaccharides [i.e. $Glc_1Man_9(GlcNAc)_2$, $Glc_2Man_9(GlcNAc)_2$ and $Glc_3Man_9(GlcNAc)_2$] became labelled. This was interpreted as suggesting that all three glucoses came from this glucosyl donor (Staneloni et al., 1980b). However, if only the first glucose was donated via dolichyl-P-[^{14}C]glucose and the other glucoses came directly from endogenous UDP-glucose, and if the $Glc_1Man_9(GlcNAc)_2$-PP-dolichol served as an obligatory precursor for the other glucosylated lipid-linked oligosaccharides, then one would expect all three oligosaccharide-lipids to become labelled during an in vitro incubation. Thyroid microsomes also incorporated glucose from dolichyl-P-glucose into glucosylated lipid-linked oligosaccharides whereas UDP-glucose was not utilised (Murphy and Spiro, 1981). However, here also, the microsomal fraction could contain endogenous UDP-glucose or UDP. Perhaps more conclusive evidence to implicate dolichyl-P-glucose in the formation of the three glucoses comes from studies using yeast mutants that are blocked at various steps of the lipid-linked saccharide pathway (i.e. alg mutants). Thus, yeast carrying the alg5 or alg6 mutation accumulate $Man_9(GlcNAc)_2$-PP-dolichol. Cells with the alg6 lesion are able to synthesise dolichyl-P-glucose, but are apparently defective in the first glucosyltransferase (i.e. the enzyme that adds the first α-1,3-linked glucose to the 2-linked mannose). On the other hand, mutants of the alg5 type are not able to synthesise dolichyl-P-glucose and thus are stopped at the $Man_9(GlcNAc)_2$-PP-dolichol step. Membrane preparations of this mutant are, however, able to produce glucosylated lipid-linked oligosaccharides when supplemented with exogenous dolichyl-P-glucose (Huffaker and Robbins, 1983; Runge et al., 1984; Runge and Robbins, 1986). Conclusive evidence for the glucosyl donor for each of the three glucoses will require solubilisation and purification of these enzymes. In plants, relatively little is currently known about the glucosylation reactions, either of the lipid-linked oligosaccharide or of the protein (if glucosylation of the protein does occur). However, the $Glc_3Man_9(GlcNAc)_2$-PP-dolichol has been isolated from alfalfa hypocotyls (Staneloni et al., 1980a), suspension-cultured Nicotiana tabacum cells (Lehle, 1981) and suspension-cultured soybean cells (Hori et al., 1982), and this oligosaccharide was partially characterised from the soybean cells (Hori et al., 1982). Thus, the final product of the 'dolichol' cycle, and presumably most, if not all, of the reactions in the formation of this product are similar in plant and animal cells. The details of these reactions will be better understood after purification and characterisation of these transferases involved in the assembly process.

C. Transfer of Oligosaccharide to Protein

The final step in the lipid-linked saccharide pathway is the transfer of oligosaccharide to the protein. Early studies that surveyed the amino acid sequences around the N-glycosylation site of various proteins indicated that the sequence Asn-X-Ser (Thr) was necessary, but not sufficient, for glycosylation to occur (Marshall, 1974). That is, all of the asparagine residues that carried an oligosaccharide were in this sequence [Asn-X-Ser(Thr)]. However, this sequence was also found in regions of the protein that were not glycosylated, or in proteins that are not glycoproteins. A more detailed study of the glycosylation site in a large number of proteins indicated that Asn-X-Thr was about three times more likely to be glycosylated than the Asn-X-Ser sequence (Struck and Lennarz, 1980). This study also suggested that if the amino acid in the X position were Cys, Trp or Pro, glycosylation was unlikely to occur. Further insight into the factors that affect or control glycosylation was the observation that unfolded proteins could serve as substrates for the enzyme (i.e. oligosaccharyl transferase) that catsalyses this final step (Pless and Lennarz, 1977; Kronquist and Lennarz, 1978). Furthermore, peptides generated from such proteins and synthetic tripeptides were able to serve in vitro as acceptors of the oligosaccharide, provided that the carboxy and amino termini were blocked (Hart et al., 1979). While these peptides could act as acceptors of the oligosaccharide in vitro, the oligosaccharyl-transferase can also recognise and catalyse the glycosylation of the completed polypeptide chain in vivo (Kronquist and Lennarz, 1978; Hart et al., 1979). That is, the studies described below indicate that glycosylation is not always cotranslational and that translation and translocation (i.e. passage through the ER membrane) are not necessary prerequisites for glycosylation. For example, N-linked glycosylation of the insulin receptor in LT-2 cells can be blocked by tunicamycin, but the protein portion of the receptor is still synthesised. When the tunicamycin is removed from these cells, the completed, but unglycosylated, polypeptide chain can then be glycosylated (Ronnett and Lane, 1981). Post-translational glucosylation has also been reported in both in vitro (Tucker and Pestka, 1977) and in vivo (Bergman and Kuehl, 1978) studies on the synthesis of an immunoglobulin κ light chain in the MOPC-46B cell system. Definitive evidence that translocation and glycosylation could occur as post-translational events was obtained by in vitro studies in a yeast system that was programmed to synthesise, translate and translocate the prepro-β-factor (Rothblatt and Meyer, 1986). Several inhibitors of the reaction have been obtained by either modifying one of the amino acids in the tripeptide sequence, Asn-X-Ser (Thr), to produce an altered peptide that could be used in the in vitro studies, or by feeding an amino acid analogue to cultured cells to produce an altered protein during the *in vivo* studies. As an example, a tripeptide was synthesised that contained the epoxy analogue of threonine, epoxy-ethylglycine, in the third position of the Asn-X-Ser (Thr) sequence. This compound was found to be an irreversible inhibitor of the oligosaccharyl-transferase *in vitro* since it became covalently bound to the enzyme in the vicinity of the active site (Bause, 1983). The amino acid analogue, fluoroasparagine, is toxic to cultured cells because they can incorporate it into protein in place of asparagine. The incorporation of this analogue into the protein prevents glycosylation (Rathod *et al.*, 1986). Only the threo isomer of fluoroasparagine was active, whereas the erythro configuration was incorporated into protein but had no effect on protein glycosylation.

In terms of the lipid-linked oligosaccharide that participates in this final reaction, the oligosaccharide containing three glucoses is the best substrate for transfer (Turco *et al.*, 1977; Spiro *et al.*, 1979), but shorter oligosaccharides, even those as small as GlcNAc-GlcNAc, were transferred to protein (Chen and Lennarz, 1977). In fact, in the alg yeast mutants, the $Man_{1-2}(GlcNAc)_2$-PP-dolichol-accumulating mutant was still able to transfer this oligosaccharide to protein although at a reduced rate (Runge *et al.*, 1984). Furthermore, while the presence of glucose on the oligosaccharide facilitates its transfer to protein, the number of mannose residues on the oligosaccharide apparently has very little effect on protein glycosylation. Thus, in animal cells the removal of mannose from the $Glc_3Man_9(GlcNAc)_2$-PP-dolichol by treatment with α-mannosidase had little effect on the rate of transfer of this oligosaccharide to protein (Spiro *et al.*, 1979). Although the oligosaccharyl transferase has been solubilised from the rough endoplasmic reticulum of hen oviduct and purified to some extent (Das and Heath, 1980), its extreme lability has made it difficult to gather much detailed information on the properties of this key enzyme of N-linked glycosylation. Recently, however, a novel approach has been taken towards identifying this protein and determining its properties. A photoaffinity reagent, $N^α$-[^{3}H]Ac-Asn-Lys($N^ε$-*p*-azidobenzoyl)-Thr-NH$_2$, was developed to label the active site of oligosaccharyl transferase from hen oviduct rough endoplasmic reticulum (Weply *et al.*, 1985). This probe indicated that the enzyme (or its subunits) had an apparent molecular weight of 60 kDa when examined by SDS gel electrophoresis. Labelling of the enzyme with this probe required the presence of Mn^{2+}, a divalent cation that is also necessary for enzymatic activity. Subsequently, the photoaffinity labelling technique was improved by the development of another active site-directed probe, ^{125}I-labelled $N^α$-3(4-hydroxyphenylpropionyl)-Asn-Lys-($N^ε$-*p*-azidobenzoyl)-Thr-NH$_2$. Since the specific activity of this probe was much greater than that with the ^{3}H-probe, much lower amounts of material were needed in the assay mixtures. As a result, the non-specific background radioactivity was reduced as well as the time necessary for autoradiography. Using this new probe, the authors were able to show that the tripeptide reacted with a 57 kDa protein of the endoplasmic reticulum that was subsequently glycosylated and converted to a 60 kDa form. Antibody prepared against the 57 kDa protein was able to immunoprecipitate the photoaffinitly labelled protein as well as to recognise it by immunoblotting. Although the 57 kDa protein immunoprecipitated with the antibody was still able to interact with the photoaffinity probe, it was inactive in catalysing the glycosylation of peptides (Kaplan *et al.*, 1988). While active enzyme has not yet been obtained, this approach should provide valuable information about the enzyme, and eventually lead to the purification of the active enzyme. The transfer of oligosaccharide from the $Glc_3Man_9(GlcNAc)_2$-PP-dolichol to protein has been demonstrated using membrane preparations of plants, but this activity has not been solubilised or further characterised (Staneloni et al., 1980).

D. Topology of the Lipid-linked Saccharides in the Endoplasmic Reticulum

A considerable amount of evidence has been obtained to support the luminal location of oligosaccharide transfer to protein acceptors. For example, when the tripeptide acceptor, acetyl-Asn-Leu-Thr-$NHCH_3$, was incubated with hen oviduct microsomal vesicles, the glycopeptide product was found to be trapped within the vesicles (Wepley et al., 1983). The results indicated that the hydrophobic peptide had crossed the membrane and been glycosylated at the luminal surface, to yield a hydrophilic glycopeptide that could not escape from the vesicles. Other studies (Hanover and Lennarz, 1980) have shown that peptides that are glycosylated in vitro are found within the microsomal vesicles. This means that glycopeptides synthesised in cell-free systems from endogenous peptides are susceptible to digestion by endoglucosaminidase H only if the vesicles are first disrupted. In experiments designed to determine the length of the newly synthesised polypeptide chain necessary to accommodate an oligosaccharide, it was concluded that glycosylation cannot occur until there is a segment of at least 32 amino acids between the acceptor side and the ribosome (Glabe et al., 1980). This segment is long enough to extend across the ER membrane into the lumen.

Several laboratories have attempted to determine the orientation of the endoplasmic reticulum enzymes that participate in the assembly of the $Glc_3Man_9(GlcNAc)_2$-PP-dolichol. Usually in these studies, intact and detergent-disrupted microsomal vesicles are treated with proteases and the effect of such treatment on the activity of the enzyme is measured. If the enzyme is not inactivated in intact vesicles but activity is lost in disrupted vesicles, then one assumes that the enzyme is likely to have a luminal orientation. Of course, various control experiments must be done to show that the protease does not penetrate the intact vesicle, or does not affect the integrity of the vesicles.

In a study with rat liver microsomes, it was found that the enzymatic activities for the synthesis of glucosyl-P-dolichol, mannosyl-P-dolichol and GlcNAc-GlcNAc-PP-dolichol were sensitive to protease digestion, regardless of whether the microsomal vesicles were disrupted with detergent or not (Snider et al., 1980). This was also true for the enzyme(s) that transfer glucose from glucosyl-P-dolichol into the oligosaccharide-lipids. These results were confirmed in hen oviduct microsomes (Hanover and Lennarz, 1982) and in calf thyroid microsomes (Spiro and Spiro, 1985). The conclusion of these experiments was that these enzymes have protease-sensitive sites that face the cyto-plasmic side of the ER membrane. However, it was not possible to determine the orientation of the catalytic sites of these enzymes from such experiments. There are a number of possible models that could explain these results. For example, the enzymes could have their catalytic sites facing the cytoplasmic side of the membrane. Or, the enzymes could have a catalytic site that is oriented on either side of the membrane (i.e. transmembrane) since these enzymes transfer a sugar from the cytoplasmic nucleoside diphosphate sugar to a luminally-oriented acceptor. It is also possible that the catalytic site faces the lumen but that a region of the enzyme is oriented towards the cytoplasm and that destruction of this cytoplasmic region inactivates the enzymes. Some enzymatic activities that were examined in these studies were resistant to proteolysis in both the intact and the disrupted vesicles. These include the synthesis of both GlcNAc-PP-dolichol and mannosyl-P-dolichol from rat liver or hen oviduct microsomes. These data indicate that these enzymes are simply not sensitive to the proteolytic conditions and therefore no conclusions about orientation can be made. In one study, the stilbene derivative, DIDS, which labels membrane proteins, was used to determine the orientation of the ER proteins. Application of DIDS to the intact calf pancreas vesicles caused inactivation of the glucosyl-P-dolichol synthase, suggesting that this enzyme has a cytoplasmic orientation, at least as far as its catalytic site is concerned (Spiro and Spiro, 1985).

In the lipid-linked saccharide pathway (Fig. 3.3), there are 16 intermediates formed on the pathway to synthesis of $Glc_3Man_9(GlcNAc)_2$-PP-dolichol. Studies from several laboratories have provided information on the orientation of some of these intermediates. Intermediates were localised on both sides of the ER membrane, suggesting that assembly of the lipid-linked oligosaccharides occurs on both the cytoplasmic and luminal faces.

One method for approaching the orientation was to use the enzyme galactosyltransferase. This enzyme could convert GlcNAc-GlcNAc-PP-dolichol to Gal-GlcNAc-GlcNAc-PP-dolichol in disrupted microsomes, but in intact vesicles, the endogenous GlcNAc-GlcNAc-PP-dolichol was inaccessible to the probe (Hanover and Lennarz, 1982). This suggested a luminal orientation of this lipid. Another probe used for localisation studies was the lectin, Concanavalin A. Oligosaccharide-lipids are normally extracted from membranes with organic solvents but when complexed with the lectin, they are not extractable. This technique showed that the $Glc_3Man_9(GlcNAc)_2$-PP-dolichol resides on the luminal side of the membrane (Snider and Robbins, 1982) since the oligosaccharide was bound by Concanavalin A (Con A) in disrupted but not in intact vesicles. On the other hand, the $Man_{3-5}(GlcNAc)_2$-PP-dolichol was bound by Con A to the extent of 50–70% in intact vesicles as well as in disrupted vesicles (Snider and Rogers, 1984). The larger species, such as $Man_{6-9}(GlcNAc)_2$-PP-dolichol, were bound in disrupted vesicles but not in intact vesicles, again supporting the luminal orientation. These studies support the hypothesis that the lipid-linked oligosaccharides up to Man_5-$(GlcNAc)_2$-PP-dolichol face the cytoplasm whereas the larger oligosaccharides face the lumen. There are, however, discrepancies, since GlcNAc-PP-dolichol also appears to face the lumen. Is it possible that some of these lipid-linked saccharides do move or 'flip-flop' in the membranes, even in the absence of a carrier or transporter?

Since some of the reactions occur on the luminal side of the ER, the sugars must somehow be able to cross the ER membrane in order to be incorporated into oligosaccharide. Recent studies by Perez and Hirschberg (1986) and Hirschberg and Snider (1987) have shown that radioactive UDP-GlcNAc and UDP-Glc are translocated into RER vesicles. These authors used a mixture of nucleotide sugars that were radiolabelled with one isotope in the sugar and a second in the nucleotide and showed that the isotope ratio of the transported solute was similar to that of the substrates added to the medium. Such data support the notion that the intact nucelotide sugars are transported. In addition, transport was temperature dependent and saturable with apparent K_m of 3–4 μM for both sugar nucleotides. On the other hand, no transport system could be shown for GDP-mannose, indicating that this nucleotide sugar does not cross the ER membrane and that the incorporation of mannose must occur at the cytoplasmic face.

E. Regulation of the Lipid-linked Saccharide Pathway

Among the interesting questions still to be answered with regard to the biosynthesis of the lipid-linked saccharides is whether or not the pathway is under any sort of regulation, and if so, how does this regulation work?

Since dolichyl-P serves as an obligatory carrier in the biosynthesis of N-linked oligosaccharides, its levels in the membrane may be one of the controlling factors in terms of the rate of synthesis of the lipid-linked oligosaccharides. In fact, as discussed earlier in this chapter, the addition of exogenous dolichyl-P to membrane fractions frequently results in a stimulation in the incorporation of the sugar from its nucleoside diphosphate derivative (i.e. GDP-mannose, UDP-GlcNAc, UDP-Glc) into lipid-soluble material. Such studies imply that the levels of dolichyl-P are limiting in the various membrane fractions. Further evidence that dolichyl-P-levels may control the rate of synthesis of the lipid-linked oligosaccharides is the finding that 25-hydroxycholesterol, an inhibitor of HMG-CoA reductase, inhibited both glycoprotein and dolichol-linked saccharide synthesis in aortic smooth muscle cells in culture (Mills and Adamany, 1978). This inhibition is presumably due to an inhibition of dolichol synthesis at the level of the HMG-CoA reductase. On the other hand, another study indicated that when mouse L cell cultures were incubated with 25-hydroxycholesterol, large fluctuations in cholesterol synthesis could occur with relatively small changes in dolichol synthesis (James and Kandutsch, 1974). These authors suggest that while both pathways (cholesterol and dolichol) share common intermediates and a regulatory enzyme, they nevertheless maintain a large degree of regulatory independence. Similar results were obtained in a detailed study with rat liver slices (Keller, 1986).

While dolichyl-P levels are undoubtedly one of the important factors that control the extent and rate of formation of the lipid-linked oligosaccharides, there do appear to be other levels of control over this pathway. For example, several inhibitors of protein synthesis have been shown to affect the synthesis of lipid-linked saccharides, indicating a close link between protein synthesis and oligosaccharide synthesis. When Madin Darby Canine Kidney (MDCK) cells were grown in the presence of cycloheximide or puromycin, the incorporation of [^{3}H]leucine into protein was markedly inhibited, as was the incorporation of [2-^{3}H]mannose into lipid-linked saccharides. However, the formation of dolichyl-P-mannose was affected only slightly (Schmitt and Elbein, 1979). Since cycloheximide did not affect mannose incorporation (from GDP-mannose) into lipid-linked saccharides in cell-free extracts, and since the levels of the glycosyltransferases of the dolichol pathway are not very much affected during the course of these experiments, the results are best explained either by a limitation in the availability of dolichyl-P, or by a feedback mechanism, perhaps involving the lipid-linked oligosaccharides.

Another study, using actinomycin D to depress levels of mRNA, or cycloheximide to inhibit protein synthesis, showed that synthesis of lipid-linked oligosaccharides was proportional to the rate of protein synthesis. The regulated step appeared to be earlier than the formation of Man$_5$(GlcNAc)$_2$-PP-dolichol, leading the authors to speculate that the likely control point was dolichyl-P-levels (Hubbard and Robbins, 1980).

However, in mouse LM cells inhibited with cycloheximide, the incorporation of mannose into lipid-linked oligosaccharides stopped, but there was no effect on the formation of dolichyl-PP-GlcNAc, dolichyl-PP-GlcNAc-GlcNAc or dolichyl-P-mannose. When exogenous dolichyl-P was added to control cells, there was a 300% increase in the incorporation of mannose into lipid-linked oligosaccharides. However, the addition of exogenous dolichyl-P to cells inhibited with cycloheximide had no effect on mannose incorporation. These results indicated that this inhibition was not due to limitations in the amount of dolichyl-P, and led the authors to propose that feedback regulation, possibly caused by elevated levels of a metabolite such as GTP, might be responsible for the inhibition (Grant and Lennarz, 1983). In MDCK cells inhibited by either cycloheximide or puromycin, it was also found that the addition of various amounts of dolichyl-P (dissolved in either methanol or dimethylsulphoxide (DMSO) to the cell cultures could not overcome the inhibition of mannose incorporation into lipid-linked oligosaccharides. However, adding dolichyl-P to uninhibited cells had a pronounced stimulatory effect on lipid-linked oligosaccharide formation (Pan and Elbein, 1989). Thus, there may be an important regulatory mechanism in terms of the rate of glycosylation, which involves the inhibition of this pathway by metabolites. It will probably require a more careful scrutiny of the individual enzymes of the lipid-linked saccharide pathway to determine the exact site and mechanism of this control.

F. Processing of the Oligosaccharide

Once the oligosaccharide is added to the protein, the oligosaccharide undergoes a number of modification reactions that may ultimately give rise to a variety of oligosaccharide structures that are referred to as high-mannose, hybrid and complex N-linked oligosaccharides. Although the details of processing are beyond the scope of this chapter, they will be briefly summarised here in the interest of completeness. As the $Glc_3Man_9(GlcNAc)_2$ is added to the growing polypeptide chain in the lumen of the ER, two ER membrane-bound enzymes remove all three of the glucose residues from the 'glycoprotein'. Glucosidase I removes the outermost α-1,2-linked glucose, whereas glucosidase II removes the next two α-1,3-linked glucoses. These reactions leave a $Man_9(GlcNAc)_2$-protein which, in the absence of any further processing, is probably the direct precursor of the high-mannose type of glycoprotein. Or, this oligosaccharide may be further processed by the removal of a number of mannose residues as the protein traverses the ER and is transported to the Golgi apparatus. Thus, an α-mannosidase activity in the ER may remove one or several of the α-1,2-linked mannoses, while Golgi mannosidase I (actually several activities referred to as mannosidase IA, IB, etc.) can remove up to four α-1,2-linked mannose residues. The results of these reactions is a $Man_5(GlcNAc)_2$-protein. Further processing reactions can lead to the addition of other sugars, such as GlcNAc, fucose, xylose, sialic acid, as well as the removal of several other mannose residues. Processing reactions have also been demonstrated in plants, both *in vivo* and *in vitro*. However, the details of these reactions are not as well understood. Nevertheless, several of the processing enzymes have been highly purified (glucosidase I, mannosidase I, GlcNAc transferase I, etc.) and some of their properties have been determined. A recent review covers our current understanding of plant glycoprotein processing (Elbein, 1988). Figure 3.5 outlines the processing reactions that have been shown to occur in animal cells that lead to the formation of the various high-mannose, hybrid and complex types of oligosaccharides.

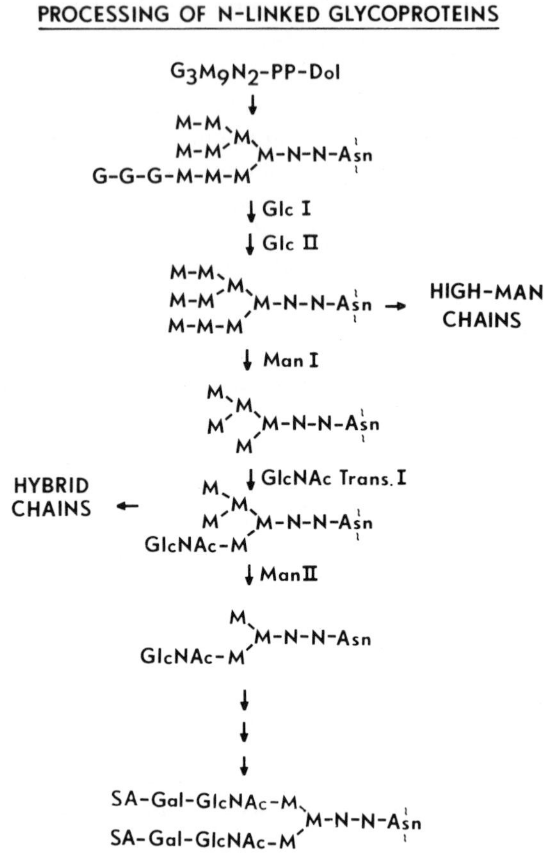

PROCESSING OF N-LINKED GLYCOPROTEINS

FIG 3.5. Series of reactions involved in the processing of the oligosaccharide chains of the N-linked glycoproteins in animal cells.

REFERENCES

Adamany, A. and Spiro, R. G. (1975). *J. Biol. Chem.* **250**, 2842–2854.

Adrian, G. S. and Keenan, R. W. (1979). *Biochim. Biophys. Acta* **575**, 431–438.

Allen, C. M., Kalin, J. R., Sach, J. and Veruzzo, P. (1978). *Biochemistry* **17**, 5020–5026.

Arakawa, H. and Moorkerjea, S. (1984). *Eur. J. Biochem.* **140**, 297–302.

Babczinski, P. (1972). *Eur. J. Biochem.* **112**, 53–58.

Bailey, D. S., DeLuca, V., Durr, M., Verma, D. P. S. and Maclachlan, G. A. (1980). *Plant Physiol.* **66**, 1113–1118.

Ballou, L., Gopal, P., Krummel, B., Tammi, M. and Ballou, C. (1986). *Proc. Natl. Acad. Sci. USA* **83**, 3081–3085.

Banerjee, D. K., Scher, M. G. and Waechter, C. J. (1981). *Biochemistry* **20**, 1561–1568.

Bause, E. (1983). *Biochem. J.* **209**, 323–330.

Baynes, J. W., Hsu, A.-F. and Heath, E. C. (1973). *J. Biol. Chem.* **248**, 5693–5704.

Beevers, L. and Mense, R. M. (1977). *Plant Physiol.* **60**, 703–708.

Behrens, N. H. and Leloir, L. F. (1970). *Proc. Natl. Acad. Sci. USA* **66**, 153–159.

Behrens, N. and Tabora, E. (1978). *Methods Enzymol.* **50**, 402–437.

Behrens, N. H., Parodi, A. J. and Leloir, L. F. (1971a). *Proc. Natl. Acad. Sci. USA* **68**, 2857–2860.

Behrens, N. H., Parodi, A. A., Leloir, L. F. and Krisman, C. (1971b). *Arch. Biochem. Biophys.* **143**, 375–380.

Bergman, L. W. and Kuehl, W. M. (1978). *Biochemistry* **17**, 5174–5180.

Brett, C. T. and Leloir, L. F. (1977). *Biochem. J.* **161**, 93–101.

Burgos, J. and Morton, R. A. (1962). *Biochem. J.* **82**, 454–456.

Butterworth, P. H. W. and Hemming, F. W. (1968). *Arch. Biochem. Biophys.* **128**, 503–513.

Caccam, J. F., Jackson, J. J. and Eylar, E. H. (1969). *Biochem. Biophys. Res. Commun.* **35**, 505–511.

Caccam, R., Hoflack, B. and Verbert, A. (1980). *Eur. J. Biochem.* **106**, 473–479.

Carlo, P. L. and Villemez, C. L. (1979). *Arch. Biochem. Biophys.* **198**, 117–123.

Chambers, J. P. and Elbein, A. D. (1975). *J. Biol. Chem.* **250**, 6904–6915.

Chambers, J. P., Forsee, W. T. and Elbein, A. D. (1977). *J. Biol. Chem.* **252**, 2498–2506.

Chapman, A., Li, E. and Kornfeld, S. (1979). *J. Biol. Chem.* **254**, 10243–10249.

Chapman, A., Fujimoto, K. and Kornfeld, S. (1980). *J. Biol. Chem.* **255**, 4441–4446.

Chen, W. W. and Lennarz, W. J. (1976). *J. Biol. Chem.* **251**, 7802–7809.

Chen, W. W. and Lennarz, W. J. (1977). *J. Biol. Chem.* **252**, 3473–3479.

Daleo, G. R. and Pont Lezica, R. (1977). *FEBS Lett.* **74**, 47–250.

Dallner, G., Behrens, N. H., Parodi, A. J. and Leloir, L. F. (1972). *FEBS Lett.* **24**, 315–319.

Das, R. and Heath, E. C. (1980). *Proc. Natl. Acad. Sci. USA* **83**, 3081–3085.

Delmer, D., Kulow, C. and Ericson, M. C. (1978). *Plant Physiol.* **61**, 25–29.

DeLuca, L. M. (1977). *Vitam. Horm.* **35**, 1–57.

DeLuca, L., Maestri, N., Rosso, G. and Wolf, G. (1973). *J. Biol. Chem.* **248**, 641–651.

Elbein, A. D. (1979). *Ann. Rev. Plant Physiol.* **30**, 239–272.

Elbein, A. D. (1980). *In* "The Biochemistry of Plants", Vol 3, (F. Loewus and W. Tanner, eds), pp. 571–587. Academic Press, New York.

Elbein, A. D. (1988). *Plant Physiol.* **87**, 291–295.

Elbein, A. D., Forsee, W. T., Schultz, J. C. and Laine, R. A. (1975). *Lipids* **10**, 427–440.

Elbein, A. D., Gafford, J. and Kang, M. S. (1979). *Arch. Biochem. Biophys.* **196**, 311–318.

Ericson, M. C. and Delmer, D. P. (1977). *Plant Physiol.* **59**, 341–347.

Ericson, M. C., Gafford, J. and Elbein, A. D. (1977). *J. Biol. Chem.* **252**, 7431–7433.

Ericson, M. C., Gafford, J. T. and Elbein, A. D. (1978). *Plant Physiol.* **61**, 274–277.

Evans, P. J. and Hemming, F. W. (1972). *FEBS Lett.* **31**, 335–338.

Forsee, W. T. and Elbein, A. D. (1973). *J. Biol. Chem.* **248**, 2858–2867.

Forsee, W. T. and Elbein, A. D. (1975). *J. Biol. Chem.* **250**, 9283–9293.

Forsee, W. T., Valkovich, G. and Elbein, A. D. (1976). *Arch. Biochem. Biophys.* **174**, 469–479.

Garcia, R. C., Recondo, E. and Dankert, M. (1974). *Eur. J. Biochem.* **43**, 93–105.

Glabe, C. G., Hanover, J. A. and Lennarz, W. J. (1980). *J. Biol. Chem.* **255**, 9236–9242.

Gold, P. and Green, M. (1983). *J. Biol. Chem.* **258**, 12967–12975.

Gough, D. P. and Hemming, F. W. (1970a). *Biochem. J.* **117**, 309–317.

Gough, D. P. and Hemming, F. W. (1970b). *Biochem. J.* **118**, 163–166.

Grange, D. K. and Adair, W. L. (1977). *Biochem. Biophys. Res. Commun.* **79**, 734–740.

Grant, W. and Lennarz, W. J. (1983). *Eur. J. Biochem.* **134**, 575–583.

Green, J. R. and Northcote, D. H. (1979). *Biochem. J.* **178**, 661–671.

Hanover, J. A. and Lennarz, W. J. (1980). *J. Biol. Chem.* **255**, 3600–3604.

Hanover, J. A. and Lennarz, W. J. (1982). *J. Biol. Chem.* **259**, 2787–2794.

Hart, G. W., Brew, K., Grant, G. A., Bradshaw, R. A. and Lennarz, W. J. (1979). *J. Biol. Chem.* **254**, 9747–9753.

Haselbeck, A. and Tanner, W. (1982). *Proc. Natl. Acad. Sci. USA* **79**, 1520–1524.

Heifetz, A. and Elbein, A. D. (1977). *J. Biol. Chem.* **252**, 3057–3063.

Heifetz, A., Keenan, R. W. and Elbein, A. D. (1979). *Biochemistry* **18**, 2186–2192.

Hemming, F. W. (1974a). *In* "The Biochemistry of Lipids", Vol. 4. (T. W. Goodwin, ed.) pp. 39–97. Butterworth, London.

Hemming, F. W. (1974b). *MTP Int. Rev. Sci. Biochem. Lipids*, 39–97.

Herscovics, A., Warren, C. D. and Jeanloz, R. W. (1975). *J. Biol. Chem.* **250**, 8079–8084.

Herscovics, A., Bugge, B. and Jeanloz, R. W. (1977a). *J. Biol. Chem.* **252**, 2271–2278.

Herscovics, A., Golovtchenko, A. M., Warren, C. D., Bugge, B. and Jeanloz, R. W. (1977b). *J. Biol. Chem.* **252**, 224–234.

Herscovics, A., Warren, C. D., Bugge, B. and Jeanloz, R. W. (1978). *J. Biol. Chem.* **253**, 160–165.

Hirschberg, C. B. and Snider, M. D. (1987). *Ann. Rev. Biochem.* **56**, 63–87.

Hopp, H. E., Romero, P. R., Daleo, G. R. and Pont Lezica, R. (1977). *Plant Physiol.* **59**, 82–88.

Hopp, H. E., Daleo, G. R., Romero, P. A. and Pont Lezica, R. (1978a). *Plant Physiol.* **61**, 248–257.

Hopp, H. E., Romero, P. A., Daleo, G. R. and Pont Lezica, R. (1978b). *Eur. J. Biochem.* **84**, 561–571.

Hori, H. and Elbein, A. D. (1982). *Plant Physiol.* **70**, 12–20.

Hori, H., James, Jr., D. W. and Elbein, A. D. (1982). *Arch. Biochem. Biophys.* **215**, 12–21.

Hou, C. T., Umemura, U., Nakamura, M. and Funahashi, S. (1968). *J. Biochem. (Tokyo)* **63**, 351–357.

Hsu, A.-F., Baynes, J. W. and Heath, E. C. (1974). *Proc. Natl. Acad. Sci. USA* **71**, 2391–2395.

Hubbard, S. C. and Ivatt, R. J. (1981). *Ann. Rev. Biochem.* **50**, 555–583.

Hubbard, S. C. and Robbins, P. W. (1980). *J. Biol. Chem.* **255**, 11782–11793.

Huffaker, T. C. and Robbins, P. W. (1983). *Proc. Natl. Acad. Sci. USA* **80**, 7466–7470.

Ito, T., Kudama, Y., Kawamura, K., Suzuki, K., Takatsuki, A. and Tamura, G. (1979). *Agric. Biol. Chem.* **43**, 1187–1195.

James, D. W., Preiss, J. and Elbein, A. D. (1985). *In* "The Polysaccharides", Vol. 3, (G. Aspinall, ed.), pp. 107–207. Academic Press, New York.

James, M. J. and Kandutsch, A. (1974). *J. Biol. Chem.* **254**, 8442–8446.

Jensen, J. W. and Schutzbach, J. S. (1981). *J. Biol. Chem.* **256**, 12899–12904.

Jensen, J. W. and Schutzbach, J. S. (1985). *Eur. J. Biochem.* **153**, 41–48.

Kang, M. S., Spencer, J. P. and Elbein, A. D. (1978a). *Biochem. Biophys. Res. Commun.* **82**, 568–574.

Kang, M. S., Spencer, J. P. and Elbein, A. D. (1978b). *J. Biol. Chem.* **253**, 8860–8866.

Kang, M. S., Spencer, J. P. and Elbein, A. D. (1979). *J. Biol. Chem.* **254**, 10037–10043.

Kaplan, H. A., Naider, F. and Lennarz, W. J. (1988). *J. Biol. Chem.* **263**, 7814–7820.

Kaushal, G. P. and Elbein, A. D. (1985). *J. Biol. Chem.* **260**, 16303–16309.

Kaushal, G. P. and Elbein, A. D. (1986a). *Plant Physiol.* **81**, 1086–1091.

Kaushal, G. P. and Elbein, A. D. (1986b). *Plant Physiol.* **82**, 748–752.

Kaushal, G. P. and Elbein, A. D. (1986c). *Arch. Biochem. Biophys.* **250**, 38–47.

Kaushal, G. P. and Elbein, A. D. (1987). *Biochemistry* **26**, 7953–7960.

Kaushal, G. P., Szumilo, T. and Elbein, A. D. (1988). *In* "The Biochemistry of Plants", Vol. 14, (J. Preiss, ed.), pp. 421–456. Academic Press, New York.

Kauss, H. (1969). *FEBS Lett.* **5**, 81–84.

Kean, E. (1977). *J. Biol. Chem.* **252**, 5622–5629.

Kean, E. (1980). *J. Biol. Chem.* **255**, 1921–1927.

Kean, E. (1982). *J. Biol. Chem.* **257**, 7952–7954.

Keenan, R. W. and Kruczek, M. E. (1976). *Biochemistry* **15**, 1586–1591.

Keller, R. K. (1986). *J. Biol. Chem.* **261**, 12053–12059.

Keller, R. K., Boon, D. Y. and Crum, F. C. (1979). *Biochemistry* **18**, 3946–3952.

Kornfeld, R. and Kornfeld, S. (1985). *Ann. Rev. Biochem.* **54**, 631–664.

Kronquist, K. E. and Lennarz, W. J. (1978). *J. Supramol. Struct.* **8**, 51–65.

Laine, R. and Elbein, A. D. (1971). *Biochemistry* **10**, 2547–2553.

Lehle, L. (1981). *FEBS Lett.* **123**, 63–66.

Lehle, L. and Bause, E. (1984). *Biochim. Biophys. Acta* **799**, 246–251.

Lehle, L. and Tanner, W. (1976). *FEBS Lett.* **71**, 167–180.

Lehle, L. and Tanner, W. (1978). *Biochim. Biophys. Acta* **539**, 218–229.

Lehle, L. and Tanner, W. (1983). *Biochem. Soc. Trans.* **11**, 568–574.

Lehle, L., Fartaczek, F., Tanner, W. and Kauss, F. (1976). *Arch. Biochem. Biophys.* **175**, 419–426.

Leloir, L. F., Staneloni, R. J., Carminatti, H. and Behrens, N. H. (1973). *Biochem. Biophys. Res. Commun.* **52**, 1285–1292.

Levy, J. A., Carminatti, H., Cantarella, A. I., Behrens, N. H., Leloir, L. H. and Tabora, E. (1974). *Biochem. Biophys. Res. Commun.* **60**, 118–125.

Marshall, R. D. (1974). *Biochem. Soc. Symp.* **40**, 17–26.
McCloskey, M. A. and Troy, F. A. (1980). *Biochemistry* **19**, 2061–2068.
McEvoy, F. A., Ellis, D. E. and Shall, S. (1977). *Biochem. J.* **164**, 273–275.
Mills, J. T. and Adamany, A. (1978). *J. Biol. Chem.* **253**, 5270–5273.
Molner, J., Chao, H. and Ikehara, Y. (1971). *Biochim. Biophys. Acta* **239**, 401–410.
Muller, T., Bause, E. and Jaenicke, L. (1981). *FEBS Lett.* **128**, 208–212.
Murphy, L. A. and Spiro, R. G. (1981). *J. Biol. Chem.* **256**, 7487–7494.
Nakaido, H. and Hassid, W. Z. (1971). *Adv. Carbohyd. Chem. Biochem.* **26**, 351–483.
Olden, K., Parent, J. B. and White, S. (1982). *Biochim. Biophys. Acta* **650**, 209–232.
Palamarczyk, G. and Hemming, F. W. (1975). *Biochem. J.* **148**, 245–251.
Palamarczyk, G., Lehle, L., Mankowski, T., Chojnacki, T. and Tanner, W. (1980). *Eur. J. Biochem.* **105**, 517–523.
Pan, Y. T. and Elbein, A. D. (1989). Unpublished observations.
Parodi, A. J. and Leloir, L. F. (1979). *Biochim. Biophys. Acta* **559**, 1–37.
Parodi, A. J., Behrens, N. H., Leloir, L. F. and Carminatti, H. (1972). *Proc. Natl. Acad. Sci. USA* **69**, 3268–3272.
Parodi, A. J., Quesada-Allue, L. A. and Cazzulo, J. J. (1981). *Proc. Natl. Acad. Sci. USA* **78**, 6201–6205.
Parodi, A. J., Lederkremer, G. Z. and Mendelzon, D. H. (1983). *J. Biol. Chem.* **258**, 5589–5595.
Parodi, A. J., Martin-Barrientos, J. and Engel, J. C. (1984). *Biochem. Biophys. Res. Commun.* **118**, 1–7.
Perez, M. and Hirschberg, C. B. (1986). *J. Biol. Chem.* **261**, 6822–6830.
Pless, D. D. and Lennarz, W. J. (1977). *Proc. Natl. Acad. Sci. USA* **74**, 134–138.
Plouhar, P. L. and Bretthauer, R. K. (1982). *J. Biol. Chem.* **257**, 8907–8911.
Pont Lezica, R., Brett, C. T., Martinez, P. R. and Dankert, M. (1975). *Biochem. Biophys. Res. Commun.* **66**, 980–987.
Pont Lezica, R., Daleo, G. R. and Dey, P. M. (1986). *Adv. Carbohyd. Chem. Biochem.* **44**, 341–385.
Prakash, C., Katial, A., Kang, M. S. and Vijay, I. K. (1984). *Eur. J. Biochem.* **139**, 87–93.
Rathod, P. K., Tasjian, A. H. Jr. and Abeles, R. H. (1986). *J. Biol. Chem.* **261**, 6461–6469.
Rearick, J. I., Chapman, A. and Kornfeld, S. (1981a). *J. Biol. Chem.* **256**, 6255–6261.
Rearick, J. I., Fujimoto, K. and Kornfeld, S. (1981b). *J. Biol. Chem.* **256**, 3762–3769.
Richards, J. B. and Hemming, F. W. (1972). *Biochem. J.* **130**, 77–83.
Richards, J. B., Evans, P. J. and Hemming, F. W. (1971). *Biochem. J.* **124**, 957–959.
Riedell, W. E. and Miernyk, J. A. (1988). *Plant Physiol.* **87**, 420–426.
Reitman, M. L. and Kornfeld, S. (1981). *J. Biol. Chem.* **256**, 4275–4281.
Romero, P. A. and Pont Lezica, R. (1976). *Acta Physiol. Latinoam.* **26**, 364–370.
Ronnett, G. V. and Lane, M. D. (1981). *J. Biol. Chem.* **256**, 4704–4707.
Rothblatt, J. A. and Meyer, D. I. (1986). *EMBO J.* **5**, 1031–1036.
Runge, K. W. and Robbins, P. W. (1986). *J. Biol. Chem. Microbiol.*, 312–316.
Runge, K. W., Huffaker, T. C. and Robbins, P. W. (1984). *J. Biol. Chem.* **259**, 412–417.
Sasak, W., Levrat, C., Warren, C. D. and Jeanloz, R. W. (1984). *J. Biol. Chem.* **259**, 332–337.
Schmitt, J. and Elbein, A. D. (1979). *J. Biol. Chem.* **254**, 12291–12294.
Schutzbach, J. S., Springfield, J. D. and Jensen, J. W. (1980). *J. Biol. Chem.* **255**, 4170–4175.
Sharma, C. B., Babczinski, P., Lehle, L. and Tanner, W. (1976). *Eur. J. Biochem.* **46**, 35–41.
Sharma, C. B., Lehle, L. and Tanner, W. (1981). *Eur. J. Biochem.* **116**, 101–108.
Sharma, C. B., Lehle, L. and Tanner, W. (1982). *Eur. J. Biochem.* **126**, 319–325.
Snider, M. (1983). *In* "Biology of Carbohydrates", Vol. 2, (V. Ginsburg and P. W. Robbins, eds), pp. 163–198. Wiley Interscience, New York.
Snider, M. D. and Robbins, P. W. (1982). *J. Biol. Chem.* **257**, 6796–6801.
Snider, M. D. and Rogers, O. C. (1984). *Cell* **36**, 753–761.
Snider, M. D., Sultzman, L. A. and Robbins, P. W. (1980). *Cell* **21**, 385–392.
Spencer, J. P. and Elbein, A. D. (1980). *Proc. Natl. Acad. Sci. USA* **77**, 2524–2527.
Spiro, M. J. and Spiro, R. G. (1985). *J. Biol. Chem.* **260**, 5808–5815.
Spiro, M. J., Spiro, R. G. and Bhoyroo, V. D. (1976). *J. Biol. Chem.* **251**, 6400–6408.
Spiro, M. J., Spiro, R. G. and Bhoyroo, V. D. (1979). *J. Biol. Chem.* **254**, 7668–7674.

Staneloni, R. J., Tolmasky, M. E., Petriella, C., Ugalde, R. A. and Leloir, L. F. (1980a). *Biochem. J.* **191**, 257–260.

Staneloni, R. J., Ugalde, R. A. and Leloir, L. F. (1980b). *Eur. J. Biochem.* **105**, 275–278.

Staneloni, R. J., Tolmasky, M. E., Petriella, C. and Leloir, L. F. (1981). *Plant Physiol.* **68**, 1175–1179.

Stoll, J., Losenberg, L., Carson, D. D., Lennarz, W. J. and Krag, S. S. (1985). *J. Biol. Chem.* **260**, 232–236.

Struck, D. K. and Lennarz, W. J. (1977). *J. Biol. Chem.* **252**, 1007–1013.

Struck, D. K. and Lennarz, W. J. (1980). *In* "The Biochemistry of Glycoproteins and Proteoglycans" (W. J. Lennarz, ed.), pp. 35–73. Plenum Press, New York.

Takatsuki, A. and Tamura, G. (1982). *In* "Tunicamycins" (G. Tamura, ed.). Japan Sci. Soc. Press, Tokyo.

Tamura, G. (ed.) (1982). "Tunicamycins" Japan Sci. Soc. Press, Tokyo.

Tanner, W. (1969). *Biochem. Biophys. Res. Commun.* **35**, 144–150.

Tanner, W. and Lehle, L. (1987). *Biochim. Biophys. Acta* **906**, 81–99.

Tkacz, J. and Lampen, J. O. (1975). *Biochem. Biophys. Res. Commun.* **65**, 248–253.

Trimble, R. B., Byrd, J. C. and Maley, F. (1980). *J. Biol. Chem.* **255**, 11892–11895.

Troy, F. A. (1979). *Ann. Rev. Microbiol.* **33**, 519–560.

Tucker, P. and Pestka, S. (1977). *J. Biol. Chem.* **252**, 4474–4486.

Turco, S., Stetson, B. and Robbins, P. W. (1977). *Proc. Natl. Acad. Sci. USA* **74**, 4411–4414.

Vijay, I. K., Perdew, G. H. and Lewis, D. E. (1980). *J. Biol. Chem.* **255**, 11210–11220.

Vijay, I. K. and Perdew, G. H. (1982). *FEBS Lett.* **139**, 321–324.

Villemez, C. L. and Clark, A. F. (1969). *Biochem. Biophys. Res. Commun.* **36**, 57–63.

Villemez, C. L. and Carlo, P. L. (1980). *J. Biol. Chem.* **255**, 8174–8178.

Vliegenthart, J. F. G., Dorland, L. and van Halbeek, H. (1983). *Adv. Carbohyd. Chem. Biochem.* **41**, 209–373.

Waechter, C. J. and Harford, J. B. (1977). *Arch. Biochem. Biophys.* **181**, 185–198.

Waechter, C. J. and Lennarz, W. J. (1976). *Ann. Rev. Biochem.* **45**, 95–112.

Waechter, C. J. and Scher, M. G. (1976). *Arch. Biochem. Biophys.* **188**, 385–393.

Waechter, C. J., Lucas, J. J. and Lennarz, W. J. (1973). *J. Biol. Chem.* **248**, 7570–7579.

Waechter, C. J., Lucas, J. J. and Lennarz, W. J. (1974). *Biochem. Biophys. Res. Commun.* **56**, 343–350.

Wepley, J. K., Shenbagamurthi, P., Lennarz, W. J. and Naider, F. (1983). *J. Biol. Chem.* **258**, 11856–11863.

Wepley, J. K., Shenbagamurthi, P., Naider, F., Park, H. R. and Lennarz, W. J. (1985). *J. Biol. Chem.* **260**, 6459–6565.

Wong, T. K. and Lennarz, W. J. (1982). *J. Biol. Chem.* **257**, 6619–6624.

Wong, T. K., Decker, C. P. and Lennarz, W. J. (1982). *J. Biol. Chem.* **257**, 6614–6618.

Zatta, P., Zakim, D. and Vessey, D. A. (1976). *Biochem. Biophys. Res. Commun.* **70**, 1014–1019.

Zatz, M. and Barondes, S. H. (1969). *Biochem. Biophys. Res. Commun.* **36**, 511–517.

4 Disaccharides

GAD AVIGAD

Department of Biochemistry, UMDNJ-Robert Wood Johnson Medical School, Piscataway, NJ 08854, USA

METHODS IN PLANT BIOCHEMISTRY Vol. 2
ISBN 0-12-461012-9

I. INTRODUCTION

There is an enormously broad range of methods available for sugar analysis, and the author is guided by his own experience and probably by some amount of a biased judgement to make the selection presented in this chapter. The difficulty of this task can be perceived from the fact that just in the cumulative Chemical Abstracts Index for 1982–1986 there are about 10 000 listings for sucrose, of which about half concern biological aspects of the sugar, and at least 3000 abstracts for enzymes related to disaccharide utilisation or transformation. In comparison, only about 600 abstracts for trehalose, 2500 for maltose, and 250 for melibiose are recorded for the same period.

The basic methods and principles for the determination of disaccharides are the same as those applied to the study of monosaccharides or of longer chain glycosides. Consequently, since a large number of the procedures for the determination of disaccharides are based on their hydrolysis to the monosaccharide components, some cases of overlap between the texts in this chapter and Chapter 1 occur. Chemical reagents, specific enzymes and chromatographic tchniques can be used effectively and sensitively for disaccharide analysis. Unfortunately, it is impossible within the space available to provide detailed protocols for each of the methods described. Accordingly, except for a small number of cases where protocols are briefly given, the reader is presented with an extensive selection of literature citations where detailed descriptions of experimental procedures can be located.

Analysis of carbohydrates in plant tissue involves the procedures of extraction, separation, structural identification and quantitation. With the analytical technologies available today, several of these steps can be executed simultaneously. This is exemplified, for example, by the application of liquid chromatography techniques when the sugars in the sample are modified chemically, separated, their structure determined, and their concentration assayed in one 'run'. Similarly both sugar identification and differential measurements can be achieved with coupled enzyme reaction systems and with radioisotope labelling.

Knowledge of the basic chemical properties of carbohydrates and understanding of the mechanism of action of enzymes involved in their transformations are essential to assure proper application of the analytical methods, and to grasp their limitations. This rule assumes even more importance in our time when commercially provided time-saving kits of ready-made, pre-packed analytical reagents are commonly used in many research and analytical laboratories, and when automated instrumentation is employed to perform the analysis.

For general literature about disaccharide analysis, consult the series of *Methods in Carbohydrate Chemistry* Vols I–VIII (1962–1980) published by Academic Press, New York, and the monographs by Lee (1980), White and Kennedy (1981), Lee and Lindley (1982), Schneider (1982), Williams (1984), and by Chaplin and Kennedy (1986). For enzymic analysis, consult Bergmeyer *et al.*'s *Methods of Enzymatic Analysis*, 3rd edition, published by Verlag Chemie (1982–1985), and the volumes on Carbohydrate Metabolism and Complex Carbohydrates in the Series of *Methods in Enzymology* (161 vols) published by Academic Press, London, since 1955. References to reviews on specific analytical methodologies (for example about chromatography) are given in the respective sections. Among the 'compact' general overviews of carbohydrate chemistry, the books by Shallenberger (1982), Kennedy (1988) and by ElKhadem (1988) are recommended. The series of *Advances in Carbohydrate Chemistry and Biochemistry* published by Academic Press, London, since 1945 is one of the richest troves of knowledge for those interested in carbohydrate chemistry and biology. The classic textbook by Browne and Zerban (1941) is an excellent source of information about methods for disaccharide analysis available before the introduction of chromatography and the use of advanced enzymic reagents.

II. HISTORY AND INVENTORY

A. History

The history of disaccharide research is intimately entangled in the development of the science of biochemistry. The endeavours to understand the structure of sugars, the process of fermentation, the nature of enzymes, and biological catalysis, the route of carbon assimilated during photosynthesis and tracing the course of metabolic pathways, often involved the study of disaccharides, primarily that of sucrose. A wealth of literature describing various aspects of the history of sugar research is available and the very short outline presented here is based on the works of Reed (1866), Tucker (1881), Effront (1902), VonLippmann (1904), Browne (1912), Czapek (1913), Armstrong (1924), Sumner and Somers (1943), Deerr (1949–1950), Levi and Purvis (1949), Gottschalk (1950), Neuberg and Mandl (1950) and Fruton (1972).

Preparations of pure crystalline cane-sugar (sucrose) were known from ancient times, but its chemical nature was a mystery until the middle of the nineteenth century. Similarly was the case with the disaccharide lactose, which was first isolated by Bartoletti in 1619. It was Sigismund Andreas Marggraf who between 1747 and 1762 succeeded in isolating crystalline sugar from beets and other plants by extraction in a hot ethanolic solution and in the presence of lime. He claimed on the basis of taste, shape of crystals and 'stability' in alkaline solutions that the material obtained from all sources is the same as cane-sugar. It took an additional period of two hundred years of reasearch finally to establish the chemical nature of this compound and to understand how it is synthesised in nature. Two of the techniques that Marggraf employed, hot alcohol extraction and use of alkaline milk of lime 'solution' in which sucrose is stable, remain as fundamental practical methods used for the extraction, refining, and analysis of sucrose to this day.

Between 1810 and 1830, the chemical composition of sugar was established by GayLussac, Berzelius, and Dumas. DeSaussure in 1819 and Payen and Presoz in 1833 discovered the 'diastase' activity of malt, and in 1847, the product of this reaction was identified as a disaccharide, maltose, by Dubrunfaut. Dumas in 1828, Presoz in 1833, Peligot in 1838, and Biot in 1842, all concluded that sucrose can be cleaved by acid or by a 'ferment' from yeast or other biological sources, to yield a 'reducing' sugar. It was Dubrunfaut in 1847 who finally established that this reaction provides glucose and a ketose, shown to be identical to the fructose obtained in inulin hydrolysates. Yeast invertase was identified as an individual catalytic entity by Berthelot in 1860. Further studies by O'Sullivan, E. Fischer, and Armstrong towards the end of the nineteenth century, and by C. S. Hudson between 1908 and 1915, transformed invertase into a commonly used agent for sucrose analysis and in sugar technology in general.

Additional disaccharides, such as trehalose, melibiose and turanose, were isolated and characterised during the middle and late nineteenth century, and together with the chemical identification of these compounds, enzymes that can cause their hydrolysis were obtained from various biological sources. Prominent scientists of that period such as Liebig, Berthelot, Bourquelot, E. Fischer, Bertrand, Willstatter, Armstrong, and others contributed extensively to these earlier studies.

Though scientists of that era envisaged the potential that these enzymes have for the specific analysis of disaccharides, their efforts in this direction were thwarted because most enzyme preparations then used were impure reagents, containing more than one hydrolase activity.

In the early nineteenth century, analysis of sucrose solutions was based on the determination of specific gravity (Baumé or Brix degrees), a technique which is still in use in the sugar industry, and which is suitable for the determination of 0.1–99% sugar solutions. Biot in 1836 and Ventzke in 1842 discovered the optical properties of sugars, and the 'inversion' of optical rotation which occurs when sucrose is hydrolysed by acids. This immediately led to the development of the polarimetric methods for the determination of sugar solutions, using polarimeters or saccharimeters. Refractive index methods for the measurement of sugar solution were also introduced during that period.

Becquerel in 1831 found that some sugar can reduce alkaline copper, and Barreswill in 1846 used this reaction to develop a reagent for the quantitation of sugar in solution. H. Fehling in 1849 improved the procedure and thereby established the permanency of both one of the most important reagents available for sugar analysis, as well as of his own name for posterity in the annals of sugar chemistry. Titrimetric and gravimetric 'Fehling' methods in numerous variations were commonly in use for over a century, and manuals for sugar analysis usually contained scores of tables suitable for the determination of the equivalent of 2–300 mg glucose per sample.

Toward the end of the nineteenth century, additional reactions for sugar analysis based primarily on the reactivity of the carbonyl carbon, such as oxidations, amination, hydrazone formation and others, were developed by Gentele, Killiani, Wohl, and E. Fischer. Among methods used for sugar isolation and purification was the formation of 'saccharate' complexes with cations such as calcium, strontium and lead salts. Specific qualitative colour tests for sugar such as reactions with various phenolic compounds in strong mineral acids were developed. During the beginning of the twentieth century, these colorimetric reactions, as well as the classical metallo-oxidation methods, were converted into quantitative spectrophotometric assays for sugars, and the sensitivity of sugar determination increased to a level of 0.1–1 µmol per sample.

Between 1928 and 1938, key biochemical reactants such as ATP, NAD^+ and $NADP^+$, hexokinase, phosphoglucoisomerase and mutase, and glucose 6-phosphate dehydrogenase were discovered by Meyerhof, Lohman, Von Euler, Warburg, Cori and others. The use of these enzymes in a spectrophotometric assay procedure for the analysis of glucose at the nmol level emerged as a bonus from these biochemical studies. Glucose oxidase as an analytical reagent also was ushered in about 1945.

The introduction of chromatography, the use of radioisotopes, and development of spectroscopic techniques in the 1940–1950 period opened the gates to unprecedented capabilities for the determination of sugars in biological studies. The increased technical sophistication in instrumentation, and the impressive effort of various commercial enterprises to provide reliable and pure biochemical reagents, brought us to where we stand now in our ability to study and analyse sugars, rapidly, accurately, and at the pmol level of sensitivity.

As for the structure of sucrose, the methylation analysis of Irvine and Haworth, between 1903 and 1927, and the polarimetric studies of Hudson in the same period, furnished the structural model of the disaccharide as we know it today. NMR spectroscopy in the 1960s determined the exact spatial conformation of the molecule. Lemieux and Huber in 1953 chemically synthesised sucrose for the first time. The enzymic synthesis of sucrose using the bacterial sucrose phosphorylase was achieved by Hassid, Doudoroff, and Barker in 1944, synthesis by the plant sucrose synthase was accomplished by Leloir and Cardini in 1955.

B. Inventory

Sucrose is the major product of carbon assimilation during photosynthesis. It is the dominant transportable organic carbon compound which is delivered to sink tissues in the plant, and it can accumulate in storage tissues in large quantities. Trehalose, also a non-reducing disaccharide, is equivalent to sucrose, being a reserve storage molecule in many species of non-photosynthetic organisms, such as fungi, yeast, actinomycetes, some bacteria, and the body fluids of insects and other invertebrata. It may appear also in some algae and ferns. In several species of algae, seaweeds and fungi, the monoglycosyl inositols and other monoglycosyl-polyols sometimes are found as reserve sugars at low levels.

A large number and variety of reducing disaccharides may be detected in plant tissue extracts. These compounds are usually present at very low levels, and their appearance is the product of catabolism of structural or reserve glycosides in the organism. Among these disaccharides are those formed from the degradation of storage oligosaccharides (such as the α-galactosides), of reserve polysaccharides (starch and fructans), of structural polysaccharides (β-glucans), and some other rare disaccharides at low levels which arise from degradation of other glycosides.

Comprehensive reviews on disaccharides in plant can be found in the summaries written by Staneck et al. (1965), Aspinall et al. (1967), Lee (1980), Avigad (1982), Dey and Dixon (1985), and Collins (1987).

1. Sucrose

In addition to its central position in agricultural and food chemistry, this disaccharide continues to be a focus of interest for research in carbohydrate metabolism and physiology in plants. A brief discussion of this topic is presented in Section III.

(a) *Sucrose esters.* A variety of polar, mono- and poly-*O*-acylated sucrose esters were recently isolated from various plants. Among these were esters isolated from *Tulipa* (Strack et al., 1981) from *Polygala* (Hamburger and Hostettmann, 1985), from *Solanum* (King et al., 1986, 1987), from *Lilliaceae* (Nakano et al., 1986), from tobacco leaves (Severson et al., 1985; Wahlberg et al., 1986; Garegg et al., 1988), and from rhubarb (Kashiwada et al., 1988). Among the acyl groups identified were acetyl-, capryl-, isobutyryl-, methylpentanoyl-, feruloyl- and galloyl- residues. These compounds, present usually in small quantities, are eluted from the plant material by organic solvents, and a significant amount of them can be found in the soluble carbohydrate pool obtained by 80% ethanol extraction. Since these compounds are both β-fructofurano-sids and esters, they are liable to hydrolysis by acid and by alkali, respectively. Detailed chromatographic, chemical, enzymic and carbon-13 NMR methods for the identifi-cation of these sucrose esters can be found in the references cited. The presence of these compounds in plants may be much more common than had been thought at first. Very little is known about the biochemical mechanisms involved in their biosynthesis (probably by esterification of sucrose) or their metabolism.

A novel structure (Agrocinopine A) containing L-arabinopyranose-2-phosphate linked as a phosphodiester to the 4-OH of the fructose residue in sucrose, has been identified in crown gall tumours (Ryder *et al.*, 1984).

(b) *Sucrose analogues*. A large number of structural analogues of sucrose have been prepared either by chemical synthesis or by specific enzymic reactions. Many of these disaccharides have been used in biochemical and physiological experimentation exploring reactions that normally involve sucrose. Among the analogues, we find several α-D-glycosyls such as various α-D-glucopyranosyl-ketofuranosides (Hassid *et al.*, 1951; Hassid and Ballou, 1957); α-D-xylosyl- and α-D-galactosyl-D-fructofuranoside (Avigad *et al.*, 1956; Feingold *et al.*, 1957; Hestrin and Avigad, 1958; Tanaka *et al.*, 1981; Nisizawa *et al.*, 1986; Cheetham *et al.*, 1989); α-D-allosyl-D-fructofuranoside (Fukui *et al.*, 1963; Gruber and Feingold, 1964; Hough and O'Brien, 1980; Binder and Robyt, 1984); 3-deoxysucrose, 4,6-dideoxysucrose, 6-deoxysucrose, 6,6'-dideoxysucrose (Chen *et al.*, 1983; Binder and Robyt, 1984, 1985, 1986; Tanriseven and Robyt, 1989); 6-thiosucrose (Binder and Robyt, 1984); 3-deoxy-3-fluorosucrose, 6-deoxy-6-fluoro-sucrose (Binder and Robyt, 1986; Card *et al.*, 1986; Eklund and Robyt, 1988); the inver-tase resistant 1'-deoxy-1'-fluorosucrose (Card and Hitz, 1984; Hitz *et al.*, 1985, 1986; Schmalstig and Hitz, 1987a,b; Damon *et al.*, 1988); 6'-deoxy-6'chlorosucrose (Hashi-moto *et al.*, 1985); 1'-azido-1'-deoxysucrose, 4'-deoxy-4'-fluorosucrose, and 4-deoxy-4-fluorosucrose (Card *et al.*, 1986). Sucrose 3',4'-epoxide was synthesised by Guthrie *et al.* (1983) and 1-thiosucrose by Defaye *et al.* (1984). Additional literature on structural analogues of sucrose is listed by Khan (1976), Jenner (1980), Binder and Robyt (1986), and Collins (1987). A sucrose 6-phosphate analogue where the phosphate is linked to the C-6 glucosyl residue, is produced by strains of *Streptococci* or *Escherichia coli* when grown on sucrose (Schmidt *et al.*, 1982; Chassy and Porter, 1982; Martin and Russell, 1987). This ester is synthesised by the phosphoenolpyruvate–phosphotransferase system of the bacteria and has also been synthesised chemically.

2. Trehalose

α,α-Trehalose is widely distributed in nature, and is found in insects, fungi, achinomyces and some bacteria (Elbein, 1974; Lee, 1980; Avigad, 1982), and it is an important osmolite in several strains of cyanobacteria (Reed and Stewart, 1988). It has been only rarely detected in photosynthetic organisms. The presence of trehalose in solutions that originated from plant tissues is almost always a biochemical marker for contribution by yeasts or fungi. Significant advances have been made in understanding the pattern and control of trehalose metabolism, particularly in yeast and other fungi (Avigad, 1982; Thevelein, 1988; Section III). Of particular interest to plant biochemists is the observation that trehalose is found in soybean root nodules (Streeter, 1982, 1985, 1987; Reibach and Streeter, 1983; Salminen and Streeter, 1986; Mellor, 1988). Since the review by Avigad (1982), information about trehalose accumulation in several bacterial strains has been published by Streeter (1985), Iwahara (1988), and by Schimz and Overhoff (1987), and in spores of several fungi described by Van Laere et al. (1987) and by Van Laere and Slegers (1987).

(a) *Trehalose analogues.* A number of structural analogues of the natural α,α-trehalose molecule are known (Elbein, 1974; Lee, 1980; Collins, 1987). Among these are several aminotrehalose isomers known to be weakly bacteriostatic reagents, the β,β- and the α,β-trehalose. The α-D-allopyranosyl-α-D-glucopyranoside analogue can be prepared via the reduction of 3-ketotrehalose (Section VII.C.7(b)), and the α-D-glucopyranoside has been enzymically synthesised by Kasumi et al. (1986). Several deoxyfluorotrehalose derivatives have been prepared by Penglis (1981), and 2-deoxy, 3-deoxy, 2,3-dideoxy-trehalose have been synthesised by Baer et al., (1985). Various α,α-dialdohexosyl and dixylosyl analogues of trehalose have been prepared by Defaye et al. (1980, 1983). Isomers of D-galactopyranosyl-D-glucopyranoside or D-galactopyranoside have been described by Nimami et al. (1985) and α-D-mannopyranosyl-α-D-mannopyranoside has been prepared by Liav and Goren (1983).

3. Maltose

This disaccharide normally appears in plant tissues only at very low levels as a product of starch degradation. This process is intensified in photosynthetic tissues during dark periods, whereas when illuminated, maltose is almost undetectable in the plastids. Maltose concentration can increase significantly during the processes of germination and growth and in senescent tissues when stored starch is being hydrolysed by amylolytic enzymes. For detailed literature, see Stitt and ApRees (1980), Tarelli (1980), Avigad (1982), Nakamura and Kozimi (1985), Manners (1985), and Steup (1988). The appearance of maltose in algae-invertebrate symbionts has been discussed by Douglas (1988); (see also Avigad, 1982).

4. Melibiose

Melibiose (6-*O*-α-D-galactopyranosyl-D-glucose) usually appears as a product of the cleavage of raffinose by invertase. Its level in tissues which contain raffinose is usually very low. Levels of disaccharide tend to increase during periods when storage carbo-hydrates are actively consumed such as during seed germination and fruit ripening. In addition to the surveys by Rathbone (1980), Avigad (1982), and Dey (1985), several recent studies related to the metabolism of α-galactoside oligosaccharide where meli-biose may have appeared have been published (Schoenwitz and Ziegler, 1982; Costello et al., 1982; Muller and Jacks, 1983; Van Den et al., 1986; Saravitz et al., 1987; Kasai et al., 1981; Madore et al., 1988; Kuo et al., 1988; Quemener, 1988).

5. Other disaccharides, glycosyl-alditols and glycosyl-inositols

Very little new information has been added recently to our knowledge about the variety of reducing disaccharides which appear in plant tissues as a result of enzymic degradation of polysaccharides or other glycosides (Lee, 1980; Avigad, 1982; Dey, 1985; Dey and Dixon, 1985).

Laminaribiose was detected during studies of UDP-glucose and sucrose metabolism by plant cell protoplasts (Preisser and Komor, 1988; Maretzki and Thom, 1988; Niemetz and Hawker, 1988). Gentiobiose and 1-O-β-glucopyranosylglycerol in *Lilium* were studied by Kaneda *et al.* (1984). Floridoside metabolism has been studied by Kremer and Kirst (1981) and Meng *et al.* (1987); β-galactosylglycerol was obtained by Nakano *et al.* (1988). Glycosylinositols such as galactinol were studied by Schweitzer and Horman (1981), Madore and Webb (1982), Nicolas *et al.* (1984), Pharr *et al.* (1985, 1987), Pharr and Sox (1984), Saravitz *et al.* (1987) and Madore *et al.* (1988). The presence of α-D-glucopyranosyl (1-2)glycerol (lilioside) as an osmolyte in cyanobacteria was discussed by Reed and Stewart (1988).

III. BIOCHEMICAL BACKGROUND

A. Sucrose

The principal reactions which are involved in sucrose synthesis and degradation in plants have been defined as the following (Akazawa and Okamoto, 1980; Avigad, 1982; Edwards and Walker, 1983):

(a) carbon dioxide fixed during photosynthesis in the plastids is in part converted to starch, and in substantial quantities transported to cytosol as triose phosphates. Here the triose phosphates are converted into hexose phosphates.

(b) In the cytosol, the UDP-glucose pyrophosphorylase (EC 2.7.7.9) catalyses the reaction:

$$\alpha\text{-D-Glucose 1-phosphate} + UTP \rightleftharpoons \text{UDP-Glucose} + PPi$$

(c) In what is believed to be the principal reaction for sucrose synthesis, sucrose phosphate synthase (EC 2.4.1.14; SPS) catalyses the reaction:

$$\text{UDP-Glucose} + \text{fructose 6-phosphate} \rightleftharpoons \text{Sucrose 6'-phosphate} + UDP$$

(d) Sucrose is released by the action of sucrose phosphatase (EC 3.1.3.24):

$$\text{Sucrose 6'-phosphate} \longrightarrow \text{Sucrose} + Pi$$

(e) An enzyme which catalyses the direct synthesis of sucrose is found in most tissues, and it is particularly abundant in storage compartments. This enzyme, sucrose synthase (EC 2.4.1.13; SS), catalyses the reaction:

$$\text{UDP-Glucose} + \text{fructose} \rightleftharpoons \text{Sucrose} + UDP$$

This reaction is readily reversible, and is thought to have an important role in the degradation of stored sucrose, providing UDP-glucose and fructose for cellular metabolism.

(f) Tissues contain two types of invertase (β-fructofuranoside). One, most likely vacuolar, has an acid pH optimum for its activity, and the second, an 'alkaline' invertase, is probably a cytosolic enzyme. The exact manner and quantitative aspects by which these two hydrolases contribute to the overall mechanisms of sucrose breakdown in the tissue, and how these activities are regulated, are far from being understood.

Several new discoveries related to sucrose metabolism and which promoted an impressive surge in research activity were made at about the time when the reviews cited above were written. The important recent developments in this area are based on the following:

(a) The discovery of pyrophosphate—fructose 6-phosphate 1-phosphotransferase (EC 2.7.1.90; PFP)—which catalyses the reaction:

PPi + fructose 6-phosphate $\rightleftharpoons$ Fructose 1,6-bisphosphate + Pi

The enzyme is abundant in plant tissues cytosol and provides an alternative pathway for synthesis and degradation of fructose 1,6-bisphosphate which does not depend on ATP or on the regulatory phosphofructokinase (EC 2.7.1.11; PFK) and fructose 1,6-bisphosphatase (Carnal and Black, 1979, 1983).

(b) The discovery of fructose 2,6-bisphosphate and the observation that it is a potent activator of PFP, as well as an inhibitor of fructose 1,6-bisphosphatase (Sabularse and Anderson, 1981; Czeke et al., 1982, 1984).

(c) The identification of highly regulated enzyme entities which are responsible for the synthesis of fructose 2,6-bisphosphate (phosphofructokinase-2; EC 2.7.1.105) and for its hydrolysis (fructose 2,6-bisphosphate; EC 3.1.3.46), both in photosynthetic and in storage tissues (Czeke and Buchanan, 1983; Avigad and Bohrer, 1984).

(d) Detailed characterisation of the kinetic and regulatory properties of enzymes involved in the scheme of sucrose metabolism, particularly those of SPS, of PFP, of PFK, of fructose 1,6-bisphosphatase and of enzymes responsible for fructose 2,6-bisphosphate synthesis and hydrolysis.

All these observations provide us with a better basis for understanding how sucrose synthesis and utilisation is regulated in plant tissues. The prominent role of triose phosphates, hexose 6-phosphate, fructose 2,6-bisphosphate, and Pi and PPi levels and light in modulating the flux of carbohydrate to and from sucrose at selected different physiological situations has been established. It should be pointed out that whereas experimental evidence has been found that the action of SS provides a major step for sucrose utilisation, the observation of Barber (1985) that SPS catalyses a reversible reaction, has not yet been integrated into the many metabolic schemes for sucrose metabolism published recently.

An abundant crop of excellent, though often repetitive, reviews on the biochemistry of sucrose in plants has appeared lately. The reader should turn to Hawker (1985), Heldt and Stitt (1987), Huber et al., (1985, 1987), Stitt (1987b) and Stitt et al. (1987a) for general discussion and working hypotheses on the subject. The articles by Huber (1986), Van Schaffingen (1987), Stitt et al. (1987b), Preiss (1987), Stitt (1987a), Black et al. (1987a,b), ApRees (1988) and Xu et al. (1989), present discussions which emphasise the role of fructose 2,6-bisphosphate and pyrophosphate in regulation. Preiss et al. (1987), Preiss (1987, 1988), Keeling et al. (1988), Hargreaves and ApRees (1988), Doehlert et al. (1988), ApRees (1988) and Tyson and ApRees (1989) all emphasise the metabolic relationship between sucrose and other compounds such as structural glycosides, starch, and the pool of soluble metabolites in the cell during various physiological states. It has been recently claimed that light (Huber *et al.*, 1989a), and phosphorylation of SPS protein (Huber *et al.*, 1989b) are probably important factors that contribute to the regulation of rate of sucrose biosynthesis.

Recent studies of the structure of the sucrose synthase genes, their sequencing and cloning, are likely in the long run to shed more light on the regulatory mechanism associated with sucrose biochemistry. In maize two genes (*Sh*1, and *Ss*2 or *Css*) proscribe the formation of two sucrose synthase isozymes, SS-1 and SS-2, respectively (McCormick *et al.*, 1982; Werr *et al.*, 1985; Springer *et al.*, 1986; McCarty *et al.*, 1986; Gupta *et al.*, 1988 and Rickers *et al.*, 1989). *Sh*1 is the structural gene for the major endosperm enzyme which is a homotetramer of 92 kDa (Hawker, 1985). The *Ss*2 gene is expressed in several tissues, whereas *Sh*1 is predominantly in the endosperm. Both genes are on chromosome 9 and have a significant sequence homology. The two isozymes are expressed differently during the course of plant development and physiological states. Synthesis of *SS*1, for example, is induced by anaerobiosis (Springer *et al.*, 1986; McElfresh and Chourey, 1988). Very similar gene structure and sucrose synthase isozyme distribution have been found in the potato (Salanoubat and Belliard, 1987), soybean nodules (Thummler and Verma, 1987) and in wheat (Marana *et al.*, 1988).

B. Trehalose

The biosynthesis of trehalose follows a pathway which is universal in all organisms known to produce this disaccharde (Elbein, 1974; Avigad, 1982; Salminen and Streeter, 1986). Two key enzymic reactions are involved in this process:

(a) UDP-Glucose + glucose   6-phosphate $\longrightarrow$ α,α-Trehalose   6-phosphate + UDP catalysed by trehalose synthase (EC 2.4.1.15)

(b) α,α-Trehalose 6-phosphate + H_2O $\longrightarrow$ α,α-Trehalose + Pi catalysed by trehalose-P phosphatase (EC 3.1.3.12)

Active catabolism of trehalose occurs in all organisms which synthesise the molecule. The reaction in this process is the hydrolysis to glucose by specific trehalase (EC 3.2.1.28) entities. The regulation of trehalose metabolism has been studied in great detail in fungi and yeast (Panek, 1985; Dellamora-Ortiz *et al.*, 1986; Thevelein, 1984, 1988; Keller *et al.*, 1988; Van Dooren *et al.*, 1988). The rate of metabolic mobilisation of trehalose depends on the activities of two independent trehalase isoenzymes. One type, a 'non-regulatory' species, has an acid pH optimum for its activity and is localised in the vacuole. A second trehalase is present in the cytoplasm, requires a neutral pH for its activity, and it is strongly regulated by a 3,5-cyclic AMP-dependent protein phosphory-lation (enzyme activation), and also by protein dephosphorylation (enzyme deactiva-tion). This regulatory mechanism is responsible for the burst of trehalose degradation which occurs at various physiological states of the organism (Van Laere *et al.*, 1987; Attfield, 1987; Van Laere and Slegers, 1987; Hottiger *et al.*, 1987; Toyoda *et al.*, 1987; Harris and Cotter, 1988).

Trehalase is also produced by many microorganisms which do not synthesise trehalose, but which can use it as a carbon source for growth. Of particular interest is the pattern of trehalose utilisation by *Escherichia coli* (Boos *et al.*, 1987). The enzyme is also produced by the intestinal and kidney epithelium of mammals (Semenza, 1986; Chen *et al.*, 1987; Yoneyama, 1987).

Trehalose phosphorylase (EC 2.4.1.64) and 6-phosphotrehalase (EC 3.2.1.93) have been found in several microorganisms, but their role in trehalose metabolism is not clear (see also Section V.C).

IV. DISACCHARIDE ANALYSIS

A. Strategy

Before applying highly specific analytical procedures it is necessary to have a good idea of which carbohydrates are present and of their concentration ranges in the solution studied. If the system is completely unknown, qualitative evaluation by thin layer chromatography and by total sugar analysis should be obtained first. Choice of specific quantitative methodology will depend on many factors:

(a) the purpose of the analysis—whether it is a survey of disaccharides level to be executed in repetitive series of many samples, a study of an isolated enzyme system, an investigation of a metabolic flux in the tissue, or a structural characterisation;
(b) the concentration range of the disaccharide, whether it is in the mmol or pmol range;
(c) the presence of carbohydrates or of other compounds which may interfere with a particular assay method for the disaccharide;
(d) the degree of accuracy required in the study;
(e) the disaccharide should not be degraded prior to or during the analysis so as to give rise to false data. This concern is particularly important with regard to sucrose, which is readily hydrolysed in mild acid conditions;
(f) the availability of the specific analytical reagents and their cost effectiveness; and
(g) the accessibility to specific instrumentation.

It should be noted that for many purposes, although perhaps taking more time, simple inexpensive techniques for disaccharide analysis can achieve excellent results—almost as good as those that can be obtained with sophisticated and costly instrumentation and with some esoteric reagents.

A critical point to remember is that characterisation of a disaccharide by one technique should always be verified by a second independent assay. Appearance of a spot or a peak on a chromatogram does not always provide a true identity. A different chemical assay, or preferably the use of a specific enzyme reagent, is highly recommended for proof of a structure.

B. Extraction of Disaccharides from Biological Material

The disaccharides are usually found in tissue extracts which contain a pool of low molecular weight metabolites. If the extraction is carried out in water or in dilute salt or buffer solutions, macromolecules will also be solubilised. Rapid deactivation of enzymes which may act on the disaccharides must be accomplished; this is achieved by rapid heating to boiling point (Streeter and Jeffers, 1979; Hoogenboom et al., 1988). A commonly used procedure for obtaining the soluble pool of metabolites, including disaccharides, is rapid extraction in hot 80% ethanol. Extracts from plant tissues may often be slightly acidic, consequently when studying β-fructofuranosides, care should be taken to neutralise the solution. If needed, a small amount of KOH, $Ba(CO_3)_2$, or preferably an organic base (e.g. triethanolamine), may be carefully added to adjust the pH to 7.0–8.0. The biological material extracted should be mechanically cut, crushed, homogenised or pulverised into small fragments. Immediate freezing of the solid sample in liquid nitrogen or dry ice/methanol, or dropping into boiling ethanol, will aid in arresting enzyme activity. Extraction, either by refluxing with water or with 80% ethanol, must be repeated two or three times to secure maximal recovery of soluble sugar, and each extraction should be of at least 20–30 min (extraction periods ranging from only 5 min up to 24 h are recorded in the literature). The volume of the extracting solution should be about 5 ml per 100 mg plant material to be extracted. Handling of very small amounts of biological samples, such as is often the case in metabolic studies and when radioisotopes are employed, is simplified by the use of small Teflon-capped reaction vials and a solid-block heater.

When the aim is to analyse only a non-reducing glycoside such as sucrose or trehalose, extraction can be carried out using a hot (100°C) 0.5 N NaOH solution and the disaccharide is then directly analysed by a colorimetric acid reagent (Rébeillé et al., 1985; Journet et al., 1986).

In some cases, for example in studying seeds, extraction with aqueous solution should be preceded by extraction with an organic phase such as petrol–ether, CCl_4–n-heptane or with chloroform–methanol mixtures. Dickson (1979), Kaiser and Bassham (1979) and Redgwell (1980) have provided detailed procedures for extraction of soluble metabolites from small samples of plant tissues, and with only small variations these methods are followed by many investigators. In the basic protocol up to 50 mg plant material is homogenised and extracted two or three times by 2.0 ml methanol–chloroform–H_2O (12:5:3; v/v). The pellet obtained by centrifugation is washed with chloroform–water (2:1; v/v), then extracted by 80% ethanol. This extract, which contains low molecular weight metabolites including the neutral disaccharides, is clarified by one or more of the following procedures: filtration through a membrane; passage through a bed of an ion-exchange resin; gel permeation; and treatment with acid-washed charcoal. Characterisation of individual components in the solution would then follow by the use of selected analytical methods.

Gerhardt and Heldt (1984) employed CCl_4–n-heptane (66:34; v/v) for the first extraction of liquid N_2 frozen and lyophilised samples of plant tissue. The dry material was then extracted with 10% $HClO_4$ at 0°C for 5 min. The supernatant obtained after centrifugation was neutralised with a 5 N KOH/1 M triethanolamine solution; it contained all the soluble metabolites, including neutral sugars and disaccharides. It should be noted that when perchloric acid solutions are used for extraction of small metabolites and for protein denaturation, solutions must be kept cold and immediate neutralisation must follow. Sucrose, as well as other metabolites such as fructose 2,6-bisphosphate, and even ATP, may otherwise be hydrolysed. Similar procedures for extraction of plant tissues (using 0.3–1.3 M $HClO_4$ at 0°C, and a subsequent neutralisation with KOH or K_2CO_3) have been described by Giersch et al. (1980), Wirtz et al. (1980), Stitt and

ApRees (1980), Gerhardt and Heldt (1984), Blunden and Wilson (1985) and Klein (1987).

Several alternative extraction methods have occasionally been described in the literature. The use of 5% trichloroacetic solution (for example Cheung and Suhadolnik, 1979; Quemener, 1988) requires the subsequent extraction of the acid with ether. It is suitable for stable glycosides such as trehalose, but it is risky for sucrose-containing solutions because of slow sugar inversion. Another method requires the addition of 0.2 ml 0.23 M $ZnSO_4$ and 0.3 ml saturated $Ba(OH)_2$ solutions to 1.0 ml of the sugar-containing solution (e.g. an enzyme reaction mixture), mixing and removal of the pellet by centrifugation. This procedure has been used for many years for protein denaturation and solution clarification in preparing samples containing sucrose or other relatively acid-labile glycosides for the cuprimetric determination of reducing sugars or for the glucose oxidase assay (Leigh et al., 1979).

Another deproteinisation reagent, often employed for sugar analysis in food, is the sequential addition and mixing of 0.025 volume of Carrez-I solution (3.6% of $K_4[Fe(CN)_6].3H_2O$ in water); additional 0.025 vol of Carrez II solution (7.2% of $ZnSO_4.7H_2O$ in water), and then about 0.05 vol of 0.1 M NaOH (adjustment to pH 7–8) to 0.9 vols of the sugar solution to be analysed (Boeringer-Mannheim, 1986). The solution is clarified by centrifugation or by filtration (using membrane filters for small samples, with or without addition of polyamide or polyvinylpolypyrrolidone powder). The filtrate obtained is ready to be analysed enzymically for sucrose, glucose and fructose as well as for other carbohydrates.

Finally, it has been recommended (Au et al., 1989) that in preparation for chromatographic analysis, very small tissue samples could be directly extracted by mixing with ice-cold acetonitrile (>95% final concentration) for 2 min. Water is added to about 45%, and after mixing the pellet obtained by centrifugation, is washed twice with 100 vols of 55% acetonitrile in water. The combined supernatants are ready for high pressure liquid chromatography (HPLC) analysis and contain excellent yields of metabolites. This method has many advantages over the perchloric acid extraction, and since it is successful for mammalian cells, it could be applicable to plant tissues as well.

V. NON-ENZYMIC METHODS FOR DISACCHARIDES

It is envisaged that there will be some overlap of techniques with Chapter 1; however, for the sake of completeness I will briefly describe the relevant methods.

A. Spectrophotometric Methods

1. Phenol–H₂SO₄ reagent

This is probably the most common procedure for the estimation of the total content of reducing sugars and glycosides (Dubois et al., 1956). A sample of 0.01–0.25 μmol sugar in 0.3 ml is mixed with 0.01 ml 10% phenol. A stream of 1.0 ml analytical grade concentrated H_2SO_4 is injected into the sugar–phenol mixture. After mixing and incubating for 30 min at room temperature, absorbance is read at 490 nm. Some non-carbohydrates, such as heavy metal cations, thio- and azo- compounds may interfere.

2. Anthrone–H₂SO₄ reagent

Though mostly displaced by phenol–H_2SO_4 as a general colorimetric reagent for sugars, anthrone is still employed by many investigators. In one version of the procedure (Loewus, 1952) 1.0 ml of the sugar sample (5–200 nmol) is mixed with 0.25 ml 2% anthrone in ethylacetate. Concentrated H_2SO_4 (2.5 ml) added and the precipitate formed is dissolved by vigorous vibromixing. Absorbance at 620 nm is read after 10 min incubation at room temperature. Another variation of the anthrone procedure uses a thiourea-containing solution and a prolonged heating period for increased sensitivity. This variation is beneficial for the assay of glycosides, as well as for non-reducing disaccharides such as trehalose (Spiro, 1966).

For the assay of a sucrose-containing solution, a 100 μl sample (0.05–0.4 μmol of the sugar) is mixed with 100 μl of 30% NaOH, and heated at 100°C for 10 min. After cooling to room temperature, 3.0 ml anthrone reagent (0.15% in 80% H_2SO_4) is added. After 15 min at 40°C, absorbance is read at 620 nm. This procedure (Cardini et al., 1955; Van Handel, 1968) is useful for the analysis of sucrose only in mixtures with monosaccharides, but not if the sample contains other glycosides that are stable to degradation by alkali.

A modified anthrone reagent suitable for the assay of ketose in the presence of a large excess of aldoses, for example for the determination of sucrose in a mixture with glucose, has been described by Boratynski (1984). In this method the anthrone–H_2SO_4 colour is developed in the presence of 4% boric acid and 2% acetone, conditions which significantly suppress the reactivity of aldoses.

3. Resorcinol reagent of ketoses

The resorcinol assay procedure is very convenient for determining the total content of free fructose and fructosides in the presence of aldoses. If a mixture of ketoses is first reduced by excess $NaBH_4$, only the fructoside will remain reactive in the acid resorcinol assay, thus permitting the determination of sucrose in the presence of fructose. The reagent of Roe et al. (1949) was modified for greater sensitivity as follows (G. Avigad, unpubl. res.):

Reagents:
 A. Resorcinol, 0.01 g; thiourea, 0.2 g, and sulphamic acid, 1 g are dissolved in 100 ml glacial acetic acid. Reagent A can be kept for several months at room temperature.
 B. HCl, analytical grade.
 C. Standard fructose (or sucrose), 0.1 mM in water.

For the day's use, mix 10 ml of Reagent A with 65 ml of Reagent B.

Procedure: In a reaction vial or a small test tube, a 0.4 ml sample containing 5–200 nmol ketose is mixed with 0.8 ml resorcinol–HCl reagent and heated for 10 min at 80°C using a water bath or a heating block. After cooling, the absorbance is measured at 515 nm.

4. Determination of uronic acids

Two procedures are suitable for both free uronic acids and glycuronosides.

Reagents:

    A. Carbazole reagent (Bitter and Muir, 1962). Uronic acid (2–20 nmol in 0.25 ml) is mixed with 1.5 ml 0.025 M sodium tetraborate in concentrated H_2SO_4 at 0°C. The mixture is heated for 10 min at 100°C, cooled on ice and then mixed with 50 µl 0.125% carbazole in ethanol. The mixture is reheated for 15 min at 100°C, cooled rapidly, and the absorbance is read at 530 nm.

    B. β-Hydroxydiphenyl reagent (Blumenkrantz and Asboe-Hansen, 1973). Uronic acid (2–50 nmol in 0.2 ml) is mixed with 1.2 ml of 0.12 M disodium tetraborate in concentrated H_2SO_4 and kept at 100°C for 5 min. After cooling on ice, 25 µl of 0.15% β-hydroxydiphenyl is added with mixing, and within 5 min absorbance is read at 520 nm against a reagent blank.

Weak interference in these colorimetric assays (1–10% on a molar basis) is contributed by the presence of non-hexuronic acid-reducing glycoses, particularly by pentoses.

5. Assay of disaccharides and other glycosides in the presence of reducing sugars

Advantage is taken of the inability of polyols to produce chromogens in most of the mineral acid colorimetric reagents for sugars. Consequently, when a mixture of sugars is first reduced by $NaBH_4$ (at pH > 7.0), only the residual glycoside will react with the acid reagents such as the phenol- and the anthrone-H_2SO_4, or with the resorcinol or the carbazol reagents for ketosides and uronosides, respectively. After their conversion to polyols, all monosaccharides and the glycose residues in oligosaccharides will be rendered inert and will not be determined by the colorimetric acid reagents. The disaccharide alditols, such as maltitol or melibiitol, still react as a simple monoglycose in the acid colorimetric reagents. Sucrose and trehalose will not be affected by the borohydride, and their reactivity with these acid reagents will not be altered.

A simple protocol (Asensio *et al.*, 1963) is as follows: 0.9 ml of the neutral sugar sample is reduced for about 15 min at room temperature with 0.1 ml of a freshly prepared 5% $NaBH_4$ in 0.01 N NaOH. Aliquots containing 2–50 nmol glycoside are assayed with the phenol-H_2SO_4 or the anthrone-H_2SO_4 reagents. Standard curves are prepared with maltose (or maltitol), methyl-α-D-glucoside, sucrose, trehalose, or any other glycoside most suitable for the sugar system studied.

6. Raybin test for sucrose

Diazouracil in alkaline solution interacts with sucrose to produce a poorly soluble unidentified olive green product (Raybin, 1933, 1937). Other β-ketofuranosyl-α-glycosides such as α-glucopyranosyl-β-xylofuranoside and raffinose react positively in this reaction (Hassid and Ballou, 1957). The test is not very sensitive (low detection level of about 2 µmol: Feingold *et al.*, 1956). The method has been used occasionally for a qualitative spot test analysis, but has defied attempts to be made into a quantitative and more sensitive assay for sucrose and its ketoside analogues.

B. Determination of Reducing Sugars

A large number of reagents for the quantitation of reducing sugars is available to the analyst. Their application and selection of the most suitable reagent for the determination of disaccharides will depend on the contents of the sugar mixture analysed. When sucrose is present, which is the most common case when studying plant systems, the reagent employed should not be acidic to prevent sucrose hydrolysis. On the other hand, if complete hydrolysis of the disaccharide by acid or by a glycohydrolase preceded reducing sugar determination, almost any of the reagents described can be employed for analysing the monosaccharides. A selection of several reagents for the determination of reducing sugars in manual procedures is presented here. Adaptations of some of these as well as of other reagents which are suitable for the automated analysis of reducing sugars resolved by chromatographic separations, will be discussed further on in this chapter.

1. Cuprimetric reagents

The classical Somogyi–Nelson colorimetric method (Nelson, 1944) is one of the most enduring and popular general assays for reducing sugars. The method is described in numerous manuals (for example, Spiro, 1966) usually for the determination of 0.05–0.60 nmol keto- and aldo-hexoses. The availability of micropipettes, small reaction vials, heating blocks and sensitive spectrophotometric instrumentation makes it possible to reduce the volume of the reactants originally prescribed and consequently increase the sensitivity of the assay several-fold, providing that the reaction vessel used has a low surface area-to-depth ratio.

A typical protocol is as follows: in a 2 ml reaction vial, mix 0.25 ml sugar solution (2–50 nmol reducing sugar) with 0.25 ml copper reagent (28 g anhydrous disodium phosphate, 40 g Na-K tartrate in 700 ml H_2O. Add 100 ml 1 N NaOH, stir in 80 ml 10% $CuSO_4.5H_2O$; dissolve in 90 g anhydrous Na_2SO_4, make up to 1.0 litre with water, and filter). Cap the vial and heat at 100°C for 20 min, cool to room temperature, add 0.2 ml Nelson's reagent (25 g ammonium molybdate in 450 ml concentrated H_2SO_4 and 3 g disodium arsenate in 25 ml H_2O; keep at 37°C in a brown bottle). Mix, add 1.0 ml H_2O, and read the absorbance at 650 nm against reaction blank. Compare with a standard curve prepared with known quantities of glucose, for the reducing disaccharide studied.

Among equivalent alternatives for the Nelson's procedure, the alkaline copper-2,2'-bicinchoninate reagent for as little as 1 nmol reducing sugars can be recommended (Mopper and Grindler, 1973; McFeeters, 1980; Waffenschmidt and Jaenicke, 1987). Another useful choice of a cuprimetric assay is that employing a neocuproin reagent (Dygert et al., 1965; Chaplin and Kennedy, 1986). The detection range in this method is increased to 0.2–1.5 nmol glucose equivalents, consequently the reaction is more likely to be sensitive to false positive contribution by non-carbohydrate-reducing contaminants in the solution. A slightly acidic copper reagent was found to be selective and particularly sensitive to hexuronic acids (Avigad, 1975b).

2. Dinitrosalicylic acid reagent

The original Bernfield procedure (Bernfield, 1955) is still a widely used assay method for disaccharide and oligosaccharide analysis, particularly for starch and cellulose hydrolysates. The method is not very sensitive; but consequently it is less influenced by other reducing contaminants in the biological samples studied. In the procedure described

(Chaplin and Kennedy, 1986) the reducing sugar sample (0.05–3 μmol) in 100 μl is mixed with 1.0 ml of the dinitrosalicylic reagent (0.25 g 3,5-dinitrosalicylic acid and 75 g K-Na tartrate are dissolved in 50 ml 2 N NaOH, and volume adjusted to 250 ml with H_2O). The assay mixture is heated at 100°C for 10 min, cooled to room temperature and absorbance is read at 570 nm. Absorbance of blanks can be lowered by prior purging the reagent solution with nitrogen or helium.

3. Alkaline ferricyanide reagent

This method (Park and Johnson, 1949; Spiro, 1966) is a simple and sensitive procedure for reducing sugar determination. However, since the stability of the Prussian Blue product is strongly influenced by the salt concentration, the nature of the cation and other solutes present in biological systems, has limited the use of this reagent. Porro *et al.* (1981) have described a modified reagent with a significantly improved precision. In this case, to 0.5 ml of the reducing sugar (5–60 nmol) solution, 1.0 ml of a ferricyanide reagent (1.5 mM potassium ferricyanide, 50 mM Na_2CO_3 and 10 mM KCN stored in the dark) is added (note: sugar solution should be neutral; presence of a strong buffer may interfere with the reaction). After 15 min heating at 100°C, 2.5 ml of 1.36 mM ferric ammonium sulphate solution in 25 mM H_2SO_4 is added. The mixture is kept 15 min at 50°C, cooled to room temperature and 0.5 ml of a 0.125 M solution of oxalic acid added with vigorous mixing. Absorbance is read at 690 nm against a reagent blank.

A modified, very sensitive ferricyanide reagent for 1–100 nmol reducing sugars, which depends on the reduction of a triazine dye, has been described by Avigad (1975a). This oxidation reduction procedure is sensitive to interference by thiols, by heavy metal ions and by chelating agents.

4. 2-Furoylhydrazine (FAH) reagent

Aroylhydrazines interact with reducing sugars in an alkaline, bismuth-containing solution to produce coloured derivatives. This reaction is suitable for the assay of reducing sugars, including oligosaccharides in the presence of sucrose or trehalose and other non-reducing glycosides (Lever, 1977; Lever *et al.*, 1984). Both 4-hydroxybenzoyl-hydrazine, and the more water-soluble 2-furoylhydrazine, are suitable for use in this reaction.

Reducing sugar (2–100 nmol) is heated for 15 min at 70° in 2.5 ml 0.2 N NaOH containing 1.0 mM bismuth sodium tartrate and 50 mM FAH. After cooling the absorbance at 410 nm is read against reagent blanks. Small differences in molar extinction values between aldoses, ketoses and reducing disaccharides have been observed with this reagent.

5. 2-Cyanoacetamide reagent

Aldose, 5–750 nmol, in 1.0 ml is mixed with 1.0 ml of 1% 2-cyanoacetamide in water and 2.0 ml of 0.1 M borate buffer, pH 9.0. After heating at 100°C for 10 min, tubes are cooled and absorbance at 276 nm is measured against reagent blanks. The method (Honda *et al.*, 1981a, 1982) is suitable for the determination of aldoses, including disaccharides, in the presence of sucrose and trehalose. Molar absorbance values of the coloured products obtained for different aldoses (mono- and disaccharides) vary significantly. The presence of UV-absorbing compounds such as nucleotides could seriously interfere by contributing high background absorption.

The same reagent has been adapted for a sensitive fluorimetric analysis of reducing sugars, including oligosaccharides (Honda *et al.*, 1981b; Crowther and Wetmore, 1987).

6. *Tetrazolium salt reagents*

The rapid reduction of tetrazolium salts by glycose in alkaline solution with the formation of a coloured formazan provides a useful technique for reducing-sugar determinination. The rate of reduction by different sugars can vary markedly. For example, ketoses react much faster than aldoses, disaccharides with substitution on carbons adjacent to the carbonyl carbon are significantly less reactive, e.g. 2-glucosyl-glucoses and 1-glycosylfructose compared to the 4- or 6-glycosylhexoses (Avigad *et al.*, 1961). The presence of strongly reducing substances such as ascorbate, ketoamines or thiols will strongly interfere in the assay. The rate of tetrazolium reduction depends also on the nature of the solvent and since the formazan product is usually poorly soluble in water, the analytical reagent used should contain an organic solvent for its solubilisation. Several procedures for the quantitation of reducing sugar (0.01–1 mM) with a tetrazolium reagent are available both for manual and automatic methods adapted for use in liquid chromatography (Avigad *et al.*, 1961; Mopper and Degens, 1972; Van Eijk *et al.*, 1984).

In a manual procedure, 0.01–0.5 nmol sugar in 0.7 ml aqueous solution (preferably 50% ethanol) is mixed first with 0.1 ml 0.2% tetrazolium salt [2,3,5-triphenyltetrazolium chloride (TTC) or *p*-anisyltetrazolium blue (BT), or nitrobluetetrazolium chloride (NBT)], and then with 0.2 ml 0.2 N NaOH containing 0.2 N sodium tetraborate. Incubation time at room or higher temperature will depend on the nature of the sugar analysed and the tetrazolium salt employed (5–10 min for ketoses, 3–60 min for aldoses or disaccharides). The reaction is terminated by the addition of 2.0 ml of 10% acetic acid in 2-propanol or 10% HCl in pyridine. Absorbance is measured at 485 nm for TTC, 520 nm for BT and at 560 nm for NBT. It should be noted that a large variety of tetrazolium salts with different solubilities and redox potential values is commercially available. Many of them should be suitable for use in reducing-sugar analysis (Altman, 1976).

7. *Reduction with* NaB^3H_4

Using tritiated borohydrides provides an ideal technique for the introduction of a radioactive label into the end group of a reducing oligosaccharide. This method has also been applied extensively to disaccharides (Conrad, 1976; Takasaki and Kobata, 1978). In the basic procedure, 20–200 nmol sugar is reduced at pH > 7.0 by about a 5 molar excess of NaB^3H_4 for about 2 h at room temperature. Hall and Patrick (1989) suggest the use of 10–20 mM NaB^3H_4 and about 15 mM NaOH at pH 10–11. This alkalinity must be maintained (use thymol blue as an indicator) to secure complete reduction of pmol levels of reducing oligosaccharides. Excess borohydride is decomposed (use a safety hood) by the careful addition of 1 M acetic acid until evolution of gas ceases. Solution is then passed through a cation-exchange column, washed with water and lyophilised. Boric acid is removed by repeated evaporation with methanol. The dry matter is dissolved in 10–100 μl water. The [^{3}H]alditol content is analysed by PC or TLC and by radioscanning. Radioactive spots can be eluted from the chromatograms and determined by counting. Alternatively, samples can be analysed by HPLC and the ^{3}H-label serves as a tag for the location, identification and the quantification of the compounds (Section VI.B.3). The tritiation procedure is useful for the analysis and tracing of very small amounts of reducing disaccharides, such as maltose or malibiose, in the presence of a large excess of sucrose or other non-reducing glycoses.

In order to apply the reduction by NaB^{3}H$_4$ for quantitative analysis, the specific activity of the reagent has to be accurately determined, since the value provided by the manufacturer is usually only an approximation. A simple way of obtaining this value is to reduce a known amount of glucose 6-phosphate (e.g. 0.01 ml of a 1–10 mM solution) accurately determined by the glucose 6-phosphate dehydrogenase assay, by adding a 50 μl solution containing about a 10 molar excess of a freshly diluted sample of the NaB^{3}H$_4$ in 0.01 N NaOH. In an alternative procedure, accurately determined amounts of a reducing sugar, such as lactose, can substitute for glucose 6-phosphate. After about 15 min at room temperature, 100 μl 1 M acetic acid is added, and the borate is removed by three evaporations, each with 3 ml methanol. The final dry matter is dissolved in 0.5 ml water and samples are measured in a scintillation counter. If a sugar such as lactose is being used for this assay, the concentration of the ^{3}H-lactitol in the final solution can be verified colorimetrically by the phenol–H$_2$SO$_4$ assay. This procedure and literature on related methods have been described by Avigad (1979a). Alternative procedures which rely on the formation of [^{3}H]NADH or [^{3}H]NADPH by NaB^{3}H$_4$ or NaCNB^{3}H$_4$ reductions have been described by Avigad (1979b) and by Murtishaw et al. (1983).

Roy et al. (1988) have found that ketoses—but not aldoses—are chemoselectively reduced by NaBH$_4$ in the presence of Ce(III) chloride in alkaline solutions. This selectivity could possibly be utilised for analytical applications and tritium labelling of a mixture of sugars in solution.

8. Reductive amination

Methods based on the use of sodium cyanoborohydride for the reduction of the Schiff base adducts produced by incubating a reducing sugar with a primary amine at about neutral pH conditions have been employed for HPLC pre-column derivatisation analysis (Section VI.B.3). NaCNBH$_3$ is more suitable than NaBH$_4$ for this purpose since it reduces the sugars to polyols only in acidic (pH 2–4) conditions, but at neutral pH it is an excellent reducer of ketoimine to ketoamine. NaBH$_4$, on the other hand, decomposes rapidly below pH 7, and in alkaline solutions reduces both the carbonyl and the ketoimine. Tritiated NaCNBH$_3$ provides a reagent by which a radioactive label can be introduced into the ketoamine produced by reductive amination.

Application of the reaction to reducing disaccharides can be carried out with a polymeric material that has free amines, either directly (Gray, 1978) or by using a 2-(4-aminophenyl)ethylamine bridge (Zopf et al., 1978; Semprevivo, 1988). This technique produces a disaccharide linked to a protein or to an insoluble polymer which can be used as the solid support for affinity chromatography separations of proteins and for immunochemical and radioimmunoassay reagents. In general it is difficult to achieve rapidly a stoichiometric derivatisation of reducing sugar by reductive amination. The method has gained only limited use in quantitative analysis, or in chromatrographic analysis (Section VI.B.3) and it is mostly applied for synthetic purposes or for semi-quantitative tagging.

9. Fluorescence assay

Several fluorimetric methods for reducing sugars (see section VI.B.3(d)) were adapted for use in HPLC procedures. O'Neill et al. (1989) used an ethanolamine-borate reagent for the manual determination of samples containing less than 500 pmol reducing sugars. The reagent is useful for following the course of glycoside hydrolysis when the initial substrate, such as sucrose or trehalose, is not reducing and it is not masked by other interfering glycoses.

10. Oxidation to aldobionic acids

The oxidation of aldoses by halogens (such as bromine) or by hypohalites (such as hypoiodite) in mildly acidic or neutral solution, will result in the formation of the corresponding aldonic acid (Browne and Zerban, 1941; Schaffer and Isbell. 1963; Isbell, 1963). Titrimetric versions of this oxidation, known as the 'Willstatter–Schudel reaction', used in the past to provide a very popular way of determining aldose concentrations in solution. Under alkaline conditions the oxidation will usually lead to further degradation processes. The method has been used extensively for synthetic purposes such as the preparation of maltobionic, melibionic or lactobionic acids and some other molecules derived from them by further chemical manipulation. The technique has been occasionally used also for verification of disaccharide structure, but it is rarely employed as an analytical tool in biochemical and physiological research. For further information about the subject consult general carbohydrate chemistry textbooks.

C. Periodate Oxidation

The formation of formaldehyde by periodate oxidation of carbohydrates can be used for the estimation of sugar concentration, and the technique has been also applied for the detection of sugar in automated GLC and HPLC (Sections VI.B.2 and VI.B.3). To improve the sensitivity of the method, reducing disaccharides should be first incubated at pH 8.0 with about a 10 molar excess of $NaBH_4$ for at least 15 min. The pH of the solution is then adjusted to 5.5–6.0 by careful addition of dilute acetic acid. When gas formation ceases, a freshly prepared solution with about a 10 molar excess of sodium meteperiodate is added. After about 1 h in the dark, the amount of formaldehyde can be assayed by a variety of reagents: fluorimetrically with cyclohexane-1,3-dione (Pesez and Bartos, 1967); photometrically by the Hantzsch reaction (Speck, 1962; Gallop et al., 1981), or by the 4-amino-3-mercapto 1,2,4-triazole (AHMT) reagent (Avigad, 1983). Hydroxymethyl groups in the disaccharide alditol will in most cases be released very rapidly as formaldehyde. Other aldehyde groups produced by periodate oxidation of the glycosyl residue will be released much more slowly from the intermediate ester structures, and this process will be facilitated by more intense acid or alkali hydrolysis. When using the periodate method as a quantitative assay for a disaccharide, a careful calibration curve for the procedure using known amounts of the individual disaccharide analysed should be prepared.

In some analytical systems, it is of interest to measure the consumption of periodate and evaluate the number of vicinal diols oxidised. Such analysis can provide useful information on the structure of the glycoside oxidised. A selection of titrimetric and spectrophotometric methods for the assay of periodate is available (see Avigad, 1969 and Honda *et al.*, 1989 for further literature). The silica high column capillary zone electrophoresis recently described by Honda *et al.* (1989) provides an additional technique for periodate analysis.

D. Physico-chemical Parameters

1. Nuclear magnetic resonance (NMR) spectroscopy of disaccharides

Proton NMR and carbon-13 NMR spectroscopy can be utilised very effectively to elucidate disaccharide structure and to gain information about their metabolic fluxes. Bock *et al.* (1984), Hoffman *et al.* (1986) and Small and McIntyre (1989) summarised a large amount of carbon-13 NMR data for most of the commonly occurring disaccharides. Additional more detailed studies describing NMR data, particularly for sucrose and trehalose, can be found in Deslauriers *et al.* (1980); Coxon (1980); Schaefer *et al.* (1980); Hall and Morris (1980); Bax *et al.* (1981); Den Hollander and Shulman (1981); Chen and Whistler (1983); Ryder *et al.* (1984); Dijkema *et al.* (1985); Martin (1985); Defaye *et al.* (1983); Martin *et al.* (1985, 1988); Wahlberg *et al.* (1986); Nakano *et al.* (1986); D'Accorso *et al.* (1986); Kashiwada *et al.* (1988); Mega and Van Etten (1988) and Keeling *et al.* (1988). In deciphering disaccharide structures, the high resolution carbon-13 and proton NMR data obtained for other complex oligosaccharides can also be helpful and informative (Barker *et al.*, 1982; Vliegenthart *et al.*, 1984; Dill *et al.*, 1985; Koerner *et al.*, 1987; Bauman *et al.*, 1988). The elegant use of carbon-13 NMR spectroscopy for the evaluation of the sucrose and starch interrelationships in wheat grain is described by Keeling *et al.* (1988).

2. Optical properties of disaccharides

Optical rotation, optical rotary dispersion and circular dichroism data provide important information about the structure and conformation of disaccharides (as well as glycosides in general). These analytical methods, some of which are used frequently by the structural carbohydrate chemist and in the sugar and food industry, are rarely employed for routine analysis of disaccharides in physiological and biochemical research and therefore will not be discussed in this chapter. This subject has been reviewed in detail in most monographs and advanced textbooks on carbohydrate chemistry.

3. Mass spectrometry (MS)

In combination with liquid chromatography methodologies, analytical techniques based on MS provide powerful information about sugar structure. Specific applications are listed in Section VI.B. Useful recent reviews on the subject are presented by Raven (1987), Santikarn *et al.* (1987), Reinhold (1987), Dell (1987) and Bierman and McGinnis (1988).

4. Isotope discrimination analysis

The $^{13}C/^{12}C$ ratio [or $\delta^{13}C$ (‰)] found for an individual sugar isolated from plant tissue (as exemplified particularly in sucrose and starch), can indicate whether the plant that produced it has a C_3, C_4, CAM, or a mixed type of metabolism. The availability of benchtop automated nitrogen carbon analyser-isotope ratio mass spectrometer instrumentation is likely to simplify and increase the applications of this sensitive and useful technique (Barrie and Lemley, 1989). For general reviews and literature citations see O'Leary (1981), Avigad (1982), Williams (1984), and Edwards and Ku (1987) and Farquhar et al. (1989). The value of the stable isotope carbon ratio can also reflect the variation in the supply of CO_2 that had occurred during photosynthesis and sugar biosynthesis and can provide information on metabolic partitioning in the tissue (Andrews and Abel, 1979; Doner and Bills, 1982; Leavitt and Long, 1985; Lipp et al., 1988; Brugnioli et al., 1988; Farquhar et al., 1988; Bowman et al., 1989; Sasakawa et al., 1989). Analysis of the $^{18}O/^{16}O$ ratio [or $\delta^{18}O$ (‰)] in sucrose and other carbohydrates is another parameter which can indicate the source and history of the sugar molecule (Sternberg and De Niro, 1983; Doner and Bills, 1982; Doner et al., 1987).

VI. CHROMATOGRAPHY OF DISACCHARIDES

A. Planar Chromatography*

1. Paper chromatography (PC)

Though largely superseded by thin layer chromatography (TLC), PC is still a useful and simple technique to resolve and analyse a mixture of sugars. This subject has been reviewed in many monographs which should be consulted for further detailed information or selection of reagents (Block et al., 1958; Hough and Jones, 1962; Churms, 1982, 1983).

A choice of useful solvent systems suitable for the separation of disaccharides is:

n-Propanol–ethyl acetate–H_2O (7:1:2, v/v);
n-Butanol–ethanol–H_2O (4:1:2, v/v);
n-Butanol–acetic acid–H_2O (4:1:5, v/v);
n-Butanol–pyridine–H_2O (6:4:3, v/v);
Ethyl acetate–pyridine–H_2O (10:4:3, v/v).

Reagents for sugar detection: see Section VI.A.3.

2. Thin layer chromatography (TLC)

Separation, detection and quantitation of sugars by TLC procedures is an economical,

---

* The solvent systems and reagents listed in Sections VI.A.1 and VI.A.2 include some of the ones most commonly used, and were selected by the author on the basis of personal experience and judgement. References to the original publications which introduced these, as well as numerous other reagents used in the planar chromatographic analysis of sugars are to be found in the extensive review literature and monographs which are cited here.

rapid and sensitive technique that is extensively used in chemical, biological and applied research. TLC is very suitable for the analysis of disaccharides found in plant material. Many detailed reviews and manuals for this method, are available (Stahl, 1969; Wing and BeMiller, 1972; Ghebregzabher *et al.*, 1976, 1979; Kirchner, 1978; Churms, 1982, 1983; Robards and Whitelow, 1986; Chaplin and Kennedy, 1986) and should be consulted for detailed information. High quality ready-made thin layer plates of different dimensions and with a wide variety of absorbent materials are today available commercially. This provides the analyst with a standardised uniform product which allows highly reproducible separations. TLC can be applied for quantitative analysis of a sugar mixture, or for qualitative evaluation by direct densitometry or after the elution of individual separated compounds. Use of plates with a thick absorbing layer can facilitate preparative separation. Recent advances in TLC instrumentation improve the use of the technique for fast quantitative analytical purposes (Touchstone, 1988). The absorbents employed for TLC of sugars are microcrystalline cellulose, Kieselguhr or silica gel. These absorbents are usually impregnated with various amounts of salts and binders such as gypsum, borate, acetate, or phosphate. These inclusions may exert a strong effect on the effectiveness of separation with different solvents. Also, their presence can critically influence the reactivity of the separated sugars with the colour detection reagents employed. Since most of these reagents require high acid concentrations, increasing the concentration of acid in the detecting reagent (orthophosphoric acid is recommended), or multiple applications of reagent to the plates, is often required to secure maximal colour development. A mixture of known amounts of sugar standards (e.g. 2 µg each of glucose, sucrose, and maltose) should always be applied to a track on the TLC plate. The separation of disaccharides, as well as higher oligosaccharides, can often be improved by repeated solvent applications, i.e. when solvent front reaches an advanced position the plates are dried, then subjected to a repeated solvent development. The second solvent can be identical to the one applied first, or of a different composition. Application of the second solvent can also be carried out in a dimension perpendicular to that of the first run.

(a)   *Solvent systems.* A choice of some useful solvent systems suitable for the separation of disaccharides is given below.

For microcrystalline cellulose plates:

Ethyl acetate–pyridine–H_2O (8:2:1; v/v).
Ethyl acetate–pyridine–acetic acid–H_2O (10:8:1:3; v/v).
n-Butanol–ethanol–H_2O (3:1:1; v/v).
n-Butanol–pyridine–H_2O (6:4:3; v/v).
n-Propanol–ethyl acetate–H_2O (6:1:2; v/v).

For silica gel and Kieselguhr plates:

n-Butanol–pyridine–H_2O (6:4:3; v/v).
n-Butanol–acetic acid–H_2O (2:1:1; v/v).
n-Butanol–methanol–H_2O (5:3:2; v/v).
n-Butanol–ethanol–H_2O (7:1:2; v/v).
n-Butanol–ethyl acetate–2-propanol–H_2O (35:10:6:3; v/v).
Ethyl acetate–n-propanol–H_2O (6:3:1; v/v).

2-Propanol–acetone–1 M lactic acid (2:2:1; v/v).
Ethyl acetate–2-propanol–H_2O (3:4:2; v/v).
Ethyl acetate–pyridine–acetic acid–H_2O (70:30:5:10; v/v).
Ethyl acetate–methanol–acetic acid–H_2O (60:10:15:10; v/v).
Ethyl methyl ketone–2-propanol–acetonitrile–0.5 M boric acid–0.25 M isopropyl-
amine in acetic acid: (40:30:20:15:0.4; v/v).
Ethyl methyl ketone–benzene–2-propanol–0.5 M isopropylammoniumbenzoate in
H_2O (30:20:40:15; v/v).

(b)   *Reagents for disaccharide detection*

H_2SO_4 charring (not for paper or cellulose plates): spray with 25% H_2SO_4 in ethanol.
Heat briefly at 110°C.

Silver nitrate: Dip plates or paper in a solution made by mixing 0.1 ml of a saturated
$AgNO_3$ solution in 20 ml acetone and adding several drops of water to facilitate
dissolution. Dry plate, and spray with 0.5 N NaOH in 90% ethanol. Spots appear in
5–10 min. Chromatogram can be sprayed with a sodium thiosulphate solution, washed
with water, then dried to preserve spots. All reducing sugars react with this reagent.
Non-reducing sugars (e.g. trehalose, polyols) react more slowly and are somewhat less
sensitive.

Diphenylamine–aniline: 2% diphenylamine and 2% aniline in acetone–85% H_3PO_4
(5:1; v/v). Can be used as a dip or a spray. Heat chromatogram 10 min at 105°C. A
general popular reagent; different sugars appear in distinct colours.

Naphthoresorcinol: 0.2% naphthoresorcinol in ethanol– 85% H_3PO_4 (8.5:1.5; v/v).
Chromatograms are heated for 10 min at 90°C; different sugars appear in distinct
colours.

p-Anisidine: 1.25% p-anisidine and 1.75% phthalic acid in 90% ethanol. Chromatograms are heated for 10 min at 105°C. A general reagent, particularly for reducing
sugars.

Urea: 3% (w/v) urea in n-butanol–ethanol–H_2O orthophosphoric acid (80:8:5:7;
v/v). Chromatogram sprayed and heated for 10 min at 105°C. Reagent highly specific
for ketoses.

Tetrazolium blue: Freshly prepared 0.04% Tetrazole Blue (p-anisyltetrazolium chloride) in 3 N NaOH in 33% ethanol. Spray and heat at 60°C for 10–20 min. Very reactive
for reducing sugars.

Triphenyltetrazolium: 2% 2,3,5-triphenyltetrazolium chloride in 0.5 N NaOH in
methanol. Chromatogram sprayed and heated in a moist environment for 10 min
at 80°C. Very sensitive (1 nmol) to reducing sugars.

Thymol: 0.5% thymol in ethanol containing 5% H_2SO_4. Plates heated at 100°C for
10 min.

Orcinol: 6% orcinol in ethanol containing 3% 1-octanol is mixed 1:10 with 1% $FeCl_3$
in 10% H_2SO_4. Heat 10 min at 100°C. Reagent specifically good for silica plates
saturated with borate.

(c)   *Radioautography.*

In metabolic studies, ^{13}C, ^{3}H or ^{32}P isotopes are often used for
labelling. After separation by planar chromatography, standard radioautography
techniques are applied to detect the labelled compounds. Non-radioactive carrier can be

added before chromatographic separation to increased accuracy if spots are to be eluted for quantitation by counting. Many examples of the use of this technique for the study of sucrose metabolism are available (see for example: Outlaw *et al.*, 1975; Mbaku *et al.*, 1978; Dickson, 1979; Redgwell, 1980; Preisser and Komor, 1988).

Sensitivity of detection and visualisation of separated beta-emitters radioactive sugars can be significantly increased when the chromatograms are treated with the flour PPC in a suitable solvent such as 2-methylnaphthalene-toluene or diethyl ether (Randerath, 1970; Bonner and Stedman, 1978). Such enhancing dipping solutions or sprays can be obtained from various suppliers (DuPont-New England Nuclear Co., and the Amersham Corporation, for example). Lucher and Lego (1989) recommend the use of sodium salicylate in 2-methylnaphthalene as the flour, achieving 150-fold amplification in the detection of ^3H-labelled material in TLC fluorography.

(d) *TLC of derivatised disaccharides.* Though most existing TLC procedures for disaccharide separations are for the unaltered native sugars, several methods which rely on chemical derivatisation prior to chromatographic separation have also been described. Most of these methods are similar and complementary to those used in gas chromatography (GC) and HPLC analysis.

The procedure described by Wang *et al.* (1983), suitable for both TLC and GC, is based on the preparation of permethylated oligosaccharide alditols. Excellent separation of disaccharides can be obtained. In this method, the sugar solution, after passage through a Biogel-P4 column, is passed through a small mixed-bed ion-exchange column and reduced in volume by flash evaporation or lyophilisation. The sugar is reduced by $NaBH_4$ (Matsuura *et al.*, 1981) and then permethylated by a modified Hakomori procedure (Stellner *et al.*, 1973; Churms, 1982; Ciucanu and Kerek, 1984; Chaplin and Kennedy, 1986; Carpita and Shea, 1988). The permethylated alditols are purified and dried by chloroform–methanol extractions. Sugars are separated on silica gel plates using benzene–methanol (16:1; v/v) as the solvent, sprayed with 0.5% orcinol in 4 N sulphuric acid and revealed by heating at 100°C for several minutes. Table 4.1 shows the separation of oligosaccharides by this methods, using both TLC and GC analysis.

Another method described by Forsythe and Feather (1989) is based on acetylation of the oligosaccharide mixture extracted from plant tissues, and the subsequent separation of chloroform soluble products by TLC or by silica-gel column chromatography.

(e) *High performance thin layer chromatography (HPTLC).* Developments in the implementation of new HPLC techniques for separation and detection of sugars which are very useful for the analysis of disaccharides have been recently described (Gauch *et al.*, 1979; Lee *et al.*, 1979; Touchstone and Dobbins, 1983; Koizumi *et al.*, 1985; Doner, 1988; Klaus and Fischer, 1988; Preisser and Komor, 1988). This methodology, which requires specific equipment for sample application and detection, permits oligosaccharides separation within very short periods of time, has excellent high resolution between disaccharides with closely related structures, and has lower detection limits than TLC. Instrumentation for programmed automated multiple development (AMD) in HPTLC greatly improves the resolution and therefore the usefulness of this technique.

Absorbents for the stationary phase in HPLC are usually bonded-phase silica or small particle silica gel impregnated with borate or phosphate and applied to aluminium

TABLE 4.1. The structures of oligosaccharides and the separation of their permethylated alditol derivatives by thin layer chromatography and by capillary gas chromatography (from Wang et al., 1983).

Oligosaccharide used		$R_{IM}{}^a$	RI[b]
Glcα1–1αGlc	Trehalose[c]	164	2281
Glcβ1–2Glc	Sophorose	117	2329
Glcβ1–3Glc	Laminaribiose	64	2306
Glcβ1–4Glc	Cellobiose	105	2292
Glcα1–4Glc	Maltose	96	2295
Glcβ1–6Glc	Gentiobiose	115	2420
Glcα1–6Glc	Isomaltose	100	2380
Xylβ1–4Xyl	Xylobiose	158	2024
Galβ1–4Glc	Lactose	71	2297
Galα1–6Glc	Melibiose	77	2400
Glcα1–2Fru	Sucrose[c]	163	2222
Glcα1–6Glcα1–6Glc	Isomaltotriose	67	3227
Glcα1–4Glcα1–4Glc	Maltotriose	66	3095
Glcβ1–3Glcβ1–3Glc	Laminaritriose	90	3171
Xylβ1–4Xylβ1–4Xyl	Xylotriose	126	2869
Galα1–6Glcα1–2Fru	Raffinose[c]	89	3052
Glcα1–(4Glcα1)$_2$–4Glc	Maltotetraose	40	3914
Glcα1–(6Glcα1)$_2$–6Glc	Isomaltotetraose	46	4017
Glcβ1–(3Glcβ1)$_2$–3Glc	Laminaritetraose	65	4049
Xylβ1–(4Xylβ1)$_2$–4Xyl	Xylotetraose	102	3742
Galα1–6Galα1–6Glcα1–2Fru	Stachyose[c]	43	3873
Glcα1–(4Glcα1)$_3$–4Glc	Maltopentaose	33	1.247[d]
Glcα1–(6Glcα1)$_3$–6Glc	Isomaltopentaose	33	1.249[d]
Glcβ1–(3Glcβ1)$_3$–3Glc	Laminaripentaose	51	1.301[d]
Xylβ1–(4Xylβ1)$_3$–4Xyl	Xylopentaose	80	1.207[d]
Acetolysates of high mannose oligosaccharides			
Manα1–3Man		105	2309
Manα1–2Man		75	2325
Manα1–2Manα1–3Man		68	2949
Manβ1–4GlcNAc		27[e]	2698
Manα1–3Manβ1–4GlcNAc		23[e]	3595
Manα1–2Manα1–3Manβ1–4GlcNAc		19[e]	4264
Manα1–2Manα1–2Manα1–3Manβ1–4GlcNAc		16[e]	1.308[d]

[a] Samples were developed twice on pre-coated silica gel 60 plates with benzene–methanol (16:1; v/v). The mobilities are expressed as relative mobilities to isomaltose (R_{IM}).
[b] Retention index (RI) on DB-1 fused-silica capillary column (0.25mm × 12 m) and helium as carrier gas for GLC at 150–330°C, 5°C min^{-1}. Flame ionisation detector used.
[c] No reaction with NaBH$_4$.
[d] Relative retention time to n-tetracosane ($C_{40}H_{82}$).
[e] Conditions are the same as in (a) except with benzene–methanol (10:1; v/v) as solvent.

sheets. Separation is facilitated by solvent systems similar to TLC. For example, for disaccharides:

n-propanol–nitromethane–water–acetic acid (5:3:2:0.1; v/v).
n-butanol–pyridine–H$_2$O (8:5:4; v/v).
n-butanol–ethanol–H$_2$O (5:4:3; v/v).

Repeating the development two or three times may increase resolution between oligosaccharides. Sugar bands are revealed by acidic colour reagents such as used in TLC. The diphenylamine-aniline, or the naphthoresorcinol reagents (Section VI.A.2) are often employed. In addition, methods based on *in situ* derivatisation of the sugar bands in a solution of lead tetra-acetate in 0.4% acetic acid–toluene (5:5:190; v/v) has been recommended as a sensitive detection method for oligosaccharides (Doner, 1988; Klaus and Fischer, 1988). HPTLC analysis is very suitable for the separation and detection by radioautography of low levels of labelled sugars isolated from very small samples of plant material.

3. *Planar electrophoresis*

Separation of sugars as well as of disaccharides, usually as their borate complexes, can be achieved by high voltage paper electrophoresis (Foster, 1962; Weigel, 1963; Whitaker, 1967). Similarly, thin layer electrophoresis of neutral and acidic sugars, including disaccharides, can be carried out in borate at pH 10 on silica gel plates covered with a thin layer of 1-octanol (Bonn, 1985; Bonn *et al.*, 1986).

B. Column Chromatography

1. *Low pressure liquid chromatography (LPLC)*

Historically, LPLC techniques developed before the advent of high performance chromatography (HPLC) and require only low hydrostatic pressure usually applied by gravity or by a peristaltic pump. Today, most LPLC procedures have been superseded by HPLC techniques; nevertheless, they still provide a useful and economical method for the analysis of sugar mixtures. LPLC can be helpful for a preliminary purification step of a disaccharide, or for a quantitative analytical separation when the more expensive HPLC instrumentation is not accessible. Automatic monitoring of the sugar peaks eluted from a LPLC column may be achieved by any of the methods described for HPLC (Section VI.B.3). Detailed literature reviews on this subject have been presented by Churms (1982, 1983), and by Robards and Whitelaw (1986).

(a) *Charcoal column.* Bulk crude separation of disaccharides from sugar mixtures can be achieved by partition on acid-washed charcoal–Celite (about 1:1; w/w) columns. Dilute ethanol solution (1–4%), applied in a gradient or batchwise, is used for elution of the disaccharide fraction (Baker *et al.*, 1979). This preparative procedure, which may serve as a first step of purification from a very crude extract, usually results in loss of some material, and the disaccharide fraction often includes some mono- and trisaccharides contaminants. The method is not suitable for separation and recovery of very small quantities of sugar on a scale usually used in metabolic studies.

(b) *Gel permeation chromatography (GPLC).* Separation of the disaccharide-containing fraction in a mixture with compounds of much larger molecular weight can be obtained by passage of the solution through cross-linked gels of defined porosity. Sephadex G-10 or G-25 (cross-linked dextran manufactured by Pharmacia) or Bio-Gel P-2 or P-4 (cross-linked polyacrylamide manufactured by BioRad Laboratories) are the gels most commonly employed. Disaccharides will usually elute from these columns much behind polymeric materials such as proteins, polysaccharides or larger oligo-saccharides, but separation from monosaccharides is not very effective. The rapid passage of a solution containing disaccharides through such a gel-permeation column is often carried out as a purification, and to some extent concentration, step prior to the subsequent application of a more discriminatory analytical assay such as HPLC. However, GPLC techniques can provide effective separations of mixtures of oligo-saccharides of different molecular sizes (Kennedy and Fox, 1980b; Churms, 1982, 1983; Robards and Whitelaw, 1986; Lin et al., 1988). Practically all of the low pressure GPLC techniques for oligosaccharides have now been replaced by HPLC and FPLC methodology (Section VI.B.3).

(c) *Ion-exchange chromatography (IEC)*. Low pressure partition chromatography of neutral sugars on columns of ion-exchange resins, usually using borate solutions for elution, has been successfully employed for many years. The method has been applied to the separation and analysis of disaccharides (for reviews see Jandera and Churáček, 1974; Kennedy and Fox, 1980a; Churms, 1982, 1983; Robards and Whitelaw, 1986). Charged disaccharides, such as molecules containing hexuronic acid or hexosamine residues, can be more sharply separated by chromatography on ion-exchange resin columns. Gradients of salt solutions such as LiCl, ammonium acetate or formate, are used for elution.

Partition chromatography of disaccharides on ion-exchange polymers (Samuelson, 1972; Churms, 1983; Robards and Whitelaw, 1986) using an ethanol–water or aceto-nitrile–water mixture for elution, has developed into some of the widely employed HPLC methodology (Section VI.B.3). For separation by IEC, it should be remembered that sucrose is very susceptible to acid hydrolysis; therefore both solid and liquid phases employed should be maintained as close to neutral as possible. Also, borate solutions used for elution are usually alkaline, and consequently can cause changes in the structure of any aldoses present in the solution analysed. If exposure to alkaline condition is prolonged, quantitative data obtained from the fractions eluted from the column will not accurately represent the concentration of the aldose in the original mixture. Non-reducing sugars such as trehalose and sucrose will not be altered by the alkaline conditions.

2. Gas liquid chromatography (GLC)

This highly sensitive analytical technique is very useful for both the quantitative and the qualitative determination of carbohydrates, disaccharides included. It is a powerful, sometimes indispensable, method for exact structural determination, particularly when coupled with mass spectroscopy (MS). Although not so commonly used for routine analysis, because of its high sensitivity GLC-MS can be helpful when only very minute quantities of material (less than 1 nmol) are available. Most of the basic GLC procedures, which were first developed for monosaccharides have been adapted for the analysis of disaccharides as well as for longer chains of glycosides. The reader should consult detailed reviews and experimental manuals describing the vast literature on the subject (Clamp, 1977; Churms, 1982, 1983; Robards and Whitelaw, 1986; Chaplin and Kennedy, 1986; Bierman and McGinnis, 1988; Elwood et al., 1988). Very valuable and specific information is usually provided by the manufacturers of GLC equipment.

Reagent kits sold by these sources significantly simplify the task of the analyst. In the present survey, only a small number of selected methods useful for the GLC analysis of disaccharides will be cited.

(a) *Trimethylsilylated (TMS) sugars.* This derivatisation (Sweely *et al.*, 1963, 1966; Laine *et al.*, 1972) in its various adaptations is one of the most commonly used for GLC of carbohydrates. The production of multiple peaks from a sugar with an unsubstituted carbonyl could result in a complex separation profile when even a simple mixture of carbohydrates is analysed by TMS derivatisation. The method has been used to analyse extracts prepared from various plants or fungi and which contained, among other carbohydrates, disaccharides such as sucrose or trehalose (Ericsson *et al.*, 1978; Prager and Miskiewic, 1979; Roberts and Mitchell, 1979; Belliardo *et al.*, 1979; Königshofer *et al.*, 1979; Streeter, 1980; Womersley, 1981; Oshima *et al.*, 1982, 1983; Cowie and Hedges, 1984; Kenyon *et al.*, 1985 and Gadd *et al.*, 1987). Laker (1979) describes in detail an excellent GLC separation of ten commonly occurring disaccharides (0.02–1 mg samples) after direct TMS derivatisation (Table 4.2).

TABLE 4.2.   Separation of several disaccharides as their TMS derivatives by GLC on OV-17 (Supelco) column (2.7 m) at 255°C. Sugar samples of 0.02–10 ng applied (from Laker, 1979).

Sugar	Relative retention[a]
Cellobiose	0.94, 1.22
Gentiobiose	1.79
Lactose	0.86, 1.13
Lactulose	0.73
Maltose	0.94, 1.22
Melibiose	1.62
Isomaltulose	1.12
Sucrose	0.81
Trehalose	1.10
Turanose	1.00 (21 min)

[a] Retention time for turanose (21 min) is taken as 1.00.

(b) *GLC of alditol acetates.* This popular procedure requires first the reduction of all carbonyls by $NaBH_4$, followed by acetylation or trifluoroacetylation of the alditol. Applied extensively, with or without MS analysis, to the determination of monosaccharides, many examples for the use of this derivatisation for the analysis of disaccharide-containing sugar mixtures, particularly sucrose, have been described (Schwind *et al.*, 1978; Doner *et al.*, 1979; Englmaier, 1986; Bierman and McGinnis, 1988). The procedure has been used to identify products of disaccharide hydrolysis (Blakeney *et al.*, 1983; Churms, 1983; Bierman and McGinnis, 1988), for uronic acid-containing glycosides (Lehrfeld, 1985; Walters and Hedges, 1988), and for amino sugars (Säämänan and Tammi, 1988). GLC-MS analyses of several disaccharide alditol acetates have been described by Klok *et al.* (1981), Olsen *et al.* (1988) and by O'Neill *et al.* (1988) and used for structural analysis of these sugars in very small samples of plant tissues. The acetylated aldononitrile derivatisation of sugars (Churms, 1983; Bierman and McGinnis, 1988) has not been applied extensively to the GLC analysis of disaccharides.

(c) *GLC of sugar oximes*. Excellent separation of the trimethylsilylated sugar *O*-alkyloximes is a highly effective method for the analysis of disaccharides such as sucrose, maltose, gentiobiose, trehalose and isomaltose in a mixture. The use of this procedure (Sweeley *et al.*, 1963) in various modifications, including the analysis of various disaccharides, has been described in detail by Mason and Slover (1971), Adam and Jennings (1975), Toba and Adachi (1977), Ferguson *et al.* (1979), Schäffler and DuBoil (1981), Li and Schuhmann (1981), Molnar-Perl *et al.* (1984), Shaw and Dickinson (1984), Streeter (1987), Pierce (1988), and Bierman and McGinnis (1988). Since the derivatisation is particularly suitable for the direct analysis of disaccharides, the basic procedure is given here as follows: a dry sugar sample (up to 5 mg, containing no more than 5 μl water) is mixed in a Teflon-coated vial with 0.5 ml of 2.5% hydroxylamine-HCl in dry pyridine at 80°C. The mixture is vigorously shaken or sonicated for 30 min. Hexamethyldisilazane (HMDS), 0.5 ml and trifluoroacetic acid, 0.05 ml are added. After an additional incubation of 30 min at 80°C samples are analysed by GLC. A variety of columns and elution programmes can be employed, as described in the references cited. To prevent hydrolysis of β-fructosides, e.g. sucrose, Schäffler and DuBoil (1981) recommend the addition of 25 μl of dimethylaminoethanol to the oximation reaction.

Extension of the oximation procedure to include MS analysis (Frank *et al.*, 1981; Chaves Das Neves *et al.*, 1982; Chaves Das Neves and Riscado, 1986) provides increased sensitivity for the analysis of reducing disaccharides. In these modifications, the aldoximes are reduced with borane or by $NaCNBH_3$ to form the corresponding aminoalditols. These products are then methanolised, trimethylsilylated and analysed by GLC and MS. Many examples of the use of the oximation reaction for the GLC-MS analysis of monosaccharides in glycoside (including disaccharide) hydrolysates can be found in the work of Pelletier and Cadieux (1982), Guerrant and Moss (1984), and Neeser (1985). Rubino (1989) describes a silylaldonitrile derivation of glycoses as a convenient procedure for their GLC-MS analysis.

(d) *GLC of methanolysed disaccharides*. Cleavage of the glycosidic bond by methanolysis is a good method for the characterisation of the nature of the glycoside structure. The methylglycoside products obtained are derivatised by TMS (Clamp, 1977; Chaplin, 1982), or by trifluoroacetylation (Tomana *et al.*, 1978; Wrann and Todd, 1978).

(e) *Permethylation of disaccharides*. As one of the fundamental methods for the study of glycoside structure, methylation analysis can be applied effectively to disaccharides. The protocol most commonly used for this method is that of Hakomori (1964) and it involves the use of methylsulphinyl carbanion as the alkoxide-forming permethylation reagent. This is achieved by the treatment of dimethylsulphoxide (DMSO) with sodium or potassium hydride (Hakomori, 1964; Phillips and Fraser, 1981). It has been claimed that improved methylation occurs when the lithium salt of DMSO is employed (Blakeney and Stone, 1985; PazParente *et al.*, 1985). Kvenheim (1988) observed that direct alkoxide formation for the methylation of 1–2 ng of sugar sample can be achieved by using butyllithium (15% in hexane)-DMSO reagent and the subsequent addition of methyliodide into the same reaction vial. The preferred starting material for carbohydrate methylation is a non-reducing glycoside, such as an oligosaccharide alditol obtained by $NaBH_4$ reduction. Ciucanu and Kerek (1984) claim that the Haworth classical reagent, using methyliodide, solid NaOH or KOH in almost dry methyl-sulphoxide, yields a completely permethylated sugar within less than 10 min.

Permethylated glycoside-alditols such as short chain oligosaccharides can be detected by TLC chromatography, and separated effectively by capillary GLC (Karkkainen, 1970, 1971; Wang et al., 1983; Table 4.1). Numerous procedures for the processing of the permethylated glycosides and their analysis by GLC-MS have been described (see for example Lindberg, 1972; Wang et al., 1983, 1984; Waege et al., 1983; Harris et al., 1984; Levery and Hakomori, 1987; Bierman and McGinnis, 1988). In the most common procedures, the permethylated glycoside is hydrolysed by acid, the monosaccharide components are reduced by $NaBH_4$, peracetylated and the partially methylated alditol acetates analysed by GLC-MS (Hakomori, 1964; Stellner et al., 1973; Kvenheim, 1988). A formic acid–trifluoroacetic acid mixture usually promotes an excellent degree of permethylated glycoside hydrolysis. As an alternative to the hydrolysis, reduction and acetylation of the permethylated glycoside, its analysis by methanolysis and acetylation has been recommended (Fournet et al., 1981; Chaves Das Neves et al., 1982; Chaves Das Neves and Riscado, 1986). The partially methylated and acetylated methylglycosides are assayed by GLC-MS.

An additional procedure available for determining the nature of the glycosidic linkage in oligosaccharides is based on a reductive cleavage of the permethylated sugar and analysis of the released anhydromethylethers by GLC (Gray, 1987).

(f) *Periodate oxidised disaccharides.* GLC analysis of fragments produced by Smith degradation (periodate oxidation, borohydride reduction and mild acid hydrolysis) of disaccharides could be applied to their structural analysis (Yamaguchi and Mukumoto, 1977; Honda et al., 1979a,b). This technique can be useful in establishing the identity of an unknown disaccharide isolated from an enzymic digest of a polymeric substrate or from a tissue extract.

3. High performance liquid chromatography (HPLC)

It is unquestionable that today the most commonly used method for the separation and analysis of sugar mixtures in biological preparations is based on HPLC procedures. HPLC separations can be carried out on a variety of stationary phase columns, such as cation-exchange resins, amine bonded microparticulate silica gels, reverse phase alkylated silica, or a hydroxylated polymeric support (see Churms, 1983; Robards and Whitelaw, 1986; Bendiak et al., 1988; Shaw, 1988; and Hicks, 1988, for detailed description). Recent developments in the instrumentation, automation, computerisation and the use of fast methods for the detection and quantification of sugar peaks separated by HPLC, have transformed the method into a versatile, sensitive, rapid and overall an economical analytical technique. In plant systems, where in most cases a relatively simple mixture of disaccharide-containing solutions has to be analysed, HPLC is particularly useful and straightforward, which explains the popularity that the method has attained. The technique can be helpful differentiating between disaccharides of similar structural features, and can also be used on a preparative scale. A large selection of HPLC instrumentation as well as individual hardware components from which to assemble a system adapted for a particular analytical mission, are now commercially available. The manufacturers and suppliers are usually eager to provide the scientist with an impressive amount of literature, technical notes and experimental protocols.

Honda (1984) presented a comprehensive summary and tabulated several hundred protocols for sugar separation, including disaccharides, by HPLC. Additional helpful technical information about this subject is reviewed by Ross (1977), Aitzetmüller (1978, 1980), Verhaar and Kuster (1981), Heyraud and Rinando (1981), Churms (1982, 1983), Hanai (1986), Robards and Whitelaw (1986); Housnell (1986), Verzele et al. (1987), Olechno et al. (1987), Lin et al. (1988) and Hicks (1988). Shaw (1988) summarised a large number of HPLC separation profiles of sugars found in food products and in some plant extracts, where sucrose and other disaccharides are commonly present. An up-to-date survey of general HPLC methodologies and detection techniques is presented by Barth et al. (1988).

In this chapter only a very general survey of HPLC methods, particularly those suitable for disaccharide analysis, will be listed. The reader should consult the literature recommended, so as to be able to select and apply the best procedure for the specific biological system studied and the instrumentation available. Investigators should not be deterred from improvising and adapting published methods which are to be used as a general guideline. Finally, a word of caution: it has been noticed by Muzika and Kovar (1987) that a low degree of sucrose hydrolysis can occur during HPLC separations on cation-exchange resins which are often in the Ca^{2+} form, and sometimes employ Ca-propionate as the elution solvent.

(a) *Sample preparation.* A most important step, before application of the sample to the HPLC column, is to clarify the sugar solution by high speed centrifugation and/or by filtration so that all material which may clog the solid phase matrix is removed (Honda, 1984; Shaw, 1988). Membrane filters (with 0.1–1.2 μm pores) are suitable for this purpose. It is usually beneficial to remove excess salt from the sugar solution, particularly if the HPLC separation is to be carried out on an ion-exchange polymer. Deionisation is most commonly achieved by passing the sugar solution through small columns of ion-exchange (such as a mixed-bed) resins. It should be remembered that very small amounts of carbohydrate can be adsorbed to the ion-exchange resin, and the loss of a neutral sugar could be noticeable when very dilute solutions are manipulated. Such is often the case in metabolic studies when radiolabelled sugar is isolated from a tissue or an enzyme system. Care should therefore be given to elute the resin with a sufficient volume of water, or with a suitable aqueous solution chosen to secure maximal recovery. Eluant at 2- to 3-fold of column's volume should be passed for this purpose (Keeling and James, 1986). When handling β-fructofuranoside-containing solutions such as sucrose, attention should be given to secure a neutral pH for all liquid phases employed for elution. Many manufacturers of HPLC instrumentation now supply ready-made cartridges for desalting and filtration, a convenience that simplifies the whole procedure. Boric acid and related boron species can be removed under diminished pressure as the volatile methyl ester. To accomplish this, small volumes of methanol are added repeatedly to the sample, and flash evaporated (three or four times) to dryness at 40–60°C. Borate can also be removed by a high capacity ion-exchange resin and a boron-selective resin (Amberlite IRA-743) has been recommended for this purpose (Hicks et al., 1986). It has also been advised that since many of the cellulose, Teflon, or nylon membrane filters used for the clarification of sugar solutions contain a wetting agent such as glycerol, membranes should be rinsed before use to avoid appearance of a misleading 'sugar' peak (Johnson and Harris, 1987).

(b) *Detection of separated sugars without derivatisation.* The simplest and ideal methods of detection are those that do not require any chemical derivatisation of the sugar. However, since the sensitivity of most of the direct methods available until very recently was not sufficiently high for the study of very dilute sugar solutions, many chemical derivatisation procedures have been developed to increase detectability. The rapid advances in the technology of chromatographic detectors is impressive and promise the availability of ever increasing sensitivity of detection (Yeung, 1989).

Refractive index (RI) detectors are the most commonly used. Most standard detectors can respond to 10–20 µg carbohydrate, but some RI detectors which have higher sensitivity have been described. Woodruff and Yeung (1982) and Bornhop et al. (1987) have developed a laser-based RI detector which responds to 1 µg sugar. Hancock and Synovec (1989) constructed an RI gradient detector which can sense 50 ng sugars separated by microbore HPLC. The application of this sensitive technique to the analysis of sugar mixtures in biological samples, has yet to be tested. Another direct method depends on the use of UV detectors, usually measuring at 190 nm. These detectors are claimed to be sensitive to 1–10 µg sugar. This method is very sensitive to interference from UV-absorbing materials present in samples from biological sources and in the solvents used (Yang et al., 1981). Other methods for direct determination, such as the mass detector, and the flame ionisation detector which can detect less than 1 µg sugar (Macrae et al., 1982; Veening et al., 1986) are technically complicated and as yet have not gained many advocates. Amperometric detectors are discussed in Section VI.B.3.

TABLE 4.3. HPLC separation of some disaccharides on Water Associates amine-bonded carbohydrate analysis column. Acetonitrile–water (83:17; v/v) used as the mobile phase (from Maretzki and Thom, 1988).

Disaccharide	Elution time (min)
Sucrose	10.86
Laminaribiose	12.20
Turanose	12.45
Maltose	13.75
Cellobiose	13.95
Trehalose	16.09
Isomaltose	16.88
Gentiobiose	19.60

Several examples, arbitrarily selected, for the use of RI detectors in HPLC analysis of disaccharides will be cited here: Maretzki and Thom (1988) studied the pool of neutral sugars in sugar cane vacuoles incubated with UDP-^{14}C-glucose, and obtained excellent resolution between several disaccharides (Table 4.3). Blanken et al. (1985) and Nikolov et al. (1985a,b) described the HPLC separation of a large number of oligosaccharides (Tables 4.4, 4.5). Separation of disaccharides in sugar cane juice is described by Wong-Chong and Martin (1979); in sucrose–maltose mixtures by Kawamoto and Okada

(1983), Picha (1985), and Van Den *et al.* (1986); in various plant extracts by McBee and Maness (1983) and Wilson and Lucas (1987); in the study of sucrose metabolism by Schnyder *et al.* (1988); in fermentation liquors by Ross and Chapital (1987); and in honey by Lipp *et al.* (1988). Numerous other examples are listed by Honda (1984), Robards and Whitelaw (1986), and by Shaw (1988).

TABLE 4.4. HPLC separation of several sugars on Lichrosorb-NH$_2$ column (Merck) using acetonitrile–H$_2$O (80:20; v/v) as the mobile phase. UV detector (from Blanken *et al.*, 1985).

Sugar		Elution time (min)
Fuc	Fucose	5.1
Man	Mannose	7.4
Glc	Glucose	8.6
Gal	Galactose	9.3
GalOH	Galactitol	9.0
GlcNAc	*N*-Acetylglucosamine	5.6
GlcNAcOH	*N*-Acetylglucosaminitol	7.5
GalNAc	*N*-Acetylgalactosamine	5.7
GalNAcOH	*N*-Acetylgalactosaminitol	6.8
Glcα1–4Glc	Maltose	22.2
Glcα1–6Glc	Isomaltose	29.3
Glcβ1–4Glc	Cellobiose	20.3
Glcβ1–6Glc	Gentiobiose	34.9
Galα1–6Glc	Mellibiose	33.1
Galα1–6GlcOH	Mellibiitol	29.2
Galβ1–4Glc	Lactose	23.8
Galβ1–4GlcOH	Lactitol	28.3
Galβ1–6Glc	Allolactose	35.9
Galβ1–3GlcNAc		13.2
Galβ1–4GlcNAc	*N*-Acetyllactosamine	13.8
Galβ1–6GlcNAc		20.7
Galβ1–3GalNAc		18.4
Galβ1–3GalNAcOH		17.3
GlcNAcβ1–2Man		19.5
GlcNAcβ1–6Man		25.1
GlcNAcβ1–3Gal		21.9
GlcNAcβ1–4Gal		19.9
GlcNAcβ1–6Gal		29.2
GlcNAcβ1–4GlcNAc	di-*N,N'*-Acetylchitobiose	12.7
Glcα1–2βFru	Sucrose	15.8
Glcα1–4Glcα1–4Glc	Maltotriose	44.0
Glcα1–6Glcα1–6Glc	Isomaltotriose	54.1
Galα1–6Glcα1–2βFru	Raffinose	46.6
Galα1–6Galα1–6Glcα1–2βFru	Stachyose	60.7

HPLC procedures adapted for the separation of sucrose and the α-galactoside series of oligosaccharides, such as melibiose, raffinose, galactinol and floridoside, have been described by Dunmire and Otto (1979), Wells and Lester (1979), Pharr *et al.* (1985, 1987), Pharr and Sox (1984), Lattanzio *et al.* (1986), Van Den *et al.* (1986), Saravitz *et al.* (1987), Meng *et al.* (1987), Kuo *et al.* (1988) and Quemener (1988). Trehalose-containing solutions were analysed by HPLC as reported by Schimz *et al.* (1985), Schimz and Overhoff (1987), Van Laere *et al.* (1987), Van Laere and Slegers (1987),

Iwahara and Miki (1988), Thevelein (1988) and Mellor (1988). Separation of ketodisaccharides, such as maltulose and cellobiulose, were obtained by Hicks *et al.* (1983), and of fructobioses as well as other oligosaccharides by Pollock (1982), Hidaka *et al.* (1988) and Muramatsu *et al.* (1988).

TABLE 4.5. HPLC separation of neutral di- and trisaccharides on (A) Supelcosil LC-NH$_2$ (Supelco), and (B) Zorbax-NH$_2$ (DuPont) columns, both 4.5 × 250 mm. A Bio-Sil-NH$_2$ (BioRad) pre-column was used with both columns. Elution at 1.0 ml min^{-1} was with 80% acetonitrile–H$_2$O in (A), and 77% and 75% acetonitrile–H$_2$O for di- and trisaccharides, respectively, in (B); RI detector. Capacity value $[k^1 = (t_R - t_0)/t_0]$ is calculated from retention time (t_R) and column dead time value (t_0) which is retention time for the water peak (about 3.5–4.0 min in these experiments) (from Nikolov *et al.*, 1985a,b).

Oligosaccharide		k^1 A	k^1 B
Xylβ1–4Xyl	Xylobiose	1.36	2.32
Glcα1–2βFru	Sucrose	2.22	3.49
Glcα1–3Fru	Turanose	2.47	3.80
Glcα1–6Fru	Palatinose (Isomaltulase)	2.50	3.89
Glcα1–4Fru	Maltulose	2.65	4.09
Glcα1–5Fru	Leucrose	2.76	4.20
Glcβ1–3Glc	Laminaribiose	2.41	3.77
Glcα1–3Glc	Nigerose	2.77	4.31
Glcβ1–4Glc	Cellobiose	2.77	4.36
Glcα1–4Glc	Maltose	2.80	4.38
Glcβ1–2Glc	Sophorose	2.92	4.50
Glcα1–1βGlc	α,β-Trehalose	2.98	4.61
Glcα1–2Glc	Kojibiose	3.14	4.85
Glcα1–1βGlc	α,α-Trehalose	3.22	4.91
Glcβ1–1αGlc	β,β-Trehalose	3.42	5.16
Glcα1–6Glc	Isomaltose	3.51	5.39
Glcβ1–6Glc	Gentibiose	3.64	5.65
Galβ1–4Fru	Lactulose	3.03	4.63
Galβ1–4Glc	Lactose	3.31	5.22
Galα1–6Glc	Melibiose	3.94	6.02
Xylβ1–4Xylβ1–4Xyl	Xylotriose	2.62	2.93
Fruβ2–1Fruβ2–2Glc	1-Kestose	—	5.84
Glcα1–3Fruβ2–1αGlc	Melezitose	4.51	5.87
Galα1–6Glcα1–2βFru	Raffinose	5.68	7.22
Glcβ1–3Glcβ1–3Glc	Laminaritriose	3.99	5.66
Glcβ1–4Glcβ1–3Glc	3-*O*-β-Cellobiosyl-D-glucose	—	6.61
Glcβ1–3Glcβ1–4Glc	4-*O*-β-Laminaribiosyl-D-glucose	—	6.67
Glcα1–4Glcα1–4Glc	Maltotriose	5.28	7.31
Glcβ1–4Glcβ1–4Glc	Cellotriose	5.53	7.42
Glcα1–6Glcα1–3Glc	3-*O*-α-Isomaltosyl-D-glucose	6.22	8.15
Glcα1–4Glcα1–4Glc	Panose	6.45	8.52
Glcα1–4Glcα1–6Glc	Isopanose	6.77	8.69
Glcα1–6Glcα1–6Glc	Isomaltotriose	8.23	10.22

(c) *Pulse-amperometric determination.* A most powerful detection method using a gold electrode for the detection of sugars in HPLC effluents has recently been developed (Hughes and Johnson, 1982; Rocklin and Pohl, 1983; Reim and Van Effen, 1986). The technique (promoted by the Dionex Corp.) is very sensitive (less than 1 ng, or 5 pmol levels of sugars can be detected) and rapid. It has already been effectively applied for the detection of a variety of carbohydrates, monosaccharides, simple or complex oligo-saccharides and sugar phosphates (Franklin, 1985; Hardy et al., 1988; Hardy and Townsend, 1988; Townsend et al., 1988; Chen et al., 1988; Pollman, 1989). The method is based on the formation of an electrometric pulse generated by the sugar in the alkaline (pH 13.0) liquid phase. Separation is on an anion-exchange solid phase column, and the special electrode is constructed so as to survive the highly corrosive alkaline solution. There is reason to believe that this detection method will also be useful in studies involving sugar analysis in plant biochemistry research. Watanabe (1985) developed an automated electrochemical detection for sugars at the pmol level using a CuII-bis(1,10-phenanthroline) as a mediator. This technique has been applied to the analysis of various mixtures of sugars (Watanabe and Inoue, 1983; Tabata and Ide, 1988). Similarly, a stable glassy carbon electrode coated by a Cu(II) layer was found to be useful in detecting sub-ng levels of mono- and disaccharides (Prabhu and Baldwin, 1989).

(d) *Derivatisation prior to column application.* Many methods in which sugar solution is modified chemically prior to separation by HPLC have been described. The modifi-cation is carried out in order to introduce a chemical tag into the sugar molecule so that sensitivity of its detection will be increased much above the RI detection level. Whereas some pre-column derivatisation techniques are simple and straightforward (for example reduction by NaB^3H_4 for reducing sugars), many are difficult to apply to quantitative analysis for several reasons. Among the problems are: different reactivity of various sugars in the modification reaction; derivatisation is not always stoichiometric; degra-dation products may be produced and consequently complicate the interpretation of the results; and finally, the specific extinction coefficient of the derivatives produced may vary for different sugars. Several pre-column derivatisation methods which have been used for disaccharides will be described here.

It is likely that the majority of the pre-column modification reagents described in the literature will not become widely used. This is because most of the requirements which necessitated their development for the determination of concentration levels of carbo-hydrates in tissue extracts can be satisfied by the improvement in sensitivity of the RI and pulse-amperometry techniques of detection. Some derivatisations which are specifi-cally helpful for structural analysis may retain their usefulness. It is important at this point to mention an ingenious and simple detection system for fluorescent compounds described by Cheng and Dovichi (1988). Laser-induced fluorescence could detect sub-fmol levels (10^{-18} mol) of fluorescent derivatised compounds (such as amino acids) separated by HPLC. It is not difficult to see that this ultra-microsensitive procedure can be expanded to the detection of properly derivatised carbohydrates.

Reduction. The solution of sugar is reduced by $NaBH_4$ (or NaB^3H_4) and the alditols acetylated or permethylated (Wells and Lester, 1979; Hull and Turco, 1985; Hall and Patrick, 1989), then separated by HPLC. Dethy et al. (1984) and Miwa et al. (1988) applied the reduction method for the analysis of hexitols and disaccharide alditols by reacting them with phenylisocyanate, and detecting the eluted sugar peaks at 240 nm. Takeuchi et al. (1987) have effectively employed NaB^3H_4 for preparing the [^{3}H]alditols obtained in glycoside, including disaccharide, hydrolysates. The [^{3}H]alditol peaks were separated and quantified automatically by RI and radioactivity counting.

Hydrazones. The conversion of reducing sugars into their hydrazones, for example by using 2,4-dinitrophenylhydrazine, has been described by Mopper and Johnson (1983), McNicholas *et al.* (1984) and Karamanos *et al.* (1987). The hydrazones are separated by HPLC. The method is designed predominantly for the detection of monosaccharides. Quantitative conversion of reducing disaccharides may not be complete, and sucrose, if present, could be partially hydrolysed during the derivatisation.

Dansylation. The use of dansyl hydrazine [1-naphthalenesulphonyl-5-(dimethyl-amino)-hydrazine] as a fluorescent tag for reducing mono- and oligosaccharides has been described by Avigad (1977). The original method which was designed for TLC separations has been applied to the HPLC procedure (Alpenfels, 1981; Takeda *et al.*, 1982; Mopper and Johnson, 1983; Hull and Turco, 1985; Eggert and Jones, 1985; Finden *et al.*, 1985; Karamanos *et al.*, 1987, 1988). The high sensitivity of the method allows determination of as little as 10 pmol of aldose. Sucrose, and other β-fructo-furanosides, will be cleaved during this derivatisation, and trehalose and other more stable glycosides will not react.

Several analogues of dansylhydrazine which are suitable for sugar derivatisation and yield very high fluorescent products have been described (Anderson, 1986; Lin and Wu, 1987). These can be adapted for both TLC and HPLC separations of reducing sugars, but may be more susceptible to interference by trace amounts of carbonyls other than sugars that are present in biological samples.

Reductive amination. Introduction of a fluorescent tag to a reducing sugar can be achieved by reductive amination procedures, based on the reduction of Schiff base adducts formed between the sugar and a suitable amine. Sodium cyanoborohydride is the preferred reductant for these reactions. It has not always been established whether a stoichiometric derivatisation of the sugars present in a mixture occurs during reductive amination reactions described in the literature, and some loss of products may occur during steps devised to remove excess reagents from the samples before application to the HPLC columns. The technique, however, was found to be useful in several cases.

One of the more publicised reagents for this method is 2-aminopyridine (Hase *et al.*, 1978, 1979, 1981; Takemoto *et al.*, 1985; Carr *et al.*, 1985; Tomiya *et al.*, 1987). Also described are the use of 7-amino-4-methylcoumarin (Prakash and Vijay, 1983), of ethanolamine (Kato and Kinoshita, 1980), 7-amino-1-naphthol (Cole *et al.*, 1985), and 4'-*N,N*-dimethylamino-4-aminobenzene (Rosenfelder *et al.*, 1985; Towbin *et al.*, 1988).

A successful application of reductive amination suitable for both TLC and HPLC analysis has been recommended by Wan *et al.* (1984) and Webb *et al.* (1988). In this method, condensation of the reducing sugars, including disaccharides, was done with *p*-aminobenzoic ethyl ester and with $NaCNBH_3$. The products, detected at 229 or 254 nm after HPLC separation, could be further analysed by GLC-fast atom bombardment MS. This method, which can be carried out in one vial, is a very powerful procedure for determining structure from as little as 0.5 nmol disaccharide (Table 4.6).

Perbenzoylation. This method, originally designed for GLC, has also been used for HPLC separations. Perbenzoylated sugars are detected by absorbance at 230 nm (Lehrfeld, 1976; Nachtman and Budna, 1977; Galensa, 1984). This method, together with methanolysis, has also been applied to the structural study of various oligosaccharides (Daniel *et al.*, 1981; Jentoft, 1985). Karamanos *et al.* (1988) used perbenzoylation and reduction (Taylor and Conrad, 1972) to characterise the structure of oligosaccharides containing uronic acids.

Miscellaneous. Oxime derivatives of reducing sugars, which are used for GLC analysis, have also been employed in HPLC procedures (Lawson and Russel, 1980; Chen and McGinnis, 1983). Oximation and perbenzoylation were used by Thompson (1978). A list of several other reagents suggested for pre-column tagging is given by Hicks (1988).

TABLE 4.6. Relative retention time of disaccharides–*p*-aminobenzoic ethyl ester (ABEE) derivatives on Lichrosorb Si-60 column using for elution a linear gradient of acetonitrile–water (15–40%) containing 0.05% 1,4-diamino(butane) (from Wang *et al.*, 1984).

Disaccharide–ABEE	Relative retention[a]
Maltose	0.66
Isomaltose	0.84
Cellobiose	1.00
Laminaribiose	0.63
Gentiobiose	1.01
Lactose	0.96
Melibiose	0.92
Sophorose	0.87

[a] Retention time for cellobiose (16 min) is taken as 1.00.

(e) *Derivatisation following column elution.* Chemical reactions to increase the detection level of the sugar in column effluents are carried out immediately after elution, and the product is detected photometrically or by a choice of other techniques. Post-column procedures are usually executed in fully automated HPLC systems, and the reagents to be used should preferably not contain highly corrosive chemicals such as mineral acids. During earlier periods in the development of HPLC techniques, reagents such as hot anthrone or phenol-H_2SO_4 were employed for analysis, providing good data for sugar detection, but promoting rapid deterioration of the instrumentation.

Periodate oxidation. Post-column oxidation of carbohydrates by periodate produces aldehydes which can be sensitively detected. Among these products, the formaldehyde formed from hydroxymethyl groups in the sugar molecule is most easily detected. To facilitate production of formaldehyde, e.g. from a reducing disaccharide, the sugar should be first reduced to alditol with $NaBH_4$. The formaldehyde (1 nmol detection level) is analysed photometrically (Speck, 1962; Hauser and Cummins, 1964; Pesez and Bartos, 1967; Gallop *et al.*, 1981; Avigad, 1983) or by a selection of other reagents (Sawicki and Sawicki, 1978; see also Section V.C). Simatupang *et al.* (1978) and Honda *et al.* (1983) used the periodate method for the colorimetric or fluorimetric detection of sugar peaks (0.2–2 nmol) in HPLC separation. Nordin (1983) has measured the consumption of periodate at 260 nm to detect the position of sugar peak separated by chromatography.

Luminescence. If glucose can be a product of the sugar eluted from a HPLC column, such as in the case of a disaccharide after enzymic or acid hydrolysis, it can be detected by reaction in the glucose oxidase system. The H_2O_2 produced is coupled to a chemiluminescence reaction (Koerner and Nieman, 1988). This and related techniques are discussed in Section VII.D on biosensors.

Copper complexes. Spectral measurements of the complex formed between sugars and cuprammonium ions were used to detect 2–5 nmol levels of monosaccharides (Grimble et al., 1983; Abbou and Sioufi, 1987; McKay et al., 1987). The method was used for analysing sucrose-containing mixtures. Mopper and Grindler (1973), Mopper (1978), McFeeters (1980) and Waffenschmidt and Jaenicke (1987) described a copper-bicinchoninate reagent for detecting sugar peaks in HPLC.

Miscellaneous. Reaction with hydrazine and fluorescamine (Avigad, 1976) or with glycamine and o-phthalaldehyde (Perini and Peters, 1982) yields fluorescent compounds from reducing sugars. Use of these reagents is limited because of high sensitivity to amine-containing compounds which contaminate most impure sugar solutions from biological sources.

Vrátny et al. (1985) and Femia and Weinberger (1987) have used the interaction with 4-aminobenzoic hydrazone in post-column detection of reducing sugars in HPLC. The alkaline 2-cyanoacetamide reagent has been extensively used by Honda and collaborators (Honda et al., 1980, 1981a,b, 1988; Schlabach and Robinson, 1983). This reagent will not react with sucrose or trehalose; it can therefore help in detecting small amounts of reducing sugar in the presence of the non-reducing glycosides. In contrast, ethylenediamine reagent (Mopper et al., 1980; Honda et al., 1986), the p-toluenesulphonic acid reagent (Villanueva et al., 1987) and the ethanolamine-boric acid reagent (Kato and Kinoshita, 1980) have been claimed to be able to detect both non-reducing and reducing disaccharides in HPLC effluents.

One of the classical reactions for measuring reducing sugars, the reduction of tetrazolium salts to coloured formazans, has also been adapted to HPLC analysis (Mopper and Degens, 1972; Thomas and Lobel, 1976; Wight and Van Niekerk, 1983; Van Eijk et al., 1984; Escott and Taylor, 1985). This method is simple and sensitive, but because of the relatively insoluble nature of the formazan in aqueous solutions, the solvent composition used for the derivatisation (e.g. addition of alcohol) has to be precisely adjusted to prevent precipitation and clogging.

C. Chromatographic Analysis of Phosphorylated Intermediates

Sugar phosphates are nucleotides and products obtained in enzymic reaction employed for disaccharide analysis (see Section VII.C). Analytical procedures based on ion-exchange chromatography can be used to determine and quantify the appearance of such products as glucose 1-phosphate, glucose 6-phosphate, UDP or UDP-glucose. Protocols and conditions for such separations by liquid chromatography are described by Giersch (1979), Geiger et al. (1980), Giersch et al. (1980), Heldt et al. (1980), Lehmann (1981), Bruut and Hokse (1983), Reiss et al. (1984), Meyer and Wagner (1985), Tjioe et al. (1985), Stikkelman et al. (1985), Henderson and Henderson (1986), Smith et al. (1988), Outlaw et al. (1988), Smrcka and Jensen (1988), Galloway et al. (1988) and Lim and Peters (1989). Some of the recent procedures described in these studies are based on HPLC methodology. They are highly effective, rapid and sensitive. The FPLC system (a reverse phase, ion-exchange and gel filtration Mono-Q column offered by Pharmacia) is also suitable for the separation of phosphate ester mixtures and requires the use of a suitable concentration gradient of a buffered salt solution for sugar elution. Planar chromatographic procedures for various phosphate esters are described by Whitaker (1967); Stahl (1969); Redgwell (1980); Preisser and Komor (1988); Ram et al. (1989) and by Feingold and Barber, Chapter 2 this volume.

VII. ENZYME REAGENTS FOR THE ANALYSIS OF DISACCHARIDES

The advantages of using enzymes for the determination of disaccharides are their high specificity, sensitivity, and versatility of the assay techniques by which they can be assayed. Most of the enzymic methods depend on the cleavage of the disaccharide by a specific hydrolase; the products of hydrolysis are determined either in a separate subsequent step, or by end point determination in coupled assay systems. These can also be analysed by a chemical assay for reducing sugar or by chromatographic analysis.

In addition to the hydrolases, several non-hydrolytic enzymic reactions specific for individual disaccharides are known and can be applied analytically. These reactions are discussed in Section VII.C.

A. Glycoside Hydrolases

Complete hydrolysis of a disaccharide can be achieved by heating it in dilute mineral acid (for example, at 100°C, sucrose will be hydrolysed in 0.1 N HCl within 10 min, whereas most other common disaccharides will hydrolyse in 1 N HCl within 1 to 2 h; see Hassid and Ballou, 1957; BeMiller, 1967; Pazur, 1970). Because of its lack of selectivity, acid hydrolysis does not provide an ideal assay method for a mixture of glycosides unless the products of hydrolysis are first separated by chromatography. In comparison, the use of a specific glycoside hydrolase provides a convenient, selective and relatively specific reagent for both the purpose of quantitative analysis and for establishing the disaccharide structure (Lee and Lindley, 1982). Hydrolases for all glycosidic linkages are known, and many which can be used as analytical reagents for disaccharides, are commercially available. Many of these enzyme preparations are from yeast and fungi, and only partially purified. Investigators should be aware of the possible presence of contaminating hydrolase activity which may complicate an analysis of a mixture of glycosides. The glycoside hydrolase activities of each new enzyme preparation obtained should be surveyed carefully on a variety of substrates.

Larner and Gillespie (1956), Avigad (1958), and Halvorson and Elias (1958) have noticed that Tris(hydroxymethyl)aminomethane is a powerful competitive inhibitor of α-glucosidases (K_i, 1–10 mM at pH 6–7). Numerous studies have since established that this inhibition is characteristic of a broad spectrum of α-gluco-disaccharide hydrolases such as maltase, trehalase and dextransucrase from many biological sources. Unless the inclusion of Tris in the reaction mixture is purposely designed to inhibit disaccharide cleavage, the reagent should not be used as a buffer if achievement of complete enzymic hydrolysis is desired. It should be noted that the plant alkaline invertase (EC 3.2.1.26) is also strongly inhibited by Tris (Morell and Copeland, 1984).

1. Sucrose

(a) *Invertase.* The use of invertase (β-fructofuranosidase, EC 3.2.1.26) for the specific assay of sucrose, has been practised for over a century and it is broadly applied in analytical chemistry. It is the most common reagent for the analysis of sucrose in biochemical studies. When sucrose is present in mixture with other reducing sugars, particularly with glucose, these can be destroyed by heating with alkali (1 N NaOH at 100°C for 1 h). Alternatively, prior to sucrose inversion, the reducing sugars can be converted to alditols by treatment with excess $NaBH_4$, or the glucose can be specifically removed by exposure to a glucose oxidase/catalase mixture. The products of sucrose hydrolysis, namely glucose and fructose, can be assayed chemically, chromatographically, or by specific enzymic reactions as described in Section VII.B. (see Bergmeyer *et al.*, 1983, and Boehringer-Mannheim, 1986, for a selection of suitable protocols). Many examples are available to show how the method is applied to analyse sucrose in plant extracts (see Rufty and Huber, 1983; Stitt *et al.*, 1985; Huber and Akazawa, 1986, as an arbitrarily limited selection). Since invertase cleaves other β-fructofuranosides (such as raffinose) which are often found in plant tissues together with sucrose, the determination of glucose in the hydrolysate will provide a true value for the sucrose originally present, whereas fructose will indicate the total amount of β-fructosides that was cleaved.

One of the most popularly used procedures for the determination of sucrose via hydrolysis by invertase is that developed by Jones *et al.* (1977) and Jones and Outlaw (1981). In their detailed protocols, pre-existing glucose and fructose are destroyed by heating in alkali. Sucrose is hydrolysed, and the reduction of $NADP^+$ is measured by a series of coupled enzymic reactions. Sensitivity of detection is from 10^{-7}–10^{-14} moles of sucrose depending on the procedure selected: spectrophotometric (5–70×10^{-9} mol), fluorimetric (0.1–5×10^{-9} mol), or recycling fluorimetric procedures (0.1–10×10^{-13} mol). Pure yeast invertase, with an optimum pH of activity of 4.5, is an inexpensive reagent which can be obtained from many commercial sources (Woodward and Wiseman, 1982). Similar to one of the plant invertase isozymes which has a neutral pH optimum are several bacterial invertases (Kunst *et al.*, 1974, 1977; Bugbee, 1984; Yamamoto *et al.*, 1986). The use of these preparations could be advantageous in a coupled enzyme assay system for the analysis of sucrose.

(b) *a-Glucosucrase*. An α-glucosidase that can hydrolyse sucrose is produced by insects and mammals. The sucrase-isomaltase (EC 3.2.1.28) made by the brush border intestinal and kidney epithelium, have been studied in great detail (Hauser and Semenza, 1983; Semenza, 1986; Hunziker *et al.*, 1986). The enzyme has two covalently linked subunits, one of which acts on sucrose and the other (EC 3.2.1.10) has the catalytic site for isomaltose. This interesting enzyme, which is easy to isolate, has the potential to be a useful analytical reagent for α-glucoside disaccharides.

An 'α- glucosaccharase' was thought to be produced by several strains of mould. It is probable that these early reports actually evaluated a mixture of invertase and α-glucosidase (maltase), rather than an individual and specific α-glucosucrase entity.

2. *Trehalose*

The specific α-glucoside hydrolase, trehalase (EC 3.2.1.28), is produced by all organisms which are able to synthesise trehalose, as well as by many microorganisms which can use trehalose as a carbon source for growth (see reviews by Avigad, 1982; Labat-Robert, 1982). Good sources for this enzyme are insects, yeast and fungi (Patterson *et al.*, 1972; Avigad, 1982; Keller *et al.*, 1982; Hehre *et al.*, 1982; Londesborough and Varimo, 1984; Thevelein, 1984; Dellamora-Ortiz *et al.*, 1986; Weiser *et al.*, 1988; Araujo *et al.*, 1989; Sumida *et al.*, 1989). It is important that the trehalase used for analysis be devoid of contaminating glycosidases which is a common occurrence. Enzyme prepared from invertase-less yeast strains is particularly recommended (Avigad, 1982; Araujo *et al.*, 1989). Trehalase is also produced in plant tissues such as soybean root nodules (Streeter, 1985; Salminen and Streeter, 1986; Mellor, 1988). At present the only commercially available trehalase is a preparation from porcine kidney (Semenza, 1986; Yoneyama, 1987). A potentially good source for trehalase could develop from future cloning studies of this enzyme from *E. coli* (Guitierrez *et al.*, 1989).

3. Maltose

(a) *Maltase.* Hydrolysis of maltose to glucose is facilitated by α-glucosidase (EC 3.2.1.20). The enzyme, which is produced by many strains of bacteria, fungi and yeasts, is usually a broad spectrum α-glucosidase which cleaves, in addition to maltose, other α-glucodisaccharides including sucrose, but at slower rates (Larner, 1960; Bucke, 1982; Chiba and Minamiura, 1988). The enzyme is strongly inhibited by Tris(hydroxy-methyl) aminomethane. Some highly purified preparations of α-glucosidase, such as that from *Candida tropicalis* (Hehre *et al.*, 1977), are devoid of sucrase activity. If the maltose analysed is in a mixture with sucrose or isomaltose, for example, the relative specificity of the α-glucosidase (maltase) preparation toward these various α-glucosides should be evaluated. Also, some impure preparations from yeast (commercially available maltase is in most cases a yeast product) may still be contaminated with invertase (β-fructofuranosidase). An α-glucosidase preparation from rice which can also serve as a maltase (Chiba *et al.*, 1983; Weiser *et al.*, 1988) is sold by several manufacturers. The glucose liberated from maltose can be estimated with glucose oxidase, or by any other procedures for glucose discussed in Section VII.B.1, or by chromatography. For protocols, see Halvorson (1966), Bergmeyer *et al.* (1983, 1984) and Boehringer-Mannheim (1986). It should be remembered that α-glucosidase hydrolases also catalyse extensive transglucosylation reactions, a property that should be considered, particularly when the substrates are present in relatively high concentrations (Nisizawa and Hashimoto, 1970; Hehre *et al.*, 1973; Kitahata *et al.*, 1981). Mammalian α-glucosidases from the kidney and intestinal mucosa have been studied extensively (Reiss and Sacktor, 1982; Semenza, 1986) and are of potential value as reagents for maltose analysis.

(b) *Amyloglucosidase.* This α-glucosidase (EC 2.3.1.3.) removes terminal α-1,4-glucosyl residues from 4-α-gluco-oligosaccharides. Maltose is also hydrolysed but usually at a slower rate than longer chain oligosaccharides. Enzyme preparations from various fungi such as *Aspergillus niger* and *Rhizopus niveus*, as well as from several bacteria, have been studied in detail and are commercially available as highly purified protein (Pazur *et al.*, 1973; Suetsugu *et al.*, 1973; Hehre *et al.*, 1980; Bucke, 1982; Tsujisaka, 1988; Saha and Zeikus, 1989). The enzyme has poor ability to hydrolyse the α-1,6-glucosyl linkage and does not act on sucrose. Many fungal amyloglucosidase preparations sold are not excessively purified, and may still be contaminated with β-fructofuranosidase (invertase). The capacity of amyloglucosidases to catalyse glucosyl transfer reactions should be taken into account when using these enzymes to study the cleavage of maltose, particularly when present in relatively high concentrations (Pazur, 1972; Pazur *et al.*, 1973; Hehre *et al.*, 1973; Kitahata *et al.*, 1981; Fujimoto *et al.*, 1988). An α-glucosidase from spinach leaves has been used to prepare [3H]-labelled maltose by exchange reaction with [3H]glucose (Shoaf *et al.*, 1979).

4. α-Galactosides

An α-galactosidase from a variety of biological sources (*Aspergillus*, *Escherichia coli*, coffee beans and other seeds, for example) is available for the analysis of α-galactosides, such as melibiose. The free galactose released can be determined as described in Section VII.B.3. Descriptions of various α-galactosidase preparations are given by Courtois and Petek (1966), Li and Li (1972), Agrawal and Bahl (1972), Lindley (1982) and Dey (1985). A thermostable bacterial α-galactosidase has been described by Mitsutomi and Ohtakara (1984, 1988) and an insolubilised enzyme has been prepared by Ohtakara and Mitsutomi (1987). As discussed elsewhere, for some analytical systems it is important to determine the degree of contamination by other glycoside hydrolases of the α-galactosidase preparations.

5. β-Glycosides

The β-galactosidase (EC 3.2.1.23) preparations which are available from *E. coli* and from yeast are usually highly purified, have a high specific activity and are relatively free from contaminating glycoside hydrolase activities (Agrawal and Bahl, 1972; Stephens and DeBusk, 1975; Nijpels, 1982; Bergmeyer *et al.*, 1983, 1984).

The classical β-glucosidase (EC 3.2.1.21) is that from almond emulsin. Some highly purified preparations are now available, but their ability to hydrolyse glycosidic bonds, including β-galactosides, should always be verified. Barring small differences in kinetic parameters, most β-glucosidases are not specific enough to distinguish between different β-glucoside disaccharides (Lee, 1972; Lindley, 1982; McCleary and Harrington, 1988).

6. β-Hexuronides

β-Glucuronidase (EC 3.2.1.31) preparations (from *E. coli, Helix* or *Abalone*) are commercially available. The degree of purity of these enzymes must be ascertained before use. This enzyme can be employed in the study of β-glucouronyl disaccharides (glucobiuronic acids) produced during the enzymic degradation of various plant polysaccharides.

A β-galactouronidase (EC 3.2.1.15) is a component of the 'pectinase' preparations (fungal or from a plant origin) that can hydrolyse β-galacturonides (pectin substances) including galactobiuronic acid disaccharides.

7. Disaccharide phosphates

(a) *Sucrose 6'-phosphate.* The product of the sucrose phosphate synthetase reaction has been usually identified by planar chromatography or electrophoresis carried out under conditions that differentiate between neutral and phosphorylated sugars (Dutton *et al.*, 1961; Hatch, 1964; Hawker, 1985). The phosphate ester can be separated from neutral sugars and from other negatively charged molecules by a variety of ion chromatographic techniques (see Section VI.C). As in the case of the sucrose synthase system, treatment with alkali (Cardini *et al.*, 1955) or reduction by $NaBH_4$ (introduced by Rorem *et al.*, 1960) will leave the sucrose-phosphate intact so that it can be determined chemically or chromatographically. Sucrose 6'-phosphate, which is resistant to cleavage by invertase, can be hydrolysed to free sucrose by incubation with bacterial alkaline phosphatase, or with sucrose phosphatase, if available (Whitaker, 1984; Cardemil and Varner, 1984; Hawker and Smith, 1984; Hawker, 1985; Hawker *et al.*, 1987). Alternatively, it can be carefully hydrolysed to glucose and fructose 6-phosphate by mild acid (30 min at 80°C at pH 2.5). Analysis of the reaction products in either procedure is carried out by standard methods.

Barber (1985) has clearly demonstrated that sucrose phosphate synthase is a reversible reaction, and that the difficulties encountered in establishing this in past investigations were due to unidentified inhibitors present in the reaction mixture. It remains to be seen whether sucrose phosphate synthase can be employed for preparative or analytical goals, and more importantly, if reversal of sucrose phosphate synthesis has any role in carbohydrate metabolic fluxes in plant tissues.

(b)   *Tehalose 6-phosphate.* Since the discovery of this product by Cabib and Leloir (1958), most of the procedures for its identification have been based on methods similar to those used for sucrose 6'-phosphate. These include the destruction by alkali of all reducing sugars in the solution and the subsequent determination of trehalose phosphate colorimetrically with the anthrone or phenol H_2SO_4 reagents. Phosphate can be removed by hydrolysis with non-specific bacterial alkaline phosphatase to release the free trehalose (Lapp *et al.*, 1972; Killick, 1979; Salminen and Streeter, 1986). Trehalose 6-phosphate is also synthesised in several strains of bacteria when grown on various α-glucosides (Marechal, 1984; Martin and Russell, 1987). The more specific trehalose phosphatase (EC 3.1.3.12), which is present in all organisms that can synthesise trehalose (for reviews and further information see Mitchell *et al.*, 1972; Avigad, 1982; Thevelein, 1988), can also be used for the same purpose, but it is not readily available. A phosphotrehalase activity which cleaves trehalose 6-phosphate to glucose and glucose 6-phosphate has been detected in a bacterium (Bhumiratana *et al.*, 1974; Marechal, 1984) and in soybean nodule bacterioids (Salminen and Streeter, 1986). Future studies are needed to characterise this enzyme further.

B.   Determination of Disaccharide Hydrolysates

1.   *Glucose*

(a)   *Assay of glucose with glucose oxidase* (EC 1.1.3.4). This reagent, which produces glucono δ-lactone and H_2O_2 from glucose, is one of the most commonly used for glucose determination in solution. It is also a method of choice for evaluating glucodisaccharides after their hydrolysis. There are numerous applications of this enzymic reagent adapted for manual use, for semi-automatic use, for autoanalysers and for continuous flow injection techniques. Immobilised glucose oxidase sensors are also available, as discussed in Section VII.D. It is advantageous to include Tris(hydroxymethyl)aminomethane, an inhibitor of α-glucosidases, at concentrations of > 20 M, in the glucose oxidase reagent (White and Subers, 1961). This will prevent contaminating α-glucosidase activity, often present in glucose oxidase preparations, from interfering in the assay of glucose when present in a mixture with α-glucoside oligosaccharides.

A pyranose 2-oxidase (EC 1.1.3.10) has been obtained from the mycelia of several *Basidiomycetes*, and has been proposed as an analytical reagent for glucose (Volc *et al.*, 1985; Volc and Eriksson, 1988). The enzyme oxidises D-glucose to D-*arabino*-2-hexulose (glucosone) and H_2O_2. D-Xylose and L-sorbose are also substrates for this oxidase, but D-galactose, D-fructose and glucosides are very poorly reactive. Assay of the reaction is based on the colorimetric determination of H_2O_2 or in any of the selection of methods used for glucose oxidase systems.

Most of the assays using glucose oxidase rely on analysis of the H_2O_2 stoichiometrically produced during oxidation of glucose (Bergmeyer et al., 1983, 1984). A variety of ready-made reagent kits for glucose determination based on the glucose oxidase reaction are commercially available. Below are brief descriptions of the principal methodologies available for H_2O_2 determination with glucose oxidase.

Colorimetry. In the presence of excess peroxidase, H_2O_2 and a suitable hydrogen acceptor, a soluble chromophore with a high specific absorbance is produced. Among the most commonly used reagents (for the determination of 0.01–1.0 mM glucose) are:

(a) Benzidine reagents such as o-dianisidine or o-tolidine (Sols and DeLaFuente, 1961; Lloyd and Whelan, 1969; Cerning-Beroard, 1975; Guilbault, 1976; Pazur and Dreher, 1980; Artiss et al., 1981; Josephy et al., 1982; Bergmeyer et al., 1983; Worthington, 1988).

(b) ABTS (2,2'-azino-di(3-ethylbenzthiazoline-6-sulphonate)) (Werner et al., 1970; Majkic-Singh et al., 1981; Bergmeyer et al., 1983; Ellis and Rand, 1987).

(c) The Trinder reagent (4-aminoantipyrine plus a phenol) (Bergmeyer et al., 1983, 1984; Blake and McLean, 1989).

(d) MBTH (3-methyl 2-benzothiazolinone hydrazone coupled with dimethylamino-benzoic acid or with formaldehyde) (Ngo and Lenhoff, 1980; Bergmeyer et al., 1983, 1984; Capaldi and Taylor, 1983; Sabin and Wesserman, 1987). This is probably the most sensitive colorimetric assay described for glucose oxidase.

(e) APC (4-aminophenazone plus chromotropic acid) (Prencipe et al., 1987).

(f) Ti-(IV). Matsubara et al. (1985a,b) used a titanium(IV)-4-(2-pyridylazo) resorcinol (Ti-PAR), and Ti(IV)-2-((5-bromopyridyl)azo) 5-(N-sulphopropylamino) phenol (Ti-PAPS) for colorimetric assay of H_2O_2 in glucose oxidase systems.

Spectrophotometry. Hydrogen peroxide can be used to oxidise NADH in a coupled reaction with NADH-peroxidase (EC 1.11.1.1) and the consumption of NADH is monitored spectrophotometrically (Avigad, 1978).

Fluorimetry. The presence of pmol levels of H_2O_2 can be detected fluorimetrically in specific peroxidase catalysed reactions. Zaitsu and Ohkura (1980), and Mendez et al., (1987) used p-hydroxyphenyl and Matsumoto et al. (1981) employed homovanillic acid as the substrate. Lazrus et al. (1985) and Miller and Kester (1988) studied in detail conditions for the fluorimetric determination of H_2O_2 using the peroxidase catalysed dimerisation of (p-hydroxyphenyl) acetic acid.

Catalase coupled system (Hantzsch reaction). In the presence of catalase, H_2O_2 oxidises methanol to formaldehyde. This product is condensed with NH_3^+ and acetylacetone to form a coloured lutidine chromogen (Bergmeyer et al., 1983). This colorimetric method is not as sensitive as the peroxidase-dye coupled assays listed above. A more sensitive variation has been described by Kobayashi and Kawai (1983) in which the formaldehyde produced by catalase is analysed by GLC as a pentafluoro-benzyloxylamine derivative.

Luminometry. In recent years significant advances have been reported in applying chemiluminescent reactions to the assay of H_2O_2 in solution. With proper reagents and

instrumentation, detection levels of less than one pmol H_2O_2 can be achieved. In these determinations either luminol or diaryloxalates serve as hydrogen donors in a reaction which can be enhanced by peroxidase. Among the many reviews and manuals recently published, and which contain detailed descriptions of the application of chemilumines-cence to the glucose oxidase reaction, are: Auses *et al.* (1975), Bostick and Hercules (1975), Puget and Michelson (1976), Seitz (1978, 1981), Whithead *et al.* (1979), Bergmeyer *et al.* (1983), Madsen and Kromis (1984), Tsuji *et al.* (1985), Thorpe *et al.* (1985), Grayeski *et al.* (1986), Imai (1986) and Campbell (1988); see also Section VII.D.
    Electrochemical determination. See Section VII.D on biosensors.

(b)   *Glucose determination via the hexokinase reaction.* One of the most common assay systems employed for glucose in biological research is the conversion of glucose to glucose 6-phosphate by Mg^{2+}-ATP and hexokinase (EC 2.7.1.1) and subsequent oxidation of the glucose 6-phosphate by $NAD(P)^+$ and glucose 6-phosphate dehydrogen-ase (EC 1.1.1.49) in a coupled system. Reagent kits for this assay are provided by many manufacturers, and the method has been adapted for autoanalyser technology. The more specific glucokinase (EC 2.7.1.2), such as the thermostable *Bacillus* product which is commercially available, can replace the hexokinase. The yeast glucose 6-phosphate dehydrogenase is specific for $NADP^+$, whereas the bacterial enzyme usually can use both NAD^+ and $NADP^+$. The NAD(P)H produced in this coupled reaction can be measured by a variety of techniques (see also Chapter 1, this volume).
    Spectrophotometry. The traditional and most common procedure is based on record-ing the increase in absorbance at 340 nm as a measure of NAD(P)H production. Ready-made reagent kits for this assay are supplied by many manufacturers. Detailed protocols are available in many publications; Bergmeyer *et al.* (1983, 1984) is probably one of the most cited and accessible sources. Additional protocols can be found in Guilbault (1976), Jones *et al.* (1977), Sturgeon (1980a,b), Jones and Outlaw (1981), Chaplin and Kennedy (1986), and Boehringer-Mannheim (1986).
    Colorimetry. Reduced pyridine nucleotides can be oxidised by phenazine metho-sulphate which, in turn, will be reoxidised by a tetrazolium salt. The coloured formazan product is measured colorimetrically (Guilbault, 1976; Bergmeyer *et al.*, 1983, 1984; Cairns, 1987). Alternatively, the NAD(P)H can be oxidised by diaphorase (EC 1.8.1.4) in the presence of a suitable tetrazolium salt as the electron acceptor (Gella *et al.*, 1981). Colorimetric assays of this design have been often used to determine NAD(P)H in analytical clinical chemistry, including the determination of glucose in the hexokinase/ glucose 6-phosphate dehydrogenase coupled reaction. It has been observed by many investigators that oxygen strongly represses tetrazolium reduction in a phenazine coupled reaction. It is, therefore, advisable (Van Noorden and Butcher, 1989) that an assay based on the formation of formazan should be carried out under N_2, in the presence of cyanide (5 mM), and with an elevated concentration of tetrazolium salt (5 mM). Bergel *et al.* (1989) evaluated several amplified diaphorase coupled reactions for the measurement of NAD(P)H. These procedures can be adapted for the sensitive determination of monosaccharides oxidised by specific pyridine nucleotide-linked dehydrogenases.
    Fluorimetry. The classical experimental protocol of Lowry and Passaneau (1972) has been extensively employed for the determination of pmol levels of NAD(P)H-produced enzyme-catalysed reaction, including the determination of glucose in the hexokinase/

glucose 6-phosphate dehydrogenase-assay systems (see also Bergmeyer *et al.*, 1983, 1984). The sensitivity of this fluorimetric method can be amplified several thousand-fold using a variety of enzymic cycling procedures (Lowry and Passaneau, 1972; Jones *et al.*, 1977; Chi *et al.*, 1978; Cox *et al.*, 1982; MacGregor and Matschinsky, 1984; Shayman *et al.*, 1987) and could be appropriately applied for the assay of specific monosaccharides present in disaccharide hydrolysates. A more specific application for glucose 6-phosphate dehydrogenase assay has also been described by Feraudi *et al.* (1983) and Harrison *et al.* (1988).

Coupling an NAD(P)H-forming system with phenazine methosulphate or with diaphorase and with resazurin results in the formation of a fluorescent resorufin. This reaction provides another analytical method for NAD(P)H produced by glucose 6-phosphate dehydrogenase, as well as in many other NAD(P)$^+$-dependent oxidations (Guilbault, 1975; Avigad, 1984; Soyama and Ono, 1987).

Luminometry. The NAD(P)H produced in the hexokinase/glucose 6-phosphate dehydrogenase coupled reaction can be monitored in a bacterial luciferase reaction. Initial rates or end-point measurement of bioluminescence make it possible to determine a range of 0.1–200 nM NAD(P)H (Kricka, 1988). For a selection of detailed descriptions and protocols of these analytical methods, particularly when applied to the assay of glucose and sucrose hydrolysates, see Whitehead *et al.* (1979), Golden and Katz (1980), Seitz (1981), Palmisano and Schwartz (1982), Feraudi *et al.* (1983), Hughes (1983), Campbell *et al.* (1985), Kurkijarvi *et al.* (1985), Weinhausen and DeLuca (1986), Idahl *et al.* (1986); Wieland *et al.* (1986); Campbell (1988) and Welch *et al.* (1989).

Chromatographic assay of glucose 6-phosphate. The glucose 6-phosphate produced in the hexokinase reaction can be identified and separated by planar or ion-exchange liquid chromatography (Smith, 1988; Smith *et al.*, 1988; see also Section VI.C). This analytical procedure is mostly employed in biochemical studies when the specific radioactivity of [^{14}C]metabolites isolated in very small quantities has to be determined, for example in studies of metabolic fluxes where the distribution of ^{14}C in the different glycosyl residues in sucrose or starch has to be assessed. One approach which facilitates detection of the phosphorylated sugar is to use very high specific activity Mg-AT^{32}P in the hexokinase reaction. The hexose phosphate peaks separated by ion-exchange chromatography can then be determined with great sensitivity.

(c) *Glucose dehydrogenase* (EC 1.1.1.47). The *Bacillus* or the yeast NAD$^+$ (in some species also NADP$^+$) dependent D-glucose dehydrogenase is rapidly becoming a popular analytical reagent for the specific assay of glucose in disaccharide hydrolysates. It can serve as a suitable substitute for the glucose oxidase or the hexokinase/glucose 6-phosphate dehydrogenase assays. It should be noted that glucose dehydrogenases from various sources very often can oxidise several other hexoses, in particular 2-deoxy-D-glucose and D-mannose (Anderson and Dahms, 1975; Avigad and Englard, 1975). Reaction with glucose dehydrogenase has to be carried out with relatively high NAD(P)$^+$ concentrations (2–5 mM) to secure maximum reduction of glucose (1–100 mM). The NAD(P)H produced can be measured by standard spectrophotometric assay or with fluorimetric techniques to increase sensitivity. Also it can be coupled to bioluminescent reactions (Welch *et al.*, 1989) or used in biosensors (Section VII.D). It should be noted that to facilitate the formation of the β-anomer of glucose, which is the substrate in the dehydrogenase reaction, aldose 1-epimerase (EC 5.1.3.3) is usually included in the coupled assay system. Protocols for the use of these enzyme reagents for the determination of glucose in various types of biological preparations can be found in Arion *et al.* (1980), Bergmeyer *et al.* (1983, 1984), Vormbrock (1984), Alegre *et al.* (1988) and Stio *et al.* (1988). Cairns (1987) has developed a sensitive colorimetric glucose dehydrogenase- and formazan-forming system which can be used to determine glucose in sucrose hydrolysates (see also Section VII.B.1). Immobilised glucose dehydrogenase electrodes have also been described (see Section VII.D).

In addition to $NAD(P)^+$-dependent glucose dehydrogenase, other analogous glucose dehydrogenase systems (EC 1.1.99.10 and EC 1.1.99.17) which are flavoproteins or pyrroloquinoline quinone proteins, respectively, that oxidise glucose, and to some extent also other aldoses, but with the participation of quinones rather than $NAD(P)^+$ as electron acceptor, have been isolated mostly from bacteria (Ameyama, 1982; Matsushita and Ameyama, 1982; Geiger and Goerisch, 1986; Dokter *et al.*, 1986; Duin and Jongejan, 1989). The use of these enzymes as analytical reagents for glucose is made redundant by the availability of glucose oxidase, and also by their lack of substrate specificity.

2. D-Fructose

The most common enzymic assay method for D-fructose in sucrose hydrolysates is based on phosphorylation in the Mg^{2+}-ATP hexokinase reaction, isomerisation with phosphoglucose isomerase, and determination of the glucose 6-phosphate produced in the glucose 6-phosphate dehydrogenase reaction which can be carried out using a variety of techniques (see Section VII.B.1).

A specific enzyme, D-fructose 5-dehydrogenase (EC 1.1.99.11), which oxidises D-fructose is produced by several strains of *Acetobacter*. It can provide a reagent to determine fructose colorimetrically in the presence of a suitable redox dye, or electrochemically with hexacyanoferrate (III) as the redox acceptor (Ameyama *et al.*, 1981; Ameyama, 1982; Matsumoto *et al.*, 1986). This interesting enzyme is not available commercially and the long-term stability of the preparations is not very good.

Sorbitol (L-iditol) dehydrogenase (EC 1.1.1.14) can be used to convert fructose to sorbitol with the consumption of NADH. To pull the reaction beyond equilibrium and to achieve a complete reduction of fructose, the system must be coupled with an efficient NADH-regenerating enzymic reaction such as the lactate or alcohol dehydrogenases. The amount of the secondary product (pyruvate or acetaldehyde) that is formed corresponds to the amount of fructose reduced (Guilbault, 1976; Bergmeyer *et al.*, 1983, 1984).

The NAD^+ mannitol 1-phosphate dehydrogenase (EC 1.1.1.17) which is produced by various strains of bacteria (Klungsoyr, 1966; Horowitz, 1966; Novotny *et al.*, 1984) and by fungi (Jennings, 1984) can be used for the determination of fructose. For this purpose, fructose 6-phosphate (K_m about 0.5 mM) produced in the hexokinase reaction is reduced with NADH. The kinetic properties of this oxidation/reduction system favours an efficient fructose 6-phosphate reduction, and end-point determination can be achieved by use of a small excess of NADH/fructose 6-phosphate in the reaction mixture. Coupling the system to an efficient NAD^+-regeneration enzyme and providing the catalytic amount of NAD^+ can promote a cycle for the quantitative conversion of fructose 6-phosphate to mannitol 1-phosphate.

3. D-Galactose

(a) *Assay with D-galactose dehydrogenase* (EC 1.1.1.48). The use of the commercially available NAD$^+$-dependent enzyme provides the simplest reagent for galactose determination in disaccharide hydrolysates. The NADP$^+$-dependent dehydrogenase (EC 1.1.1.120) has not been used extensively for analytical purposes. A large number of protocols for this assay have been published (Wallenfels and Kurz, 1975; Sturgeon, 1980a; Maier and Kurtz, 1982; Bergmeyer *et al.*, 1983, 1984; Orfanos *et al.*, 1986). Most of these methods depend on a spectrophotometric determination of the NADH produced. Other more sensitive fluorimetric and automated techniques have also been described (Urbanowski *et al.*, 1980; Fujimura *et al.*, 1981: Kulski and Buehring, 1982). Coupled with a bacterial luciferase reaction, the NADH produced can be measured luminometrically, as has been described for glucose (Arthur *et al.*, 1989).

(b) *Assay of galactose with galactose oxidase* (EC 1.1.3.9). This Cu(II)-pyrroloquinoline quinone enzyme oxidises free galactose, as well as a large number of galactosides, with the production of H_2O_2. The reaction can be monitored by any of the assay systems designed for H_2O_2 analyses in the glucose oxidase reaction described above (see Section VII.B.1). The kinetic properties of galactose oxidase and its relative lack of specificity have restricted the use of galactose oxidase as an analytical reagent for galactose. However, it is a useful enzyme with which to obtain information about galactoside structure and for the introduction of various markers, such as tritium labelling, into galactosyl residues, disaccharides included (Avigad *et al.*, 1962; Avigad, 1978; Guilbault, 1976; Pazur and Dreher, 1980; Bergmeyer *et al.*, 1984; Hatton and Regoeczi, 1982; Ellis and Rand, 1987; Turner, 1988, and Worthington, 1988).

(c) *Galactokinase* (EC 2.7.1.6) This enzymic reaction is not a convenient choice for the routine assay of galactose. Nonetheless, it can be of some analytical use as a reaction to 'trap' small amounts of ^{14}C-labelled galactose, or of non-radioactive galactose when AT$_\gamma$^{32}P is used as the substrate. The product of the reaction, galactose 1-phosphate, can be adsorbed to small disks of DEAE paper (Whatman DE-81), or detected by other techniques of ion-exchange chromatography (Blume and Beutler, 1975; Wilson and Schell, 1982; Gross and Pharr, 1982).

4. Determination of alditols

Glycerol, inositol isomers and occasionally hexitols are released from various dimeric glycosides such as the floridosides, galactinol and others (Dey, 1980, 1985; Avigad, 1982). The simplest way to determine the polyol released is by using chromatographic procedures.

Since in some analytical methods for reducing disaccharides reduction by borohydride is employed as a first step, free D-sortibol (D-glucitol) or both D-sorbitol and D-mannitol will be released when reduced glycosyl-glucoses or glycosyl-fructoses are hydrolysed by acid or enzymically. In most situations the hexitols released can be easily determined by chromatography. In some cases it may be convenient to quantify them in an enzymic assay. NAD$^+$ sorbitol (L-iditol) dehydrogenase (EC 1.1.1.14) and NAD$^+$

mannitol dehydrogenase (EC 1.1.1.67), obtained from mammalian tissues or from various microorganisms, can be used for such determinations (Kersters and DeLey, 1966; Barnett, 1968; Malone *et al.*, 1980; Liao *et al.*, 1980; Liessing and McGinnis, 1982; O'Brien *et al.*, 1983; Bergmeyer *et al.*, 1984; Tani and Vongsuvanlert, 1987; Vongsuvanlert and Tani, 1988). It should be noted that the polyol dehydrogenases from various sources may vary in the degree of specificity for mannitol and sorbitol (Kulbe and Chmiel, 1988). The substrate specificity of each enzyme preparation should be verified carefully before it is used in analysis.

5. Miscellaneous. The NADPH-aldose reductase (EC 1.1.1.21) is a non-specific enzyme that can reduce many aldehydes and aldohexoses such as glucose, galactose, and xylose. The reaction is enhanced if coupled with a NADPH-regenerating system. The mammalian aldose reductase isolated from extrahepatic tissues, such as lens, muscle and red blood cells, can be used for this analysis (Wermuth and Von Wartburg, 1982; Branlant, 1982; Srivastava *et al.*, 1985; Cromlish and Flynn, 1983). NADPH-D-glucuronic acid reductase (EC 1.1.1.19) and *myo*-inositol oxygenase (EC 1.13.99.1) have been isolated from mammalian tissues (Reddy *et al.*, 1981). These enzymes provide useful reagents for the determination of glucuronic acid in aldobiuronic acid hydrolysates, and of inositol in galactinol hydrolysates. Specific D-xylose dehydrogenase (EC 1.1.1.175), D-aldohexose dehydrogenase which also acts on D-mannose (EC 1.1.1.119), glycerol dehydrogenase (EC 1.1.1.72), L-arabinose dehydrogenase (EC 1.1.1.46), as well as a specific fungal glycerol oxidase are all enzymes which can be employed for the determination of monomeric products in disaccharide hydrolysates. Description of these enzymes can be found in several sections of *Methods in Enzymology*, Vol. 41 (1975) and Vol. 89 (1982).

C. Enzymic Determination of Disaccharides using Non-hydrolytic Reactions

A selection of highly specific enzymes which act on disaccharide substrates can be used for their analysis. Some of these enzymes are commercially available; others can be relatively easily prepared. Among the reactions involved are disaccharide phosphorolysis, oxidation, and transglycosylation. When products of the reaction are phosphate esters, such as sugar phosphates or nucleotides, analysis can be accomplished by chromatographic procedures (see Section VI.C) or by coupling to a spectrophotometric enzyme system (see Section VII.B). Non-phosphorylated sugar products can be analysed by many of the chemical, chromatographic or enzymic assay methods described in this chapter. Detection and identification of the phosphate esters chromatographically separated from enzymic systems can be greatly facilitated by using radioisotopes. This can be achieved, for example, employing [^{32}P]orthophosphate, AT$_\gamma^{32}$P, UD^{32}P or [^{14}C]UDP as substrates, and by adding very small amounts of highly radioactive [^{14}C]- or [^{3}H]disaccharide to tag the non-radioactive glycobiose which is to be analysed.

1. Reversal of the sucrose synthase reaction

The ability of sucrose synthase to catalyse an easily reversible reaction (Cardini *et al.*, 1955) has been used extensively as one of the procedures for the determination of this enzyme activity. For this purpose, the assay is usually conducted at pH 6.5–7.0 in the presence of high sucrose concentrations (such as 0.2 M), and the UDP-dependent rate of sucrose cleavage is evaluated. The products of the reaction, UDP-glucose and fructose, can be determined by a variety of techniques, either in a two-step assay, or in a continuous system when coupled to an auxilliary enzyme reaction. Several of these methods, based on chemical or spectrophotometric analysis, have been described by Avigad (1964) and Avigad and Milner (1966), and variations of these procedures are found in most of the published studies on sucrose synthase (see Avigad, 1982; Hawker, 1985). Some useful detailed protocols have been presented by Delmer (1972), Su and Preiss (1978), Cheung and Suhadolnik (1979), Burrows and Cintron (1983), Salerno et al. (1979), Morell and Copeland (1985), Huber and Akazawa (1986), Hargreaves and ApRees (1988); Keller et al. (1988) and Moriguchi and Yamaki (1988).

Because of the simplicity and availability of the invertase-dependent methods and because of the high K_m of the disaccharide for the sucrose synthase, this enzyme has rarely been used for quantitative determination of sucrose. It has been useful, however, as a mechanism by which to synthesise or manipulate specifically radiolabelled sucrose or to convert it directly to sugar nucleotides such as UDP-glucose and ADP-glucose. For example, small quantities of [^{14}C]sucrose isolated in metabolic studies can be incubated in a UDP-sucrose synthetase system, and the UDP-[^{14}C]glucose readily isolated by simple chromatographic procedures (Delmer, 1972; Su and Preiss, 1978; see also Section VI.C).

2. Sucrose phosphorylase (EC 2.4.1.7)

This enzyme is produced by several bacteria such as *Leuconostoc mesenteroides* and *Pseudomonas saccharophila* (Doudoroff, 1961; Mieyal, 1972; Vandamme et al., 1987). Known for many years, and occasionally used for the identification of sucrose or for its specific ^{14}C-labelling (Hassid et al., 1951; Doudoroff, 1961; Avigad, 1964; Silverstein et al., 1967; Chassy and Krichevsky 1972), the interest in the analytical application of this enzyme has been revived recently since it is now commercially available (DeLaporte et al., 1982; Birnberg and Brenner, 1984; Stikkelman et al., 1985; Waldmann et al., 1986).

Incubation of sucrose in the presence of excess orthophosphate (usually 10 mM) and in the presence of sucrose phosphorylase will release α-D-glucose 1-phosphate and fructose. The reaction can be coupled with phosphoglucomutase and NAD(P)$^+$/glucose 6-phosphate dehydrogenase and measured spectrophotometrically or fluorimetrically. When the reaction is completed, the formation of NAD(P)H corresponds to the amount of sucrose consumed. Other variations of the assay can be based on the use of arsenate to replace the orthophosphate (Doudoroff, 1961). In this case final products will be glucose and fructose, and these can be assayed by any reagent used to determine sucrose via the invertase reaction, such as the versatile glucose oxidase detection systems (see Section VII.B.1). The development of sensitive HPLC for sugar phosphates also provides a useful procedure to analyse sucrose cleavage by sucrose phosphorylase (see Section VI.C). Using ^{32}Pi as a substrate will facilitate the analysis of the phosphorolytic cleavage of sucrose.

3. Other disaccharide phosphorylases

Several highly specific, reversible, and quite efficient (low K_m) diglucoside phosphorylases have been characterised. Since they all release glucose when they act on the disaccharide, coupling the reaction with glucose oxidase (in any variation as described in Section VII.B.1) provides a convenient assay for the disaccharide and 'pulls' the reaction from equilibrium toward complete cleavage. As in the case of sucrose phosphorylase, in the presence of arsenate, rather than phosphate, the net result of the reaction will be the appearance of two moles glucose per mole disaccharide. The reversibility of the phosphorylase reactions also provides a mechanism by which a radiolabelled disaccharide can be synthesised. Label can be in the glucosyl or the glucose (reducing) residue, depending on whether the aldose acceptor or the glucose 1-phosphate is the radioactive substrate.

It is surprising that relatively little research has been carried out on these enzymes and their application as analytical reagents.

(a) *Cellobiose* A bacterial cellobiose phosphorylase (EC 2.4.1.20) that produces α-D-glucose 1-phosphate from cellobiose is a highly specific Mg^{2+}-dependent enzyme (Doudoroff, 1961; Alexander, 1972; Sasaki *et al.*, 1983; Sasaki, 1988). In the reverse reaction, the enzyme can use α-D-glucose, α-2-deoxyglucose or D-xylose as the glucosyl acceptor to produce the β-glucosyl-aldose disaccharide. Ng and Zeikus (1986) used this enzyme to produce selectively ^{14}C-labelled cellobiose in either of its glucose residues.

(b) *Trehalose.* Trehalose phosphorylase (EC 2.4.1.64) which produces β-D-glucose 1-phosphate by phosphorolysis of α,α-trehalose has been found in *Euglena* (Marechal and Belcopitow, 1972; Miyatake *et al.*, 1984). The study of this enzyme and its use for trehalose determination has not been pursued in great detail. Production of the β- rather than the α-isomer of glucose 1-phosphate prevents coupling the reaction with phosphoglucomutase (EC 5.4.2.2). A specific phosphoglucomutase (EC 5.4.2.6) is probably responsible for conversion of the β-isomer to glucose 6-phosphate (Ben Zvi and Schramm, 1961).

(c) *Laminaribiose.* A phosphorolytic cleavage of (laminaribiose) 3-β-D-glucopyranosyl-D-glucose, which yields α-D-glucose 1-phosphate and glucose, is catalysed by a laminari-biose phosphorylase (EC 2.4.1.31) found in *Euglena* by Goldemberg *et al.*, (1966), in *Astasia* by Manners and Taylor (1967), and in *Ochromonas* by Albrecht and Kauss (1971). The enzyme is probably distinguished from a 1,3-β-oligoglucan phosphorylase (EC 2.4.1.30) by being more active on short chain oligoglucosides. This enzyme is useful for the identification of β-1–3-glucosides from β-glucoside oligosaccharides with different linkages, which appear as degradation products of plant β-glucans.

(d) *Maltose.* A highly specific maltose phosphorylase (EC 2.4.1.8) has been obtained from strains of *Neisseria* and *Lactobacilli* (Doudoroff, 1961; Wood and Rainbow, 1961; Kamogawa *et al.*, 1973; Martin and Russell, 1987). The substrate for this enzyme is the α-maltose isomer (Tsumuraye *et al.*, 1984) and the products of phosphorolysis are β-D-glucose 1-phosphate and glucose. By coupling with trehalose phosphorylase (see (b) above), maltose could be converted in high yields to trehalose in the presence of maltose phosphorylase and catalytic amounts of orthophosphate (Murao *et al.*, 1985). The maltose phosphorylase reaction can be easily monitored with a glucose oxidase system, with a glucokinase reaction (Kondo *et al.*, 1988), or with any other test for glucose. The β-glucose 1-phosphate is inert with yeast or muscle phosphoglucomutase. It can be hydrolysed with alkaline phosphatase, and alternatively if the maltose cleavage is carried out with arsenate, the disaccharide will be converted to glucose (Kamogawa *et al.*, 1974). The enzyme does not act on maltitol, isomaltose, trehalose, or sucrose. It provides an ideal system for specific maltose determination and hopefully will attract more attention in the future.

4. *Detection of sucrose with glycansucrases.*

The bacterial enzymes which use sucrose as the glycosyl donor for the formation of a specific homopolymeric glycan (Hehre, 1951) can be of some use for the detection of this disaccharide. Dextransucrase (EC 2.4.1.5) produces from sucrose an α-glucanan (dextran) and free fructose (Hehre, 1951; Robyt, 1980; Robyt and Martin, 1983; Nisizawa *et al.*, 1986; Mayer, 1987). Levansucrase (EC 2.4.1.10), produces a β-fructofuranan (levan) and free glucose (Hestrin *et al.*, 1956; Hestrin and Avigad, 1958; Dedonder, 1966; Tanaka *et al.*, 1981; Imai *et al.*, 1984). Since the dominant polymerisation reactions catalysed by these enzymes are accompanied by a small degree of sucrose inversion, the use of these systems for quantitation of sucrose based only on polymer synthesis is not simple. On the other hand, the formation of a very high molecular weight polysaccharide can be simply and sensitively determined by planar chromatography (for example, Feingold and Avigad, 1956; Nisizawa *et al.*, 1986; Miller and Robyt, 1986). Sensitivity of detection is increased if the sucrose analysed is ^{14}C-labelled. Levansucrase and dextransucrase can also catalyse low levels of transglycosylation reactions which lead to the formation of short chain oligosaccharides. This, however, starts to be noticeable only at relatively high substrate concentrations. In addition, levansucrase also catalyses fructosyl exchange between sucrose and glucose or other aldoses, making it possible to use the reaction for a specific radioactive labelling of the sucrose molecule or the synthesis of sucrose analogues (Hestrin and Avigad, 1958; Tanaka *et al.*, 1981; Cheetham *et al.*, 1989).

Amylosucrase (EC 2.4.1.4) is an enzyme produced by strains of the bacterium *Neisseria perflava* (Hehre *et al.*, 1949; Okada and Hehre, 1974; MacKenzie *et al.*, 1977) and uses sucrose as the substrate for the synthesis of α-1,4-glucan. Application of amylosucrase as an analytical reagent for sucrose has been limited (Avigad *et al.*, 1956; Feingold *et al.*, 1957; Okada and Hehre, 1974; Tao *et al.*, 1988). The enzyme catalyses reversible α-glucosyl transfer between sucrose or α-amylodextrins and [14C]fructose (Okada and Hehre, 1974).

5. *Miscellaneous transglycosylations with sucrose*

(a) *Detection of sucrose with sucrose/sucrose fructosyl transferase* (EC 2.4.1.99). This enzyme is found in many plant tissues which can produce fructofuranans, usually of the inulin type. It exhibits a transferase activity which can synthesise short chain oligofructosides from sucrose (reviewed by Pontis and Del Campillo, 1985; Wiemken *et al.*, 1986; Pollock and Chatterton, 1988). A most common product of this reaction is the trisaccharide isokestose (1[F]-fructosylsucrose; Kandler and Hopf, 1982; Shiomi, 1989). The action of this β-fructosyltransferase on sucrose, and the formation of typical oligosaccharides, can be followed by TLC or HPLC methods (Frehner *et al.*, 1984;

Shiomi, *et al.*, 1985; Wagner and Wiemken, 1987; Housley *et al.*, 1989; Incoll *et al.*, 1989). Cairns (1987) employed a coupled hexokinase/glucose 6-phosphate dehydrogenase system with phenazine methosulphate and a tetrazolium salt to follow the degree of sucrose cleavage in such fructosyltransferase reaction.

(b) *Conversion of sucrose to isomaltulose.* A very efficient and specific conversion of sucrose to isomaltulose (6-*O*-D-glucopyranosyl-D-fructose) as well as a small measure of trehalulose (1-*O*-α-D-glucopyranosyl-D-fructose) can be catalysed by several micro-organisms (Nakajima, 1988), such as *Erwinia rhapontici* (Cheetham *et al.*, 1982; Cheetham, 1984) and *Protaminobacter rubrum* (Weidenhagen and Lorenz, 1957; Munir *et al.*, 1987). Although the transglucosylase reaction responsible for this synthesis has not been characterised in great detail, it is of potential interest for future analytical applications.

6. Amylomaltase

This enzyme (EC 2.4.1.25) is defined today as a 1,4-α-D-glucan:1,4-α-D-glucan 4-α-D-glycosyltransferase (or the disproportionating D-enzyme) since it can catalyse glucosyl transfer not only from maltose, but also from longer α-1,4-gluco-oligosaccharides. The product of the reaction with maltose and particularly short chain maltodextrins as the substrate is long chain maltodextrins and free glucose, with equilibrium reached at about 60% conversion. Pulling the reaction with glucose oxidase will complete the cleavage of all the malto-oligosaccharides presented as the substrate. The enzyme was found first in *E. coli* (Monod and Torriani, 1950; Hassid *et al.*, 1951) where it is induced by growth on maltose. The enzyme is also produced by other bacteria strains such as *Pseudomonas*, *Streptococcus* and *Aeromonas* (Barker *et al.*, 1965; Wober, 1973; Kitahara, 1988). The reaction it catalyses had been studied in detail by Palmer *et al.* (1973) and by Szmelcman and Schwartz (1976). The enzyme was rarely applied as an analytical reagent for maltose and the potential for its broadened use for this specific purpose is questionable (Kitahara, 1988).

7. Disaccharide oxidoreductases

(a) *1-Oxidoreductases.* A limited number of enzymes that can modify disaccharides in an oxidation–reduction reaction have been identified and applied for use as analytical reagents.

Cellobiose dehydrogenase (quinone 1-oxidoreductase, EC 1.1.5.1) is produced by several fungi and can convert cellobiose, as well as lactose, to the corresponding glycobionic acid (Dekker, 1980, 1988; Coudray *et al.*, 1982; Sadana and Patil, 1985). For an assay in a two-step reaction, the glycobionic acid is oxidised by periodate to produce glyoxylic acid, which is assayed spectrophotometrically with NADH and lactate dehydrogenase (Ayers and Eriksson, 1982). The simpler assay for the cellobiose oxidase reaction is based on the photometric determination of a suitable quinone serving as the electron acceptor (Coudray *et al.*, 1982; Westermark and Eriksson 1988).

A very similar enzyme, cellobiose dehydrogenase (acceptor 1-dehydrogenase, EC 1.1.99.18) of fungal origin, can use several redox dyes as the electron acceptors (Dekker, 1980, 1988; Coudray *et al.*, 1982). Assay of cellobiose in the presence of glucose (but also of lactose and other short chain β-glucosides) can be monitored by the reduction of 2,4-dichlorophenolindophenol (Holm, 1986; Dekker, 1988; Sadana and Patil, 1988; Canevascini, 1988). These cellobiose oxidases are stable enzymes which seem to be very useful for the study of cellulose degradation and metabolism. A fungal cellobiose oxidase (EC 1.1.3.25) which utilises O_2 and yields aldobionic acid and H_2O_2 (or O_2) has been described by Ayers and Eriksson (1982) and Morpeth (1985). The three cellobiose oxidoreductase reagents described have high affinity to cellobiose and have been found to be helpful for monitoring the process of cellulose degradation by cellulolytic enzymes (Canevascini, 1985; Kelleher et al., 1987).

Several partially purified enzyme preparations defined as 'maltose dehydrogenase' from bacteria have been described. From *Corynebacterium* Kobayashi and Horikoshi (1980a,b) and Kobayashi et al. (1982) obtained NAD^+-dependent dehydrogenase that oxidised maltose and malto-oligosaccharides to the corresponding glucosyl-gluconates. Unfortunately for the analyst, this preparation also oxidised glucose, lactose, and cellobiose. Another preparation from the same cell was an NAD^+-dehydrogenase that oxidised glucose, gentiobiose, and cellobiose, but not maltose.

A maltose oxidoreductase (acceptor 1-dehydrogenase) was obtained from *Staphylococcus* (Ishikawa et al., 1985) and from *Serratia* (Kido et al., 1986). D-Glucose, D-galactose, and D-xylose were also efficiently oxidised by this enzyme preparation, suggesting that it belongs to the aldose dehydrogenase (EC 1.1.99.17) group of quinoproteins. The proton acceptor in the reaction could be phenazine-tetrazolium salt, but not $NAD(P)^+$. The usefulness of these enzymes for the analysis of maltose-containing solutions has yet to be evaluated.

The bacterial glucose dehydrogenase (acceptor) (EC 1.1.99.17) can oxidise maltose (as well as other aldobioses such as cellobiose and melibiose) at rates somewhat slower than that of glucose (Hauge, 1966; Ameyama, 1982; Dokter et al., 1986). It has also been claimed that $NAD(P)^+$ aldehyde reductase (EC 1.1.1.21), which is obtained from yeast and other microorganisms, can reduce disaccharides such as maltose and lactose, leading to the production of the corresponding glycosorbitol. The rate of this reaction is very slow, but in coupling with an efficient $NAD(P)^+$-regeneration system (such as the formate dehydrogenase), reduction of the disaccharides is accelerated significantly (Kulbe and Chmiel, 1988).

(b) *D-Glycoside 3-dehydrogenase* (EC 1.1.99.13). This oxidoreductase produced by strains of *Agrobacterium* and *Flavobacterium* converts a large number of glycosides such as maltose, lactose, melibiose and sucrose to the corresponding 3-ulosides (Fukui et al., 1963; Van Beeumen and DeLey, 1968, 1975a,b; Takeuchi et al., 1986, 1988). The oxidation reaction can be determined amperometrically or in combination with a suitable dye as the electron acceptor. This relatively non-specific enzymic reaction has been useful for the preparation of several D-allosyl analogues of the D-glucosyl substrates oxidised, and has also been used for the formation of 3-[^{3}H] D-glucosyl disaccharides.

(c) *Determination with galactose oxidase* (EC 1.1.3.9). Many galactosides with a terminal α- or β-galactosyl residue can be oxidised by this enzyme. Among these substrates are lactose, melibiose, melibi-itol, galactinol and floridoside (Avigad et al., 1962; Tressel and Kosman, 1982; Avigad, 1985). Because of its properties, assay of galactosides with this enzyme is not too sensitive and usually relies on measurement of first-order kinetics. However, the use of galactose oxidase has been helpful for the preparation of 6-^{3}H-labelled or fluorescent-tagged galactosyls in disaccharides and other carbohydrates (Avigad, 1967, 1985; Wilcheck and Bayer, 1987).

8. Disaccharide binding proteins

A periplasmic protein which has high affinity to maltose is produced by *E. coli* and other bacteria. Binding of [^{14}C]maltose between 0.25 and 75 μM can be measured by techniques of equilibrium dialysis (Kellermann and Fereuci, 1982; Reizer and Peter-kofsky, 1987; Bankaitis *et al.*, 1987).

There is extensive experimental evidence that sucrose-binding proteins which are involved in the transport of the disaccharide exist in some strains of bacteria, yeast, and in plant cell membranes (Hitz and Giaquinta, 1987). Characterisation of these proteins must await future research.

D. Electrochemical Biosensors for Disaccharides

Very useful and specific techniques for disaccharide analysis evolve from the construc-tion of specific electrodes. These methods are based on further developments of the standard glucose sensor which is a Clark oxygen electrode covered with immobilised glucose oxidase. This electrode is broadly used to measure glucose, particularly as an 'on-line' probe in automated analytical systems in clinical chemistry and in the food and fermentation industries. Detection of H_2O_2 (0.1–50 mM) produced by an equivalent amount of glucose is conveniently accomplished by potentiometric or polarographic techniques. In some experimental systems, the glucose oxidase electrode is coupled to a fluorimetric or luminometric assay system, thereby increasing the sensitivity of H_2O_2 detection several-fold (see Section VII.B.1). In addition to the more familiar sensors for H_2O and O_2 based on the Clark electrode, other sensors, based on electroactive mediators which shuttle electrons betweeen the enzyme reaction products and the electrode, have been developed. Among these are electrodes that can electrochemically measure the NAD(P)H produced in enzyme-catalysed reactions. For detailed reviews and descriptions of the rapidly developing biosensor techniques, particularly as applied for glucose and some related carbohydrates, see Guilbault, (1976, 1982, 1984), Schneider (1982), Bergmeyer *et al.* (1983), Lowe (1985), Gorton (1986), Frew and Hill (1987), Turner *et al.* (1987), Matsumoto *et al.* (1988), Taylor (1988), Ho (1988), Turner (1988), Shichiri *et al.* (1988) and Sternberg *et al.* (1988).

Some of the concepts which are used to develop commercial scale coenzyme–carbohydrate conversions in immobilised enzyme reactors could be applied for future development of analytical sensors (Kulbe and Chmiel, 1988). For the analysis of disaccharides, a simple and very rapid two-step assay is usually required. First, the disaccharide is completely hydrolysed, preferably by enzyme catalysis and then, one of the monosaccharides produced is measured with a suitable electrochemical sensor. Since glucose is the most common product released by hydrolysis of naturally occurring disaccharides, it can be monitored with the glucose oxidase electrode. Such sensor systems for disaccharides based on the use of specific hydrolases (e.g. invertase or α-glucosidase) have been described by several authors (Bergmeyer *et al.*, 1983; Guilbault,

1984; Scouten, 1987; Fukui *et al.*, 1987; Rahni *et al.*, 1987; Mascini *et al.*, 1988; Yoda, 1988; Matsumoto *et al.*, 1988, and Monsan and Combers, 1988). In a different arrangement, the glucose liberated from the disaccharide is oxidised by hexacyano-ferrate(III) (Mattos *et al.*, 1988) or in a hexokinase/glucose 6-phosphate dehydrogenase NAD(P)H-producing reaction (Schubert *et al.*, 1986; Gorton, 1986; Ho, 1988) and the products of the reaction are electrochemically measured. A similar principle was used to develop pen- or needle-sized glucose sensors (Hicks, 1985; Iwai and Akihama, 1986; Lomen *et al.*, 1986; Matthews *et al.*, 1987; Karube *et al.*, 1988; Turner and Swain, 1988; Jackson and Phillips, 1988; Yamasaki *et al.*, 1989), tools which could greatly facilitate the determination of sugars in very small volumes.

The sensitivity of detection can be increased many fold by co-immobilising a suitable bacterial luciferase and NAD(P)H oxidoreductase onto a glucose sensor constructed with an immobilised NAD(P)H-producing enzyme system, such as the NAD^+-glucose dehydrogenase, or with hexokinase/glucose 6-phosphate dehydrogenase. The NAD(P)H produced in amounts equimolar to the glucose present is measured by luminometry (Kricka *et al.*, 1984; Weinhausen and DeLuca, 1986; Weinhausen *et al.*, 1987; Girotti *et al.*, 1986; Ugarova *et al.*, 1988; Kricka, 1988). Chemiluminescence assay can also be employed to detect the H_2O_2 produced when using the glucose oxidase electrode. The most suitable reaction for this purpose is the luminol chemiluminescence reaction. It has been successfully applied to the detection of $0.1\ \mu\text{M}{-}1\ \mu\text{M}$ glucose (Pilosof and Nieman, 1982; Malavolti *et al.*, 1984; Koerner and Nieman, 1988; Petersson *et al.*, 1986). This method has been used effectively for the assay of low concentrations of disaccharides, such as sucrose and maltose (Swindelhurst and Nieman, 1988). Another way of determining glucose at the μM level, also based on the glucose oxidase sensor is by measuring the H_2O_2 in a peroxyoxalate chemiluminescence reaction (Abdel Latif and Guilbault, 1988).

For galactose in disaccharide hydrolysates, electrodes based on immobilised galactose oxidase have been described (Guilbault, 1984; Olsson *et al.*, 1985; Matsumoto *et al.*, 1988; Turner, 1988; Taylor, 1988; Kiba *et al.*, 1989). Though suitable for the determi-nation of free galactose, galactose oxidase also acts on galactosides (see Section VII.B.3). A galactose oxidase sensor could, therefore, be used to detect galactoside disaccharides without prior hydrolysis, but would be therefore less helpful for the determination of a mixture of galactosides and galactose.

Fructose, such as in a mixture produced by the hydrolysis of sucrose, could be measured with a sensor prepared by the immobilisation of fructose isomerase (EC 5.3.1.5) on a glucose electrode (Swindelhurst and Nieman, 1988), or by an immobilised fructose 5-dehydrogenase (EC 1.1.99.11) sensor using hexacyanoferrate(III) as the redox acceptor (Matsumoto *et al.*, 1986, 1988).

The availability of electrochemical detectors for the direct assay of NAD(P)H (Gorton, 1986; Schubert *et al.*, 1986; Ho, 1988) may provide sensors for the specific amperometric and potentiometric determination of any monosaccharide which can be oxidised by a coupled enzyme system that produces a reduced pyridine nucleotide. A technique similar to the assay of glucose (and fructose) by the hexokinase/glucose 6-phosphate dehydrogenase system described above has been used by Marko Varga (1987; 1989) for determining glucose in the NAD^+-glucose dehydrogenase (EC 1.1.1.118) sensor. A very effective sensor recently developed by McNeil *et al.* (1989) is based on the immobilisation of a thermostable NADH-oxidase (EC 1.11.1.1) which allows the amperometric determination of H_2O_2 produced from NADH oxidation. Coupling of this probe to any specific enzymic reaction that forms NADH provides a simple assay for the substrate which was oxidised by the dehydrogenase.

Employing the same basic techniques which were developed for monosaccharides, construction of several experimental sensors which contain immobilised disaccharide hydrolases have recently been described. Rahni *et al.* (1987) and Xu *et al.*, (1989) cross-linked invertase, mutarotase and glucose oxidase to an α-chymotrypsinogen-treated intestinal membrane (or onto other artificial membranes) and mounted on an oxygen electrode. Sucrose between 0.01 and 1.5 mM was amperometrically determined by initial rate or steady-state assays. This sensor was relatively stable and was used to measure sucrose in a variety of plant and food preparations. Matsumoto *et al.* (1988) also constructed an immobilised invertase reactor suitable for determination of 0.02–1.9 mM sucrose. A similar approach has been employed for the preparation of invertase bound to poly(ethylene-co-vinyl) alcohol hollow fibre membrane (Shiomi *et al.*, 1988). Immobilised disaccharidases for the assay of sucrose, lactose, or maltose have also been studied by several other investigators (Gestrelius and Mosbach, 1987; Monsan and Comber, 1988; Fukui *et al.*, 1987; Scouten, 1987; Ellis and Rand, 1987; Swindelhurst and Nieman, 1988). Yoda *et al.* (1988) devised a multi-enzyme electrode system which contained an anti-interference layer specifically constructed for the determination of sucrose and other α-glucosides such as maltose and starch, in the presence of excess free glucose.

The long-term stability of the disaccharide sensors described to date is not sufficient to provide a stable product suitable for commercial marketing. The ingenious sensor technology which is being actively explored and developed for sugars, and particularly for disaccharide analysis, has an attractive and promising future.

ACKNOWLEDGEMENTS

I thank Dr David S. Feingold for reading the manuscript and for his valuable comments, and Miss Karen Goudie for her skilful and attentive typing.

REFERENCES

Abbou, M. and Sioufi, A. M. (1987). *J. Liq. Chromatogr.* **10**, 95–106.
Abdel-Latif, M. S. and Guilbault, G. G. (1988). *Anal. Chem.* **60**, 2671–2674.
Adam, S. and Jennings, W. G. (1975). *J. Chromatogr.* **115**, 218–221.
Agrawal, K. M. L. and Bahl, Om. P. (1972). *Methods Enzymol.* **28**, 720–728.
Aitzetmüller, K. (1978). *J. Chromatogr.* **156**, 354–358.
Aitzetmüller, K. (1980). *Chromatographia* **13**, 432–436.
Akazawa, T. and Okamoto, K. (1980). *In* "The Biochemistry of Plants", Vol. 3 (J. Preiss, ed.), pp. 199–220. Academic Press, New York.
Albrecht, G. J. and Kauss, H. (1971). *Phytochemistry* **10**, 1293–1298.
Alegre, M., Cuidad, C. J., Fillat, C. and Guinovart, J. J. (1988). *Anal. Biochem.* **173**, 185–189.
Alexander, J. K. (1972). *Methods Enzymol.* **28**, 944–948.
Alpenfels, W. F. (1981). *Anal. Biochem.* **114**, 153–157.
Altman, F. P. (1976). *Progr. Histochem. Cytochem.* **9**, 1–51.
Ameyama, M. (1982). *Methods Enzymol.* **89**, 20–29.

Ameyama, M., Shinagawa, E., Matsushita, K. and Adachi, O. (1981). *J. Bacteriol.* **145**, 814–823.
Anderson, J. M. (1986). *Anal. Biochem.* **152**, 146–153.
Anderson, K. L. and Dahms, A. S. (1975). *Methods Enzymol.* **41**, 147–150.
Andrews, T. J. and Abel, K. M. (1979). *Plant Physiol.* **63**, 650–656.
ApRees, T. (1988). *In* "The Biochemistry of Plants, Carbohydrates", Vol. 14 (J. Preiss, ed.), pp. 1–33. Academic Press, San Diego.
Araujo, P. S., Panek, A. C., Ferreira, R. and Panek, A. D. (1989). *Anal. Biochem.* **176**, 432–436.
Arion, W., Lange, A., Wells, H. E. and Ballas, L. M. (1980). *J. Biol. Chem.* **255**, 10396–10406.
Armstrong, E. F. (1924). "Carbohydrates and the Glucides". Longman, Green & Co., London.
Arthur, P. G., Kent, J. C. and Hartmann, P. E. (1989). *Anal. Biochem.* **176**, 449–456.
Artiss, J., Thibert, R. J., McIntosh, J. M. and Zak, B. (1981). *Microchem. J.* **26**, 487–505.
Asensio, C., Avigad, G. and Horecker, B. L. (1963). *Arch. Biochem. Biophys.* **103**, 299–309.
Aspinall, G. O., Percival, E., Rees, D. A. and Rennie, M. (1967). *In* "Rodd's Chemistry of Carbon Compounds", Vol. 1F (S. Coffey, ed.), pp. 596–631. Elsevier, Amsterdam.
Attfield, P. V. (1987). *FEBS Lett.* **225**, 259–263.
Au, J. L.-S., Su, M.-H. and Wientjes, M. G. (1989). *Clin. Chem.* **35**, 48–51.
Aurand, L. W., Woods, A. E. and Wells, M. R. (1987). "Food Composition and Analysis". Avi Books, Van Nostrand Reinhold, New York.
Auses, J. P., Cook, S. L. and Maloy, J. T. (1975). *Anal. Chem.* **244**, 244–249.
Avigad, G. (1958). *Bull. Res. Council Isr.* **7A**, 112.
Avigad, G. (1964). *J. Biol. Chem.* **239**, 3613–3618.
Avigad, G. (1967). *Carbohydr. Res.* **3**, 430–434.
Avigad, G. (1969). *Carbohydr. Res.* **11**, 119–123.
Avigad, G. (1975a). *Methods Enzymol.* **41**, 27–29.
Avigad, G. (1975b). *Methods Enzymol.* **41**, 29–31.
Avigad, G. (1976). *J. Carbohydr. Nucleosides. Nucleotides.* **3**, 307–313.
Avigad, G. (1977). *J. Chromatogr.* **139**, 343–347.
Avigad, G. (1978). *Anal. Biochem.* **86**, 470–476.
Avigad, G. (1979a). *Anal. Chim. Acta.* **111**, 315–319.
Avigad, G. (1979b). *Biochim. Biophys. Acta.* **571**, 171–174.
Avigad, G. (1982). *In* "Encyclopedia of Plant Physiology, New Series, Carbohydrates I" (F. A. Loewus, and W. Tanner, eds), Vol. 13A, pp. 217–347. Springer, Berlin and Heidelberg.
Avigad, G. (1983). *Anal. Biochem.* **134**, 499–504.
Avigad, G. (1984). *Anal. Lett.* **17B**, 371–384.
Avigad, G. (1985). *Arch. Biochem. Biophys.* **239**, 531–537.
Avigad, G. and Bohrer, P. J. (1984). *Biochim. Biophys. Acta* **798**, 317–324.
Avigad, G. and England, S. (1975). *Methods Enzymol.* **41**, 142–147.
Avigad, G. and Milner, Y. (1966). *Methods Enzymol.* **8**, 341–345.
Avigad, G., Feingold, D. S. and Hestrin, S. (1956). *Biochim. Biophys. Acta.* **20**, 129–134.
Avigad, G., Zelikson, R. and Hestrin, S. (1961). *Biochem. J.* **80**, 57–61.
Avigad, G., Amaral, D., Asensio, C. and Horecker, B. L. (1962). *J. Biol. Chem.* **237**, 2736–2743.
Avigad, G., Ziv, O. and Neufeld, E. (1965). *Biochem. J.* **97**, 715–722.
Ayers, A. R. and Eriksson, K.-E. (1982). *Methods Enzymol.* **89**, 129–135.
Baer, H. H., Mekarska, M. and Boucher, F. (1985). *Carbohydr. Res.* **136**, 335–345.
Baker, J. K., Skelton, R. E. and Ma, C. Y. (1979). *J. Chromatogr.* **168**, 417–427.
Bankaitis, V. A., Altman, E. and Emr, S. D. (1987). *In* "Bacterial Outer Membranes as Model Systems" (M. Inouye, ed.), pp. 75–116. John Wiley and Sons, New York.
Barber, G. A. (1985). *Plant Physiol.* **79**, 1127–1128.
Barker, S. A. and Farisi, M. A. (1965). *Carbohydr. Res.* **1**, 97–105.
Barker, R., Nuñez, H. A., Rosevear, P. and Serianni, A. S. (1982). *Methods Enzymol.* **83**, 58–69.
Barnett, J. A. (1968). *J. Gen. Microbiol.* **52**, 131–159.
Barrie, A. and Lemley, M. (1989). *Amer. Lab.* **21**, 54–63.
Barth, H. G., Barber, W. E., Lochmuller, C. H., Majors, R. E. and Regnier, F. E. (1988). *Anal. Chem.* **60**, 387R–435R.
Bauman, H., Jansson, P. E. and Kenne, L. (1988). *J. Chem. Soc. Perkin Trans. I.* 209–217.
Bax, A., Freeman, R., Frenkiel, T. A. and Levitt, M. H. (1981). *J. Magn. Reson.* **43**, 478–483.

Belliardo, F., Buffa, M., Patetta, A. and Manino, A. (1979). *Carbohydr. Res.* **71**, 335–338.
BeMiller, J. N. (1967). *Adv. Carbohydr. Chem.* **22**, 25–108.
Bendiak, B., Orr, J., Brockhausen, I., Vella, G. and Phoebe, C. (1988). *Anal. Biochem.* **175**, 96–105.
Ben Zvi, R. and Schramm, M. (1961). *J. Biol. Chem.* **236**, 2186–2189.
Bergel, A., Shouppe, J. and Comtat, M. (1989). *Anal. Biochem.* **179**, 382–388.
Bergmeyer, H. U., Bergermeyer, J. and Grassl, M. (1983). "Methods of Enzymatic Analysis", 3rd edn, Vol. 1, Fundamentals; Vol. 3, Oxidoreductases. Verlag Chemie, Weinheim and New York.
Bergmeyer, H. U., Bergmeyer, J. and Grassl, M. (1984). "Methods of Enzymatic Analysis", 3rd edn, Vol. 6, Carbohydrates. Verlag Chemie, Weinheim and New York.
Bernfield, P. (1955). *Methods Enzymol.* **1**, 149–158.
Bhumiratana, A., Anderson, A. L. and Costilow, R. N. (1974). *J. Bacteriol.* **119**, 484–493.
Bierman, C. J. and McGinnis, G. D. (1988). "Analysis of Carbohydrates by GLC and MS". CRC Press, Boca Raton, FL.
Binder, T. P. and Robyt, J. F. (1984). *Carbohydr. Res.* **132**, 173–177.
Binder, T. P. and Robyt, J. F. (1985). *Carbohydr. Res.* **140**, 9–20.
Binder, T. P. and Robyt, J. F. (1986). *Carbohydr. Res.* **147**, 149–154.
Birnberg, P. R. and Brenner, M. L. (1984). *Anal. Biochem.* **142**, 556–561.
Bitter, T. and Muir, H. M. (1962). *Anal. Biochem.* **4**, 330–334.
Black, C. C., Mustardy, L., Sung, S. S., Kormanik, P. P., Xu, D. P. and Paz, N. (1987a). *Physiol. Plant* **69**, 387–394.
Black, Jr. C. C., Xu, D.-P., Mustardy, L., Paz, N. and Kormanik, P. P. (1987b). *In* "Phosphate Metabolism and Cellular Regulation in Microorganisms" (A. Torriani-Gorini, F. G. Rothman, S. Silver, A. Wright, and E. Yagil, eds), pp. 264–268. American Soc. Microbiol., Washington, D.C.
Blake, D. A. and McLean, N. V. (1989). *Anal. Biochem.* **177**, 156–160.
Blakeney, A. B. and Stone, B. A. (1985). *Carbohydr. Res.* **140**, 319–324.
Blakeney, A. B., Harris, P. J., Henry, R. J. and Stone, B. A. (1983). *Carbohydr. Res.* **113**, 291–299.
Blanken, W. M., Bergh, M. L. E., Koppen, P. L. and VanDenEijnden, D H. (1985). *Anal. Biochem.* **145**, 322–330.
Block, R. J., Durrum, E. L. and Zweig, G. (1958). "Paper Chromatography and Paper Electrophoresis", 2nd edn. Academic Press, New York.
Blume, K. G. and Beutler, E. (1975). *Methods Enzymol.* **42**, 47–53.
Blumenkrantz, H. and Asboe-Hansen, G. (1973). *Methods Enzymol.* **54**, 484–489.
Blunden, C. A. and Wilson, M. F. (1985). *Anal. Biochem.* **151**, 403–408.
Bock, K., Pedersen, C. and Pedersen, H. (1984). *Adv. Carbohydr. Chem. Biochem.* **42**, 193–225.
Boehringer-Mannheim, GmbH. (1986). "Methods of Biochemical Analysis and Food Analysis using Test Combinations". Boehringer-Mannheim Biochemica, Mannheim, FRG.
Bonn, G. (1985). *J. Chromatogr.* **322**, 411–424.
Bonn, G., Grunwald, M., Scherz, H. and Bobleter, O. (1986). *J. Chromatogr.* **370**, 485–493.
Bonner, W. M. and Stedman, J. P. (1978). *Anal. Biochem.* **89**, 247–256.
Boos, W., Ehmann, U., Bremer, E., Middendorf, A. and Postma, P. (1987). *J. Biol. Chem.* **262**, 13212–13218.
Boratynski, J. (1984). *Anal. Biochem.* **137**, 528–532.
Bornhop, D. J., Nolan, T. G. and Dovichi, N. J. (1987). *J. Chromatogr.* **384**, 181–187.
Bostick, D. T. and Hercules, D. M. (1975). *Anal. Chem.* **47**, 447–452.
Bowman, W. D., Hubick, K. T., Von Caemmerer, S. and Farquhar, G. D. (1989). *Plant Physiol.* **90**, 162–166.
Branlant, G. (1982). *Eur. J. Biochem.* **129**, 99–104.
Browne, C. A. (1912). "A Handbook of Sugar Analysis". John Wiley and Sons, New York.
Browne, C. A. and Zerban, F. W. (1941). "Physical and Chemical Methods of Sugar Analysis". John Wiley and Sons, New York.
Brugnioli, E., Hubick, K. T., Von Caemmerer, S., Wong, S. C. and Farquhar, G. D. (1988). *Plant Physiol.* **88**, 1418–1424.
Bruut, K. and Hokse, H. (1983). *J. Chromatogr.* **268**, 131–137.

Bucke, C. (1982). *In* "Developments in Food Carbohydrate", Vol. 3 (C. K. Lee and M. G. Lindley, eds), pp. 49–80. Applied Science Publishers, London.
Bugbee, W. M. (1984). *Can. J. Microbiol.* **30**, 1326–1329.
Buhmiratana, A., Anderson, R. L. and Costilow, R. N. (1974). *J. Bacteriol.* **119**, 484–493.
Burrows, R. B. and Cintron, C. (1983). *Anal. Biochem.* **130**, 376–378.
Cabib, E. and Leloir, L. F. (1958). *J. Biol. Chem.* **231**, 259–275.
Cairns, A. J. (1987). *Anal. Biochem.* **167**, 270–278.
Campbell, A. K. (1988). "Chemiluminescence, Principles and Applications in Biology and Medicine". Verlag-Chemie and Ellis Horwood Ltd., Weinheim FDR, and Chichester, England.
Campbell, A. K., Hallet, M. B. and Weeks, I. (1985). *Methods Biochem. Anal.* **31**, 317–416.
Canevascini, G. (1985). *Anal. Biochem.* **147**, 419–427.
Canevascini, G. (1988). *Methods Enzymol.* **160**, 443–448.
Capaldi, D. J. and Taylor, K. F. (1983). *Anal. Biochem.* **129**, 329–336.
Card, P. J. and Hitz, W. D. (1984). *J. Am. Chem. Soc.* **106**, 5348–5550.
Card, P. J., Hitz, W. D. and Ripp, K. G. (1986). *J. Am. Chem. Soc.* **108**, 158–161.
Cardemil, L. and Varner, J. E. (1984). *Plant Physiol.* **76**, 1047–1054.
Cardini, C. E., Leloir, L. F. and Chiriboga, J. (1955). *J. Biol. Chem.* **214**, 149–155.
Carnal, N. W. and Black, Jr. C. C. (1979). *Biochem. Biophys. Res. Commun.* **86**, 20–26.
Carnal, N. W. and Black, Jr. C. C. (1983). *Plant Physiol.* **71**, 150–155.
Carpita, N. C. and Shea, E. M. (1988). *In* "Analysis of Carbohydrates by GLC and MS" (C. J. Bierman, and G. D. McGinnis, eds), pp. 157–216. CRC Press, Inc., Boca Raton, FL.
Carr, S. A., Reinhold, V. N., Green, B. N. and Hass, J. R. (1985). *Biomed. Mass Spectrom.* **12**, 288–295.
Cerning-Beroard, J. (1975). *Cereal Chem.* **52**, 431–438.
Chaplin, M. F. (1982). *Anal. Biochem.* **123**, 336–341.
Chaplin, M. F. and Kennedy, J. F. (1986). "Carbohydrate Analysis, A Practical Approach". IRL Press, Oxford.
Chassy, B. M. and Krichevsky, M. I. (1972). *Anal. Biochem.* **49**, 232–239.
Chassy, B. M. and Porter, E. V. (1982). *Methods Enzymol.* **90**, 556–559.
Chaves Das Neves, H. J. and Riscado, A. M. V. (1986). *J. Chromatogr.* **367**, 135–143.
Chaves Das Neves, H. J., Bayer, E., Blos, G. and Frank, H. (1982). *Carbohydr. Res.* **99**, 70–74.
Cheetham, P. S. J. (1982). *In* "Developments in Food Carbohydrate" (C. K. Lee and M. G. Lindley, eds), Vol. 3, pp. 107–140. Applied Science Publishers, London.
Cheetham, P. S. J. (1984). *Biochem. J.* **220**, 213–220.
Cheetham, P. S. J., Imber, C. E. and Isherwood, J. (1982). *Nature (London)* **299**, 628–631.
Cheetham, P. S. J., Hacking, A. J. and Vlitos, M. (1989). *Enzyme Microb. Technol.* **11**, 212–219.
Chen, C.-C. and McGinnis, G. D. (1983). *Carbohydr. Res.* **122**, 322–326.
Chen, C.-C. and Whistler, R. L. (1983). *Carbohydr. Res.* **117**, 318–321.
Chen, C.-C., Whistler, R. L. and Daniel, J. R. (1983). *Carbohydr. Res.* **117**, 318–321.
Chen, C.-C., Guo, W. J. and Isselbacher, K. J. (1987). *Biochem. J.* **247**, 715–724.
Chen, L.-M., Yet, M.-G. and Shao, M. C. (1988). *FASEB J.* **2**, 2819–2824.
Cheng, Y. F. and Dovichi, N. J. (1988). *Science* **242**, 562–564.
Cheung, C. P. and Suhadolnik, R. J. (1979). *Plant Physiol.* **63**, 146–148.
Chi, M. M. Y., Lowry, C. V. and Lowry, O. H. (1978). *Anal. Biochem.* **89**, 119–129.
Chiba, S. and Minamiura, N. (1988). *In* "Handbook of Amylases and Related Enzymes" (The Amylase Research Society of Japan, ed.), pp. 104–116. Pergamon Press, Oxford.
Chiba, S., Kimura, A. and Matsui, H. (1983). *Agric. Biol. Chem.* **47**, 1741–1746.
Chiba, S., Brewer, C. F., Okada, G., Matsui, H. and Hehre, E. J. (1988). *Biochemistry* **27**, 1564–1569.
Churms, S. C. (1982). "CRC Handbook of Chromatography: Carbohydrates" (G. Zweig and J. Sherma eds), Vol. I. CRC Press, Boca Raton, FL.
Churms, S. C. (1983). *In* "Chromatography, Part B; Applications" (E. Heftman, ed.), *J. Chromatogr.* Library, Vol 22B, pp. 223–286. Elsevier, Amsterdam.
Ciucanu, I. and Kerek, F. (1984). *Carbohydr. Res.* **131**, 209–217.
Clamp, J. R. (1977). *Biochem. Soc. Trans.* **5**, 1693–1695.

Cole, E., Reinhold, V. N. and Carr, S. A. (1985). *Carbohydr. Res.* **139**, 1–11.
Collins, P. M. (ed.) (1987). "Carbohydrates". Chapman and Hall, London.
Conrad, H. E. (1976). *Methods Carbohydr. Anal.* **1**, 71–75.
Costello, L. R., Bassham, J. A. and Calvin, M. (1982). *Plant Physiol.* **69**, 77–82.
Coudray, M. R., Canevascini, G. and Meier, H. (1982). *Biochem. J.* **203**, 277–284.
Courtois, J. E. and Petek, F. (1966). *Methods Enzymol.* **8**, 565–571.
Cowie, G. L. and Hedges, J. I. (1984). *Anal. Chem.* **56**, 497–504.
Cox, C., Camus, P., Buret, J. and Duvivier, J. (1982). *Anal. Biochem.* **119**, 185–193.
Coxon, B. (1980). *In* "Developments in Food Carbohydrates", Vol. 2 (C. K. Lee, ed.), pp. 351–390. Applied Science Publishers, London.
Cromlish, J. A. and Flynn, T. G. (1983). *J. Biol. Chem.* **258**, 3416–3424.
Crowther, R. S. and Wetmore, R. F. (1987). *Anal. Biochem.* **163**, 170–174.
Czapek, F. (1913). "Biochemie der Pflanzen". Gustav Fischer, Jena.
Czeke, C. and Buchanan, B. B. (1983). *FEBS Lett.* **155**, 139–142.
Czeke, C., Weeden, N. F., Buchanan, B. B. and Uyeda, K. (1982). *Proc. Natl. Acad. Sci. USA* **79**, 4322–4326.
Czeke, C., Balogh, A., Wong, J. H., Buchanan, B. B., Stitt, M., Herzog, B. and Heldt, H. W. (1984). *TIBS* **9**, 533–535.
D'Accorso, N. B., Vasquez, I. M. and Thiel, M. E. (1986). *Carbohydr. Res.* **156**, 207–213.
Damon, S., Hewitt, J. H., Nieder, M. and Bennett, A. B. (1988). *Plant Physiol.* **87**, 731–736.
Daniel, P. F., De Fendis, D. F., Lott, I. T. and McLuer, R. H. (1981). *Carbohydr. Res.* **97**, 161–180.
Dedonder, R. (1966). *Methods Enzymol.* **8**, 500–505.
Deerr, N. (1949–1950). "The History of Sugar", 2 vols. Chapman and Hall, London.
Defaye, J., Driquez, H. and Henrissat, B. (1980). *In* "Mechanisms of Saccharide Polymerization and Depolymerization" (J. J. Marshall, ed.), pp. 331–353. Academic Press, New York.
Defaye, J., Driguez, H. and Henrissat, B. (1983). *Carbohydr. Res.* **124**, 262–273.
Defaye, J., Driquez, H., Ponce, S. and Chambert, R. (1984). *Carbohydr. Res.* **130**, 299–315.
Dekker, R. F. H. (1980). *J. Gen. Microbiol.* **120**, 309–316.
Dekker, R. F. H. (1988). *Methods Enzymol.* **160**, 454–463.
DeLaporte, A., DeValk, L. and Vandamme, E. J. (1982). *Antonie van Leuwenhoek* **48**, 516–519.
Dell, A. (1987). *Adv. Carbohydr. Chem. Biochem.* **45**, 19–72.
Dellamora-Ortiz, G. M., Ortiz, C. H. D., Maia, J. C. C. and Panek, A. D. (1986). *Arch. Biochem. Biophys.* **251**, 205–214.
Delmer, D. P. (1972). *Plant Physiol.* **50**, 469–472.
Den Hollander, J. A. and Shulman, R. G. (1981). *Tetrahedron* **39**, 3529–3538.
Deslauriers, R., Jarrel, H. C., Byrd, R. A. and Smith, I. C. P. (1980). *FEBS Lett.* **118**, 185–190.
Dethy, J. M., Callaert-Deveen, B., Janssens, M. and Lenaers, A. (1984). *Anal. Biochem.* **143**, 119–124.
Dey, P. M. (1980). *Adv. Carbohydr. Chem. Biochem.* **37**, 284–372.
Dey, P. M. (1985). *In* "Biochemistry of Storage Carbohydrates in Green Plants" (P. M. Dey and R. A. Dixon, eds), pp. 85–129. Academic Press, London.
Dey, P. M. and Dixon, R. A. (eds) (1985). "Biochemistry of Storage Carbohydrates in Green Plants" Academic Press, Inc., London.
Dickson, R. E. (1979). *Plant Physiol.* **45**, 480–488.
Dijkema, C., Kester, H. C. M. and Visser, J. (1985). *Proc. Natl. Acad. Sci. USA* **82**, 14–18.
Dill, K., Berman, E. and Pavia, A. A. (1985). *Adv. Carbohdyr. Chem. Biochem.* **43**, 1–49.
Doehlert, D. C., Kuo, T. M. and Felker, F. C. (1988). *Plant Physiol.* **86**, 1013–1019.
Dokter, P., Frank, Jr. J. and Duine, J. A. (1986). *Biochem. J.* **239**, 163–167.
Doner, L. W. (1988). *Methods Enzymol.* **160**, 176–180.
Doner, L. W. and Bills, D. D. (1982). *J. Assoc. Off. Anal. Chem.* **65**, 608–615.
Doner, L. W., White, J. W. Jr. and Phillips, J. G. (1979). *J. Assoc. Off. Anal. Chem.* **62**, 182–189.
Doner, L. W., Ajie, H. O., Sternberg, L. daS. L., Milburn, J. M., DeNiro, M. J. and Hicks, K. B. (1987). *J. Agric. Food Chem.* **35**, 610–612.
Doudoroff, M. (1961). "The Enzymes", 2nd edn, Vol. 5 (P. D. Boyer, H. Lardy and K. Myrbäck, eds), pp. 229–236. Academic Press, New York.

Douglas, A. E. (1988). *In* "Biochemistry of the Algae and Cyanobacteria" (L. J. Rogers and J. R. Gallon, eds), pp. 297–309. Clarendon Press, Oxford.

Dubois, M., Gilles, K. A., Hamilton, J. K., Rebers, D. A. and Smith, F. (1956). *Anal. Chem.* **28**, 350–356.

Duin, J. A. and Jongejan, J. A. (1989). *Ann. Rev. Biochem.* **58**, 403–426.

Dunmire, D. L. and Otto, S. E. (1979). *J. Assoc. Off. Anal. Chem.* **62**, 176–185.

Dutton, J. V., Carruthers, A. and Oldfield, J. F. T. (1961). *Biochem. J.* **82**, 266–272.

Dygert, S., Li, L. H., Florida, D. and Thoma, J. A. (1965). *Anal. Biochem.* **13**, 367–374.

Eades, D. M., Williamson, J. R. and Sherman, W. R. (1989). *J. Chromatogr.* **490**, 1–8.

Edwards, G. E. and Ku, M. S. B. (1987). *In* "The Biochemistry of Plants" (M. D. Hatch and N. K. Boardman, eds), Vol. 10, pp. 275–325. Academic Press, San Diego.

Edwards, G. E. and Walker, D. (1983). "C_3, C_4: Mechanisms and Cellular and Environmental Regulation of Photosynthesis". University of California Press, Berkeley, CA.

Effront, J. (1902). "Enzymes and Their Applications". John Wiley and Sons, New York.

Eggert, F. M. and Jones, M. (1985). *J. Chromatogr.* **333**, 123–131.

Eklund, S. H. and Robyt, J. F. (1988). *Carbohydr. Res.* **178**, 253–258.

Elbein, A. D. (1974). *Adv. Carbohydr. Chem. Biochem.* **30**, 227–256.

ElKhadem, H. S. (1988). "Carbohydrate Chemistry, Monosaccharides and their Oligomers". Academic Press, San Diego.

Ellis, P. C. and Rand, A. G. R. (1987). *J. Assoc. Off. Anal. Chem.* **70**, 1063–1068.

Ellwood, P. C., Reid, W. K., Marcell, P. D., Allen, R. A. and Kohouse, J. F. (1988). *Anal. Biochem.* **175**, 202–211.

Englmaier, P. (1986). *Z. Anal. Chem.* **324**, 338–339.

Ericsson, A., Hansen, J. and Delgaard, L. (1978). *Anal. Biochem.* **86**, 552–560.

Escott, R. E. A. and Taylor, A. F. (1985). *J. High Resolut. Chromatogr. Commun.* **8**, 290–292.

Esteban, N. V., Liberato, D. J., Sidburg, J. B. and Yergey, A. L. (1987). *Anal. Chem.* **59**, 1674–1677.

Farquhar, G. D., Hubick, K. T., Condon, A. G. and Richards, R. A. (1988). *In* "Application of Stable Isotope Ratios to Ecological Research" (P. W. Rundel, J. R. Ehleringer and K. A. Nagy, eds), pp. 21–40. Springer, Heidelberg.

Farquhar, G. D., Ehleringer, J. R. and Hubick, K. T. (1989). *Ann. Rev. Plant Physiol.* **40**, 503–507.

Feingold, D. S. and Avigad, G. (1956). *Biochim. Biophys. Acta* **22**, 196–197.

Feingold, D. S., Avigad, G. and Hestrin, S. (1956). *Biochem. J.* **64**, 351–361.

Feingold, D. S., Avigad, G. and Hestrin, S. (1957). *J. Biol. Chem.* **224**, 295–307.

Femia, R. A. and Weinberger, R. (1987). *J. Chromatogr.* **402**, 127–134.

Feraudi, M., Gärtner, C., Kolb, J. and Weicker, H. (1983). *J. Clin. Chem. Clin. Biochem.* **21**, 193–197.

Ferguson, J. E., Dickson, D. B. and Rhodes, A. M. (1979). *Plant Physiol.* **63**, 416–420.

Finden, D. A. S., Fysh, R. R. and White, P. C. (1985). *J. Chromatogr.* **342**, 179–185.

Forsythe, K. L. and Feather, M. S. (1989). *Carbohydr. Res.* **185**, 315–319.

Foster, A. B. (1962). *Methods in Carbohydr. Chem.* **1**, 51–58.

Fournet, B., Strecker, G., Leroy, Y. and Montreuil, J. (1981). *Anal. Biochem.* **116**, 489–502.

Frank, H., Chaves Das Neves, H. J. and Bayer, E. (1981). *J. Chromatogr.* **207**, 213–220.

Franklin, G. D. (1985). *Am. Lab.* **6**, 65–69.

Frehner, M., Keller, F. and Wiemken, A. (1984). *J. Plant Physiol.* **116**, 198–208.

Frew, J. E. and Hill, H. A. O. (1987). *Anal. Chem.* **59**, 933A–944A.

Friedman, S. (1966). *Methods Enzymol.* **8**, 600–603.

Fruton, J. S. (1972). "Molecules and Life". Wiley-Interscience, New York.

Fujimoto, H., Nishida, H. and Ajisaka, K. (1988). *Agric. Biol. Chem.* **52**, 1345–1351.

Fujimura, Y., Ishii, S., Kawamura, M. and Naruse, H. (1981). *Anal. Biochem.* **117**, 187–195.

Fukui, S., Hochster, R. M., Durbin, R., Grebner, E. E. and Feingold, D. S. (1963). *Bull. Res. Coun. Isr. Sec. IIa.* **4**, 262–268.

Fukui, S., Sonomoto, K. and Tanaka, A. (1987). *Methods Enzymol.* **135**, 230–252.

Gadd, G. M., Chalmers, K. and Reed, R. H. (1987). *FEMS Microbiol. Lett.* **48**, 249–254.

Galensa, R. (1984). *Z. Lebensmitt. Unters. Forsch.* **178**, 199–202.

Gallop, P. M., Fluckiger, R., Hanneken, A., Mininsohn, M. M. and Gabbay, K. H. (1981). *Anal. Biochem.* **117**, 427–432.
Galloway, C. M., Dugger, W. M. and Black, Jr. C. C. (1988). *Plant Physiol.* **88**, 980–982.
Gaudreault, P. R. and Webb, J. A. (1986). *Plant Sci.* **45**, 71–75.
Garegg, P. J., Oscarson, S. and Ritzen, H. (1988). *Carbohydr. Res.* **181**, 89–96.
Gauch, R., Leuenberger, U. and Baumgartner, E. (1979). *J. Chromatogr.* **174**, 195–200.
Geiger, O. and Goerisch, H. (1986). *Biochemistry* **25**, 6043–6048.
Geiger, P. J., Ahn, S. and Bessman, S. P. (1980). *Methods Carbohydr. Chem.* **8**, 21–32.
Gella, F. J., Olivella, M. T., Pegueroles, F. and Gener, J. (1981). *Clin. Chem.* **27**, 1686–1689.
Gerhardt, R. and Heldt, H. W. (1984). *Plant Physiol.* **75**, 542–547.
Gestrelius, S. and Mosbach, K. (1987). *Methods Enzymol.* **136**, 353–356.
Ghebregzabher, M., Rufini, S., Monaldi, B. and Lato, M. (1976). *Chromatogr. Rev., J. Chromatogr.* **127**, 133–162.
Ghebregzabher, M., Rufini, S., Sapia, G. M. and Lato, M. (1979). *J. Chromatogr.* **180**, 1–16.
Giersch, C. (1979). *J. Chromatogr.* **172**, 153–161.
Giersch, C., Heber, U., Kaiser, G., Walker, D. A. and Robinson, S. P. (1980). *Arch. Biochem. Biophys.* **205**, 246–259.
Girotti, S., Roda, A., Ghini, S., Piacentini, A. L., Carrea, G. and Bovara, R. (1980). *Anal. Chim. Acta* **183**, 187–196.
Goldemberg, S. H. and Marechal, L. R. (1972). *Methods Enzymol.* **28**, 953–960.
Goldemberg, S. H., Marechal, L. R. and DeSuza, B. C. (1966). *J. Biol. Chem.* **241**, 45–50.
Golden, S. and Katz, J. (1980). *Biochem. J.* **188**, 799–805.
Goldstein, A and Lampen, J. O. (1975). *Methods Enzymol.* **42**, 504–511.
Gorton, L. (1986). *J. Chem. Soc. Faraday Trans. I* **82**, 1245–1258.
Gottschalk, A. (1950). *In* "The Enzymes", Vol. I, Part I (J. B. Sumner and K. Myrback, eds), pp. 551–582. Academic Press, New York.
Gray, G. R. (1978). *Methods Enzymol.* **50**, 155–160.
Gray, G. R. (1987). *Methods Enzymol.* **138**, 26–38.
Grayeski, M. L., Woolf, E. J. and Helly, P. J. (1986). *Anal. Chim. Acta* **183**, 207–215.
Grimble, G. K., Barker, H. M. and Taylor, R. H. (1983). *Anal. Biochem.* **128**, 422–428.
Gross, K. C. and Pharr, D. M. (1982). *Plant Physiol.* **69**, 117–121.
Gruber, E. E. and Feingold, D. S. (1965). *Biochem. Biophys. Res. Commun.* **19**, 37–42.
Guerrant, G. O. and Moss, C. W. (1984). *Anal. Chem.* **56**, 633–638.
Guilbault, G. E. (1975). *Methods Enzymol.* **41**, 53–56.
Guilbault, G. E. (1976). "Handbook of Enzymatic Methods of Analysis". Marcel Dekker, New York.
Guilbault, G. E. (1982). *Appl. Biochem. Biotechnol.* **1**, 85–98.
Guilbault, G. E. (1984). "Handbook of Immobilized Enzymes". Marcel Dekker, New York.
Guittierez, C., Ardourel, M., Bremer, E., Middendorf, A. and Boos, W. (1989). *Mol. Gen. Genet.* **217**, 347–354.
Gupta, M., Chourey, P. S., Burr, B. and Still, P. E. (1988). *Plant Mol. Biol.* **10**, 215–224.
Guthrie, R. D., Jenkins, I. D., Thang, S. and Yamasaki, R. (1983). *Carbohydr. Res.* **121**, 107–109.
Hakomori, S. (1964). *J. Biochem. (Tokyo)* **55**, 205–208.
Hall, L. D. and Morris, G. A. (1980). *Carbohydr. Res.* **82**, 175–184.
Hall, N. A. and Patrick, A. D. (1989). *Anal. Biochem.* **178**, 378–384.
Halvorson, H. (1966). *Methods Enzymol.* **8**, 559–565.
Halvorson, H. and Elias, L. (1958). *Biochim. Biophys. Acta* **30**, 28–40.
Hamburger, M. and Hostettmann, K. (1985). *Phytochemistry* **28**, 1793–1797.
Hanai, T. (1986). *Adv. Chromatogr.* **25**, 279–307.
Hancock, D. O. and Synovec, R. E. (1989). *J. Chromatogr.* **464**, 83–91.
Hardy, M. R. and Townsend, R. R. (1988). *Proc. Natl. Acad. Sci. USA* **85**, 3289–3293.
Hardy, M. R., Townsend, R. R. and Lee, Y. C. (1988). *Anal. Biochem.* **170**, 54–62.
Hargreaves, J. A. and ApRees, T. (1988). *Phytochemistry* **27**, 1621–1629.
Harris, P. J., Henry, R. J., Blakeney, A. B. and Stone, B. A. (1984). *Carbohydr. Res.* **127**, 59–73.
Harris, S. D. and Cotter, D. A. (1988). *Can. J. Microbiol.* **34**, 835–838.

Harrison, J., Hodson, A. W., Skillen, A. W., Stappenbeck, R., Agius, L. and Alberti, K. G. M. M. (1988). *J. Clin. Chem. Clin. Biochem.* **26**, 141–146.

Hase, S., Ikenaka, T. and Matsushima, Y. (1978). *Biochem. Biophys. Res. Commun.* **85**, 257–263.

Hase, S., Hara, S. and Matsushima, Y. (1979). *J. Biochem. (Tokyo)* **85**, 217–220.

Hase, S., Ikenaka, T. and Matsushima, Y. (1981). *J. Biochem. (Tokyo)* **90**, 407–414; 1275–1279.

Hashimoto, H., Sekiguchi, M. and Yoshimura, J. (1985). *Carbohydr. Res.* **144**, C6–C8.

Hassid, W. Z. and Ballou, C. E. (1957). *In* "The Carbohydrates" (W. Pigman, ed.), pp. 478–535. Academic Press, New York.

Hassid, W. Z., Doudoroff, M. and Barker, H. A. (1951). *In* "The Enzymes", Vol. I, Part 2 (J. B. Sumner and K. Myrbäck, eds), pp. 1014–1039. Academic Press, New York.

Hatch, M. D. (1964). *Biochem. J.* **93**, 521–526.

Hatton, M. W. C. and Regoeczi, E. (1982). *Methods Enzymol.* **89**, 172–176.

Hauge, J. E. (1966). *Methods Enzymol.* **9**, 92–98; 107–111.

Hauser, T. R. and Cummins, R. L. (1964). *Anal. Chem.* **36**, 679–681.

Hauser, H. and Semenza, G. (1983). *CRC Crit. Rev. Biochem.* **14**, 319–345.

Hawker, J. S. (1985). *In* "Biochemistry of Storage Carbohydrates in Green Plants" (P. M. Dey and R. H. Dixon, eds), pp. 1–51. Academic Press, London.

Hawker, J. S. and Smith, G. M. (1984). *Phytochemistry* **23**, 245–249.

Hawker, J. S., Smith, G. M., Phillips, H. and Wiskich, J. T (1987). *Plant Physiol.* **84**, 1281–1285.

Hehre, E. J. (1951). *Adv. Enzymol.* **11**, 297–337.

Hehre, E. J. and Okada, G. (1974). *J. Biol. Chem.* **249**, 126–135.

Hehre, E. J., Hamilton, D. M. and Carlson, A. S. (1949). *J. Biol. Chem.* **177**, 267–279.

Hehre, E. J., Okada, G. and Genghof, D. S. (1973). *In* "Carbohydrates in Solution", *Adv. in Chem. Ser.* **117** (H. S. Isbell, ed.), pp. 309–333. American Chemical Society, Washington, D.C.

Hehre, E. J., Genghof, D. S., Sternlicht, H. and Brewer, C. F. (1977). *Biochemistry* **16**, 1780–1786.

Hehre, E. J., Brewer, C. F., Uchiyama, T., Schlesselmann, P. and Lehmann, J. (1980). *Biochemistry* **19**, 3557–3564.

Hehre, E. J., Sawai, T., Brewer, C. F., Nakano, M. and Kanda, T. (1982). *Biochemistry* **21**, 3090–3096.

Heldt, H. W. and Stitt, M. (1987). *In* "Progress in Photosynthesis Research", Vol. 3 (J. Biggins, ed.), pp. 10.675–10.684. Martinus Nijhoff, Dordrecht.

Heldt, H. W., Portis, A. R., McClilley, R., Mosbach, A. and Chon, C. J. (1980). *Anal. Biochem.* **101**, 278–287.

Henderson, S. K. and Henderson, D. E. (1986). *J. Chromatogr. Sci.* **24**, 198–203.

Hestrin, S. and Avigad, G. (1958). *Biochem. J.* **69**, 388–398.

Hestrin, S., Feingold, D. S. and Avigad, G. (1956). *Biochem. J.* **64**, 340–351.

Heyraud, A. and Rinando, M. (1981). *J. Liq. Chromatogr.* **4** (Suppl. 2), 175–293.

Hicks, J. M. (1985). *Clin. Chem.* **31**, 1391–1395.

Hicks, K. B. (1988). *Adv. Carbohydr. Chem. Biochem.* **48**, 17–72.

Hicks, K. B., Symanski, E. V. and Pfeffer, P. E. (1983). *Carbohydr. Res.* **112**, 37–50.

Hicks, K. B., Simpson, G. L. and Bradbury, A. G. N. (1986). *Carbohydr. Res.* **147**, 39–48.

Hidaka, H., Hirayama, M. and Sumi, N. (1988). *Agric. Biol. Chem.* **52**, 1181–1187.

Hitz, W. D. and Giaquinta, R. T. (1987). *BioEssays* **6**, 217–221.

Hitz, W. D., Schmitt, M. R., Card, P. J. and Giaquinta, R. T. (1985). *Plant Physiol.* **77**, 291–295.

Hitz, W. D., Card, P. J. and Ripp, K. G. (1986). *J. Biol. Chem.* 11986–11991.

Ho, M. H. (1988). *Methods Enzymol.* **137**, 271–287.

Hodge, J. E. and Hofreiter, B. T. (1962). *Methods Carbohydr. Chem.* I, 380–399.

Hoffman, R. E., Christofides, J. C., Davies, D. B. and Lawson, C. J. (1986). *Carbohydr. Res.* **153**, 1–16.

Holm, K. A. (1986). *Anal. Chim. Acta* **188**, 285–288.

Honda, S. (1984). *Anal. Biochem.* **140**, 1–47.

Honda, S., Takai, Y. and Kakehi, K. (1979a). *Anal. Chim. Acta.* **105**, 153–161.

Honda, S., Kakehi, K. and Kubono, Y. (1979b). *Carbohydr. Res.* **75**, 61–70.

Honda, S., Matsuda, Y., Takahashi, M., Kakehi, K. and Ganno, S. (1980). *Anal. Chem.* **52**, 1079–1082.

Honda, S., Takahashi, M., Kakehi, K. and Ganno, S. (1981a). *Anal. Biochem.* **113**, 130–138.
Honda, S., Takahashi, M., Nishimura, Y., Kakehi, K. and Ganno, S. (1981b). *Anal. Biochem.* **118**, 162–167.
Honda, S., Nishimura, Y., Takahashi, M., Chiba, H. and Kakehi, K. (1982). *Anal. Biochem.* **119**, 194–199.
Honda, S., Takahashi, M., Shimada, S., Kakehi, K. and Ganno, S. (1983). *Anal. Biochem.* **128**, 419–437.
Honda, S., Enami, K., Konishi, S., Susuki, S. and Kakehi, K. (1986). *J. Chromatogr.* **361**, 321–329.
Honda, S., Kakehi, K., Fukjikawe, K., Oka, Y. and Takahashi, M. (1988). *Carbohydr. Res.* **183**, 59–69.
Honda, S., Suzuki, K. and Kakehi, K. (1989). *Anal. Biochem.* **177**, 62–66.
Hoogenboom, G., Huck, M. G., Peterson, C. M. and Goli, A. (1988). *J. Assoc. Off. Anal. Chem.* **71**, 844–848.
Horikoshi, T., Koga, T. and Hamada, S. (1987). *J. Chromatogr.* **416**, 353–356.
Horowitz, S. B. (1966). *Methods Enzymol.* **9**, 150–166.
Hottiger, T., Boller, T. and Wiemken, A. (1987). *FEBS Lett.* **220**, 113–115.
Hough, L. and Jones, J. K. N. (1962). *Methods Carbohydr. Chem.* **1**, 21–31.
Hough, L. and O'Brien, E. (1980). *Carbohydr. Res.* **84**, 95–102.
Housley, T. L., Kanabus, J. and Carpita, N. C. (1989). *J. Plant Physiol.* **134**, 192–195.
Housnell, E. F. (1986). *In* "HPLC of Small Molecules; A Practical Approach" (C. K. Lim, ed.), pp. 49–68. IRL Press, Oxford.
Huber, S. C. (1986). *Ann. Rev. Plant Physiol.* **37**, 233–246.
Huber, S. C. and Akazawa, T. (1986). *Plant Physiol.* **81**, 1008–1013.
Huber, S. C., Kerr, P. S. and Kalt-Torres, W. (1985). *In* "Regulation of Carbon Partitioning in Photosynthetic Tissues" (R. L. Heath and J. Preiss, eds), pp. 199–214. Waverly Press, Baltimore, MD.
Huber, S. C., Kalt-Torres, W., Usada, H. and Bickett, M. (1987). *In* "Progress in Photosynthesis Research", Vol. 3 (J. Biggins, ed.), pp. 10.717–10.724. Martinus Nijhoff, Dordrecht.
Huber, S. C., Nielsen, T. H., Huber, J. L. A. and Pharr, D. M. (1989a). *Plant Cell Physiol.* **30**, 277–285.
Huber, J. L. A., Huber, S. C. and Nielsen, T. H. (1989b). *Arch. Biochem. Biophys.* **270**, 681–690.
Hughes, R. J. (1983). *Anal. Biochem.* **131**, 318–323.
Hughes, S. and Johnson, D. C. (1982). *J. Agric. Food. Chem.* **30**, 712–714.
Hull, S. R. and Turco, S. J. (1985). *Anal. Biochem.* **146**, 143–149.
Hunziker, W., Spiess, M., Semenza, G. and Lodish, H. (1986). *Cell* **46**, 227–234.
Idahl, L. A., Sandström, P. E. and Sehlin, J. (1986). *Anal. Biochem.* **155**, 177–181.
Imai, K. (1986). *Methods Enzymol.* **133**, 435–449.
Imai, S., Takeuchi, K., Shibata, S., Yoshikawa, S., Kitahata, S., Okada, S., Araya, S. and Nisizawa, T. (1984). *J. Dent. Sci. Res.* **63**, 1293–1297.
Incoll, L. D., Bonnett, G. D. and Gott, B. (1989). *J. Plant Physiol.* **134**, 196–202.
Isbell, H. S. (1963). *Methods Carbohydr. Chem.* **2**, 13–15.
Ishikawa, H., Matsuura, K. and Misaki, H. (1985). *Ger. Offen.* DE3,442,856; CA 103, 69813.
Iwahara, S. and Miki, S. (1988). *Agric. Biol. Chem.* **52**, 867–868.
Iwai, H. and Akihama, S. (1986). *Chem. Pharm. Bull.* **34**, 3471–3474.
Jackson, J. A. and Phillips, R. N. (1988). *Amer. Clin. Lab.*, July, pp. 18–21.
Jaenchen, D. E. (1988). *Amer. Lab.*, March, pp. 63–73.
Jandera, P. and Churáćek, J. (1974). *J. Chromatogr.* **98**, 79–104.
Jenner, M. R. (1980). *In* "Developments in Food Carbohydrate", Vol. 2 (C. K. Lee, ed.), pp. 91–143. Applied Science, London.
Jennings, D. H. (1984). *Adv. Microb. Physiol.* **25**, 150–193.
Jentoft, N. (1985). *Anal. Biochem.* **148**, 424–433.
Johnson, J. M. and Harris, C. H. (1987). *J. Chromatogr. Sci.* **25**, 267–269.
Jones, M. G. K. and Outlaw, Jr. W. H. (1981). "Techniques in Carbohydrate Metabolism", B302, pp. 1–8. Elsevier-North Holland, Ireland.
Jones, M. G. K., Outlaw, Jr. H. and Lowry, O. H. (1977). *Plant Physiol.* **60**, 379–383.

Josephy, P. D., Eling, T. and Mason, R. P. (1982). *J. Biol. Chem.* **257**, 3669–3675.
Journet, E. P., Bligny, R. and Douce, R. (1986). *J. Biol. Chem.* **261**, 3193–3199.
Kaiser, W. M. and Bassham, J. A. (1979). *Z. Pflanzenphysiol.* **94**, 377–385.
Kamogawa, A., Yokobayashi, K. and Fukui, T. (1973). *Agric. Biol. Chem.* **37**, 2813–2819.
Kamogawa, A., Yokobayashi, K. and Fukui, T. (1974). *Anal. Biochem.* **57**, 303–305.
Kandler, O. and Hopf, H (1982). *In* "Encyclopedia of Plant Physiol", New Series, Vol. 13A, Carbohydrates I (F. A. Loewus and W. Tanner, eds), pp. 348–383. Springer, Berlin and Heidelberg.
Kaneda, M., Kobayashi, K., Nishida, K. and Katsuta, S. (1984). *Phytochemistry* **23**, 795–798.
Karamanos, N. K., Tsegenidis, T. and Antonopoulos, G. A. (1987). *J. Chromatogr.* **405**, 221–228.
Karamanos, N. K., Hjerpe, A., Tsegenidis, T., Engfeldt, B. and Antonopoulos, C. A. (1988). *Anal. Biochem.* **172**, 410–419.
Karkkainen, J. (1970). *Carbohydr. Res.* **14**, 27–33.
Karkkainen, J. (1971). *Carbohydr. Res.* **17**, 11–18.
Karube, I., Tamiya, E., Sode, K., Yokoyama, K., Suzuki, H. and Hattori, S. (1988). *Ann. N.Y. Acad. Sci.* **542**, 470–479.
Kasai, T., Fujita, O. and Kawamura, S. (1981). *Kagawa Daigaku Nogakubu Gakujutsu Hokoku* **32**, 103–109; CA 98, 157849.
Kashiwada, Y., Nonaka, G. I. and Nishioka, I. (1988). *Phytochemistry* **27**, 1469–1472.
Kasumi, T., Brewer, C. F., Reese, E. T. and Hehre, E. J. (1986). *Carbohydr. Res.* **146**, 39–49.
Kato, T. and Kinoshita, T. (1980). *Anal. Biochem.* **106**, 238–243.
Kawamoto, T. and Okada, E. (1983). *J. Chromatogr.* **258**, 284–288.
Keeling, P. L. and James, P. (1986). *J. Liquid Chromatogr.* **9**, 983–992.
Keeling, P. L., Wood, J. R., Tyson, R. H. and Bridges, I. G. (1988). *Plant Physiol.* **87**, 311–319.
Kelleher, T., Montenecourt, B. S. and Eveleigh, D. E. (1987). *Appl. Microbiol. Biotechnol.* **27**, 299–305.
Keller, F., Schellenberg, M. and Wiemken, A. (1982). *Arch. Microbiol.* **131**, 298–301.
Keller, F., Frehner, M. and Wiemken, A. (1988). *Plant Physiol.* **88**, 239–241.
Kellermann, O. K. and Fereuci, T. (1982). *Methods Enzymol.* **90**, 459–463.
Kennedy, J. F. (ed.) (1988). "Carbohydrate Chemistry". Clarendon Press, Oxford.
Kennedy, J. F. and Fox, J. E. (1980a). *Methods Carbohydr. Chem.* **8**, 3–12.
Kennedy, J. F. and Fox, J. E. (1980b). *Methods Carbohydr. Chem.* **8**, 13–19.
Kenyon, W. H., Severson, R. F. and Black, Jr. C. C. (1985). *Plant Physiol.* **77**, 183–189.
Kersters, K. and DeLey, J. (1966). *Methods Enzymol.* **9**, 170–179.
Khan, R. (1976). *Adv. Carbohydr. Chem. Biochem.* **33**, 235–294.
Khayat, E. (1987). *Carbohydr. Res.* **162**, 329–332.
Kiba, N., Shitara, K. and Furasawa, M. (1989). *J. Chromatogr.* **463**, 183–187.
Kido, Y., Yano, J., Mijagawa, E., Mimura, H. and Motoki, Y. (1986). *Agric. Biol. Chem.* **50**, 1641–1642.
Killick, K. (1979). *Arch. Biochem. Biophys.* **196**, 121–133.
Kinchi, K., Mutoh, T. and Naoi, M. (1984). *Anal. Biochem.* **140**, 146–151.
King, R. R., Pelletier, Y., Singh, R. P. and Calhoun, L. A. (1986). *J. Chem. Soc., Chem. Commun.*, pp. 1078–1079.
King, R. R., Singh, R. P. and Calhoun, L. A. (1987). *Carbohydr. Res.* **166**, 113–121.
Kirchner, J. G. (1978). "Techniques of Chemistry", 2nd edn, Vol. XIV, pp. 508–535. John Wiley and Sons, New York.
Kitahara, S. (1988). *In* "Handbook of Amylases and Related Enzymes" (The Amylase Research Society of Japan, ed.), pp. 169–172. Pergamon Press, Oxford.
Kitahata, S., Brewer, C. F., Genghof, D. S., Sawai, T. and Hehre, E. J. (1981). *J. Biol. Chem.* **256**, 6017–6026.
Klaus, R. and Fischer, W. (1988). *Methods Enzymol.* **160**, 159–175.
Klein, U. (1987). *Plant Physiol.* **85**, 892–897.
Klok, J., Nieberg Van Velzen, E. H., De Leeuw, J. W. and Schenck, P. A. (1981). *J. Chromatogr.* **207**, 273–275.
Klungsoyr, L. (1966). *Biochim. Biophys. Acta* **128**, 55–62.
Kobayashi, Y. and Horikoshi, K. (1980a). *Agric. Biol. Chem.* **44**, 2261–2269.

Kobayashi, Y. and Horikoshi, K. (1980b). *Biochim. Biophys. Acta* **614**, 256–265.
Kobayashi, Y. and Kawai, S. (1983). *J. Chromatogr.* **275**, 394–399.
Kobayashi, Y., Ueyama, H. and Horikoshi, K. (1982). *Agric. Biol. Chem.* **46**, 2139–2142.
Koerner, P. J. and Nieman, T. A. (1988). *J. Chromatogr.* **449**, 217–228.
Koerner, T. A. W., Prestegard, J. H. and Yu, R. K. (1987). *Methods Enzymol.* **138**, 38–59.
Koizumi, K., Utamura, T. and Okada, Y. (1985). *J. Chromatogr.* **321**, 145–157.
Kondo, H., Shiraishi, T., Nagata, K. and Tomita, K. (1988). *Clin. Chim. Acta* **172**, 131–140.
Königshofer, H., Albert, R. and Kinzel, H. (1979). *Z. Pflanzenphysiol.* **92**, 449–453.
Kremer, B. P. and Kirst, G. O. (1981). *Plant. Sci. Lett.* **23**, 349–357.
Kricka, L. J. (1988). *Anal. Biochem.* **175**, 14–21.
Kricka, L. J., Stanley, P. E., Thorpe, G. H. G. and Whithead, T. P. (1984). "Analytical Applications of Bioluminescence and Chemiluminescence", pp. 111–139. Academic Press, London.
Kulbe, K. D. and Chmiel, H. (1988). *Ann. N.Y. Acad. Sci.* **542**, 444–464.
Kulski, J. K. and Buehring, G. G. (1982). *Anal. Biochem.* **119**, 341–350.
Kunst, F., Pascal, M. Lepesant, J. A., Walle, J. and Dedonder, R. (1974). *Eur. J. Biochem.* **42**, 611–620.
Kunst, F., Steinmetz, M., Lepesant, J. A. and Dedonder, R. (1977). *Biochimie* **59**, 287–292.
Kuo, T. M., VanMiddlesworth, J. F. and Wolf, W. J. (1988). *J. Agric. Food Chem.* **36**, 32–36.
Kurkijarvi, K., Rannio, R., Lavi, J. and Lovgren, T. (1985). *In* "Bioluminescence and Chemiluminescence, Instruments and Applications", Vol. II (K. VanDyke, ed.), pp. 167–184. CRC Press, Boca Raton, FL.
Kvenheim, A. L. (1988). *Acta Chem. Scand.* **B41**, 150–152.
Labat-Robert, J. (1982). *In* "Developments in Food Carbohydrate", Vol. 3 (C. K. Lee and M. G. Lindley, eds), pp. 81–106. Applied Science Publishers, London.
Laine, R. A., Esselman, W. J. and Sweeley, C. C. (1972). *Methods Enzymol.* **28**, 159–164.
Laker, M. F. (1979). *J. Chromatogr.* **163**, 9–18.
Lapp, D. F., Patterson, B. W. and Elbein, A. D. (1972). *Methods Enzymol.* **28**, 515–519.
Larner, J. (1960). *In* "Enzymes", 2nd edn, Vol. 4A (P. D. Boyer, H. Lardy, and K. Byrback, eds), pp. 369–378. Academic Press, New York.
Larner, J. and Gillespie, R. E. (1956). *J. Biol. Chem.* **223**, 709–725.
Lattanzio, V., Bianco, V. V., Miccolis, V. and Linsalata, V. (1986). *Food Chem.* **22**, 17–25.
Lawson, M. A. and Russell, G. F. (1980). *J. Food Sci.* **45**, 1256–1258.
Lazrus, A. L., Kok, G. L., Gitlin, S. N., Lind, J. A. and McLaren, S. E. (1985). *Anal. Chem.* **57**, 917–922.
Leavitt, S. W. and Long, A. (1985). *Plant Physiol.* **78**, 427–429.
Lee, C. K. (ed.) (1980). "Developments in Food Carbohydrate Vol. 2. Disaccharides" Applied Science Publishers, London.
Lee, C. K. and Lindley, M. G. (eds) (1982). "Developments in Food Carbohydrate", Vol. 3, Dissacharidases. Applied Science Publishers, London.
Lee, K. W., Nurk, D. and Zlatkis, A. (1979). *J. Chromatogr.* **174**, 187–193.
Lee, Y. C. (1972). *Methods Enzymol.* **28**, 699–702.
Lehmann, J. (1981). *Z. Pflanzenphysiol.* **102**, 415–424.
Lehrfeld, J. (1976). *J. Chromatogr.* **120**, 141–147.
Lehrfeld, J. (1985). *Anal. Chem.* **57**, 346–348.
Leigh, R. A., ApRees, T., Fuller, W. A. and Banfield, J. (1979). *Biochem. J.* **178**, 539–547.
Lever, M. (1977). *Anal. Biochem.* **81**, 21–27.
Lever, M., Walmsley, T. A., Visser, R. S. and Ryde, S. J. (1984). *Anal. Biochem.* **139**, 205–211.
Levery, S. B. and Hakomori, S.-I. (1987). *Methods Enzymol.* **138**, 13–25.
Levi, I. and Purvis, C. B. (1949). *Adv. Carbohydr. Chem.* **4**, 1–35.
Li, Y. T. and Li, S. C. (1972). *Methods Enzymol.* **28**, 714–720.
Li, B. W. and Schuhmann, P. J. (1981). *J. Food. Sci.* **46**, 425–427.
Liao, J. I., Rountree, M., Good, R. and Punko, C. (1988). *Clin. Chem.* **34**, 2327–2330.
Liav, A. and Goren, M. B. (1983). *Carbohydr. Res.* **123**, C22–C24.
Liessing, N. and McGinnis, E. T. (1982). *Methods Enzymol.* **89**, 135–140.
Lim, C. K. and Peters, T. V. (1989). *J. Chromatogr.* **461**, 259–266.

Lin, J. K. and Wu, S. S. (1987). *Anal. Chem.* **59**, 1320–1326.
Lin, J. K., Jacobson, J., Pereira, A. N. and Ladish, M. R. (1988). *Methods Enzymol.* **160**, 145–159.
Lindberg, B. (1972). *Methods Enzymol.* **28**, 178–195.
Lindley, M. G. (1982). *In* "Developments in Food Carbohydrate", Vol. 3 (C. K. Lee and M. G. Lindley, eds), pp. 141–165. Applied Science Publishers, London.
Lipp, J., Ziegler, H. and Conrady, E. (1988). *Z. Lebensmitt. Unters. Forsch.* **187**, 334–338.
Lloyd, J. B. and Whelan, D. (1969). *Anal. Biochem.* **30**, 467–470.
Lomen, C. E., deAlwis, U. and Wilson, W. S. (1986). *J. Chem. Soc. Faraday Trans. I* **82**, 1265–1270.
Londesborough, J. and Varimo, K. (1984). *Biochem. J.* **219**, 511–518.
Lowe, C. R. (1985). *Biosensors* **1**, 3–16.
Lowry, O. H. and Passaneau, J. V. (1972). "A Flexible System of Enzymatic Analysis." Academic Press, New York.
Loewus, F. A. (1952). *Anal. Chem.* **24**, 219–220.
Lucher, L. A. and Lego, T. (1989). *Anal. Biochem.* **178**, 327–330.
MacGregor, L. C. and Matschinsky, F. M. (1984). *Anal. Biochem.* **141**, 382–389.
MacKenzie, C. R., Johnson, K. G. and McDonald, I. J. (1977). *Can. J. Microbiol.* **23**, 1303–1307.
Macrae, R. and Dick, J. (1981). *J. Chromatogr.* **210**, 138–145.
Macrae, R., Trugo, L. C. and Dick, J. (1982). *Chromatographia* **15**, 476–478.
Madore, M. and Webb, J. A. (1982). *Can. J. Bot.* **60**, 126–130.
Madore, M. A., Mitchell, D. E. and Boyd, C. M. (1988). *Plant Physiol.* **87**, 588–591.
Madsen, B. C. and Kromis, M. S. (1984). *Anal. Chem.* **56**, 2849–2850.
Maier, E. and Kurz, G. (1982). *Methods Enzymol.* **89**, 176–181.
Majkic-Singh, N., Stojanov, M., Spastic, S. and Berkes, I. (1981). *Clin. Chim. Acta* **116**, 117–123.
Malavolti, N. L., Pilosof, D. and Nieman, T. A. (1984). *Anal. Chem.* **56**, 2191–2195.
Malone, J. I., Knox, G., Benford, S. and Tedesco, T. A. (1980). *Diabetes* **29**, 861–864.
Manners, D. J. (1985). *In* "Biochemistry of Storage Carbohydrates in Green Plants" (P. M. Dey and R. A. Dixon, eds), pp. 149–203. Academic Press, London.
Manners, D. J. and Taylor, D. C. (1967). *Arch. Biochem. Biophys.* **121**, 443–451.
Marana, C., Garcia-Olmedo, F. and Carbonero, P. (1988). *Gene* **63**, 253–260.
Marechal, L. R. (1984). *Arch. Microbiol.* **137**, 70–73.
Marechal, L. R. and Belcopitow, E. (1972). *J. Biol. Chem.* **247**, 3223–3228.
Maretzki, A. and Thom, M. (1988). *Plant Physiol.* **88**, 266–269.
Marko-Varga, G. A. (1987). *J. Chromatogr.* **408**, 157–170.
Marko-Varga, G. A. (1989). *Anal. Chem.* **61**, 831–838.
Martin, F. (1985). *Physiol. Veg.* **23**, 463–490.
Martin, S. A. and Russell, J. B. (1987). *Appl. Environ. Microbiol.* **53**, 2388–2393.
Martin, F. D., Canet, D. and Marchal, J. P. (1985). *Plant Physiol.* **77**, 499–502.
Martin, F., Ramstedt, M., Soderhall, K. and Canet, D. (1988). *Plant Physiol.* **86**, 935–940.
Mascini, M., Moscone, D., Palleschi, G. and Pillotou, R. (1988). *Anal. Chim. Acta.* **213**, 101–111.
Mason, B. S. and Slover, H. T. (1971). *J. Agric. Food Chem.* **19**, 551–554.
Matsubara, C., Iwamoto, T., Nishikawa, Y., Takamura, K., Yano, S. and Yoshikawa, S. (1985a). *J. Chem. Soc. Dalton Trans. I*, pp. 81–84.
Matsubara, C., Kudo, K., Kawashita, T. and Takamura, K. (1985b). *Anal. Chem.* **57**, 1107–1109.
Matsumoto, K., Hamada, O., Okeda, H. and Osajima, Y. (1986). *Anal. Chem.* **58**, 2732–2734.
Matsumoto, K., Kamikado, H., Matsubara, H. and Osajima, Y. (1988). *Anal. Chem.* **60**, 147–151.
Matsumoto, T., Suzuki, O., Katsumata, Y., Oya, M., Nimura, Y., Akita, M. and Hattori, T. (1981). *Clin. Chim. Acta* **112**, 141–148.
Matsushita, K. and Ameyama, A. (1982). *Methods Enzymol.* **89**, 149–154.
Matsuura, F., Nuñez, H. A., Grabowski, G. A. and Sweeley, C. C. (1981). *Arch. Biochem. Biophys.* **207**, 337–352.
Matthews, D. R., Bowen, E., Watson, A., Holman, R. R., Steemson, J., Hughes, S. and Scott, D. (1987). *Lancet* **i**, 778–779.
Mattos, I. L., Zagatto, E. A. G. and Jacintho, A. O. (1988). *Anal. Chim. Acta* **214**, 247–257.
Mayer, R. M. (1987). *Methods Enzymol.* **138**, 649–661.
Mbaku, S. B., Fritz, G. J. and Bowes, G. (1978). *Plant Physiol.* **62**, 510–515.

McBee, G. G. and Maness, N. O. (1983). *J. Chromatogr.* **264**, 474–478.
McCarty, D. R., Shaw, J. R. and Hannah, L. C. (1986). *Proc. Natl. Acad. Sci. USA* **83**, 9099–9103.
McCleary, B. V. and Harrington, J. (1988). *Methods Enzymol.* **160**, 575–583.
McCormick, S., Mauvais, J. and Fedoroff, N. (1982). *Mol. Gen. Genet.* **187**, 494–500.
McElfresh, K. C. and Chourey, P. S. (1988). *Plant Physiol.* **87**, 542–546.
McFeeters, R. F. (1980). *Anal. Biochem.* **103**, 302–306.
McKay, D. B., Tanner, G. P., Maclean, D. J. and Scott, K. J. (1987). *Anal. Biochem.* **165**, 392–398.
McNeil, C. J., Spoors, J. A., Cocco, D., Cooper, J. M. and Bannister, J. V. (1989). *Anal. Chem.* **61**, 25–29.
McNicholas, P. A., Balley, M. and Redmond, J. W. (1984). *J. Chromatogr.* **315**, 451–456.
Mega, T. L. and Van Etten, R. L. (1988). *J. Am. Chem. Soc.* **110**, 6372–6376.
Mellor, R. B. (1988). *J. Plant Physiol.* **133**, 173–177.
Mendez, A. J., Cabeza, C. and Hsia, S. L. (1987). *Anal. Biochem.* **156**, 386–389.
Meng, J., Rosell, K. G. and Srivastava, L. M. (1987). *Carbohydr. Res.* **161**, 171–180.
Meyer, R. and Wagner, K. G. (1985). *Anal. Biochem.* **148**, 269–276.
Mieyal, J. J. (1972). *Methods Enzymol.* **28**, 935–943.
Miller, A. W. and Robyt, J. F. (1986). *Anal. Biochem.* **156**, 357–363.
Miller, W. L. and Kester, D. R. (1988). *Anal. Chem.* **60**, 2711–2715.
Mitchell, M., Matula, M. and Elbein, A. D. (1972). *Methods Enzymol.* **28**, 520–522.
Mitsutomi, M. and Ohtakara, A. (1984). *Agric. Biol. Chem.* **48**, 3153–3155.
Mitsutomi, M. and Ohtakara, A. (1988). *Agric. Biol. Chem.* **52**, 2305–2311.
Miwa, I., Kanbara, M., Wakazono, H. and Okuda, J. (1988). *Anal. Biochem.* **173**, 39–44.
Miyatake, K., Kuramoto,Y. and Kitakoa, S. (1984). *Biochem. Biophys. Res. Commun.* **122**, 906–911.
Molnar-Perl, I., Pintér-Szakaes, M., Kovago, A. and Petroczy, J. (1984). *J. Chromatogr.* **295**, 433–443.
Monod, J. and Torriani, A. M. (1950). *Am. Inst. Pasteur* **78**, 65–77.
Monsan, P. and Comber, D. (1988). *Methods Enzymol.* **137**, 584–598.
Mopper, K. (1978). *Anal. Biochem.* **87**, 162–168.
Mopper, K. and Degens, E. T. (1972). *Anal. Biochem.* **45**, 147–153.
Mopper, K. and Gindler, E. M. (1973). *Anal. Biochem.* **56**, 440–442.
Mopper, K. and Johnson, L. (1983). *J. Chromatogr.* **256**, 27–38.
Mopper, K., Dawson, R., Liebezeit, G. and Hansen, H. P. (1980). *Anal. Chem.* **52**, 2018–2022.
Morell, M. and Copeland, L. (1984). *Plant Physiol.* **74**, 1030–1034.
Morell, M. and Copeland, L. (1985). *Plant Physiol.* **78**, 149–154.
Moriguchi, T. and Yamaki, S. (1988). *Plant Cell Physiol.* **29**, 1301–1366.
Morpeth, F. F. (1985). *Biochem. J.* **228**, 557–564.
Muller, L. L. and Jacks, T. J. (1983). *Plant Physiol.* **71**, 703–704.
Munir, M., Schneider, B. and Schiweck, H. (1987). *Carbohydr. Res.* **164**, 477–485.
Muramatsu, M., Kainuma, S., Miwa, T. and Nakakuki, T. (1988). *Agric. Biol. Chem.* **52**, 1303–1304.
Murao, S., Nagano, H., Ogura, S. and Nishino, T. (1985). *Agric. Biol. Chem.* **49**, 2113–2118.
Murtiashaw, M. H., Thorpe, S. R. and Baynes, J. W. (1983). *Anal. Biochem.* **135**, 443–446.
Muzika, K. and Kovar, J. (1987). *J. Liquid Chromatogr.* **10**, 2291–2304.
Nachtmann, F. and Budna, K. W. (1977). *J. Chromatogr.* **136**, 279–287.
Nakajima, Y. (1988). In "Handbook of Amylases and Releatead Enzymes" (The Amylase Research Society of Japan, ed.), pp. 230–231. Pergamon Press, Oxford.
Nakamura, Y. and Kozimi, Y. (1985). *J. Chromatogr.* **333**, 83–92.
Nakano, H., Takenishi, S. and Watanabe, Y. (1988). *Agric. Biol. Chem.* **52**, 1913–1921.
Nakano, K., Murakami, K., Takaishi, Y. and Tomimatsu, T. (1986). *Chem. Pharm. Bull.* **34**, 5005–5010.
Neeser, J. R. (1985). *Carbohydr. Res.* **138**, 189–198.
Nelson, N. (1944). *J. Biol. Chem.* **153**, 375–380.

Neuberg, C. and Mandl, I. (1985). *In* "The Enzymes", Vol. I, Part I (J. B. Sumner and K. Myrback, eds), pp. 527–550. Academic Press, New York.

Ng, T. and Zeikus, J. G. (1986). *Appl. Environ. Microbiol.* **52**, 902–904.

Ngo, T. T. and Lenhoff, H. M. (1980). *Anal. Biochem.* **105**, 389–397.

Nicolas, P., Gertsch, I. and Parisod, C. (1984). *Carbohydr. Res.* **131**, 331–334.

Niemietz, C. and Hawker, J. S. (1988). *Am. J. Plant Physiol.* **15**, 359–366.

Nijpels, H. H. (1982). *In* "Developments in Food Carbohydrate" Vol. 3 (C. K. Lee and M. G. Lindley, eds), pp. 23–48. Applied Science Publishers, London.

Nikolov, Z. L., Meagher, M. M. and Reilly, P. J. (1985a). *J. Chromatogr.* **319**, 51–57.

Nikolov, Z. L., Meagher, M. M. and Reilly, P. J. (1985b). *J. Chromatogr.* **321**, 393–399.

Nimami, Y., Yazawa, K., Nakamura, K. and Tamura, Z. (1985). *Chem. Pharm. Bull.* **33**, 710–714.

Nisizawa, K. and Hashimoto, Y. (1970). *In* "The Carbohydrates", 2nd edn, Vol. IIA (W. Pigman and D. Horton, eds), pp. 241–300. Academic Press, New York.

Nisizawa, T., Takeuchi, K., Imai, S., Kitahata, S. and Okada, S. (1986). *Carbohydr. Res.* **147**, 135–144.

Nordin, P. (1983). *Anal. Biochem.* **131**, 492–498.

Novotny, M. J., Reizer, J., Esch, F. and Saier, M. (1984). *J. Bacteriol.* **159**, 986–990.

O'Brien, M. M., Schofield, P. J. and Edwards, M. R. (1983). *Biochem. J.* **211**, 81–90.

Ohtakara, A. and Mitsutomi, M. (1987). *J. Ferment. Technol.* **65**, 493–498.

Okada, G. and Hehre, E. J. (1974). *J. Biol. Chem.* **249**, 126–135.

O'Leary, M. H. (1981). *Phytochemistry* **20**, 553–567.

Olechno, J. D., Carter, S. R., Edwards, W. T. and Gillen, D. G. (1987). *Am. Biotechnol. Lab.* **5**(5), 38–50.

Olson, B. (1982). *Anal. Chim. Acta* **136**, 113–119.

Olson, A. C., Gray, G. M., Chin, M. C., Betschart, A. A. and Turnlund, J. R. (1988). *J. Agric. Food Chem.* **36**, 300–304.

Olsson, B., Lundback, H. and Johansson, G. (1985). *Anal. Chim. Acta* **167**, 123–136.

O'Neill, R. A., White, A. R., York, W. S., Darvill, A. G. and Albersheim, P. (1988). *Phytochemistry* **27**, 329–333.

O'Neill, R. A., Darvill, A. and Albersheim, P. (1989). *Anal. Biochem.* **177**, 11–15.

Orfanos, A. P., Jinks, D. C. and Guthrie, R. (1986). *Clin. Biochem.* **19**, 225–228.

Oshima, R., Yamanuchi, Y. and Kumanotani, J. (1982). *Carbohydr. Res.* **107**, 169–176.

Oshima, R., Kumanotani, J. and Watanabe, C. (1983). *J. Chromatogr.* **259**, 159–163.

Outlaw, Jr. W. H., Fisher, D. B. and Christy, A. C. (1975). *Plant Physiol.* **55**, 704–711.

Outlaw, Jr. W. H., Hite, D. R. C. and Fiore, G. B. (1988). *Anal. Biochem.* **171**, 104–107.

Palmer, T. N., Wober, G. and Whelan, W. J. (1973). *Eur. J. Biochem.* **39**, 601–612.

Palmisano, J. and Schwartz, J. H. (1982). *Anal. Biochem.* **126**, 409–413.

Panek, A. D. (1985). *J. Biotechnol.* **3**, 121–130.

Park, J. T. and Johnson, M. J. (1949). *J. Biol. Chem.* **181**, 149–151.

Patterson, B. W., Ferguson, A. H., Matula, M. and Elbein, A. D. (1972). *Methods Enzymol.* **28**, 998–1000.

PazParente, J., Cardon, P., Leroy, Y., Montreuil, J. and Fournet, B. (1985). *Carbohydr. Res.* **141**, 41–47.

Pazur, J. H. (1970). *In* "The Carbohydrates, Chemistry and Biochemistry", Vol. IIA (W. Pigman and D. Horton, eds), pp. 69–137. Academic Press, New York.

Pazur, J. H. (1972). *Methods Enzymol.* **28**, 931–934.

Pazur, J. H. and Dreher, K. L. (1980). *Methods Carbohydr. Chem.* **8**, 138–144.

Pazur, J. H., Dropkin, D. J. and Hetzler, C. E. (1973). *In* "Carbohydrates in Solution", *Adv. in Chem. Ser.* **117** (H. S. Isbell, ed.), pp. 374–386. American Chemical Society, Washington, D.C.

Pelletier, O. and Cadieux, S. (1982). *J. Chromatogr.* **231**, 225–235.

Penglis, A. A. E. (1981). *Adv. Carbohydr. Chem. Biochem.* **38**, 195–285.

Perini, F. and Peters, B. P. (1982). *Anal. Biochem.* **123**, 357–363.

Pesez, M. and Bartos, J. (1967). *Talanta* **14**, 1097–1108.

Petersson, B. A., Hansen, E. H. and Ruzicka, J. (1986). *Anal. Lett.* **19**, 649–665.

Pharr, D. M. and Sox, H. N. (1984). *Plant Sci. Lett.* **35**, 187–193.

Pharr, D. M., Huber, S. C. and Sox, H. N. (1985). *Plant Physiol.* **77**, 104–108.

Pharr, D. M., Hendrix, D. L., Robbins, N. S., Gross, K. C. and Sox, H. N. (1987). *Plant Sci.* **50**, 21–26.

Phillips, L. R. and Fraser, B. A. (1981). *Carbohydr. Res.* **90**, 149–152.

Picha, D. H. (1985). *J. Food Sci.* **50**, 1189–1190.

Pierce (1988). "Handbook and General Catalog", p. 156. Pierce, Rockford, IL.

Pilosof, D. and Nieman, T. A. (1982). *Anal. Chem.* **54**, 1698–1701.

Pollman, R. M. (1989). *J. Assoc. Off. Anal. Chem.* **72**, 425–428.

Pollock, C. J. (1982). *Phytochemistry* **21**, 2461–2465.

Pollock, C. J. and Chatterton, N. J. (1988). *In* "The Biochemistry of Plants", Vol. 15, Carbohydrates' (J. Preiss, ed.), pp. 109–140. Academic Press, San Diego.

Pontis, H. G. and Del Campillo, E. (1985). *In* "Biochemistry of Storage Carbohydrates in Green Plants" (P. M. Dey and R. A. Dixon, eds), pp. 205–227. Academic Press, London.

Porro, M., Viti, S., Antoni, G. and Neri, P. (1981). *Anal. Biochem.* **118**, 301–306.

Prabhu, S. V. and Baldwin, R. P. (1989). *Anal. Chem.* **61**, 852–856.

Prager, M. J. and Miskiewicz, M. A. (1979). *J. Assoc. Off. Anal. Chem.* **62**, 262–265.

Prakash, C. and Vijay, I. K. (1983). *Anal. Biochem.* **128**, 41–46.

Preiss, J. (1987). *Physiol. Plant* **69**, 373–376.

Preiss, J. (1988). *In* "The Biochemistry of Plants" Vol. 14, Carbohydrates (J. Preiss, ed.), pp. 181–245. Academic Press, San Diego.

Preiss, J., Morell, M., Bloom, M., Knowles, V. L. and Liu, T. P. (1987). *In* "Progress in Photosynthesis Research", Vol. 3 (J. Biggins, ed.), pp. 10.693–10.704. Martinus Nijhoff, Dordrecht.

Preisser, J. and Komor, E. (1988). *Plant Physiol.* **88**, 259–265.

Prencipe, L., Iaccheri, E. and Manzati, C. (1987). *Clin. Chem.* **33**, 486–489.

Puget, K. and Michelson, A. M. (1976). *Biochimie* **58**, 757–758.

Quemener, B. (1988). *J. Agric. Food Chem.* **36**, 754–759.

Rahni, M. A. N., Lubrano, G. J. and Guilbault, G. E. (1987). *J. Agric. Food Chem.* **35**, 1001–1004.

Ram, P. A., Fung, M., Millette, C. F. and Armant, D. R. (1989). *Anal. Biochem.* **178**, 421–426.

Randerath, K. (1970). *Anal. Biochem.* **34**, 188–205.

Rathbone, E. B. (1980). *In* "Developments in Food Carbohydrates", Vol. 2 (C. K. Lee, ed.), pp. 145–185. Applied Science Publishers, London.

Raven, J. A. (1987). *In* "The Biochemistry of Plants", Vol. 13, Methodology (D. D. Davies, ed.), pp. 127–180. Academic Press Inc., Orlando, FL.

Raybin, H. W. (1933). *J. Am. Chem. Soc.* **55**, 2603–2604.

Raybin, H. W. (1937). *J. Am. Chem. Soc.* **59**, 1402–1403.

Rébeillé, F., Bligny, R., Martin, J. P. and Douce, R. (1985). *Biochem. J.* **226**, 679–684.

Reddy, C. C., Swan, J. S. and Hamilton, G. A. (1981). *J. Biol. Chem.* **256**, 8510–8518.

Redgwell, R. J. (1980). *Anal. Biochem.* **107**, 44–50.

Reed, R. H. and Stewart, W. D. P. (1988). *In* "Biochemistry of the Algae and Cyanobacteria" (L. V. Rogers and J. R. Gallon, eds), pp. 217–231. Clarendon Press, Oxford.

Reed, W. (1866). "The History of Sugar and Sugar Yielding Plants". Longman, Green & Co., London.

Reibach, P. H. and Streeter, J. G. (1983). *Plant Physiol.* **72**, 634–640.

Reim, R. E. and Van Effen, R. M. (1986). *Anal. Chem.* **58**, 3203–3207.

Reinhold, V. N. (1987). *Methods Enzymol.* **138**, 59–94.

Reiss, P. D., Zuurendonk, P. F. and Veech, R. L. (1984). *Anal. Biochem.* **140**, 162–171.

Reiss, U. and Sacktor, B. (1982). *Biochim. Biophys. Acta* **704**, 422–426.

Reizer, J. and Peterkofsky, A. (1987). "Sugar Transport and Metabolism in Gram Positive Bacteria". Ellis Horwood, Halsted Press, New York.

Rickers, J., Cushman, J. C., Michalowski, C. B., Schmitt, J. M. and Bohnert, H. J. (1989). *Mol. Gen. Genet.* **215**, 447–454.

Robards, K. and Whitelaw, M. (1986). *J. Chromatogr.* **373**, 81–110.

Roberts, S. M. and Mitchell, D. T. (1979). *New Phytol.* **83**, 494–508.

Robyt, J. F. (1980). *In* "Mechanisms of Saccharide Polymerization and Depolymerization" (J. J. Marshall, ed.), pp. 43–54. Academic Press, New York.

Robyt, J. F. and Martin, P. J. (1983). *Carbohydr. Res.* **113**, 301–315.

Rocklin, R. D. and Pohl, C. A. (1983). *J. Liquid Chromatogr.* **6**, 1577–1590.

Roe, J. H., Epstein, J. H. and Goldstein, N. P. (1949). *J. Biol. Chem.* **178**, 839–845.

Rorem, E. S., Walker, H. G. and McCready, R. M. (1960). *Plant Physiol.* **35**, 269–272.

Rosenfelder, G., Morgelin, M., Chang, J. Y., Schonenberger, C. A., Braun, D. E. and Towbin, H. (1985). *Anal. Biochem.* **147**, 156–165.

Ross, M. S. F. (1977). *J. Chromatogr.* **141**, 67–119.

Ross, L. F. and Chapital, D. C. (1987). *J. Chromatogr. Sci.* **25**, 112–117.

Roy, R., Gervais, S., Gamian, A. and Boratynski, J. (1988). *Carbohydr. Res.* **177**, C5–C8.

Rubino, F. M. (1989). *J. Chromatogr.* **473**, 125–133.

Rufty, T. W. and Huber, S. C. (1983). *Plant Physiol.* **72**, 474–480.

Ryder, M. H., Tate, M. E. and Jones, G. P. (1984). *J. Biol. Chem.* **259**, 9704–9710.

Säämänen, A. M. and Tammi, M. (1988). *Glycoconjugate Res.* **5**, 235–243.

Sabin, R. D. and Wasserman, B. P. (1987). *J. Agric. Food Chem.* **35**, 649–651.

Sabularse, D. C. and Anderson, R. L. (1981). *Biochem. Biophys. Res. Commun.* **103**, 848–855.

Sadana, J. C. and Patil, R. V. (1985). *J. Gen. Microbiol.* **131**, 1917–1923.

Sadana, J. C. and Patil, R. V. (1988). *Methods Enzymol.* **160**, 448–454.

Saha, B. C. and Zeikus, J. G. (1989). *Starch.* **41**, 57–64.

Salanoubat, M. and Belliard, G. (1987). *Gene* **60**, 47–56.

Salerno, G. L., Gamundi, S. S. and Pontis, H. G. (1979). *Anal. Biochem.* **93**, 196–199.

Salminen, S. O. and Streeter, J. G. (1986). *Plant Physiol.* **81**, 538–541.

Salminen, S. O. and Streeter, J. G. (1987). *Plant Physiol.* **83**, 535–540.

Samuelson, O. (1972). *Methods Carbohydr. Chem.* **6**, 65–75.

Santikarn, S., Her, G. R. and Reinhold, V. N. (1987). *J. Carbohydr. Chem.* **6**, 141–154.

Saravitz, D. M., Pharr, D. M. and Carter, T. E. (1987). *Plant Physiol.* **83**, 185–189.

Sasakawa, H., Sugiharto, B., O'Leary, M. H. and Sugiyama, T. (1989). *Plant Physiol.* **90**, 582–585.

Sasaki, T. (1988). *Methods Enzymol.* **160**, 468–472.

Sasaki, T., Tanaka, T., Nakagawa, S. and Kainuma, K. (1983). *Biochem. J.* **209**, 803–807.

Sawicki, E. and Sawicki, C. R. (1978). "Aldehyde Photometric Analysis", Vol. 5. Academic Press, London.

Schaefer, J. S., Kier, J. D. and Stejskal, E. O. (1980). *Plant Physiol.* **65**, 254–259.

Schaffer, R. and Isbell, H. S. (1963). *Methods Carbohydr. Chem.* **2**, 11–12.

Schäffler, K. J. and DuBoil, P. G. M. (1981). *J. Chromatogr.* **207**, 221–229.

Scheller, F. W., Renneberg, R. and Schubert, F. (1988). *Methods Enzymol.* **137**, 29–43.

Schimz, K. L. and Overhoff, B. (1987). *FEMS Microbiol. Lett.* **40**, 333–337.

Schimz, K. L., Irrang, K. and Overhoff, B. (1985). *FEMS Microbiol. Lett.* **30**, 105–169.

Schlabach, T. D. and Robinson, J. (1983). *J. Chromatogr.* **282**, 169–177.

Schmalstig, J. E. and Hitz, W. D. (1987a). *Plant Physiol.* **85**, 407–412.

Schmalstig, J. E. and Hitz, W. D. (1987b). *Plant Physiol.* **85**, 902–905.

Schmidt, K., Schupfner, M., and Schmitt, R. (1982). *J. Bacteriol.* **151**, 68–76.

Schneider, F. ed. (1982). "Sugar Analysis, International Commission for Uniform Methods of Sugar Analysis" (ICUMSA). Dublin, Ireland.

Schnyder, H., Ehses, U., Bestajovsky, J., Mehrhoff, R. and Kuhbauch, W. (1988). *J. Plant Physiol.* **132**, 333–338.

Schoenwitz, R. and Ziegler, H. (1982). *Can. J. Plant Sci.* **63**, 415–420.

Schubert, F., Kirsten, D., Scheller, F., Abraham, M. and Boross, L. (1986). *Anal. Lett.* **19**, 2155–2167.

Schweizer, T. F. and Horman, I. (1981). *Carbohydr. Res.* **95**, 61–71.

Schweizer, T. F., Horman, I. and Wursch, P. (1978). *J. Sci. Food Agric.* **29**, 148–154.

Schwind, H., Scharbert, F., Schmidt, R. and Kattermann, R. (1978). *J. Clin. Chem. Clin. Biochem.* **16**, 145–149.

Scouten, W. H. (1987). *Methods Enzymol.* **135**, 30–65.

Seitz, W. R. (1978). *Methods Enzymol.* **57**, 445–462.

Seitz, W. R. (1981). *CRC Critical Rev. Anal. Chem.* **13**, 1–58.
Semenza, G. (1986). *Ann. Rev. Cell. Biol.* **2**, 355–412.
Semprevivo, L. H. (1988). *Carbohydr. Res.* **177**, 222–227.
Severson, R. F., Arrendale, R. F., Chortyk, O. T., Green, C. R., Thomas, F. A., Stewart, J. L. and Johnson, A. W. (1985). *J. Agric. Food Chem.* **33**, 870–875.
Shallenberger, R. S. (1982). "Advanced Sugar Chemistry". Avi Publishing, Westport, CT.
Shaw, J. R. and Dickinson, D. B. (1984). *Plant Physiol.* **75**, 207–211.
Shaw, P. E. (1988). "Handbook of Sugar Separations in Food by HPLC". CRC Press, Boca Raton, FL.
Shayman, J. A., Morrison, A. R. and Lowry, O. H. (1987). *Anal. Biochem.* **162**, 562–568.
Shichiri, M., Kawamori, R. and Yamasaki, Y. (1988). *Methods Enzymol.* **137**, 326–337.
Shiomi, N., Kido, H. and Kiriyama, S. (1985). *Phytochemistry* **24**, 695–698.
Shiomi, T., Tohyama, M., Satoh, M. and Imai, K. (1988). *Biotechnol. Bioeng.* **32**, 664–668.
Shoaf, C. R., Caplow, M. and Heizer, W. D. (1979). *Anal. Biochem.* **96**, 12–20.
Silverstein, R., Voet, J., Reed, D. and Abeles, R. H. (1967). *J. Biol. Chem.* **242**, 1338–1346.
Simatupang, M. H., Sinner, M. and Dietrichs, H. H. (1978). *J. Chromatogr.* **155**, 446–449.
Small, G. W. and McIntyre, K. (1989). *Anal. Chem.* **61**, 666–674.
Smith, R. E. (1988). "Ion Chromatography Applications". CRC Press, Boca Raton, FL.
Smith, R. E., Howell, S., Yourtee, D., Premkumar, N., Pond, T., Sun, G. Y. and MacQuarrie, R. A. (1988). *J. Chromatogr.* **439**, 83–92.
Smrcka, A. V. and Jensen, R. G. (1988). *Plant Physiol.* **86**, 615–618.
Sols, A. and DeLaFuente (1961). *Methods Med. Res.* **9**, 302–309.
Soyama, K. and Ono, E. (1987). *Clin. Chem. Acta* **168**, 259–260.
Speck, Jr. J. C. (1962). *Methods Carbohydr. Chem* **1**, 441–445.
Spiro, R. E. (1966). *Methods Enzymol.* **8**, 3–26.
Springer, B., Werr, W., Starlinger, P., Bennett, D. C., Zokolica, M. and Freeling, M. (1986). *Mol. Gen. Genet.* **205**, 461–468.
Srivastava, S. K., Hair, G. A. and Das, B. (1985). *Proc. Natl. Acad. Sci. USA* **82**, 7222–7226.
Stahl, E. (1969). "Thin Layer Chromatography". Springer, New York.
Stanek, J., Cerney, M. and Pacak, J. (1965). "The Oligosaccharides". Academic Press, New York.
Stellner, K., Saito, H. and Hakomori, S. (1973). *Arch. Biochem. Biophys.* **155**, 464–472.
Stephens, R. and DeBusk, A. G. (1975). *Methods Enzymol.* **42**, 497–503.
Sternberg, L. and DeNiro, M. J. (1983). *Science* **220**, 947–948.
Sternberg, R., Bindra, D. S., Wilson, G. S. and Thevenot, D. R. (1988). *Anal. Chem.* **60**, 2781–2786.
Steup, M. (1988). *In* "The Biochemistry of Plants", Vol. 14, Carbohydrates" (J. Preiss, ed.), pp. 255–296. Academic Press, San Diego.
Stikkelman, R. M., Tjioe, T. T., VanDerWiel, J. P. and VanRantwijk, F. (1985). *J. Chromatogr.* **322**, 220–222.
Stio, M., Vanni, P. and Pinzautti, G. (1988). *Anal. Biochem.* **174**, 32–37.
Stitt, M. (1987a). *Plant Physiol.* **84**, 201–204.
Stitt, M. (1987b). *In* "Progress in Photosynthesis Research", Vol. 3 (J. Biggins, ed.), pp. 10.685–10.692. Martinus Nijhoff, Dordrecht.
Stitt, M. and ApRees, T. (1980). *Biochim. Biophys. Acta* **627**, 131–143.
Stitt, M., Wirtz, W., Gerhardt, R., Heldt, H. W., Spencer, C., Walker, P. and Foster, C. (1985). *Planta* **166**, 354–364.
Stitt, M., Huber, S. and Kerr, P. (1987a). *In* "The Biochemistry of Plants", Vol. 10 (M. D. Hatch and N. K. Boardman, eds), pp. 327–409. Academic Press, San Diego.
Stitt, M., Gerhardt, R., Wilke, I. and Heldt, H. W. (1987b). *Plant Physiol.* **69**, 377–386.
Strack, D., Sachs, G., Römer, A. and Wiermann, R. (1981). *Z. Naturforsch.* **36c**, 721–723.
Streeter, J. G. (1980). *Plant Physiol.* **66**, 471–476.
Streeter, J. G. (1982). *Planta* **155**, 112–115.
Streeter, J. G. (1985). *J. Bacteriol.* **164**, 78–84.
Streeter, J. G. (1987). *Plant Physiol.* **85**, 768–773.
Streeter, J. G. and Jeffers, D. L. (1979). *Crop Sci.* **19**, 729–734.
Sturgeon, R. J. (1980a). *Methods Carbohydr. Chem.* **8**, 131–133.

Sturgeon, R. J. (1980b). *Methods Carbohydr. Chem.* **8**, 135–137.

Su, J. C. and Preiss, J. (1978). *Plant Physiol.* **60**, 389–393.

Suetsugu, N., Hirooka, E., Yasui, H., Hiromi, K. and Ono, S. (1973). *J. Biochem. (Tokyo)* **73**, 1223–1232.

Sumida, M., Ogura, S., Miyata, S., Arai, M. and Murao, S. (1989). *J. Ferment. Bioeng.* **67**, 83–86.

Sumner, J. B. and Somers, G. F. (1943). "Chemistry and Methods of Enzymes". Academic Press, New York.

Sweeley, C. C., Bentley, R., Makita, M. and Wells, W. W. (1963). *J. Am. Chem. Soc.* **85**, 2497–2507.

Sweeley, C. C., Wells, W. W. and Bentley, R. (1966). *Methods Enzymol.* **8**, 95–108.

Swindlehurst, C. A. K. and Nieman, T. A. (1988). *Anal. Chim. Acta* **205**, 195–205.

Szmelcman, S. and Schwartz, M. (1976). *Eur. J. Biochem.* **65**, 13–19.

Tabata, S. and Ide, T. (1988). *Carbohydr. Res.* **176**, 245–251.

Takasaki, S. and Kobata, A. (1978). *Methods Enzymol.* **50**, 50–52.

Takeda, M., Maeda, M. and Tsuji, A. (1982). *J. Chromatogr.* **244**, 347–355.

Takemoto, H., Hase, S. and Ikenaka, T. (1985). *Anal. Biochem.* **145**, 245–250.

Takeuchi, M., Ninomiya, K., Kawebata, K., Asano, N., Kameda, Y. and Matsui, K. (1986). *J. Biochem. (Tokyo)* **100**, 1049–1055.

Takeuchi, M., Takasaki, S., Inoue, N. and Kobata, A. (1987). *J. Chromatogr.* **400**, 207–213.

Takeuchi, M., Asano, N., Kameda, Y. and Matsui, K. (1988). *Agric. Biol. Chem.* **52**, 1905–1912.

Tanaka, T., Yamamoto, S., Oi, S. and Yamamoto, T. (1981). *J. Biochem. (Tokyo)* **90**, 521–526.

Tani, Y. and Vongsuvanlert, V. (1987). *J. Ferment. Technol.* **65**, 405–411.

Tanriseven, A. and Robyt, J. F. (1989). *Carbohydr. Res.* **186**, 87–94.

Tao, B. Y., Reilly, P. J. and Robyt, J. F. (1988). *Carbohydr. Res.* **181**, 163–174.

Tarelli, E. (1980). *In* "Developments in Food Carbohydrates", Vol. 2 (C. K. Lee, ed.), pp. 187–227. Applied Science Publishers, London.

Taylor, R. L. and Conrad, H. E. (1972). *Biochemistry* **11**, 1383–1388.

Taylor, R. T. (1988). *In* "Biosensors" (D. L. Wise, ed.). CRC Press, Boca Raton, FL.

Thevelein, J. M. (1984). *Microbiol. Rev.* **48**, 42–59.

Thevelein, J. M. (1988). *Exp. Mycol.* **12**, 1–12.

Thomas, J. and Lobel, L. H. (1976). *Anal. Biochem.* **73**, 222–226.

Thompson, R. M. (1978). *J. Chromatogr.* **166**, 201–212.

Thorpe, G. H. G., Kricka, L. J., Moseley, S. B. and Whithead, T. P. (1985). *Clin. Chem.* **31**, 1335–1341.

Thummler, F. and Verma, D. P. S. (1987). *J. Biol. Chem.* **262**, 4730–4736.

Tjioe, T. T., VanDerWiel, J. P., Stikkelman, K. M., Straathof, A. J. J. and VanRantwijk, F. (1985). *J. Chromatogr.* **330**, 412–414.

Toba, T. and Adachi, S. (1977). *J. Chromatogr.* **135**, 411–417.

Tolmasky, M. E., Staneloni, R. J., Ugalde, R. A. and Leloir, L. F. (1980). *Arch. Biochem. Biophys.* **203**, 358–364.

Tomana, M., Niedermeier, W. and Spivey, C. (1978). *Anal. Biochem.* **89**, 110–118.

Tomiya, N., Kurono, M., Ishihara, H., Tejima, S., Endo, S., Arata, Y. and Takahashi, N. (1987). *Anal. Biochem.* **163**, 489–499.

Touchstone, J. C. (1988). *J. Chromatogr. Sci.* **26**, 645–649.

Touchstone, J. C. and Dobbins, M. L. (1983). "Practice of Thin Layer Chromatography", 2nd edn. John Wiley & Sons, New York.

Towbin, H., Schoenenberger, C. A., Braun, D. G. and Rosenfelder, G. (1988). *Anal. Biochem.* **173**, 1–9.

Townsend, R. R., Hardy, M. R., Hindsgaul, O. and Lee, Y. C. (1988). *Anal. Biochem.* **174**, 459–470.

Toyoda, Y., Fujii, H., Miwa, I., Okuda, J. and Sy, J. (1987). *Biochim. Biophys. Res. Commun.* **143**, 212–217.

Tressel, P. S. and Kosman, D. J. (1982). *Methods Enzymol.* **89**, 163–171.

Tsuji, A., Maeda, M. and Arakaw, H. (1985). *In* "Bioluminescence and Chemiluminescence Instruments and Applications", Vol. 1 (K. Van Dyke, ed.) pp. 185–202. CRC Press, Boca Raton, FL.

Tsujisaka, Y. (1988). *In* "Handbook of Amylases and Related Enzymes" (The Amylase Research Society of Japan, ed.), pp. 117–120. Pergamon Press, Oxford.
Tsumuraya, Y., Brewer, C. F. and Hehre, E. J. (1984). *Am. Chem. Soc.* 188th Meeting. Abst. CARB-25.
Tucker, J. H. (1881). "A Manual of Sugar Analysis", 2nd edn. Van Nostrand, New York.
Turner, A. P. F. (1988). *Methods Enzymol.* **137**, 90–103.
Turner, A. P. F., Karube, I. and Wilson, G. S. (1987). "Biosensors, Fundamentals and Applications". Oxford University Press, London.
Turner, A. P. F. and Swain, A. (1988). *Am. Biotechnol. Lab.* Nov., pp. 10–18.
Tyson, R. H. and ApRees, T. (1989). *Plant Sci.* **59**, 71–76.
Ugarova, N. N., Lebedeva, O. V. and Frumkina, I. G. (1988). *Anal. Biochem.* **173**, 221–227.
Urbanowski, J. C., Wunz, T. M. and Dain, J. A. (1980). *Anal. Biochem.* **105**, 461–467.
Van Beeumen, J. and DeLey, J. (1968). *Eur. J. Biochem.* **6**, 331–343.
Van Beeumen, J. and DeLey, J. (1975a). *Methods Enzymol.* **41**, 22–27.
Van Beeumen, J. and DeLey, J. (1975b). *Methods Enzymol.* **41**, 153–158.
Vandamme, E. J., Van Loo, J., Machtelinckx, L. and DeLaporte, A. (1987). *Adv. Appl. Microbiol.* **32**, 163–201.
Van Den, T., Bierman, C. J. and Marlett, J. A. (1986). *J. Agric. Food. Chem.* **34**, 421–425.
Van Dooren, J., Scholte, M. E., Postma, P. W., Van Driel, R. and Van Dam, K. (1988). *J. Gen. Microbiol.* **134**, 785–890.
Van Eijk, H. G., Van Noort, W. L., Dekker, C. and Van der Heul, C. (1984). *Clin. Chem. Acta* **139**, 187–193.
Van Handel, E. (1968). *Anal. Biochem.* **22**, 280–283.
Van Laere, A. J. and Slegers, L. K. (1987). *FEMS Microbiol. Lett.* **41**, 247–252.
Van Laere, A., Francois, A., Overloop, K., Verbeke, M. and Van Gerven, L. (1987). *J. Gen. Microbiol.* **133**, 239–245.
Van Noorden, C. J. F. and Butcher, R. G. (1989). *Anal. Biochem.* **176**, 170–174.
Van Schaffingen, E. (1987). *Adv. Enzymol.* **59**, 316–395.
Veening, H., Tock, P. P. H., Kraak, J. C. and Poppe, H. (1986). *J. Chromatogr.* **352**, 345–350.
Verhaar, L. A. Th. and Kuster, B. F. M. (1981). *J. Chromatogr.* **220**, 313–328.
Verzele, M., Simoens, G. and VanDamme, F. (1987). *Chromatographia* **23**, 292–300.
Villanueva, V. R., LeGoff, M. Th., Mardon, M. and Moncelon, F. (1987). *J. Chromatogr.* **393**, 115–121.
Vliegenthart, J. F. G., Dorland, L. and Van Halbeck, (1983). *Adv. Carbohydr. Chem. Biochem.* **41**, 209–374.
Volc, J. and Eriksson, K. E. (1988). *Methods Enzymol.* **161**, 316–322.
Volc, J., Nerud, F. and Musilek, J. (1985). *Folia Microbiol.* **30**, 141–147.
Vongsuvanlert, V. and Tani, Y. (1988). *Agric. Biol. Chem.* **52**, 419–426.
Von Lippmann, E. O. (1904). "Die Chemie der Zuckerarten", 2 Aufl. Vieweg u. Sohn, Braunschweig.
Vormbrock, R. (1984). *In* "Methods of Enzymatic Analysis", Vol. VI (H. U. Bergmeyer, J. Bergmeyer and M. Grassi, eds), pp. 172–178. Verlag Chemie, Weinheim and New York.
Vrátny, P., Brinkman, J. A. T. and Frei, R. W. (1985). *Anal. Chem.* **57**, 224–229.
Waege, T. J., Darvill, A. G., McNeil, M. and Albersheim, P. (1983). *Carbohydr. Res.* **123**, 281–304.
Waffenschmidt, S. and Jaenicke, L. (1987). *Anal. Biochem.* **165**, 337–340.
Wagner, W. and Wiemken, A. (1987). *Plant Physiol.* **85**, 706–710.
Wahlberg, I., Walsh, E. B., Forsblom, I., Oscarson, S., Enzell, C. R., Ryhage, R. and Isaksson, R. (1986). *Acta Chem. Scand.* **B40**, 724–730.
Waldmann, H., Gygax, D., Bednarski, M. D., Shangraw, W. R. and Whitesides, G. M. (1986). *Carbohydr. Res.* **157**, C4–C7.
Wallenfels, K. and Kurz, G. (1975). *Methods Enzymol.* **9**, 112–116.
Walters, J. S. and Hedges, J. I. (1988). *Anal. Chem.* **60**, 988–994.
Wang, W. T., Matsuura, F. and Sweeley, C. S. (1983). *Anal. Biochem.* **134**, 398–405.
Wang, W. T., LeDonne, Jr. N. C., Ackerman, B. and Sweeley, C. C. (1984). *Anal. Biochem.* **141**, 366–381.

Watanabe, N. (1985). *J. Chromatogr.* **330**, 333–338.
Watanabe, N. and Inoue, M. (1983). *Anal. Chem.* **55**, 1016–1019.
Watanabe, Y., Kibesaki, Y., Takenishi, S., Sakai, K. and Tsujisaka, Y. (1979). *Agric. Biol. Chem.* **43**, 943–950.
Webb, J. W., Jiang, K., Gillece-Castro, B. L., Tarentino, A. L., Plummer, T. H., Byrd, J. C., Fisher, S. J. and Burlingame, A. L. (1988). *Anal. Biochem.* **169**, 337–349.
Weidenhagen, R. and Lorenz, S. (1957). *Angew. Chem.* **69**, 641.
Weigel, H. (1963). *Adv. Carbohydr. Chem.* **18**, 61–97.
Weinhausen, G. and DeLuca, M. (1986). *Methods Enzymol.* **133**, 198–208.
Weinhausen, G., Kricka, L. J. and DeLuca, M. (1987). *Methods Enzymol.* **136**, 82–93.
Weiser, W., Lehman, J., Chiba, S., Matsui, H., Brewer, C. F. and Hehre, E. J. (1988). *Biochemistry* **27**, 2294–2300.
Welch, C., Bryant, A. and Kealey, T. (1989). *Anal. Biochem.* **176**, 228–233.
Wells, G. B. and Lester, R. L. (1979). *Anal. Biochem.* **97**, 184–190.
Wermuth, B. and Von Wartburg, J.-P. (1982). *Methods Enzymol.* **89**, 181–186.
Werner, W., Rey, H. G. and Wielinger, H. (1970). *Z. Anal. Chem.* **252**, 224–228.
Werr, W., Frommer, W. B., Maas, S. and Starlinger, P. (1985). *EMBO J.* **4**, 1373–1380.
Westermark, U. and Eriksson, K. E. (1988). *Methods Enzymol.* **160**, 463–468.
Whitaker, J. R. (1967). *In* "Paper Chromatography and Electrophoresis", Vol. I (G. Zweig and J. R. Whitaker, eds), pp. 232–279. Academic Press, New York.
Whitaker, D. P. (1984). *Phytochemistry* **23**, 2429–2430.
White, C. A. and Kennedy, J. F. (1981) *In* "Techniques in Carbohydrate Metabolism", B312 1–64. Elsevier-North Holland, Ireland.
White, J. W. and Subers, M. H. (1961). *Anal. Biochem.* **2**, 380–384.
Whithead, T. P., Kricka, L. J., Carter, T. J. N. and Thorpe, G. H. G. (1979). *Clin. Chem.* **25**, 1531–1546.
Wieland, E., Wilder-Smith, E. and Kather, H. (1986). *J. Clin. Chem. Clin. Biochem.* **24**, 399–403.
Wiemken, A., Frehner, M., Keller, F. and Wagner, W. (1986). *Curr. Top. Plant Biochem. Physiol.* **5**, 17–35.
Wight, A. W. and Van Niekerk, P. J. (1983). *Food. Chem.* **10**, 211–224.
Wilcheck, M. and Bayer, E. A. (1987). *Methods Enzymol.* **138**, 429–442.
Williams, S. ed. (1984). *In* "Official Methods of Analysis, Assoc. Off. Anal. Chem.", 14th edn., Sections 13, 130–31, 161. Arlington, VA.
Wilson, D. B. and Schell, M. A. (1982). *Methods Enzymol.* **90**, 30–35.
Wilson, C. and Lucas, W. J. (1987). *Plant Physiol.* **84**, 1088–1095.
Wing, R. E. and BeMiller, J. N. (1972). *In* "Methods in Carbohydrate Chemistry", Vol. VI, pp. 42–59. Academic Press, New York.
Wirtz, W., Stitt, M. and Heldt, H. W. (1980). *Plant Physiol.* **66**, 187–193.
Wober, G. (1973). *Hoppe Seyler's Z. Physiol. Chem.* **354**, 75–82.
Womersley, C. (1981). *Anal. Biochem.* **112**, 182–189.
Wong, C. G., Sung, S. S. J. and Sweeley, C. C. (1980). *Methods Carbohydr. Chem.* **8**, 55–65.
Wong-Chong, J. and Martin, F. A. (1979). *J. Agric. Food Chem.* **27**, 927–932.
Wood, B. J. B. and Rainbow, C. (1961). *Biochem. J.* **78**, 204–209.
Woodruff, S. D. and Yeung, E. S. (1982). *Anal. Chem.* **54**, 2124–2125.
Woodward, J. and Wiseman, A. (1982). *In* "Developments in Food Carbohydrate" Vol. 3 (C. K. Lee and M. G. Lindley, eds), pp. 1–21. Applied Science Publishers, London.
Worthington, C. C. (1988). "Worthington Enzyme Manual". Worthington, Biochemical, Co., Freehold, NJ.
Wrann, M. M. and Todd, C. W. (1978). *J. Chromatogr.* **147**, 309–316.
Xu, D. P., Sung, S. J. S., Loboda, T., Kormanik, P. P. and Black, C. C. (1989). *Plant Physiol.* **90**, 635–642.
Xu, Y., Guilbault, G. E. and Kuan, S. S. (1989). *Anal. Chem.* **61**, 782–784.
Yamaguchi, H. and Makumoto, T. (1977). *J. Biochem. (Tokyo)* **82**, 1673–1680.
Yamamoto, K., Kitamoto, Y., Ohata, N. and Isshiki, S. (1986). *J. Ferment. Technol.* **64**, 285–291.
Yamasaki, Y., Ueda, N., Nao, K., Sekija, M., Kawamori, R., Shichiri, M. and Kamada, T. (1989). *Clin. Chim. Acta* **93**, 93–98.

Yang, M. T., Milligan, L. P. and Mathison, G. W. (1981). *J. Chromatogr.* **209**, 316–322.
Yeung, E. S. (1989). *Liq. Chromatogr. Gas Chromatogr.* **7**, 118–126.
Yoda, K. (1988). *Methods Enzymol.* **137**, 61–68.
Yoneyama, Y. (1987). *Arch. Biochem. Biophys.* **255**, 168–175.
Zaitsu, K. and Ohkura, Y. (1980). *Anal. Biochem.* **109**, 109–113.
Zopf, D. A., Tsai, C. M. and Ginsburg, V. (1978). *Methods Enzymol.* **50**, 163–169.

5 Oligosaccharides

P. M. DEY

Department of Biochemistry, Royal Holloway and Bedford New College, University of London, Egham, Surrey TW20 0EX, UK

I. INTRODUCTION

The D-galactose-containing oligosaccharides that occur in free form will be the main subject of discussion in this chapter. Of these, the α-galactosyl derivatives of sucrose are

METHODS IN PLANT BIOCHEMISTRY Vol. 2
ISBN 0-12-461012-9

the most commonly occurring members in the plant kingdom. A general survey reveals that these are the most abundant soluble sugars in plants and rank only second to sucrose. These sugars are the primary oligosaccharides and are synthesised *in vivo*. In contrast, the secondary oligosaccharides are those which are derived by enzymic degradation of higher homologous oligosaccharides, heterosides and polysaccharides.

One physiological function of the galactosyl oligosaccharides is as reserve carbohydrates in storage organs such as seeds and tubers. They are mobilised during the early stages of germination (see Dey, 1985). They are therefore synthesised and deposited in these organs during the maturation process. It is also conceivable that a fraction of these oligosaccharides arrive at this location as translocated material. However, in some plants they constitute the major soluble carbohydrates that are translocated from leaves, the site of their biosynthesis (Webb and Gorham, 1964; Trip *et al.*, 1965; Senser and Kandler, 1967a,b; King and Zeevart, 1974; Peel, 1975; Turgeon and Webb, 1975; Ziegler, 1975; Zimmerman and Ziegler, 1975; Dickson and Nelson, 1982; Hendrix, 1982; Kandler and Hopf, 1980, 1982). In winter-hardy plants, the oligosaccharides are known to act as frost resistance factors (see Alden and Hermann, 1971; Levitt, 1972; Larcher *et al.*, 1973; Santarius and Milde, 1977; Dey, 1980a; Lineberger and Steponkus, 1980; Kandler and Hopf, 1982). As D-galactose toxicity in plants is well recognised (Ordin and Bonner, 1957; Baker and Ray, 1965; Ernst, 1967; Malca *et al.*, 1967; Göering *et al.*, 1968; Ernst *et al.*, 1971; Roberts *et al.*, 1971; Maretzki and Thom, 1978; Kapista and Evdonina, 1981; Dey, 1985), it is also possible that the formation of galactose-containing oligosaccharides serves as a detoxifying process.

The *in vivo* synthesis of D-galactosyl sucrose derivatives involves UDP-D-galactose as the galactosyl donor. However, galactinol [O-α-D-galactopyranosyl-($1 \rightarrow 1$)-*myo*-inositol] serves this function solely in the biosynthesis of the raffinose family of oligosaccharides (see Dey, 1985). The main pathway of degradation of these oligosaccharides is via α-galactosidase-catalysed cleavage of D-galactosyl groups (Dey and Pridham, 1972; Pridham and Dey, 1974; Dey, 1980a, 1985). The resulting D-galactose is readily phosphorylated and further metabolised (Dey, 1983, 1984, 1985). The ultimate product, sucrose, then enters various metabolic pathways. The D-fructosyl group of the sucrose moiety can be cleaved by both acid and alkaline invertases, but this pathway is unlikely to operate *in vivo*. The relative rate of D-fructosyl removal from the oligosaccharides is very much lower as compared to sucrose (see Dey and Del Campillo, 1984).

It has been generally observed that biosynthetic and/or degrading enzymes of galactosyl oligosaccharides are also found in the tissues that are used for isolating such sugars. It is, therefore, important to take appropriate precautions so that the action of endogenous enzymes does not produce secondary oligosaccharides and other products. For the same reason extremes of pH and high temperatures should also be excluded from the isolation protocol. Some measures that should be included are: (a) incorporation of enzyme inhibitors; (b) freezing the plant material in liquid nitrogen prior to blending and homogenisation; and (c) using boiling solvent for alcohol extraction. When dealing with lipid-rich plant sources, it is advisable to remove major lipids by treating and washing the tissue homogenate with chloroform. A chloroform–methanol (3:1; v/v) mixture may also be used for removing lipids from dry tissue powders; however, some sugars may also be solubilised and lost with the washings if the water content of the tissues is high enough ($> 8\%$). Finally, it is also important to deionise the sugar extract by passing it through a mixed bed of anion- and cation-exchange resins. This treatment allows an improved chromatographic resolution of the sugars.

The quantitative determination of oligosaccharides in an extract can be carried out via three possible routes: (a) enzymic method; (b) paper chromatographic separation followed by estimation of the pure oligosaccharide fractions; and (c) high performance liquid chromatographic (HPLC) determination. It should be remembered, however, that authentic samples of the sugars must be used as markers. In a homologous series of oligosaccharides, if the higher homologue markers are not available, a rough identification can be made from their paper chromatographic mobilities (R_f). French and Wild (1953) demonstrated a distinct correlation of structure with R_f. A straight line is obtained when the logarithm of the partition function, $\alpha' = R_f/(1 - R_f)$, for each oligosaccharide is plotted against molecular size. Increasing the size of an oligosaccharide by one monosaccharide unit generally decreases the logarithm of the partition function by a fraction which is dependent on the type of the monosaccharide unit being added and also on the type of linkage by which it is attached.

The enzymic method of determining α-D-galactosyl oligosaccharides will involve incubation with α-galactosidase followed by measuring the liberated D-galactose using either a chemical method or D-galactose dehydrogenase (see Chapter 1, this volume, for details). A useful colorimetric method for determining reducing carbohydrates has been recently described by Honda et al. (1982). Alternatively, the total oligosaccharide content can be expressed as sucrose units which would be the end-product of α-galactosidase action. Sucrose can be readily estimated in the same digest by first reacting with invertase followed by the colorimetric assay of glucose using glucose oxidase reagent (see Chapter 1, this volume). However, the α-galactosidase method is not adequate for determining an individual oligosaccharide in a mixture containing other homologous sugars.

Thin layer (TLC) or paper chromatographic methods are satisfactory, although time-consuming, and give good separation of oligosaccharides. The separated zones can be eluted with water and total carbohydrate quantified by a colorimetric method. Standard curves should be constructed from eluates obtained after similarly processing known concentrations of the marker oligosaccharides. The HPLC method, on the other hand, provides a rapid and efficient method of separating and quantifying the oligosaccharides with the aid of a refractive index monitoring system (Hicks, 1988). (The techniques relevant to this chapter can also be found in Chapters 1, 2, 4 and 6 of this volume.)

II. TRISACCHARIDES

The trisaccharides discussed in this section are the α-D-galactosides of sucrose in which the D-galactosyl group can be linked to either D-glucose or D-fructose moiety as shown in Table 5.1. Of the six known trisaccharides only three have been well investigated, i.e. umbelliferose, raffinose and plantenose.

A. Umbelliferose

This trisaccharide [α-D-galactopyranosyl- $(1 \rightarrow 2)$-α-D-glucopyranosyl-$(1 \leftrightarrow 2)$-β-D-fructofuranoside] was found in the roots of various umbellifers (Boerheim-Svendsen, 1954, 1958; Wickström and Boerheim-Svendsen, 1956) and its structure established (Fig. 5.1) as an isomer of raffinose. This sugar was further isolated from other plant organs as shown in Table 5.2. Umbelliferose was also detected in the families Araliaceae and

TABLE 5.1. Mono-*O*-α-D-galactosylsucrose derivatives.

α-D-Galactosyl linkage with	Product
C-2 of D-glucose	Umbelliferose
C-3 of D-glucose	Unnamed
C-6 of D-glucose	Raffinose
C-1 of D-fructose	Unnamed
C-3 of D-fructose	Unnamed
C-6 of D-fructose	Planteose

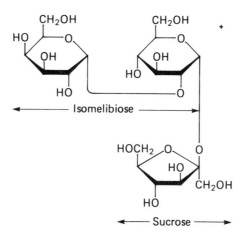

FIG. 5.1. Umbelliferose: α-D-galactopyranosyl-(1→2)-α-D-glucopyranosyl-(1↔2)-β-D-fructofuranoside.

Pittosporaceae, which have similarities with Umbelliferae (Hopf, 1973; Kandler and Hopf, 1982).

Umbelliferose was extracted by Boerheim-Svendsen (1958) by macerating 10–50 g fresh plant material with 150–250 ml ethanol plus 1–2 g CaCO$_3$ and Amberlite IR 120 (H$^+$) and Amberlite IRA 400 (OH$^-$), respectively. After concentrating the solution, it was passed through an alumina column to obtain a colourless solution. The trisaccharide gave a $R_{sucrose}$ value of 0.7 by paper chromatography with the solvent system isopropanol–pyridine–water (7:7:6; v/v). Crowden *et al.* (1969) found $R_{glucose}$ values of 0.46 with butanol–acetic acid–water (4:1:5; v/v), 0.62 with butanol–benzene–pyridine–water (4:1:1:3; v/v) and 0.7 with butanol–ethanol–pyridine–water (4:1:2:2; v/v).

The pathway of umbelliferose biosynthesis was established as follows using an enzyme preparation from the leaves of *Aegopodium podagraria* (Hopf and Kandler, 1974):

$$\text{UDP-D-}[^{14}\text{C}]\text{Galactose} + \text{sucrose} \rightleftharpoons [^{14}\text{C}]\text{umbelliferose} + \text{UDP}$$

TABLE 5.2. Distribution of umbelliferose and sucrose in the leaves, storage organs (rhizomes and roots), and fruits of various unbellifers.[a]

Species	Umbelliferose			Sucrose		
	Leaf	Storage organ	Fruit	Leaf	Storage organ	Fruit
Aethusa cynapium	1.0	—	2.1	38.1	—	5.5
Aegopodium podagraria	15.9	6.6	1.1	17.9	19.1	6.3
Anethum graveolens	—	—	10.5	—	—	2.1
Anthriscus cerefolium	6.0	1.6	11.6	50.6	25.2	4.2
Apium graveolens	2.6	0.9	7.4	29.6	12.5	5.2
Archangelica officinalis	7.4	9.5	27.4	17.0	9.2	26.1
Astrantia mairor	0.8	1.3	10.2	51.0	34.2	39.4
Azorella trifurcata	1.5	3.1	—	14.6	19.0	—
Bifora radians	—	—	8.3	—	—	3.5
Bowlesia incana	27.2	2.1	—	7.4	12.8	—
Bupleurum fruticosum	0.6	0.6	21.9	34.1	6.4	2.2
Carum carvi	1.1	3.0	5.6	32.1	29.0	10.3
Chaerophyllum bulbosum	19.0	5.3	6.0	45.0	11.8	18.2
Circuta virosa	—	—	10.2	—	—	8.2
Conium maculatum	0.8	0.8	2.8	51.5	24.3	13.2
Coriandrum sativum L.	—	—	11.1	—	—	6.5
Cryptotaenia japonica	—	—	10.7	—	—	20.2
Daucus carota	—	0.8	3.5	—	34.3	8.3
Dorema ammoniacum	20.9	8.0	15.5	43.6	17.2	22.6
Eryngium giganteum	1.3	1.4	4.4	44.1	6.8	19.3
Falcaria vulgaris	1.2	2.6	—	31.5	24.1	—
Foeniculum vulgare	1.0	0.7	6.5	50.3	28.6	6.7
Hacquetia epictis	1.0	0.7	—	52.9	21.9	—
Heracleum sphondylium	1.2	1.4	20.5	50.8	9.6	11.0
Hydrocotyle dissecta	—	—	0.6	—	—	4.4
Laserpitium siler	3.1	3.3	9.9	18.2	17.0	6.3
Ligusticum mutellina	1.8	3.4	12.1	42.6	10.9	20.1
Levisticum officinale	6.8	5.3	9.4	6.0	7.9	10.5
Myrrhis odorata	18.2	—	9.7	50.8	—	1.4
Oenanthe crocata	1.3	0.7	9.8	42.8	36.2	14.9
Orlaya grandiflora	—	—	8.8	—	—	5.7
Pastinaca sativa	1.2	—	8.3	52.7	—	10.7
Petroselinum crispum	1.5	—	8.3	55.4	—	7.2
Peucedanum ostruthium	25.6	11.4	9.8	34.2	6.8	8.2
Pimpinella siifolia	4.6	4.7	10.7	47.5	8.0	10.4
Pleurospermum austriacum	23.4	8.8	24.8	48.9	5.6	30.8
Scandix pecten-veneris	—	—	6.7	—	—	1.1
Seseli glaucum	0.7	1.5	8.3	26.8	24.5	11.9
Silaus teniufolius	1.1	1.5	—	47.6	42.4	—
Siler trilobus	2.5	—	—	52.4	—	—
Sium sisarum	2.9	1.3	12.1	41.6	37.5	24.3
Smyrnium olusatrum	1.7	0.7	4.8	52.7	26.4	15.5
Trinia kitaibellii	5.1	1.6	—	28.6	16.4	—

[a] Expressed as mg per g fresh weight of tissue; —, not examined. Data taken from Hopf and Kandler (1976).

The enzyme showed a pH optimum of ~ 7.5 and was stable at $-20°C$. One important step for the enzyme preparation was the removal of endogenous plant phenolics which caused loss of enzymic activity. The preparation also displayed other activities such as α-galactosidase, invertase, sucrose synthase and UDP-D-galactose 4'-epimerase which caused labelling of sucrose on prolonged incubation of the enzyme preparation with UDP-D-[^{14}C]galactose and sucrose. Under the pH conditions of the incubation there was no significant degradation of umbelliferose by the contaminating hydrolases. Table 5.3 shows that the galactosyl donor specificity of the enzyme is much restricted to UDP-D-galactose. The enzyme is also unable to catalyse an exchange reaction between [^{14}C]sucrose and umbelliferose.

TABLE 5.3. The acceptor and donor specificities of UDP-D-galactose:sucrose 2-α-D-galactosyl-transferase (umbelliferose synthase) from the leaves of *Aegopodium podagraria*.[a]

Acceptor	Donor	Newly synthesised labelled oligosaccharide
Cellobiose	UDP-D-[^{14}C]Galactose	None
D-Galactose	UDP-D-[^{14}C]Galactose	None
D-Glucose	UDP-D-[^{14}C]Galactose	None
Lactose	UDP-D-[^{14}C]Galactose	None
Maltose	UDP-D-[^{14}C]Galactose	None
Melibiose	UDP-D-[^{14}C]Galactose	None
Raffinose	UDP-D-[^{14}C]Galactose	None
Trehalose	UDP-D-[^{14}C]Galactose	None
Umbelliferose	UDP-D-[^{14}C]Galactose	None
Sucrose	UDP-D-[^{14}C]Galactose	Umbelliferose
[^{14}C]Sucrose	UDP-D-Galactose	Umbelliferose
[^{14}C]Sucrose	Galactinol	None
[^{14}C]Sucrose	D-Galactosyl phosphate	None
[^{14}C]Sucrose	Raffinose	None
[^{14}C]Sucrose	Umbelliferose	None

[a] After Hopf and Kandler (1974).

Hopf and Kandler (1976) showed that photosynthetic fixation of $^{14}CO_2$ in *A. podagraria* resulted in equal distribution of radioactivity in sucrose and umbelliferose. As compared to sucrose, the trisaccharide is not translocated and has a slower turnover. There was no ^{14}C incorporation into galactinol in mature leaves; however, it did occur in immature leaves. But the latter also contained raffinose, and galactinol is a recognised galactosyl donor for its biosynthesis. The *in vivo* utilisation of umbelliferose takes place by the action of α-galactasidase yielding sucrose and D-galactose. As far as the action of invertase is concerned, umbelliferose is fairly resistant to both acid and alkaline forms of the enzyme (Sömme and Wickström, 1965; see also Kandler and Hopf, 1980).

B. Raffinose

This trisaccharide (Fig. 5.2) was first isolated in a crystalline form from *Eucalyptus manna* (Johnston, 1843) and since then its occurrence in the plant kingdom has been

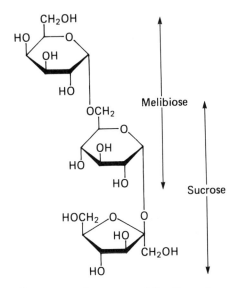

FIG. 5.2. Raffinose: α-D-galactopyranosyl-(1→6)-α-D-glycopyranosyl-(1↔2)-β-D-fructofuranoside.

TABLE 5.4. Amounts of α-D-galactosides of sucrose in some legume seeds (% of dry matter).[a]

Legume seeds	Galactinol	Sucrose	Raffinose	Stachyose	Verbascose
Chick pea					
(*Cicer Arietinum*)	0.39	2.69	0.45	1.72	0.10
			(1.10)	(2.50)	
Cow pea					
(*Vigna unguiculata*)	0.12	2.64	0.41	4.44	0.48
			(0.40)	(4.80)	(0.50)
Faba bean					
(*Vicia faba*)	0.22	2.00	0.22	0.67	1.45
Field bean					
(*Phaseolus vulgaris*)	0.15	3.01	0.26	2.16	0.03
			(0.20)	(1.20)	(4.00)
Lentil					
(*Lens culinaris*,	Traces	3.36	0.31	1.47	0.47
L. esculenta)			(0.90)	(2.70)	(1.40)
Lima bean					
(*Phaseolus lunatus*)	0.10	18.5	0.46	2.76	0.31
Mung bean					
(*Vigna radiata*)	0.19	0.96	0.23	0.95	1.83
Lupin					
(*Lupinus albus*)	0.23	2.63	0.82	4.11	0.48
Pea					
(*Pisum sativum*)	0.17	1.85	0.60	1.71	2.30
			(0.60; 0.30)	(1.90; 1.70)	(2.20)
Soybean					
(*Glycine max*)	Traces	6.35	1.15	2.85	—
			(0.80)	(5.40)	

[a] Data are taken from Sosulski *et al.* (1982), except that those presented in parentheses are from Cristofaro *et al.* (1974).

proved ubiquitous. The abundance of its distribution stands probably next to sucrose amongst the soluble sugars. The detection of the enzyme α-galactosidase, which degrades raffinose into sucrose and D-galactose, in any plant or its organs can be taken as a rough guide to predicting the *in vivo* presence of raffinose. Legume seeds are a particularly good source of this sugar (see Table 5.4); however, it coexists with other members of its family. The higher homologues are generally richer in the storage organs of plants (Table 5.5). Plant leaves synthesise this trisaccharide and it is then translocated to distant parts. Thus, radioactively labelled raffinose can be prepared by photosynthetic fixation of $^{14}CO_2$ (see Table 5.6).

TABLE 5.5. The distribution of galactinol, sucrose, and the raffinose family of oligosaccharides in various organs of some plant species.[a]

Plant species	Galactinol	Sucrose	Raffinose	Stachyose	Verbascose
Andromeda japonica					
Leaf	0.34	2.26	0.55	1.16	—
Stem	0.28	2.69	0.64	2.76	—
Buddleia davidii					
Leaf	0.18	3.68	0.22	0.16	—
Catalpa bignonioides					
Leaf, young	1.24	2.58	0.29	0.55	—
Leaf, old	2.32	5.67	0.86	3.61	—
Lamium maculatum					
Leaf	0.48	1.15	0.66	0.68	—
Root and rhizome	0.99	2.07	0.89	9.62	1.36
Lycopus europaeus					
Leaf	2.81	3.43	2.49	2.12	0.82
Root	2.12	3.60	3.27	7.70	5.50
Marrubium vulgare					
Leaf	0.57	2.00	0.65	1.34	—
Root and hypocotyl	0.22	2.65	0.33	1.84	0.62
Oenothera pumila					
Leaf	0.88	4.75	0.42	0.87	—
Stem	0.10	7.32	0.23	0.15	—
Origanum vulgare					
Leaf	0.97	8.28	0.78	1.40	—
Root and rhizome	0.88	4.86	0.87	1.67	0.92
Prunella grandiflora					
Leaf	0.72	2.47	0.52	1.88	—
Root and rhizome	0.68	3.20	0.93	2.78	1.42

[a] The quantities are expressed in mg per g fresh weight of tissue. After Senser and Kandler (1967a).

The effective extraction of raffinose and other oligosaccharides is generally done with 80% boiling ethanol (Knudsen, 1986), either by blending wet tissue material or dry powder in this solvent. A refluxing of approximately 1 h is required before filtering and concentrating the extract. The separation of the sugars can be achieved either by paper chromatography (see Table 5.7) or HPLC (Knudsen, 1986). It is advisable to deproteinate the extract before injecting it into the separation column. This is achieved by mixing the extract with acetonitrile (1:2; v/v) and leaving the mixture for 1 h at 5°C followed by filtration (Millipore, 0.45 μm). The HPLC method is also particularly useful

TABLE 5.6. The distribution (%) of ^{14}C among oligosaccharides during foliar photosynthesis in the presence of $^{14}CO_2$ in some plant species.

Plant species	Duration of photosynthesis (min)	Galactinol	Sucrose	Raffinose	Stachyose
Aristolochiales					
Aristolochiaceae					
Aristolochia clematis	60	10.3	86.7	2.3	0.1
Asarum europaeum	30	1.3	98.5	0.1	—
Celastrales					
Celastraceae					
Evonymus alatus	30	18.5	64.6	4.8	11.8
E. hamiltonianus	30	5.4	81.0	6.3	7.2
E. phellomanus	30	31.2	50.7	6.9	11.0
E. verrugineus	30	13.2	77.6	2.5	6.6
Cistales					
Violaceae					
Viola odorata	60	2.3	95.4	2.0	0.2
V. canadensis	60	0.6	98.4	0.9	0.1
Ericales					
Ericaceae					
Calluna vulgaris	30	2.5	96.3	1.0	0.2
Erica carnea	60	21.0	70.9	6.7	1.2
Rhododendron russatum	30	6.7	87.9	4.2	0.6
Vaccinium myrtillus	30	0.8	97.5	1.4	0.2
Gentianales					
Buddleiaceae					
Buddleia davidii	60	25.4	58.6	5.6	10.1
Oleaceae					
Forstiera neomexicana	30	6.9	88.9	0.7	3.5
Syringa villosa	30	26.3	54.4	3.1	16.2
Lamiales[a]					
Lamiaceae					
Ajuga reptans	60	13.3	52.3	8.6	24.1
Calamintha illyrica	30	6.2	75.7	8.9	8.5
C. vulgaris	30	8.5	65.5	4.0	19.5
Dracocephalum iberica	60	6.3	78.8	6.9	7.4
D. moldavica	60	12.1	57.0	8.1	21.4
D. sibiricum	30	32.5	35.6	5.8	23.9
Elscholtzia stauntonii	30	13.5	77.2	6.8	2.2
Galeopsis dubia	60	10.9	47.5	19.5	20.4
Hyssopus officinalis	60	20.1	53.7	7.6	16.8
Lamium album	60	8.5	46.2	14.3	29.4
L. maculatum	60	6.9	55.6	11.0	22.6
Lavendula latifolia	60	13.8	70.4	13.2	2.0
Lycopus europaeus	30	4.9	73.2	9.9	12.0
Marrubium vulgare	30	8.5	65.5	4.0	19.5
Mentha longifolia	30	6.2	76.5	7.5	9.2
Origanum vulgare	30	8.4	77.8	7.7	5.4
Phlomis tuberosa	60	8.4	65.3	6.7	19.6
Prunella grandiflora	60	12.2	55.3	14.1	17.7
Salvia cruinea	30	18.9	64.8	6.9	9.0
Salvia farinnacea	30	11.5	48.9	5.7	32.4
Scutellaria alpina	30	18.5	54.0	9.5	17.3

Continued

TABLE 5.6—continued.

Plant species	Duration of photosynthesis (min)	Galactinol	Sucrose	Raffinose	Stachyose
S. baicalensis	60	12.5	69.3	5.2	12.7
Stachys officinalis	60	16.6	58.2	6.4	17.1
Thymus villosa	60	20.6	54.3	9.4	14.1
Teucrium montanum	60	6.6	81.5	4.3	7.5
Verbenaceae					
Verbena hybridum (var. compactum)	30	7.0	60.1	9.4	21.3
Myrtalis					
Onagracea					
Epilobium palustre	30	7.1	70.4	14.9	7.4
Jussieua elegans	60	3.6	67.5	9.8	19.1
Oenothera tetraptera	30	1.4	93.7	4.7	0.2
Punicaceae					
Punica granata	30	11.0	65.4	4.5	19.1
Trapaceae					
Trapa natans					
Surface leaves	30	0.8	98.7	0.3	0.1
Submerged leaves	30	0.1	99.6	0.2	0.1
Papaverales					
Papaveraceae					
Dicentra spactabilis	30	0.7	99.0	0.2	—
Glaucium flavum	30	1.1	98.2	0.4	—
Papaver nudicaulis	30	0.6	99.2	0.1	—
Passifloralis					
Cucurbitaceae					
Cucurbita pepo	30	3.4	81.8	2.8	12.0
C. sativus	60	9.6	84.8	1.8	3.8
Ecballium elaterum	60	8.0	47.5	13.0	31.2
Thladiantha dubia	60	18.0	35.9	13.2	32.8
Trichosanthes japonica	60	13.5	58.8	9.2	18.5
Rosales					
Rosaceae					
Agrimonia leucantha	30	3.0	96.2	0.8	—
Alchemilla mollis	30	8.6	84.9	5.3	0.1
Filipendula hexapetala	60	2.2	94.2	3.0	0.1
Fragaria vesca	30	4.9	94.3	0.6	0.1
Rosa fendleri	30	6.6	90.5	1.3	0.1
Rutales					
Rutaceae					
Citrus trifoliatus	30	0.7	99.1	0.1	0.1
Dictamnus alba	30	2.6	92.7	2.7	2.0
Ruta graveolens	30	1.7	95.4	2.6	0.1
Scrophulales					
Scrophulariaceae					
Verbascum longifolium (var. pannosum)	30	13.7	79.8	1.5	4.1
Wulfenia carinthiaca	60	0.8	97.0	1.5	—

[a] Radioactivity in verbascose accounted for up to 3% in a number of species. After Senser and Kandler (1967a).

TABLE 5.7. Paper chromatographic mobilities ($R_{sucrose}$) of various α-D-galactosylsucrose derivatives (Whatman No. 1 paper at 25°C) using different solvent systems.[a]

Oligosaccharides	Solvent systems[b]			
	1	2	3	4
Trisaccharide				
Planteose	0.70	0.55	0.48	0.48
Raffinose	0.76	0.58	0.50	0.63
Umbelliferose	0.90	0.62	0.58	0.66
Tetrasaccharide				
Isolychnose	0.76	0.47	0.25	0.54
Lychnose	0.48	0.43	0.20	0.34
Sesamose	0.48	0.43	0.20	0.34
Stachyose	0.48	0.43	0.20	0.34
Pentasaccharide				
Verbascose	0.32	0.37	0.10	0.24
Hexasaccharide				
Ajugose	0.20	0.19	0.04	0.14

[a] After Kandler and Hopf (1982).
[b] 1. 88% Phenol–acetic acid–EDTA–water (840:10:1:160; v/v). 2. Solution A–Solution B (1:1; v/v) [Solution A = *n*-butanol–water (750:50; v/v); Solution B = propionic acid–water (352:448; v/v)]. 3. *n*-Butanol–pyridine–water–acetic acid (60:40:30:3; v/v). 4. *n*-Butanol–ethylacetate–acetic acid–water (40:30:25:40; v/v).

for quantitating the oligosaccharides; it is of course important to calibrate the column with authentic sugar samples. Alternatively, a biochemical analysis kit is available (Boehringer-Mannheim, GmbH) for raffinose determination. This technique is based upon the following reactions whereby the formation of NADH is measured spectro-photometrically:

$$\text{Raffinose} + \text{H}_2\text{O} \xrightarrow{\text{α-galactosidase}} \text{D-galactase} + \text{sucrose}$$

$$\text{D-Galactose} + \text{NAD}^+ \xrightarrow{\text{galactose dehydrogenase}} \text{galatonolactone} + \text{NADH} + \text{H}^+$$

It should be noted that the presence of other α-D-galactosyl sugars would interfere with raffinose estimation.

The well established pathway of raffinose biosynthesis (Senser and Kandler, 1967b; see also Kandler, 1967; Kandler and Hopf, 1980, 1982; Dey, 1980a, 1985) is as follows:

$$\text{UDP-D-Galactose} + myo\text{-inositol} \longrightarrow \text{galactinol} + \text{UDP}$$

$$\text{Galactinol} + \text{sucrose} \longrightarrow \text{raffinose} + myo\text{-inositol}$$

The D-galactosyl donor, galactinol [O-α-D-galactopyranosyl-(1→1)-*myo*-inositol], is of common occurrence in plants that also contain the raffinose family of oligosaccharides (see Tables 5.4, 5.5 and 5.6) and is synthesised by galactinol synthase (see review by Dey, 1985). This enzyme has been isolated and partially purified (Pharr *et al.*, 1981; Webb, 1982). An up-dated method for the extraction and purification of galactinol from *Cucumis sativus* has been described by Pharr *et al.* (1987). Raffinose synthase (galactinol: sucrose 6^{glu}-α-D-galactosyltransferase) which forms raffinose from galactinol and sucrose, was purified from the seeds of *Vicia faba* (Lehle and Tanner, 1972, 1973; see also Dey, 1985). The enzyme also catalyses the following exchange reaction:

$$\text{Raffinose} + [^{14}C]\text{sucrose} \longrightarrow [^{14}C]\text{raffinose} + \text{sucrose}$$

The activity–pH profile shows a broad pH dependence (pH 5.5–8.0) with an optimum at 6.8. At pH 7.2 the K_m values for galactinol and sucrose are 7×10^{-3} M and 9×10^{-4} M, respectively. The enzyme has a rigid requirement for sucrose as the glycosyl acceptor whereas galactinol, *p*-nitrophenyl-α-D-galactopyranoside and raffinose work as significant D-galactosyl donors.

Raffinose and related oligosaccharides generally accumulate during seed maturation (Holl and Vose, 1980). However, the rise in the levels of higher homologues of raffinose, e.g. stachyose and verbascose, is rapid. This is presumably because raffinose acts as precursor of these oligosaccharides. All of these oligosaccharides serve as storage carbohydrates and also as factors responsible for frost resistance in some winter-hardy plants. These sugars are mobilised during seed germination with the *in vivo* action of α-galactosidase.

C. Planteose

Planteose (Fig. 5.3) was first isolated from the seeds of *Plantago major* and *P. ovata* (Wattiez and Hans, 1943). It also occurs in plant species of the orders Gentianalis (Apocynaceae, Asclepiadaceae, Buddleiaceae, Loganiaceae), Hippuridales (Hippuridaceae),Lamiales (Lamiaceae, Verbenaceae), Loasales (Loasaceae;), Oleales (Oleaceae), Scrophulariales (Bignoniaceae, Pedaliaceae, Plantagenaceae, Scrophulariaceae) and Solanales (Boraginaceae, Convolvulaceae, Nolanaceae, Polemoniaceae, Solanaceae). Jukes and Lewis (1974) have suggested that this oligosaccharide may be of chemotaxonomic importance.

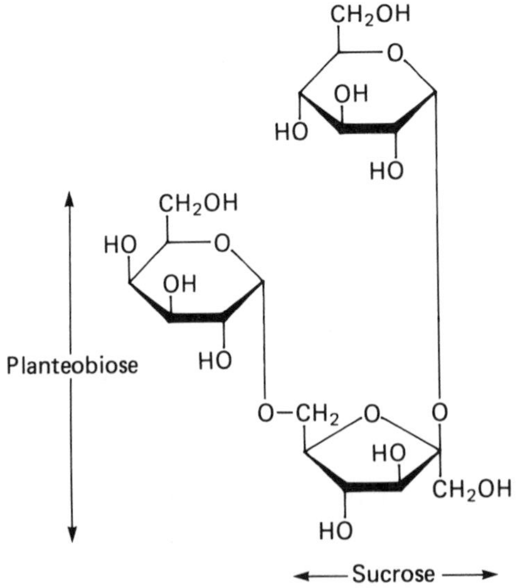

FIG. 5.3. Planteose: α-D-galactopyranosyl-(1→6)-β-D-fructofuranosyl-(2↔1)-α-D-glucopyranoside.

For isolation of planteose, defatted seedpowder of *P. ovata* was extracted with boiling 70% ethanol. Sugars in the extract were converted to their barium complexes and filtered, the complexes were treated with H_2SO_4 and precipitated barium salts removed, the solution was evaporated and syrup extracted with 85% ethanol. Planteose dihydrate crystallised out from this solution. As sucrose occurs in high concentration in mature seeds, this should be removed by yeast fermentation prior to complexing with barium. Paper chromatographic separation of planteose can be achieved by using the solvent systems described in Table 5.7. Hopf *et al.* (1984a) successfully used the solvent system ethylacetate–pyridine–water (10:4:3; v/v); paper electrophoresis in 0.05 M $Na_4B_2O_7$–NaOH buffer, pH 10.0 (2 kV, 25°C, 90 min) also gave good separation. Alkaline $AgNO_3$ (Trevelyan *et al.*, 1950) was used for visualising the sugar. Dey (1980b) used *p*-aminobenzoic acid as the spray reagent and the planteose spot gave a specific fluorescence when viewed under UV light.

The *in vitro* biosynthesis of planteose was first demonstrated by Dey (1980b) using the enzyme preparation from *Sesamum indicum*. The enzyme preparation from 10-week-old seeds (following anthesis) showed a preference for UDP-D-galactose over galactinol as D-galactosyl donor. Hopf *et al.* (1984a) later partially purified planteose synthase (UDP-D-galactose: sucrose 6^{fru}-α-D-galactosyltransferase) from the mature seeds of *S. indicum*. It catalyses the following reaction:

$$\text{UDP-D-Galactose} + \text{sucrose} \longrightarrow \text{planteose} + \text{UDP}$$

The enzyme was optimally active at pH 6.2 in 0.1 M Tris-acetate buffer, however, Imidazole-HCl and MES-NaOH were significantly inhibitory. Sodium citrate buffer caused nearly total inhibition. The enzyme was highly specific for sucrose as glycosyl acceptor and UDP-D-galactose as galactosyl donor; the K_m values for these substrates were respectively, 3.6–6.0 mM and 0.2–0.5 mM. At a concentration of 10 mM, UMP caused 55% inhibition of the activity whereas UDP, UDP-D-glucose D-galactose 1-phosphate, D-glucose and D-galactose had no effect at 1 mM. However, Mn^{2+} and Co^{2+} (at 10 mM) increased the enzymic activity, probably by inhibiting the contaminating UDP-D-glucose 4-epimerase; this enzyme competes for UDP-D-galactose.

In seeds, planteose concentration increases during the development and maturation periods (Amuti and Pollard, 1977; Dey, 1980b). It is interesting to note that in some species sucrose, raffinose and stachyose are present at high concentrations in roots and stems while planteose is concentrated in seeds. It therefore seems that planteose is synthesised *de novo* in the seeds. The *in vivo* function of planteose is probably as a storage carbohydrate which is utilised during seed germination via the action of α-galactosidase and invertase.

D. Other Trisaccharides

As noted in Table 5.1, substitutions of α-Dgalactosyl group at C-2 and C-6 of the D-glucose moiety of sucrose yield umbelliferose and raffinose, respectively. Such substitutions at C-3 and C-4 are also possible. The C-3 substituted isomer, α-D-galactopyranosyl-(1→3)-α-D-glucopyranosyl-(1↔2)-β-D-fructofuranoside, has been isolated from the seeds of *Lolium* and *Festuca* (MacLeod and McCorquodale, 1958a,b; Sömme and

Wickström, 1965). Both species belong to the family Graminae. The C-4 substituted isomer is not known to occur in a free form.

Of the products of α-D-galactosyl substitution on the D-fructose moiety of sucrose, planteose (substitution at C-6) is well characterised (see Section II.C). The substitution products at C-1 and C-3 are known to occur in the roots of *Silene inflata* (Davy and Courtois, 1965, 1966). The C-4 substituted isomer has not yet been detected.

III. TETRASACCHARIDES

The oligosaccharides described in this section are mainly stachyose, lychnose, isolychnose and sesamose, which are the higher homologues of the trisaccharides discussed in Section II. The tetrasaccharide based upon the structure of umbelliferose has not yet been characterised, although Kandler and Hopf (1982) detected it in *Aegopodium podagraria*.

A. Stachyose

Stachyose, a homologue of raffinose, was first isolated from the rhizomes of *Stachys tuberifera* by von Planta and Schulz (1890). Like raffinose, this sugar is abundantly distributed in the plant kingdom, and is a major oligosaccharide in several plant species (see Fig. 5.4, Tables 5.4, 5.5, 5.6).

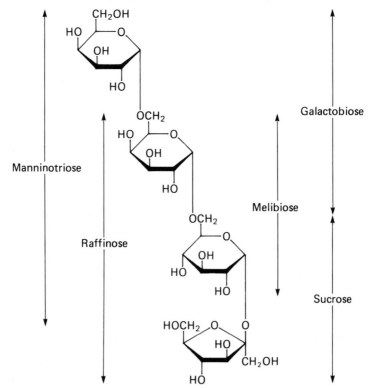

FIG. 5.4.   Stachyose: α-D-galactopyranosyl-(1→6)-α-D-galactopyranosyl-(1→6)-α-D-glucopyranosyl-(1↔2)-β-D-fructofuranoside.

As stachyose coexists with members of its homologous series and other sugars, earlier isolation and purification methods were lengthy (French, 1954). However, French *et al.* (1953) described a simpler method of isolation from *S. tuberifera*. The juice extracted from 256 g rhizomes was diluted with water to 1 litre and passed through a charcoal–celite (1:1; w/w) column (48 mm × 85 mm). The decolorised solution was evaporated and the syrup treated with glacial acetic acid and butyl alcohol. Seeding with stachyose hydrate crystals resulted into 11 g crystalline stachyose. This was recrystallised by dissolving in water and adding a mixture of methyl and butyl alcohol. The addition of acetic acid enhanced the first crystallisation.

Stachyose can be best detected by paper chromatography, TLC or HPLC with the help of authentic sample as a reference (see Table 5.7; Knudsen, 1986; Madore *et al.*, 1988). The quantitative determination of the oligosaccharide can be achieved with the use of α-galactosidase and D-galactose dehydrogenase as described for raffinose. As the release of D-galactose may occur in two steps.

$$\text{Stachyose} \xrightarrow{\text{α-Galactosidase}} \text{raffinose} + \text{D-galactose}$$

$$\text{Raffinose} \xrightarrow{\text{α-Galactosidase}} \text{sucrose} + \text{D-galactose}$$

The reaction should be allowed to run to its completion so that both the galactosyl groups are cleaved off. Under the conditions of partial hydrolysis, as when determining K_m of α-galactosidase with stachyose, the reaction product may also contain raffinose in addition to sucrose and D-galactose. Thus, it would be erroneous to consider the release of two D-galactose moieties from one stachyose molecule. The best approach will, therefore, be the use of HPLC method for D-galactose determination with appropriate correction for raffinose content in the reaction mixture.

The enzyme responsible for the synthesis of stachyose (galactinol : raffinose 6^{gal}-α-D-galactosyltransferase) was isolated by Tanner and Kandler (1966, 1968) from the seeds of *Phaseolus vulgaris* (see also Kandler, 1967; Tanner, 1969; Lehle and Tanner, 1972). Galactinol was the D-galactosyl donor

$$[^{14}C]\text{Galactinol} + \text{raffinose} \longrightarrow [^{14}C]\text{stachyose} + myo\text{-inositol}$$

and raffinose acted as the only efficient D-galactosyl acceptor. Stachyose synthase has also been isolated from the seeds of *Vicia faba* and the leaves of *Cucurbita pepo*; a comparative account of these enzymes is presented in Table 5.8. The enzymes are unable to synthesise raffinose, verbascose, or higher homologues which generally accumulate in the storage organs. In a typical assay, the enzyme is incubated with $[^{14}C]$galactinol (0.015 M) and raffinose (0.1 M) in 0.5 M phosphate buffer, pH 7.0 at 32°C. The labelled stachyose is separated by paper chromatography and the radioactive areas cut out and measured in a scintillation counter. In the event of using unlabelled reactants in the assay, the liberated *myo*-inositol can be measured with *myo*-inositol dehydrogenase (Weissbach, 1958). The enzyme from *P. vulgaris* is also able to catalyse the following exchange reactions (Tanner and Kandler, 1968):

$$myo\text{-}[^{14}C]\text{Inositol} + \text{galactinol} \longrightarrow [^{14}C]\text{galactinol} + myo\text{-inositol}$$

$$[^{14}C]\text{Raffinose} + \text{stachyose} \longrightarrow [^{14}C]\text{stachyose} + \text{raffinose}$$

TABLE 5.8. Comparison of some of the properties of galactinol:raffinose 6-α-D-galactosyltransferase from various sources.

Properties	Source of the enzyme		
	Cucurbita pepo[a]	*Phaseolus vulgaris*[b]	*Vicia faba*[c]
K_m (mM)			
Raffinose	4.6	0.84	0.85
Melibiose	5.2	12.00	3.90
Stachyose	—	—	3.30
Galactinol	7.7	7.30	11.00
V_{max} (%)			
Raffinose	—	100	100
Melibiose	—	46	80
Stachyose	—	—	35
Galactinol	—	—	—
pH Optimum	6.5–6.9	6.0–7.0	6.3
Inhibitors	Melibiose, *myo*-inositol, Tris, Zn^{2+}, Mn^{2+}, Ni^{2+}, Mg^{2+}, Ca^{2+}, Cu^{2+}, Ag$^+$	Sulphydryl compounds	Raffinose, stachyose
Specificity			
Donor	Galactinol and *p*-nitrophenyl-α-D-galactoside; UDP-D-galactose and melibiose do not act as donors	Galactinol; UDP-D-galactose and ADP-D-galactose do not act as donors	Galactinol
Acceptor	Raffinose, melibiose (poor, forms manninotriose), D-galactose (poor, two unidentified products); D-fructose, D-glucose, cellobiose, gentiobiose, maltose, melizitose, sucrose, trehalose, maltotriose, manninotriose, and stachyose do not act as acceptors	Raffinose, D-glucose (poor, forms melibiose), D-galactose (poor), lactose (poor); glycerol, D-fructose, sucrose, maltose, cellobiose, trehalose, gentiobiose, melizitose, and stachyose do not act as acceptors	Stachyose, raffinose, melibiose, D-galactose (poor), D-glucose (poor), lactose (poor); glycerol, D-fructose D-glucose 1-phosphate, D-glucose 6-phosphate, sucrose, maltose, cellobiose, trehalose, gentiobiose, melizitose and verbascose do not act as acceptors
Reversibility of the reaction	Freely reversible	Freely reversible $$\frac{[\text{stachyose}][myo\text{-inositol}]}{[\text{raffinose}][\text{galactinol}]} = 4$$	—

Exchange reactions	myo-[^{14}C]Inositol + galactinol $\rightleftharpoons$ [^{14}C]galactinol + myo-inositol; [^{14}C]Galactinol + stachyose $\rightleftharpoons$ [^{14}C]stachyose + galactinol [^{14}C]Raffinose + stachyose $\rightleftharpoons$ [^{14}C]stachyose + raffinose	myo-[^{14}C]Inositol + galactinol $\rightleftharpoons$ [^{14}C]galactinol + myo-inositol [^{14}C]Raffinose + stachyose $\rightleftharpoons$ [^{14}C]stachyose + raffinose	—
Stability	For 1 month at 4°C in the presence of the 20 mM 2-mercaptoethanol	For several months at 0°C	—

[a] Gaudreault and Webb (1981).
[b] Tanner and Kandler (1968), Lehle and Tanner (1973a).
[c] Tanner et al. (1967), Lehle and Tanner (1972).

The synthesised stachyose in leaves is transported to various organs where it can be stored or converted to other α-D-galactosyl oligosaccharides. For example, in roots and seeds, where endogenous synthesis may also occur, stachyose is deposited for further utilisation during germination periods. On the other hand, stachyose synthesised in mature leaves can be metabolised by immature leaves, thus serving as a ready source of energy (Webb and Gorham, 1964, 1965). α-Galactosidase and invertase are the main enzymes responsible for the breakdown of stachyose into its constituents. The immature leaves are unable to synthesise this oligosaccharide (Turgeon and Webb, 1975; Gaudreault and Webb, 1981). Stachyose is also known to provide frost-hardiness to winter-hardy plants.

B. Lychnose

The structure of lychnose resembles that of raffinose where a second α-D-galacto pyranosyl group is attached to the C-1 of the D-fructose moiety (Fig. 5.5). This tetrasaccharide is found in the vegetative storage organs of all species of the Caryophyllaceae; however, the seeds contain the oligosaccharides of the raffinose series (Schwarzmaier, 1973; see also Kandler and Hopf, 1980, 1982). Lychnose was first isolated from the roots of *Lychnis dioica* (Archambault *et al.*, 1956a,b; Courtois, 1957; Wickström *et al.*, 1958a,b) and later from other species, for example *Cucubalus baccifera* (Courtois and Ariyoshi, 1960), *Dianthus caryophyllus* (Courtois and Ariyoshi, 1962a,b), *D. lumnitzeri* (Königshofer *et al.*, 1979), *L. alba* (Paquin, 1958), *Silene inflata* (Davy and Courtois, 1965) and *Cerastium arvense* (Hopf *et al.*, 1984b).

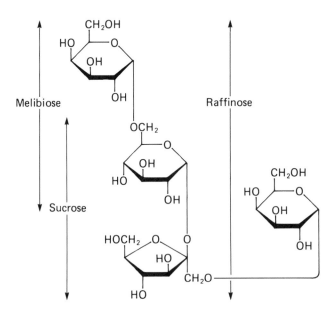

FIG. 5.5. Lychnose: α-D-galactopyranosyl-(1→6)-α-D-glucopyranosyl-(1↔2)-β-D-fructofuranosyl-(1←1)-α-D-galactopyranoside.

The basic procedure for isolating lychnose consists of homogenising the plant material in ethanol, concentrating the extract, redissolving the syrup in 70% ethanol, decolorising with activated charcoal, and finally concentrating and resolving by paper chromatography or HPLC. The data on the selection of chromatographic solvent systems and relative mobilities are presented in Table 5.7.

Schwarzmaier (1973) studied the labelling kinetics of oligosaccharides in several species of Caryophyllaceae during foliar photoassimilation of $^{14}CO_2$ and demonstrated rapid labelling of sucrose, galactinol and raffinose. Lychnose required a longer period of photoassimilation for labelling. This indicates raffinose working as the precursor of lychnose; UDP-D-galactose, galactinol, D-galactose 1-phosphate and stachyose failed to act as D-galactosyl donors in *in vitro* experiments (Schwarzmaier, 1973; Hopf *et al.*, 1984b). An enzyme preparation from leaves of winter-hardened *Cerastium arvense*, when incubated with [U-^{14}C]raffinose, produced labelled lychnose and sucrose. The pH optimum was 6.0 (Hopf *et al.*, 1984b). No higher homologues were formed, thus lychnose did not act as a D-galactosyl acceptor.

$$\text{Raffinose + raffinose} \longrightarrow \text{lychnose + sucrose}$$

In whole plant experiments lychnose accumulation was demonstrated during late autumn and winter and the sugar was consumed in spring. Thus, the oligosaccharide provides winter-hardiness to the plant (Hopf *et al.*, 1984b).

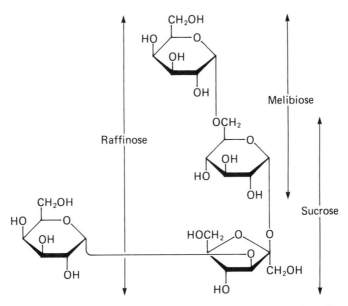

FIG. 5.6. Isolychnose: α-D-galactopyranosyl-(1→6)-α-D-glucopyranosyl-(1↔2)-β-D-fructofuranosyl-(3←1)-α-D-galactopyranoside.

C. Isolychnose

This tetrasaccharide is an isomer of lychnose where, instead of C-1, the C-3 of D-fructose moiety is linked to the α-D-galactosyl group (Fig. 5.6). This sugar was first isolated from the roots of *Lychnis dioica* (Wickström *et al.*, 1959) and further shown to coexist with lychnose in various roots and leaves (see Section III.B). Isolation and chromatographic separation are similar to that of lychnose (see also Table 5.7).

The enzyme preparation from *Cerastium arvense* which was shown to synthesise lychnose from two molecules of raffinose also synthesised isolychnose (Hopf *et al.*, 1984b).

$$\text{Raffinose} + \text{raffinose} \longrightarrow \text{isolychnose} + \text{sucrose}$$

However, it is not clear whether the same enzyme catalyses the formation of both oligosaccharides. The specificity for the D-galactosyl donor and acceptor and the optimum pH for synthesis are the same for both isomers. The physiological conditions under which isolychnose is synthesised by the plant parallel those of lychnose, i.e. the oligosaccharides are produced and accumulate most efficiently when the plants have acquired frost-hardiness. However, isolychnose occurs in quantities of only 5–10% that of lychnose (Hopf *et al.*, 1984b).

D. Sesamose

This tetrasaccharide (Fig. 5.7) is the next higher homologue of planteose (Fig. 5.3) with an additional α-D-galactosyl group. It was first isolated from the mature seeds of *Sesamum indicum* by Hatanaka (1959). It coexists along with the planteose; the extraction procedure and separation methods are therefore similar to those described in Section II.C.

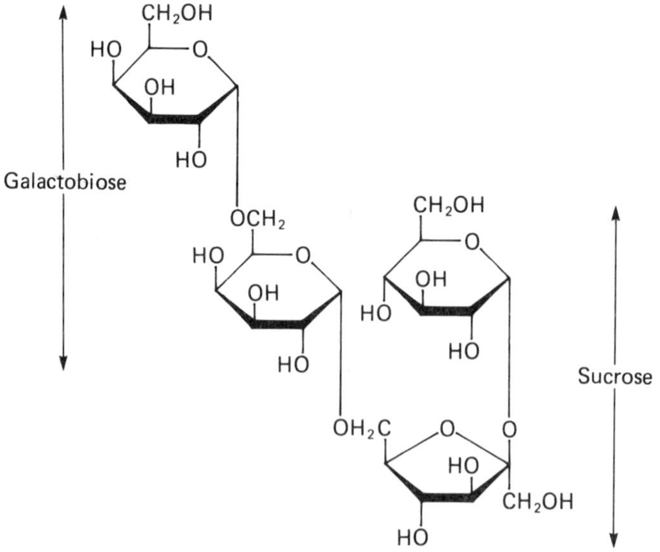

FIG. 5.7. Sesamose: α-D-galactopyranosyl-(1→6)-α-D-galactopyranosyl-(1→6)-β-D-fructopyranosyl-(2↔1)-α-D-glucopyranoside.

The synthesis of sesamose in *S. indicum* seeds occurs during the maturation period (Dey, 1980b). A crude enzyme preparation from this source was able to synthesise the tetrasaccharide from UDP-D-galactose and sucrose. However, according to Hopf *et al.* (1984a), who used the purified enzyme preparation, planteose failed to act as an α-D-galactosyl acceptor. Thus, it is not clear whether there are two distinct enzymes responsible for the synthesis of planteose and sesamose, respectively.

E. Other Tetrasaccharides

Kato *et al.* (1979) isolated two tetrasaccharides, I (Fig. 5.8) and II (Fig. 5.9), from cotton seeds by 80% methanol extraction. These sugars, together with small amounts of verbascose, were eluted from an activated carbon column with 15–20% ethanol and located between raffinose and stachyose on paper chromatograms; however, they had identical R_f values. When submitted to Dowex-1(borate form) column chromatography, I and II were eluted at 8 mM and 20 mM borate buffer, respectively. The tetrasaccharide I was detected earlier in *Triticum* seeds (White and Secor, 1953; Saunders and Walker, 1969; Saunders *et al.*, 1975). The biosynthetic details of the two sugars are not known (see also Dey, 1985).

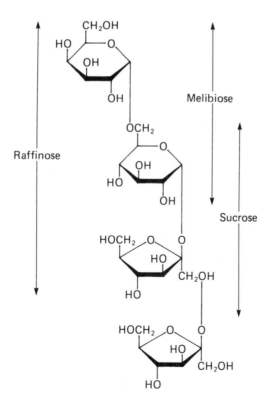

FIG. 5.8. Tetrasaccharide I: α-D-galactopyranosyl-(1→6)-α-D-glucopyranosyl-(1↔2)-[β-D-fructofuranosyl-(2→1)]-β-D-fructofuranoside.

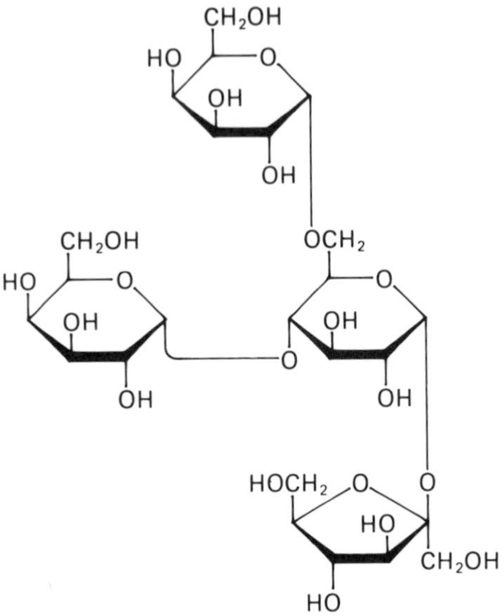

FIG. 5.9. Tetrasaccharide II: α-D-galactopyranosyl-(1→6)-[α-D-galactopyranosyl-(1→4)]-α-D-glucopyrano-syl-(1↔2)]-β-D-fructofuranoside.

A tetrasaccharide, α-D-galactopyranosyl-(1→4)-α-D-galactopyranosyl-(1→3)-α-D-glucopyranosyl-(1↔2)-β-D-fructofuranoside, was isolated from *Festuca rubra* by Mor-genlie (1970). The related trisaccharide with one fewer α-D-galactopyranosyl group (see Section II.D) also occurs in this source.

A tetrasaccharide, homologous to umbelliferose (see Section II.A), was reported to occur in *Aegopodium podagraria* (Kandler and Hopf, 1982).

IV. HIGHER HOMOLOGUES

Oligosaccharides belonging to the homologous series stachyose, sesamose, lychnose and isolychnose, are discussed below.

A. Stachyose Series

The first member of this series, the pentasaccharide verbascose, is formed by linking a new α-D-galactosyl group to C-6 of the non-reducing α-D-galactose moiety of stachyose (Fig. 5.4). Successive additions of further α-D-galactosyl groups produce the higher homologues.

Verbascose was first isolated by Bourquelot and Bridel (1910) from the roots of *Verbascum thapsus*. Other members of the series are also present in this source and can be separated by column chromatography (Hérissey *et al.*, 1954a), or by HPLC (Knudsen, 1986; see also Madore *et al.*, 1988). The paper chromatographic mobilities are shown in Table 5.7. In order to isolate the pentasaccharide in a crystalline form, 3 kg fresh roots of *V. thapsus* was extracted with boiling ethanol, vacuum-concentrated and the sugar precipitated with three volumes of 95% ethanol. The sugar was purified by barium complexing, removal of barium and finally precipitation with ethanol. The vacuum-dried product was extracted with 80% boiling methanol, filtered and mixed with an equal volume of absolute ethanol. On refrigeration, needle-shaped crystal clusters of verbascose were deposited (French, 1954).

Verbascose occurs in most leguminous plants along with other members of the series (Kandler and Hopf, 1982), but the highest concentrations are found in the storage organs (see Tables 5.4 and 5.5). The concentration of this sugar can be high in various seeds such as: *Cajanus cajan, Cicer arietinum, Dolichos uniflorus, Lens esculenta, Lupinus angustifolius, L. luteus, Phaseolus vulgaris, Pisum sativum, Vicia faba, V. sativa* and *Vigna radiata* (Cristofaro *et al.*, 1974; Cerning-Béroard and Filiartre, 1977, 1979, 1980; Holl and Vose, 1980; Kandler and Hopf, 1982; Sosulski *et al.*, 1982; see also Amuti and Pollard, 1977).

The formation of verbascose was demonstrated during foliar photoassimilation of $^{14}CO_2$ (Kandler and Hopf, 1982; Madore *et al.*, 1988); however, *in vitro* biosynthesis using leaf enzyme preparation has not so far been shown. On the other hand, an enzyme preparation from the seeds of *Vicia faba* catalysed the following two reactions (Tanner *et al.*, 1967; Lehle and Tanner, 1972):

$$\text{Galactinol} + \text{raffinose} \longrightarrow \text{stachyose} + myo\text{-inositol}$$

$$\text{Galactinol} + \text{stachyose} \longrightarrow \text{verbascose} + myo\text{-inositol}$$

A strong inhibition of the reactions was observed when both the substrates were present in the incubation mixture. This indicates that the same enzyme is responsible for the synthesis of both stachyose and verbascose.

The hexasaccharide of this series, ajugose, was first detected in the roots of *Ajuga nipponensis* (Murakami, 1942). The roots of *Verbascum thapsus* (Hérissey *et al.*, 1954a) and *Salvia pratensis* (Courtois *et al.*, 1956) and various legume seeds contain this sugar in small quantities (Courtois *et al.*, 1956; Tharanathan *et al.*, 1976; Amuti and Pollard, 1977; Cerning-Béroard and Filiartre, 1977, 1980). The higher homologues with a degree of polymerisation up to 15 have also been detected in some plants (Hérissey *et al.*, 1954b; Hattori and Hatanaka, 1958; Cerning-Béroard and Filiartre, 1977, 1980).

The biosynthesis of ajugose involves galactinol as the D-galactosyl donor and verbascose as the acceptor. The enzyme preparations from *Pisum sativum* and *Vicia sativa* were able to catalyse the synthesis of this hexasaccharide (Kandler and Hopf, 1982); however, verbascose does not work as an acceptor for the enzyme preparation from *V. faba* (Lehle and Tanner, 1972). A general pathway for the biosynthesis of the stachyose series of oligosaccharides is depicted in Fig. 5.10.

B. Sesamose Series

As in the stachyose series, successive additions of α-D-galactosyl groups to the C-6 of the existing D-galactose moiety of oligosaccharide (e.g. sesamose; see Fig. 5.7) produce the members of this series. The pentasaccharide (unnamed) was first isolated by Hatanaka (1959) from the seeds of *Sesasum indicum*; planteose and sesamose, along with raffinose and stachyose, were also detected in this source (Wankhede and

212   P. M. DEY

Tharanathan, 1976). However, Dey (1980b) demonstrated the absence of raffinose and its family of oligosaccharides. The isolation (Wankhede and Tharanathan, 1976) of sesamose and its penta- and hexasaccharides from *S. indicum* involved extraction of the defatted flour with 70% ethanol, passage through ion-exchanged resins [Dowex-(Cl$^-$) and Dowex-50 (H$^+$)], concentration and separation of the sugars on a charcoal–celite column using aqueous ethanol (increasing from 2–25%). Paper chromatographic separation and detection have been demonstrated by Dey (1980b).

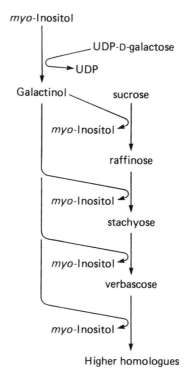

FIG. 5.10.   Pathway for the biosynthesis of the raffinose family of oligosaccharides.

In mature seeds of *S. indicum* the concentrations of the oligosaccharides decrease with increasing degree of polymerisation (Dey, 1980b). The biosynthetic pathway of the pentasaccharide was shown to involve UDP-D-galactose as the D-galactosyl donor. However, the enzyme preparation from 8-week-old seeds (following anthesis) produced no apparent synthesis of the oligosaccharide, while the 10-week-old seed extract gave positive results (Dey, 1980b).

C.   Lychnose Series

The oligosaccharides of this series (Fig. 5.11) occur in the vegetative storage organs of plant species belonging to Caryophyllacae, for example in the roots of *Cucubalus baccifera, Dianthus caryophyllus* and *Lychnis dioica* (Courtois *et al.*, 1958, 1960; Wickström *et al.*, 1959; Courtois and Ariyoshi, 1960, 1962a,b). Members of both the lychnose and isolychnose series coexist in the same plant sources (Courtois and Davy, 1965; Courtois and Percheron, 1965), and oligosaccharides containing 8-12 α-D-galactosyl groups have been identified (Courtois et al., 1960).

The lower homologues can be extracted from the roots of L. dioica with 70% ethanol; however, the higher homologues are extracted with water from the residual tissue (Courtois et al., 1960). The penta- and the hexasaccharides are more abundant than the higher homologues.

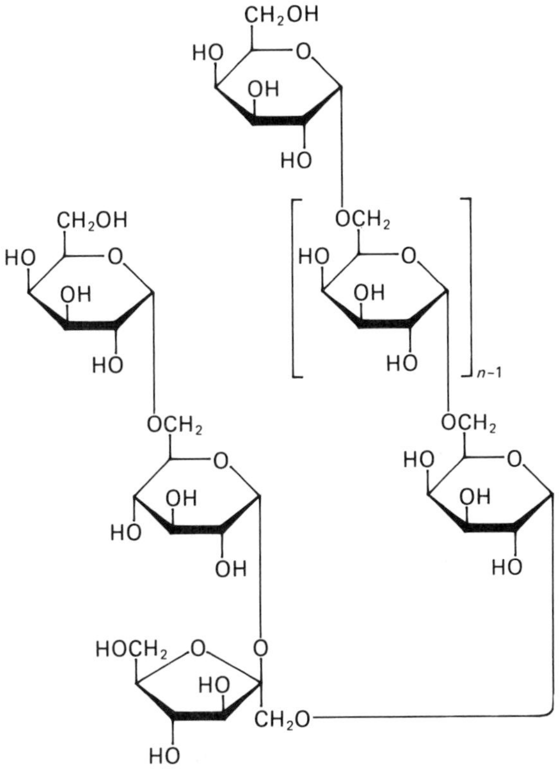

FIG. 5.11.   Lychnose series of oligosaccharides: higher homologues contain n number of α-D-galactopyrano-syl groups.

The function of the lychnose and isolychnose series is considered to be as reserve carbohydrates. They are formed in the plant in late autumn and reach their peak level during winter (Courtois, 1960; Schwarzmaier, 1973; Hopf et al., 1984b), thus also providing frost-hardiness. It is presumed that the pathway of biosynthesis of the oligosaccharides is analogous to those of lychnose and isolychnose (see Sections III.B and III.C).

D. Isolychnose Series

The members of this series (Fig. 5.12) have been isolated from the roots of plant species belonging to the family Sileneae (Courtois and Percheron, 1965; Davy and Courtois, 1966). These oligosaccharides frequently coexist in the same plant organ with those of lychnose series. However, in *Lychnis dioica*, isolychnose oligosaccharides dominate in the spring season (Courtois *et al.*, 1958; Wickström, 1959).

Members of the isolychnose series were separated by column chromatography using either cellulose or activated charcoal (Davy and Courtois, 1966). However, if the oligosaccharides of both series are present in the extract, it is difficult to resolve the isomers of identical molecular weight.

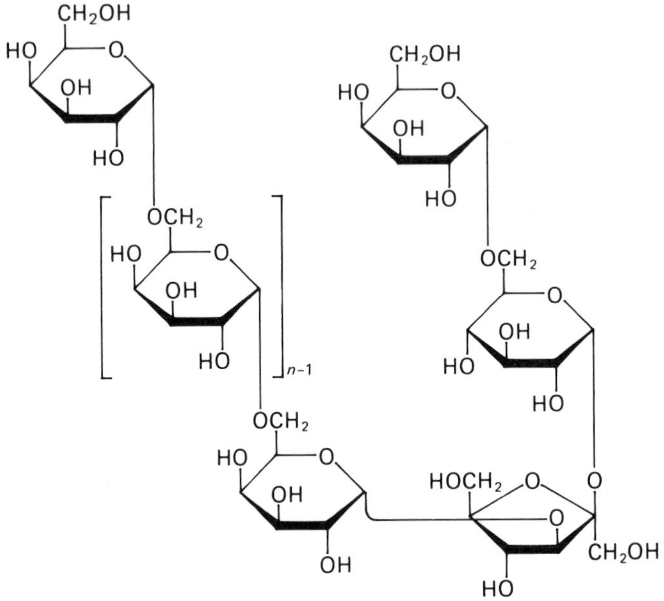

FIG. 5.12. Isolychnose series of oligosaccharides: higher homologues contain *n* number of α-D-galacto-pyranosyl groups.

V. CONCLUDING REMARKS

The types of freely occurring sugars in plants, other than sucrose, are fairly limited. Sucrose-based oligosaccharides and their homologues, especially those constituted with D-galactose residues, are the most common of all. As shown in Table 5.1, there still remain undiscovered members of this class of oligosaccharide, thus necessitating further work. These sugars function as storage carbohydrates and also impart frost-hardiness to many plants. In addition to these functional significances, they can also be considered as chemotaxonomic markers. The lack of a systematic survey in this area does not allow one to draw generalised conclusions. There is also a dearth of information on the cellular location of the D-galactosyl sucrose oligosaccharides. As far as the enzymology of the oligosaccharides is concerned, much is known of the enzymes that degrade them, i.e. mainly α-galactosidase and invertase. However, the biosynthetic enzymes have not been isolated in purified forms. It is important to study the properties of these enzymes, their control mechanisms, and their cellular location.

The general isolation techniques that have been employed for the oligosaccharides described in this chapter have in common the use of aqueous ethanol. No better alternative has thus far been developed in this respect. However, much progress has been made in the field of purification and resolution of oligosaccharides, most importantly on the microscale. Developments in HPLC techniques, coupled with more sensitive monitoring methods, are of special interest. These methods provide compactness, speed and improved resolution when compared with conventional chromatographic separations. Ion-exchange and ion-suppression amine adsorption HPLC methods are useful in determining charge and size characteristics of oligosaccharides. The use of lectins in fractionation and structural assessment of complex sugars is becoming widely accepted (Osawa and Tsuji, 1987). The lectin affinity HPLC technique (Green and Baenziger, 1989) should find increasing application in this respect. In the field of structural determination, although the basic chemical approach is of great value, the application of physical techniques, such as NMR, mass spectrometry and FAB-mass spectrometry have offered increased convenience and accuracy (Sweeley and Nunez, 1985; Dell, 1987).

The oligosaccharides which occur in combined form, for example in glycoproteins, find increasing importance in the field of cell recognition, differentiation, malignant transformation, protein targetting and expression in different cell types (Bacic and Clarke, 1987; Ferguson and Williams, 1988; Rademacher *et al.*, 1988). On the other hand, cell wall-derived oligosaccharides are recognised to work as 'hormones' and 'second messengers' and are implicated in a wide range of physiological functions (Albersheim *et al.*, 1983; Darvill and Albersheim, 1984; West *et al.*, 1984, 1985; Albersheim and Darvill, 1985; Robertsen, 1986; Ryan, 1987; Bacic *et al.*, 1988). These oligosaccharides and their molecular mode of action will undoubtedly attract much attention in the future.

REFERENCES

Alden, J. and Hermann, R. K. (1971). *Bot. Rev.* **37**, 37–142.
Albersheim, P. and Darvill, A. G. (1985). *Sci. Am.* (September), 44–50.
Albersheim, P., Darvill, A. G., McNeil, M., Valent, B., Sharp, J. K., Nothnagel, E. A., Davis, K. R., Yamazaki, N., Gollin, D. J., York, W. S., Dudman, W. F., Darvill, J. E. and Dell, A. (1983). *In* "Structure and Function of Plant Genomes" (O. Ciferri and L. Dure, III, eds), pp. 293–312. Plenum Press, New York.
Amuti, K. S. and Pollard, C. J. (1977). *Phytochemistry* **16**, 533–537.
Archambault, A., Courtois, J. E., Wickström, A. and Le Dizet, P. (1956a). *Bull. Soc. Chim. Biol.* **38**, 1121–1131.
Archambault, A., Courtois, J. E., Wickström, A. and Le Dizet, P. (1956b). *C.R. Hebd. Seances Acad. Sci.* **242**, 2875–2877.
Bacic, A. and Clarke, A. E. (1987). *Recent Adv. Phytochem.* **22**, 61–81.
Bacic, A., Harris, P. J. and Stone, B. A. (1988). *In* "Biochemistry of Plants" (P. K. Stumpf and E. E. Conn, eds.), Vol. 14, pp. 297–371. Academic Press, New York.
Baker, D. B. and Ray, P. M. (1965) *Plant Physiol.* **40**, 360–368.
Boerheim-Svendsen, A. (1954). Ph.D. Thesis, University of Oslo.
Boerheim-Svendsen, A. (1958). *Medd. Nor. Farm. Selsk.* **20**, 1–8.
Bourquelot, E. and Bridel, M. (1910). *C.R. Hebd. Seances Acad. Sci.* **151**, 760–762.
Cerning-Béroard, J. and Filiartre, A. (1977). *Comm. Eur. Commun.* [*Rep.*] *EUR.* **5686**, 65–79.
Cerning-Béroard, J. and Filiartre, A. (1979). *Lebensm. Wiss. Technol.* **12**, 273–280.

Cerning-Béroard, J. and Filiartre, A. (1980). *Z. Lebensm.-Unters.-Forsch.* **171**, 281–285.
Courtois, J. E. (1957). *Biokhimiya* **22**, 248–258.
Courtois, J. E. (1960). *Bull. Soc. Chim. Biol.* **42**, 1451–1466.
Courtois, J. E. and Ariyoshi, U. (1960). *Bull. Soc. Chim. Biol.* **42**, 737–751.
Courtois, J. E. and Ariyoshi, U. (1962a). *Bull. Soc. Chim. Biol.* **44**, 23–30.
Courtois, J. E. and Ariyoshi, U. (1962b). *Bull. Soc. Chim. Biol.* **44**, 31–37.
Courtois, J. E. and Davy, J. (1965). *C.R. Hebd. Seances Acad. Sci.* **261**, 3483–3485.
Courtois, J. E. and Percheron, F. (1965). *Mem. Soc. Bot. Fr.*, 29–39.
Courtois, J. E., Archambault, A. and Le Dizet, P. (1956). *Bull. Soc. Chim. Bio.* **38**, 1117–1119.
Courtois, J. E., Le Dizet, P. and Wickström, A. (1958). *Bull. Soc. Chim. Biol.* **40**, 1059–1065.
Courtois, J. E., Le Dizet, P. and Davy, J. (1960). *Bull. Soc. Chim. Biol.* **42**, 351–364.
Cristofaro, E., Mottu, F. and Wuhrmann, J. J. (1974). *In* "Sugars in Nutrition" (H. L. Sipple and K. W. McNutt, eds), pp. 313–336. Academic Press, New York.
Crowden, R. K., Harborne, J. B. and Heywood, V. H. (1969). *Phytochemistry* **8**, 1963–1984.
Darvill, A. G. and Albersheim, P. (1984). *Ann. Rev. Plant Physiol.* **35**, 243–275.
Davy, J. and Courtois, J. E. (1965). *C.R. Hebd. Seances Acad. Sci.* **261**, 3483–3485.
Davy, J. and Courtois, J. E. (1966). *Medd. Nor. Farm. Selsk.* **28**, 197–210; *Chem. Abstr.* **67**, 91053s (1967).
Dell, A. (1987). *Adv. Carbohydr. Chem. Biochem.* **45**, 19–72.
Dey, P. M. (1983). *Eur. J. Biochem.* **136**, 155–159.
Dey, P. M. (1984). *Phytochemistry* **23**, 729–732.
Dey, P. M. (1980a). *Adv. Carbohydr. Chem. Biochem.* **37**, 283–372.
Dey, P. M. (1980b). *FEBS Lett.* **114**, 153–156.
Dey, P. M. (1985). *In* "Biochemistry of Storage Carbohydrates in Green Plants" (P. M. Dey and R. A. Dixon, eds), pp. 53–129. Academic Press, London and New York.
Dey, P. M. and Del Campillo, E. (1984). *Adv. Enzymol.* **56**, 141–249.
Dey, P. M. and Pridham, J. B. (1972). *Adv. Enzymol.* **36**, 91–130.
Dickson, R. E. and Nelson, E. A. (1982). *Physiol Plant.* **54**, 393–401.
Ernst, R. (1967). *Am. Orchid Soc. Bull.* **36**, 1068–1073.
Ernst, R., Arditti, J. and Healey, P. L. (1971). *Am. J. Bot.* **58**, 827–835.
Ferguson, M. A. J. and Williams, A. F. (1988). *Ann. Rev. Biochem.* **57**, 285–320.
French, D. (1954). *Adv. Carbohydr. Chem.* **9**, 149–184.
French, D. and Wild, G. M. (1953). *J. Am. Chem. Soc.* **75**, 2612–2616.
French, D., Wild, G. M. and James, W. J. (1953). *J. Am. Chem. Soc.* **75**, 3664–3666.
Gaudreault, P.-R. and Webb, A. J. (1981). *Phytochemistry* **20**, 2629–2633.
Göering, H., Recklin, E. and Kaiser, R. (1968). *Flora (Jena)* **159**, 82–103.
Green, E. D. and Baenziger, J. U. (1989). *Trends Biochem. Sci.* **14**, 168–172.
Hatanaka, S. (1959). *Arch. Biochem. Biophys.* **82**, 188–194.
Hattori, S. and Hatanaka, S. (1958). *Bot. Mag.* **71**, 417–423.
Hendrix, J. E. (1982). *Plant. Sci. Lett.* **25**, 1–7.
Hérissey, H., Fleury, P., Wickström, A., Courtois, J. E. and Le Dizet, P. (1954a). *Bull. Soc. Chim. Biol.* **36**, 1507–1518.
Hérissey, H., Fleury, P., Wickström, A., Courtois, J. E. and Le Dizet, P. (1954b). *Bull. Soc. Chim. Biol.* **36**, 1519–1524.
Hicks, K. B. (1988). *Adv. Carbohydr. Chem. Biochem.* **46**, 17–72.
Holl, F. B., and Vose, J. R. (1980). *Can. J. Plant Sci.* **60**, 1109–1114.
Honda, S., Nishimura, Y., Takahashi, M., Chiba, H. and Kakehi, K. (1982). *Anal. Biochem.* **119**, 194–199.
Hopf, H. (1973). Ph.D. Thesis, University of Munich.
Hopf, H. and Kandler, O. (1974). *Plant Physiol.* **53**, 13–14.
Hopf, H. and Kandler, O. (1976). *Biochem. Physiol. Pflanz.* **169**, 5–36.
Hopf, H., Spanfelner, M. and Kandler, O. (1984a). *Z. Pflanzenphysiol. Bd.* **114**, 485–492.
Hopf, H., Gruber, G., Zinn, A. and Kandler, O. (1984b). *Planta* **162**, 282–288.
Johnston, J. F. W. (1843). *Phil. Mag.* **23**, 14–18.
Jukes, C. and Lewis, D. H. (1974). *Phytochemistry* **13**, 1519–1521.

Kandler, O. (1967). *In* "Harvesting the Sun: Photosynthesis in Plant Life" (A. S. Pietro, F. A. Greer and T. J. Army, eds), pp. 131–152. Academic Press, New York.

Kandler, O. and Hopf, H. (1980). *In* "Biochemistry of Plants" (J. Preiss, ed.), Vol. 3, pp. 221–270. Academic Press, New York.

Kandler, O. and Hopf, H. (1982). *In* "Encyclopedia of Plant Physiology, New Series" (F. A. Loewus and W. Tanner, eds), Vol. 13A, pp. 348–383. Springer, Berlin and New York.

Kapista, O. S. and Evdonina, L. V. (1981). *Genetika (Moscow)* 17, 424–436.

Kato, K., Abe, M., Ishiguro, K. and Ueno, Y. (1979). *Agric. Biol. Chem.* 43, 293–297.

King, R. W. and Zeevart, J. A. D. (1974). *Plant Physiol.* 53, 96–103.

Knudsen, I. M. (1986). *J. Sci. Food Agric.* 37, 560–566.

Königshofer, H., Albert, R. and Kinzel, H. (1979). *Z. Pflanzenphysiol.* 92, 449–453.

Larcher, W., Heber, V. and Santarius, K. A. (1973). *In* "Temperature and Life" (H. J. Precht, H. Christopherson, H. Hansel and W. Larcher, eds), pp. 195–292. Springer, Berlin and New York.

Lehle, L. and Tanner, W. (1972). *In* "Methods in Enzymology" (V. Ginsburg, ed.), Vol 28, pp. 522–530. Academic Press, New York.

Lehle, L. and Tanner, W. (1973). *Eur. J. Biochem.* 38, 103–110.

Levitt, J. (1972). *In* "Responses of Plants to Environmental Stresses". Academic Press, New York.

Lineberger, R. D. and Steponkus, P. L. (1980). *Plant Physiol.* 65, 298–304.

MacLeod, A. M. and McCorquodale, H. (1958a). *New Phytol.* 57, 168–182.

MacLeod, A. M. and McCorquodale, H. (1958b). *Nature (London)* 182, 815–817.

Madore, M., Mitchell, D. E. and Boyd, C. M. (1988). *Plant Physiol.* 87, 588–591.

Malca, I., Endo, R. M. and Long, M. R. (1967). *Phytopathology* 57, 272–278.

Maretzki, A. and Thom, M. (1978). *Plant Physiol.* 61, 544–548.

Morgenlie, S. (1970). *Acta Chem. Scand.* 24, 2149–2155.

Murakami, S. (1942). *Acta Phytochim.* 13, 37–56.

Ordin, L. and Bonner, J. (1957). *Plant Physiol.* 32, 212–215.

Osawa, T. and Tsuji, T. (1987). *Ann. Rev. Biochem.* 56, 21–42.

Paquin, R. (1958). M.Sc. Thesis, University of Montreal.

Peel, A. J. (1975). *In* "Encyclopedia of Plant Physiology, New Series" (M. H. Zimmermann and J. A. Milburn, eds), Vol. 1, pp. 171–196. Springer, Berlin and New York.

Pharr, D. M., Sox, H. N., Locy, R. D. and Huber, S. C. (1981). *Plant Sci. Lett.* 23, 25–33.

Pharr, D. M., Hendrix, D. L., Robbins, N. S., Gross, K. C. and Sox, H. N. (1987). *Plant Sci.* 50, 21–26.

Pridham, J. B. and Dey, P. M. (1974). *In* "Plant Carbohydrate Biochemistry" (J. B. Pridham ed.), pp. 83–96. Academic Press, London and New York.

Rademacher, T. W., Parekh, R. B. and Dwek, R. A. (1988). *Ann. Rev. Biochem.* 57, 783–838.

Roberts, R. M., Heiseman, A. and Wicklin, A. (1971). *Plant Physiol.* 48, 36–42.

Robertsen, B. (1986). *In* "Biology and Molecular Biology of Plant–Pathogen Interactions" (J. Bailey, ed.), pp. 177–183. Springer, Berlin.

Ryan, C. A. (1987). *Recent Adv. Phytochem.* 22, 163–179.

Santarius, K. A. and Milde, H. (1977). *Planta* 136, 163–166.

Saunders, R. M. and Walker, H. B., Jr. (1969). *Cereal Chem.* 45, 472–478.

Saunders, R. M., Betschart, A. A. and Lorenz, K. (1975). *Cereal Chem.* 52, 472–478.

Schwarzmaier, G. (1973). Ph.D. Thesis, University of Munich.

Senser, M. and Kandler, O. (1967a). *Phytochemistry* 6, 1533–1540.

Senser, M. and Kandler, O. (1967b). *Z. Pflanzenphysiol.* 57, 376–388.

Sömme, R. and Wickström, A. (1965). *Acta Chem. Scand.* 19, 537–540.

Sosulski, F. W., Elkowicz, L. and Reichert, R. D. (1982). *J. Food. Sci.* 47, 498–502.

Sweeley, C. C. and Nuñez, H. A. (1985). *Ann. Rev. Biochem.* 54, 765–801.

Tanner, W. (1969). *Ann. N.Y. Acad. Sci.* 165, 726–742.

Tanner, W. and Kandler, O. (1966). *Plant Physiol.* 41, 1540–1542.

Tanner, W. and Kandler, O. (1968). *Eur. J. Biochem.* 4, 233–239.

Tanner, W., Lehle, L. and Kandler, O. (1967). *Biochem. Biophys. Res. Commun.* 29, 166–171.

Tharanathan, R. N., Wankhede, D. B. and Rao, R. R. (1976). *J. Food. Res.* **41**, 715–716.
Trevelyan, W. E., Procter, D. D. and Harrison, L. S. H. (1950). *Nature (London)* **166**, 444–445.
Trip, P., Nelson, C. D. and Krotkov, G. (1965). *Plant Physiol.* **40**, 740–747.
Turgeon, R. and Webb, J. A. (1975). *Planta* **123**, 53–62.
von Planta, A. and Schulz, E. (1890). *Ber. Dtsch. Chem. Ges.* **23**, 1692–1699.
Wankhede, D. B. and Tharanathan, R. N. (1976). *J. Agric. Food Chem.* **24**, 655–659.
Wattiez, N. and Hans, M. (1943). *Bull. Acad. R. Med. Belg.* **8**, 386–396.
Webb, J. A. (1982). *Can. J. Bot.* **60**, 1054–1059.
Webb, J. A. and Gorham, P. R. (1964). *Plant Physiol.* **39**, 663–672.
Webb, J. A. and Gorham, P. R. (1965). *Can. J. Bot.* **43**, 97–103.
Weissbach, A. (1958). *Biochem. Biophys. Acta.* **27**, 608–611.
West, C. A., Bruce, R. J. and Jin, D. F. (1984). *In* "Structure, Function and Biosynthesis of Plant Cell Walls" (W. M. Dugger and S. Bartnicki-Garcia, eds), pp. 359–380. Waverly Press, Baltimore.
West, C. A., Moesta, P., Jin, D. F., Lois, A. F. and Wickham, K. A. (1985). *In* "Cellular and Molecular Biology of Plant Stress" (J. L. Key and T. Kosuge, eds), pp. 335–350. Alan R. Liss, New York.
White, L. M. and Secor, G. E. (1953). *Arch. Biochem. Biophys.* **44**, 244–245.
Wickström, A. and Boerheim-Svendsen, A. (1956). *Acta Chem. Scand.* **10**, 1199–1207.
Wickström, A., Courtois, J. E., Le Dizet P. and Archambault, A. (1958a). *Bull Soc. Chim Fr.* pp. 1410–1415.
Wickström, A., Courtois, J. E., Le Dizet, P. and Archambault (1958b). *C.R. Hebd. Seances Acad. Sci.* **246**, 1624–1626.
Wickström, A., Courtois, J. E., Le Dizet, P. and Archambault, A. (1959). *Bull. Soc. Chim. Fr.* pp. 871–878.
Ziegler, H. (1975). *In* "Encyclopedia of Plant Physiology, New Series" (M. H. Zimmermann and J. A. Milburn, eds), Vol. 1, pp. 59–100. Springer, Berlin and New York.
Zimmermann, M. H. and Ziegler, H. (1975). *In* "Encyclopedia of Plant Physiology, New Series" (M. H. Zimmermann and J. A. Milburn, eds), Vol. 1, pp. 480–505. Springer, Berlin and New York.

6   Cyclitols

FRANK A. LOEWUS

*Institute of Biological Chemistry, Washington State University, Pullman, WA
99164-6340, USA*

I.   INTRODUCTION

Although cyclitols represent only a small group among the many polyhydroxylated
compounds found in nature, their status has grown considerably in recent years as
newly discovered functional roles for these substances have been identified in living
cells. The trivial term *cyclitol* is used to identify polyhydroxycycloalkanes and their

METHODS IN PLANT BIOCHEMISTRY Vol. 2
ISBN 0-12-461012-9

derivatives, while the more selective term *inositol* singles out those which are cyclo-hexanehexols bearing one hydroxyl group on each of three or more ring atoms. Because most analytical concern focuses on the latter due to their vital role in cellular metabolism, particular attention will be given to the inositol group of compounds while broader implications will be noted where appropriate.

II. CONFIGURATIONAL CONSIDERATIONS

The stereochemical qualities of inositols place special demands on nomenclature which are not usually encountered in the general rules of organic chemistry. To this end, recommendations were published by IUPAC-IUB (1973) which dealt with stereochemical assignments for polyhydroxylated cycloalkanes containing only one kind of substituent. The central role of *myo*-inositol in cellular metabolism of inositol phospholipids has attracted particular attention to stereochemical features which are peculiar to cyclitols and these are discussed in recent publications (Parthasarathy and Eisenberg, 1986; Irvine, 1986; Michell and Berridge, 1988). For present purposes it is sufficient to illustrate stereoisomeric configurations for each of the eight diasteriomeric forms of inositol (Fig. 6.1). Since *chiro*-inositol is enantiomeric, a total of nine structures is possible. A detailed discussion of the effect of substitution on molecular asymmetry is beyond the scope of this chapter, but a simple introduction to the stereochemistry of the biological parent of naturally occurring inositols is in order. *myo*-Inositol contains a plane of symmetry that divides the molecule into prochiral halves. Substitution at C-1 or C-3 will create one set of enantiomers, while substitution at C-4 or C-6 will create another set. To illustrate with specific examples, there are six monomethyl ethers of *myo*-inositol (Fig. 6.2), two pairs of which are enantiomeric. Of the six methyl ethers, four occur naturally, 1D-1-*O*-methyl-*myo*-inositol [D-(−)-bornesitol], 1L-1-*O*-methyl-*myo*-inositol [L-(+)-bornesitol], 1-D-4-*O*-methyl-*myo*-inositol [D-(+)-ononitol] and 5-methyl-*myo*-inositol (sequoyitol). A numeral preceding the chiral designation identifies the numbered ring atom, clockwise (L) or counterclockwise (D) as shown in Fig. 6.1.

III. SOURCES IN THE LITERATURE

A comprehensive treatise on the chemistry and biochemistry of cyclitols prepared by T. Posternak (1962), later revised and translated into English (Posternak, 1965), provides an excellent source of information on this subject. Although it was prepared prior to the recommended changes in cyclitol nomenclature, these rules are readily inserted where needed. Additional details on the physical, chemical and biochemical properties of cyclitols are reviewed by Angyal and Anderson (1959) and Anderson (1972). Another sourcebook, limited to the inositol phosphates but comprehensive in scope, has been prepared by Cosgrove (1980). Many of the methodologies mentioned in these sources are still practised, and readers of this chapter are urged to consult these references in this regard. Reviews of a more limited nature on metabolic processes of cyclitols and their physiological roles in plants have appeared (Loewus and Loewus, 1980; Hoffmann-Ostenhof and Pittner, 1982; Loewus and Dickinson, 1982; Loewus and Loewus, 1983; Drew, 1984; Mórré *et al.*, 1990).

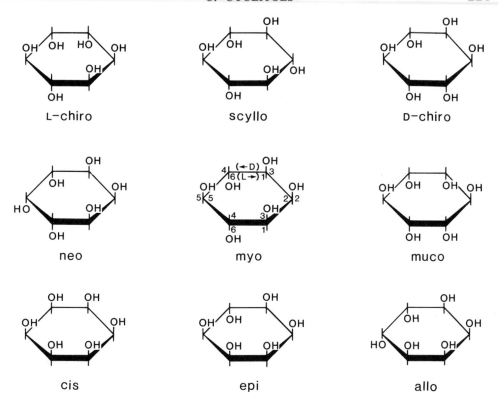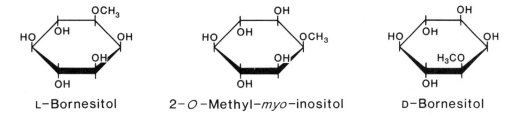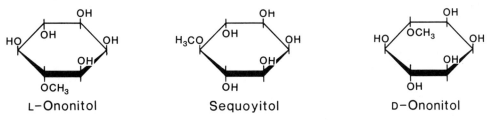

FIG. 6.1. Stereoisomers of inositol.

FIG 6.2. Isomers of monomethyl-*myo*-inositol.

A review on polyols in fungi and plants discusses methods of detection and quantitative estimation (Lewis and Smith, 1967). Although this review specifically excludes cyclitols from consideration, it provides useful information on earlier methods used to analyse polyhydric compounds as a class.

Recently, Mahadevappa and Holub (1987) reviewed various chromatographic and allied methods used for analysis of phosphoinositides and their breakdown products in activated blood platelets and neutrophils. Their review addresses various methodologies of extraction, separation, identification and quantitation used to resolve phospholipids, molecular species of phosphatidylinositol, inositol phosphates, acylglycerols and phosphatidic acids. This document is a rich source of information on current techniques, especially those involving thin layer chromatography of lipophylic structures, and has immediate application to plant as well as animal systems.

IV. SEPARATION OF ISOMERIC FORMS OF INOSITOL

Of the nine stereoisomeric forms of inositol, six have been identified as natural constituents in one or more plant species. They include *myo*-, *scyllo*-, D-*chiro*-, L-*chiro*-, *muco*-, and *neo*-inositol. *muco*-Inositol is a constituent of certain seagrasses (Drew, 1984) and *neo*-inositol has been reported to be present in an euphorbiacious plant, *Croton celtidifolius* (Mukherjee and Axt, 1984). *Epi*-, *cis*- and *allo*-inositols have not been found in plants. The separation of these isomers and their *O*-methyl ethers has offered a challenge to each succeeding development in the chromatography of natural products.

A. Cellulose Powder and Paper Chromatography

Cellulose powder chromatography, one of the earliest methods applied to the separation of inositol isomers, is still considered a practical method for large-scale resolution and purification in certain cases. *scyllo*- and *myo*-Inositol are readily resolved on a well-packed microcrystalline cellulose column (100×2 cm) by elution with acetone–water (4:1; v/v). Generally speaking, R_f values obtained when using paper sheets correspond to values obtained upon separation in packed columns. Posternak (1965, pp. 21, 45–49) has listed a number of solvent systems and methods of detection along with R_f values for all isomers except *muco*-inositol. A practical method for detection of non-reducing saccharides including cyclitols has been developed. In this system, the paper is first sprayed with 1% sodium periodate (in 50% aqueous acetone), then a preparation of silver nitrate (1 ml saturated silver nitrate in 100 ml of 95% aqueous acetone), and finally 1% sodium hydroxide. When the spots have developed, the paper is dipped in 5% sodium thiosulphate and dried to preserve the record (Yamada *et al.*, 1975).

B. Gas–Liquid Chromatography

The extremely simple procedure of trimethylsylation for preparing volatile derivatives of the cyclitols (Sweeley *et al.*, 1963) opened a new approach to qualitative and quantitative analysis that is still widely used. A wide choice of column packings allows the separation of all of the inositol isomers as well as their methyl ethers (Loewus and

Shah, 1972). Ford (1985) developed methods for separating three inositol methyl ethers and their parent inositols which occur in grain and forage legumes.

Refinements such as capillary columns and the use of mass spectrometry to analyse products after separation have enhanced the use of these derivatives (Niwa *et al.*, 1983; Rittenhouse, 1987). Systematic analysis of *O*-methyl-inositols has provided optimal conditions for the identification and separation of pinitol, ononitol, bornesitol and *chiro*-inositol in reference to *myo*-inositol (Binder and Haddon, 1984). Hexa*kis*trimethyl-silyl-*myo*-inositol is widely used as an internal standard in experiments involving trimethylsilylation of hydrolysates from plant cell walls and other plant polysaccharides. Since both hexa*kis*trimethylsilyl-*myo*-inositol and -*scyllo*-inositol can be prepared as crystalline solids, these inositol derivatives offer a practical approach to quantitation (Loewus, 1966).

An interesting example of the use of pinitol as a natural marker to detect adulteration of cocoa powder by carob powder has recently appeared (Baumgartner *et al.*, 1986). GC-MS analysis of oxime trimethylsilyl derivatives of fermented carob extract revealed the presence of pinitol, sorbitol plus *chiro*-inositol, sequoyitol plus ononitol plus bornesitol and *myo*-inositol in four separate peaks. The major cyclitol was pinitol, which is absent from cocoa powder.

Other volatile forms of cyclitols which are useful in gas chromatography include hexa*kis*-*O*-acetyl, hexa*kis*-*O*-trifluoroacetyl and butaneboronyl ester derivatives (Laker, 1980). Irving (1981) successfully resolved the hexa*kis*-*O*-acetyl esters of all five naturally occurring isomers of inositol. This derivative has been used successfully as the hexa*kis*-*O*-acetyl-*myo*-[2-^{2}H]inositol to determine the content of *myo*-inositol in mammalian tissues and fluids by GC-MS (Anderson *et al.*, 1982, Swahn, 1985). Similar applications in plant-derived extracts are indicated.

Hexa*kis*-*O*-trifluoroacetyl esters have higher volatility than the corresponding *O*-acetyl derivatives and are substantially more stable than the corresponding *O*-trimethyl-silyl derivatives. *O*-Trifluoroacetyl esters of several cyclitols have been resolved in less than 20 min on a high-temperature-resistant stationary phase of polycarbonane siloxane (Dexsil) (Englmaier, 1985). The trifluoroacetyl derivatives are well suited to mass analysis and have been used to compare *myo*-inositol with its hexadeuterated counterpart (Turk *et al.*, 1986). Butaneboronyl esters have more limited use due to the bifunctional reagent, butaneboronic acid, which selects compounds with even numbers of reactive groups (hydroxyl, amino or carboxyl) (Eisenberg, 1972). Since it is also quite useful as a means of achieving baseline resolution of mannitol, glucitol and galactitol derivatives, this method has value in studies that are concerned with glycosylated inositols.

chiro-Inositol is the only diastereomer which in its unsubstituted form has enantiomers. Leavitt and Sherman (1982a,b) have successfully resolved these enantiomers into D and L forms on a Chirosil-Val capillary column as the hexa*kis*(heptafluoro-butyric) esters.

C. Thin Layer (Planar) Chromatography

Systems employing this technique are often dedicated to separation of compounds such as phospholipids which are readily separated through two-dimensional development. A disadvantage of this approach is the limit of one sample per plate. Introduction of pre-adsorbent silica gel high performance thin layer chromatographic plates enables this separation to be made in one dimension (Korte and Casey, 1982; Dugan, 1985). Less attention has been given to general applications that involve water-soluble compounds, although some useful procedures have been developed such as separation of *myo*-inositol and its monophosphates on silica gel glass-fibre sheets (Hokin-Neaverson and Sadeghian, 1976). Relatively inexpensive supplies and equipment are required for this approach to separation of cyclitols, which suggests that a greater effort should be made to exploit current knowledge regarding derivatives and their properties as gained from gas–liquid chromatography and apply these findings to simple systems. Observations such as those reported by Tate and Bishop (1962) on the separation of the acetyl esters of *myo*- and *chiro*-inositols are readily refined with the aid of recently developed technology.

D. Electrophoresis

Paper electrophoresis of cyclitols in dilute borate solution is an effective way of resolving many isomers and their derivatives (Angyal and McHugh, 1957). In 0.012 M sodium tetraborate, all isomers except *neo*- and *myo*-inositol were separated. Increasing the borate concentration to 0.15 M separated the latter pair. Zweig and Whitaker (1967, see pp. 264–265) have assembled a useful table in which they compared the electrophoretic mobilities of cyclitols in nine different complexing systems, including borate, molybdate, germanate, stannate, sulphonated phenylboronic acid, arsenite and basic lead acetate. The introduction of plastic-backed, cellulose powder, thin layer plates should greatly improve resolution, speed and sensitivity with these techniques.

E. High Performance Liquid Chromatography (HPLC)

When access to equipment required for HPLC is practical, this approach to separation and analysis of cyclitols and their derivatives has distinct advantages. A wide selection of column packings, mobile phases and detectors provides convenient answers to practically every analytical problem. Honda (1984) has prepared a comprehensive listing of such information. Generally speaking, procedures developed for sugars and polyols apply equally well to cyclitols (Hendrix *et al.*, 1981). In an elegant demonstration of the value of HPLC as applied to cyclitols, Sasaki *et al.* (1988) separated all eight isomers of inositol on a calcium-form, cation-exchange resin (Bio-Rad HPX-87C) with just water as the eluting solvent. In an earlier study, Angyal *et al.* (1979) demonstrated the practical use of this type of packed column as a means of separating a number of sugars, inositols and inositol methyl ethers. The acid form of this resin (HPX-87H) effectively separated *scyllo*-inositol and *myo*-inositol (Sasaki and Loewus, 1980), a practical matter when one wishes to prepare *myo*-[2-^{3}H]inositol and *scyllo*-[R-^{3}H]inositol by reduction of *myo*-inosose-2 with sodium borotritide (Reymond, 1957). When mixtures of *scyllo*- and *myo*-inositol are too large to handle by HPLC, an ion-exchange column packed with Dowex 1 resin in the borate form is an effective method (Spector, 1978). Pinitol, sequoyitol and *chiro*-inositol, forms often encountered together in coniferous plants, were readily separated on a Microsil column with acetonitrile–water, 78:22; v/v (Ghias-Ud-Din *et al.*, 1981). It should be noted that the HPLC mobile phases used in the examples just described are volatile aqueous systems or water which are readily removed to give separated components for further use or study in underivatised form.

V. DETECTION AND QUANTITATIVE ANALYSIS OF *myo*-INOSITOL

Although *myo*-inositol is considered to occur ubiquitously in plants, its presence was often overlooked when conventional carbohydrate-detecting reagents were employed in plant analysis. Historically, Scherer's reaction which converts cyclitols to polyoxygenated quinones (oxidation with nitric acid, evaporation to dryness, addition of 5% calcium chloride and a drop of concentrated ammonia to be followed by further drying to produce a red colour; see Posternak, 1965, pp. 44–45) offered the most sensitive test but was non-specific for individual inositols. Specificity for *myo*-inositol was achieved by use of microbiological procedures with yeasts or *Neurospora* in which *myo*-inositol is a growth factor (see Posternak, 1965, pp. 78–81). While microbiological methods are highly specific and very sensitive, they are usually tedious, slow, and exacting as regards conditions and reagents. Chromatographic methods, especially GC-MS and HPLC, have largely supplanted such methods.

When the need for specificity and sensitivity is pre-eminent, enzymic analysis provides these qualities (Loewus, 1990). Two enzymic reactions have been used successfully; *myo*-inositol oxygenase, EC 1.13.99.1, and *myo*-inositol 2-dehydrogenase, EC 1.1.1.18. The oxygenase was used as an assay by Kean and Charalampous (1959). The reaction is shown below:

$$\textit{myo}\text{-Inositol} + O_2 \longrightarrow \text{D-Glucuronate} + H_2O$$

Subsequent improvements in purification, characterisation and use of this enzyme for the assay of *myo*-inositol have been reported (Channa Reddy *et al.*, 1981a,b; Naber *et al.*, 1986), but the enzyme is unavailable commercially.

The dehydrogenase, on the other hand, is available commercially (Sigma Chemical Co., St. Louis, MO) as a highly purified enzyme from *Enterobacter aerogenes* (Larner, 1962). It is also prepared from *Streptomyces hygroscopicus forma glebosus* ATCC 14607 (*S. glebosus*) when this organism is grown with *myo*-inositol as a major energy source (Walker, 1975). The reaction is shown below:

$$\textit{myo}\text{-Inositol} + NAD^+ \rightleftharpoons \textit{myo}\text{-Inosose-2} + NADH + H^+$$

As originally used for assay of *myo*-inositol, the reaction does not proceed to completion, although maximum NADH formation is proportional to the concentration of *myo*-inositol (Weissbach, 1984). This problem has been circumvented by coupling the reaction with malate dehydrogenase, EC 1.1.1.37, and fluorimetric assay for malate, which is stoichiometric to *myo*-inositol, (MacGregor and Matschinsky, 1984). This method will detect *myo*-inositol in concentrations ranging from 10 to 80 μM when the macro-assay is employed. A micro-assay variant will detect 1–18 pmol of *myo*-inositol in dried samples. Recently, Dolhofer and Wieland (1987) provided a simpler version in which the dehydrogenase-inhibiting NADH accumulation is avoided by using NADH generated in the oxidation of *myo*-inositol to reduce Fe^{3+}-bathophenanthroline disulphonic acid to Fe^{2+}-bathophenanthroline disulphonic acid. The latter is measured spectrophotometrically with phenazine methosulphate at 546 nm. Linearity between the amount of *myo*-inositol and absorbance was obtained in the range of 0.5–3 nmol of *myo*-inositol per assay. Although this method was developed for determination of *myo*-inositol in serum, its application to plant analysis should find no major obstacles.

VI. ANALYSIS OF INOSITOL GLYCOSIDES

Interest in this class of substituted cyclitols has grown significantly since the first discovery of 1L-1-*O*-α-D-galactopyranosyl-*myo*-inositol (trivial term: galactinol) as a constituent of sugar beet sap (Brown and Serro, 1953). In that original study, *myo*-inositol and galactinol were resolved on a charcoal column by elution of the former with water followed by 2% ethanol to recover the latter. Galactinol is the galactosyl donor in the biosynthesis of raffinose and its higher homologues which are important constituents of the phloem (Zimmermann and Ziegler 1975; see also Chapter 5, this volume). It is readily isolated from cucumber leaves with hot 80% ethanol and recovered after deionisation by paper chromatography in a solvent of *n*-propanol–ethylacetate–water (7:1:2; v/v). Further separation by HPLC on a Waters Sugar-Pak I column with water as solvent resolves stachyose, raffinose, galactinol and sucrose (Pharr and Sox, 1984). Alternatively, these oligosaccharides separate on a reverse phase C_{18} column with 0.06 M phosphoric acid (adjusted to pH 2.5 with ammonium hydroxide) (Handley *et al.*, 1983).

Another inositol galactoside that is found in seeds of several leguminous species is 1D-2-*O*-(α-D-galactopyranosyl)-4-*O*-methyl-*chiro*-inositol (trivial term: pinitol galactoside) (Beveridge *et al.*, 1977). A positional isomer, 1D-5-*O*-(α-D-galactopyranosyl)-4-*O*-methyl-*chiro*-inositol, and the demethylated form of pinitol galactoside, 1D-2-*O*-(α-D-galactopyranosyl)-*chiro*-inositol are also present in certain legumes, including soybean. These three galactosides were resolved by passage through a column of charcoal–Celite (1:1; w/w) and eluting first with 2% and then 5% aqueous ethanol (Schweitzer and Horman, 1981). A pinitol digalactoside, ciceritol, is also widely distributed among legumes (Quemener and Brillouet, 1983). Retention times for the pinitol galactosides and sugars of the raffinose series with HPLC and GC (trimethylsilylated forms) are found in this study.

Conjugates of indole-3-acetic acid include at least two structures in which the ester, indole-3-acetyl-2-*O*-*myo*-inositol, is glycosylated at carbon 5 of *myo*-inositol with either arabinose or galactose (Cohen and Bandurski, 1982). Quantitative analysis and identification of these structures was recently reviewed by Bandurski and Ehmann (1986). Mass spectral fragmentation patterns for the ester and its two glycosides are given.

VII. ANALYSIS OF INOSITOL PHOSPHATES

Phosphoric esters of inositol constitute one of the most abundant forms of organic phosphate in nature. This is especially true of phytate, the hexa*kis*phosphate of *myo*-inositol, which is the major phosphate reserve in seeds and storage organs of higher plants. The role of *myo*-inositol is central to all aspects of inositol phosphate metabolism, but other isomers (*scyllo*-, D-*chiro*- and *neo*-) are present as soil phosphates, possibly the result of bacterial action on phytate from seeds and vegetation (Cosgrove, 1980). Cosgrove's monograph on the inositol phosphates summarised virtually all information then available on that subject. Together with Posternak's treatise (1965), it provides a reliable record on methods and procedures in use prior to 1980.

The virtual explosion of interest in inositol phosphates in the present decade due to the discovery of phosphatidylinositol-linked signal transduction (Berridge, 1987) and membrane-bound anchoring structures (Low and Saltiel, 1988), and a fuller appreciation of inositol phosphate metabolism (Majerus *et al.*, 1988), has brought with it a flood of new methods, many of which depend upon new forms of instrumentation and analysis. Much of this new material is found in biomedical literature, and the reader is urged to consult this source for applications which can be adapted to studies in plant biology. Here, attention will focus on plant studies.

A. *myo*-Inositol Monophosphate

Although numerous methods for separating and analysing *myo*-inositol monophosphates were developed in conjunction with studies on its occurrence and chemistry (Cosgrove, 1980), it was the discovery of *myo*-inositol 1-phosphate synthase, the key biosynthetic step between D-glucose 6-phosphate and 1L-*myo*-inositol 1-phosphate, that prompted the need for a selective and specific assay. When labelled substrate is employed, the assay developed by Chen and Charalampous (1965), as modified by Loewus *et al.* (1984), proved effective. It involves an initial treatment with alkaline phosphatase to convert *myo*-inositol monophosphate to free *myo*-inositol, a boiling of the reaction mixture in the presence of barium hydroxide to oxidise glucose, passage through cationic and anionic exchange resins to remove salts and sugar acids, and paper or thin layer chromatography to recover pure *myo*-inositol for radioassay. An enzymic assay with *myo*-inositol monophosphatase, EC 3.1.1.25, in which inorganic phosphate is determined, is useful when the substrate is completely free of interfering phosphates (Eisenberg and Parthasarathy, 1984). Caution is recommended since this enzyme also acts on a number of phosphorylated substances, although poorly or not at all on hexose phosphates (Loewus and Loewus, 1982; Gumber *et al.*, 1984). A colorimetric assay based on rapid oxidation of *myo*-inositol monophosphate relative to hexose phosphate by periodic acid (Barnett *et al.*, 1970) is widely used in studies with animal tissues, but extension to plant tissues is not recommended unless high levels of *myo*-inositol monophosphate are present.

Leavitt and Sherman (1982c) have described the detection and separation of *myo*-inositol monophosphates as per(trimethylsilyl) and trimethylsilyl ether–dimethylphosphate derivatives on several column packings which enabled them to resolve the 1, 2, 4, 5, and cyclic 1, 2 forms. Since sugar phosphates are also readily detected, this procedure is very useful if the instrumentation is available.

Instances where the chirality of enantiomeric forms of *myo*-inositol 1-phosphate must be determined are readily met by the use of chiral gas–liquid chromatography (Leavitt and Sherman, 1982a,b). This method was especially useful for determining the chirality of *myo*-inositol 1-phosphate obtained from *myo*-inositol 1-phosphate synthase and *myo*-inositol kinase in plants (Loewus *et al.*, 1982). Its usefulness should increase as new situations arise where the origin of metabolically derived *myo*-inositol 1-phosphate, i.e. phosphoinositide, *de novo* synthesis or free *myo*-inositol, is sought. Although chiral columns are also used in HPLC, applications involving separation of chiral forms of *myo*-inositol 1-phosphate from plants have not been reported.

B. Higher Order Phosphoric Esters of *myo*-Inositol

Standard methods for the separation of these esters by paper chromatography and electrophoresis are found in Cosgrove's monograph (1980). Two useful buffer systems for electrophoretic separation of products of phytic acid hydrolysis are 0.1 M ammonium formate, pH 3.8 and 0.1 M sodium oxalate, pH 1.58 (Roberts and Loewus, 1968). While they separate mixtures into congeneric groups containing the same number of phosphates, they fail to resolve isomeric forms within a group. The same comment applies to ion-exchange column chromatography which is quite useful for separation of larger quantities of inositol phosphates. Wilson and Harris (1966) used a Dowex 1 (formate) column with a gradient of formic acid/ammonium formate to separate the six groups of phosphates. Phillippy *et al.* (1987) have successfully separated gram quantities of phytate hydrolysate on Dowex 1 resin in the chloride form after treatment of phytate with wheat phytase or HCl. Partial resolution of penta*kis*-, tetra*kis*-, and trisphosphates allowed tentative assignment of substituted positions. Recently, Wreggett and Irvine (1987) introduced a technique for separating inositol phosphates on the basis of their phosphate content which determined strength of binding to a commercial anion-exchange cartridge (Waters ACCELL QMA SEP-PAK). The procedure is empirical and caution is advised if this method is chosen since the conditions may vary with different batches of cartridges. For broad-based separations prior to HPLC, this approach may be useful (see Wreggett and Irving, 1989).

To achieve resolution of isomeric phosphates within each group, HPLC has been the method of choice. Binder *et al.* (1985) used a Waters μBondapak amine column with a gradient of ammonium acetate/acetic acid, pH 4 to separate *myo*-inositol mono-(1,2 cyclic, 1 and 2=4)-, bis-(1,4 and 4,5)-, and tris-(1,4,5)-phosphate with near baseline efficiency over a 120 min period. Meek and Nicoletti (1986) chose a strong anion-exchange resin-packed column (Pharmacia, Mono Q HR 5/5 or a Bio-Rad Aminex A-27) and ran their gradient from weak to strong sulphate concentration over a period of 20–30 min to separate bis- and trisphosphates. Detection involved on-line enzymic hydrolysis and assay of resulting inorganic phosphate. A somewhat similar anion-exchange HPLC employing a Whatman Partisil SAX column and a citrate-buffered mobile phase gradient was used to separate *myo*-inositol-1,4,5-trisphosphate from its hydrolysis products, 1,4-bisphosphate and 1-phosphate (Mathews *et al.*, 1988). One serious problem encountered in the use of strong anion-exchangers is on-column hydrolysis. Another problem is poor resolution of lower congeners when working with hydrolysates of phytic acid. Minear *et al.* (1988) circumvent these difficulties by including tetrasodium ethylenediaminetetraacetate (EDTA) in the mobile phase of a NaCl gradient. A significantly simpler set-up involving isocratic elution with ion-pair chromatography from a Water μBondapack C_{18} column provided base-line resolution of *myo*-inositol 1-, 1,4-, 1,4,5- and 1,3,4-phosphates within a 20 min period (Shayman and BeMent, 1988).

In contrast to these procedures, a highly sophisticated complexometric dye- and transition metal-based post-column detection system for polyanions (referred to as metal-dye detection) has been developed that will measure inositol phosphates in the picomolar range when separated on a Pharmacia, Mono Q HR 5/5 column in an HCl gradient containing a tervalent transition metal (Ho and Y gave the highest detector sensitivity). Detection involved complex formation of the cation-specific dye, 4-(2-pyridylazo)resorcinol, with a polyanion such as inositol phosphate (Mayr, 1988).

C. Phytic Acid

Phytic acid, the hexa*kis*phosphoric ester of *myo*-inositol, usually exists in plants as a complex salt with calcium, magnesium and potassium as the major inorganic cations. Strong acid extraction solubilises phytic acid, but in the presence of EDTA, milder conditions may be used (Sasaki and Loewus, 1980). Since phytic acid forms an insoluble stable complex with ferric ion in dilute acid solution, presumably the only inositol phosphate with this property, this analytical step has been widely adopted (see Kikunaga *et al.*, 1985 for additional references), at least as a first step, in assays of phytic acid. Caution is advised, since the ferric ion method is not specific (Frolich *et al.*, 1986). Numerous reviews address the various quantitative and qualitative aspects of phytate analysis (Posternak, 1965; Cosgrove, 1980; Maga, 1982; Reddy *et al.*, 1982; Oberleas and Harland, 1986; Harland and Oberleas, 1987).

As in the case of *myo*-inositol, phytic acid has a plane of symmetry through C-2 and C-5. Removal of a single phosphate results in one of four diastereomeric forms. Since two of these are enantiomeric, there are six stereoisomers of *myo*-inositol penta*kis*phosphate as shown in Fig. 6.3. One of these, *myo*-inositol-1,3,4,5,6-penta*kis*phosphate occurs in avian erythrocytes and was mistakenly considered to be 'phytic acid' until Johnson and Tate (1969) identified it as the penta*kis*phosphate through the use of ^{31}P NMR. With the use of fully proton decoupled ^{31}P NMR, phytic acid is clearly identified by its four-peak spectrum (δ 3.87, 4.04, 4.38 and 5.22), as has been demonstrated in the detection of phytic acid in pollens (Jackson *et al.*, 1982). This analytical tool is especially useful in identifying specific isomeric forms of the congeneric inositol phosphates (Frolich *et al.*, 1986; Lindon *et al.*, 1986; Cerdan *et al.*, 1986; Phillippi, 1989).

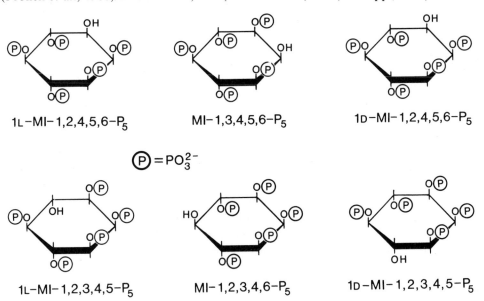

FIG. 6.3.   Isomers of *myo*-inositol penta*kis*phosphate.

As in the case of lower order polyphosphates, identification and quantitative analysis of phytic acid is amenable to HPLC. Indeed, most of the procedures for inositol phosphates noted earlier also adapt to phytate. One relatively rapid method, useful when the major component is phytate, employs a Waters IC-PAK-A anion-exchange column and 0.09 M sodium nitrate–0.036 M nitric acid (1:1; v/v) as mobile phase (Cilliers and van Niekerk, 1986). The phytic acid peak has a retention time of 5–8 min.

D. Phosphatidylinositol and Its Phosphates

This rapidly evolving area of phospholipid metabolism has created a bewildering array of analytical choices, largely applied to studies involving animal tissues. Many are described in a methods manual (Conn and Means, 1987; see Section III) and recent reviews (Bleasdale et al., 1983; Jungalwala, 1985; Putney, 1986; Irvine, 1986; Mahadevappa and Holub, 1987). Only one, directed primarily toward phospholipid biosynthesis, focuses on plant systems (Moore, 1987). Individual projects in plant biology that focus on the putative role of phosphatidylinositol in second messenger formation and signal transmission rely heavily on methods developed for animal tissues. While this is expedient (Irving et al., 1989), it presents certain pitfalls for the plant scientist, in that myo-inositol-linked events in plants encompass numerous metabolic processes not encountered in animals such as cell-wall polysaccharide biosynthesis, galactosyl transfer reactions, phytic acid formation and breakdown, and the production of complex inositol-containing lipids unlike those found in animal tissues (Loewus and Loewus, 1983). Notable examples of such encounters are found in several reports. Coté et al. (1987) detected anomalous radiolabelled products during HPLC separation of inositol phosphates from [2-³H]- inositol-labelled Samanea pulvini. Their study included direct comparison with inositol phosphates that had been obtained by treatment of red blood cell ghosts with endogenous membrane-localised, phosphoinositide-specific phosphodiesterase (phospholipase C) (Downes et al., 1982). In a separate study on the phototransducing role of inositolphospholipid in Samanea pulvini, Satter's group states (Morse et al., 1987):

> In plants, inositol functions not only in the inositolphospholipid cycle but also as an intermediate in both phytate and cell wall synthesis. Its multiple use does not preclude its function in signal transduction, although it adds a number of complexities to its metabolic regulation.

In another study on inositol-containing lipids in suspension-cultured tomato cells, several ill-characterised, lipid-extractable compounds containing both inositol and phosphate moieties were isolated which might easily have been confused with phosphatidylinositol phosphates had it not been for the careful analytical constraints imposed by the investigators (Drøbak et al., 1988). With these considerations in mind, one can be directed to a number of useful studies in which phosphatidylinositol and its proximal metabolites have been analysed with respect to physiological responses in plants. In a study of the effect of hypo-osmotic shock on changes in polyphosphoinositide metabolism, Einsphar et al. (1988) separated inositol phospholipids from the unicellular green alga, Dunaliella salina, by thin layer chromatography, which was followed by further paper chromatography of hydrolysis products. A similar approach was used by Ettlinger and Lehle (1988) to follow rapid changes in polyphosphoinositides induced by auxin in suspension culture of Catharanthus roseus. The latter used HPLC involving a

Whatman Partisil-SAX anion-exchange column, rather than paper chromatography, to identify inositol phosphates. A survey of plant-related studies involving *myo*-inositol, phosphatidylinositol and their phosphates will appear shortly (Morré *et al.*, 1990).

REFERENCES

Anderson, J. R., Larsen, E., Barbo, H., Bertelsen, B., Christensen, J. E. J. and Gregersen, G. (1982). *Biomed. Mass. Spectrom.* **9**, 135–140.

Anderson, L. (1972). *In* "The Carbohydrates", 2nd edn (W. Pigman and D. Horton, eds), Vol. 1A, pp. 519–579. Academic Press, New York.

Angyal, S. J. and Anderson, L. (1959). *Adv. Carbohydr. Chem.* **14**, 135–212.

Angyal, S. J. and McHugh, D. J. (1957). *J. Chem. Soc.*, 1423–1431.

Angyal, S. J., Bethell, G. S. and Beveridge, R. H. (1979). *Carbohydr. Res.* **73**, 9–18.

Bandurski, R. S. and Ehmann, A. (1986). *In* "Gas Chromatography/Mass Spectrometry, Modern Methods of Plant Analysis" (H. F. Linskens and J. F. Jackson, eds), Vol. 3, pp. 189–213. Springer, Berlin.

Barnett, J. E. G., Brice, R. E. and Corina, D. L. (1970). *Biochem. J.* **119**, 183–186.

Baumgartner, S., Genner-Ritzmann, R., Haas, J., Amadò, R. and Neukom, H. (1986). *J. Agric. Food. Chem.* **34**, 827–829.

Berridge, M. J. (1987). *Ann. N. Y. Acad. Sci.* **494**, 39–48.

Beveridge, R. J., Ford, C. W. and Richards, G. N. (1977). *Aust. J. Chem.* **30**, 1583–1590.

Binder, H., Weber, P. C. and Siess, W. (1985). *Anal. Biochem.* **148**, 220–227.

Binder, R. G. and Haddon, W. F. (1984). *Carbohydr. Res.* **129**, 21–32.

Bleasdale, J. E., Eichberg, J. and Hauser, G. (eds) (1983). "Inositol and Phosphoinositides: Metabolism and Regulation", 698 pp. Humana Press, Clifton, NJ.

Brown, R. J. and Serro, R. F. (1953). *J. Am. Chem. Soc.* **75**, 1040–1042.

Channa Reddy, C., Swain, J. S. and Hamilton, G. A. (1981a). *J. Biol. Chem.* **256**, 8510–8518.

Channa Reddy, C., Pierzchala, P. A. and Hamilton, G. A. (1981b). *J. Biol. Chem.* **256**, 8519–8524.

Cerdan, S., Hansen, C. A., Johanson, R., Inubushi, T. and Williamson, J. R. (1986). *J. Biol. Chem.* **261**, 14 676–14 680.

Chen, I. W. and Charalampous, F. C. (1965). *J. Biol. Chem.* **246**, 3507–3512.

Cilliers, J. J. L. and van Niekerk, P. J. (1986). *J. Agric. Food. Chem.* **34**, 680–683.

Cohen, J. D. and Bandurski, R. S. (1982). *Ann. Rev. Plant Physiol.* **33**, 403–430.

Conn, P. M. and Means, A. R. (1987). *Methods Enzymol.* **141**, Part B, 83–271.

Cosgrove, D. J. (1980). "Inositol Phosphates", 191 pp. Elsevier, Amsterdam.

Coté, G. G., Morse, M. J., Crain, R. C. and Satter, R. L. (1987). *Plant Cell Rep.* **6**, 352–355.

Dolhofer, R. and Wieland, O. H. (1987). *J. Clin. Chem. Clin. Biochem.* **25**, 733–736.

Downes, C. P., Mussat, M. C. and Michell, R. H. (1982). *Biochem. J.* **203**, 169–177.

Drew, E. A. (1984). *In* "Storage Carbohydrates in Vascular Plants" (D. H. Lewis, ed.), pp. 133–155. Cambridge University Press, Cambridge.

Drøbak, B. K., Ferguson, I. B., Dawson, A. P. and Irvine, R. F. (1988). *Plant Physiol.* **87**, 217–222.

Dugan, E. A. (1985). *Liquid Chromatogr.* **3**, No. 2.

Einsphar, K. J., Peeler, T. C. and Thompson, G. A. Jr. (1988). *J. Biol. Chem.* **263**, 5775–5779.

Eisenberg, F. (1972). *Methods Enzymol.* **28**, 168–178.

Eisenberg, F. Jr. and Parthasarathy, R. (1984). *In* "Methods of Enzymatic Analysis, 3rd edn, Vol. 6, Metabolites 1: Carbohydrates", (H. U. Bergmeyer, ed.), pp. 371–375. Verlag Chimie, Weinheim.

Englmaier, P. (1985). *Carbohydr. Res.* **144**, 177–182.

Ettlinger, C. and Lehle, L. (1988). *Nature (London)* **331**, 176–178.

Ford, C. W. (1985). *J. Chromatogr.* **333**, 167–170.

Frolich, W., Drakenberg, T. and Asp, N.-G. (1986). *J. Cereal Sci.* **4**, 325–334.

Ghias-Ud-Din, M., Smith, A. E. and Phillips, D. V. (1981). *J. Chromatogr.* **211**, 295–298.

Gumber, S. C., Loewus, M. W. and Loewus, F. A. (1984). *Plant Physiol.* **76**, 40–44.
Handley, L. W., Pharr, D. M. and McFeeters, R. F. (1983). *J. Am. Soc. Hort. Sci.* **108**, 600–605.
Harland, B. F. and Oberleas, D. (1987). *Wld. Rev. Nutr. Diet.* **52**, 235–259.
Hendrix, D. L., Lee, R. E. Jr and Baust, J. G. (1981). *J. Chromatogr.* **210**, 45–53.
Hoffmann-Ostenhof, O. and Pittner, F. (1982). *Can. J. Chem.* **60**, 1863–1871.
Hokin-Neaverson, M. and Sadeghian, K. (1976). *J. Chromatogr.* **120**, 502–505.
Honda, S. (1984). *Anal. Biochem.* **140**, 1–47.
Irvine, R. F. (1986). In "Phosphoinositides and Receptor Mechanisms" (J. W. Putney, ed.), pp. 89–107. Alan R. Liss, New York.
Irvine, R. F., Letcher, A. J., Lander, D. J., Drøbak, D. J., Dawson, A. P. and Musgrave, A. (1989). *Plant Physiol.* **89**, 888–892.
Irving, G. C. J. (1981). *J. Chromatogr.* **205**, 460–463.
IUPAC-IUB (1973). *Pure. Appl. Chem.* **37**, 285–297.
Jackson, J. F., Jones, G. and Linskens, H. F. (1982). *Phytochemistry* **21**, 1255–1258.
Johnson, L. F. and Tate, M. E. (1969). *Can. J. Chem.* **47**, 63–73.
Jungalwala, F. B. (1985). In "Phospholipids in Nervous Tissues" (J. Eichberg, ed.), pp. 1–44. John Wiley & Sons, New York.
Kean, E. L. and Charalampous, F. C. (1959). *Biochim. Biophys. Acta* **36**, 1–3.
Kikunaga, S., Takahashi, M. and Huzisige, H. (1985). *Plant Cell Physiol.* **26**, 1323–1330.
Korte, K. and Casey, M. L. (1982). *J. Chromatogr.* **232**, 47–53.
Laker, M. F. (1980). *J. Chromatogr.* **184**, 457–470.
Larner, J. (1962). *Methods Enzymol.* **5**, 326–328.
Leavitt, A. L. and Sherman, W. R. (1982a). *Carbohydr. Res.* **103**, 203–212.
Leavitt, A. L. and Sherman, W. R. (1982b). *Methods Enzymol.* **89**, 3–8.
Leavitt, A. L. and Sherman, W. R. (1982c). *Methods Enzymol.* **89**, 9–18.
Lewis, D. H. and Smith, D. C. (1967). *New Phytol.* **66**, 184–204.
Lindon, J. C., Baker, D. J., Farrant, R. D. and Williams, J. M. (1986). *Biochem. J.* **233**, 275–277.
Loewus, F. A. (1966). *Carbohydr. Res.* **3**, 130–133.
Loewus, F. A. (1990). In "Methods in Carbohydrate Chemistry" Vol. 10. (D. J. Manners and R. J. Sturgeon, eds). Academic Press, Orlando. (In press).
Loewus, F. A. and Dickinson, D. B. (1982). In "Encyclopedia of Plant Physiology, New Series" Vol. 13A (F. A. Loewus and W. Tanner, eds), pp. 194–216. Springer, Berlin.
Loewus, F. A. and Loewus, M. W. (1980). In "The Biochemistry of Plants, Vol. 3, Carbohydrates: Structure and Function" (J. Preiss, ed.), pp. 43–76. Academic Press, New York.
Loewus, F. A. and Loewus, M. W. (1983). *Ann. Rev. Plant Physiol.* **34**, 137–161.
Loewus, F. A. and Shah, R. H. (1972). *Methods Carbohydr. Chem.* **6**, 14–20.
Loewus, M. W. and Loewus, F. A. (1982). *Plant Physiol.* **70**, 760–766.
Loewus, M. W., Bedgar, D. L. and Loewus, F. A. (1984). *J. Biol. Chem.* **259**, 7644–7647.
Loewus, M. W., Sasaki, K., Leavitt, A. L., Munsell, L., Sherman, W. R. and Loewus, F. A. (1982). *Plant Physiol.* **70**, 1661–1663.
Low, M. G. and Saltiel, A. R. (1988). *Science* **239**, 268–275.
MacGregor, L. C. and Matschinsky, F. M. (1984). *Anal. Biochem.* **141**, 382–389.
Maga, J. A. (1982). *J. Agric. Food Chem.* **30**, 1–9.
Majerus, P. W., Connolly, T. M., Bansal, V. S., Inhorn, R. C., Ross, T. S. and Lips, D. L. (1988). *J. Biol. Chem.* **263**, 3051–3054.
Mahadevappa, V. G. and Holub, B. J. (1987). In "Chromatography of Lipids in Biochemical Research and Clinical Diagnosis" (A. Kuksis, ed.), pp. 225–265. Elsevier, Amsterdam.
Mathews, W. R., Guido, D. M. and Huff, R. M. (1988). *Anal. Biochem.* **168**, 63–70.
Mayr, G. W. (1988). *Biochem. J.* **254**, 585–591.
Meek, J. L. and Nicoletti, F. (1986). *J. Chromatogr.* **351**, 303–311.
Michell, R. H. and Berridge, M. J. (1988). *Phil. Trans. R. Soc. London* **B320**, 237–238.
Minear, R. A., Segars, J. E., Elwood, J. W. and Mulholland, P. J. (1988). *Analyst* **113**, 645–649.
Moore, T. S. Jr. (1987). *Methods Enzymol.* **148**, 585–596.
Morré, D. J., Boss, W. and Loewus, F. A. (eds) (1990). "Inositol Metabolism in Plants", Alan R. Liss. New York. (In press).
Morse, M. J., Crain, R. C. and Satter, R. L. (1987). *Proc. Natl. Acad. Sci. USA* **84**, 7075–7078.

Mukherjee, R. and Axt, E. M. (1984). *Phytochemistry* **23**, 2682–2684.
Naber, N. I., Swan, J. S. and Hamilton, G. A. (1986). *Biochemistry* **25**, 7201–7207.
Niwa, T., Yamamoto, N., Maeda, K. and Yamada, K. (1983). *J. Chromatogr.* **277**, 25–39.
Oberleas, D. and Harland, B. F. (1986). *In* "Phytic Acid: Chemistry and Applications" (E. Graf, ed.), pp. 77–100. Pilatus Press, Minneapolis, MN.
Parthasarathy, R. and Eisenberg, F. Jr. (1986). *Biochem. J.* **235**, 313–322.
Pharr, D. M. and Sox, H. N. (1984). *Plant Sci. Lett.* **35**, 187–193.
Phillippi, B. Q. (1989). *J. Agric. Food Chem.* **37**, 1261–1265.
Phillippi, B. Q., White, K. D., Johnston, M. R., Tao, S.-H. and Fox, M. R. S. (1987). *Anal. Biochem.* **162**, 115–121.
Posternak, T. (1962). "Les Cyclitols". Hermann, Paris.
Posternak, T. (1965). "The Cyclitols", 431 pp. Holden-Day, San Francisco.
Putney, J. W. (ed.) (1986). "Phosphoinositides and Receptor Mechanisms", 400 pp. Alan R. Liss, New York.
Quemener, B. and Brillouet, J.-M. (1983). *Phytochemistry* **22**, 1745–1751.
Reddy, N. R., Sathe, S. K. and Salunkhe, D. K. (1982). *Adv. Food Res.* **28**, 1–92.
Reymond, D. (1957). *Helv. Chim. Acta* **40**, 492–494.
Rittenhouse, S. E. (1987). *Methods Enzymol.* **141** (Cell Reg., Part B), 143–149.
Roberts, R. M. and Loewus, F. A. (1968). *Plant Physiol.* **43**, 1710–1716.
Sasaki, K. and Loewus, F. A. (1980). *Plant Physiol.* **66**, 740–745.
Sasaki, K., Hicks, K. B. and Nagahashi, G. (1988). *Carbohydr. Res.* **183**, 1–9.
Schweitzer, T. F. and Horman, I. (1981). *Carbohydr. Res.* **95**, 61–71.
Shayman, J. A. and BeMent, D. M. (1988). *Biochem. Biophys. Res. Commun.* **151**, 114–122.
Spector, R. (1978). *J. Neurochem.* **31**, 1113–1115.
Swahn, C.-G. (1985). *J. Neurochem.* **45**, 331–334.
Sweeley, C. C., Bentley, R., Makita, M. and Wells, W. W. (1963). *J. Am. Chem. Soc.* **85**, 2497–2507.
Tate, M. E. and Bishop, C. T. (1962). *Can. J. Chem.* **40**, 1043–1048.
Turk, J., Wolf, B. A. and McDaniel, M. L. (1986). *Biomed. Environ. Mass Spectrom.* **13**, 237–244.
Walker, J. B. (1975). *Methods Enzymol.* **43**, 433–439.
Weissbach, A. (1984). *In* "Methods of Enzymatic Analysis, 3rd edn, Vol. 6, Metabolites 1: Carbohydrates" (H.U. Bergmeyer, ed.), pp. 366–370. Verlag Chimie, Weinheim, FRG.
Wilson, A. M. and Harris, G. A. (1966). *Plant Physiol.* **41**, 1416–1419.
Wreggett, K. A. and Irvine, R. F. (1987). *Biochem. J.* **245**, 655–660.
Wreggett, K. A. and Irvine, R. F. (1989). *Biochem. J.* **262**, 997–1000.
Yamada, T., Hisamatsu, M. and Taki, M. (1975). *J. Chromatogr.* **103**, 390–391.
Zimmerman, M. H. and Ziegler, H. (1975). *In* "Transport in Plants. I. Phloem Transport" (M. H. Zimmermann and J. A. Milburn, eds), Vol. I, Encycl. Plant Physiol., New Series, pp. 480–503. Springer, Berlin.
Zweig, G. and Whitaker, J. R. (1967). "Paper Chromatography and Electrophoresis", 420 pp. Academic Press, New York.

7 Branched-chain Sugars and Sugar Alcohols

ERWIN BECK[1] and HERBERT HOPF[2]

[1]*Lehrstuhl Pflanzenphysiologie, Universität Bayreuth, Universitätsstrasse 30, D 8580 Bayreuth, FRG*
[2]*Botanisches Institut, Universität München, Menzinger Strasse 67, D 8000 München 19, FRG*

I. INTRODUCTION

Branched-chain monosaccharides and sugar alcohols represent two groups of carbohydrates which, apart from their interconvertibility by reduction and oxidation,

METHODS IN PLANT BIOCHEMISTRY Vol. 2
ISBN 0-12-461012-9

respectively, do not share many chemical features. From the viewpoint of a plant biochemist they are usually filed under the 'exotics', holding an intermediate position between the primary and secondary plant constituents. Whereas there are only three branched-chain monosaccharides which, although widespread in the plant kingdom, are accumulated by only a few species, a great number of sugar alcohols of plant origin are known, several of which have attained considerable importance, e.g. in the food industry. On the other hand, the branched-chain sugars, because of their involvement in a variety of different biochemical reactions, deserve much attention by biochemists and biologists. Consequently, the various topics in the chapters 'branched-chain sugars' and 'sugar alcohols' are covered with differing emphasis.

Modern analytical methods have started a triumphant advance into the biochemical laboratories and therefore classical methods, such as paper chromatography (PC), paper electrophoresis (PE) and thin layer chromatography (TLC), may seem to be remnants of a bygone era. However, not all laboratories have access to gas chromatography-mass spectroscopy (GC-MS), nuclear magnetic resonance (NMR) and fast atom bombardment-mass spectroscopy (FAB-MS) instruments, and for many studies, especially those including preparative steps, the classical methods are still irreplaceable. Furthermore, much of the research into branched-chain sugars and sugar alcohols has been performed using the classical analytical procedures and therefore, in many cases, the corresponding NMR, MS and other spectroscopical data are simply not available. Consequently, both the classical and the modern techniques had to be considered in this chapter.

II. BRANCHED-CHAIN MONOSACCHARIDES

A. Occurrence, Biochemistry, Physiology and Chemistry of D-Apiose, D-Hamamelose and Aceric acid

D-Hamamelose and D-apiose are the only branched-chain true carbohydrates yet found in the plant kingdom. By contrast, in microorganisms and fungi approximately 25 branched-chain monosaccharides have been detected which are predominantly deoxy-sugars. These compounds will not be treated here and for further information the reader is directed to a detailed review by Grisebach (1978) and to the more recent compilations by Williams and Wander (1980) and by Yoshimura (1984).

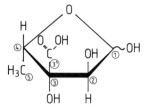

FIG. 7.1.   3-C-carboxy-5-deoxy-L-Xylose: Aceric acid.

Recently, a branched-chain acidic monosaccharide, namely 3-C-carboxy-5-deoxy-L-xylose (Fig. 7.1), has been detected as a minor constituent of plant cell walls and has been termed 'aceric acid' (Spellman *et al.*, 1983). Research into this compound has so far been concentrated on structure elucidation using exclusively modern spectroscopic methods. Classical chromatographic data for this compound are not available and therefore this compound will be discussed in Section II.E.4 together with the application of modern analytical methods.

D-Apiose and D-hamamelose are glycosidic components of secondary plant substances. Apiin (Fig. 7.2) and hamamelitannin (Fig. 7.3) are the main constituents of aqueous extracts from medicinal plants and have been in therapeutic use for centuries. The decoction of parsley leaves (*Petroselinum crispum* (MILL) A. W. HILL), containing apiin, was recommended by Dioscorides. The knowledge of the utilisation of an aqueous bark extract from witch hazel (*Hamamelis virginiana* L.) can be traced back to about 1840 and to the Indians of the Oneida tribe, located in central New York State (Shafizadeh, 1956). Because of its astringency the *Hamamelis* extract, rich in tannins such as hamamelitannin, was applied for treating burns, boils and all kinds of wounds. It is still included as an ingredient of lotions for cutaneous use (Gaudy *et al.*, 1988), of ointments for the therapy of haemorrhoids and of hair lotion preparations (Oku, 1988).

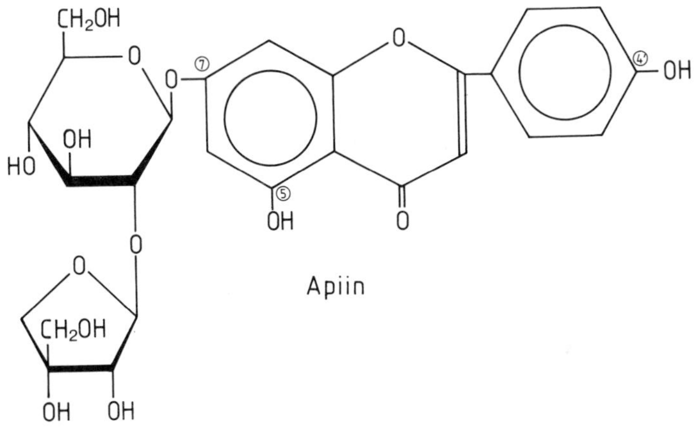

FIG. 7.2. Apiin.

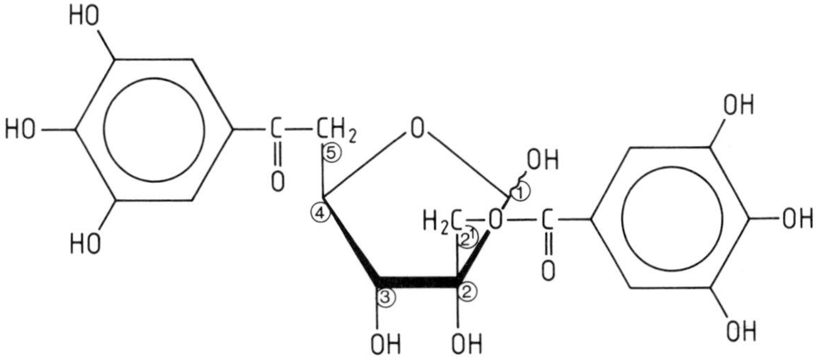

FIG. 7.3. Hamamelitannin.

In nature, D-apiose has never been found in the free form. Hence, the identification of an apiitol-dehydrogenase in the homogenate of two microorganisms, *Aerobacter aerogenes* (Neal and Kindel, 1970) and a strain of *Micrococcus* (Hanna *et al.*, 1973), appears somewhat puzzling. However, those bacteria have been isolated from parsley seedlings and being presumptive inhabitants of this plant, may have developed the capability of metabolising the branched-chain sugar. By contrast free D-hamamelose, mainly by ^{14}C-labelling of the products of photosynthesis, has been detected as a more or less ubiquitous constituent of higher plants (Sellmair *et al.*, 1977). Since a physiological role was difficult to assign to both branched-chain monosaccharides, they have been looked upon as curiosities of plant biochemistry.

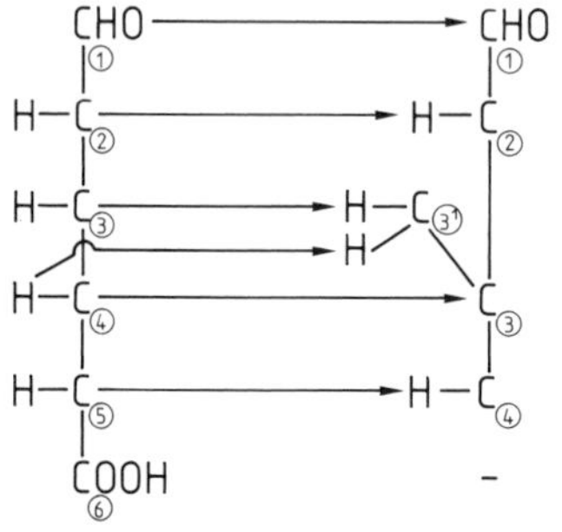

FIG. 7.4. Biogenetic relation between the carbons and attached H-atoms of D-glucuronic acid and D-apiose.

Knowlege of the widespread occurrence of apiose-containing low molecular weight secondary plant constituents, as well as of its contribution to pectic cell-wall polysaccharides, has considerably increased over the past 30 years and several reviews have been published (Watson and Orenstein, 1975; Grisebach, 1980; Beck, 1982) covering predominantly the biochemical features, such as biosynthesis and transfer to the final acceptors. Since these are not the focus of this chapter, they will only be summarised here: UDP-D-apiose is generated from UDP-D-glucuronic acid by an enzyme which even in the homogenous state simultaneously produces UDP-D-xylose (Matern and Grisebach, 1977). The enyzme has been isolated from parsley cell cultures (Matern and Grisebach, 1977) and from duckweed (Mendicino and Abou-Issa, 1974; Gustine *et al.*, 1975) and has been given various names (see Table 7.4), UDP-apiose/UDP-xylose synthase being the most appropriate term. It is clearly different from UDP-D-glucuronic acid decarboxylase (EC 4.1.1.35) which is also present in both plant sources (Wellmann *et al.*, 1971; Gustine *et al.*, 1975). After decarboxylation of UDP-D-glucuronic acid, UDP-D-apiose forms by an intramolecular rearrangement of the carbon chain by which carbon-3 is excluded to produce carbon-3[1] (Fig. 7.4). This type of ring contraction requires cleavage of the bond between carbons-2 and -3 followed by an aldol condensation of carbon-4 with carbon-2 (Grisebach, 1980). A particulate preparation of *Lemna minor* has been shown to transfer the D-apiosyl group from UDP-D-apiose to pectins, thus producing the D-apiosyl- and D-apiobiosyl side chains of the galacturonans (Pan and Kindel, 1977; Mascaro and Kindel, 1977). Another apiosyltransferase, UDP-apiose-flavone apiosyltransferase (EC 2.4.2.25), catalyses the transfer of D-apiose to

7-O-(β-D-glucosyl)-apigenin (Watson and Kindel, 1970; Ortmann *et al.*, 1972). Apart from these and the above-mentioned reduction to apiitol, no further biochemical reactions of D-apiose have been reported. *Lemna* as well as the quoted strain of a *Micrococcus* are capable of quantitative oxidation of free D-apiose to CO_2 (Hanna *et al.*, 1973). Although apiose-containing compounds are constituents of mammalian food (e.g. celery, parsley, chickpeas), the importance of this sugar in animal metabolism is not known (Watson and Orenstein, 1975).

Chemical synthesis of D-apiose has attracted much attention (see Williams and Wander, 1980) for various reasons connected with research into carbohydrate chemistry, but not for the production of large quantities of this sugar. Several routes have been developed starting either from a sugar or the respective di-O-isopropylidene derivative (see Williams and Wander, 1980; Araki *et al.*, 1981; Yoshimura, 1984; Koos and Mosher, 1986; Snyder and Serianni, 1987) or from non-carbohydrate precursors (Williams and Wander, 1980). By the former routes the final carbon-3^1 is added to C-2 and the molecule has to be inverted so that the original carbon-1 becomes carbon-4. Special problems originating from stereochemistry cannot be discussed here and the reader is directed in that respect to the papers quoted above. Although not meeting the central interest of plant biochemists, the more recent of these papers provide method-ological details, especially regarding the various techniques of NMR and mass spec-troscopy, which nevertheless may be of importance for several aspects of their work. Except for one report (Ovodova *et al.*, 1968) which mentions crystalline D-apiose in passing, neither natural nor synthetic apiose has been obtained in crystalline form.

By contrast, both synthetic and natural D-hamamelose can be crystallised (Overend and Williams, 1965; Gilck *et al.*, 1975). While all apiose-containing compounds (except apiitol which is not of plant origin) are glycosides (apiosides), a natural hamameloside has hitherto not been detected. The tannins known so far (hamamelitannin and monogalloyl-hamamelose) as well as the phosphates are esters, and the reducing group of D-hamamelose is not involved in the ester linkages. Consequently, biosynthesis of D-hamamelose does not occur at the stage of a nucleoside diphosphate-activated form. As stated by Beck (1982), activation by nucleoside diphosphate of the precursor of D-apiose (D-glucuronic acid), as well as of the precursors of the branched monosaccharides of microorganisms, for chemical reasons is not required in order to accomplish branching of the carbon chain. Rather this type of activation has to be interpreted to enable ready transfer of the sugar by providing moiety to phenolic compounds, polysaccharides or antibiotics by providing the energy for the production of the glycosidic bond.

Biosynthesis of free D-hamamelose takes place by the reaction sequence:

$$\text{D-Fructose-1,6-BP} \longrightarrow \text{D-hamamelose-}2^1\text{,5-BP} \longrightarrow (\text{D-hamamelose-}2^1\text{-P}$$
$$+ \text{D-hamamelose-5-P}) \longrightarrow \text{D-hamamelose}$$

where the letters BP stand for bisphosphate, and P for monophosphate (Gilck and Beck, 1974; Beck and Knaupp, 1974). Branching of the hexose skeleton is accomplished by an intramolecular rearrangement following the introduction of a second carbonyl function at carbon-5 of fructose-BP (carbon-4 of hamamelose-BP). Ring contraction by aldol condensation causes expulsion of carbon-3 concomitantly with its oxidation to the carbonyl group and the reduction of the original carbonyl group at position 2 to a tertiary alcohol group (Fig. 7.5). The enzyme catalysing the branching reaction has been localised to the chloroplasts while subsequent dephosphorylation, at least of the second phosphate group, may take place outside the plastids (Gilck and Beck, 1974).

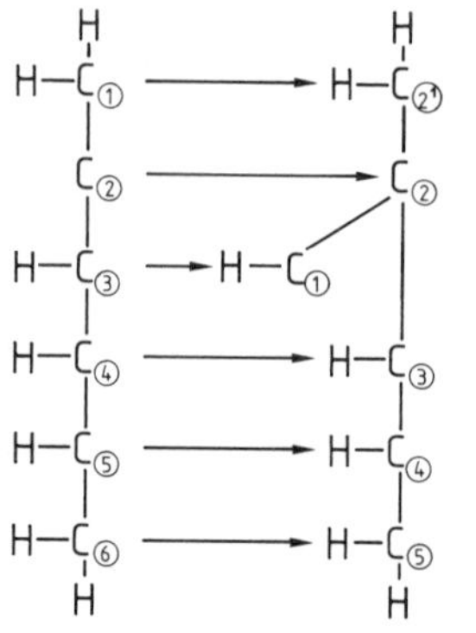

D-Fructose   D-Hamamelose

FIG. 7.5. Biogenetic relation between the carbons and attached H-atoms of D-fructose and D-hamamelose.

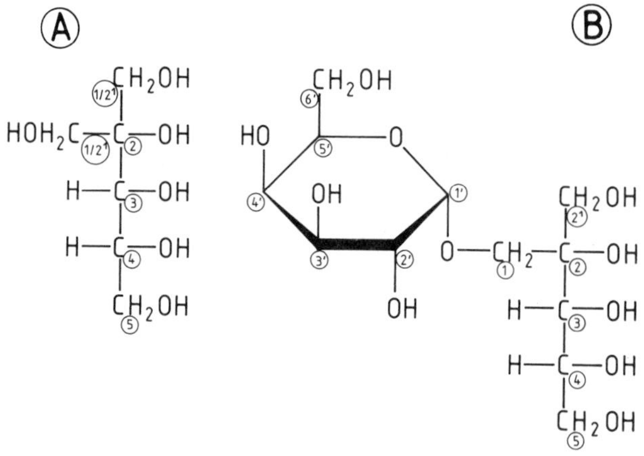

FIG. 7.6. (A) D-Hamamelitol: note the exchangeability of the hydroxymethyl-groups of carbon-1 and -2[1]. (B) Clusianose: 1-0-α-D-galactopyranosyl-hamamelitol.

In the case of D-hamamelose, the oxidation product, viz. D-hamamelonic acid (Thanbichler *et al.*, 1971), as well as the related alditol, D-hamamelitol (Sellmair *et al.*, 1968) have been identified as natural products. Substantial amounts of the latter, as well as of its derivative clusianose (1-*O*-α-D-galactosyl-hamamelitol: Beck, 1969; see Fig. 7.6) have been detected in Primulaceae and could be used as chemotaxonomic markers for several subgenera of *Primula* (Sellmair *et al.*, 1977). Catabolism of free D-hamamelose could not be demonstrated in plants. The free sugar and its derivatives hamamelitol and clusianose, once produced, accumulate in the plant tissue. Since low temperature treatment enhanced the accumulation of both hamamelitol and clusianose, these compounds are thought to be cryoprotectants (Sellmair and Kandler, 1970; Sellmair *et al.*, 1969). From the environment of a hamamelose-accumulating primrose (*Primula clusiana* TAUSCH) two strains of bacteria were isolated which could utilise free D-hamamelose. While a strain of *Pseudomonas* isolated from the natural root-bed could only oxidise D-hamamelose to D-hamamelonic acid (Thanbichler *et al.*, 1971), *Kluyvera citrophila*, collected from the leaves of *Primula clusiana*, under aerobic conditions completely metabolised the sugar to CO_2. Anaerobic dissimilation, however, resulted in mixed acid fermentation producing lactate, succinate, formate, acetate and ethanol (Thanbichler and Beck, 1974). The first steps of dissimilation were identified as phosphorylation at C-2^1 (a D-hamamelose-2^1-kinase has been purified from *Kluyvera citrophila*: Beck *et al.*, 1980), rearrangement to fructose-1-P and phosphorylation to fructose-1,6-BP. The mechanism of the rearrangement has not yet been elucidated.

As with D-apiose, several routes of chemical synthesis of D-hamamelose have been reported. The classical synthesis (Burton *et al.*, 1965) starting from methyl-3,4-*O*-isopropylidene-β-D-erythro-pentopyranoside-2-ulose (which is readily obtained from D-arabinose) used diazomethane as the source for the branch carbon. The isomers obtained by alkaline hydrolysis could be separated by crystallisation, and demethylation finally resulted in D-hamamelose and D-epihamamelose. Making use of the stereochemical influence of a neighbouring isopropylidene group on nucleophilic addition to a keto group, diazomethane could be replaced by the 1,3-dithiane-anion, the ring of which can easily be cleaved after addition to the isopropylidene product of the methylpentuloside (Paulsen *et al.*, 1972, 1980). Crystalline D-hamamelose was obtained from a concentrated methanol solution (Overend and Williams, 1965).

B. The Nomenclature of Branched-chain Monosaccharides and its Relevance to Recent Research in Photosynthesis

The structure of branched-chain monosaccharides is not covered by the rules of carbohydrate nomenclature as published in 1970 (Pigman and Horton, 1970). When confining ourselves to the three representatives of the plant kingdom there should not be any confusion because the trivial names D-hamamelose, D-apiose and aceric acid have been widely accepted. In spite of this, as will be shown, there has been unexpected ambiguity with nomenclature for a derivative of one of these compounds and hence the rules of carbohydrate nomenclature as well as the general principles of established organic nomenclature should be applied as far as possible. According to rules 6 and 9 (Pigman and Horton, 1970), D-hamamelose must be addressed as 2-C-(hydroxymethyl)-D-ribose while the systematic name of D-apiose is 3-C-(hydroxymethyl)-D-glycero-aldotetrose. The stereoepimer corresponding to D-hamamelose [2-C-(hydroxymethyl)-D-arabinose] has been termed D-epihamamelose (Overend and Williams, 1965) and must not be confused with L-hamamelose (Burton *et al.*, 1965). Note that the cyclic form of apiose exhibits an additional asymmetric carbon at position 3 which has been demonstrated by Hulyalkar *et al.* (1965) to adopt the D-configuration (Fig. 7.7). The presence of the D-apio-D-furanose form in all naturally occurring apiosides was explained by Grisebach (1980) in terms of specific transferases transferring the erythro- but not the threofuranose from UDP-D-apiose to the various acceptors. While the cyclic form of apiose necessarily adopts the furanose ring, the formation of the semiacetal of D-hamamelose may result in the furanose as well as the pyranose ring. As revealed by proton- and carbon-13 NMR spectroscopy (Schilling and Keller, 1977), D-hamamelose

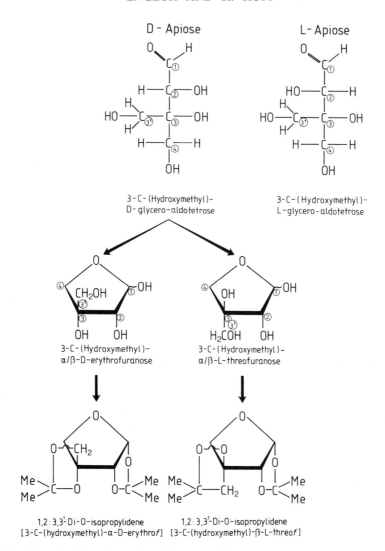

FIG. 7.7. Stereochemistry of D- and L-apiose as established by the di-*O*-isopropylidene derivatives.

exists in solution as a mixture of the two furanoses and the two pyranoses (Fig. 7.8). The ratio of furanose to pyranose is approximately 7:3 and the pyranoses prefer the $^1C_4$-conformation which is stabilised by an intramolecular H-bridge. In the recently discovered monogalloyl hamameloses, a similar equilibrium of hamamelofuranoses (predominant) and pyranoses was demonstrated by carbon-13 NMR spectroscopy (Schilling and Keller, 1986). The occurrence of several stereoisomers of D-hamamelose, even in a secondary plant constituent, is conceivable since the hemiacetal group of the cyclic form is not involved in the ester bond(s) to gallic acid. Hamamelitannin has been described as $2^1,5$-digalloyl-hamamelofuranose (Fig. 7.3; Mayer *et al.*, 1965; Ezekiel *et al.*, 1969). Since in the 1960s carbon-13 NMR spectroscopy was not used in structural elucidation, a small portion of digalloyl-hamamelopyranose would have been easily overlooked.

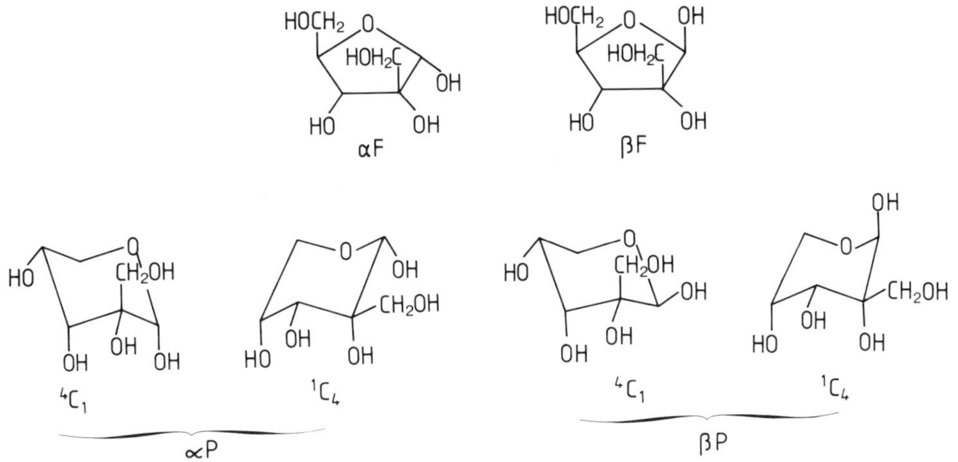

FIG. 7.8. The possible stereoisomers of cyclic D-hamamelose in aqueous solution. F, furanose; P, pyranose.

In numbering the carbon atoms of D-hamamelose and D-apiose the proposal of Shafizadeh (1956) is recommended: the straight-chain carbons are consecutively numbered, starting with the most oxidised carbon atom. The carbon of the side chain is identified by the number of the branching carbon in the straight chain plus a superscript 1 (see Figs. 7.1, 7.7 and 7.9).

Discussion of the nomenclature of branched-chain monosaccharides prompts us to return to history again and to recall the excited attention which plant physiologists paid to D-hamamelose and its derivatives in those days. In order to differentiate between the CN^--sensitive oxidative phosphorylation and the rather insensitive photophosphorylation Kandler (1958) supplied KCN to $^{14}CO_2$-assimilating *Chlorella* suspensions whereupon $^{14}CO_2$-fixation stopped and a compound accumulated which was identified as the phosphate ester of D-hamamelonic acid. Obviously in these experiments at least the bulk of the branched-chain aldonic acid had arisen from ribulose 1,5-bisphosphate by cyanohydrine synthesis (Kandler and Gibbs, 1959). Fifteen years later, Beck *et al.* (1971) showed that isolated spinach chloroplasts actually produce hamamelose-2^1,5-BP upon CO_2 fixation. The conversion of D-fructose-BP to D-hamamelose-BP has been demonstrated with a particulate protein fraction of isolated chloroplasts (Gilck and Beck, 1974) resulting in an equilibrium at a ratio of 10 : 1 (Beck, 1982). However, a specific physiological role of the branched hexose-BP could not be identified. Recently, in leaves of several *Phaseolus* species, soybean, *Vigna* and *Cucumis* (Seemann *et al.*, 1985) as well as in tobacco (Servaites, 1985) and potato leaves (Gutteridge *et al.*, 1986), a potent inhibitor of ribulose 1,5-bisphosphate carboxylase (Rubisco) was detected which preferentially binds to the activated form of the enzyme, although preserving that form greatly abolishes Rubisco activity (Seemann *et al.*, 1985). Having a carboxyl as well as a phosphate ester group the inhibitor is of anionic nature, and NMR spectroscopy and comparison with synthesised reference substances has identified it as 2-carboxy-arabinitol-1-P (Gutteridge *et al.*, 1986; Berry *et al.*, 1987). If the correct nomenclature had been applied by the authors, the inhibitor would have been termed D-hamamelonic acid-2^1-P! Figure 7.9 shows the identity of both molecules. D-Hamamelo-

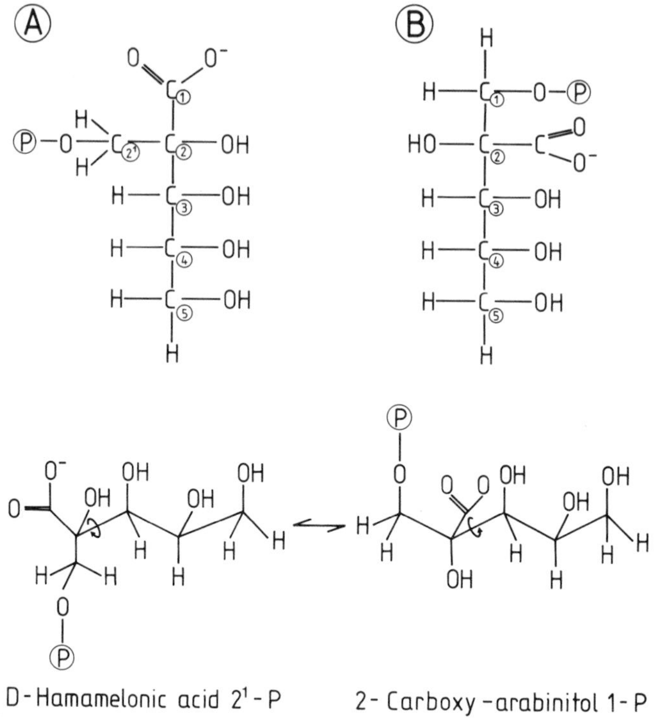

FIG. 7.9. Illustration of the identity of D-hamamelonic acid 2^1-P (A) and 2-carboxy-D-arabinitol-1-P (B).

nic acid-2^1-P closely resembles the enol form of 3-keto-hamamelonic acid 2^1,5-BP ('2-carboxy-3-keto-arabinitol-1,5-BP'), which is the intermediate of the carboxylase reaction adn therefore inhibits the active site of Rubisco (Gutteridge *et al.*, 1986; Schloss, 1988). It accumulates during the night, reaching a stromal concentration of 2–4 mM (Servaites, 1985; Seemann *et al.*, 1985). At those concentrations it binds to Rubisco at a ratio of 0.5 moles inhibitor per mole of active site (Servaites, 1985). The K_d of the Rubisco–inhibitor complex was found to be smaller than 0.1 μM (Gutteridge *et al.*, 1986) and a K_i with respect to ribulose-BP of 0.8 μM has been reported (Seemann *et al.*, 1985). Treatment of the Rubisco-bound inhibitor with alkaline phosphatase resulted in Rubisco activation (Berry *et al.*, 1987; Seemann *et al.*, 1985) which was explained in terms of dephosphorylation of that portion of the inhibitor which dissociates from the complex. An enyzme catalysing the dissociation of the Rubisco–inhibitor complex has been described and has been termed Rubisco activase (Robinson and Portis, 1988). This enzyme did not, however, reverse the inhibition of Rubisco by D-hamamelonic acid 2^1,5-BP, nor did it metabolise the inhibitor. Reductive inactivation of D-hamamelonic acid 2^1-P has been demonstrated with a soluble protein fraction from tobacco chloroplasts (Salvucci *et al.*, 1988). The reaction product has not yet been identified, but, for theoretical reasons, it could be D-hamamelose 2^1-P. The enzyme concerned appears to be activated by light and thiols.

C. Preparation and Chemical Degradation of D-Apiose, D-Hamamelose, Aceric Acid and Related Compounds

Although convenient procedures of chemical synthesis of D-apiose and D-hamamelose are available, both monosaccharides have been isolated from plant material rather than synthesised whenever substantial amounts have been required. Carbohydrate chemists, however, may need special forms, e.g. position-labelled sugars or the epimers or L-stereoisomers, and hence a short summary of the literature of chemical synthesis will be provided. This chapter concentrates on methods of isolation from natural sources.

1. D-Apiose

(a) *Isolation from natural sources.* In spite of its widespread occurrence in the plant kingdom (see Table 7.1), only four plant species appear as productive sources for the isolation of D-apiose, viz. *Petroselinum hortense* HOFFM. (parsley), *Lemna minor* L. (duckweed), *Zostera marina* L. (seaweed) and *Posidonia australis* L. (the source of the marine fibre). In parsley plants and seeds, several flavones have been demonstrated bearing an O-β-D-apiofuranosyl(1 → 2)β-D-glucosyl moiety at carbon-7: apiin (Fig. 7.2; Nordström *et al.*, 1953) which is the predominant compound, petroselinin (also termed graveolbioside A, Nordström *et al.*, 1953) and the corresponding flavone from chryso-eriol (graveolbioside B; Grisebach and Bilhuber, 1967). From these compounds, apiose can be selectively liberated by hydrolysis with $0.025 \, \text{N} \, H_2SO_4$ at 100°C for 1 h (Grisebach and Bilhuber, 1967), $0.1 \, \text{N} \, \text{HCl}$ at 100°C for 20 min (Mendicino and Picken, 1965) or $0.7 \, \text{N} \, H_2SO_4$ at 100°C for 30 min (Beck and Kandler, 1966).

Dried parsley leaves, as available from the spice market, as well as parsley seeds, have been used as starting material. Crystalline crude apiin was obtained in 1% yield from dried leaves (Beck and Kandler, 1966) while from seeds and using the lead acetate precipitation procedure according to Gupta and Seshadri (1952), a yield of 0.8% was reported (Grisebach and Bilhuber, 1967). No further purification of crude apiin is necessary if it is utilised as starting material for the preparation of D-apiose. After hydrolysis, the solution is kept for another 12 h at 5°C whereupon the aglycons and flavone glucosides precipitate (Grisebach and Döbereiner, 1966; Beck and Kandler, 1966). From the neutralised supernatant D-apiose has been isolated either by prepara-tive PC (Mendicino and Picken, 1965) or by column chromatography (Beck and Kandler, 1966; Roberts *et al.*, 1967). The yield of pure D-apiose was 5.3% (by weight) as related to crude apiin (Beck and Kandler, 1966). Otherwise D-apiose has been extracted from the syrup by preparation of crystalline derivatives, such as the phenylosazone, the N-benzyl-N-phenylhydrazone, di-O-isopropylidene derivative or the corresponding aldonit apiitol. Duckweed has also been proven as a productive source of D-apiose which is present as D-apiosyl and D-apiobiosyl side chains in several galacturonans (Beck, 1967; Hart and Kindel, 1970a,b; Mascaro and Kindel, 1977). Hydrolysis at 100°C with 0.1 N HCl (30 min) is sufficient to release essentially all of the apiose from the polysaccharide. Upon subsequent isolation of apiose by PC a yield of 0.4% as related to plant dry matter was reported (Picken and Mendicino, 1967). The same yield was obtained by a similar procedure starting with seaweed (Williams and Jones, 1964; see also Ovodova *et al.*, 1968, who reported the isolation of crystalline D-apiose from

TABLE 7.1a. Structure and occurrence of D-apiose-containing secondary plant constituents.

Compound	Chemical structure	Plant source	References
Apiosides			
Frangulin B	6-*O*-Apiofuranosyl-1,6,8-trihydroxy 3-methyl-anthraquinone	*Rhamnus frangula* L.	Wagner and Demuth (1972, 1974)
Iridoid-glycoside	6'-*O*-Apiosylebuloside	*Sambucus ebulus*	Gross *et al.* (1987)
Apiosylglycosides			
Apiosyl-2-glycosides:			
Apiin	Apigenin-7-β-D-apiosyl-β-D-2-glucoside	*Petroselinum crispum*,	Grisebach and Bilhuber (1967), Nordström *et al.* (1953)
		Cell cultures of *Petroselinum*,	Kreuzaler and Hahlbrock (1973)
		Apium graveolens	Farooq *et al.* (1957, 1958), Rahman (1958)
		Chrysanthemum spp.	Wagner and Kirmayer (1957)
		Cuminum cyminum	Chakraborti (1959)
		Bellis perennis	Wagner and Kirmayer (1957)
		Anthemis nobilis	Wagner and Kirmayer (1957)
		Matricaria spp.	Wagner and Kirmayer (1957)
		Centaurea spp.	Wagner and Kirmayer (1957)
		Serratula coronata	Wagner and Kirmayer (1957)
		Echinops gmelini	Wagner and Kirmayer (1957)
		Crotalaria anagyroides	Subramanian and Nagarajan (1970)
		Digitalis purpurea	Watson and Orenstein (1975)
		Vicia hirsuta	Nakaoki *et al.* (1955)
		Capsicum spp.	Rangoonwala and Friedrich (1967)
Graveobioside A (= Petroselinin)	Luteolin-7-apiosyl-β-D-2-glucoside	*Petroselinum crispum*	Nordström *et al.* (1953)
		Apium graveolens	Farooq *et al.* (1953)
Graveobioside B	Chrysoeriol-7-apiosyl-β-D-2-glucoside	*Petroselinum crispum*	Grisebach and Bilhuber (1967)
		Luffa echinata	Seshadri and Vydeeswaran (1971)
—	Quercetin-3-apiosyl-β-D-2-glucoside	*Securidaca diversifolia*	Hamburger *et al.* (1985)
—	Quercetin-3-apiosyl-β-D-2-galactoside	*Securidaca diversifolia*	Hamburger *et al.* (1985)
—	Quercetin-3-apiosyl-β-D-xylanoside	*Securidaca diversifolia*	Hamburger *et al.* (1985)

Compound	Plant species	Reference
— Quercetin-3-apiosyl-α-L-arabinoside	*Securidaca diversifolia*	Hamburger et al. (1985)
— Kaempferol-3-apiosyl-β-D-glucoside	*Securidaca diversifolia*	Hamburger et al. (1985)
—	*Cicer arietinum*	Hösel and Barz (1970)
Disaccharide-Dibenzoate 3'-Benzoyl-2-O-D-apiosyl-D-benzoyl-glucoside	*Daviesia latifolia*	Hansson et al. (1966)
Apiosyl-3-glycosides		
Onjisaponin A Presenegine-3-β-D-glc$p$,-28-β-D-Fuc$p$, [2 ← α-L-Rham$p$ 4 ← β-D-Xyl$p$ 4 ← β-D-Gal$p$ 3 ← β-D-Api$f$], [3 ← α-L-Rham$p$] [4 ← 4'-methoxy-cinnamic acid]	*Polygala tenuifolia*	Sakuma and Shoji (1982)
Onjisaponin F Presenegine-3-β-D-glc$p$,-28-β-D-Fuc$p$, [2 ← α-L-Rham$p$ 4 ← β-D-Xyl$p$ 3 ← α-L-Ara$p$ 3 ← β-D-Api$f$], [4 ← 3',4',5'-trimethoxy-cinnamic acid]	*Polygala tenuifolia*	Sakuma and Shoji (1981)
Onjisaponin G Presenegine-3-β-D-glc$p$,-28-β-D-Fuc$p$, [2 ← α-L-Rham$p$ 4 ← β-D-Xyl$p$ 3 ← β-D-Api$f$], [4 ← 3',4',5'-trimethoxycinnamic acid]	*Polygala tenuifolia*	Sakuma and Shoji (1981)
Apiosyl-6-glucosides		
Furcatin p-Vinylphenol-D-apiosyl-6-glucoside	*Viburnum furcatum*	Hattori and Imaseki (1959)
Apiopaeonoside Paeonol-D-apiosyl-6-glucoside	*Paeonia suffruticosa*	Yu et al. (1986)
Apiosyl-glucosides of unknown linkage		
Lanceolarin Biochanin-A-7-apiosylglucoside	*Dalbergia lanceolaria*	Malhotra et al. (1965)
Homoflavoyadorinin B	*Viscum album*	Ohta and Tagishita (1970)
— Quercetin-3-O-(2^{G}β-D-apiosyl-rutinoside)	*Solanum glaucophyllum*	Rappaportt et al. (1977)

TABLE 7.1b. Occurrence of D-apiose as a cell-wall constituent.

Compound	Composition	Plant source	References
Apiogalacturonan	Apiosyl and apibiosyl-(D-Apif/3^1-D-Apif) side chains of polygalacturonic acid	*Lemna minor* (*Lemna gibba*)	Hart and Kindel (1970a,b), Beck (1967)
Apio-xylo-galacturonan	Apiosyl and xylosyl side chains of polygalacturonic acid	*Lemna minor*	Beck (1967)
Zosterine	Polygalacturonic acid + gal, ara, xyl, O-Me-xyl, api.	*Zostera marina*	Ovodova *et al.* (1968), Ovodov *et al.* (1971)
		Zostera pacifica	Ovodova *et al.* (1968)
		Phyllospadix spp.	Ovodova *et al.* (1968)
		Phyllospadix torreyi	Woolard and Jones (1978)
Rhamnogalacturonan II	RGII contains (60%) two heptasaccharides (A, B)	*Acer pseudoplatanus*-cell suspension cultures	Darvill *et al.* (1978), York *et al.* (1985)
		Pisum sativum	McNeil *et al.* (1984)
		Lycopersicum esculentum	
		Pinto bean	
		Avena sativa	
		Pseudotsuga spp.	
Pectic cell-wall fraction	Unknown	*Tilia vulgaris*	Bacon and Cheshire (1971)
		Posidonia australis	Bell *et al.* (1954)
		Wolffia arrhiza	Duff (1965)

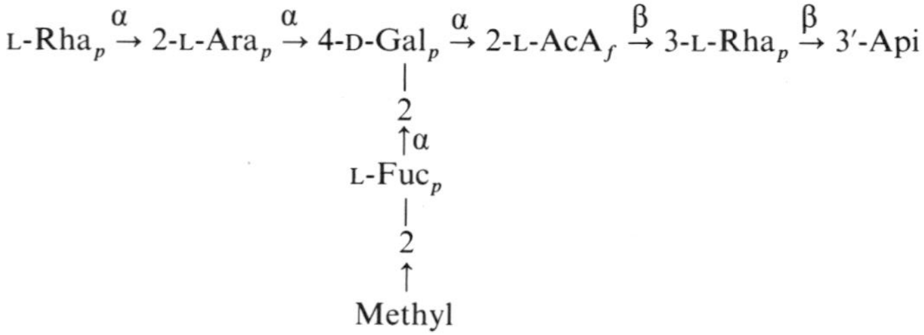

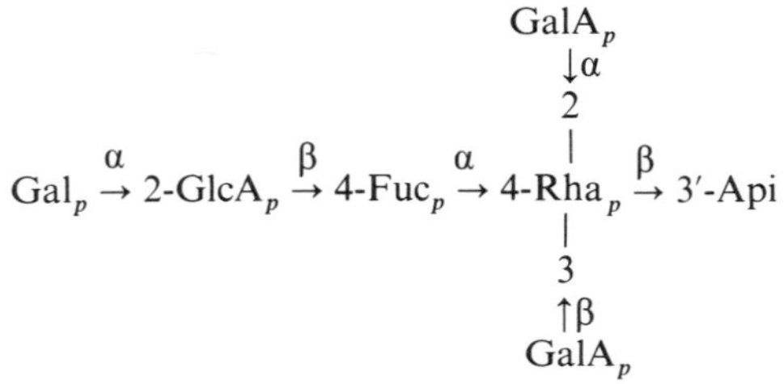

(A)

$$\text{L-Rha}_p \overset{\alpha}{\rightarrow} \text{2-L-Ara}_p \overset{\alpha}{\rightarrow} \text{4-D-Gal}_p \overset{\alpha}{\rightarrow} \text{2-L-AcA}_f \overset{\beta}{\rightarrow} \text{3-L-Rha}_p \overset{\beta}{\rightarrow} 3'\text{-Api}$$

$$\begin{array}{c} | \\ 2 \\ \uparrow\alpha \\ \text{L-Fuc}_p \\ | \\ 2 \\ \uparrow \\ \text{Methyl} \end{array}$$

$$\begin{array}{c} \text{GalA}_p \\ \downarrow\alpha \\ 2 \\ | \end{array}$$

(B)

$$\text{Gal}_p \overset{\alpha}{\rightarrow} \text{2-GlcA}_p \overset{\beta}{\rightarrow} \text{4-Fuc}_p \overset{\alpha}{\rightarrow} \text{4-Rha}_p \overset{\beta}{\rightarrow} 3'\text{-Api}$$

$$\begin{array}{c} | \\ 3 \\ \uparrow\beta \\ \text{GalA}_p \end{array}$$

TABLE 7.1c. List of plants containing substantial amounts of D-apiose in unknown linkage which have not been mentioned in Tables 7.1a and 7.1b. An ample screening (175 plant species from 100 families: Duff and Knight, 1963) by hydrolysis of dried plant material and subsequent PC was performed by R. B. Duff who died before the detailed results were published. Watson and Orenstein (1975) gave a summary of Duff's data. (D) refers to that compilation. Species containing only traces or moderate amounts of D-apiose are not listed here. Van Beusekom (1967) screened the monocotyledoneous group of the Helobiae for apiose: (B) refers to that work.

Aegopodium podagraria (D), *Conopodium majus* (D), *Cymodocea nodosa* (B), *Datus aucuparia* (D), *Desfontania spinosa* (D), *Hevea brasiliensis* (Patrick, 1956), *Ilex aquifolia* (D), *Lavatera annua* (D), *Menyanthes trifoliata* (D), *Nerium oleander* (Duff and Knight, 1963), *Posidonia oceanica* (B), *Potamogeton pectinatus* (B), *Ruppia spiralis* (B), *Spirodela polyrhiza* (B), *Taraxacum kok-saghyz* (Chrastil, 1956), *Thalassia hemprichii* (B), *Vinca minor* (D).

Zostera marina). The classical procedure utilises *Posidonia* leaves or marine fibres as plant source which are hydrolysed with 0.4 N H_2SO_4 at 100°C for 1 h (Bell, 1962).

For the preparation of smaller amounts of D-apiose, apiin can be purchased from Carl Roth GmbH (Karlsruhe, FRG). Di-*O*-isopropylidene-apiose is available from Pfanstiehl Labs (Wankegan, Illinois, USA). The free sugar can be liberated from these products by hydrolysis with H_2SO_4, neutralisation and decolorisation with charcoal (Bell, 1962). Usually the yields are given in terms of the more or less water-free syrup. For exact determination of D-apiose see Section II.E.3.

Uniformly ^{14}C-labelled D-apiose has been obtained in reasonable yields from duckweed after 24 h photosynthesis in air plus 1% [14]CO_2 (Beck and Kandler, 1965).

(b)   *Synthesis and position-labelling of apiose and its analogues.* For special purposes, chemical synthesis of apiose may be advantageous and several procedures have been reported in the literature for the synthesis of D-apiose (Gorin and Perlin, 1958; Khalique, 1962; Williams and Jones, 1964; Ezekiel *et al.*, 1969; Ho, 1979; Koos and Mosher, 1986), L-apiose (Schaffer, 1959; Overend *et al.*, 1970; Ho, 1979) and DL-apiose (Raphael and Roxburgh, 1955; Araki *et al.*, 1981). Position-labelling is accomplished either by supplying the correspondingly labelled precursors (e.g. [14]C- or ^{3}H-labelled glucose or glucuronic acid) to parsley seedlings (Grisebach and Döbereiner, 1966; Grisebach and Sandermann, 1966) or to sterile cultures of duckweed (Beck and Kandler, 1966; Picken and Mendicino, 1967) or by incorporation of stable isotopes upon chemical synthesis according to Snyder and Serianni (1987). In addition to free D-apiose, the following analogues have been synthesised: D-apiose aldal (Nachman *et al.*, 1986), 3-deoxy-D-apiose (and various derivatives: Ball *et al.*, 1969) and apiitol (Neal and Kindel, 1970).

(c)   *Synthesis of derivatives of D-apiose.* Except for one unconfirmed report (Ovodova, *et al.*, 1968), D-apiose has never been obtained in crystalline form. Hence derivatives were prepared whenever a crystalline product was required. In earlier work these derivatives, in addition to chromatography, were used to confirm the presence of D-apiose. The derivatives usually prepared are the *N*-benzyl-*N*-phenylhydrazone, the phenylosazone, *p*-bromophenylosazone and the di-isopropylidene-apiose. The latter can easily be purified by sublimation. However, care must be taken with respect to synthesis because upon treatment with acetone containing 5% (v/v) of concentrated sulphuric acid the furanose ring opens and reforms again, resulting in a mixture of the β-L-threofuranose and the α-D-erythrofuranose, both present as 1,2 : 3,3[1]-di-*O*-isopropylidene derivatives (Ball *et al.*, 1969, 1971; Fig. 7.7). If the reaction is terminated by neutralisation before equilibrium is reached both derivatives will be isolated, otherwise only the di-*O*-isopropylidene-β-L-threofuranose is obtained. The melting point and optical rotation data of the various derivatives are presented in Table 7.2.

(d)   *UDP-D-apiose and D-apiose phosphates.* Following the route of biosynthesis, UDP-D-apiose has been prepared in good yield (about 60%) from UDP-D-glucuronic acid with UDP-apiose/UDP-xylose synthase purified from duckweed (Kindel and Watson, 1973) and from suspension cultures of parsley (Baron *et al.*, 1973). The presence of UDP-D-apiose in parsley plants has also been demonstrated (Sandermann and Grisebach, 1968). Apiosyl-cyclic-1,2-phosphate was prepared by alkaline treatment of UDP-

D-apiose and apiose-2-P was produced by acid hydrolysis from it (Kindel and Watson, 1973). From a mixture of β-D-apiose tetra-acetates α-D-apiosyl 1-P and α-L-apiosyl 1-P have been isolated in 20% yield (Mendicino and Hanna, 1970). The tetra-acetates of the two ring forms were separated by PC and TLC. Apiosyl nucleosides have been synthesised, the structures of which are given by Watson and Orenstein (1975).

(e) *Chemical degradation of D-apiose.* Free D-apiose can easily be degraded by periodate oxidation producing formic acid, formaldehyde and glycolic acid as shown in Fig. 7.10a. To distinguish between carbons 1 and 2 and then in turn between C-3^1 and C-3, an aliquot of the starting material was reduced to apiitol, periodate oxidation of which generates the same products, but from other carbons (Fig. 7.10b). For degradation studies crystalline starting material is usually desirable. Hence crystalline derivatives of D-apiose have also been degraded with periodate: upon oxidation of the N-benzyl-N-phenylhydrazone C-1 plus C-2 were obtained as glyoxal-N-benzyl-N-phenylhydrazone, C-3 as CO_2 (after decarboxylation of glycolic acid with cerium-IV-salt) and C-3^1 plus C-4 were collected as formaldehyde (Beck and Kandler, 1966). C-3^1 and the hydrogen atoms attached to it could be selectively recovered by IO_4^--oxidation of apiin (Kelleher and Grisebach, 1971): Position-^{14}C- or ^{3}H-labelled UDP-glucuronic acid was enzymatically converted to UDP-D-apiose which was then enzymatically transferred to 7-glucosyl-apigenin. Another method of isolation of C-3^1 by periodate oxidation started from apiofuranosyl 1,2-cyclic phosphate which was prepared from UDP-D-apiose by treatment with alkali (Mendicino and Abou-Issa, 1974).

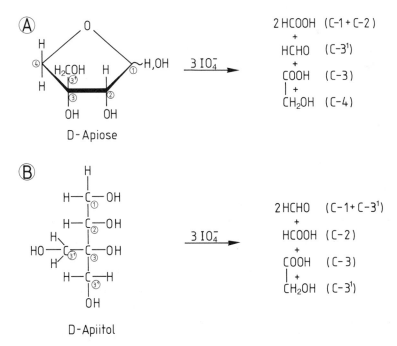

FIG. 7.10. Periodate oxidation of D-apiose (A) and D-apiitol (B) according to Picken and Mendicino (1967).

TABLE 7.2. Characterisation of apiose, apibiose and their frequently used derivatives.

Compound	Melting point (°C, not always corrected)	Optical rotation $[\alpha]_D$	Reference
D-Apiose (Syrup)	—	$+9.1 \pm 0.5°$ (H_2O)	Bell (1962)
	—	$+6.4°$ (c 1.0, H_2O)	Gorin and Perlin (1958)
	—	$+5.2°$ (c 1.1, H_2O)	Ho (1979)
	—	$+5.1°$ (c 1.1, H_2O)	Koos and Mosher (1986)
	—	$+4.8°$ (H_2O)	Mendicino and Picken (1965)
		$+4.2°$ (c 0.48, H_2O)	Khalique (1962)
(Crystalline)	167	$+5°$ (c 3, H_2O)	Ovodova et al. (1968)
D-Apiose-phenylosazone	158	—	Ovodova et al. (1968)
	155–157	—	Bell (1962)
	154–156	—	Chrastil (1956), Patrick (1956)
	148–157	—	Duff (1965)
D-Apiose-p-bromophenylosazone	209.5–210.5		Bell (1962)
	290 (perhaps printing error, should read 209)		Chrastil (1956)
	206–208		Raphael and Roxburgh (1955)
D-Apiose-N-benzyl-N-phenyl-hydrazone	140	$-29.5°$ (c 1.1, MeOH)	Beck and Kandler (1966)
	139	$-68°$ (c 1.3, Pyridine)	Koos and Mosher (1986)
	137–139	$-29°$ (c 1.1, MeOH)	Williams and Jones (1964)
	—	$-75°$ (c 1.7, Pyridine)	Williams and Jones (1964)
	137		Ho (1979)
	135		Bell (1962)
L-Apiose-N-benzyl-N-phenyl-hydrazone	194–195.5	—	Gorin and Perlin (1958)
	189–190	—	Williams and Jones (1964)
	188.5–190.5	—	Bell (1962)

1,2:3,3¹-Di-O-isopropylidene-D-apiose	81–83	+56° (EtOH)	Bell (1962)
	81–83	+55.5° (c 1.1, EtOH)	Khalique (1962)
	81–82	−61°	Hansson et al. (1966)
	80–82	+58° (c 1.17, EtOH)	Williams and Jones (1964)
	—	+54 ± 1° (c 0.55, EtOH)	Duff (1965)
Di-O-isopropylidene-L-apiose	56–58	−16° (c 1.14, EtOH)	Williams and Jones (1964)
D-Apibiose	—	−69.1° (c 5, H₂O)	Hart and Kindel (1970a)
D-Apibiose-phenylosazone	91–93	—	Hart and Kindel (1970a)
D-Apibiose-phenylosotriazol	116.5–117.0	−80° (c 5, H₂O)	Hart and Kindel (1970a)

2. D-Hamamelose

(a) *Preparation from hamamelitannin.* For the preparation of substantial quantities of D-hamamelose, isolation of the free sugar from plant material is not recommended for two reasons: (1) the plants rich in D-hamamelose, namely the primroses and related genera, are rare species; and (2) D-hamamelose is accompanied by the corresponding alditol (D-hamamelitol: Sellmair *et al.*, 1977) which can be separated from the aldose only via the preparation of chemical derivatives of the latter, e.g. the phenylhydrazone. In those species which are not capable of reducing D-hamamelose to the sugar alcohol, the free sugar could only be detected after ^{14}C-labelling and hence these species are not utilisable as sources for its preparation. A convenient procedure for the preparation of this monosaccharide follows the classical route via the isolation of hamamelitannin (Freudenberg and Blümmel, 1924). Hydrolysis of this compound with 1 N HCl and purification of the sugar from the neutralised solution (from which crystalline gallic acid is removed by centrifugation) by column chromatography yields pure hamamelose. The whole procedure, starting with the bark of witch hazel (*Hammamelis virginiana*) which is available from the herb shop, has been described in detail by Gilck *et al.* (1975). Since D-hamamelose crytallises readily from a solution in dry (!) methanol saturated with petroleum ether, synthesis of crystalline derivatives is not as essential as in the case of D-apiose.

For the preparation of small reference samples of D-hamamelose hamamelitannin is available from Carl Roth GmbH (Karlsruhe, FRG).

(b) *Preparation by chemical synthesis and labelling of D-hamamelose.* As outlined in Section II.A, various methods of chemical synthesis of D-hamamelose (Paulsen *et al.*, 1972; Overend and Williams, 1965), methyl-β-D-hamameloside (Paulsen *et al.*, 1980; Depezay and Duréault, 1978), D-epihamamelose (Overend and Williams, 1965), methyl-D-epihamamelose (Depezay and Duréault, 1978) and L-hamamelose (Burton *et al.*, 1965) have been developed, which at present seem only moderately important for biochemical and physiological work. For U-^{14}C-labelling of D-hamamelose ^{14}CO$_2$-fixation with primroses (Sellmair *et al.*, 1977) or *Soldanella* (which produces only insignificant amounts of hamamelitol) is recommended. ^{14}C-D-Hamamelose can be extracted from the leaves and purified by PC and paper electrophoresis (if hamamelitol is present) as described by Beck *et al.* (1968) and Sellmair *et al.* (1977). Position-labelled D-hamamelose may be obtained from the same plant species after administering a solution of ^{14}C- or ^{3}H-position-labelled glucose via the petiole. Specifically C-1-labelled D-hamamelose can be prepared by cyanohydrine synthesis with D-ribose and labelled cyanide (Serianni *et al.*, 1982).

D-Hamamelose cannot produce an osazone. However, crystalline *p*-nitrophenyl-hydrazone has been prepared in good yield (70%; m.p. 162–165°C: Beck *et al.*, 1968; m.p. 165–166°C, $[\alpha]_D^{20} = +144°$ (c 4%, pyridine: Freudenberg and Blümmel, 1924). Less recommendable from the point of view of yield is the synthesis of the *p*-toluene-hydrazone (m.p. 155°C, $[\alpha]_D^{21} = +76.1°$ (c 6%, pyridine: Freudenberg and Blümmel, 1924).

(c) *Preparation of naturally occurring compounds related to D-hamamelose.* In addition to the classical source of D-hamamelose, viz. hamamelitannin, a few naturally occurring derivatives of D-hamamelose have been isolated and identified. Table 7.3 provides a summary of the members of the 'D-hamamelose family'.

Hamamelitannin. The preparation of hamamelitannin has been described in context with the isolation of D-hamamelose. It should be mentioned that this compound has also been synthesised from methyl-β-D-hameloside and tri-*O*-benzylgalloyl chloride (Ezekiel *et al.*, 1969). Monogalloyl-hamamelose which turned out to be a mixture of 2^1 and of 5-galloyl-hamamelose presumably originates *in vivo* by deacylation of hamamelitannin (Schilling and Keller, 1986). Due to its poor solubility in ethyl acetate it could be separated from the digalloyl compound. Isolation of one of the monogalloyl-hamameloses was not achieved.

D-*Hamamelitol.* The preparation of D-hamamelitol was mainly performed by chemical reduction of D-hamamelose mentioned in Section III.E.10 (see also Sellmair *et al.*, 1968).

D-*Hamamelonic acid.* D-Hamamelonic acid is produced from D-hamamelose by a strain ('H1') of *Pseudomonas* which has been isolated from the soil around roots of *Primula clusiana*, a primrose producing substantial quantities of hamamelose and hamamelitol. Usually D-hamamelonic acid is obtained from the medium as a mixture of the lactone and the anion, thus producing two spots on PCs and paper pherograms (Thanbichler *et al.*, 1971).

D-*Hamamelonic acid-2^1P.* D-Hamamelonic acid 2^1-P is the recently detected natural inhibitor of Rubisco. It has been extracted from darkened potato leaves as a Rubisco–inhibitor complex which was precipitated with polyethylene glucol (PEG). After further purification of that complex, D-hamamelonic acid 2^1-P was separated from the protein by denaturation of the latter with MeOH. It was purified by reversed phase HPLC and ion-exchange chromatography (Gutteridge *et al.*, 1986). Another procedure (Berry *et al.*, 1987), starting with ion-exchange chromatography of a bean leaf extract, also made use of affinity-binding to Rubisco as a powerful purification step of the inhibitor. The compound has also been synthesised via cyanohydrine synthesis with ribulose-BP (RuBP) and cyanide, partial dephosphorylation with phosphatase and separation of the monophosphates by binding of D-hamamelonic acid 2^1-P to Rubisco (Berry *et al.*, 1987). Several branched-chain carboxypentitol-bisphosphates and 4-ketocarboxypentitol-bisphosphates have been synthesised and used to study the intermediate of the RuBP carboxylase reaction (Pierce *et al.*, 1980; Schloss, 1988).

D-*Hamamelose $2^1,5$-BP.* U-^{14}C-Hamamelose-BP was obtained from isolated spinach chloroplasts which had been allowed to fix $^{14}CO_2$ for 15–20 min (Beck *et al.*, 1971; Beck and Knaupp, 1974) or to which U-^{14}C-fructose 1,6-BP had been administered in the light for 15 min (Gilck and Beck, 1974). Anion-exchange chromatography of the soluble chloroplast extract yielded the fraction of the sugar-bisphosphates. Fructose-BP together with hamamelose-BP were separated from sedoheptulose-BP, glucose-BP, 6-P-gluconate and 3-P-glyceric acid by PE. Finally fructose-BP was dephosphorylated specifically with fructose-bisphosphatase whereupon hamamelose-BP could be isolated by PE (Beck and Knaupp, 1974). The enyzme catalysing the intramolecular rearrangement of fructose-BP and hamamelose-BP is obviously associated with the membrane fraction of the chloroplast (Beck, 1982).

D-*Hamamelose 2^1-P.* D-Hamamelose 2^1-P is the first intermediate in the catabolic route from D-hamamelose to CO_2 as accomplished by *Kluyvera citrophila* strain 627 which belongs to the *Enterobacteriaceae* (Thanbichler and Beck, 1974). The enzyme

TABLE 7.3. Naturally occurring members of the D-hamamelose family.

Compound	Structure	Melting point (°C)	Optical rotation $[\alpha]_D$	Occurrence	References
D-Hamamelose	2-C-(Hydroxymethyl)-D-ribose	110	$-6.7 \pm 0.5°$ (H₂O)	Ubiquitous	Gilck et al. (1975)
		110–111	$+7.7$ $-7.0°$	Synthetic	Sellmair et al. (1977)
					Overend and Williams (1965)
D-Hamamelitol (syrup)	2-C-(Hydroxymeythyl)-D-ribitol	—	$+23°$ (c 1, MeOH)	Primroses	Sellmair et al. (1968)
Clusianose D-hamamelitol	1-O-α-D-Galactopyranosyl-	180–190	$+131°$ (H₂O)	Primroses	Sellmair et al. (1977)
					Beck (1969)
D-Hamamelonic acid	2-C-(Hydroxymethyl)-D-ribonic acid (2-carboxy-D-arabinitol)	116–118	—	Pseudomonas spp.	Sellmair et al. (1977)
					Thanbichler et al. (1971)
Hamamelitannin	2^1,5-Digalloyl-D-hamamelose	144–147	$+35.7°$ (c 2, H₂O)	Hamamelis virginiana	Meyer et al. (1965)
				Hamamelis virginiana	Freudenberg and Blümmel (1924)
		—		Castanea sativa	Mayer and Kunz (1959)
		—		Quercus rubra	Mayer et al. (1965)
				Synthetic	Ezekiel et al. (1969)
Monogalloyl-D-hamamelose	2^1- and 5-Galloyl-D-hamamelose	145–147.5	$+31.3°$ (c 1.5, H₂O)	Hamamelis virginiana	Schilling and Keller (1986)
		—	$+10.4$ (c 2, H₂O)		
—	D-Hamamelose 2^1,5-BP	—		Spinach chloroplasts	Beck et al. (1971)
				Primrose leaves	Gilck and Beck (1974)
—	D-Hamamelose 2^1-P	—		Kluyvera citrophila	Thanbichler and Beck (1974)
					Beck et al. (1980)
—	D-Hamamelonic acid 2^1-P (2-carboxy-D-arabinitol-1P)	—	—	Phaseolus vulgaris	Berry et al. (1987)
—		—	—	Solanum tuberosum	Gutteridge et al. (1986)

catalysing this step, a specific hamamelose-kinase, has been purified (Beck *et al.*, 1980) and used to prepare sufficient hamamelose 2^1-P for structure elucidation (Thanbichler and Beck, 1974).

(d)   *Chemical degradation of D-hamamelose.* Two procedures for the degradation of D-hamamelose by periodate oxidation have been reported, starting either with the *p*-nitrophenylhydrazone or with crystalline D-hamamelose. The latter (easier) procedure allows specific determination of H(T) at C-1, but does not differentiate between C-1 and C-2; C-2^1 and C-5, and between C-3 and C-4 (and of H attached to these carbons), respectively (Gilck and Beck, 1974). Degradation of D-hamamelose-*p*-nitrophenyl-hydrazine proceeds as shown in Fig. 7.11 (Beck *et al.*, 1968). The resulting glyoxylic acid-*p*-nitrophenylhydrazone can be easily decarboxylated in order to differentiate between C-1 and C-2.

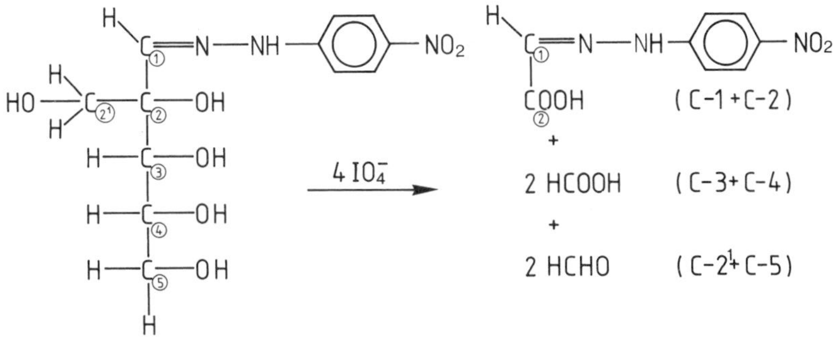

FIG. 7.11.   Periodate oxidation of D-hamamelose-*p*-nitrophenylhydrazone according to Beck *et al.* (1968).

3.   Aceric acid (Spellman et al., 1983)

Aceric acid has been isolated from purified rhamnogalacturonanII (RGII), which also contains D-apiose. The RGII was prepared from primary cell walls of suspension-cultured sycamore (*Acer pseudoplatanus*) cells. Aceric acid (25–50 µg) was obtained by anion-exchange chromatography after hydrolysis (1 h, 120°C) of 3 g RGII with 2 M trifluoroacetic acid. The aceric acid content of RGII is 7%. A reasonable quantity (60 mg) of aceric acid was prepared by a similar procedure from 600 g of Pectinol AC (Corning, Inc.), a commercial preparation of *Aspergillus niger* exoenzymes. If the fungus is grown on plant cell-wall material Pectinol AC contains a polysaccharide similar to RGII at a proportion of 1.5%. For elucidation of the chemical structure, aceric acid was reduced to the corresponding carboxy-deoxy-pentitol, the lactone of which could then be acetylated.

D.   Enzymes Catalysing Biosynthesis and Metabolism of Apiose, Hamamelose and Related Compounds

While enzymological work in the field of D-hamamelose is almost completely lacking, a remarkable number of enzymes related to the biochemistry of D-apiose have been

TABLE 7.4. Enzymes catalysing the biosynthesis and metabolism of D-hamamelose, D-apiose and related compounds.

Enzyme	Catalysed reaction	Properties				Source	Reference
		MW (kDa)	k_M	Specific activity (U)	Optimum pH		
D-Hamamelose-kinase	D-Hamamelose + ATP → D-Ham-2'P + ADP	21	$k_{M\ Ham}$:3 mM $k_{M\ ATP}$:2.5 mM	6		*Kluyvera citrophila* 627	Beck *et al.* (1980)
UDP-Apiose/UDP-xylose-synthase	UDP-GlucUA → UDP-Api + UDP-Xyl	2 × 34 + 2 × 44	$k_{M\ UDPGlucUA}$: 2 μM (Tris) 60 μM^a (PO_4^{3-}) 30 μM^b (PO_4^{3-})	6.7×10^{-3}	7.5 8	Parsley cell suspension cultures	Grisebach (1980) Matern and Grisebach (1977)
UDP-D-Glucuronate cyclase+carboxy-lyase I	UDP-GlucUA → UDP-Api + UDP-Xyl	—	$k_{M\ UDPGlucUA}$: 9.8 μM^a (PO_4^{3-}) 8.3 μM^b (PO_4^{3-})	180×10^{-3}	7.7-8.1	*Lemna minor*	Gustine *et al.* (1975)
UDP-D-Apiose-synthase	UDP-GlucUA → UDP-Api + UDP-Xyl	110 ± 10	$k_{M\ UDPGlucUA}$: 2 μM (Tris)	0.3×10^{-3}	7-8.5	*Lemna minor*	Mendicino and Abou-Issa (1974)
UDP-Apiose/UDP-xylose-synthase	UDP-GlucUA → UDP-Api + UDP-Xyl	—	$k_{M\ UDPGlucUA}$: 5 μM (Tris)	0.02×10^{-3}	7-8	*Lemna minor*	Wellmann and Grisebach (1971)
UDP-Apiose: flavone apiosyltransferase (EC 2.4.2.25)	UDP-Apiose + Flavone-β-D-glycoside → Flavone-glycobioside + UDP	50	$k_{M\ UDPApi}$: 66 μM	2.2	7.0	Parsley cell suspension cultures	Ortmann *et al.* (1972)
Apiin-synthetase	UDP-Api + Apigenin-7-O-glucoside → Apiin + UDP	—	$k_{M\ UDPApi}$: 6 μM $k_{M\ Flavone\ βgluc}$: 70 μM	—	7.6-8.4	Parsley plants *Digitalis purpurea*	Watson and Kindel (1970) Watson and Orenstein (1975)

Enzyme	Reaction		Km		pH optimum	Source	Reference
UDP-D-Apiose: acceptor D-apiosyltransferase (+ D-xylosyltransferase)	UDP-Api + Galacturonan or Apiogalacturonan → Apiosylgalacturonan or Apibiosylgalacturonan + UDP	Particulate enzyme	$k_{M\ UDPApi}$: 4.9 μM	—	5.7	*Lemna minor*	Pan and Kindel (1977) Mascaro and Kindel (1977)
D-Apiose reductase (EC 1.1.1.114)	D-Apiose + NADH → D-Apiitol + NAD$^+$	—	$k_{M\ Apiose}$: 20 mM $k_{M Apiitol}$: 10 mM	1.4	7.5–10.5	*Aerobacter aerogenes*	Neal and Kindel (1970)
D-Apiitol dehydrogenase	D-Apiitol + NAD$^+$ → D-Apiose + NADH	110	$k_{M\ Apiose}$: 71 mM $k_{M\ Apiitol}$: 11.6 mM $k_{M\ NAD}$: 0.35 mM $k_{M\ NADH}$: 15 μM	0.1	7.5–10	*Micrococcus*	Hanna *et al.* (1973)
β-Glycosidase (2 enzymes)	Biochanin-A-7-β-apioglucoside → Biochanin A + Apiosyl 1 → 2 glucose	120–140 (both)	$k_{M\ Lanceolarin}$: 0.15 mM	10	5.5	*Cicer arietinum*	Hösel (1976)
Furcatinase	p-Vinylphenol-D-apiosyl-6-glucoside → p-Vinylphenol + Apiosyl 1 → 6 glucose	—	$k_{M\ Furcatin}$: 3.6 mM	—	6.0	*Viburnum furcatum*	Imaseki and Yamamoto (1961)

[a] Synthesis of UDP-api.
[b] Synthesis of UDP-xylose.

demonstrated, purified and characterised. Rather than confronting the reader with a verbose description of the enzymology, a compilation of these enzymes is provided in Table 7.4. With reference to the scope of this article, it should be mentioned that, with the exceptions of hamamelose kinase and apiitol dehydrogenase, the convenient type of spectrophotometric assay could not be applied. ^{14}C-Labelled substrates were required (which in most cases have to be prepared), either because of low enzyme activities (μ units!) or because differentiation between more than one reaction product was necessary (e.g. D-apiose and D-xylose). The latter problem, as well as the preparation of substrates such as ^{14}C-UDP-D-apiose, not infrequently demands time consuming chromatographic procedures (e.g. PC for 5 days: Ortmann et al., 1972) and re-examination of the separated compounds via dephosphorylation, PC and/or PE.

E. Analytical Methods

1. Paper chromatography, paper electrophoresis, gas chromatography of branched-chain sugars

Neither D-hamamelose and its derivatives, nor D-apiose and its related compounds, are encountered in the files of the *CRC Handbook of Chromatography* (Churms, 1982). Except for very few data, mainly on GLC of chemical derivatives of apiose, neither does the *Handbook of Biochemistry and Molecular Biology* (Fasman, 1975) provide chromatographic data of these monosaccharides. The data compiled in Tables 7.5 and 7.6 have therefore been collected from the original papers. However, in many papers characteristic values such as R_f values are not shown or the lack of data for general reference substances, e.g. glucose, render those chromatographic results less useful. PC and PE have been widely employed for the identification of both branched-chain monosaccharides and the data are shown in Tables 7.5 and 7.6. TLC has successfully been employed in the separation of the apiose-containing flavones (Egger, 1967; Ortmann et al., 1972) and also for the separation of hamamelitannin and its chemical derivatives (Mayer et al., 1965; Ezekiel et al., 1969). However, TLC data of hamamelose and apiose and of their other natural derivatives have not been reported. A solvent system which by the pattern of separation could also be useful for TLC of both monosaccharides makes use of the complexing capability of boric acid (Götz et al., 1978). GLC on packed columns of TMS-hamamelose (Beck et al., 1971), of apiitol acetates (Darvill et al., 1978; Hanna et al., 1973), of O-methyl-ethers (Hulyalkar et al., 1965), of apiose and of TMS-D-apiose and D-apiitol (Roberts et al., 1967) combined with FID has been reported. GLC in combination with MS will be reviewed in Section II.E.4.

Separation of UDP-D-apiose from other nucleoside diphosphate sugars by PC or PE has not been achieved (Sandermann and Grisebach, 1968). Therefore, the sugar-nucleotide fraction has consistently been analysed after liberation of the sugar moieties by acid or enzymatic (phosphodiesterase) hydrolysis. Likewise, satisfactory separation of the individual sugar-bisphosphates and -monophosphates, respectively, could not be attained with those methods (for a compilation of the $R_{picric\ acid}$ values, see Beck and Knaupp, 1974). Again, analysis of the sugar-bisphosphates and -monophosphates required previous dephosphorylation.

TABLE 7.5. $R_{Glucose}$ values of branched-chain monosaccharides and related compounds as compared to other sugars upon PC.

Compound	Solvent									
	I	II	III	IV[a]	V	VI	VII	VIII	IX	X
D-Hamamelose	1.51	1.50	1.27	—	1.18	1.53	1.27	1.37	1.30	2.28
D-Hamamelitol	1.55	1.54	1.20	—	1.20	1.50	1.17	1.45	1.35	2.08
D-Hamamelonic acid	0.48	0.91	0.24	—	—	—	—	—	—	—
D-Hamamelono-lactone	1.45	1.89	1.83	—	—	—	—	—	—	—
Hamamelitannin	—	—	—	—	—	0.26[b]	—	—	—	—
Clusianose	1.22	0.54	0.55	—	0.81	—	—	—	—	—
D-Apiose	1.42	1.58	1.70	1.52–1.57	1.63	1.68	—	—	—	—
D-Apiitol	—	—	—	1.27	—	—	—	—	—	—
D-Galactose	1.11	0.94	0.87	—	0.93	0.95	1.06	1.31	1.13	—
D-Fructose	1.37	1.28	1.21	—	1.21	1.33	1.23	1.37	1.35	—
D-Xylose	1.25	1.35	1.30	—	1.26	1.55	1.45	1.51	1.35	2.27
D-Ribose	1.56	1.67	1.50	—	1.53	1.68	—	—	1.50	3.46
D-Rhamnose	1.61	1.82	1.89	—	1.75	1.79	—	—	—	—

Usually descending PC on Whatman No. 1 paper was performed.

[a] R_f value.

[b] Whatman No. 3 MM.

Solvents (v/v).

I: 88% (w/w) phenol–water–acetic acid–1 M Na_2EDTA (480:160:10:1).

II: butanol–water (Solution 1): propionic acid–water (Solution 2) (1:1); Solution 1: butanol–water (750:50); Solution 2: propionic acid–water (352:448).

III: butanol–pyridine–water–acetic acid (60:40:30:3).

IV: butanol–pyridine–water (60:40:30).

V: butanol–ethyl acetate–acetic acid–water (4:3:2.5:4).

VI: butanol–acetic acid–water (4:1:5; upper phase).

VII: ethyl acetate–pyridine–water (140:70:30).

VIII: isopropanol–water–acetic acid (75:15:10).

IX: butanol–ethanol–water (5:1:4).

X: butanol–benzene–formic acid–water (100:19:10:25).

The data were compiled from Beck and Kandler (1965); Sellmair *et al.* (1968); Beck (1969); Neal and Kindel (1970) and Beck *et al.* (1971). A variety of other solvent systems have been employed, especially in research into apiose and its derivatives. However, in many cases R_f values are not quoted or the solvent was used for a specific purpose. For the separation of PC of stereoisomers of apiose, the reader is directed to the paper of Sandermann and Grisebach (1968) and for chromatographic data of apibiose and its derivatives the article by Hart and Kindel (1970a) should be consulted.

TABLE 7.6. $R_{\text{picric acid}}$ values of branched-chain monosaccharides and general reference substances upon paper electrophoresis.

Compound	R_{picrate} value in buffer		
	I	II	III
Hamamelose	2.00	0.81	0.92
Hamamelitol	1.87	0.83	1.23
Clusianose	1.20	1.00	0.78
Xylose	1.85	0.38	0.95
Ribose	1.03	0.77	0.93
Ribulose	1.76	0.84	1.42
Glucose	0.00	0.24	1.51

The data were collected from Beck (1969) and Beck et al. (1971).
Buffer systems and conditions:
    I: 0.1 M sodium-molybdate, pH 5.0, 20 V cm^{-1}, 8–10°C, 6 h.
    II: 0.05 M sodium-germanate, pH 10.7, 15 V cm^{-1}, 8–10°C, 8 h.
    III: 0.1 M sodium-borate, pH 9.2, 20 V cm^{-1}, 8–10°C, 6 h.
All pherograms were run on Whatman No. 1 paper.

2. Detection of branched-chain sugars on paper chromatograms and pherograms

A general and very sensitive method for the detection of D-hamamelose and its related compounds (including hamamelitannin) as well as of D-apiose and apibiose is the treatment of the chromatograms and pherograms with the so-called 'alkaline silver-nitrate reagent' (Trevelyan et al., 1950), as described in detail in the *CRC Handbook of Chromatography* (Churms, 1982). By exposing the finished chromatograms to steam from boiling water non-reducing carbohydrates such as apiitol or hamamelitol can also be visualised. This additional treatment is especially useful if the sensitivity of detection is decreased due to the impregnation of the paper with non-volatile buffer substances, e.g. sodium germanate. Treatment with alkaline silver-nitrate reagent cannot be recommended for paper pherograms developed in sodium molybdate. In that case the reaction with periodate and benzidine (Churms, 1982) may be employed; however, the molybdate–sugar complexes must be destroyed first, e.g. by spraying with glacial acetic acid.

The reaction with silver nitrate produces blackish spots which do not allow differentiation of the compounds by a specific colour. Since apiose (as well as aceric acid: Spellman et al., 1983) does not produce furfuraldehyde when treated with acid (Patrick, 1956) a characteristic colour with the benzidine-trichloroacetic acid or aniline-phthalate spray reagents (Churms, 1982) cannot be observed. Faint yellowish spots appear with these reagents which produce a white to light-blue fluorescence in UV light. Although the reaction is not sensitive, a considerable advantage results from the application of aniline-phthalate, for example: since apiose is not decomposed to yield CO_2 the spots can be visualised and subsequently used for radioactivity counting (Gebb et al., 1975).

Hamamelose, on the other hand, apparently produces furfuraldehyde and thus can be detected by generating a brown colour with aniline-phthalate or aniline-trichloroacetic acid. Again, the reaction is much less sensitive than with other sugars such as glucose. Hamamelitannin, owing to its content of gallic acid, can be detected with a dilute (approximately 2%) aqueous solution of $FeCl_3$ which produces a dark violet colour.

UDP-apiose can be localised on paper by the UV-fluorescence quenching characteristic of nucleotides (Mendicino and Abou-Issa, 1974).

3. Quantification of D-apiose and D-hamamelose

Since the anthrone reagent is not applicable to apiose, the Dische test for the determination of tetroses (with fructose, H_2SO_4 and cystein; Sandermann and Grisebach, 1968) has been employed for its quantification. In addition, the Somogy and Nelson test (Somogy, 1945), based upon the reducing capacity of the free sugar has been used (Picken and Mendicino, 1967). From paper chromatograms, apiose has been quantitated with the tetrazolium chloride method (Linskens, 1955; Beck, 1967). However, this method is susceptible to substantial errors if, for example, the resulting colour is exposed to daylight (note: one must work in a dark room).

For the quantitative determination of hamamelose the reducing power of the carbonyl group has been utilised (Sellmair *et al.*, 1968) in the method described by Folin (1929). Hamamelitol, as well as hamamelose, can also be quantitated with the periodate–chromotropic acid procedure (Lambert and Neish, 1950; Sellmair *et al.*, 1968). Clusianose is easily determined via its component galactose.

4. Proton and Carbon-13 NMR, GLC-MS and FAB-MS of branched-chain monosaccharides and their derivatives

Proton and carbon-13 NMR have been employed successfully for the determination of the conformation of branched-chain sugars, and together with electron impact mass spectroscopy (EIMS) in the structural analysis of aceric acid and hamamelonic acid 2^1-P. While the resolution of the proton decoupled chemical shift of the carbon-13 signals in D_2O is optimal between 60 and 100 MHz, a magnetic field of at least 300 MHz is required for a satisfactory separation of the proton signals. Naturally, explanations for the size and the assignment of the signals are a matter of discussion and the interpretation of spin-coupling data with respect to twisting of the several ring forms still requires further investigation. The first interpretation of the proton NMR spectra of β-D-apio-D- and β-D-apio-L-furanosyl tetra-acetate was elaborated by Mendicino and Hanna (1970). From the absence of a proton coupling of the protons at C-1 and C-2 ($J_{1,2} = 0$) and from signals of protons at C-4 and C-3[1] which were interpreted as resulting from long-range coupling, a twisted conformation of the furanose ring was concluded. A similar study was performed with free β-D-apio-D- and β-D-apio-L-furanosyl-1-P and -1,2-cyclic P. Proton NMR data of 2,3-O-isopropylidene-β-D-apiose were established by Ho (1979). The most comprehensive NMR study on DL-apiose in aqueous solution was performed by Snyder and Serianni (1987) using selective carbon-13 and proton enrichment and establishing 2 dimensional and $^{13}C\{^{13}C\}$ NMR spectra. In order to investigate the influence of the configuration of the furanose ring on the rates of ring opening these authors chose DL-apiose which forms the D- as well as the L-ring. In addition, the effect of an hydroxymethyl side chain on the anomerisation rate could also be studied with apiose. In the quoted article carbon-13 and proton chemical shift data for the four apio-furanoses dissolved in D_2O were presented as well as the ^1H–^1H, ^{13}C–^{13}C and ^{13}C–^1H spin coupling constants. In a similar manner, the conformation of D-hamamelose and of the hamamelose moiety of monogalloyl-hamamelose in aqueous (D_2O and $[D_6]$ DMSO) solutions has been studied (Schilling and Keller, 1977, 1986). In both investigations the carbon-13 and proton chemical shift data have been recorded at 20 MHz and with dioxan as an internal standard. The results of these studies have already been described in Section II.B.

Structure elucidation of the Rubisco-inhibitor hamamelonic acid 2^1-P and of aceric acid provide excellent examples of the successful combination of GC-MS and NMR spectroscopy. Trimethylsilyl (TMS) derivatives were prepared from an incubation mixture of the inhibitor with alkaline phosphatase. Upon GLC TMS_3-PO_4 and TMS_6-hamamelonic acid and TMS_4-hamamelono lactone separated as distinct peaks. The mass spectrum of TMS_6-hamamelonic acid together with an interpretation of the ion masses is shown in Fig. 7.12A (Berry et al., 1987). Although this spectrum is typical of 2-carboxy-pentitols it does not allow differentiation between the 2-carboxy-derivatives of the four possible pentitols. Identification, however, is easily achieved by proton NMR spectroscopy. According to the assignment of the proton signals of various carboxy-pentitol-bisphosphates (Pierce et al., 1980) the proton NMR spectrum of the Rubisco-inhibitor could be clearly identified as resulting from 2-carboxy-arabinitol (hama-melonic acid, Fig. 7.12B; Gutteridge et al., 1986).

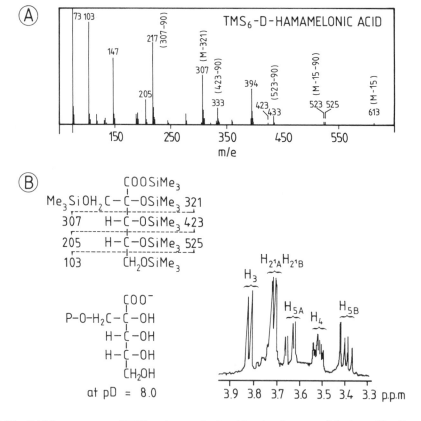

FIG. 7.12. (A) Mass spectrum of TMS_6-D-hamamelonic acid and assignment of the ions. After Berry et al. (1987), modified. (B) Proton NMR-spectrum of free D-hamamelonic acid and assignment of the protons according to Gutteridge et al. (1986), modified.

GLC-MS analysis requires derivatisation of the substance under investigation, either by trimethylsilylation or by peracetylation (preferentially of the corresponding alditol). If only the molecular weight and a preliminary 'fingerprint' for comparative purposes is required, FAB-MS of a few micrograms of the unsubstituted compound, dissolved in a polar solvent, is a powerful technique. This method has been employed, together with GLC-MS, NMR spectroscopy and X-ray crystallography, in the structure elucidation of aceric acid (Spellmann *et al.*, 1983). Since a FAB-mass spectrum of a branched-chain carbohydrate has not been published we have produced the anionic FAB spectrum of hamamelitannin which is shown in Fig. 7.13. In addition to the ion corresponding to the molecular weight (484 − 1), the fragments of gallic acid (153 and 169) can be detected.

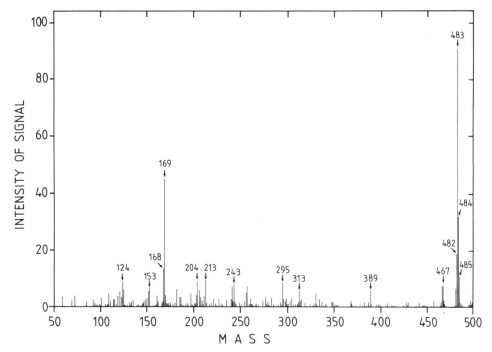

FIG. 7.13. FAB(SIMS) Mass spectrum of hamamelitannin (original). A Finnigan MAT 8500 mass spectrometer was used and the negative ions were produced with a Cs$^+$ beam of 4.6 kV. The compound was dissolved in glycerol.

III. SUGAR ALCOHOLS

A. Introduction

According to rule 23 of carbohydrate nomenclature (IUPAC-IUB, Pigman and Horton, 1970) sugar alcohols (alditols) are defined by the general formula H—$(CHOH)_{n+3}$—H where 'n' ranges from 0 to 7, yielding the groups of sugar alcohols of triitol, tetritols, pentitols, hexitols, heptitols, octitols, nonitols and decitols (Table 7.7). The terms of the sugar alcohols are derived from the names of the corresponding monosaccharides by changing the suffix '-ose' to '-itol', e.g. D-glucose to D-glucitol. Many of the sugar alcohols have common names (Table 7.7), frequently referring to the plant from which they were first isolated, e.g. D-dulcitol = D-sorbitol (from *Sorbus aucuparia*, Rosaceae). Since some of the trivial names of the sugar alcohols are established in literature (e.g. sorbitol) they will be used in this article instead of the systematic chemical term. The sugar alcohols listed in Table 7.7 are all of the D-configuration with exception of L-iditol

TABLE 7.7. The sugar alcohols of plants.

Sugar alcohol	Common name (synonym)	Source	Reference
Triitol			
Glycerol	—	*Olea europaea*	Scheele (1779)
Tetritols			
meso-Erythritol	Erythrol, Erythrite	*Rocella montagnei*	Stenhouse (1848)
2-C-methyl-Erythritol	—	*Convolvulus glomeratus*	Anthonsen *et al.* (1976)
D-Threitol	—	*Armellaria melleae*	Kratzl *et al.* (1963)
Pentitols			
meso-Ribitol	Adonitol, Adonite	*Adonis vernalis*	Merck (1893)
2-C-hydroxymethyl-Ribitol	Hamamelitol	*Primula clusiana*	Sellmair *et al.* (1968)
D-Arabinitol	D-Arabitol, D-Lyxitol	Reduction of arabinose	Kiliani (1887)
L-Arabinitol		*Lobaria pulmonaria*	Nolan and Keane (1933)
meso-Xylitol	—	Reduction of xylose	Bertrand (1891)
	—	*Pullularia pullulans*	Kiessling *et al.* (1962)
Hexitols			
meso-Allitol	Allodulcitol	Synthesis	Lespieau and Weimann (1932)
		Itea ilicifolia	Plouvier (1959)
D-Glucitol	Sorbitol, Sorbite	*Sorbus aucuparia*	Boussingault (1872)
	L-Gulitol		
1,5-anhydro-D-Glucitol	Polygalitol, Aceritol	*Polygala amara*	Chodat (1988)
D-Mannitol	Mannite	*Fraxinus* sp.	Proust (1806)
1,5-anhydro-D-Mannitol	Styracitol	*Styrax obassia*	Asahina (1907)
meso-Galactitol	Dulcitol, Dulcite	*Melampyrum nemorosum*	Hünefeld (1857)
L-Iditol	Sorbierite	Synthesis	Fischer and Fay (1895)
		Sorbus aucuparia	Vincent and Meunier (1898)
Heptitols			
D-glycero-D-gluco-Heptitol	β-Sedoheptitol	Synthesis	Merill *et al.* (1947)

Alditol	Common name	Source	Reference
D-glycero-D-manno-Heptitol	Volemitol α-Sedoheptitol β-Mannoheptitol	*Sedum* sp. *Lactarius volema*	Charlson and Richtmyer (1960) Bourquelot (1895)
D-glycero-D-galacto-Heptitol	Perseitol	*Persea gratissima*	Muntz and Marcano (1884)
D-glycero-D-ido-Heptitol	α-Mannoheptitol —	*Pichia miso*	Onishi and Perry (1965)
1-deoxy-D-glycero-D-manno-Heptitol	Siphulitol	*Siphula* sp.	Lindberg and Meier (1962)
Octitol			
D-erythro-D-galacto-Octitol	—	*Persea gratissima*	Charlson and Richtmyer (1960)

and L-arabinitol and of optically inactive *meso*-sugar alcohols. The simplest of the sugar alcohols is glycerol; as it is widespread only in its bound form in lipids and phospholipids it will not be considered further in this Section. Table 7.7 in addition to the L- and *meso*-alditols lists three tetritols, five pentitols, seven hexitols, five heptitols and one octitol, as well as two anhydro-hexitols, two branched-chain sugar alcohols and one deoxy-alditol, all of which have been isolated from plant material. So far nonitols and decitols are not known to occur in plants. Derivatives of sugar alcohols which have been detected in plant material are glycosides and phosphate esters (for details, see Bieleski, 1982). During the last 50 years a series of review articles about sugar alcohols have been published. They should be consulted by anyone who wants more general information about this subject (Barker, 1955; Lohmar, 1957; Bourne, 1958; Touster and Shaw, 1962; Plouvier, 1963; Stanek *et al.*, 1963; Lewis and Smith, 1967; Percival and McDowell, 1967; Brimacombe and Webber, 1972; Stacey, 1974; Bieleski, 1982; Lewis, 1984).

The reviews by Lewis and Smith (1967), Bieleski (1982) and by Lewis (1984) are the most comprehensive and important ones and should be read in conjunction with this chapter.

B. Discovery of Sugar Alcohols

The desire to find sweetening agents other than honey and cane sugar stimulated the study of new sources in the nineteenth century. The first sugar alcohol isolated was mannitol from manna of ash (*Fraxinus* sp.) by Proust in 1806 (Table 7.7). Most of the sugar alcohols were first isolated from plant material, but some were first prepared by reduction of monosaccharides and were detected later in plant material. The sugar alcohols were found to be sweet-tasting, readily soluble in water, insoluble in most organic solvents, stable against acids and alkali and easy to crystallise.

C. Occurrence of Sugar Alcohols in the Plant Kingdom

The occurrence of sugar alcohols in the plant kingdom is shown in Table 7.7. Only three of them, i.e. sorbitol, mannitol and galactitol, are widespread. Many of the sugar alcohols have a very restricted occurrence, sometimes only known from one species (e.g. allitol and iditol). The group of plants showing the greatest variety of sugar alcohols are the fungi.

As pointed out by Bieleski (1982), our knowledge of the distribution of sugar alcohols in the plant kingdom may change upon extended screening using modern techniques for the detection of small quantities and further work on carbohydrate enzymology; it also may change when major plant groups such as ferns, mosses, gymnosperms and tropical angiosperms are better investigated for sugar alcohols. The distribution of sugar alcohols in the plant world has been reviewed in detail by Plouvier (1963), Lewis and Smith (1967), Brimacombe and Webber (1972), Bieleski (1982), and has been listed in detail in Hegnauer's *Chemotaxonomie der Pflanzen* (1962–1986) and in Karrer's *Konstitution und Vorkommen der organischen Pflanzenstoffe* (1958–1985).

D. Physiology of Sugar Alcohols

The physiology of sugar alcohols is excellently reviewed by Bieleski (1982) and Lewis (1984). In the following sections the main aspects are briefly summarised.

1. Biosynthesis and degradation of sugar alcohols

In higher plants the main site of sugar alcohol formation is the leaf where sugar alcohols together with sucrose represent end products of photosynthesis. Sugar alcohols can also be synthesised in leaves from exogenously applied hexoses, e.g. sorbitol from fructose by apple leaves (Grant and apRees, 1981). The site of their biosynthesis is supposed to be the cytoplasm; evidence which favours the extrachloroplastic site is summarised by Lewis (1984). Redgwell and Bieleski (1978) studied the labelling kinetics of hexose and hexose-phosphates in apricot leaves and postulated the following pathway of bio-synthesis and degradation of sorbitol:

Biosynthesis: Glucose 6-phosphate $\xrightarrow{(1)}$ sorbitol 6-phosphate

Sorbitol 6-phosphate $\xrightarrow{(2)}$ sorbitol

Degradation: Sorbitol $\xrightarrow{(3)}$ fructose

The enzymes involved in this pathway are: (1) glucose 6-phosphate reductase (glucose 6-phosphate: sorbitol 6-phosphate oxidoreductase, NADPH-dependent); (2) sorbitol 6-phosphate phosphatase; and (3) sorbitol dehydrogenase (sorbitol: fructose oxido-reductase, NAD^+-dependent). Glucose 6-phosphate reductase and sorbitol dehydro-genase were characterised by Negm and Loescher (1979, 1981). The mechanism of biosynthesis of mannitol may be the same as for sorbitol because Rumpho *et al.* (1983) demonstrated mannose 6-phosphate reductase in leaves of celery. In fungi (Lewis and Smith, 1967) and in algae (Ikawa *et al.*, 1972) the same mechanism for mannitol biosynthesis was suggested. The biosynthesis of pentitols in fungi appears to occur only by a direct reduction from pentose to pentitol (Suzuki and Onishi, 1975).

Little or nothing is known so far about other pathways of sugar alcohol biosynthesis and degradation (for a few details, see Bieleski, 1982).

2. Translocation of sugar alcohols

In higher plants sugar alcohols are mainly synthesised and accumulated in leaves. From there some of the sugar alcohols are translocated to the stem, root, fruits and seeds. Zimmermann and Ziegler (1975) have listed those plant species which contain sugar alcohols in their sieve elements. They belong to the *Celastraceae* (galactitol), *Combreta-ceae* (mannitol), *Oleaceae* (mannitol) and *Rosaceae* (except *Rosoideae*; sorbitol). The sieve tube sap of these species also contains sucrose and varying amounts of raffinose and stachyose.

3. Storage of sugar alcohols

Like sucrose and oligosaccharides (Kandler and Hopf, 1982), sugar alcohols are soluble storage carbohydrates which in special cases accumulate to make up more than 1% of the dry weight of the plant material. Most sugar alcohols in leaves turn over very slowly and show no significant diurnal fluctuations, for example, sorbitol in plum (Anderson *et al.*, 1961). In contrast, marked seasonal variations have been demonstrated for many sugar alcohols showing highest concentrations during the winter. These observations have led to the suggestion that they are involved in plant cold- and frost-hardiness (see Kandler and Hopf, 1982). Sugar alcohols may act in a biophysical role rather than in a biochemical role as demonstrated, for example, hamamelitol (Sellmair and Kandler, 1970). The sugar alcohol content in plants may also correlate with osmotic stress caused by desiccation during summer and winter or due to high salinity (for details, see Bieleski, 1982).

E. Methods of Analysis

1. General remark on the literature

Methods to detect and to quantify sugar alcohols in plant extracts as developed up to 1966 have been reviewed by Lewis and Smith (1967).

Numerous methods are described in detail in Volumes I–VIII of *Methods of Carbohydrate Chemistry* (Whistler and Wolfrom, 1962–1979), in Volumes IA, IB, IIA, IIB of *The Carbohydrates, Chemistry and Biochemistry* (Pigman and Horton, 1970–1972) and in Volumes A–E 'Carbohydrate Metabolism' as well as in Volumes A–E 'Complex Carbohydrates' of *Methods in Enzymology* (Colowick and Kaplan, 1955–1988). A brief survey of modern methods is also given by Bieleski (1982).

2. Extraction of the plant material

Sugar alcohols should be extracted immediately from freshly harvested plant material to avoid degradation by enzymes. Air-dried plant material and even material stored at deep freeze temperatures ($-25°C$) over a long period should be avoided for the same reason. The plant material must be free of microbial infection and contamination because microorganisms (e.g. fungi) may contain large quantities of sugar alcohols (see Bieleski, 1982). For extraction, 70% boiling ethanol is usually used. The resulting solution is concentrated under reduced pressure or by lyophilisation. Other suitable extraction procedures have been discussed by Lewis and Smith (1967) and Bieleski (1982).

3. Fractionation of the plant extract

Extracts of plant material frequently contain substances which interfere with some of the methods of detection and estimation of sugar alcohols. Percolation of the aqueous extract with petroleum ether removes lipophilic substances. Decolorisation is achieved with charcoal. Proteins and mucilages can be removed by precipitation with metal ions (e.g. lead acetate, barium hydroxide, calcium hydroxide) as described by Bell (1955). Usually ion-exchange resins (H^+, OH^-) are used to remove inorganic salts, organic acids, amino acids and phosphate esters which interfere with the chromatographic procedures at concentrations greater than 5%. However, the resin may also bind varying amounts of sugar alcohols which are eluted from the resin only by prolonged washing. Subsequent samples may therefore be contaminated.

4. Isolation of sugar alcohols by their benzylidene derivatives

Sugar alcohols may be isolated from fractioned plant extracts by virtue of their benzylidene derivatives (Meunier, 1888; Fischer, 1895). Owing to their insolubility in water these derivatives can easily be crystallised and purified from adhering matter by washing with water. As described by the authors mentioned above, a concentrated extract is dissolved in 50% sulphuric acid and benzaldehyde is added (1 : 1; v/v). After standing at room temperature for 24 h the crystalline sugar alcohol benzylidene acetal is separated; after boiling the derivative in 5% sulphuric acid for 1 h, extraction of the benzaldehyde with ether and neutralisation of the acid with barium hydroxide, the sugar alcohol can be crystallised from ethanolic solutions.

5. Separation methods: chromatography and electrophoresis

Chromatographic and electrophoretic methods are used for the isolation and identification of sugar alcohols. Simultaneously, information about homogeneity, molecular size and chemical structure can be obtained. The theoretical background of these methods with examples of application has been provided by Horowitz (1980).

TABLE 7.8. Separation of monosaccharides and sugar alcohols by PC, TLC and PE.

Compound	A ($R_{mannitol} \times 100$)	B ($R_f \times 100$)	C ($R_{glucose} \times 100$)	D ($R_{cis\text{-}inositol} \times 100$)
Allitol	—	—	90	9
D-Arabinitol	152	47	87	13
L-Arabinose	78	37	96	—
Erythritol	193	64	75	8
Fructose	65	34	—	—
Galactitol	96	25	97	17
Galactose	39	16	—	—
D-Glucitol	—	—	—	20
D-Glucose	39	28	100	—
Glycerol	—	73	—	—
L-Iditol	—	—	81	24
D-Mannitol	100	30	91	14
D-Mannose	59	—	72	—
Perseitol	—	13	98	—
Ribitol	165	53	85	8
D-Ribose	251	47	77	—
β-Sedoheptitol	—	—	88	—
D-Sorbitol	116	21	83	—
L-Threitol	—	61	75	12
Volemitol	—	18	93	—
Xylitol	156	38	79	18
D-Xylose	106	55	100	—

A: PC (Lewis and Smith, 1967). Solvent, ethyl-methyl-ketone–acetic acid–saturated aq. soln of boric acid (9:1:1; v/v); paper, Whatman No. 1; room temperature.
B: TLC (Kremer, 1978a,b). Solvent, isopropanol–acetone–0.2 M lactic acid (60:30:10; v/v); layer, Silicagel 60, impregnated with 0.5 M NaH$_2$PO$_4$.
C: PE (Frahn and Mills, 1959). Buffer, 0.05 M borate, pH 9.2; paper, Whatman No. 4; 1.5 h at 500 V (15 V cm^{-1}).
D: PE (Angyal et al., 1974). Buffer, 0.2 M soln of Ca-acetate in 0.2 M acetic acid; paper, Whatman No. 1; 2 h at 20 V cm^{-1}; 25°C.

(a) *Paper chromatography of sugar alcohols.* With most of the solvent systems used in paper chromatography for the separation of monosaccharides and sugar alcohols, the distinction between these two groups on the one hand and between the stereoisomeric sugar alcohols on the other may be very difficult owing to the similar R_f values. The former problem is overcome by using specific detection reagents for sugar alcohols, e.g. benzidine-sodium periodate (Beau and Porter, 1959) or by reacting the monosaccharides with triphenyltetrazolium chloride prior to paper chromatography (Steiner and Maas, 1957). One procedure widely used in the past has been the removal of sugars by fermentation with yeasts leaving the sugar alcohol behind; this method suffers from the fact that yeasts themselves are able to form sugar alcohols (Bieleski 1982). Chromatographic separation of the stereoisomeric sugar alcohols was achieved using solvents allowing the formation of ionic complexes of sugar alcohols with borate. Out of a number of such solvent systems (see Lewis and Smith, 1967) one is given in Table 7.8 showing reasonable separation of monosaccharides and sugar alcohols. The combination of the carbon-14 isotope technique with paper chromatography is still one of the best methods, not only for the detection of sugar alcohols, but also for the determination of their turnover and elucidation of their biosynthesis. The solvent systems used by Calvin's group (Benson *et al.*, 1950) furnishes a complete separation of sorbitol from all the other ^{14}C-labelled water-soluble compounds. Further data for solvent systems and R_f values are listed by Hais and Macek (1963), Lewis and Smith (1967) and Churms (1982). Reagents for detection of sugar alcohols are described later.

(b) *Thin layer chromatography of sugar alcohols.* With respect to velocity, TLC is superior to PC and the quality of separation is similar in both techniques. TLC separation of sugar alcohols using different solvent systems and materials has been reviewed by Lewis and Smith (1967) and by Churms (1982). Distinction of sugar alcohols from similarly migrating monosaccharides can be achieved as described for PC. Kremer (1978) reported an improved method for thin layer chromatography of monosaccharides and sugar alcohols which allows a complete separation of the respective compounds even when using unfractioned plant extracts in ethanol (Table 7.8).

(c) *Paper electrophoresis of sugar alcohols.* Plant extracts containing sugar alcohols cannot be used for separation of the latter by PE since many compounds can drastically change migration of these substances in the electric field. Hence, prior to PE separation of the sugar alcohols into tetritols, pentitols etc. by PC or TLC is recommended. The compounds are eluted from the chromatograms with water and subjected to PE in a complexing buffer such as borate, germanate, molybdate, etc. (for details, see Weigel, 1963). Mobilities of sugar alcohols in borate buffer are given in Table 7.8. To distinguish between sugars and sugar alcohols molybdate buffer (Bourne *et al.*, 1961) is recommended where sugars show almost no mobility. Data for PE of sugar alcohols are listed by Weigel (1963), Lewis and Smith (1967) and Churms (1982).

(d) *Detection of sugar alcohols on paper chromatograms and pherograms.* A method has been developed by Beau and Porter (1959) for differentiating between sugar alcohols and monosaccharides on the same chromatogram. First, the monosaccharides are located by using benzidine; subsequently the chromatogram is treated with metaperiodate whereby the sugar alcohols appear as white to yellow spots on a blue background. A widely used and very sensitive detection reagent is 'alkaline silver nitrate' (Trevelyan et al., 1950) yielding brown to black spots with both monosaccharides and sugar alcohols on a white to yellow background. The detection is inhibited when borate was used in the solvent system. Frahn and Mills (1959) recommend the addition of 4.5% pentaerythritol to the sodium hydroxide reagent when using the silver nitrate technique. Special methods have been developed to detect sugar alcohols on paper pherograms in the presence of non-volatile buffer substances which seriously interfere with the detection methods described above (Lewis and Smith, 1967).

(e) *Gas–liquid chromatography of sugar alcohols.* The qualitative and quantitative analysis of sugar alcohols by GLC is an efficient and accurate method. The volatile derivatives of sugar alcohols commonly used are trimethylsilyl ethers and acetates which are separated on different column packings. Reviews of these techniques are given by Holligan (1971), Holligan and Drew (1971), Laine et al. (1972) and Churms (1982). As Sloneker (1972) pointed out, sugar alcohol acetates are better resolved than other derivatives. A recommendable method for acetylation of sugar alcohols was described by Blankeney et al. (1983); the separation of sugar alcohol acetates by GLC is shown in Table 7.9. Since derivatives of monosaccharides, oligosaccharides and other plant constituents may show similar retention times, a further characterisation by mass spectrometry (Lönngren and Svensson, 1974) is recommended.

TABLE 7.9. Separation of sugar alcohol peracetates by gas–liquid chromatography on a Silar 10C glass-capillary column. The temperature was kept at 190°C for 4 min and then increased to 230°C at a rate of $4°C min^{-1}$ (Blankeney et al., 1983).

Peracetylated sugar alcohol	Retention time (min)
Glycerol	0.91
Erythritol	4.66
Ribitol	9.12
D-Arabinitol	9.68
Xylitol	11.40
Allitol	13.00
D-Mannitol	14.20
Galactitol	15.10
D-Sorbitol	16.20
L-Iditol	18.60

(f) *Column chromatography of sugar alcohols.* Column chromatography using charcoal, Fuller's earth, celite, cellulose and other adsorbents have been employed to isolate sugar alcohols from plant extracts in larger quantities. Details of these techniques are reviewed by Binkley (1955). The separation of sugar and sugar alcohols on cation-exchange resins in the calcium or barium form was discovered by Felicetta et al. (1959). Jones and Wall (1960), using a column of Dowex 50W X-8 (200–400 mesh, Ba^{2+}), could separate mannitol, sorbitol and xylitol. Angyal et al. (1979) demonstrated that chromatography on a column of Dowex 450 W X-2 (200–400 mesh, Ca^{2+}) is a convenient method for the separation of sugar alcohols. The main advantages are as follows:

separation is rapid and clean, the recovery is nearly quantitative, the solvent is water only, the capacity of the column is large, the column is stable for a long time and need not be regenerated and the sequence of emergence of sugar alcohols from the column corresponds to their mobility in PE (Angyal *et al.*, 1974; Table 7.10). This method is employed in industrial preparation (US, British and German patents—for details, see Fong *et al.*, 1975; Angyal *et al.*, 1979). The separated sugar alcohols are detected by the same methods as are described for the HPLC technique. Barker *et al.* (1968) showed the separation of isomeric sugar alcohols (erythritol and threitol) and of sugars using an anion-exchange resin (AG 1-X8) in the molybdate form. The analysis of the separated compounds was performed automatically whereby sugars were determined by reaction with cystein–sulphuric acid reagent and sugar alcohols by acetyl acetone to determine the formaldehyde released on periodate oxidation. Further examples of automatic sugar alcohol analysis are given by Samuelson (1972) and Larsson and Samuelson (1976).

TABLE 7.10.   Relative retention times of sugar alcohols and monosaccharides using Aminex Q 15-S Ca^{2+} column (Conrad and Palmer, 1976). Elution with water at $1\ ml\ min^{-1}$ and $80°C$ from a $60 \times 0.7\ mm$ column. Glucose retention time was about 15 min. Detection by refractive index.

Compound	Relative retention time
Sucrose	0.84
Glucose	1.00
Galactose	1.10
Mannose	1.12
Fructose	1.18
Arabinose	1.19
Mannitol	1.35
Glycerol	1.48
Sorbitol	1.63

(g) *HPLC of sugar alcohols.* HPLC provides rapid separation of sugar alcohols for analytical purposes. This technique was introduced to carbohydrate analysis about 1975. Since then, an abundant literature has accumulated which has recently been reviewed by Honda (1984). Separations may be accomplished using adsorption chromatography, straight or reversed phase partition chromatography and ion-exchange (cation and anion) chromatography. Most useful for the separation of sugar alcohols are cation-exchange resins in calcium form such as Aminex Q 15-S (Bio-Rad), µSpherogel Carbohydrate (Beckman), Sugar-Pak I (Waters) and Unigel CHO (Kratos). As shown in Table 7.11, HPLC with Aminex Q 15-S Ca^{2+} column packing separates glycerol, sorbitol and mannitol from each other and from sucrose, glucose and fructose. Another example of a separation of sugar alcohols is given by Honda *et al.* (1983) with an anion-exchange resin and elution of the sugar alcohols with borate buffers (Table 7.8). Plant extracts normally cannot be used directly in HPLC as substances such as proteins, lipids and salts rapidly may inactivate the column packing; in any case the special instructions for purification of plant extracts prior to chromatography of the respective column producers must be followed. Sugar alcohols can be detected by refractive index change or by post-column labelling. Refractivity measurement is not specific for sugar alcohols (carbohydrates and other organic compounds interfere) and therefore should be used only in special cases, e.g. in the routine analysis of foods or fruit juices if the composition is known (Conrad and Palmer, 1976). A specific detection of sugar alcohols is achieved after cleavage by periodate. The resulting formaldehyde is reacted with acetylacetone to yield lutidine which in turn can be determined photometrically at 412 nm (Samuelson and Strömberg, 1966). Under controlled conditions sugar alcohols may be determined without significant interference by reducing carbohydrates (Honda et al., 1983). Dethy et al. (1984) reported a pre-column derivatisation of sorbitol and galactitol (reaction with phenylisocyanate to yield UV-absorbing derivatives absorbing UV-light at 240 nm) followed by a separation of the products on a Nucleosil $5 C_{18}MN$ column.

For routine analysis carbohydrate analysers are commercially available; their capabilities have been compared by Honda (1983).

TABLE 7.11. Separation of sugar alcohols by HPLC with Hitachi 2633 anion-exchange resin. Elution with 0.5 M borate buffer (pH 7.1), 0–35 min; 0.3 M borate buffer (pH 8.0), 35–65 min; 0.5 M borate buffer (pH 10.5), 65–90 min. Detection of sugar alcohols photometrically after post-column derivatisation (Honda et al., 1983).

Sugar alcohol	Approximate elution time (min)
Xylitol	35
Ribitol	37
Arabinitol	43
Sorbitol	58
Mannitol	78
Galactitol	85

6. *Chemical and physical characterisation of sugar alcohols*

Sugar alcohols isolated from plant material may be characterised by their derivatives as well as by PC and PE and comparison of the R_f values with authentic substances. The chemical derivatives commonly prepared include acetates, acyclic and cyclic acetals and ketals which are, in turn identified by their physical constants. Elemental analysis (C, H, O) need be performed only when a hitherto unknown sugar alcohol is described. Physical characterisation of crystalline sugar alcohols as well as of their derivatives comprise the determination of molecular weight, optical rotation (in aqueous solution as well as in borax or molybdate) and melting point. In addition, IR, NMR and MS techniques, if available, are useful. For details of methods see Volumes I–VIII of *Methods in Carbohydrate Chemistry* (Whistler and Wolfrom, 1962–1979).

7. *Preparation of sugar alcohols from monosaccharides*

Sugar alcohols can be prepared by reduction of aldoses or ketoses; while aldoses yield only one alditol, reduction of ketoses results in the formation of two sugar alcohols (e.g. fructose yields sorbitol and mannitol). The most convenient method of preparing sugar alcohols from monosaccharides is achieved by reduction with sodium borohydride (Wolfrom and Thompson, 1963). Commonly, borohydride is added to a solution of the respective monosaccharide in water which is then kept at room temperature (1–2 h) until a droplet from it no longer reduces Fehling's solution. The excess of borohydride is then destroyed with acetic acid. The sugar alcohol may be recovered by deionisation of the solution with cation/anion-exchange resin, or by repeated evaporation of the borate as methanol ester under reduced pressure. Reduction by other metal hydrides (e.g. lithium or aluminium hydride) as described by Lewis *et al.* (1963) is usually less satisfactory. Monosaccharides can also be reduced by catalytic hydrogenation with Raney nickel (Karabinos, 1963) or by electrolytic reduction. The latter two procedures are widely used in industrial production of sorbitol, mannitol and xylitol.

8. Preparation of radioactively labelled sugar alcohols

Uniformly ^{14}C-labelled sugar alcohols can be isolated by chromatography from leaves after photosynthesis in $^{14}CO_2$ (Benson *et al.*, 1950; Scherpenberg *et al.*, 1965). Position-labelled sugar alcohols can be prepared by biological or chemical reduction of the corresponding monosaccharides (for details, see Hough and Richardson, 1972). Monosaccharides can be reduced by borotritide to yield ^{3}H-labelled sugar alcohols (Natowicz and Baenzinger, 1980). Radioactively labelled sugar alcohols have been separated by HPLC and detected by scintillation counting (Barr and Nordin, 1980).

9. Quantitative determination of sugar alcohols

The quantitative determination of sugar alcohols has become less of a problem since GLC and HPLC became available. Among the many chemical procedures listed by Lewis and Smith (1967), the periodate oxidation of sugar alcohols followed by the determination of the liberated formaldehyde with chromotropic acid (Lambert and Neish, 1950; Speck, 1962) has been the most widely used procedure. The major difficulty in the application of this technique to plant extracts is the presence of interfering sugars; a variety of methods has been developed for overcoming this problem (see Lewis and Smith, 1967). Quantification of sorbitol, mannitol and xylitol, three sugar alcohols which are widely used as sweeteners in foods and beverages, is achieved by enzymatic methods (for details, see Bergmeyer, 1984). These methods are superior to GLC and HPLC as sugar alcohols can be determined directly in the aqueous extract. In general, a sugar alcohol is oxidised to the respective monosaccharide by a dehydrogenase and the amount of NAD(P)H formed is measured photometrically at 339, 334 or 366 nm. The amount of NAD(P)H is stoichiometric with the sugar alcohol. For the quantitative determination of D-mannitol, mannitol dehydrogenase is used (Horikoshi, 1984):

$$\text{D-Mannitol} + \text{NADP} \xrightarrow{\text{Mann-DH}} \text{fructose} + \text{NADPH}$$

The increase in NADPH is proportional to the amount of mannitol in the sample.

Quantitative determination of D-sorbitol in a mixture of sugar alcohols requires a composed assay because sorbitol dehydrogenase is not specific for sorbitol (Beutler, 1984a):

(a)   D-Sorbitol + NAD$^+$ $\xrightarrow{\text{SDH}}$ fructose + NADH

(b)   Fructose + ATP $\xrightarrow{\text{HK}}$ fructose 6-P + ADP

(c)   Fructose 6-P $\xrightarrow{\text{PGI}}$ glucose 6-P

(d)   Glucose 6-P + NADP$^+$ $\xrightarrow{\text{G6PDH}}$ gluconate 6-P + NADPH

(e)   NADH + pyruvate $\xrightarrow{\text{LDH}}$ NAD$^+$ + lactate

SDH = sorbitol dehydrogenase, HK = hexokinase, PGI = phosphoglucose isomerase, G6PDH = glucose 6-P-dehydrogenase, LDH = lactate dehydrogenase.
   For quantitation of xylitol an enzymic colour reaction is used (Beutler, 1984b):

$$\text{Xylitol} + \text{NAD}^+ \xrightarrow{\text{SDH}} \text{xylulose} + \text{NADH}$$

$$\text{NADH} + \text{INT} \xrightarrow{\text{D}} \text{NAD}^+ + \text{formazan}$$

SDH = sorbitol dehydrogenase, INT = iodonitrotetrazoliumchloride, D = diaphorase.

Xylitol can be determined only in samples which do not contain other sugar alcohols, as sorbitol dehydrogenase is not specific for xylitol. The second enzymatic reaction with diaphorase is necessary to yield quantitative reduction of xylitol.

10.   Characteristic data of the individual sugar alcohols

Data of chemical and physical properties are listed and information about productive sources for isolation, industrial production and utilisation is provided.

(a)   *Tetritols*

meso-Erythritol (Fig. 7.14, **1**)

(Syn. Erythrol, Erythrite)
$C_4H_{10}O_4$; mol. wt 122.12; tetragonal prisms; twice as sweet as sucrose; m.p. 121.5°C; very soluble in water; slightly soluble in ethanol; insoluble in ether; tetra-acetate m.p. 89–90°C; dibenzylidene acetal m.p. 202°C.
   Main sources. Fungi (Smith *et al.*, 1969), green algae (Smith, 1974), lichens (Smith *et al.*, 1969), grasses (Hofmann, 1874), roots of *Primula officinalis* (Regbie and Richtmyer, 1966), further ref. see, Karrer No. 144, Merck Index No. 3618.

Industrial production. Isolation from *Aspergillus niger* or *Penicillium herquei.*

Producers. BDH, Ega, Fluka, Koch-Light, Merck, Riedel, Senn, Serva.

Use. Coronary vasodilator, also erythritol tetranitrate. Erythritol anhydride is used to prevent microbial spoilage.

2-C-methyl-D-Erythritol (Fig. 7.14, **2**)

$C_5H_{12}O_4$; mol. wt 124.1; m.p. 82°C; $[\alpha]_D = + 23.7°C$ (H_2O).

Sources. Convolvulus glomeratus (Anthonsen *et al.*, 1976), *Liriodendron tulipifera* (Dittrich and Angyal, 1988).

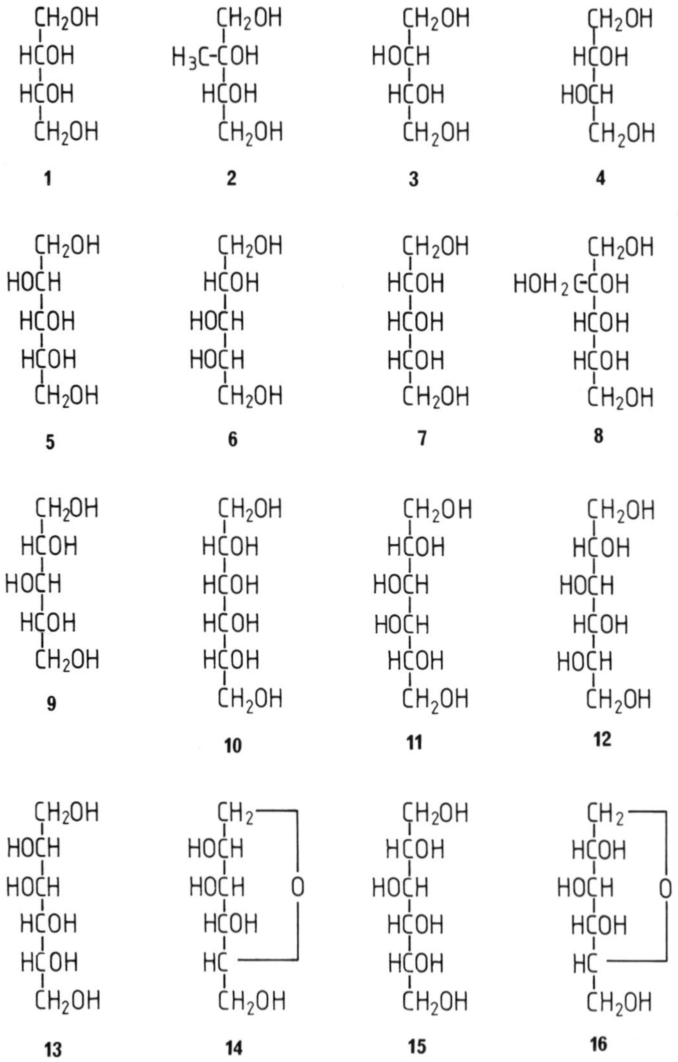

FIG. 7.14.   Structures of some sugar alcohols. **1**, *meso*-Erythritol; **2**, 2-C-Methyl-D-Erythritol; **3**, D-Threitol; **4**, L-Threitol, **5**, D-Arabinitol; **6**, L-Arabinitol; **7**, *meso*-Ribitol; **8**, D-Hamamelitol; **9**, *meso*-Xylitol; **10**, *meso*-Allitol; **11**, *meso*-Galactitol; **12**, L-Iditol; **13**, D-Mannitol; **14**, 1,5 anhydro-D-Mannitol; **15**, D-Glucitol; **16**, 1,5-anhydro-D-Glucitol.

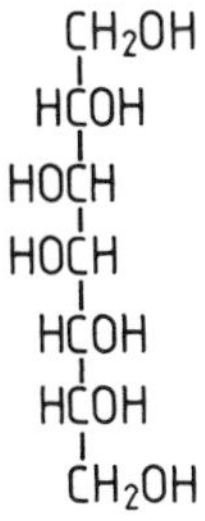

17

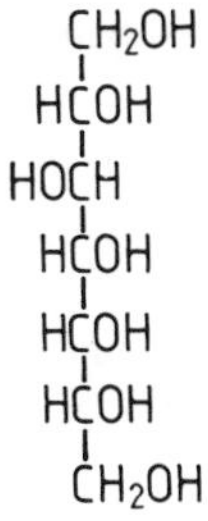

18

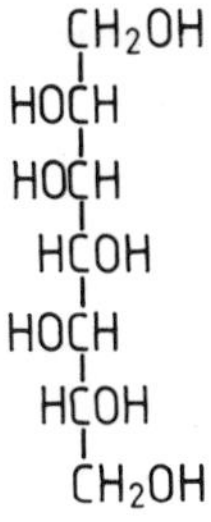

19

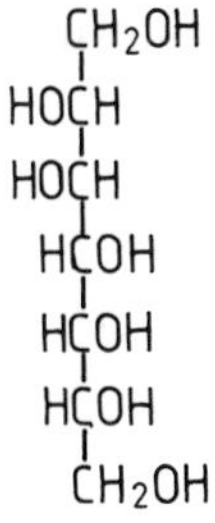

20

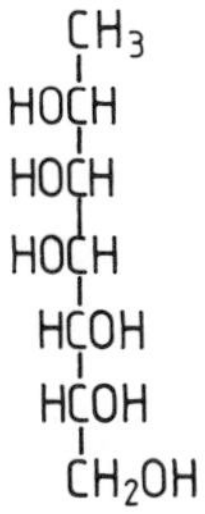

21

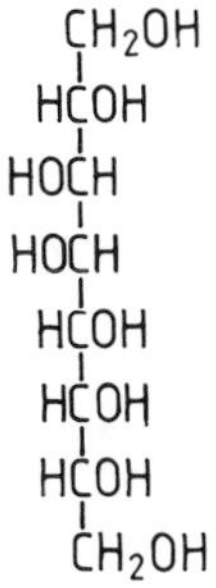

22

FIG. 7.14. (Continued). **17**, D-glycero-D-galacto-Heptitol; **18**, D-glycero-D-gluco-Heptitol; **19**, D-glycero-D-ido-Heptitol; **20**, D-glycero-D-manno-Heptitol; **21**, 7-deoxy-D-glycero-D-manno-Heptitol = 1-deoxy-D-glycero-D-talo-Heptitol; **22**, D-erythro-D-galacto-Octitol.

D-Threitol (Fig. 7.14, **3**)

$C_4H_{10}O_4$; mol. wt 122.12; m.p. 88°C; $[\alpha]_D = 4.3°$ (H_2O); dibenzylidene acetal m.p. 231°C; tetrabenzoate m.p. 97°C.

Sources. Armillaria mellea (Kratzl *et al.*, 1963), other fungi (Lewis and Smith, 1967); further ref. see Karrer No. 145.

L-Threitol (Fig. 7.14, **4**)

Has not been reported to occur naturally but has been isolated from plants that have been fed with L-sorbose (McComb and Rendig, 1963).

(b) *Pentitols*

D-Arabinitol (Fig. 7.14, **5**)

(Syn. D-Arabitol, D-Lyitol)
$C_5H_{12}O_5$; mol. wt 152.15; big prismatic crystals; sweet taste; m.p. 102°C; $[\alpha]_D = 7.8°$ (borax); penta-acetate m.p. 86°C.

Main sources. Fungi (Frèrejacque, 1939), lichens (Lindberg *et al.*, 1953), avocado seeds (Richtmyer, 1970a), *Fabriana imbricata* (Richtmyer, 1970b); further ref. see Karrer No. 147, Merck Index No. 784.

Industrial production. Fermentation from molasses using *Saccharomyces rouxii* and *Saccharomyces mellis.*

Producers. Baker, BDH, Fluka, Koch-Light, Roth, Senn.

L-Arabinitol (Fig. 7.14, **6**)

$C_5H_{12}O_5$; mol. wt 152.15; wart-like crystals; sweet taste; freely soluble in water; m.p. 102°C; $[\alpha]_D = -7.2°$ (Borax); penta-acetate m.p. 72–73°C.

Source. Supposed to be a minor component in fungi and angiosperms (see Bieleski, 1982); further ref. see Karrer No. 5402, Merck Index No. 784.

meso-Ribitol (Fig. 7.14, **7**)

(Syn. Adonitol, Adonite)
$C_5H_{12}O_5$; mol. wt 152.15; large crystals; freely soluble in water; m.p. 102°C; dibenzylidene acetal m.p. 164–165°C.

Main sources. Ribitol is the major pentitol in plants as it is a constituent of riboflavin; it is also a constituent in the cell walls of some bacteria (ribitol teichoic acids). Algae (Maruo *et al.*, 1965), lichens (Smith, 1974), *Adonis vernalis* (Merck, 1893), *Bupleurum falcatum* (Wessely and Wang, 1938); further ref. see Karrer No. 146, Merck Index No. 156.

Producers. Baker, BDH, Ega, Fluka, Koch-Light, Merck, Riedel, Roth, Senn, Serva.

2-C-hydroxymethyl-D-Ribitol (Fig. 7.14, **8**)

(Syn. D-Hamamelitol)
$C_6H_{14}O_6$; mol. wt 182.11; syrup; tribenzylidene acetal m.p. 213°C; hexabenzoate m.p. 136°C.

Source. *Primula clusiana* (Sellmair *et al.*, 1968). Constituent of clusianose (1-*O*-α-D-galactopyranosyl-hamamelitol: Beck, 1969).

meso-Xylitol (Fig. 7.14, **9**)

$C_5H_{12}O_5$; mol. wt 152.15; prisms from ethanol; stable form m.p. 93–94.5°C; soluble in water; relative sweetness 1.2–1.3 (as compared with sucrose); penta-acetate m.p. 62–63°C.

Main sources. Fungi (Barnett, 1976), occurrence in higher plants may be artifacts (see Bieleski, 1982); further ref. see Karrer No. 5401, Merck Index No. 8994.

Industrial production. Produced by hydrogenation of xylose obtained from xylan-containing plants by acid hydrolysis.

Producers. Barker, BDH, Fluka, Koch-Light, Roth, Senn.

Uses. As oral or intravenous nutrient, as sugar substitute in diabetic food, used in the prevention of dental caries (as sweetening agent in chewing gum), for conservation of moisture content in foods. In mammalian organisms xylitol is oxidised by sorbitol dehydrogenase to xylulose which is metabolised in the pentose phosphate cycle.

(c) *Hexitols*

meso-Allitol (Fig. 7.14, **10**)

(Syn. Allodulcitol)
$C_6H_{14}O_6$; mol. wt 182.17; sweet-tasting crystals from ethanol; very soluble in water; m.p. 150°C; hexa-acetate m.p. 61°C.

Source. *Itea* spp. (Plouvier, 1959); further ref. see Karrer No. 3202.

meso-Galactitol (Fig. 7.14, **11**)

(Syn. Dulcitol, Dulcite, Dulcose, Melampyrite, Melampyrin)
$C_6H_{14}O_6$; mol. wt 182.17; prisms; slightly sweet-tasting; slightly soluble in cold water; soluble in hot water; m.p. 188.5–189°C; hexa-acetate m.p. 168–169°C.

Main sources. Fungi (Barnett, 1976), algae (Kremer, 1976), *Melampyrum nemorosum*, Madagascar manna (Hünefeld, 1857), *Evonymus atropurpureus* (Plouvier, 1949); further ref. see Karrer No. 156, Merck Index No. 4203.

Industrial production. Catalytic isomerisation of D-sorbitol.

Producers. Baker, BDH, Chenopol, Eastman, Ega, Fluka, Koch-Light, Riedel, Roth, Senn, Serva.

Uses. In bacteriology; galactitol hexanitrate (nitrodulcitol) as explosive.

L-Iditol (Fig. 7.14, **12**)

(Syn. Sorbierite)
$C_6H_{14}O_6$; mol. wt 182.17; hygroscopic prisms from ethanol; m.p. 75.7–76.7°C. $[\alpha]_D = -3.5°$ (H_2O); hexa-acetate m.p. 121.5°C.

Source. *Sorbus aucuparia* (Vincent and Meunier, 1898); further ref. see Karrer No. 155.

D-Mannitol (Fig. 7.14, **13**)

(Syn. Mannite, Manna sugar)
Trade names. Manicol, Mannidex, Diosmol, Osmitrol, Osmosal.
$C_6H_{14}O_6$; mol. wt 182.17; orthorhombic needles from ethanol; relative sweetness 0.7 (as compared with sucrose); soluble in water; m.p. 166–168°C; hexa-acetate m.p. 126°C.

Source. Most abundant sugar alcohol in nature (for details see Bieleski, 1982); further ref. see Karrer No. 148, Merck Index No. 5569.

Industrial production. Epimerisation of glucose under alkaline conditions followed by electrolytic or catalytic reduction.

Producers. Baker, BDH, Carrous, Eastman, Ega, Fluka, Hefti, Helm, Klöckner-Chemie, Koch-Light, Merck, Riedel, Roth, Senn, Serva, Ventron.

Uses. As sweetening agent, in foods as anti-caking and free-flow agent, used in the manufacture of electrolytic condensers, in making artificial resins and plasticisers. Used as osmoticum for isolation of plant protoplasts (Ruesink, 1971) and of plant mitochondria (Laties, 1974). Mannitol hexanitrate used as explosive and vasodilator.

1,5 anhydro-D-Mannitol (Fig. 7.14, **14**)

(Syn. Styracitol)
$C_6H_{12}O_5$; mol. wt 164.16; prisms from ethanol; sweet-tasting with bitter aftertaste; soluble in water; m.p. 155°C. $[\alpha]_D = -44.9°$ (H_2O); tetrabenzoate m.p. 142°C.

Source. *Styrax obasia* (Asahina, 1907); further ref. see Karrer No. 152.

D-Sorbitol (Fig. 7.14, **15**)

(Syn. Sorbite, D-Glucitol, L-Gulitol)
Trade names. Sorbol, Sorbicolan, Sorbo, Sorbostyl, Nivitin, Cholaxine, Karion, Sionit, Sionon, Sorbilande, Diakarmon.
$C_6H_{14}O_6$; mol. wt 182.17; needles from ethanol; relative sweetness 0.5 (as compared

with sucrose); soluble in water; m.p. 97°C (stable form); $[\alpha]_D = -1.8°$ (H_2O); hexa-acetate m.p. 99°C.

Main sources. *Sorbus aucuparia* (Boussingault, 1872), in fruits of *Rosaceae* (Wallaart, 1980), in algae, fungi, lichens (see Bieleski, 1982); further ref. see Karrer No. 153, Merck Index 8569.

Industrial production. High pressure hydrogenation, electrolytic or catalytic reduction of glucose.

Producers. Baker, BDH, Carroux, Finn sugar, Hefti, Helm, Koch-Light, Merck, Riedel, Roth, Ventron.

Uses. Sweetening agent, substitute for diabetics (sorbitol is converted to fructose by sorbitol dehydrogenase and further metabolised to glycogen). Prevents caries. Used for preparation of sorbose and ascorbic acid, as moisture conditioner in foods and industrial products (textiles, leather, printing rolls). Used as osmoticum for isolation of plant protoplasts (Ruesink, 1971), of plant cells (Servaites and Ogren, 1980), of chloroplasts (Schwartzbach *et. al.*, 1979) and of mitochondria (Nedergaard and Cannon, 1979). Also used in zonal centrifugation (Cline and Ryel, 1971) and in isoelectric focusing of proteins (Vesterberg, 1971).

1,5 anhydro-D-sorbitol (Fig. 7.14, **16**)

(Syn. Polygalitol, Aceritol)
$C_6H_{12}O_5$; mol. wt 164.16; prisms or needles from ethanol; soluble in water; m.p. 142.7°C; $[\alpha]_D = +39°$ (H_2O); tetra-acetate m.p. 74°C.

Sources. *Polygala amara* (Chodat, 1888), *Acer ginnala* (Perkin and Uyeda, 1922); further ref. see Karrer No. 154.

Industrial Production. Heating of sorbitol in conc. sulphuric acid at 140°C.

Producers. Atlas, Baker, Fluka, Riedel, Roth, Serva.

Uses. Sorbitans, Mono-, di- and triesters of sorbitol anhydrides, used as emulsifiers and dispersing agents (e.g. Tween 80, Span), see Merck No. 7455.

(d)   *Heptitols*

D-glycero-D-galacto-Heptitol (Fig. 7.14, **17**)

(Syn. Perseitol, α-Mannoheptitol)
$C_7H_{16}O_7$; mol. wt 212.2; needles; m.p. 188°C; $[\alpha]_D = -1.04°$ (H_2O); hepta-acetate m.p. 119°C.

Sources. *Persea gratissima* (Richtmyer, 1963), *Fabiana imbricata* (Richtmyer, 1970b); further ref. see Karrer No. 157.

D-glycero-D-gluco-Heptitol (Fig. 7.14, **18**)

(Syn. β-Sedoheptitol)
$C_7H_{16}O_7$; mol. wt 212.2; m.p. 128–129°C; $[\alpha]_D = -0.75°$ (H_2O).
Source. *Sedum* spp. (Charlson and Richtmyer, 1960); further ref. see Karrer No. 3203.

D-glycero-D-ido-Heptitol (Fig. 7.14, **19**)

$C_7H_{16}O_7$; mol. wt 212.2; small plates; m.p. 128–129°C; $[\alpha]_D = +1.20°$ (H_2O); hepta-benzoate m.p. 181–182°C.

Sources. *Pichia miso* (Onishi and Perry, 1965); also production of *meso*-glycero-ido-heptitol; further ref. see Karrer Nos. 5403, 5404.

D-glycero-D-manno-Heptitol (Fig. 7.14, **20**)

(Syn. Volemitol, α-Sedoheptitol, β-Mannoheptitol)
$C_7H_{16}O_7$; mol. wt 212.2; needles; sweet-tasting; m.p. 153°C; $[α]_D = + 2.15°$ (H_2O); hepta-acetate m.p. 63–64°C.

Main sources. Algae (Kremer, 1976), fungi (Onishi and Perry, 1972), lichens (Lindberg *et al.*, 1953), liverworts (Lewis, 1971), *Primula* spp. (Regbie and Richtmyer, 1966; Kremer, 1978a); further ref. see Karrer No. 158.

7-deoxy-D-glycero-D-manno-Heptitol (Fig. 7.14, **21**)

(Syn. Siphulitol, 7-deoxy Volemitol)
$C_7H_{16}O_6$; mol. wt 196.14.
Source. Siphula spp. (Lindberg and Meier, 1962).

(e) *Octitol*

D-erythro-D-galacto-Octitol (Fig. 7.14, **22**)

$C_8H_{18}O_8$; mol. wt 242.14; m.p. 153°C; $[α]_D = +2.4°$ (H_2O); octa-acetate m.p. 88–89°C.
Source. Persea gratissima (Charlson and Richtmyer, 1960).

REFERENCES

Anderson, J. B., Andrews, P. and Hough, L. (1961). *Biochem. J.* **81**, 149–154.
Angyal, S. J., Greeves, D. and Mills, J. A. (1974). *Aust. J. Chem.* **27**, 1447–1456.
Angyal, S. J., Bethell, G. S. and Beveridge, R. J. (1979). *Carbohydr. Res.* **73**, 9–18.
Anthonsen, T., Hagen, S., Kasi, M. A., Shah, S. W. and Tagar, S. (1976). *Acta. Chem. Scand.* **30**, 91.
Araki, Y., Nagasawa, J. and Ishido, Y. (1981). *Soc. Perkin I*, 12–23.
Archibald, A. R. and Baddiley, J. (1966). *Adv. Carbohydr. Chem.* **21**, 354.
Asahina, Y. (1907). *Arch. Pharm.* **245**, 325.
Bacon, J. S. D. and Cheshire, M. V. (1971). *Biochem. J.* **124**, 555–562.
Ball, D. H., Carey, F. A., Klundt, I. L. and Long, L. Jr. (1969). *Carbohydr. Res.* **10**, 121–8.
Ball, D. H., Bissett, F. H., Klundt, I. L., Long, L. Jr. (1971). *Carbohydr. Res.* **17**, 165–174.
Barker, S. A. (1955). In "Modern Methods of Plant Analysis" (K. Paech and M. V. Tracey, eds), Vol. II, pp. 55–63. Springer, Berlin, Göttingen and Heidelberg.
Barker, S. A., How, M. J., Peplow, P. J. and Somers, P. J. (1968). *Anal. Biochem.* **26**, 219–230.
Barnett, J. A. (1976). *Adv. Carbohydr. Chem. Biochem.* **32**, 125–234.
Baron, D., Streitberger, U. and Grisebach, H. (1973). *Biochim. Biophys. Acta* **293**, 526–533.
Barr, J. and Nordin, P. (1980). *Anal. Biochem.* **108**, 313–319.
Beau, R. C. and Porter, G. G. (1959). *Anal. Chem.* **31**, 1929.
Beck, E. (1967). *Z. Pflanzenphysiol.* **57**, 444–461.
Beck, E. (1969). *Z. Pflanzenphysiol.* **61**, 360–366.
Beck, E. (1982). "Encyclopedia of Plant Physiology, New Series" (F. A. Loewus and W. Tanner, eds), Vol. 13A, pp. 124–157. Springer, Berlin, Heidelberg and New York.
Beck, E. and Kandler, O. (1965). *Z. Naturforsch.* **20b**, 62–67.
Beck, E. and Kandler, O. (1966). *Z. Pflanzenphysiol.* **55**, 71–84.
Beck, E. and Knaupp, I. (1974). *Z. Pflanzenphysiol.* **72**, 141–147.
Beck, E., Sellmair, J. and Kandler, O. (1968). *Z. Pflanzenphysiol.* **58**, 434–451.
Beck, E., Stransky, H. and Fürbringer, M. (1971). *FEBS Lett.* **13**, 229–234.
Beck, E., Wieczorek, J. and Reinecke, W. (1980). *Eur. J. Biochem.* **107**, 485–489.
Bell, D. J. (1955). In "Modern Methods of Plant Analysis" (K. Paech and M. V. Tracey, eds), Vol. II, pp. 1–55. Springer, Berlin, Göttingen and Heidelberg.

Bell, D. J. (1962). *In* "Methods in Carbohydrate Chemistry" (R. L. Whistler, M. L. Wolfrom, J. N. BeMiller and J. Shafizadeh, eds), Vol. I, pp. 260–263. Academic Press, New York and London.

Bell, D. J., Isherwood, F. A. and Hardwick, N. E. (1954). *J. Chem. Soc.*, 3702–3706.

Benson, A. A., Bassham, J. A., Calvin, M., Goodale, T. C., Haas, V. A. and Stepka, W. (1950). *J. Am. Chem. Soc.* **72**, 1710–1718.

Bergmeyer, H. U. (ed.) (1984). "Methods of Enzymatic Analysis", Vol. VI. Verlag Chemie, Weinheim, Deerfield Beech, FL and Basel.

Berry, J. A., Lorimer, G. H., Pierce, J., Meek, J. and Freas, S. (1987). *Proc. Natl. Acad. Sci. USA* **84**, 734–738.

Bertrand, G. (1891). *Bull. Soc. Chim.* **5**, 554, 740.

Beusekom, C. F. Van (1967). *Phytochemistry* **6**, 573–576.

Beutler, H. O. (1984a). *In* "Methods of Enzymatic Anlaysis" (H. U. Bergmeyer, ed.), Vol. VI, pp. 356–362, Verlag Chemie, Weinheim, Deerfield Beech, FL and Basel.

Beutler, H. O. (1984b). *In* "Methods of Enzymatic Analysis" (H. U. Bergmeyer, ed.), Vol. VI, pp. 484–490. Verlag Chemie, Weinheim, Deerfield Beech, FL and Basel.

Bieleski, R. L. (1982). *In* "Encyclopedia of Plant Physiology, New Series" (F. A. Loewus and W. Tanner, eds), Vol. 13A, pp. 158–193. Springer, Berlin, Heidelberg and New York.

Binkley, W. (1955). *Adv. Carbohydr. Chem.* **10**, 55.

Blankeney, A. B., Harris, P. J., Henry, R. J. and Stone, B. A. (1983). *Carbohydr. Res.* **113**, 291–299.

Bourne, E. J. (1958). *In* "Encyclopedia of Plant Physiology" (W. Ruhland, ed.), Vol. VI, pp. 345–362. Springer, Berlin, Göttingen and Heidelberg.

Bourne, E. J., Hutson, D. H. and Weigel, H. (1961). *J. Chem. Soc.*, 35.

Bourquelot, E. (1895). *J. Pharm. Chim.* **2**, (6), 285.

Boussingault, J. (1872). *Ber. Deutsch. Chem. Gesellsch.* **5**, 325.

Brimacombe, J. S. and Webber, J. M. (1972). *In* "The Carbohydrates, Chemistry and Biochemistry" (W. Pigman and D. Horton, eds), Vol. IA, pp. 479–518. Academic Press, London and New York.

Burton, J. S., Overend, W. G. and Williams, N. R. (1965). *J. Chem. Soc.*, 3433–3445.

Chakraborti, S. K. (cit.) (1959). *Chem. Abstr.* 22602a.

Charlson, A. J. and Richtmyer, N. K. (1960). *J. Am. Chem. Soc.* **82**, 3428.

Chodat, M. (1888). *Arch. Sci. Phys. Nat.* **20**, 593.

Chrastil, J. (1956). *Chem. Listy* **50**, 163–164. (cit.) (1956) *Chem. Abstr.* **50**, 4456.

Churms, S. C. (1982). "CRC Handbook of Chromatography Carbohydrates" (G. Zweig and J. Sherma, eds). CRC Press, Boca Raton, FL.

Cline, G. B. and Ryel, R. B. (1971). *In* "Methods in Enzymology" (S. P. Colowick and N. O. Kaplan, eds), Vol. 22, pp. 168–204. Academic Press, New York and San Francisco.

Colowick, S. P. and Kaplan, N. O. (eds) (1955–1988). "Methods in Enzymology", Vols 1–158. Academic Press, New York and San Francisco.

Conrad, E. C. and Palmer, J. K. (1976). *Food Technol.* **26**, 84–92.

Darvill, A. G., McNeill, M. and Albersheim, P. (1978). *Plant Physiol.* **62**, 418–422.

Depezay, J.-C. and Duréault, A. (1978). *Tetrahedron Lett.* **32**, 2869–2872.

Dethy, J. M., Callaert-Deveen, B., Janssens, M. and Lenaers, A. (1984). *Anal. Biochem.* **143**, 119–124.

Dittrich, P. and Angyal, S. (1988). *Phytochemistry* **27**, 935.

Duff, R. B. (1965). *Biochem. J.* **94**, 768–772.

Duff, R. B. and Knight, A. H. (1963). *Biochem. J.* **88**, 33–34.

Egger, K. (1967). *In* "Dünnschichtchromatographie" (E. Stahl, ed.), pp. 655–673. Springer, Berlin, Heidelberg and New York.

Ezekiel, A. D., Overend, W. G. and Williams, N. R. (1969). *Carbohydr. Res.* **11**, 233–239.

Farooq, M. D., Gupta, S. R., Kiamuddin, M., Rahman, W. and Seshadri, T. R. (1953). *J. Sci. Ind. Res.* **12B**, 400–407.

Farooq, M. D., Varshney, J. P. and Rahman, W. (1957). *Naturwissenschaften* **44**, 444.

Farooq, M. D., Varshney and J. P. Rahman, W. (1958). *Naturwissenschaften* **45**, 265.

Fasman, G. D. (1975). "Handbook of Biochemistry and Molecular Biology", 3rd edn, pp. 182–225. CRC Press, Cleveland, Ohio.
Felicetta, V. F., Lung, M. and McCarthy, J. L. (1959). *Tappi* **42**, 496–502.
Fischer, E. (1895). *Ber. Deutsch. Chem. Gesellsch.* **28**, 1973.
Fischer, E. and Fay, J. W. (1895). *Ber. Deutsch. Chem. Gesellsch.* **28**, 1975.
Folin, O. (1929). *J. Biol. Chem.* **82**, 83–93.
Fong, D. H. T., Bodkin, C. L., Lang, M. A. and Garnett, J. L. (1975). *Aust. J. Chem.* **28**, 1981–1991.
Frahn, J. L. and Mills, J. H. (1959). *Aust. J. Chem.* **12**, 65.
Frèrejacque, M. (1939). *Compt. Rend.* **208**, 1123.
Freudenberg, K. and Blümmel, F. (1924). *XVII. Hamameli-tannin III. Ann. Chem.* **440**, 45–59.
Gaudy, D., Lassignardie, F., Jacob, M. and Puech, A. (1988). *Pharma* **4**, 31–36.
Gebb, C., Baron, D. and Grisebach, H. (1975). *Eur. J. Biochem.* **54**, 493–498.
Gilck, H. and Beck, E. (1974). *Z. Pflanzenphysiol.* **72**, 395–409.
Gilck, H., Thanbichler, A., Sellmair, J. and Beck, E. (1975). *Carbohydr. Res.* **39**, 160–161.
Götz, W., Sachs, A., Wimmer, H. (1978). "Dünnschichtchromatographie", pp. 82–84. Gustav Fischer, Stuttgart.
Gorin, P. A. J. and Perlin, A. S. (1958). *Can. J. Chem.* **36**, 480–485.
Grant, C. R. and apRees, T. (1981). *Phytochemistry* **20**, 1505–1511.
Grisebach, H. (1978). *Adv. Carbohydr. Chem.* **35**, 81–126.
Grisebach, H. (1980). In "The Biochemistry of Plants" (J. Preiss, ed.), Vol. III, pp. 171–197. Academic Press, New York.
Grisebach, H. and Bilhuber, W. (1967). *Z. Naturforsch.* **22b**, 746–751.
Grisebach, H. and Döbereiner, U. (1966). *Z. Naturforsch.* **21b**, 429–435.
Griesbach, H. and Sandermann, H. Jr. (1966). *Biochem. Z.* **346**, 322–327.
Gross, G.-A., Sticher, O. and Anklin, C. (1987). *Helv. Chim. Acta* **70**, 91–101.
Gustine, D. L., Yuan, D. H.-F. and Kindel, P. K. (1975). *Arch. Biochem. Biophys.* **170**, 82–91.
Gutteridge, S., Parry, M. A. J., Burton, S., Keys, A. J., Mudd, A., Feeney, J., Servaites, J. C. and Pierce, J. (1986). *Nature* **324**, 274–276.
Hais, I. M. and Macek, K. (1963). "Handbuch der Papierchromatographie", Vols I–III. Fischer, Jena.
Hamburger, M., Gupta, M. and Hostettmann, K. (1985). *Phytochemistry* **24**, 2689–2692.
Hanna, R., Picken, M. and Mendicino, J. (1973). *Biochim. Biophys. Acta* **315**, 259–271.
Hansson, B., Johansson, I. and Lindberg, B. (1966). *Acta. Chem. Scand.* **20**, 2358–2362.
Hart, D. A. and Kindel, P. K. (1970a). *Biochemistry* **9**, 2190–2196.
Hart, D. A. and Kindel, P. K. (1970b). *Biochem. J.* **116**, 569–579.
Hattori, S. and Imaseki, H. (1959). *J. Am. Chem. Soc.* **81**, 4424–4427.
Hegnauer, R. (1962–1986). "Chemotaxonomie der Pflanzen", Vols I–VII. Birkhäuser, Basel and Stuttgart.
Ho, P.-T. (1979). *Can. J. Chem.* **57**, 381–383.
Hösel, W. (1976). *Hoppe-Seyler's Z. Physiol. Chem.* **357**, 1673–1681.
Hösel, W. and Barz, W. (1970). *Phytochemistry* **9**, 2053–2056.
Hofmann, A. W. (1874). *Ber.* **7**, 508.
Holligan, P. M. (1971). *New Phytol.* **70**, 239–269.
Holligan, P. M. and Drew, E. A. (1971). *New Phytol.* **70**, 271–297.
Honda, S. (1983). *Farumashia* **19**, 1259–1266.
Honda, S. (1984). *Anal. Biochem.* **140**, 1–47.
Honda, S., Takahashi, M., Shimada, S., Kakehi, K. and Ganno, S. (1983). *Anal. Biochem.* **128**, 429–437.
Horikoshi, K. (1984). In "Methods of Enzymatic Analysis" (H. U. Bergmeyer, ed.), Vol. VI, pp. 271–276. Verlag Chemie, Weinheim, Deerfield Beech, FL and Basel.
Horowitz, M. I. (1980). In "The Carbohydrates, Chemistry and Biochemistry" (W. Pigman and D. Horton, eds), Vol. IB, pp. 1445–1470. Academic Press, London and New York.
Hough, L. and Richardson, A. C. (1972). In "The Carbohydrates, Chemistry and Biochemistry" (W. Pigman and D. Horton, eds), Vol. IA, pp. 154–158. Academic Press, London and New York.

Hulyalkar, R. K., Jones, J. K. N., Perry, M. B. (1965). *Can. J. Chem.* **43**, 2085–2091.

Hünefeld, A. (1857). *Just. Lieb. Ann. Chem.* **24**, 241.

Ikawa, T., Watcombe, T. and Nisizawa, K. (1972). *Plant Cell Physiol.* **13**, 1017–1029.

Imaseki, H. and Yamamoto, T. (1961). *Arch. Biochem. Biophys.* **92**, 467–474.

Jones, J. K. N. and Wall, R. A. (1960). *Can. J. Chem.* **38**, 2290–2294.

Kandler, O. (1958). *Arch. Biochem. Biophys.* **73**, 38–42.

Kandler, O. and Gibbs, M. (1959). *Z. Naturforsch.* **14b**, 8.

Kandler, O. and Hopf, H. (1982). *In* "Encyclopedia of Plant Physiology, New Series" (F. A. Loewus and W. Tanner, eds), Vol. 13A, pp. 348–383. Springer, Berlin, Heidelberg and New York.

Karabinos, J. V. (1963). *In* "Methods in Carbohydrate Chemistry" (R. L. Whistler and R. L. Wolfrom, eds), Vol. II, pp. 77–79. Academic Press, New York and London.

Karrer, W. (1958–1985). "Konstitution und Vorkommen der organischen Pflanzenstoffe" (4 vols). Birkhäuser, Basel and Stuttgart.

Kelleher, W. J. and Grisebach, H. (1971). *Eur. J. Biochem.* **23**, 136–142.

Khalique, A. (1962). *J. Chem. Soc.*, 2515–2516.

Kiessling, H., Lindberg, B. and McKay, J. (1962). *Acta Chem. Scand.* **16**, 1858.

Kiliani, B. (1887). *Ber. Deutsch. Chem. Gesellsch.* **20**, 1233.

Kindel, P. K. and Watson, R. R. (1973). *Biochem. J.* **133**, 227–241.

Kitsuta, K., Gupta, S. R. and Seshadri, R. T. (1952). *Proc. Indian Acad. Sci.* **35A**, 242–248.

Koos, M. and Mosher, H. S. (1986). *Carbohydr. Res.* **146**, 335–341.

Kratzl, K., Silbernagl, H. and Bässler, K. H. (1963). *Naturwissenschaften* **50**, 154.

Kremer, B. P. (1976). *Planta* **129**, 63–67.

Kremer, B. P. (1978a). *J. Chromatogr.* **166**, 355–357.

Kremer, B. P. (1978b). *Z. Pflanzenphysiol.* **86**, 453–461.

Kreuzaler, F. and Hahlbrock, K. (1973). *Phytochemistry* **12**, 1149–1152.

Laine, R. A., Esselman, W. J. and Sweely, C. C. (1972). *In* "Methods in Enzymology" (S. P. Colowick and N. O. Kaplan, eds), Vol. 28, pp. 159–167. Academic Press, New York and San Francisco.

Lambert, M. and Neish, A. C. (1950). *Can. J. Res.* **28B**, 83.

Larsson, K. and Samuelson, O. (1976). *Carbohydr. Res.* **50**, 1–8.

Laties, G. G. (1974). *In* "Methods in Enzymology" (S. P. Colowick and N. O. Kaplan, eds), Vol. 31, pp. 589–600. Academic Press, New York and San Francisco.

Lespieau, R. and Wiemann, J. (1932). *Compt. Rend.* **195**, 886.

Lewis, B. A., Smith, F. and Stephen, A. M. (1963). *In* "Methods in Carbohydrate Chemistry" (R. L. Whistler and M. L. Wolfrom, eds), Vol. II, pp. 68–75. Academic Press, New York and London.

Lewis, D. H. (1971). *Trans. Br. Bryol. Soc.* **6**, 108–113.

Lewis, D. H. (1984). "Storage Carbohydrates in Vascular Plants". Cambridge University Press.

Lewis, D. H. and Smith, D. C. (1967). *New Phytol.* **66**, 143–204.

Lindberg, B. and Meier, H. (1962). *Acta Chem. Scand.* **16**, 543.

Lindberg, B., Missiony, A. and Wachtmeister, C. A. (1953). *Acta Chem. Scand.* **7**, 591–595.

Linskens, H. F. (1955). "Papierchromatographie in der Botanik", 69 pp. Springer, Berlin, Göttingen and Heidelberg.

Lohmar, R. L. (1957). *In* "The Carbohydrates" (W. Pigman, ed), pp. 241–298. Academic Press, New York.

Lönngren, J. and Svensson, S. (1974). *Adv. Carbohydr. Chem. Biochem.* **29**, 41–106.

Malhotra, A., Murti, V. V. S. and Seshadri, T. R. (1965). *Tetrahedron Lett.* **36**, 3191–3196.

Maruo, B., Hattori, T. and Takahashi, H. (1965). *Agr. Biol. Chem.* **12**, 1084.

Mascaro, L. J. Jr, and Kindel, P. K. (1977). *Arch. Biochem. Biophys.* **183**, 139–148.

Matern, U. and Grisebach, H. (1977). *Eur. J. Biochem.* **74**, 303–312.

Mayer, W. and Kunz, W. (1959). *Naturwissenschaften* **46**, 206–207.

Mayer, W., Kunz, W. and Loebich, F. (1965). *Ann. Chem.* **688**, 232–238.

McComb, E. A. and Rendig, V. V. (1963). *Arch. Biochem. Biophys.* **103**, 84.

McNeil, M., Darvill, A. G., Fry, S. C. and Albersheim, P. (1984). *Ann. Rev. Biochem.* **53**, 625–663.

Mendicino, J. and Picken, J. M. (1965). *J. Biol. Chem.* **240**, 2797–2805.
Mendicino, J. and Abou-Issa, H. (1974). *Biochem. Biophys. Acta* **364**, 159–172.
Mendicino, J. and Hanna, P. (1970). *J. Biol. Chem.* **245**, 6113–6124.
Merck, E. (1893). *Arch. Pharm.* **231**, 129.
Merck Index (1983). "An Encyclopedia of Chemicals, Drugs and Biologicals" (M. Windholz, ed.). Merck + Co. Inc. Rahway, NY.
Merill, A. T., Haskins, W. T., Hann, R. M. and Hudson, C. S. (1947). *J. Am. Chem. Soc.* **69**, 70.
Meunier, J. (1888). *Compt. Rend.* **106**, 1425, 1732.
Muntz, A. and Marcano, V. (1884). *Compt. Rend.* **99**, 38.
Nachman, R. J., Hönel, M., Williams, T. M., Halaska, R. Ch. and Mosher, H. S. (1986). *J. Org. Chem.* **51**, 4802–4806.
Nakaoki, T., Morita, N., Motosune, H., Hiraki, A. and Takeuchi, T. (1955). *Pharm. Soc. Jpn* **75**, 172–176.
Natowicz, M. and Baenzinger, J. U. (1980). *Anal. Biochem.* **105**, 159–164.
Neal, D. L. and Kindel, P. K. (1970). *J. Bacteriol.* **101**, 910–915.
Needergaard, J. and Cannon, B. (1979). *In* "Methods in Enzymology" (S. P. Colowick and N. O. Kaplan, eds), Vol. 55, pp. 3–28. Academic Press, New York and San Francisco.
Negm, F. B. and Loescher, W. H. (1979). *Plant Physiol.* **64**, 69–73.
Negm, F. B. and Loescher, W. H. (1981). *Plant Physiol.* **67**, 139–142.
Nolan, T. J. and Keane, J. (1933). *Nature* **132**, 281.
Nordström, C. G., Swain, R. and Hamblin, A. J. (1953). *Chem. Ind.*, 85.
Ohta, N. and Tagishita, K. (1970). *Agr. Biol. Chem.* **34**, 900–907.
Oku, M. (1986–87). (cit.) *Chem. Abstr.* (1988); 108, 351.
Onishi, H. and Perry, M. B. (1965). *Can. J. Microbiol.* **11**, 929.
Onishi, H. and Perry, M. B. (1972). *Can. J. Microbiol.* **18**, 925–927.
Ortmann, R., Sutter, A. and Grisebach, H. (1972). *Biochim. Biophys. Acta* **289**, 293–302.
Overend, W. G. and Williams, N. R. (1965). *J. Chem. Soc.*, 3446–3448.
Overend, W. G., White, A. C. and Williams, N. R. (1970). *Carbohydr. Res.* **15**, 185–195.
Ovodov, Y. S., Ovodova, R. G., Bondarenko, O. D. and Krasikova, I. N. (1971). *Carbohydr. Res.* **18**, 311–318.
Ovodova, R. G., Vaskovsky, V. E., Ovodov, Y. S. (1968). *Carbohydr. Res.* **6**, 328–332.
Pan, Y.-T. and Kindel, P. K. (1977). *Arch. Biochem. Biophys.* **183**, 131–138.
Patrick, A. D. (1956). *Nature (London)* **178**, 216.
Paulsen, H., Sinnwell, V. and Stadler, P. (1972). *Chem. Ber.* **105**, 1978–1988.
Paulsen, H., Sinnwell, V. and Thiem, J. (1980). *Meth. Carbohydr. Chem.* **8**, pp. 185–194.
Percival, E. and McDowell, R. H. (1967). "Chemistry and Enzymology of Marine Algal Polysaccharides". Academic Press, London and New York.
Perkin, A. G. and Uyeda, Y. (1922). *J. Chem. Soc.* **121**, 66.
Picken, J. M. and Mendicino, J. (1967). *J. Biol. Chem.* **242**, 1629–1634.
Pierce, J., Tolbert, N. E. and Barker, R. (1980). *Biochemistry* **19**, 934–942.
Pigman, W. and Horton, D. (1970). "The Carbohydrates", Vol. 2B, Chemistry and Biochemistry, pp. 809–834. Academic Press, New York, San Francisco and London.
Pigman, W. and Derek, H. (1970–1972). "The Carbohydrates, Chemistry and Biochemistry," Vols 1A, 1B, 2A, 2B. Academic Press, New York and London.
Plouvier, V. (1949). *Compt. Rend.* **228**, 1886.
Plouvier, V. (1959). *Compt. Rend.* **249**, 2828.
Plouvier, V. (1963). *In* "Chemical Plant Taxonomy" (T. Swain, ed.), pp. 313–336. Academic Press, New York and London.
Pringsheim, H. and Krüger, D. (1932). *In* "Handbuch der Pflanzenanalyse" (G. Klein, ed.), Vol. II, 1, pp. 764–770. Springer, Wien.
Proust, M. (1806). *Ann. Chim. Phys.* **57**, 144.
Rangoonwala, R. and Friedrich, H. (1967). *Naturwissenschaften* **54**, 368.
Rahman, A. U. (1958). *Z. Naturforsch.* **13b**, 201–202.
Raphael, R. A. and Roxburgh, C. M. (1955). *J. Chem. Soc.*, 3405–3408.
Rappaportt, J., Giacopello, D., Seldes, A. M., Blanco, M. C. and Deulofeu, V. (1977). *Phytochemistry* **16**, 1115–1116.

Redgwell, R. S. and Bieleski, R. L. (1978). *Phytochemistry* **17**, 407–409.
Regbie, R. and Richtmyer, N. K. (1966). *Carbohydr. Res.* **2**, 272.
Richtmyer, N. K. (1963). *In* "Methods in Carbohydrate Chemistry" (R. L. Whistler and M. L. Wolfrom, eds), Vol. II, pp. 90–91. Academic Press, New York and London.
Richtmyer, N. K. (1970a). *Carbohydr. Res.* **12**, 135–138.
Richtmyer, N. K. (1970b). *Carbohydr. Res.* **12**, 233–239.
Roberts, R. M., Shah, R. H. and Loewus, F. (1967). *Plant Physiol.* **42**, 659–666.
Robinson, S. P. and Portis, A. R. Jr. (1988). *FEBS Lett.* **233**, 413–416.
Ruesink, A. W. (1971). *In* "Methods in Enzymology" (S. P. Colowick and N. O. Kaplan, eds), Vol. 23, pp. 197–209. Academic Press, New York and London.
Rumpho, M. E., Edwards, G. E. and Loescher, W. H. (1983). *Plant. Physiol.* **73**, 869–873.
Sakuma, S. and Shoji, J. (1981). *Chem. Pharm. Bull.* **29**, 2431–2441.
Sakuma, S. and Shoji, J. (1982). *Chem. Pharm. Bull.* **30**, 810–821.
Salvucci, M. E., Holbrook, G. P., Anderson J. C. and Bowes, G. (1988). *FEBS Lett.* **231**, 197–201.
Samuelson, O. (1972). *In* "Methods in Carbohydrate Chemistry" (R. L. Whistler and J. N. BeMiller, eds), Vol. VI, pp. 65–75. Academic Press, New York and London.
Samuelson, O. and Strömberg, H. (1966). *Carbohydr. Res.* **3**, 88–96.
Sandermann, H. Jr and Grisebach, H. (1968). *Eur. J. Biochem.* **6**, 404–410.
Schaffer, R. (1959). *J. Am. Chem. Soc.* **81**, 5452–5454.
Scheele, K. W. (see Karrer, W.) (1958). "Konstitution und Vorkommen der organischen Pflanzenstoffe," Vol. 1, Ref. 139. Birkhäuser, Basel and Stuttgart.
Scherpenberg, H. van, Gröbner, W. and Kandler, O. (1965). "Festschrift K. Mothes". Fischer, Jena.
Schilling, G. and Keller, A. (1977). *Ann. Chem.* **1977**, 1475–1479.
Schilling, G. and Keller, A. (1986). *Z. Naturforsch.* **41c**, 253–257.
Schloss, J. V. (1988). *J. Biol. Chem.* **263**, 4145–4150.
Schwartzbach, S. D., Freyssinet, G., Schiff, J. A., Hecker, L. I. and Barnett, W. E. (1979). *In* "Methods in Enzymology" (S. P. Colowick and N. O. Kaplan, eds), Vol. 59, pp. 434–437. Academic Press, New York and London.
Seemann, J. R., Berry, J. A., Freas, S. M. and Krump, M. A. (1985). *Proc. Natl. Acad. Sci. USA.* **82**, 8024–8028.
Sellmair, J. and Kandler, O. (1970). *Z. Pflanzenphysiol.* **63**, 65–83.
Sellmair, J., Beck, E. and Kandler, O. (1968). *Z. Pflanzenphysiol.* **59**, 70–79.
Sellmair, J., Beck, E. and Kandler, O. (1969). *Z. Pflanzenphysiol.* **61**, 338–342.
Sellmair, J., Beck, E., Kandler, O. and Kress, A. (1977). *Phytochemistry* **16**, 1201–1204.
Serianni, A. S., Nuñez, H. A., Hayes, M. L. and Barker, R. (1982). *In* "Methods in Enzymology" (S. P. Colowick and N. O. Kaplan, eds), Vol. 89, pp. 64–73. Academic Press, New York and London.
Servaites, J. C. (1985). *Plant Physiol.* **78**, 839–843.
Servaites, J. C. and Ogren, W. L. (1980). *In* "Methods in Enzymology" (S. P. Colowick and N. O. Kaplan, eds), Vol. 69, pp. 642–648. Academic Press, New York and London.
Seshadri, T. R. and Vydeeswaran, S. (1971). *Phytochemistry* **10**, 667–669.
Shafizadeh, F. (1956). *Adv. Carbohdyr. Chem.* **11**, 263–283.
Sloneker, J. H. (1972). *In* "Methods in Carbohydrate Chemistry" (R. L. Whistler, and J. N. BeMiller, eds), Vol. VI, pp. 20–24. Academic Press, New York and London.
Smith, D. C. (1974). *Symp. Soc. Exp. Biol.* **28**, 485–520.
Smith, D., Muscatine, L. and Lewis, D. H. (1969). *Biol. Rev.* **44**, 17–90.
Snyder, J. R. and Serianni, A. S. (1987). *Carbohydr. Res.* **166**, 85–99.
Somogy, M. (1945). *J. Biol. Chem.* **160**, 61–68.
Speck, J. C. (1962). *In* "Methods in Carbohydrate Chemistry" (R. L. Whistler and M. L. Wolfrom, eds), Vol. I., pp. 441–445. Academic Press, New York and London.
Spellman, M. W., McNeil, M., Darvill, A. G., Albersheim, P. and Henrick, K. (1983). *Carbohydr. Res.* **122**, 115–129.
Stacey, B. E. (1974). *In* "Plant Carbohydrate Biochemistry" (J. B. Pridham, ed.), Vol. X, pp. 47–59. Academic Press, New York and London.

Stanek, J., Cerny, M., Kocourek, J. and Pacak, J. (1963). "The Monosaccharides". Academic Press, New York and London.

Steiner, M. and Maas, E. (1957). *Naturwissenschaften* **44**, 90–91.

Stenhouse, J. (1848). *Just Lieb. Am. Chem.* **66**, 55, 78.

Subramanian, S. S. and Nagarajan, S. (1970). *Phytochemistry* **9**, 2581–2584.

Suzuki, T. and Onishi, H. (1975). *Agric. Biol. Chem.* **39**, 2389–2397.

Thanbichler, A. and Beck, E. (1974). *Eur. J. Biochem.* **50**, 191–196.

Thanbichler, A., Gilck, H. and Beck, E. (1971). *Z. Naturforsch.* **22b**, 912–915.

Touster, O. and Shaw, D. R. D. (1962). *Physiol. Rev.* **42**, 181–225.

Trevelyan, W. E., Procter, D. P. and Harrison, J. S. (1950). *Nature* **166**, 444–445.

Vesterberg, O. (1971). *In* "Methods in Enzymology" (S. P. Colowick and N. O. Kaplan, eds), Vol. 13, pp. 389–412. Academic Press, New York and London.

Vincent, C. and Meunier, J. (1898). *Compt. Rend.* **127**, 760.

Wagner, H. and Demuth, G. (1972). *Tetrahedron Lett.* **49**, 5013–5014.

Wagner, H. and Demuth, G. (1974). *Z. Naturforsch.* **29c**, 204–208.

Wagner, H. and Kirmayer, W. (1957). *Naturwissenschaften* **44**, 307.

Wallaart, R. A. M. (1980). *Phytochemistry* **19**, 2603–2610.

Watson, R. R. and Kindel, P. K. (1970). *Plant Physiol* **46S**, 27.

Watson, R. R. and Orenstein, N. S. (1975). *Carbohydr. Chem. Biochem.* **31**, 135–184.

Wehner, C., Thier, W. and Hadders, M. (1932). *In* "Handbuch der Pflanzenanalyse" (G. Klein, ed.), Vol. II, 1, pp. 764–770. Springer, Wien.

Weigel, H. (1963). *Adv. Carbohydr. Chem.* **18**, 61.

Wellmann, E. and Grisebach, H. (1971). *Biochim. Biophys. Acta* **235**, 389–397.

Wellmann, E., Baron, D. and Grisebach, H. (1971). *Biochim. Biophys. Acta* **244**, 1–6.

Wessely, F. and Wang, S. (1938). *Monatsh. Chem.* **72**, 168.

Whistler, R. L. and Wolfrom, M. L. (eds) (1962–1979). "Methods in Carbohydrate Chemistry", Vols I–VIII. Academic Press, New York and London.

Williams, D. T. and Jones, J. K. N. (1964). *Can. J. Chem.* **42**, 69–72.

Williams, N. R. and Wander, J. D. (1980). *In* "The Carbohydrates" (W. Pigman, D. Horton and J. D. Wander, eds), Vol. IB, pp. 778–798. Academic Press, New York, San Francisco and London.

Wolfrom, M. L. and Thompson, A. (1963). *In* "Methods in Carbohydrate Chemistry" (R. L. Whistler and M. L. Wolfrom, eds), Vol. II, pp. 65–68. Academic Press, New York and London.

Woolard, G. R. and Jones, J. K. N. (1978). *Carbohydr. Res.* **63**, 327–332.

York, W. S., Darvill, A. G., McNeil, M., Stevenson, T. T. and Albersheim, P. (1985). *In* "Methods in Enzymology" (S. P. Colowick and N. O. Kaplan, eds), Vol. 118, pp. 3–40. Academic Press, New York and London.

Yoshimura, J. (1984). *Adv. Carbohydr. Chem. Biochem.* **42**, 69–134.

Yu, J., Lang, H.-Y. and Xiao, P.-G. (1986). *Acta Pharm. Sin.* **21**, 191–197.

Ziegler, H. (1975). *In* "Encyclopedia of Plant Physiology, New Series" (M. H. Zimmermann and J. A. Milburn, eds), Vol. 1, pp. 59–136. Springer, Berlin, Heidelberg and New York.

Zimmermann, M. H. and Ziegler, H. (1975). *In* "Encyclopedia of Plant Physiology, New Series" (M. H. Zimmermann and J. A. Milburn, eds), Vol. 1, pp. 480–503. Springer, Berlin, Heidelberg and New York.

8 Cellulose

GERHARD FRANZ and WOLFGANG BLASCHEK

Institute of Pharmacy, University of Regensburg, D 8400 Regensburg, FRG

I. INTRODUCTION

In the early years of polysaccharide research the term cellulose, coined by Payen in 1838, was used for several carbohydrate polymers, since techniques to differentiate

METHODS IN PLANT BIOCHEMISTRY Vol. 2
ISBN 0-12-461012-9

between individual sugars were rather limited. Schulze in 1891 proposed that the designation of cellulose should be restricted to constituents of plant cell walls resistant to dilute acid and alkali, soluble in ammoniacal copper solutions and yielding glucose after hydrolysis.

The crystallinity of cellulose was detected some years earlier by von Nägeli using the polarising microscope. But these findings were contested at that time and final proof had to wait until the development of X-ray analysis for the determination of crystalline structures. The crystallographic unit cell of cellulose was first postulated by Polyani (1921), who pointed out that this unit could accommodate two cyclic or two linear disaccharides of polysaccharide chains. These early findings were substantiated by Meyer and Mark (1928), who showed that the X-ray structure is, in fact, compatible with long chains built up from cellobiose residues.

The linear concept of cellulose was not very clear during the early days of cellulose research, since the chemists failed to detect reducing end groups. Cellulose was considered to be associated as colloidal particles in order to account for the high molecular weight. Credit is due to Staudinger (1932), who provided the final proof of the linear macromolecular structure of cellulose, and further to Haworth (1932), who analysed the nature of the bonds between the glucose units, i.e. the 1,4-β-linked polyanhydro-D-glucopyranose polymer.

From X-ray crystallography studies and later from NMR investigations it was confirmed that the β-D-glucose units are present as pyranose rings in the chair formation, which is the lowest energy confirmation for the glucose units (Brown and Ley, 1968).

Since cellulose is known to be the most abundantly available carbohydrate, the industrial demand for this biopolymer with unique chemical and physical properties is still increasing. Many estimations have been made for the actual volume of existing cellulosic resources given as 26.5×10^{10} tonnes of cellulosic material, which is continuously replenished by the complex photosynthesis process.

Every year about 10^{15} kg of cellulose are synthesised and in part broken down by microbial attack or industrial utilisation. It has become clear that with the depletion of easily available fossil carbon sources, the industrial importance of cellulose and related products as a renewable energy source will certainly increase in the next decade. At the moment large-scale replacement of fossil energy sources is not possible because the technology is lacking. Many industrial innovations are urgently needed to resolve the problems of biomass and mainly cellulose utilisation as a constant renewable energy source.

The wide range of industrial applications of cellulose and the respective derivatives is still increasing. The derivatisation of cellulose has opened up a new range of industrial research.

Examples of industrial utilisation of cellulose include the classical field of pulp and paper (Aho, 1983), the production of glucose and its transformation into ethanol and methanol (Waymann, 1986) as a renewable energy source (Strub et al., 1983), and finally a series of derivatives with an enormous range of possible industrial applications (Quenin, 1985).

Today, many industrial companies and many academic research groups are devoted to cellulose chemistry. It is not surprising that there has been a proliferation of publications dealing with the fundamentals of cellulose chemistry and biogenesis, the physical and mechanical properties of cellulose and its derivatives as well as its biological and chemical degradation. From the enormous number of publications, reviews and specialised books about cellulose, only a few being of general interest can be cited (Brown, 1982; Wilke *et al.*, 1983; Burchard, 1985; Kennedy *et al.*, 1985; Nevell and Zeronian, 1985; Young and Powell, 1986; Fan *et al.*, 1987).

II. IMPORTANCE AND SOURCES OF CELLULOSE

The occurrence of cellulose is remarkably general in terrestrial plants showing always the same structure. No other cell-wall polysaccharide with an identical structure is known to be so widespread. The chemistry and structure of pectins and hemicelluloses is known to be variable between the different plant species.

Cellulose is an important component of several fungi and some bacteria (Aronson, 1965) and marine animals such as tunicates (Wardrop, 1970). It may even be synthesised as an abnormal constituent in animal tissue (Toriumi *et al.*, 1972).

The cellulose-containing cell-wall structures are dynamic in shape, in their specific composition and in their properties, which may constantly change in response to endogenous or exogenous events (Franz, 1972).

It is well documented that cellulose is deposited as microfibrils, these being the main structural elements of a cell wall. Besides the mechanical strength of the plant cell, cellulose is a protective component against external attack by mechanical forces or microorganisms.

The different sources of cellulose in various biological materials have determined the specific uses of this biopolymer. Textile fibres in general are not isolated from woody tissue. On the other hand, it is not economical to produce ordinary paper from cotton fibres, which are one of the rare biological sources of almost pure cellulose. Among the few bacteria that synthesise relatively large amounts of cellulose, *Acetobacter xylinum* has been studied intensively (Lin *et al.*, 1985). In the future, industrially produced bacterial cellulose might become a reality (Brown, 1987a). The great advantage of *Acetobacter* cellulose is the high purity of the product and the fact that cellulose is secreted into the medium in the form of microfibrillar aggregates, which are not accompanied by hemicelluloses or polyphenols.

TABLE 8.1.   Sources of cellulose for industrial use.

Source	Industrial importance
Bacteria	None
Cotton (lint)	Textile
Cotton (linters)	Non-textile, chemical derivatives
Fibres: Sisal	
Jute	
Flax	Mainly textile
Ramie	
Abaca	
Wood: softwood	Pulp and paper
hardwood	Pulp and paper
Straw and other biomass	Degradation by acids or enzymes

The most common sources of cellulose are wood pulp and cotton fibres. Other sources are grasses or cereal straws with a lesser content of cellulose but, as a crop residue, being a relatively economic source of cellulose.

For industrial use (see Table 8.1) the major source of cellulose is still wood pulp, even if the purity of the α-cellulose obtained from this raw material is not very high. Cellulose derivatives used for pharmaceuticals, for paints and explosives are often prepared from cotton linters.

However, alternative sources of cellulose are required, if cellulose is to be used as matrix material from which other industries can develop. Biomass from different sources which is frequently treated as a waste, should be seriously studied for this purpose.

According to Nevell and Zeronian (1985), in 1981 the world production of cellulose from pulp was 130×10^6 tonnes, most of which was used for the production of paper and board. Only a small percentage of the wood pulp was converted into chemically pure cellulose.

The world production of raw cotton in 1981 was estimated as 15.4×10^6 tonnes, most of this material being produced in the USSR, China and North and South America.

In comparing the composition of several wood types, there is some difference in the α-cellulose content. In general, wood is about half cellulose (40–50%). It is found primarily in the thickened secondary wall of tracheids where it is mixed with lignin and hemicelluloses.

The cellulose content of hardwood is generally higher than that of softwood. Furthermore, the cellulose content varies in relation to growth or differentiation of the plant cell wall or the type of organism (Fengel and Wegener, 1984).

In general, primary cell walls contain 10–20% cellulose, secondary cell walls up to 50% and only specialised walls such as cotton fibres contain up to 98% (Huwyler *et al.*, 1979). The newly regenerated cell wall of protoplasts only contains 1–5% of cellulosic material, which might increase up to 20% after a prolonged culture period (Franz and Blaschek, 1985). Cell walls from tissue cultures and suspension cultured plant cells show cellulose contents of about 20%. This depends upon the culture conditions and the plant origin, but represents more or less the amount of cellulose found in a typical primary cell wall (Blaschek *et al.*, 1981).

III. ISOLATION AND DETERMINATION OF CELLULOSE

A. The Cell Wall

Most of the cellulose used for industrial purposes originates from wood and cotton seed hair. Therefore a short survey of the complex polymer network of a plant cell wall, in which cellulose is only one of several components, is advisable.

The plant cell wall surrounds the protoplast of a cell. During mitosis, particularly the calcium salts of highly esterified pectins are deposited in the cell plate forming the middle lamella (M), which cements the cells together. In a second step the primary cell wall (P) originating from both sides of the daughter cells is layered onto the middle lamella. It is composed of additional pectic material, hemicelluloses, cellulose, some glycoproteins and small amounts of phenolics. The primary wall is highly hydrated (about 60% water/fresh weight) (Roelofsen, 1965) and it mainly consists of polysaccharides (80–90% of dry weight) with a cellulose content of only 20–30% (Albersheim, 1974). During cell elongation new cell wall material is added, e.g. by multi-net growth.

According to the Albersheim model (Darvill *et al.*, 1980a,b) of the primary cell wall, the cellulose microfibrils are surrounded by xyloglucans linked to cellulose by H-bonds. A single layer of xyloglucan molecules, on the other side, is bound to rhamnogalacturonans via arabinogalactans. Each rhamnogalacturonan can be connected to various arabinogalactans, thereby forming a linkage to different cellulose molecules. By this complex system the cellulose microfibrils are interlinked and embedded into an amorphous matrix of pectins and hemicelluloses like a steel fabric woven into concrete. The concrete, however, is a fully hydrated, viscoelastic mass. Moreover, the glycoprotein extensin is interwoven and linked to rhamnogalacturonans via arabinogalacturonans. Most of the linkages between the different polymers seem to be the result of H-bonds, ionic bonds and calcium-bridges; however, some covalent linkages may also be involved in the fixation of the cell wall network (Darvill *et al.*, 1980a,b). The relative amount and the detailed structure of the different polymers varies between plant species and the type of tissue analysed.

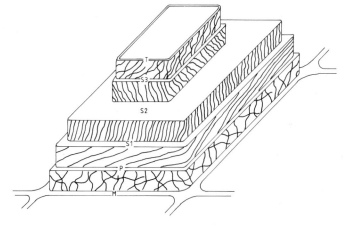

FIG. 8.1. The various cell-wall layers of a plant cell including the arrangement of cellulose microfibrils. M, middle lamella; P, primary cell-wall layer; S1–S3, secondary cell-wall layers; T, tertiary cell-wall layer.

After cessation of cell growth, secondary (S) and in some cases tertiary (T) cell-wall material is deposited in layers, the thickest wall layer usually being S2 (see Fig. 8.1; Fengel and Wegener, 1984). The cellulose content of the secondary wall increases (40–70% of dry weight), whereas the water content decreases (25–40% of fresh weight), resulting in a higher physical stability of the wall (Fengel and Wegener, 1984). The cellulose fibrils, forming an irregular network in the primary wall (Wardrop, 1962) or being deposited approximately transversely (Roelofsen, 1965), in secondary walls are oriented parallel, surrounding the cell helically at a steep angle to the longitudinal cell axis (Fengel and Wegener, 1984). Moreover, lamellae with a slightly different orientation of the microfibrils often exist (Preston, 1981). Compared to the primary wall, the diversity of polysaccharides is reduced. In angiosperms mainly xylans and in gymno-sperms glucomannans and galactoglucomannans are deposited (Kristen, 1985; Fan *et al.*, 1987). Lignification begins in the primary wall region and then extends inward to the secondary wall, making the cell wall rigid and resistant to microbial attack. Lignin seems to form covalent linkages with hemicelluloses and constitutes 25–35% of the wood structure (Northcote, 1965; Kristen, 1985).

In cotton seed hairs secondary wall deposition begins about 20 days post-anthesis and lasts for about 30 days (Beasly *et al.*, 1974). The secondary wall consists of almost pure cellulose (about 95% of dry weight) (Beasly *et al.*, 1974). The outer surface of the wall is cutinised. After opening of the cotton boll, the cell wall dehydrates, the cells collapse and develop about 50 twists per cm (Cook, 1960).

Besides the above-mentioned constituents, the cell-wall material may, in addition, contain enzymes and non-enzymatic proteins, lipophilic material (waxes, terpenes, fats, oils) and inorganic ssubstances (potassium, calcium, magnesium or silicon). In order to isolate a pure α-cellulose fraction all these 'impurities' have to be removed by various extraction procedures.

B. Isolation and Determination of Cellulose

Methods for the isolation and quantification of cellulose from various origins differ in the required precision depending on utilisation in scientific research or in industrial production. By many isolation procedures cellulose is obtained as a more or less crude preparation generally called α-cellulose (Cross and Bevan, 1912), a term created for the insoluble residue in a strong sodium hydroxide solution. After neutralisation of the alkali-soluble material the β-cellulose precipitates, whereas γ-cellulose remains in solution. Both the so-called β- and γ-celluloses consist mainly of hemicelluloses. According to a standard method, α-, β- and γ-celluloses are determined by testing the solubility of pulps in 17.5% NaOH (Tappi Standard T 203 os-74). In another procedure (Wise *et al.*, 1946) repeated extraction with 5% and 24% KOH under nitrogen is performed, resulting in an α-cellulose with less contamination by hemicelluloses and lignin. The degree of polymerisation (DP) and yield of cellulose, however, is lowered, which seems to be a general problem in all extraction procedures with alkali. Another problem is the fact that alkali treatment can change the crystalline lattice of the native cellulose I to cellulose II, resulting in a cellulose which is different from natural cellulose in its crystalline structure (for a more detailed discussion, see Section IV).

Besides NaOH and KOH, other alkaline solvents for a stepwise extraction are also used in various concentrations, e.g. lithium and calcium hydroxide, quaternary ammonium hydroxides or liquid ammonia. Alkaline extractions often are performed under nitrogen or in the presence of $NaBH_4$ in order to avoid 'peeling' of polysaccharides by β-elimination. Addition of boric acid or borates increases the dissolving power especially for mannose-containing polysaccharides. Prior to alkali treatment, extraction with hot water and dimethyl sulphoxide (DMSO) (removal of starch and heavily acetylated polysaccharides) has been found to be useful (more details in Ward and Morak, 1962; Browning, 1967; Fengel and Wegener, 1984; Fry, 1988).

For lignified cell walls a delignification is necessary prior to cellulose isolation yielding the so-called 'holocellulose'. Absolute delignification cannot be achieved by any known method, and losses of polysaccharides and depolymerisation of cellulose always take place.

The common methods for delignification are chlorination including alternating extraction with hot alcoholic solutions of organic bases (ASTM Standard D-1104-56), treatment with an acidified solution of sodium chlorite (Wise *et al.*, 1946) or extraction with dilute peracetic acid (Leopold, 1961). Numerous modifications have been applied concerning the reaction time and temperature, pH value and concentration of reagents (Fry, 1988).

For the isolation of cellulose from living tissues a first step is the purification of the cell walls. This should be performed without heating at pH-values between 4 and 7. Prolonged extraction in aqueous solutions at room temperature should be avoided because of the danger of autolysis by cell-wall-bound enzymes. By extraction of homogenised tissue with 70% ethanol for several hours at 0°C sugars, amino acids, organic acids, inorganic salts and lipophilic compounds are removed yielding an alcohol-insoluble residue called AIR (Fry, 1988). Instead of ethanol, mixtures of ethanol with benzene, cyclohexane or toluene may also be used followed by hot water extraction (Fengel and Wegener, 1984). The cytoplasmic fraction of living tissues can also be removed by homogenisation in glycerol at temepratures below 0°C (Huwyler *et al.*, 1979) or in cold aqueous buffers with optional addition of detergents, calcium salts and reducing agents (Fry, 1988). Removal of residual protein can be performed by treatment with phenol–acetic acid–water (2:1:1; w/v/v), which does not extract extensin, or by pronase digestion (Selvendran *et al.*, 1985). Non-covalently bound proteins also can be extracted from cell walls by salt solutions, sodium dodecyl sulphate (SDS), urea or guanidinium thiocyanate. The last three aqueous chaotropic agents may also dissolve some hemicelluloses, especially mannose-containing polysaccharides (Selvendran *et al.*, 1985). Extensin can be extracted incompletely by cleaving isodityrosine cross-links with acidified $NaClO_2$ (O'Neill and Selvendran, 1980). Aqueous chelating agents can release a proportion, but not all, of the pectins from the wall by complexing calcium.

It should be recalled that by all extraction procedures the remaining, insoluble cellulose may be more or less degraded, altered in crystallinity and not absolutely pure. In the past such impurities as proteins and sugars other than glucose led to speculation as to whether cellulose perhaps might be a glycoprotein or a glucan with trace amounts of other sugars located primarily in the amorphous regions of the fibril.

After all extraction procedures the cellulose content of a given cell-wall material is determined gravimetrically. The danger of losses of cellulose during extraction by degradation, however, must be considered. Due to impurities the absolute cellulose content in the remaining insoluble α-cellulose fraction is somewhat lower. It can be estimated by complete hydrolysis of α-cellulose with sulphuric acid in a two-step procedure (72%, 1 h, 25°C; 3%, 1 h, 120°C) (Adams, 1965; Saeman *et al.*, 1954) with subsequent determination of the released monosaccharides after acetylation by gas–liquid chromatography (GLC) (Blakeney *et al.*, 1983). Contamination by hemicelluloses can be calculated after hydrolysis from the sugar spectrum. Hydrolysis of α-cellulose can also be performed with 100% trifluoracetic acid (TFA) in the first stage of hydrolysis with subsequent dilution steps (Fengel and Wegener, 1979). Assays for non-cellulosic cell wall components in α-cellulose fractions may be as follows:

(a) for proteins, Coomassie Blue binding (Read and Northcote, 1981) as well as the Lowry method (Layne, 1957) or the Biuret assay (Layne, 1957);
(b) for pentoses, the orcinol method (Dische, 1962);

(c) for uronic acids, the biphenylol assay (Blumenkrantz and Asboe-Hansen, 1973);
(d) for phenols, the reaction with Folin/Ciocalteau's reagent (Forest and Bendall, 1969);
(e) for lignin, the acetyl bromide method (Johnson et al., 1961);
(f) for amorphous silicone, the assay with molybdenum blue (Volk and Weintraub, 1958); and
(g) for hydroxyprolin-rich extensin, the Kivirikko/Liesmaa method (Kivirikko and Liesmaa, 1959) after protein hydrolysis.

Other extraction methods are alcoholic nitration of cell-wall material in ethanol after KOH treatment (Kürschner and Popik, 1962) or refluxing of cell walls in an HCl-acidified mixture of acetyl-acetone and dioxane (Seifert, 1960) resulting in relatively pure, but degraded cellulose preparations.

Methods of cellulose determination without extraction procedures can be performed in the following manner. In the Updegraff method (Updegraff, 1969) the cell-wall samples are boiled in acetic–nitric reagent (acetic acid–H_2O–HNO_3: 8:2:1; v/v). The insoluble residue represents α-cellulose, as under these strong hydrolytic conditions all other polysaccharides are degraded and can be removed by centrifugation or filtration on glass fibre filters. Quantification can be performed gravimetrically or photometrically by the anthrone method (Dische, 1962). A selective hydrolysis of nearly all non-cellulosic polysaccharides can also be achieved by treatment of cell walls with 2 M TFA at 120°C for 1 h. Sequential Saeman-hydrolysis (Saeman et al., 1954) of the insoluble residue releases glucose from cellulose. If both hydrolysates are analysed by GLC, a first, helpful insight into the cellulose content of a cell wall and the sugar composition of its non-cellulosic polysaccharides may be obtained simultaneously (Blaschek et al., 1981). By nitration of cell-wall material (Spencer and Maclachlan, 1972) cellulose nitrate is obtained, being normally used for DP-determinations of cellulose by viscosity measurements. Polymers other than cellulose, such as starch, pectins, hemicelluloses and proteins, are not carried through this procedure (Spencer and Maclachlan, 1972). From the yield of cellulose nitrate the cellulose content is calculated or it is determined after denitration by sugar analysis.

IV. THE STRUCTURE OF CELLULOSE

A. The Chemical Structure of Cellulose

The description of cellulose seems to be rather simple, as it is composed of just one type of sugar residue joined in one repeating linkage. However, there still exist some open questions concerning its biosynthesis, its molecular weight distribution, its supra-molecular organisation and its arrangement in the cell wall.

It is generally accepted that cellulose is an unbranched polymer of β-1,4-linked D-anhydroglucopyranose units (see Fig. 8.2). With respect to the mean plane of the pyranose ring all OH- and both the CH_2OH-group and the glycosidic bond are equatorial, while the hydrogen atoms are in the axial positions, resulting in a $^4C_1$- (or D-C1-) chair conformation (Krässig, 1985). Fewer than 2% of the glucose units may be present in the boat or skew form (Goebel et al., 1976). As the surface of the glucan

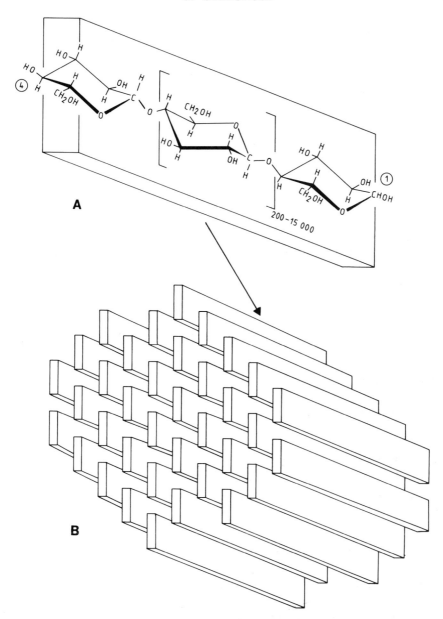

FIG. 8.2. (A) A β-1,4-glucan chain forming a flat ribbon with a reducing C-1 and a non-reducing C-4 end. (B) A cellulose microfibril of about 3.5 nm diameter composed of about 36 glucan chains, which are held together by hydrogen bonds.

chains consists mainly of hydrogen atoms, the molecule becomes hydrophobic. The β-1,4-linkage forces the alternate chain units to be rotated 180° around the main axis. Therefore the repeating unit of cellulose is an anhydrocellobiose of 1.03 nm length. The extended molecule thus forms a flat ribbon, which is further stiffened by intramolecular H-bonds, especially between the hydroxyl group at C-3 and the ring oxygen atom of the neighbouring unit (Blackwell and Marchessault, 1971). Another intramolecular H-bond between the hydroxyls on C-2 and C-6 may exist (see Fig. 8.3; Blackwell *et al.*, 1977). Intramolecular bonding stabilises the repeating distance of 1.03 nm and determines some of the chemical properties of cellulose, for example the relative reactivity of the hydroxyl groups at C-2, C-3 and C-6 for substitution normally is lowest for that at C-3 (Krässig, 1985). The glucan molecule exhibits a C-1 end with a reducing aldehyde hydrate group and a C-4 end with a non-reducing alcoholic hydroxyl group.

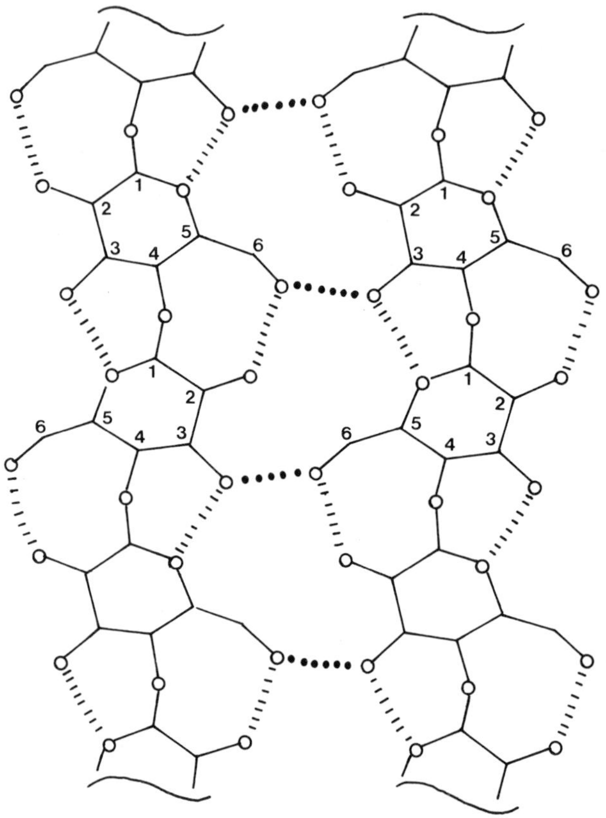

FIG. 8.3. Intramolecular (|||) and intermolecular (···) hydrogen bonds of two adjoining glucan chains in cellulose.

B. The Molecular Weight of Cellulose

The glucan chain length in cellulose can be described in different ways. The molecular weight of cellulose varies between 5×10^4 and 2.5×10^6 depending on its origin (Fengel and Wegener, 1984).

By dividing by the M_w of one repeating unit (160 Da) the degree of polymerisation (DP), ranging from 300–15 000, is obtained. This corresponds to theoretical chain lengths of 0.15–7.2 µm. Although it is assumed that the DP of cellulose at a certain stage of cell wall formation is uniform, the isolated α-cellulose fraction of a tissue is polydisperse, as (a) in different cell-wall layers cellulose of various chain lengths is deposited, (b) during ageing of a tissue the DP of cellulose decreases, and (c) cellulose isolation often causes random chain cleavage. Therefore, by DP determination methods such as viscosimetry, light-scattering or ultracentrifugation the weight average DP = $\overline{DP}$ or $\overline{P}_w$ is estimated, whereas using osmometry the number average DP = $\overline{P}_n$ describing the number of glucan chains of a certain DP can be calculated. The ratio of $\overline{P}_w$ to $\overline{P}_n$ describes the degree of polydispersity of a cellulose sample (Fengel and Wegener, 1984).

Before a less laborious, relative DP-determination method such as viscosimetry can be used, calibration with absolute methods such as osmometry, ultracentrifugation, or light-scattering is necessary (Houwink, 1940; Schulz and Mark, 1954; Claesson *et al.*, 1959). A general problem in all methods is the non-destructive dissolution of cellulose. Of all available metal complexing solvents for cellulose (Jayme, 1971; Fengel and Wegener, 1984), the cadmium complex 'cadoxene' (Jayme, 1971, 1978) and the iron tartaric sodium complex 'EWNN' (Jayme, 1971; Achwal and Chaugule, 1975) are most often used for DP determinations. Other solvents used for DP measurements are mixtures of hydrazine with water or DMSO (Kolpak *et al.*, 1977) as well as mixtures of DMSO, chloral and triethylamine (Okajima, 1978/79).

By conversion of cellulose to esters or ethers, derivatives are obtained which are soluble in acetone, propanone, ethyl acetate, water or sodium hydroxide (see Section VII). If such derivatives are used for DP determinations of cellulose, their preparation has to be performed under non-degrading conditions and their degree of substitution (DS) must be known exactly. Of these derivatives, cellulose tri-nitrate is most often used for viscosity measurements. It is prepared in a mixture of HNO_3–H_3PO_4–P_2O_5 (65:26:10; w/w) (Alexander and Mitchell, 1949; Marx-Figini, 1962; Spencer and Maclachlan, 1972) and its nitrogen content is determined (Merz, 1968). The distribution of the DP of a cellulose can be estimated by fractional dissolution (Elliott, 1967) or precipitation (Kotera, 1967; Spencer and Maclachlan, 1972; Asamizu *et al.*, 1977; Blaschek *et al.*, 1982) of cellulose or cellulose derivatives resulting in fractions of a small DP range. Fractionation is also possible by gel permeation chromatography (Minor, 1979) or counter-current distribution (Cantow, 1967). Size exclusion chromatography (SEC) can also be combined on-line with low angle laser light scattering (LALLS) for the determination of the molecular weight distribution of cellulose derivatives (Lauriol *et al.*, 1987; Eigner *et al.*, 1987).

In general, the DP of cellulose from algae and secondary walls of vascular plants is above 8000–10 000 whereas that derived from *Acetobacter xylinum* and primary walls of vascular plants is in the range of 1000–4000 (Fengel and Wegener, 1984; Delmer, 1987). Distribution curves of cellulose derived from growing primary walls of plants showed the predominance of two DP fractions (2500–4000 and 250–500). By pulse chase experiments it has been shown that the low DP fraction is a structural component of the cell wall (Blaschek *et al.*, 1982). Highest known DP values of 44 000 have been reported for the alga *Valonia* (Palma *et al.*, 1976); in higher plants cellulose from closed cotton capsules with a DP of 15 000 seems to possess the maximum chain length, whereas in opened capsules the DP is already reduced to 8000–10 000 (Goring and Timell, 1962; Fengel and Wegener, 1984). The DP-distribution curves for cotton cellulose show the presence of small amounts (about 10%) of one or two low M_w fractions in the DP range of 1000 to 6000 as well as the high M_w cellulose fraction (Marx-Figini and Schulz, 1966). Wood celluloses exhibit average DP values of 6000–10 000 with multiple peaks in the corresponding distribution curves (Goring and Timell, 1962; Fengel and Wegener, 1984). The production of pulp from wood results in a considerable decrease in DP values (500–1 000) and changes in the DP-distribution curve (Fengel and Wegener, 1984).

C. The Supramolecular Structure of Cellulose

The description of the cellulose macromolecule as a β-1,4-glucan is not sufficient. It is the ordered arrangement of the glucan chains that creates the physical and chemical characteristics of cellulose. Besides the already mentioned intramolecular linkages, intermolecular H-bonds between adjoining glucan chains in the same as well as in neighbouring lattice planes exist (see Fig. 8.3). These bonds produce a regular crystal-line arrangement of the glucan molecules resulting in distinct X-ray diffraction patterns (Meyer and Misch, 1937) and control the swelling and reactivity of cellulose. Hydrogen bonds are proposed between the hydroxyls on C-(6) and C-3″ (Marchessault and Liang, 1960) as well as probably C-6 and C-2″ (Blackwell *et al.*, 1977) within one lattice plane and the hydroxyl on C-6 and glucosidic oxygen atom O-4″ in the adjacent lattice plane (Marchessault and Liang, 1960; Blackwell *et al.*, 1977) depending on the crystalline form of cellulose. Connections of planes can also occur by van der Waals forces (Blackwell *et al.*, 1977).

TABLE 8.2. Various lattice conformations of cellulose within the parallel and anti-parallel structural families.

	Chain structure	
Parallel		Anti-parallel
Cellulose I	$\xrightarrow{1}$	Cellulose II
2↓↑3		2↓↑3
Cellulose III$_I$		Cellulose III$_{II}$
4↓		4↓
Cellulose IV$_I$		Cellulose IV$_{II}$

Treatments for lattice confirmation changes: 1 = NaOH, 20°C/H_2O; 2 = Liq. NH_3/evaporation; 3 = H_2O; 4 = 280°C.

As already indicated above, cellulose can exist in various polymorphous forms (e.g. celluloses I, II, III, IV; Table 8.2), which exhibit a different arrangement of the glucan chains and thereby produce different X-ray or electron diffraction patterns depending on the origin of the material or chemical and physical pre-treatment (Fengel and Wegener, 1984; Delmer, 1987). The most important forms are native cellulose I, which after swelling as a result of alkali treatment turns into Na-cellulose I with an expanded lattice structure. After removal of the Na-ions linked to the hydroxyl groups, cellulose II is formed (Fengel and Wegener, 1984). The linkage of interplane H-bonds in cellulose II is still a matter of discussion (Marchessault and Liang, 1960; Blackwell *et al.*, 1977; Sarko, 1978), but there is general agreement that cellulose II is more densley packed and more strongly interbound compared to cellulose I (Fengel and Wegener, 1984; Krässig, 1985). This more compact structure may be responsible for a lower reactivity of cellulose II during derivatisation (Krässig, 1985). The crystal structure of cellulose I results only from biosynthesis, and the conversion of cellulose I to II is not reversible (Sarko, 1985). A further important difference between both cellulose forms is that native cellulose I has parallel, whereas mercerised and regenerated cellulose II has anti-parallel arrangement. For a long time this was the object of controversy (French, 1985). Parallel arrangement of the glucan chains in cellulose I has been proven by X-ray and electron diffraction techniques (Sarko, 1985; French, 1985). Further proof was that electron-dense staining is restricted to one end of the microfibrils, namely the reducing end (Hieta *et al.*, 1984), and that specific cellobiohydrolases degrade only one end of the microfibrils, namely the non-reducing end (Chanzy and Henrissat, 1985). A possible mechanism of conversion of cellulose I to cellulose II might be that in native cellulose each microfibril has parallel arrangement of the glucan chains, but adjacent microfibrils are oriented anti-parallel and during mercerisation the chains of different microfibrils intermix (Sarko, 1985; French, 1985).

The main technique for the determination of the supramolecular structure of cellulose is X-ray diffraction. Related methods are electron diffraction studies performed in electron microscopes, which yield more diffraction spots than do X-ray patterns, and neutron radiation studies, which are limited by the rather diffuse scattering caused by the hydrogen atoms being also visualised. The results obtained by these techniques may be complemented by solid state carbon-13 NMR spectroscopy as well as by infrared and Raman absorption providing information on crystallinity and intra- and intermolecular relationships (the various sophisticated techniques are discussed by Sarko, 1985, 1986; French, 1985; Sterk *et al.*, 1987).

The best known proposal for the crystalline lattice of cellulose evaluated by X-ray diffraction describes a pseudomonoclinic unit cell with dimensions $a = 0.835$ nm b (fibre axis) $= 10.3$ nm $c = 0.79$ nm and an angle of $q = 84°$ between a- and c-axes (see Fig. 8.4) (Meyer and Mark, 1929; Meyer and Misch, 1937). The central glucan chain in the unit cell lies upside down with reference to the corner chains (anti-parallel arrangement). Other unit cell constants for cellulose I have been determined more recently with slightly varying unit cell parameters of $a = 0.814$–0.8247 nm, $b = 0.778$–0.786 nm, $c = 1.033$–1.038 nm and $q = 96.38$–$97°$ (since about 1970 the fibre axis has been designated the crystallographic c-axis) (French, 1985). For cellulose II unit cells with the dimensions of $a = 0.801$ nm, $b = 0.904$ nm, $c = 1.036$ nm and $q = 117.1°$ are described (Zugenmaier, 1985).

The main lattice planes of a unit cell, which are represented by peaks of different intensities in the diagrams of X-ray or electron diffraction are termed 1 0 1, 1 0 1̄ and 0 0 2 (Fig. 8.4). Different cellulose polymorphs of course exhibit different crystalline lattices; e.g. the appropriate plane distances are described to be 0.594 nm between 1 0 1-planes, 0.535 nm between 1 0 1̄-planes and 0.39 nm between 0 0 2-planes in cellulose I and 0.725 nm, 0.443 nm and 0.401 nm, respectively, in cellulose II (Schurz *et al.*, 1987). Other models for the crystalline lattice of cellulose are unit cells with parallel arrange-ment of the glucan chains especially for cellulose I (Gardner and Blackwell, 1974) and with extension of a- and c-axes resulting in four-chain (Fisher and Mann, 1960; Ruben and Bokelman, 1987) or even eight-chain (Preston, 1986) unit cells with dimensions $a = 1.634$ nm, $b = 1.572$ nm, $c = 1.038$ nm and $q = 97°$ (Zugenmaier, 1985).

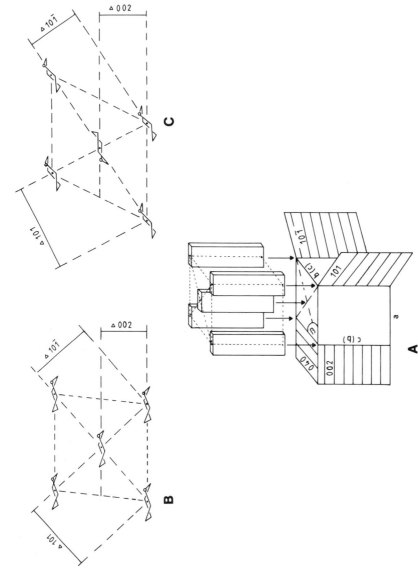

FIG. 8.4. (A) The important lattice planes (1 0 1, 1 0 1̄, 0 0 2 and 0 4 0) of the unit cell of cellulose. The crystal axes *a*, *b* and *c* and the non-90° angle β determine the figure of the unit cell. Five glucan chains (the flat ribbons of Fig. 8.2) enter the unit cell. The unit cell contains four glucose residues: from the central chain one glucose residue is positioned in the centre of the unit cell and half a glucose residue is placed at the middle of the top and the bottom, respectively; from the corner chains one-quarter of each glucose residue in the eight corners contributes to the unit cell (the corner glucose residues are shared by each of the four unit cells meeting at the corners). (B, C) View onto the 0 4 0 plane of the unit cell of cellulose I (B) and cellulose II (C): package of the (simplified) glucan chains.

D. Degree of Crystallinity of Cellulose

Cellulose aggregates are composed of highly ordered, crystalline regions and less to non-crystalline or amorphous portions. The degree of crystallinity (state of order, crystallinity index) varies from 70–80% for cotton, over 60–70% for wood, to 40–50% for regenerated cellulose (Fengel and Wegener, 1984). The imperfect crystal structure of cellulose is already indicated by its diffuse X-ray pattern (Knolle and Jayme, 1965), from which by background subtraction the degree of crystallinity can be calculated (Schurz et al., 1987; Knolle and Jayme, 1965). Other methods for the determination of the crystallinity of celluloses are IR-spectroscopy (Ferrús and Pagés, 1977), solid state carbon-13 NMR (Sterk et al., 1987), dark-field electron diffraction (Bourret et al., 1972) and deuterium exchange studies (Fengel and Wegener, 1984; Krässig, 1985).

The calculated dimensions of the highly ordered regions (crystallites) vary from 2.4 nm (wood), 5–6 nm (cotton), 7.5 nm (Acetobacter) up to 14 nm (Valonia) width and 10 nm (regenerated cellulose) up to 100 nm (Valonia) length (Fengel and Wegener, 1984; Haigler, 1985; Delmer, 1987).

The amorphous regions of cellulose are more susceptible to heterogeneous hydrolysis with dilute mineral acids. By such a treatment the crystalline regions with a certain DP, called levelling-off-DP (LDP or DP_L), depending on the dimensions of the crystallites, are left. DP_L values of 200–350 for cotton, 150–300 for wood and 15–50 for regenerated cellulose have been reported (Fengel and Wegener, 1984) corresponding to theoretical crystallite lengths of 7.5 nm at least and 175 nm at most, assuming that the glucan chains within the crystallites are unfolded. The length of hydrolysis-resistant particles of cotton cellulose measured by electron microscopy, was about 40 nm (Krässig, 1976). Hydrolysis results in a remarkable increase in the degree of crystallinity from 60–70% to more than 90% accompanied by only minor losses of weight (10%) (Jayme and Raffael, 1969).

For the distribution of crystalline and amorphous regions in cellulose fibrils, various models have been proposed (Schafizadeh and McGinnis 1971; Fengel and Wegener, 1984; Haigler, 1985). The basic concepts are: (a) individual fibrils, in which continuous chains pass through sequences of crystalline and amorphous regions; (b) longitudinally arranged glucan chains changing between individual fibrils from one crystalline region to an adjacent one, thereby producing less ordered regions at the transition areas and at the chain ends (fringed micellar concept); (c) fibrillar units of folded glucan chains with amorphous regions at the turning points.

The more recently developed 'fringe fibrilar model' (Hearle, 1963), which seems to agree best with the physical and chemical properties of cellulose, describes a network of elementary fibrils and their aggregations. The fibrils are composed of strands of elementary crystallites, which are held together along the fibre axis by glucan chains slipping from one crystallite into another, whereas the lateral cohesion between crystallites in the fibril aggregations is mainly due to secondary valence bonds.

Beside the above-mentioned amorphous regions, the fibrillar elements are surrounded by a sheath of less ordered cellulose molecules (Krässig, 1985). These two areas of lower order seem to be responsible for the physical and chemical characteristics of cellulose in heterogeneous systems. Chemical and physical processes such as adsorption, exchange or substitution seem to start on the less crystalline surface of the elementary crystallites or fibrils and in the interlinking regions (Krässig, 1985).

The fibrils measured by electron microscopy are not uniform in diameter, thus giving rise to contradictory opinions. Summarising, the term elementary fibril is usually reserved for 3.5 nm-wide fibrillar units (Mühlethaler, 1965; Heyn, 1969; Fengel and Wegener, 1984; Haigler, 1985). The elementary fibrils seem to aggregate to form microfibrils of varying diameter (10–30 nm), and microfibrils may be joined to larger 60–360 nm wide macrofibrils (discussed in Fengel and Wegener, 1984; Haigler, 1985; Krässig, 1985; Delmer, 1987). Sub-elementary fibrils of 1–3 nm, however, have also been detected in primary walls of maize coleoptiles and cambial cell walls of woods, some algae and bacteria (Fengel and Wegener, 1984). Recently by high-resolution Pt–C replication it was shown that 3.68 nm microfibrils can be composed of three 1.78 nm sub-microfibrils twisted in a left-handed fashion around the fibril axis (Ruben and Bokelman, 1987). In this model lateral fasciation along corresponding crystallographic planes of microfibrils to form thicker fibrils is not feasible.

V. CELLULOSE DEGRADATION

Biomass degradation and utilisation of the organic substances or energy fixed therein has been one of the very active areas of research during the last decade (Sahm, 1979). Only 1–2% of the biomass which is produced per year is actually utilised as foodstuff, for the pulp and paper industry or as a source for organic compounds. Since cellulose is the most plentiful organic compound of the annually produced biomass, most of the research in this field is concentrated on general cellulose utilisation and primarily on cellulose degradation. In order to convert the cellulose polymer to other useful materials, it must first be degraded to low molecular weight entities, ideally monomeric sugar units. In theory, polysaccharide conversion can be obtained by either chemical or enzymatic methods, but the latter way is preferred owing to the highly selective mode of action, thereby eliminating the formation of unwanted by-products.

A. Enzymatic Degradation

In recent years some excellent reviews have appeared which exclusively cover the field of cellulolysis, i.e. enzymatic degradation of this polymer by cellulases (Wilke *et al.*, 1983; Sahm, 1985; Finch and Roberts, 1985; Rollings, 1985; Wood, 1985; Young and Rowell, 1986; Schurz, 1986; Fan *et al.*, 1987).

The term cellulase refers to the group of enzymes that contributes to the degradation of cellulose to its monomer glucose. In most cellulolytic organisms several cellulase components form a cellulase complex which synergistically hydrolyses the polymer substrate. Microorganisms, producing enzymes capable of attacking even modified cellulose, are widely found in nature, whereas enzymes which can catalyse the breakdown of closely packed native cellulose fibrils are more restricted.

Cellulose originating from highly lignified tissues, such as wood or straw, are not ideal substrates, since the less crystalline regions are rich in lignin and hemicelluloses. The presence of these matrix substances has a retarding effect on the rate of enzyme attack, i.e. the cellulase activity.

Cellulose-degrading microorganisms include bacteria, actinomycetes and higher fungi. The latter are mostly utilised for both general studies about cellulose degradation and industrial utilisation in biotechnological processes. An excellent overview of these microorganisms and their specific abilities to degrade cellulosic materials is given by Fan *et al.* (1987). Most of the recent work which has been done with cellulases originating from three fungi, i.e. *Trichoderma reesii* (*T. viride*), *Trichoderma konigii* and *Sporotrichum pulverulentum*. Mainly from the wild-type of *T. reesii*, a series of mutants has been developed. Of particular interest today are the strains of QM-9414 and Rut-C-30 (Wilke *et al.*, 1983).

It is generally accepted that at least three enzymes are involved in the enzymatic degradation of cellulose (Sahm, 1985).

1. Endo-β-1,4-glucanases which split the native macromolecular chains, producing shorter units with a lower degree of crystallinity and a series of new chain ends (EC 3.2.1.91).
2. Exo-β-1,4-glucanases (cellobiohydrolases) which act from the non-reducing end of the shorter chains by continuously producing cellobiose in a synergistic way with the endoglucanase (EC 3.2.1.4).
3. β-1,4-glucosidases which are responsible for the hydrolysis of cellobiose to glucose (EC 3.2.1.21).

Most of the organisms produce a series of different hydrolytic endo- and exoenzymes. However, it is still uncertain whether these are real isoenzymes or enzyme modifications produced by protease activities (Sahm, 1985). A general scheme for the fractionation of cellulases, which includes a series of chromatographic procedures was proposed by Wilke *et al.* (1983).

The rate and degree of enzymatic hydrolysis of cellulose is dependent on the effective cellulose surface exposed to the enzyme in the heterogeneous cellulose/cellulase system. A pre-treatment, in order to enhance the amorphous regions and the surface area of the substrate, is considered to be essential (Fan *et al.*, 1987). The pre-treatments may be physical methods, such as pyrolysis, high-pressure steaming, milling and grinding, or chemical methods, such as treatment with alkali, acids or oxidising agents.

In the case of insoluble substrates, loss in weight or production of reducing sugar units is normally used to determine cellulase activity. Change of turbidity of a cellulose suspension or loss in tensile strength are methods which are no longer used. Enzymic attack on soluble cellulose derivatives can be measured by an increase of reducing sugar content or by a reduction in viscosity of the solution. Soluble substrates are more readily attacked by the enzymes, and consequently the estimation of enzyme activity can be carried out much faster.

An important limitation to the practical usage of cellulases is the product inhibition which may be competitive or non-competitive (Moriyama and Saida, 1986). The degradation products, cellobiose and glucose, are inhibitory even at low concentrations. They, therefore, should be eliminated from the reaction, mainly in large-scale reactions, where ultrafiltration membranes can be used successfully (Ohlson *et al.*, 1984).

B. Degradation by Acids

Acid hydrolysis as an alternative to enzymatic breakdown has long been studied for analytical as well as industrial purposes. It is well known that cellulose poses special problems in the case of acid hydrolysis due to the physical and chemical properties of cellulose (see Section IV). Physical pre-treatment such as grinding, milling or steam-explosion is essential in order to allow optimal accessibility of the acid to the fibrous material.

From the chemical viewpoint, cellulose is homogeneous, but in the biological environment, i.e. plant tissues, it is always accompanied by non-cellulosic materials. It is obvious that all these non-cellulosic materials have to be eliminated by non-destructive methods (see Section III.B).

For industrial purposes the so-called autohydrolysis procedure is of increasing importance. Autohydrolysis prior to acid hydrolysis is a steam-cracking followed by a rapid quenching, a process suitable for the elimination of hemicelluloses. In addition, much of the lignin becomes soluble afterwards in dilute alkali or in organic solvents. Extraction of lignin results in a relatively pure cellulose for hydrolysis (Wayman, 1985).

The classical cellulose hydrolysis is carried out by the Saeman method (Saeman et al., 1945) with 72% H_2SO_4. Hydrochloric acid (41%) and H_3PO_4 (85%) can be used in the same way (Wayman, 1986). The action of the three concentrated acids is to dissolve the α-cellulose at low temperatures (4–20°C). Complete hydrolysis to the corresponding monomer is then done by a careful dilution of the acids to a 3–6% concentration and subsequent heating at 100–120°C for a period of 30–360 min. Increase in temperature and pressure enhance the rate of hydrolysis and the yield of glucose (Fagan, 1971). The higher the temperature, the more effective is the hydrolysis. A detailed description of the various parameters for this important process is given by Fan et al. (1987).

The hydrolysis of cellulose with HCl in the presence of different cations has been studied in detail by Bayat-Makovi and Goldstein (1985). At elevated HCl concentrations (40%) in the presence of certain cations (Li^+, Ca^{2+} and Zn^{2+}), cellulose hydrolysis proceeds to completion under less stringent conditions. The glucose liberated in this way shows much better stability than in the case of H_2SO_4 hydrolysis. With 12 M HCl in the presence of Li^+ at 59°C, complete hydrolysis occurs in less than 10 min. Hydrolysis at weaker HCl concentrations (20%) is possible, but increases in salt concentrations and temperature are required to maintain the rate and extent of hydrolysis.

Recently cellulose hydrolysis using hydrofluoric acid has been proposed (Fan et al., 1987), whereby wood cellulose is treated with anhydrous hydrogen fluoride for 60 min in a vacuum distillation apparatus. The reaction yields glucose fluoride, which subsequently is converted into glucose and hydrogen fluoride via a reaction with water. The acid can be removed and re-used. By dilution with water, the water-soluble sugars are obtained, and the insoluble lignin fraction is precipitated. Sugar yields are reported to range from 45–95% of the theoretical value. However, during this process some of the sugars recombine to form oligomers, which have to be re-hydrolysed by dilute sulphuric acid.

VI. THE BIOSYNTHESIS OF CELLULOSE

It is important to realise that the mechanism by which cellulose is synthesised is still very badly understood. In this context only some relevant problems and results can be discussed. For further details recently published reviews are recommended (Colvin, 1980; Delmer, 1983, 1987; Maclachlan, 1983; Haigler, 1985).

There is general agreement that in most algae and in higher plants cellulose synthesis takes place at the plasma membrane by membrane-integrated and/or -associated enzyme complexes commonly called 'cellulose synthases'. Considerable efforts have been made to achieve *in vitro* synthesis of cellulose with (more or less purified) membrane preparations from various organisms (reviewed by Delmer, 1987). UDP-glucose is considered to be the glucosyl donor for cellulose synthesis. Membrane preparations usually are incubated in the presence of various concentrations of UDP-^{14}C-glucose in buffers optionally supplemented with divalent cations such as Mg^{2+} or Ca^{2+}, β-linked disaccharides and detergents.

One major problem is the precise product analysis (Franz *et al.*, 1983; Delmer, 1983, 1987; Blaschek *et al.*, 1983). In the past the inadequate realisation of this demand led to misinterpretations of numerous *in vitro* studies. The problem is due to the fact that membrane preparations, in addition to cellulose synthases, may contain enzymes catalysing the synthesis of β-1,3-glucans, mixed linkage β-1,3/1,4-glucans, short chain β-1,4-glucans and xyloglucan backbones. Product analysis therefore should include:

(a) linkage determination by methylation analysis (evaluated by radio-GLC) (Franz *et al.*, 1983; Blaschek *et al.*, 1983; Harris *et al.*, 1984; Fink *et al.*, 1987); periodate oxidation (cleaving 1,4-, but not 1,3-linkages) (Goldstein *et al.*, 1965; Heiniger and Delmer, 1977), and digestion with specific glucan hydrolases (being, however, seldom pure; Delmer, 1987);

(b) DP-determination of the *in vitro*., synthesised glucans (e.g. by viscosity measurements of tri-nitrate derivatives in the presence of unlabelled carrier cellulose (Blaschek *et al.*, 1982, 1983);

(c) verification of the cellulosic nature of the product by electron microscopy and X-ray diffraction (Lin *et al.*, 1985; by definition cellulose is the association of β-1,4-glucans in a certain crystalline arrangement).

In higher plants and fungi, almost all attempts of *in vitro* production of cellulose resulted either in the formation of β-1,3-glucans or in only a very limited synthesis of β-1,4-glucans without the proof of characteristic crystallinity. These β-1,4-glucans particulary are built at low ($< 50 \mu M$) UDP-glucose concentrations in the presence of Mg^{2+} (reviewed by Delmer, 1987).

It is remarkable that callose synthesis is usually latent in intact cells producing at the same time large amounts of cellulose. After perturbation of the physiological equilibrium (e.g. for membrane isolation), the cellulose synthesis stops, while callose formation is initiated. Thus, one enzyme tends to be active when the other is inactive (Delmer, 1987; Kauss, 1987). However, it is not known how these two glucan synthase activities are controlled. Either some unknown effectors activate or suppress alternately each of these enzymes or there exists only one glucosyl transferase, which can be switched e.g. by conformation changes between the formation of both linkage types depending on the requirements of a cell. As β-1,3-glucan synthesis is more easily achieved by *in vitro* assays, 'callose synthase' is better characterised. This enzyme is markedly activated at low UDP-glucose levels ($< 100 \mu M$), by micromolar Ca^{2+} and by millimolar β-linked disaccharide concentrations (Eigner *et al.*, 1987; Hayashi *et al.*, 1987). The digitonin-solubilised enzyme forms soluble β-1,3-glucans. Magnesium in addition to Ca^{2+} and β-glucoside stimulates the production of high M_w, alkali-insoluble, fibrillar β-1,3-glucans probably due to conformation changes of the enzyme (Hayashi *et al.*, 1987). The digitonin-solubilised enzyme resembles in size the terminal complexes seen in plasma membranes of plants, which are thought to be responsible for cellulose biosynthesis (Hayashi *et al.*, 1987).

The possible involvement of activated glucolipids and glucoproteins in the process of cellulose biosynthesis has been proposed (Franz, 1978; Hopp *et al.*, 1978; Blaschek *et al.*, 1985), but definitive evidence is lacking. There also exist indications for a need for primers (Blaschek *et al.*, 1983) as well as membrane potentials (Delmer *et al.*, 1982) in glucan synthesis. Probably the conditions for a successful *in vitro* synthesis of cellulose include membrane preparations prepared by more gentle methods than those used until now, in order to avoid the disintegration of the synthase complexes and losses of activators and primers, and the establishment of a membrane potential and precisely controlled concentrations of divalent cations and substrates.

Indications for these requirements come from the more successful work with *Acetobacter xylinum* (Ross *et al.*, 1986, 1987). Membranes of this bacterium prepared in the presence of PEG can be stimulated by GTP to synthesise large amounts of β-1,4-glucans with UDP-glucose as substrate. The specific procedure of membrane preparation prevented the loss of a protein factor, which could be identified to be a guanyl cyclase converting two molecules of GTP to bis-(3′,5′)-cyclic diguanylic acid. This newly detected compound acts as an activator of cellulose synthase, and a Ca^{2+}-inhibited membrane-bound enzyme system can degrade the activator in a regulatory system. All attempts, however, to utilise this system for higher plants have failed (Delmer, 1987). The mechanism of cellulose biosynthesis in prokaryotic and eukaryotic organisms seems to proceed in different ways.

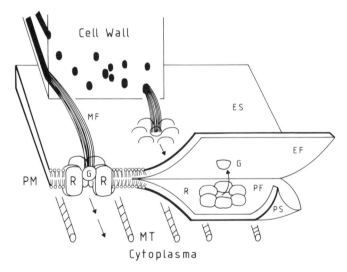

FIG. 8.5. Model of the cellulose synthase complexes of plants: the membrane faces of the plasma membrane (PM) revealed by freeze-etching are: protoplasmic surface (PS) and face (PF) as well as exoplasmic surface (ES) and face (EF). Rosettes (R) and globules (G) probably form complexes, which are thought to be involved in the biosynthesis of about 36 glucan chains associating into a microfibril (MF). The direction of microfibril deposition may be controlled by microtubules (MT).

Looking at the putative cellulose-synthesising complexes, there also seem to exist differences in eukaryotic organisms. These complexes can be made visible by the freeze etching technique in electron microscopy (Giddings *et al.*, 1980; Delmer, 1987). By this method the membrane lipid bilayer is split revealing two additional fracture faces termed the exoplasmic (EF) and the protoplasmic (PF) faces besides the exoplasmic (ES) and the protoplasmic (PS) surfaces (Fig. 8.5). In algae [e.g. *Oocystis* (Robinson and Quader, 1981) or *Valonia* (Itoh and Brown, 1984)], with large microfibrils of high crystallinity, linear terminal complexes are seen which usually fracture with the EF, but perhaps may be transmembrane proteins (Itoh and Brown, 1984). The number of subunits in these complexes is more or less in agreement with the calculated number of glucan chains in the microfibrils of these algae (Itoh *et al.*, 1984; Brown, 1987b). In other algae [e.g. *Micrasterias* (Giddings *et al.*, 1980) or *Spirogyra* (Herth, 1983)] or lower plants [e.g. *Funaria* (Reiss *et al.*, 1984) or *Adiantum* (Wada and Staehelin, 1981)] and higher plants [e.g. *Glycine* (Herth and Weber, 1984), *Avena* (Mueller and Brown, 1980) or *Lepidium* (Herth, 1985)], possessing microfibrils with elementary fibril characteristics, rosettes and/or globule-type complexes are found (see Fig. 8.5). The rosettes may be hexagonal arrays of particles found in the PF. Occasionally by double-replica fracturing single particles, called globules, as well as microfibril impressions, are seen in the complementary EF. There exist many indications that these structures are involved in cellulose biosynthesis, since they are often restricted to cell areas with high rates of cellulose biosynthesis (Herth, 1985). The ultimate proof, however, that these complexes are identical with the glucan synthases, is missing (Delmer, 1987).

For a long time it was a matter of discussion as to how the aggregation of glucan chains into elementary fibrils or microfibrils and ultimately the DP of genuine cellulose macromolecules is controlled. By experiments with dyes associating with β-1,4-glucans (e.g. Calcofluor White or Congo Red), the crystallisation of nascent chains could be prevented by intercalation of these compounds, while the rate of glucan synthesis increased (Haigler *et al.*, 1980; Benziman *et al.*, 1980; Herth, 1985). From these experiments it was concluded that the polymerisation step in biosynthesis is enzymatically controlled, whereas the subsequent crystallisation (the rate-limiting step) seems to be a physical process of self-association, in which the size of the fibrils produced might be regulated by the number of cooperating synthesising enzymes in an enzyme complex (of terminal globules or rosettes). The life-time of active rosettes is probably limited (Schnepf *et al.*, 1985; Rudolph and Schnepf, 1988). Perhaps the disintegration and reorganisation of such complexes controls the DP of cellulose, or at least the rate of cellulose formation. Changes in the organisation of such complexes also might be responsible for the generation of alternating crystalline and amorphous regions in a microfibril.

Except in the primary wall of higher plants, the deposition of microfibrils is oriented. In cellulose-synthesising cells, microtubules are often oriented parallel to the direction of the recently formed cellulose layer (Robinson and Quader, 1982). Distribution of microtubule organisation (e.g. by colchicine) results in a random deposition of microfibrils (Robinson and Quader, 1981, 1982). Therefore, microtubules might play an important role in the oriented deposition of cellulose in various cell-wall layers. According to one of various models (Robinson and Quader, 1981; Mueller and Brown, 1982; Brown, 1985), the synthases swimming in the plasma membrane are pushed forward by the process of polymerisation and the microtubules canalise the direction of movement.

VII. CELLULOSE DERIVATIVES

A. General Importance

The investigation and use of cellulose derivatives for a wide range of industrial applications has fascinated many researchers for many years. The almost unlimited supply of cheap cellulosic raw material is still an encouragement for academic research and industrial innovations.

The area 'cellulose derivatives' was opened up in 1846 by Schönbein, who was the first to synthesise cellulose nitrate. Based upon these findings the first plastic material was created by mixing cellulose nitrate with camphor to give a product which was named celluloid. At the same time cellulose acetate was discovered, and at the turn of the century, cellulose was converted to xanthate which could be regenerated to fibrous material. Considerable efforts to discover new cellulose derivatives have been made ever since and various products have been modified for the changing needs of industrial applications.

The field of cellulose derivatives has been reviewed very frequently during the last decade. (Wadsworth and Daponte, 1985; Nicholson and Meritt, 1985; Balser, 1985; Kennedy *et al.*, 1985; Ekman *et al.*, 1986; Reuben, 1986; Klausmeier, 1986). It should be expected that this field of cellulose research due to industrial demand will considerably increase in the future, since there are still many possibilities for applications.

The largest proportion of cellulose used for chemical modifications is obtained from wood pulp and from cotton (linters). However, the purest cellulose, which in the future may be used for derivatisation, comes from bacterial origin. As has been pointed out by Brown (1987a) microbial cellulose might be the best cellulose factory of the future.

Cellulose has three reactive hydroxyl groups per anhydroglucose repeating unit that form inter- and intramolecular hydrogen bonds. The bonds strongly influence the chemical reactivity and solubility of cellulose. The successful preparation of derivatives, mostly ethers and esters, requires a special approach due to their insolubility in almost all common solvent systems. All derivatisation reactions have to be carried out in heterogeneous systems. After an appropriate activation of the reactive groups, new reaction centres are created, which results in a complete reactivity of the fibrous material and the formation of a homogeneous reaction system.

B. Cellulose Esters

The alcoholic groups of cellulose undergo esterification with acids in the presence of dehydrating agents or by reaction with acid chlorides. The reaction mechanism consists in a formation of an oxonium ion of cellulose, followed by a nucleophilic substitution.

$$\text{Cell—OH} + \text{H}^+ \rightleftharpoons \text{Cell—O}^+ \overset{\text{H}}{\underset{\text{H}}{}}$$

$$\text{X}^- + \text{Cell—O}^+\overset{\text{H}}{\underset{\text{H}}{}} \rightleftharpoons \left[\text{X}^- \longrightarrow \text{Cell} \longrightarrow \text{O}^+\overset{\text{H}}{\underset{\text{H}}{}}\right] \rightleftharpoons \text{X—Cell} + \text{H}_2\text{O}$$

The resulting cellulose esters differ in their physical and chemical properties from the original polysaccharide. As a consequence they are soluble in a wide range of solvents. By controlling the degree of substitution (DS), a whole variety of derivatives, differing in their physico-chemical properties, can be obtained. In principle, esterification is possible with all inorganic and organic acids, but only a few have reached economic importance (Table 8.3).

TABLE 8.3. Cellulose esters and their utilisation (Revely, 1985).

Product	Property	Applications
Nitrate	Film formation	Smokeless powders
	Plasticiser	Propellants
	DP-determination	Membranes
Acetate	Film formation	Fibres
	Spinnable	Plastics
	Plasticiser	Films
	DP-determination	Textiles
Mixed esters	Plasticiser	Coatings
Viscose	Film formation	Fibres
	Spinnable	Textiles
	DP-determination	Sponges
		Cellophane

The most important of the inorganic esters is cellulose nitrate. Cellulose sulphate, cellulose phosphate and cellulose nitrite have almost no technical importance.

1. Cellulose nitrate

The importance of this compound stems from its ease of production with a high yield, its solubility in different organic solvent systems, and the film-forming capacity of its solutions.

Since the esterification is an equilibrium reaction, even with water-free nitric acid only a partly derivatised product (DS 2) can be obtained. For a higher degree of substitution (DS 3) the addition of sulphuric acid, phosphoric acid or acetic anhydride is necessary. The optimal molar ratio of the reactants is as follows: H_2SO_4–HNO_3–H_2O; 2:1:1 (v/v) (Balser, 1985).

In the industrial process, cellulose and the nitrating mixture are fed simultaneously into the reaction vessel. After nitration the cellulose derivative is washed with progressively weaker aqueous acid and finally with alkaline water. Solubility of the cellulose nitrate depends on the DS of the product (Balser, 1985; see Table 8.4).

TABLE 8.4. Solubility criteria of cellulose nitrates (Balser, 1985).

Degree of substitution	Solvent
1.8–2.0	Aromatic hydrocarbons
2.0–2.2	Ethanol, methanol, esters
2.2–2.3	Esters, ketones
2.3–2.8	Acetone

Conventionally, nitrocellulose prepared from cotton linters has applications in the explosives industry. Also other celluloses such as ramie and wood pulp are used for nitration, but purified cotton cellulose is still the preferred substrate.

Nitrocellulose forms complexes with a variety of organic compounds such as camphor (Celluloid®), ethylacetanilide, alkylphthalate and cyclohexanone. These complexes are in part used as plastic materials.

2. Cellulose acetate

Since the esterification of cellulose occurs in heterogeneous systems, a pre-swelling of cellulose is very important prior to the reaction. For acetylation, cellulose might be activated by a pre-treatment with acetic acid. This reduces intramolecular hydrogen bonding, and allows a better diffusion of the reactants. The addition of sulphuric acid as a catalyst serves to reduce the molecular weight of cellulose, resulting in a higher degree of esterification. The acetylating agent, together with the catalyst and a solvent, which is normally acetic acid or methylene chloride, are mixed in a reactor, which then is controlled on a temperature profile. This results in a viscous solution of cellulose triacetate, which can be separated to lower the DS. Precipitation in water or diluted acetic acid is then carried out to separate the cellulose ester from the liquid phase. The amorphous material is collected, washed and dried.

As in the case of nitrocellulose, the DS of cellulose acetate determines the solubility criteria. With decreasing degree of acetylation the solubility in polar solvents increases (Balser, 1985).

3. Mixed cellulose esters

Esters of cellulose with propionic acid or butyric acid are difficult to produce and have no advantage over cellulose acetate (Balser, 1985). However, the mixed esters of cellulose with acetic acid and propionic or butyric acid have some importance mainly in the plastics industry. The synthesis is carried out in a homogeneous system, in which the reactivity of the aliphatic fatty acids decreases with increasing chain length. The common cellulose acetate–propionate contains 5% of bound acetic acid (DS 0.25) and 58% propionic acid (DS 2.4) (Balser, 1985). On the other hand, the cellulose acetate–butyrate contains varying amounts of both acetic acid (2–45%) and butyric acid (25–65%). The different acid substitution patterns strongly influence the specific properties of cellulose derivatives. The utilisation of these compounds is comparable to the industrial application of cellulose acetates.

C. Cellulose Ethers

According to Felcht (1985), cellulose ethers are amongst the most important commercially produced water-soluble polymers (see Table 8.5). This concerns the quantity produced, the variety of applications and the technological requirements in production as well. Cellulose ethers in general are water-soluble polymers, mostly used as thickeners of aqueous media, which can be classified according to: (a) different types of substituents; (b) degree of substitution; (c) varying viscosities; (d) physical behaviour; (e) dissolving behaviour.

The most important products are: MC = methyl cellulose, EC = ethyl cellulose, PC = propyl cellulose, and CMC = carboxymethyl cellulose.

The formation of ether derivatives is comparable to ester formation, i.e. the oxonium intermediate is subsequently substituted according to the following reaction sequence.

$$\text{Cell—OH} + \text{H}^+ \rightleftharpoons \text{Cell—}\overset{\text{H}}{\underset{\text{H}}{\text{O}}}{}^+$$

$$\overset{R}{\underset{R}{\text{O}}} + \text{Cell—}\overset{\text{H}}{\underset{\text{H}}{\text{O}}}{}^+ \rightleftharpoons \left[\overset{R}{\underset{R}{\text{O}}}\!\!\longrightarrow\!\!\text{Cell}\longrightarrow\overset{\text{H}}{\underset{\text{H}}{\text{O}}}{}^+\right] \xrightarrow{-\text{HOH}} +\overset{R}{\underset{R}{\text{O}}}\!\!-\!\!\text{Cell} \xrightarrow{-\text{H}^+} \text{R—O—Cell}$$

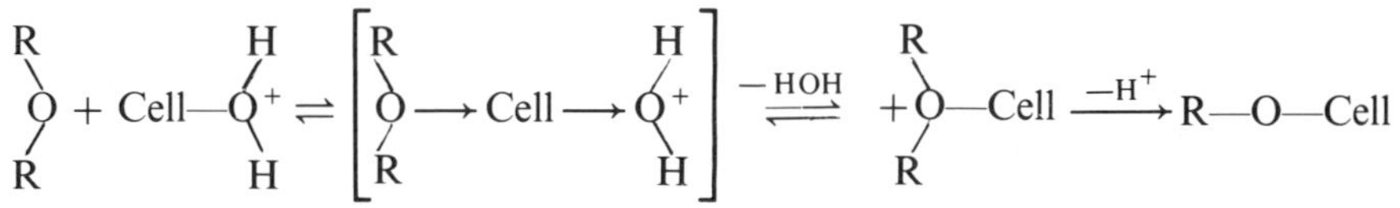

Since the OH groups of cellulose are only slightly acidic and react with the alcoholic component to only a minor degree, this mechanism results in an insufficient DS. For this reason normally pre-activated alkali-cellulose is used which leads to a higher degree of ether substituents.

$$\text{Cell—OH} + \text{NaOH} + \text{Hal—R} \longrightarrow \text{Cell—OR} + \text{NaHal} + \text{H}_2\text{O}$$

In the case of methylcellulose, the reaction is as follows:

$$\text{Cell—OH} + \text{NaOH} + \text{CH}_3\text{Cl} \longrightarrow \text{Cell—OCH}_3 + \text{NaCl} + \text{H}_2\text{O}$$

The usual DS range obtained by this reaction is 1.2–2.3. The crude cellulose ether obtained is purified by extraction of the residual salts and other products. Since commercial application of pure methylcellulose is very rare, a mixed etherification is normally carried out by adding ethylenoxide or propylenoxide to the methylation reaction, resulting in methylhydroxyethyl-cellulose or methylhydroxypropyl-cellulose, respectively. Hydroxyethyl-cellulose, which is the second important non-ionic cellulose ether, is prepared in a similar type of reaction.

By far the most important ionic cellulose ether is carboxymethyl-cellulose. The reaction is very easy and routinely carried out by reaction of alkali-cellulose with monochloracetic acid. In order to obtain a high degree of uniformly substituted products, the process is carried out in isopropanol. The commercial products have a DS range of 0.5–1.2.

TABLE 8.5. Solubility criteria of cellulose ethers.

Product	Degree of substitution-dependent solubility characteristics		
	4% NaOH	H$_2$O	Organic solvents
Methylcellulose (MC)	0.4–0.6	1.3–2.6	2.5–3.0
Ethylcellulose (EC)	0.5–0.7	0.8–1.3	2.3–2.6
Hydroxyethylcellulose (HEC)	0.5	0.5–1.0	
Carboxymethylcellulose (CMC)	0.5	0.5–1.2	

All the cellulose ethers are very stable products, which, due to their solubility in aqueous solvents, possess many interesting physico-chemical characteristics (Balser, 1985). Most important is the ability of these compounds to produce aqueous solutions with high viscosity, in which the chain length of cellulose ranging from 40 to 2000 controls the viscosity. The water retention capability of these hydrocolloids increases in parallel with increasing viscosity. This specific property is the basis for utilisation of these compounds as emulsifying and suspending agents in pharmaceutical and food industries.

D. Regenerated Cellulose

1. Cuprammonium process (CUOXAM)

In the cuprammonium process, linters or high quality wood pulp is dissolved in a copper complex: $Cu(NH_3)_4^{2+}OH_2^{2-}$. The method varies slightly if the formation of cell-glass (Cellophane®) or copper silk is required. To prepare copper silk, cellulose is dissolved in a 25% ammonia solution with 40% copper sulphate and 8% NaOH. After filtering and degassing the solution should be stored in the absence of light and air. This process is easier to carry out than the viscose method but is more expensive. To prepare cell-glass (cellulose hydrate films), it is necessary to use higher concentrations of cellulose. Otherwise the freshly prepared film would contain too much solvent, thereby changing the physical properties of the cellulose derivative.

2. Viscose method

In the viscose method, cellulose is converted into fibres or films via the xanthate (Balser, 1985):

$$Cell\text{—}OH \xrightarrow{NaOH} Cell\text{—}ONa \xrightarrow{CS_2} Cell\text{—}O\text{—}\overset{\displaystyle S}{\underset{\displaystyle S^-Na^+}{C}} \xrightarrow{H_2SO_4} Cell\text{—}OH + CS_2 + NaHSO_4$$

As in the CUOXAM method, the cellulose concentration of the reaction has to be higher for the production of films than for fibres. Sulphite wood pulp or linters are used as cellulose raw material.

In this process cellulose is first converted into alkali cellulose by a treatment with caustic soda, which is readily soluble in CS_2. The xanthate degree is low at a 10% alkali concentration. A higher degree of xanthate formation can be obtained by an 18–20% NaOH pre-treatment. By this alkali treatment, the pulp or linters is freed from residual hemicelluloses. Xanthate is a water-soluble cellulose derivative which, however, is hardly used due to its instability and high viscosity. Commercial xanthates contain 27% cellulose, 14% NaOH, 8% CS_2 and 51% water. To produce viscose silk (Rayon®), the alkaline solution is acidified by a treatment with 12% H_2SO_4 and 23% Na_2SO_4. The resulting viscose is washed to remove the acid, bleached with NaOCl or H_2O_2 and dried.

In contrast to genuine cotton, the viscose fibre has a strongly diminished stability, due to the lower molecular weight and the poorer orientation of the cellulose macro-molecular chains.

VIII. CONCLUSION

The more that contributions from different scientific disciplines add to our knowledge of cellulose, the better we shall be able to utilise this natural source for different technological needs. Not only shall we know better how to process and apply conventional sources of celluloses such as vegetable fibres, cotton and wood, but also it may be possible in the near future to utilise directly highly pure cellulose from microbial sources such as *Acetobacter xylinum* (Brown, 1987a). This may become very important, e.g. for cellulosic materials used in haemodialysis, where traces of non-cellulosic impurities can be the cause of severe side-effects. For the production of pharmaceutical cellulose derivatives such highly pure cellulose is needed.

A greater understanding of the supramolecular and morphological structure of cellulose, which decisively influence its chemical behaviour, will be necessary in order to develop further cellulose derivatisation products. For example, the search for new cellulose solvents allowing the substitution of cellulose in homogeneous reaction processes may result in some novel types of cellulose derivatives with better controlled, uniform substitution patterns and interestingly different properties.

As well as promoting the direct use of cellulose and its derivatives, the hydrolytic processes of cellulose should be further developed in order to compete with starch hydrolysis products. From an economic point of view, the products of a cellulose-based chemistry at the moment cannot compete with petrochemistry. The cost of the starting material such as wood and crop residues is not the limiting factor. There is still a need for more economic technologies in order to obtain cellulosic material of the high purity which is essential for both derivatisation and breakdown products.

ACKNOWLEDGEMENT

Part of the work was carried out at the CNRS-CERMAV, University of Grenoble, where one of us (G. F.) had the privilege to be active as invited professor. We are grateful for the encouraging discussions with J. P. Joseleau, Professor at the above-mentioned department.

REFERENCES

Achwal, W. B. and Chaugule, D. A. (1975). *Angew. Makromol. Chemie* **42**, 77–89.
Adams, G. A. (1965). *In* "Methods in Carbohydrate Chemistry" (R. L. Whistler, ed.), pp. 269–275. Academic Press, New York.
Aho, W. O. (1983). *In* "Process in Biomass Conversion" (G. L. Tillman and E. C. Jahn, eds), pp. 149–181. Academic Press, New York.
Albersheim, P. (1974). *In* "Tissue Culture and Plant Science" (H. E. Street, ed.), pp. 379–400. Academic Press, London.
Alexander, W. J. and Mitchell, F. H. (1949). *Anal. Chem.* **21**, 1497–1500.

318 G. FRANZ AND W. BLASCHEK

Aronson, J. M. (1965). *In* "The Fungi" (G. C. Ainsworth and A. S. Sussman, eds), pp. 49–55. Academic Press, New York.

Asamizu, T., Tanaka, K., Takebe, J. and Nishi, A. (1977). *Physiol. Plant.* **40**, 215–218.

Balser, K. (1985). *In* "Polysaccharide, Eigenschaften und Nutzung" (E. Burchard, ed.), pp. 84–110. Springer, Berlin, Heidelberg, New York and Tokyo.

Bayat-Makori, F. and Goldstein, J. S. (1985). *In* "Cellulose and its Derivatives" (J. F. Kennedy, A. O. Phillips, D. J. Wedlock and P. A. Williams, eds), pp. 135–142. Ellis Horwood, Chichester.

Beasly, C. A., Ting, I. P., Linkins, A. E., Birnbaum, E. H. and Delmer, D. P. (1974). *In* "Tissue Culture and Plant Science" (H. E. Street, ed.), pp. 169–192. Academic Press, London.

Benziman, M., Haigler, C. H., Brown, R. M. Jr, White, A. R. and Cooper, K. M. (1980). *Proc. Natl. Acad. Sci. USA* **77**, 6678–6682.

Blackwell, J. and Marchessault, R. H. (1971). *In* "Cellulose and Cellulose Derivatives" (N. M. Bikales and L. Segal, eds), pp. 1–37. Wiley-Interscience, New York.

Blackwell, J., Kolpak, F. J. and Gardner, H. (1977). *In* "Cellulose Chemistry and Technology" (J. C. Arthur, ed.), pp. 42–55. Adv. Chem. Series Symp. Series No. 48, American Chemical Society, Washington.

Blakeney, A. B., Harris, P. J., Henry, R. J. and Stone, B. A. (1983). *Carbohydr. Res.* **113**, 291–299.

Blaschek, W., Haaß, D., Koehler, H. and Franz, G. (1981). *Plant Sci. Lett.* **22**, 47–57.

Blaschek, W., Koehler, H., Semler, U. and Franz, G. (1982). *Planta* **154**, 550–555.

Blaschek, W., Haaß, D., Koehler, H., Semler, U. and Franz, G. (1983). *Z. Pflanzenphysiol* **111**, 357–364.

Blaschek, W., Semler, U. and Franz, G. (1985). *J. Plant Physiol.* **120**, 457–470.

Blumenkrantz, N. and Asboe-Hansen, G. (1973). *Anal. Biochem.* **54**, 484–489.

Bourret, A., Chanzy, H. and Lazaro, R. (1972). *Biopolymers* **11**, 893–898.

Brown, R. M. Jr (1982). "Cellulose and Other Natural Polymer Systems". Plenum Press, New York and London.

Brown, R. M. Jr (1985). *J. Cell Sci.* (Suppl. 2), 13–32.

Brown, R. M. Jr (1987a). *Tappi Proc.* 2.

Brown, R. M. Jr (1987b). *Food Hydrocolloids* **1**, 345–351.

Brown, R. M. Jr. and Ley, H. A. (1968). *Science* **147**, 1038.

Browning, B. L. (1967). "Methods of Wood Chemistry", Vol. II. Interscience, New York and London.

Burchard, W. (1985). "Polysaccharide, Eigenschaften und Nutzing". Springer, Heidelberg, New York and Tokyo.

Cantow, M. J. R. (1967). *In* "Polymer Fractionation" (M. J. R. Cantow, ed.), pp. 461–466. Academic Press, New York and London.

Cartier, N. (1986). Thèse d'état: Les polysaccharides de la paroi primaire des cellules de Rubus fruticosus cultivée en suspension. Université de Grenoble.

Chanzy, H. and Henrissat, B. (1985). *FEBS Lett.* **184**, 285–288.

Claesson, S., Bergmann, W. and Jayme, G. (1959). *Svensk Papperstid.* **62**, 141–155.

Clermont, L. P. and Bender, F. (1961). *Pulp. Pap. Mag. Can.* **62**, No. 1, T 28.

Colvin, J. R. (1980). *In* "Plant Biochemistry" (J. Priess, ed.), Vol. III, pp. 543–570. Academic Press, New York.

Cook, G. J. (1960). "Handbook of Textile Fibres". W. S. Cowell, Ipswich, UK.

Cross, C. F. and Bevan, E. J. (1912). "Researches on Cellulose", Vol. III. Longmans, Green, Co., London.

Darvill, A. G., McNeil, M., Albersheim, P. and Delmer, D. P. (1980a). *In* "The Biochemistry of Plants" (N. E. Tolbert, ed.), Vol. I, pp. 91–161. Academic Press, New York.

Darvill, A. G., McNeil, M., Albersheim, P. and Tolbert, N. E. (1980b). "The Plant Cell", pp. 91–162. Academic Press, New York.

Delmer, D. P. (1983). *Adv. Carbohydr. Chem. Biochem.* **41**, 105–153.

Delmer, D. P. (1987). *Ann. Rev. Plant Physiol.* **38**, 259–290.

Delmer, D. P., Benziman, M. and Padan, E. (1982). *Proc. Natl. Acad. Sci. USA* **79**, 5282–5286.

Dische, Z. (1962). *In* "Methods in Carbohydrate Chemistry" (R. L. Whistler and M. L. Wolfrom, eds), Vol. 1, pp. 475–514. Academic Press, New York.

Eigner, W. D., Billiani, J. and Huber, A. (1987). *Das Papier* **12**, 680–684.

Ekman, K., Eklund, U., Fors, J., Huttunen, J. I., Selin, J. F. and Turunen, O. T. (1986). *In* "Cellulose: Structure, Modification and Hydrolysis" (R. A. Young and R. M. Rowell, eds), pp. 131–148. John Wiley, New York.

Elliot, J. H. (1967). *In* "Polymer Fractionation" (M. J. R. Cantow, ed.), pp. 67–93. Academic Press, New York and London.

Fagan, R. D. (1971). *Environ. Sci. Technol.* **5**, 545–549.

Fan, L. T., Gharpuray, M. M. and Lee, Y. H. (1987). *In* "Cellulose Hydrolysis". Springer, Berlin, Heidelberg, New York, London, Paris and Tokyo.

Felcht, U. H. (1985). *In* "Cellulose and its Derivatives" (J. F. Kennedy, G. O. Phillips, D. J. Wedlock and P. A. Williams, eds), pp. 273–284. Ellis Horwood, Chichester.

Fengel, D. and Wegener, G. (1979). *In* "Hydrolysis of Cellulose: Mechanism of Enzymatic and Acid Catalysis" (J. R. D. Brown and L. Jurasek, eds), pp. 145–158. Adv. Chem. Ser. No. 181, American Chemical Society, Washington.

Fengel, D. and Wegener, G. (1984). "Wood: Chemistry, Ultrastructure, Reactions". Walter de Gruyter, Berlin and New York.

Ferrús, L. and Pagés, P. (1977). *Cell. Chem. Technol.* **11**, 633–637.

Finch, P. and Roberts, S. C. (1985). *In* "Cellulose Chemistry and its Applications" (T. P. Nevell and S. H. Zeronian, eds), pp. 312–343. Ellis Horwood, Chichester.

Fink, J., Jeblick, W., Blaschek, W. and Kauss, H. (1987). *Planta* **171**, 130–135.

Fisher, D. G. and Mann, J. (1960). *J. Polymer Sci.* **42**, 189–194.

Forest, G. J. and Bendall, D. S. (1969). *Biochem. J.* **113**, 741–755.

Franz, G. (1972). *Planta* **102**, 334–347.

Franz, G. (1978). *Appl. Polymer Symp.* **28**, 611–621.

Franz, G. and Blaschek, W. (1985). *In* "The Physiological Properties of Plant Protoplasts" (P. E. Pilet, ed.), pp. 171–183. Springer, Berlin and Heidelberg.

Franz, G., Blaschek, W., Haaß, D. and Koehler, H. (1983). *J. Appl. Polymer Sci.: Appl. Polymer Symp.* **37**, 145–155.

French, A. D. (1985). *In* "Cellulose Chemistry and its Applications" (T. P. Nevell and S. H. Zeronian, eds), pp. 84–111. Ellis Horwood, Chichester.

Fry, S. C. (1988). "The Growing Plant Cell Wall: Chemical and Metabolic Analysis". Longman, Harlow, Essex.

Gardner, K. H. and Blackwell, J. (1974). *Biopolymers* **13**, 1975–2001.

Giddings, T. H. Jr, Brower, D. L. and Staehelin, L. A. (1980). *J. Cell. Biol.* **84**, 327–339.

Goebel, K. D., Harvie, C. E. and Brant, D. A. (1976). *Appl. Polymer Symp.* **28**, 671–691.

Goldstein, I. J., Hay, G. W., Lewis, B. A. and Smith. F. (1965). *In* "Methods of Carbohydrate Chemistry" (R. C. Whistler and J. N. BeMiller, eds), Vol. V, pp. 361–370. Academic Press, New York.

Goring, D. A. I. and Timell, T. E. (1962). *Tappi* **45**, 454–460.

Haigler, C. H. (1985). *In* "Cellulose Chemistry and its Applications" (T. P. Nevell and S. H. Zeronian, eds), pp. 30–83. Ellis Horwood, Chichester.

Haigler, C. H., Brown, R. M. Jr and Benziman, M. (1980). *Science* **210**, 903–906.

Harris, P., Henry, R. J., Blakeny, A. B. and Stone, A. B. (1984). *Carbohydr. Res.* **127**, 59–73.

Hashimoto, A. G. (1986). *Biotechnol. Bioeng.* **28**, 1857–1865.

Haworth, W. N. (1932). *Helv. Chim. Acta* **11**, 534–539.

Hayashi, T., Read, S. M., Bussell, J., Thelen, M., Lin, F. C., Brown, M. Jr and Delmer, D. P. (1987). *Plant Physiol.* **83**, 1054–1062.

Hearle, J. W. S. (1963). *J. Appl. Sci.* **7**, 1175–1192.

Heiniger, U. and Delmer, D. P. (1977). *Plant Physiol.* **59**, 719–723.

Herth, W. (1983). *Planta* **159**, 347–356.

Herth, W. (1985). *Planta* **164**, 12–21.

Herth, W. and Weber, G. (1984). *Naturwissenschaften* **71**, 153–154.

Heyn, A. N. J. (1969). *J. Ultrastruct. Res.* **26**, 52–68.

Hieta, K., Kuga, S. and Usuda, M. (1984). *Biopolymers* **23**, 1807–1810.

Hopp, H. E., Ronsero, P. A. and Pont Lezica, R. (1978). *FEBS Lett.* **86**, 259–262.

Houwink, R. H. (1940). *J. Prakt. Chem.* **265**, 15–18.

Huwyler, H. R., Franz, G. and Meier, H. (1979). *Planta* **146**, 635–642.
Itoh, T. and Brown, R. M. Jr (1984). *Planta* **160**, 372–381.
Itoh, T., O'Neil, R. M. and Brown, R. M. Jr (1984). *Protoplasma* **123**, 174–183.
Jayme, G. (1971). In "Cellulose and Cellulose Derivatives" (N. M. Bikales and L. Segal, eds), pp. 381–410. Wiley-Interscience, New York, London, Sydney and Toronto.
Jayme, G. (1978). *Papier* **32**, 145–149.
Jayme, G. and Raffael, E. (1969). *Papier* **23**, 1–7.
Johnson, D. B., Moore, W. F. and Zankl, C. (1961). *Tappi* **44**, 793–798.
Kauss, H. (1987). *Ann. Rev. Plant. Physiol.* **38**, 47–72.
Kennedy, J. F., Phillips, G. O., Wedlock, D. J. and Williams, P. A. (1985). "Cellulose and its Derivatives". Ellis Horwood, Chichester.
Kivirikko, K. J. and Liesmaa, M. (1959). *Scand. J. Clin. Lab. Invest.* **11**, 128–133.
Klausmeier, W. H. (1986). In "Cellulose: Structure, Modification and Hydrolysis" (R. A. Young and R. M. Rowell, eds), pp. 187–201. John Wiley, New York.
Knolle, H. and Jayme, G. (1965). *Papier* **19**, 106–110.
Kolpak, F. J., Blackwell, J. and Litt, M. H. (1977). *J. Polymer Sci.* **15**, 655–658.
Kotera, A. (1967). In "Polymer Fractionation" (M. J. R. Cantow, ed.), pp. 43–66. Academic Press, New York and London.
Krässig, H. (1976). *Appl. Polymer Symp.* **28**, 777–790.
Krässig, H. (1985). In "Cellulose and its Derivatives: Chemistry, Biochemistry and Applications" (J. F. Kennedy, G. O. Phillips, D. J. Wedlock and P. A. Williams, eds), pp. 3–25. Ellis Horwood, Chichester.
Kristen, U. (1985). In "Progress in Botany", Vol. 47, pp. 1–18. Springer, Berlin and Heidelberg.
Kürschner, K. and Popik, M. G. (1962). *Holzforschung* **16**, 1–11.
Lauriol, J. M., Comtat, J., Froment, P., Pla, F. and Robert, A. (1987). *Holzforschung* **41**, 165–169.
Layne, E. (1957). *Methods Enzymol.* **111**, 447–454.
Leopold, B. (1961). *Tappi* **44**, 230–235.
Lin, F. C., Brown, R. M. Jr, Cooper, J. B. and Delmer, D. P. (1985). *Science* **230**, 822–825.
Maclachlan, G. (1983). In "New Frontiers in Plant Biochemistry" (T. Asahi and H. Imasaki, eds), pp. 83–91. Japanese Science Society Press, Tokyo.
Maclachlan, G. (1987). *Food Hydrocolloids* **1**, 365–369.
Marchessault, R. H. and Liang, C. Y. (1960). *J. Polymer Sci.* **43**, 71–84.
Marchessault, R. H. and Sarko, A. (1967). *Adv. Carbohydr. Chem.* **22**, 421–482.
Marx-Figini, M. (1962). *Macromol. Chem.* **54**, 107–118.
Marx-Figini, M. and Schulz, G. V. (1966). *Naturwissenschaften* **53**, 466–474.
Meier, H., Buchs, L., Buchala, A. J. and Homewood, T. (1981). *Nature* **289**, 821–822.
Merz, W. (1968). *Z. Anal. Chem.* **237**, 272–279.
Meyer, K. H. and Mark, H. F. (1928). *Berichte* **61**, 593–612.
Meyer, K. H. and Mark, H. F. (1929). *Z. Phys. Chem.* **B2**, 115–145.
Meyer, K. H. and Misch, L. (1937). *Helv. Chim. Acta* **20**, 232–244.
Minor, J. L. (1979). *J. Liq. Chromatogr.* **2**, 309–318.
Moriyama, J. and Saida, T. (1986). In "Cellulose: Structure, Modification and Hydrolysis" (R. A. Young and R. M. Rowell, eds), pp. 323–336. John Wiley and Sons, New York, Chichester and Toronto.
Mühlethaler, K. (1965). In "Cellular Ultrastructure of Woody Plants" (W. A. Côté, ed.), pp. 191–198. Syracuse University Press, Syracuse, NY.
Mueller, S. C. and Brown, R. M. Jr (1980). *J. Cell Biol.* **84**, 315–326.
Mueller, S. C. and Brown, R. M. Jr (1982). *Planta* **154**, 489–515.
Nevell, T. P. and Zeronian, S. H. (eds) (1985). "Cellulose Chemistry and its Applications". Ellis Horwood, Chichester.
Nicholson, M. D. and Merrit, F. M. (1985). In "Cellulose Chemistry and its Applications" (T. P. Nevell and S. H. Zeronian, eds), pp. 363–383. Ellis Horwood, Chichester.
Northcote, D. H. (1965). *Methods Carbohydr. Chem.* **5**, 201–210.
Ohlson, J., Träggardh, G. and Hahn-Hägerdahl, B. (1984). *Biotechnol. Bioeng.* **26**, 647–659.
Okajima, K. (1978/79). *Chemica Scripta* **13**, 102–107; 113–119; 120–122.

O'Neill, M. A. and Selvendran, R. R. (1980). *Biochem. J.* **187**, 53–63.

Palma, A., Büldt, G. and Jovanović, S. M. (1976). *Macromol. Chem.* **177**, 1063–1072.

Payen, A. (1838). *Compt. Rend.* **7**, 1052.

Polanyi, M. (1921). *Naturwissenschaften* **9**, 288–292.

Preston, R. D. (1981). *In* "Xylem Cell Development" (J. R. Barnett, ed.), pp. 1–13. Academic Press, London.

Preston, R. D. (1986). *In* "Cellulose: Structure, Modification and Hydrolysis" (R. A. Young and R. M. Rowell, eds), pp. 3–27. John Wiley and Sons, New York.

Quenin (1985). Précipitation du cellulose à partir de solutions dans les oxides d'amines. Thèse d'état, Université de Grenoble.

Read, S. M. and Northcote, D. H. (1981). *Anal. Biochem.* **116**, 53–64.

Reiss, H. D., Schnepf, E. and Herth, W. (1984). *Planta* **160**, 428–435.

Reuben, J. (1986). *In* "Cellulose: Structure, Modification and Hydrolysis" (R. A. Young and R. M. Rowell, eds), pp. 149–157. John Wiley and Sons, New York.

Revely, A. (1985). *In* "Cellulose and its Derivatives" (J. F. Kennedy, G. O. Phillips, D. J. Wedlock and P. A. Williams, eds), pp. 211–226. Ellis Horwood, Chichester.

Robinson, D. G. and Quader, H. (1981). *Eur. J. Cell. Biol.* **25**, 278–288.

Robinson, D. G. and Quader, H. (1982). *In* "The Cytoskeleton in Plant Growth and Development" (C. W. Lloyd, ed.), pp. 109–126. Academic Press, London.

Roelofsen, P. A. (1965). *Adv. Bot. Res.* **2**, 69–149.

Rollings, J. (1985). *Carbohydr. Polymers* **5**, 37–48.

Ross, P., Aloni, Y., Weinhouse, H., Michaeli, D., Weinberger-Ohana, P., Mayer, R. and Benziman, M. (1986). *Carbohydr. Res.* **149**, 101–117.

Ross, P., Weinhouse, H., Aloni, Y., Michaeli, D., Weinberger-Ohana, P., Mayer, R., Braun, S., deVroom, E., van der Marel, G. A., van Boom, J. H. and Benziman, M. (1987). *Nature* **325**, 279–281.

Ruben, G. C. and Bokelman, G. H. (1987). *Carbohydr. Res.* **160**, 434–443.

Rudolph, U. and Schnepf, E. (1988). *Protoplasma* **143**, 63–73.

Saeman, J. F., Buhl, J. L. and Harris, E. F. (1945). *Ind. Eng. Chem. Anal. Ed.* **17**, 35–37.

Saeman, J. F., Moore, W. E., Mitchell, R. L. and Millett, M. A. (1954). *Tappi* **37**, 336–342.

Sahm, H. (1979). *Forum Microbiol.* **2**, 177–185.

Sahm, H. (1985). *In* "Polysaccharide" (W. Burchard, ed.), pp. 54–65. Springer, Berlin, Heidelberg and New York.

Sarko, A. (1978). *Tappi* **61**, (No. 2), 59–61.

Sarko, A. (1985). *In* "New Developments in Industrial Polysaccharides" (V. Crescenzi, I. C. M. Dea and S. S. Stivala, eds), pp. 87–112. Gordon and Breach, New York, London, Paris, Montreux and Tokyo.

Sarko, A. (1986). *In* "Cellulose: Structure, Modification and Hydrolysis" (R. A. Young and R. M. Rowell, eds), pp. 29–49. J. Wiley and Sons, New York.

Schafizadeh, F. and McGinnes, G. D. (1971). *Adv. Carbohydr. Chem. Biochem.* **26**, 297–349.

Schnepf, E., Witte, O., Rudolph, U., Deichgräber, G. and Reiss, H.-D. (1985). *Protoplasma* **127**, 222–229.

Schulz, G. V. and Marx, M. (1954). *Macromol. Chem.* **14**, 52–95.

Schurz, J. (1986). *Holzforschung* **40**, 225–235.

Schurz, J., Jánosi, A. and Zipper, P. (1987). *Das Papier* **12**, 673–679.

Seifert, K. (1960). *Papier* **14**, 104–106.

Selvendran, R. R., Stevens, B. J. H. and O'Neill, M. A. (1985). *In* "Biochemistry of Plant Cell Walls" (C. T. Brett and J. R. Hillman, eds), pp. 39–78. Cambridge University Press, Cambridge.

Spencer, F. S. and Maclachlan, G. A. (1972). *Plant Physiol.* **49**, 58–63.

Staudinger, H. (1932). *In* "Die hochmolekularen organischen Verbindungen—Kautschuk und Cellulose". Springer, Berlin, Göttingen and Heidelberg.

Sterk, H., Sattler, W., Janosi, H., Paul, D. and Esterbauer, H. (1987). *Das Papier* **12**, 664–668.

Strub, A., Chartier, G. and Schleser, G. (1983). *In* "Energy from Biomass", 2nd Conference, pp. 122–209. Applied Science, London and New York.

Toriumi, J., Skrivartara, H. and Sano, K. (1972). *Acta Pathol. Japan* **22**, 591–599.

Updegraff, D. M. (1969). *Anal. Biochem.* **32**, 420–424.

Volk, R. J. and Weintraub, R. L. (1958). *Anal. Chem.* **30**, 1011–1019.

Wada, M. and Staehelin, L. A. (1981). *Planta* **151**, 462–468.

Wadsworth, L. C. and Daponte, D. (1985). *In* "Cellulose Chemistry and its Applications" (T. P. Nevell and S. H. Zeronian, eds), pp. 344–362. Ellis Horwood, Chichester.

Ward, K. Jr and Morak, A. J. (1962). *In* "Proceedings: Wood Chemistry Symposium", pp. 77–89. Butterworth, Montreal and London.

Wardrop, A. B. (1962). *Bot. Rev.* **28**, 241–285.

Wardrop, A. B. (1970). *Protoplasma* **70**, 73–77.

Warwicker, J. O. (1985). *In* "Cellulose and its Derivatives" (J. F. Kennedy, C. O. Phillips, K. J. Wedlock and P. A. Williams, eds), pp. 371–386. Ellis Horwood, Chichester.

Wayman, M. (1986). *In* "Cellulose: Structure, Modifications and Hydrolysis" (R. A. Young and R. M. Rowell, eds), pp. 265–279. John Wiley and Sons, New York.

Wilke, C. R., Maiorella, D., Sciamannia, A., Tangnu, T., Wiley, D. and Wong, H. (1983). "Enzymatic Hydrolysis of Cellulose". Noyen Data Corporation, Park Ridge, NY.

Wise, L. E., Murphy, M. and D'Addieco, A. A. (1946). *Pap. Trade J.* **122**, (No. 2), 35–43.

Wood, T. M. (1985). *In* "Cellulose and its Derivatives" (J. F. Kennedy, C. O. Phillips, D. J. Wedlock and P. A. Williams, eds), pp. 173–188. Ellis Horwood, Chichester.

Young, R. and Rowell, R. M. (eds) (1986). "Cellulose: Structure, Modification and Hydrolysis". John Wiley and Sons, New York.

Zugenmaier, P. (1985). *In* "Polysaccharide" (W. Burchard, ed.), pp. 260–279. Springer, Berlin, Heidelberg, New York and Tokyo.

9 Starch

WILLIAM R. MORRISON and JOHN KARKALAS

Department of Bioscience and Biotechnology, University of Strathclyde, James P. Todd Building, 131 Albion Street, Glasgow G1 1SD, UK

METHODS IN PLANT BIOCHEMISTRY Vol. 2
ISBN 0-12-461012-9

I. INTRODUCTION

Amylose (AM) and amylopectin (AP), the two major polysaccharides of which starch is composed, were originally characterised by mostly chemical and physical techniques (Whistler, 1964; Banks and Greenwood, 1975; French, 1975; Banks et al., 1979; Banks and Muir, 1980). Subsequent developments in chromatography, specific enzymic assays and physical chemistry have given new insights into the structural organisation and physical properties of starch granules, and the molecular architecture and rheological properties of AM and AP in solution. This chapter covers those developments of particular importance in plant biochemistry. With our better understanding of starch biochemistry (Manners, 1979, 1985a–d; Duprat et al., 1980; Nelson, 1980; Preiss and Levi, 1980; Preiss, 1982; Shannon and Garwood, 1984; Lineback, 1986) we are now moving from the relatively unsophisticated use of the major plant starches as raw materials for the food and other industries to using biotechnology to tailor starch precisely for particular end-uses, and the chapter concludes with some thoughts on this topic.

A. Occurrence of Starch

Starch, the major reserve polysaccharide of green plants, is widely distributed and has long been recognised by the characteristic microscopic appearance and iodine staining properties of the granules (Section II.B). Starch synthesised in amyloplasts and stored in the major depots of seeds, tubers and roots is very stable, it is a source of energy and carbon for the developing plant, and it is the principal food of many animals including man. Starch granules synthesised in amyloplasts, chloroplasts and chloroamyloplasts of other tissues generally have a transient existence or are present in trace amounts (e.g. in pollen), their shapes are less characteristic, and comparatively little is known about their composition and properties.

B. Granule Structures and Shapes

In starch granules the AM and AP molecules are radially oriented with their single reducing end-groups towards the centre or hilum, and synthesis is by apposition at the outer non-reducing ends (Nikuni, 1978; Blanshard, 1979; French, 1984). Small quantities of protein, identified as starch synthase, are bound within the granule (Shure et al., 1983; Sano et al., 1985; Villareal and Juliano, 1986; Schofield and Greenwell, 1987) and have been detected in tangential layers of wheat starch by fluorescent antibody staining (Skerritt, 1988). The other enzymes of starch synthesis (Section I.D) are readily detached from the granules, as is the amyloplast membrane. Normally AM and AP are accumulated together during granule development, but the percentage of AM increases in developing cereal and tuber starches (Williams and Duffus, 1977; Duffus and Murdoch, 1979; Shannon and Garwood, 1984; Morrison and Gadan, 1987), and there are gradients of AM, AP, lipids and crystallinity from the hilum to the periphery of the large A-type granules in wheat and in maize (Kassenbeck, 1978; Morrison and Gadan, 1987). Striations caused by alternating amorphous and crystalline zones may be visible by light microscopy, but they are seen best by scanning electron microscopy (SEM) of sections of granules partially digested with amylases or mineral acids which preferentially attack the amorphous regions (Evers and McDermott, 1970; Gallant *et al.*, 1972; Fuwa *et al.*, 1979; Yamaguchi *et al.*, 1979; Duprat *et al.*, 1980; Gallant and Bouchet, 1986; Galliard and Bowler, 1987). Similar zones on an even smaller scale have been revealed by high-magnification transmission electron microscopy (TEM) of thin sections (Kassenbeck, 1978; French, 1984; Gallant and Bouchet, 1986), and the smallest crystallites are of the same magnitude as the A- and B-chain segments of AP that are capable of forming crystalline parallel-stranded double helices (Section I.C). Ordering of crystallites within the starch granule causes optical birefringence (the Maltese Cross effect seen under the polarising microscope), and has been studied by X-ray, neutron, and light scattering (Section IV.A).

C. Amylose and Amylopectin

AM and AP are high molecular weight biopolymers that have many of the general properties of chemically synthesised polymers (Banks and Greenwood, 1975). Of these, the most important thing to bear in mind when reading this chapter is that neither is homogeneous. The sizes finally attained by individual primer molecules at the end of biosynthesis are determined by biochemical probability considerations so that both AM and AP cover a range of molecular weights with normal or skewed distributions, and number-averaged values are very different from weight-averaged values (Tables 9.1 and 9.2). This is discussed more fully in Section III.E.

AM has the properties of an essentially linear α-(1,4)-glucan, but in most starches one-third to two-thirds of the AM fraction has secondary chains attached through occasional α-(1,6) branch points, and in some species AM has a few phosphate groups, probably at C-6 of glucose residues (Table 9.1). The branch chains are usually moderately long, but a few may be as small as glucosyl to maltotetraosyl (G_1–G_4).

Using the chain nomenclature normally applied to AP (below), AM can be entirely C-chains (unbranched AM) or C-chains with occasional A-chains and possibly some B-chains (branched AM). Depending on source, AM has an average of 2–11 branch points and therefore 3–12 non-reducing chain ends per reducing end (Table 9.1), but the extent of branching will be greater in the branched fraction if the unbranched AM is taken into account (Takeda *et al.*, 1987b).

Digestion with β-amylase proceeds from the non-reducing ends of the glucan chains to within one to three glucose residues from α-(1,6) branch points or ester phosphate groups (Banks and Greenwood, 1975; Takeda *et al.*, 1983; Robyt, 1984; Takeda, 1987) leaving β-limit dextrins. AM leached from granules immersed in water just above their gelatinisation temperature (GT) has a lower molecular weight and higher β-amylolysis limit (90–100%) than AM leached at higher temperatures (β-amylolysis limit 70–80%) which is more extensively branched (Banks and Greenwood, 1975; Takeda *et al.*, 1984).

In dimethylsulphoxide (DMSO) and alkaline solutions AM probably has an expanded coil configuration, while in acidic or near neutral solution it is a more condensed random coil, possibly with short loose helical segments (Banks and Greenwood, 1975; Whistler and Daniel, 1984). Concentrated aqueous solutions gel readily, but at low concentrations AM retrogrades (crystallises) forming very insoluble crystals melting at 150–160°C (Stute and Konieczny-Janda, 1983). Retrogradation is particularly fast when chain length (CL) = 75–100 (Pfannenmüller *et al.*, 1971; Kodama *et al.*, 1978; Gidley *et al.*, 1986; Pfannemüller, 1986) and is retarded if crystallisation nuclei are

TABLE 9.1. Properties of some amyloses.[a]

Property	Chestnut	Kuzu	Lily	Maize	Nagaimo yam	Potato	Rice	Sweet potato	Tapioca	Water chestnut	Wheat
Iodine binding capacity (g 100 g^{-1})	19.9	20.0	20.0	20.0	19.9	20.5	20.0–21.1	20.2	20.0	20.2	19.9
[η] at 22.5°C in M KOH (ml g^{-1})	242	228	312	169		384	180–216	324	384		
β-amylolysis limit	86	76	89	84	86	80	73–84	73	75	95	82
DP$_w$, range	440–14900	480–12300	360–18900	390–13100	800–20000	840–21800	210–12900	840–19100	580–22400	160–8090	
DP$_w$, mean	4020	3220	5010	2550	6300	6360	2750–3320	5430	6680	4210	
DP$_n$, mean	1690	1540	2310	960	2000	4920	920–1110	4100	2660	800	1290
DP$_w$/DP$_n$	2.38	2.09	2.17	2.66	3.15	1.29	2.64–3.39	1.31	2.51	5.76	
CL	375	320	475	305	525	670	250–370	380	340	420	270
Chain number	4.6	4.8	4.9	3.1	3.8	7.3	2.5–4.3	11	7.8	1.9	4.8
Unbranched AM (mol. %)	66	47	61	52	71		69	30	58	89	73
P (μg g^{-1})	10	10	trace		2	3	None	6	7		

[a] From Hizukuri et al. (1983b, 1988), Hizukuri and Takagi (1984), Suzuki et al. (1981, 1986), C. Takeda et al. (1983, 1987), Y. Takeda et al. (1983, 1984, 1986a,b, 1987a).

TABLE 9.2. Properties of some amylopectins.[a]

Property	Chestnut	Kuzu	Maize	Potato	Rice			Sweet potato		Tapioca	Water chestnut
					Japonica	Indica	Waxy (jap.)	Normal	ML fraction		
Iodine binding capacity (g 100 g^{-1})	0.51	0	1.25		0.39–0.87	1.62–2.57		0.38–0.44	0.83–1.04		0.41
[η] at 22.5°C in M KOH (ml g^{-1})	176	160	168	224	134–137	150–170		175–193	187–200		
β-amylolysis limit (%)	55	57	60	55	58–59	56–59		55–56	57–58	57	59
DP$_n$ (10^3)			10.2		8.2–12.8	4.7–5.8					12.6
CL av.[b]	22.5	20.4–21.1	22	22.0–23.9	19–20	21–22	17.5–18.3	20.8–22.4	21.4–22.9	21.2	22
P (total) (μg g^{-1})	61	158	15–27	604	8–13	11–29		117–135	120–121		44
P (gluc-6-PO$_4$) (μg g^{-1})	43	121	3–5		8–13	9–28		95–115	100–102		24
Component chains,[c] DP$_n$ (wt %)											
B$_4$ F$_1$	140 (4)	119 (0.2)	91 136 (11)	104 (2.3)	120–180 (6–9)	85–130 (14–20)	101 (1)	77–162 (6–8)	210 (7)	115 (1)	130 (1)[d]
B$_3$ F$_2$	64 (9)	70 (1.5)	51 47 (19)	75 (9)	41–44 (17–19)	42–44 (19)	69 (4)	69–70 (5–8)	77 (7)	69 (5)	65 (6)
B$_2$ F$_3$	43 (25)	42 (9)	16 18 (70)	48 (26)	16–17 (74–75)	16–17 (61–67)	42 (19)	44–45 (20–23)	46 (21)	42 (23)	43 (22)
B$_1$ F$_4$	15 (62)	20 (42)		24 (35)			22 (26)	15 (62–67)	15 (65)	21 (32)	19 (44)
A		13 (47)		16 (28)			13 (50)			12 (39)	10 (24)
A:B molar ratio by GP-HPLC		0.89		0.79			2.2			0.89	0.82

[a] From Hizukuri (1986), Hizukuri et al. (1983a, 1988), Suzuki et al. (1981, 1985), C. Takeda et al. (1987), Y. Takeda et al. (1986b, 1987a, 1988a).

[b] By Smith degradation and by isoamylolysis.

[c] By isoamylase debranching and by GP-HPLC (fractions F$_1$–F$_4$) or HPLC (A, B$_1$–B$_4$). F$_1$ contains glucan-PO$_4$ with shorter CL than shown for normal chains. In papers showing bimodal distributions (lower column resolution) F$_1$ = B$_2$ + B$_3$ + B$_4$ and F$_2$ = A + B$_1$. Also 3% long-chain material included in F$_1$ of maize, rice and sweet potato.

removed by ultrafiltration. The A- and B-crystal polymorphs of AM (Sarko and Wu, 1978; Wu and Sarko, 1978a,b; Wild and Blanshard, 1986) give essentially the same X-ray diffraction patterns as the AP crystallites in native starches (Imberty et al., 1987a,b) described below.

In solutions containing suitable guest molecules segments of AM complex cooperatively to form single left-handed V-type helices with a hydrophobic cavity c. 0.5 nm in diameter. In I_2/KI solution the guest molecules are polyiodide ions, which, depending on solute concentrations, may be up to I_{13}^- but are mostly I_3^- or I_5^- (Teitelbaum et al., 1978; Knutson et al., 1982; Saenger, 1984; Moulik and Gupta, 1986; Yajima and Nishimura, 1987). At high concentrations some iodine appears to be bound outside the helix cavities. The colour and λ_{max} of the complexes vary with chain length and analytical conditions (see Sections III.C and III.E). Inclusion complexes with lower alcohols such as butan-1-ol and thymol are insoluble and heat-labile, and have been used to isolate AM (French et al., 1963; Whistler, 1964). Inclusion complexes with long-chain monoacyl lipids (e.g. fatty acids, soaps, monoglycerides and certain surfactants) are also insoluble, and dissociate on heating in water at 90–120°C (Morrison, 1988b; Raphaelides and Karkalas, 1988).

AP molecules are very large flattened disks in shape (Lelievre et al., 1986; Callaghan and Lelievre, 1987) consisting of α-(1,4)-glucan chains joined by numerous α-(1,6)-branch points (Table 9.2). There are typically five or six glucose residues between branch points in AP compared with two to four in glycogen (Banks and Greenwood, 1975; Manners, 1985a,b,d). The single reducing end-group is on a C-chain, from which all other chains ultimately depend (Peat et al., 1952). The A-chains carry no branch points, and are attached to the B-chains which have one or more branch points and are themselves attached to other B-chains or to the C-chain. Of the various models for the structure of AP proposed in recent years (French, 1972, 1975, 1984; Robin et al., 1974; Nikuni, 1978; Hizukuri, 1986) those shown in Fig. 9.1 seem most satisfactory. The A:B chain ratio, which has been the subject of some debate (Marshall, 1974; French, 1975; Manners 1979, 1985b,d; Atwell et al., 1980b; Manners and Matheson, 1981; Enevoldsen, 1985), is typically from 0.8 to 2.2 on a molar basis or from 0.4 to 1.0 on a weight basis (Hizukuri, 1986).

Linear chains in AP form red to purple polyiodide complexes (λ_{max} 530–585 nm) in $I_2/$KI solution, but normally do not complex appreciably with alcohols or monoacyl lipids. AM impurities (1–2%) can increase λ_{max} by 20–30 nm. Chains released by debranching (see Section III.E) can be resolved by gel permeation chromatography (GPC, Section III.F) into two to four populations (Table 9.2), and the shortest CL fraction may correspond to the A-chains in the parent molecule (Hizukuri, 1986). Chains carrying phosphate groups have been distinguished (Suzuki et al., 1981; Takeda et al., 1983).

There are three atypical types of AP which bind more iodine and give higher λ_{max} values than normal AP, so that they give erroneous values in the iodometric determination of AM in starches (Section III.C). In amylomaize the first type is high molecular weight AP with A and B chains 5–15 glucosyl residues longer than normal, and the second is a comparatively small AP molecule (anomalous AP) with similarly extended A and B chains that elute from GPC columns with AM (Baba et al., 1987a,b). The third type, found in maize and high-AM rices, is an otherwise normal AP in which there are extra-long chains (CL 85–180) with infrequent branching (Hizukuri, 1986; Takeda et al., 1987a, 1988a). The first two types also occur in several legume and tuber starches

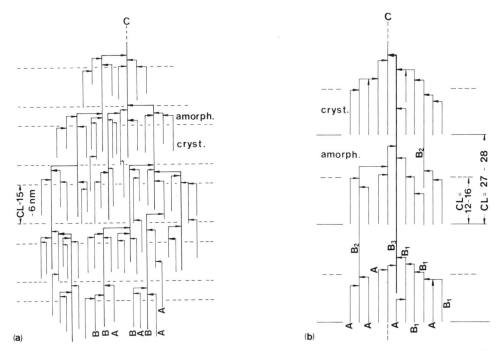

FIG. 9.1. Models for the arrangement of linear α-(1,4)-glucan chains in AP joined through α-(1,6)-glycosidic bonds (arrowheads). Left: redrawn from Robin *et al.* (1974) showing amorphous branched regions and potentially crystalline clusters of A- and B-chains; C = reducing end of single C-chain. Right: redrawn from Hizukuri (1986) showing A- and B₁-, B₂- and B₃-chains, but not superlong B₄-chains (cf. Table 9.2).

(Banks and Greenwood, 1975), and the third type has been reported in sweet potato starches (Takeda *et al.*, 1986b).

The branched regions of AP are amorphous, but the short linear chains beyond branch-points can form left-handed parallel-stranded double helices (Imberty *et al.*, 1987a,b), similar to those described for crystalline AM. The crystallinity, gelatinisation and swelling properties of starch granules are largely determined by their AP. Native granules exhibit characteristic X-ray diffraction patterns determined by CL distribution (Hizukuri, 1985), helix packing and water of crystallisation. The A-pattern is given by waxy and normal cereal starches and the B-pattern by amylomaize, root, and tuber starches, while the C-pattern of a few other starches is essentially an intermediate stage or mixture (Zobel, 1964, 1988a; Nikuni, 1978). Starches giving the B-pattern convert to the A-pattern if the plants are grown at higher temperatures or if the isolated starches are partially dehydrated (Nikuni, 1978).

D. Synthesis and Degradation of Starch

The biochemistry of starch synthesis and degradation is a major topic (Manners, 1979, 1985a,c; Duprat *et al.*, 1980; Nelson, 1980; Preiss and Levi, 1980; Preiss, 1982; Robyt, 1984) and only the salient features are discussed here. All pathways use glucose, from the action of pyrophosphorylase on sucrose and ATP or UTP, or from glycolysis. Plant phosphorylases can catalyse the transfer of glucose to glucose 1-phosphate to give

α-(1,4)-glucans, but it is now thought that they act primarily in the reverse direction to degrade glucans.

AM and AP chains are made by two types of starch synthase which catalyse the irreversible transfer of glucose to suitable acceptors such as maltosaccharides. Insoluble granule-bound starch synthase utilises UDP-glucose or ADP-glucose and makes long chains for AM, while the soluble type of starch synthase uses only ADP-glucose and makes chains for AP. Branching enzymes, closely linked with soluble synthase, cleave chains at specific points and transfer the free reducing ends (carbon-1) to carbon-6 of glucose residues in other chains to create α-(1,6)-linked branch chains with free non-reducing ends for AP. Presumably branching enzymes with specificity for much longer chains (sometimes also for much shorter chains) introduce occasional branching into some AM molecules.

Starch polysaccharides are degraded by many enzymes *in vivo*. In plants it is probable that the initial attack on native granules is always initiated by α-amylases (Preiss, 1982) and complete degradation then results with the combined actions of the other enzymes (Table 9.3). The optimum pH for most of these enzymes is about 5.0–6.5 (Fogarty and Kelly, 1979; Fogarty, 1983). The analytical applications and limitations of these enzymes are discussed in Section III.E.

II. THE INTACT STARCH GRANULE

A. Isolation and Purification

Anhydrous starch has a relative density of 1.55–1.60 and air-equilibrated starch (9–15% moisture, wet basis) has a density of *c.* 1.4 which decreases to 1.35 when fully hydrated at *c.* 30–35% moisture, corresponding to a volume increase of 70% (Dengate *et al.*, 1978; Donovan, 1979). Damaged granules absorb two to three times their weight of water, and if present in a sample they lower the hydrated density and alter the granule sedimentation characteristics. Hydrated granules are moderately permeable to small molecules which, if retained, are artifacts. Because of their high density, starch granules break through the amyloplast envelope on centrifugation and co-sediment with high-density proteinaceous components of macerated tissue. Smaller particles sediment much more slowly than large particles of the same density (Stoke's law), hence very small starch granules are invariably most contaminated with proteinaceous matter or are discarded with it when separating large granules, and great care is necessary to obtain quantitative recoveries of large and small granules together.

Intact amyloplasts (envelope plus starch granules) have been isolated from protoplasts of maize, sycamore and soybean (MacDonald and apRees, 1983; Echeverria *et al.*, 1985; Macherel *et al.*, 1985), potato tuber (Fishwick and Wright, 1980) and immature wheat (Entwhistle *et al.*, 1988). Cellulases and pectinases are used to degrade cell walls, and protoplasts are released by gentle maceration and sedimentation under gravity through a dense medium (e.g. sucrose, Ficol, Percoll, Nicodenz) to remove less dense cellular material. Sedimentation of granules through dense non-aqueous solvents can be used if water-soluble enzymes and metabolites are to be retained with the granules (Liu and Shannon, 1981a,b; Rijven, 1984).

Isolation of starch granules alone is less difficult, and the degree of purity and

TABLE 9.3. Types of enzymes that hydrolyse starch polysaccharides, and their final products[a].

Enzyme Commission (EC) number	Common name	Action	Glycosidic bond specificity	Final products
2.4.1.1	Phosphorylase	exo	α-(1,4)-glucosyl	Glucose 1-phosphate
3.2.1.1	α-Amylase(i)	endo	α-(1,4)-glucosyl	Linear and branched dextrins
	α-Amylase(ii)	endo	α-(1,4)-glucosyl[b]	Glucose or specific maltosaccharides (G_2–G_6)
3.2.1.2	β-Amylase	exo	α-(1,4)-glucosyl	Maltose and β-limit dextrins
3.2.1.3	Glucoamylase (amyloglucosidase)	exo	α-(1,4)- and α-(1,6)-glucosyl	Glucose
3.2.1.41	Pullulanase	endo	α-(1,6) with interval $> G_3$	Linear α-(1,4)-glucan chains, not glucose
3.2.1.68	Isoamylase	endo	α-(1,6) with	Linear α-(1,4)-glucan chains, not glucose or maltose
3.2.1.20	α-Glucosidase	exo	α-(1,2), α-(1,3) and α-(1,4)	Glucose

[a] From Manners (1985b), Fogarty and Kelly (1979), Fogarty (1983), Robyt (1984), and Fuwa et al. (1980).
[b] Bacterial α-(1,4)-glucanohydrolases with $\alpha(1,6) \rightarrow \alpha(1,4)$ transferase activity.
[c] Also 1 mol glucose from each odd-numbered linear chain (Banks and Greenwood, 1968).

precautions required during isolation depend on the analyses to be carried out. The problems encountered are most complex with cereal endosperm starches (McDonald and Stark, 1988; Morrison, 1988a). Milling of dry tissues to reduce particle size causes mechanical damage to granules and makes them very susceptible to attack by amylases. Cereal grains and other hard seeds should therefore be crushed gently to facilitate penetration of water so that the tissue can be softened; cautious disintegration will then release starch granules without undue breakdown of cellular material, most of which can be removed by passing the starchy slurry through a sieve of suitable aperture.

To prevent amylolysis during steeping, tissue can be brought to pH 2 briefly, and then adjusted to neutrality before proceeding (Meredith, 1970; Morrison et al., 1984). Alternatively, acetate buffer (pH 6.5) containing 0.01 M $HgCl_2$ may be used to inactivate enzymes (Adkins and Greenwood, 1966), and this will also prevent microbial growth, as does 0.01% thiomersal. However, $HgCl_2$ could precipitate proteins onto granule surfaces, and residues of thiomersal in the final starch will partially decolorise amylose–polyiodide complexes and interfere with iodometric assays (Section III.C).

It is always advisable to work at 2–4°C to minimise residual enzyme activity and, particularly with cereal starches, to minimise absorption of free fatty acids and other surface lipid artifacts (Morrison, 1981; Morrison et al., 1984). Steeping in warm water (5–10°C below gelatinisation temperature) causes annealing or reorganisation of AP crystallites so that starch gelatinisation characteristics are altered (Lorenz and Kulp, 1982; French, 1984; Krueger et al., 1987), and other undesirable changes are also likely to occur. If proteases are used to facilitate removal of protein the enzyme should be passed first through a column of starch to adsorb any amylases which may be present (McDonald and Stark, 1988). Protease treatment removes granule surface proteins but does not affect the integral proteins (Section III.H). Polyphenoloxidase in commercial proteases may discolour the starch unless sulphite is added to the steep liquor.

On the smallest scale described for isolating starch from single kernels of wheat and barley (Morrison and Gadan, 1987; South and Morrison, 1990), a dilute suspension of macerated tissue is layered above 80% CsCl and centrifuged at 38 000 × g to obtain a sediment of pure starch. Passage through CsCl removes some surface proteins from wheat starches, and the rest can be removed by washing with 1–2% sodium dodecylsulphate (SDS) at 20°C or with sodium laurate (Section III.H). Shaking aqueous slurries of starch with toluene to remove proteins (Adkins and Greenwood, 1966) is effective but tedious, and small granules are easily lost by entrainment in emulsified material at the toluene–water interface.

Finally, starches should be washed thoroughly with water to remove solutes. If the starch is rinsed twice with acetone and air-dried a free-flowing powder suitable for size analysis is obtained (Morrison and Scott, 1986), but freeze-drying can alter crystallinity and gelatinisation characteristics (Ahmed and Lelievre, 1978). Air-drying is slow and safe, but the granules are liable to clump together and they will be damaged even by mild grinding and sieving.

Granules can be separated into size fractions by sedimentation in water (Decker and Höller, 1962; Meredith et al., 1978; Cluskey et al., 1980; Morrison and Gadan, 1987). Spherical granules obey Stoke's law and diameter is inversely proportional to the square root of sedimentation time, but asymmetric granules (e.g. wheat A-type) sediment faster than predicted. For granules sedimenting through 15–18 cm water, suitable time intervals are 0.25, 0.5, 1, 1.5, 2, 5 and 16 h. Granules have also been fractionated by non-equilibrium centrifugation through layers of sucrose solution of increasing density (Williams and Duffus, 1977; Duffus and Murdoch, 1979).

Starch granules are superficially robust structures, but in practice they can be damaged quite easily in ways that affect some of their properties. In the dry state merely brushing through a fine mesh wire sieve causes measurable damage, which can be quite severe after vigorous rubbing, shearing or dry milling. Prolonged magnetic stirring of a suspension of large potato granules can also cause considerable damage. On hydration damaged granules swell two- or three-fold and appear much larger under the microscope, and they are much more susceptible to enzymic attack (Section III.I). Physical damage to wet granules causes increased cold leaching of AM, while damage to dry granules produces small fragments of AP molecules (Stark and Yin, 1986; Yin and Stark, 1988).

B. Microscopy

Classical light microscopy of starch in tissue sections and of isolated starch is still occasionally useful (Reichert, 1913; Badenhuizen, 1969, 1971; Czaja, 1971, 1978, 1980; Moss, 1976; Bechtel, 1983), and the presence of damaged granules is conveniently detected by their ability to absorb dye from very dilute solutions of Congo Red (Evers and Stevens, 1985) or Hessian Bordeau (R. G. Fulcher, pers. commun.). Normal granules stain blue-black with dilute I_2/KI (Section III.C), while waxy granules containing little or no AM stain purple, but they sometimes appear darker and this test is not reliable. Native granules and granules stained with I_2/KI viewed in polarised light show ordered and amorphous regions (Blanshard, 1987; Chandrashekar et al., 1987).

SEM can provide invaluable information on the shapes of granules, heterogeneity of populations, the presence of protein bodies and other foreign materials with similar particle sizes, and on the sites of amylolytic digestion. Care is necessary to avoid artifacts with native granules and on drying partially gelatinised granules (Bowler et al., 1987). For this reason, some authors advocate freeze-etch SEM (Leonard and Sterling, 1972; Chabot et al., 1978). TEM of starch granules requires much skill because the embedded granules have to be sectioned and handled in non-aqueous solvents to avoid hydration and swelling, and they readily fold and crease under the cutting knife to give prominent artifacts (Gallant and Bouchet, 1986). Acid hydrolysis of amorphous regions and selective oxidation in ultrathin sections have been used to reveal the distribution of crystalline regions in starch granules (Kassenbeck, 1978; Yamaguchi et al., 1979) and to detect crystallites as small as 5–6 nm which appear to correspond to the external chains in AP models (Fig. 9.1) that form double helices.

C. Size Analysis

Stereology, the technique of quantitative three-dimensional light microscopy, has been used to measure starch granules in sections of wheat (Briarty et al., 1979) but it is exceptionally tedious. Image-analysis techniques overestimate the dimensions of isolated non-spherical granules which will always tend to lie flat on the microscope slide so that only their longer axes are seen. Obtaining size distributions by image analysis is less satisfactory than by Coulter and laser techniques because it is nearly impossible to prepare random fields from very heterogeneous populations, and many fields have to be counted.

Coulter counters and similar instruments measure granule volumes, and other dimensions are then calculated assuming that the granules are spherical. Instruments counting in 16 or 32 channels (size ranges) give useful data (Mäkela et al., 1982; Mäkela and Laasko, 1984; Snyder, 1984), and modern instruments counting 10^5 to 10^6 granules divided over 100 or 256 channels are eminently suitable (Morrison and Scott, 1986).

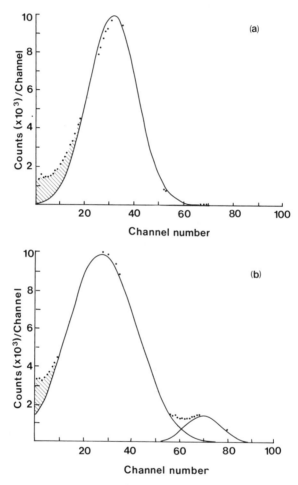

FIG. 9.2. Size distribution profiles of starch granules measured with a Coulter counter and 100-channel analyser, showing experimental points (...) fitted with normal distributions (—) and shaded areas at left discarded from calculations. (a) Rice starch: channel 1 = 2.6 μm³, 26 = 26.1 μm³, 51 = 260.7 μm³, mean volume = 62.8 μm³, mean diameter = 4.5 μm, surface = 0.795 m² g⁻¹. (b) Wheat starch: channels as for rice, mean volume B = 68.6 μm³, mean diameter = 4.4 μm, surface = 0.738 m² g⁻¹; mean volume A = 2131 μm³, mean diameter = 15.3 μm, surface = 0.257 m² g⁻¹. B granules = 26.0% by weight or volume.

Granules (typically 4–8 mg) suspended in an electrolyte, usually 0.9% saline, are disaggregated by sonication and are then drawn through a tube of suitable aperture for counting. Recommended apertures are 70 μm for small granules (oat, rice, barley),

100 μm for intermediate sizes (wheat, maize) and 120 μm or 170 μm for large granules (bean, potato, and other root and tuber starches). The channels give a logarithmic scale of granule volume and unimodal populations can be described by a normal distribution (Fig. 9.2a). The typical bimodal distributions of the A- and B-type granules in *Triticea* starches can be described by two normal distributions (Fig. 9.2b). Discarded material (shaded areas at left) is mostly cell debris, and is greater in starch isolated from finely milled flour than from cracked and macerated grain. Weight or number distributions, mean volumes, mean diameters, and specific surface areas can be determined with good precision and reproducibility (Morrison and Scott, 1986), and, given starch content (Section III.B), granules per seed or kernel can be calculated (Morrison *et al.*, 1986; Morrison and Gadan, 1987).

Counters based on the principle of diffracting a laser beam usually have 32 or 64 detector diode arrays and are comparable in precision to a Coulter counter with the same number of channels. Laser instruments can also measure dry particles in a stream of air.

III. CHEMICAL AND ENZYMIC ANALYSIS OF POLYSACCHARIDES AND MINOR COMPONENTS

This section is concerned with current methods for the analysis of starch which are largely biochemical. The classical chemical and physical procedures are well described elsewhere (Whistler, 1964; Radley, 1976a). The majority of the methods described below require starch in solution, and the importance of complete dissolution of granules without degradation of AM and AP cannot be overemphasised, particularly for structural analysis and physical chemistry.

A. Dissolving Starch

Granules do not dissolve readily in cold water, unless severely damaged by milling, and even at 100°C many do not disperse totally without strong shearing forces. Autoclaving can cause considerable degradation, and gentle boiling under nitrogen is preferred if heating is necessary (Takeda *et al.*, 1986a). Starches generally dissolve in 1 M KOH or NaOH at room temperature, but cereal starches give gels which sediment on centrifugation (Adkins *et al.*, 1970). Some degradation of AM and AP seems unavoidable with these methods even under anaerobic conditions.

Best results are usually obtained with dimethylsulphoxide (DMSO) containing 5–15% (v/v) water (Leach and Schoch, 1962; Adkins *et al.*, 1970; Wolf *et al.*, 1970; Kurtzman *et al.*, 1973; French, 1984). Most starches will dissolve with prolonged stirring at ambient temperature, but heating (< 100°C) under nitrogen may be necessary. A useful mild procedure is to pre-soak tared starch in cold water, centrifuge and decant the supernatant, then add sufficient DMSO to obtain an 85% solution with the water in the starch. Adding 1 vol. urea (6 M) to 9 vols DMSO and heating (< 10 mg starch per ml solvent) at 100°C for 90 min is also effective and quick (Morrison and Laignelet, 1983). Whichever method is used, final starch solutions must not yield precipitates or gelatinous sediments (which are transparent!) on centrifuging at $3000 \times g$.

If starch is precipitated from DMSO solution with several volumes of ethanol any lipids present remain in solution (Adkins *et al.*, 1970; Banks *et al.*, 1974; Morrison and Laignelet, 1983; Takeda *et al.*, 1986a) and the amorphous precipitate redissolves readily in cold DMSO solvents, dilute NaOH or KOH. Alternatively, the precipitate can be washed with diethyl ether to remove ethanol and residues of DMSO, and air-dried. AM and debranched glucans can retrograde rapidly in aqueous solutions and buffers, hence DMSO or alkaline solvents are often preferred for chromatography although they are not always compatible with some HPLC detectors or column packings (Section III.F).

B. Total α-Glucan

Preferred methods for the determination of starch polysaccharides or α-glucan involve complete hydrolysis to glucose with glucoamylases and colorimetric determination of glucose by an enzymic or chemical method. If starch is being determined directly in tissue, cell-wall polysaccharides in which glucose is always β-glycosidically linked do not pose a problem. Free glucose, maltosaccharides and other sugars containing α-glycosidically linked glucose, which would interfere, are removed by washing the sample with 80% (v/v) ethanol or 75% (v/v) propan-2-ol, ideally at elevated temperature to denature any native enzymes present.

Dry samples of starch or washed tissue are dispersed in water, α-amylase from *Bacillus licheniformis* (stable at 85–95°C) is added, and the suspension is heated at 85°C to cause rapid swelling of the starch with complete hydrolysis to soluble oligosaccharides. After cooling and buffering to pH 4.6 these are completely hydrolysed with glucoamylase (from *Aspergillus niger*) to glucose, which is then determined with a specific glucose oxidase–peroxidase–chromogen system (Karkalas, 1985; Holm *et al.*, 1986) or the hexokinase–glucose 6-phosphate dehydrogenase method (Kunst *et al.*, 1984) or by a non-specific chemical method for reducing sugars (Nelson, 1944; Park and Johnson, 1949; Somogyi, 1952).

Alternatively, α-glucan may be determined in solution. Tissues (free from interfering sugars) are usually extracted with hot DMSO (70–90%), which must be diluted to < 40% before enzymic determination of α-glucan (Libby, 1970; Knutson, 1986; Carpita and Kanabus, 1987) or debranching (Mercier and Kainuma, 1975; Sargeant, 1982) for chromatographic analysis and determination of eluted α-glucan chains (Sections III.E and III.F). The same enzymic methods as above may be used, or glucose content can be determined by non-specific colorimetric methods using phenol–sulphuric acid (Dubois *et al.*, 1956), anthrone–sulphuric acid (Koehler, 1952; Dische, 1955; Ough, 1964), or cysteine–sulphuric acid (Sargeant, 1982) calibrated with appropriate malto-saccharides. Perchloric acid has also been used to extract starch and mixed linkage β-glucans which were then determined as glucose (hexokinase–glucose 6-phosphate dehydrogenase method) after hydrolysis with amyloglucosidase and cellulase from *Penicillium funiculosum*, respectively (Ahluwalia and Ellis, 1984).

C. Iodometric Methods for Determining Amylose

The iodine-binding capacities (IBC) of starch polysaccharides under standard conditions of I_2/KI concentration, pH and temperature are 20–21 g I_2 per 100 g AM and

0.3–2.6 g I_2 per 100 g AP (Tables 9.1 and 9.2; see also Banks *et al.*, 1971b, 1974; Duprat *et al.*, 1980). Thus, AM content can be determined from the IBC of starch by measuring free and bound I_2 potentiometrically (Colborne and Schoch, 1964; Banks *et al.*, 1971b, 1974; Ullman, 1973), amperometrically (Larson *et al.*, 1953; Williams *et al.*, 1970; Ullman, 1973; Voss and Bergtheler, 1982) or photometrically (Ullman, 1973). Alternatively, the absorption maximum of the AM–polyiodide complex can be measured at λ_{max} in the region 630–680 nm (Gilbert and Spragg, 1964; Williams *et al.*, 1970; Wolf *et al.*, 1970; Morrison and Laignelet, 1983; C. Takeda *et al.*, 1983; Knutson, 1986; Chrastil, 1987). Iodometric methods give a measure of medium to long chain α-(1,4)-glucan content, and the small amount of iodine bound by normal AP is allowed for in calculations or by calibrating colorimetric assays with AM–AP mixtures. Morrison and Laignelet (1983) calibrated their colorimetric method against AM content of starches measured by debranching and GPC (Sections III.E and III.F).

In many high-AM cereal and legume starches the AP fractions have longer than normal A- and B-chains that bind more iodine and they give anomalously high results (Section I.C and Table 9.4). Conversely, if monoacyl lipids are present they will complex with some of the AM when the starch is dissolved and prevent iodine binding so that low apparent AM contents are obtained (Banks *et al.*, 1974; Bolling and El Baya, 1975; Morrison and Laignelet, 1983). Lipid interferences can be prevented by first precipitating the starch from DMSO solution with ethanol (Section III.A) or by complete extraction of lipids from the native granules (Section III.G).

Many colorimetric methods, including the widely-used method of Gilbert and Spragg (1964), suffer from instability of the coloured complex, and several factors must be controlled (Morrison and Laignelet, 1983). Although the natural pH of the final solution is weakly acidic, controlled acidification with trichloracetic acid has been recommended (Chrastil, 1987). Alkaline solutions have been used to minimise lipid interference and AP absorption, but they are not recommended because some iodine is converted to hypoiodite with a corresponding loss of AM–polyiodide colour.

D. Separation of Amylose and Amylopectin

The classical method for separating AM and AP, which depends on selective precipitation and recrystallisation of AM as an inclusion complex with butan-1-ol or other small guest molecules (Whistler, 1964; Banks and Greenwood, 1975; Hizukuri *et al.*, 1981) is still widely used. The pure AM-butanol complex is readily isolated from potato starch, but six or more recrystallisations are required with some other starches, and ultracentrifugation is necessary to remove traces of AP for cereal AM (Takeda *et al.*, 1986a). Intermediate material (mostly anomalous AP) in many starches greatly complicates starch fractionation (Whistler, 1964; Banks and Greenwood, 1975). AP from amylomaize starches precipitates with the AM–butanol complex while anomalous low molecular weight AP remains in solution (Baba *et al.*, 1987a,b).

An alternative approach is to use concanavalin A as a precipitant for AP (Matheson and Welsh, 1988). This protein binds to the non-reducing ends of α-glucans (and mannans), hence each molecule of AP binds many molecules of ligand and precipitates, while AM binds very little and remains in solution. This method has not been widely tested.

TABLE 9.4. Composition of some maize starches (dry weight basis).[a]

	Waxy	Normal	Sugary-2	Dull	Sugary-waxy	Dull-Sugary-2	Amylose extender
Amylose (%)							
Apparent, colorimetric[b]	0	23.7	40.4	32.0	3.4	51.8	59.4
Total, colorimetric[c]	0	25.2	43.7	36.9	3.0	55.0	63.2
Undebranched, GPC[d]	0	25.4	38.6	45.0	47.6	53.3	76.5
Debranched, GPC[e]	0	25.6	40.9	35.5	0.6	51.1	42.5
Lipid (mg %)							
Free fatty acids	6	336	635	375	46	671	410
Lysophospholipids	4	240	262	434	19	362	494
Phosphorus (mg %)							
Lipid	0.2	14.7	16.1	26.6	1.2	22.2	30.3
Non-lipid	2.8	3.2	2.9	5.2	7.1	3.0	9.8
Nitrogen (mg %)							
Lipid	0.1	6	7	11	0.5	9	12
Integral protein[f]	105	49	48	70	73	61	102

[a] From J. B. South, W. R. Morrison and O. E. Nelson (in prep.).
[b] Measured in presence of lipid.
[c] Measured on lipid-free starch.
[d] GPC on Sepharose CL-2B.
[e] GPC on Sepharose CL-6B.
[f] Nitrogen × 6.25 of Na laurate-washed starch, corrected for lipid N.

Ultracentrifugation is still used occasionally for analytical and preparative separations of AM and AP, but intermediate material thought to be anomalous AM and anomalous AP (possibly also incompletely separated normal AM and AP) often contaminates the main fractions (Banks and Greenwood, 1975).

GPC (Section III.F) on a preparative scale can provide adequate quantities of material for modern methods of analysis, and has been used in conjunction with butanol precipitation to study complex mixtures (Baba et al., 1987a,b).

E. Structural Analysis, Debranching, DP and CL

1. Molecular weights, degree of polymerisation

AM and AP, and the long and short linear glucans of which they are composed, are all heterogeneous or polydisperse and cover a range of molecular weights, whereas synthetic α-(1,4)-glucans and dextrans used to calibrate GPC columns are ideally of uniform molecular weight or monodisperse. The terminology used to describe these molecules can be confusing, and is explained below.

For a polydisperse population DP and molecular weights (M) can be expressed on a number-average or on a weight-average basis, and it is important to distinguish these clearly (MacGregor and Greenwood, 1980). The number-average molecular weight (M_n) is the sum of the weights of each monodisperse species divided by the number of moles (*n*) in the population:

$$\bar{M}_n = \frac{n_1 M_1 + n_2 M_2 + \ldots n_i M_i}{n_1 + n_2 + \ldots n_i} = \frac{\Sigma n_i M_i}{\Sigma n_i} \tag{1}$$

or, using concentration (*c*) or weight fraction (*w*)

$$\bar{M}_n = c/\Sigma(c_i/M_i) = 1/\Sigma(w_i/M_i) \tag{2}$$

and, except for very short chains (low CL values)

$$\overline{DP}_n = \bar{M}_n/162 \tag{3}$$

Using the weight fraction of a monodisperse species ($n_i M_i$) instead of the mole fraction n_i in Equation (1) gives the weight-average molecular weight

$$\bar{M}_w = \frac{\Sigma n_i M_i^2}{\Sigma n_i M_i} = \frac{\Sigma c_i M_i}{c} = \Sigma w_i M_i \tag{4}$$

and, similarly

$$\overline{DP}_w = \frac{\Sigma n_i DP_i^2}{\Sigma n_i DP_i} = \bar{M}_w/162 \tag{5}$$

In practice $\bar{M}_w > \bar{M}_n$ and $\overline{DP}_w > \overline{DP}_n$ for polymers that are not monodisperse, and the spread or polydispersity of a population is measured by the ratio DP_w/DP_n which is very high for AP, in the range 1.3–6 for AM (Table 9.1), and c. 1.1 (ideally 1.0) for monodisperse standards used to calibrate GPC columns (Section III.F).

The Z-average molecular weight (M_Z) obtained by analytical ultracentrifugation exaggerates the contribution of higher molecular weight components even more than in M_w determinations, since

$$M_Z = (\Sigma n_i M_i^3)/(\Sigma n_i M_i^2) \tag{6}$$

2. Enzymic hydrolysis

The enzymes commonly used in the analysis of starch polysaccharides in solution are highly purified forms of the debranching enzymes isoamylase and pullulanase, and the chain-shortening enzyme β-amylase (occasionally phosphorylase), each of which has specificities (Table 9.3) that can be used to elucidate structural features.

Isoamylase (from *Bacillus amyloderamosa*, *Pseudomonas* spp. or *Cytophaga* spp.) can be used in buffered aqueous solution (pH 3.5 or 5.5, respectively) or in dilute DMSO (<40%) to effect complete debranching of AP, but normally it does not release very short (G_2 and G_3) branch chains or stubs (Kainuma *et al.*, 1978). Pullulanase (from *Aerobacter aerogenes*) in aqueous buffer (pH 5.2) or dilute DMSO (< 20%) also gives complete debranching of AP and does release G_2 and G_3 branch chains but not α-(1,6)-glucosyl residues or stubs (Mercier and Kainuma, 1975).

β-Amylase from sweet potato (*Ipomoea batatas*, opt. pH 4.8) releases maltosyl residues from non-reducing ends of chains, but stops before α-(1,6)-branch points leaving β-limit dextrins with short branch chains or stubs. The β-limit dextrin from AP is presumed to have equimolar G_2 and G_3 stubs from odd- and even-numbered A-chains, respectively, and G_1 and G_2 stubs from odd- and even-numbered B-chains (Robyt, 1984). β-Amylase treatment of phosphorylase limit dextrin from AP gives a φβ-limit dextrin in which the stubs are uniformly G_2. The β-amylolysis limit (100 × maltose released/total glucan expressed as maltose) is used as an index of branching (Tables 9.1 and 9.2). Since values of 100% are only given by linear α-(1,4)-glucans this can be used as a criterion of complete debranching. Failure to obtain 100% values after thorough debranching with isoamylase and pullulanase can be caused by retrogradation of linear glucans in aqueous solution, very short G_1 or G_2 branched residues in some types of AM, or phosphate ester groups at C-6 or C-3 of glucose residues in AP (and less commonly AM) from some sources. Phosphate ester groups are also a barrier to hydrolysis by porcine and pancreatic α-amylases and *Aspergillus niger* glucoamylase, each giving characteristic phosphorylated oligosaccharide residues (Abe *et al.*, 1982; Y. Takeda *et al.*, 1983; Takeda, 1987).

Digestion of AM with β-amylase and determination of β-limit dextrins can be used to determine the molar proportions of linear and branched AM (Y. Takeda *et al.*, 1987b).

3. Analytical methods

Detailed structural analysis of starch polysaccharides requires data from appropriate and complementary methods of analysis. Most of the methods are described in this section, determination of total glucan is mentioned in Section II.B, and chromatographic methods are discussed in Section III.F.

(a)   *Permethylation*. For a highly branched polysaccharide such as AP it is possible to determine proportions of branch points and non-reducing ends, average chain lengths (CL), external chain lengths (ECL) beyond branch points, and internal chain lengths (ICL) between branch points largely from permethylation data (Whistler, 1964; Harris *et al.*, 1984). Gas chromatography (ideally with mass spectrometry) or HPLC of hydrolysis products gives the proportions of branch points from 2,3-di-*O*-methylglucose, mid-chain residues from 2,3,6-tri-*O*-methylglucose, and non-reducing ends from 2,3,4,6-tetra-*O*-methylglucose. This method is not suitable for AM which has $<1\%$ branched and terminal residues.

(b)   *Reducing power*. Several procedures depend on sensitive methods for determining reducing power, such as the modified Park and Johnson method (Hizukuri *et al.*, 1981; Y. Takeda *et al.*, 1987a) or the older methods of Nelson (1944) and Somogyi (1952) as recommended by Robyt and Whelan (1968).

(c)   *Non-reducing ends*. Non-reducing ends (= branch points + 1) in AP can be determined directly by permethylation (see (a) above) or by periodate oxidation which yields one mole of formic acid per end group (Whistler, 1964; Wharton and McCarty, 1972). A modified Smith degradation with enzymic determination of the glycerol that is produced was developed for AP (Hizukuri and Osaki, 1978) and later modified for AM (Hizukuri *et al.*, 1981; Takeda *et al.*, 1984). Non-reducing ends and branch points can also be determined from reducing power before and after debranching, usually with isoamylase (Gunja-Smith *et al.*, 1971; Hizukuri *et al.*, 1981; Suzuki *et al.*, 1981).

If a sample of AP has 4.55% branched residues, it follows that average CL = 22. If the β-amylolysis limit is 55%, and given that

$$ECL = (CL \times \beta/100) + 2 \tag{7}$$

$$\text{and } ICL = CL - ECL - 1 \tag{8}$$

it then follows that ECL = 14 and ICL = 7.

(a) DP_w, M_w. Viscometry is relatively simple and is widely used for determining M_w (Greenwood, 1964), but inspection of the data in Table 9.1 shows that the relationship $DP_w = 7.5\ [\eta]$, which is reliable for some types of AM (Takeda *et al.*, 1984) is not always accurate. Light scattering methods (Foster, 1964; Banks and Greenwood, 1975) give slightly higher values, and when used in conjunction with GP-HPLC they allow simultaneous determination of M_w and related values for small samples (Hizukuri and Takagi, 1984; Takagi and Hizukuri, 1984; Yu and Rollings, 1987).

(e) DP_n, M_n. In principle, M_n can be determined by methods based on colligative properties, such as osmometry (Greenwood, 1960; Jorgensen and Jorgensen, 1960). More commonly, DP_n is determined from the reducing power (see (b) above) of a known weight of glucan. The modified Park and Johnson method (Hizukuri *et al.*, 1981; Y. Takeda *et al.*, 1987b) is particularly suitable for AM. A simple enzymic method for *linear* amylose and maltodextrins, which depends on there being equal numbers of odd and even chain lengths, involves complete hydrolysis to maltose and glucose (one residue per odd-numbered chain) with β-amylase and determination of glucose with glucose oxidase (Banks and Greenwood, 1968).

(f) *CL*. Short chain lengths in AP have been estimated by permethylation (see (a) above), but the more usual methods are to measure non-reducing ends of a known weight of glucan using periodate oxidation or the modified Smith degradation procedure developed for AM (Hizukuri *et al.*, 1981; Takeda *et al.*, 1984). CL can also be calculated from reducing power after complete hydrolysis with isoamylase (Takeda *et al.*, 1987a).

CL of linear glucans in fractions eluted from GPC columns (Section III.F) can be *estimated* from λ_{max} of their polyiodide spectra prepared under standard conditions, but the method is only useful in the range CL = 10 to 80. The most important variables to control are temperature and concentrations of I_2 and KI (Banks *et al.*, 1971a; Knutson *et al.*, 1982; Yamamoto *et al.*, 1982; Morrison and Laignelet, 1983; Yajima *et al.*, 1987). Banks *et al.* (1971a) used 1.6 mg glucan, 1.6 mg I_2 and 40 mg KI per 100 ml at 20.4°C and derived the relationship

$$1/\lambda_{max} = 1.558 \times 10^{-3} + 102.5 \times 10^{-4}(1/DP_n) \tag{9}$$

Fales (1980) used data given by several authors (working in the range 1.6–4 mg I_2 and 40 mg KI per 100 ml at 20°C) to obtain an almost perfect linear correlation with

$$\lambda_{max} = 635 - (3290/CL) \tag{10}$$

It follows that λ_{max} of AM = 635 nm and for smaller linear glucans

$$CL = 3290/(635 - \lambda_{max}) \tag{11}$$

This method is useful if $\lambda_{max} < 590$ nm.

(g) *A:B chain ratio in AP*. The A:B chain molar ratio in AP can be calculated by comparing reducing groups released by isoamylase and by pullulanase from β-limit dextrins prepared from the AP; ϕβ-dextrins can also be used (Gunja-Smith *et al.*, 1970, 1971; Marshall and Whelan, 1974; Lii and Lineback, 1977; Manners and Matheson, 1981; Enevoldsen, 1985). Enzyme and substrate concentrations are critical, and cumulative small errors in reducing valves can give substantial errors in the final ratio (Atwell *et al.*, 1980b; Manners and Matheson, 1981; Manners, 1985b,d).

High resolution GP-HPLC columns give polymodal chain distributions from debranched AP, and the shortest fraction (CL 10–16) has been equated with A-chains and other fractions with B-1, B-2 and B-3 chains as shown in Fig. 9.1b (Hizukuri, 1986). Weight and molar chain ratios and average chain lengths for each fraction are obtained by this method.

(h) *Chain number, branch points*. Chain number (CN) is the ratio of non-reducing ends to free reducing ends, generally determined by the modified Smith and Park and Johnson methods, respectively (Tables 9.1 and 9.2), which give results in good agreement with the isoamylase method. CN also equals DP_n/CL and the number of branch points is $CN - 1$.

F. Chromatography of Intact and Debranched Polysaccharides

With the advent of cross-linked agarose and dextran gels of suitable pore size it is now possible to obtain good separations of the full range of starch-derived saccharides by gel permeation chromatography (GPC, also called size-exclusion chromatography, SEC) on analytical or preparative columns, and by high performance liquid chromatography (GP-HPLC). Separations are based primarily on molecular size and shape, and several gels are necessary to cover the full molecular weight range (1.5×10^3 to 5×10^8). In normal practice single gels are used for particular applications, but two and three GP-HPLC columns in series have been used for high resolution work (Y. Takeda et al., 1984, 1986a, 1987a; Kobayashi et al., 1985; Hizukuri, 1986). Glucose and the lower maltosaccharides (G_1–G_{10}) can be resolved as individual peaks in chromatograms, but longer α-glucans (e.g. by debranching AP) overlap, while AM and AP emerge as broad peaks due to the molecular size range (polydispersity) within each and the limitations of the gel columns. Overlapping peaks on chromatograms are commonly divided vertically at lowest points, but this is less satisfactory than deconvolution or manually separating the peaks into components with normal distributions.

For separation of AM and AP, amorphous starch, precipitated from DMSO (Section III.A) and dissolved in water, can be eluted from GPC columns with water apparently without losses due to retrogradation of AM, but many workers use water containing 10–40% DMSO or 0.01–0.05 M KOH to avoid the risk of retrogradation (Praznik, 1986), particularly with the shorter linear α-glucans obtained by debranching.

The starch polysaccharides do not have suitable chromophores for direct detection by absorption in the UV-visible range. Refractive index (RI) detectors are non-destructive and respond to mass of glucan in the eluate. They cannot be used with strong solutions of DMSO which have a high RI, or with strong alkali which damages detector surfaces and degrades some column packings. Post-column reactions (destructive) are commonly applied to collected fractions, and autoanalysers have been used to measure total glucan by the phenol–sulphuric acid (Robyt and Bemis, 1967) and cysteine–sulphuric acid methods (Sargeant, 1982; Morrison et al., 1984). Monitoring the absorbance of polyiodide complexes of undebranched starch polysaccharides at 620–650 nm has been used, in parallel with total glucan measurements, to distinguish AM and AP. However, this can be thoroughly misleading if anomalous material is present, and it is better to scan the spectra of fractions to determine λ_{max} in the region 530–680 nm.

With GPC and GP-HPLC columns log M_w, log M_n and log DP_n are proportional to elution volume or time. Columns have been calibrated with dextrans and linear glucans prepared by synthesis from primers or by controlled hydrolysis of polysaccharides and fractionation of the products (Praznik and Ebermann, 1979; Kuge et al., 1984; Kobayashi et al., 1985; Praznik, 1986; Praznik et al., 1986, 1987). For low DP_n glucans λ_{max} of polyiodide complexes is convenient for comparative work (Section III.E), but if adequate fractions can be collected the direct determination of DP_n is preferable. Low-angle laser light scattering detectors respond to the molecular volume of the glucans and allow calculation of DP_w values (Hizukuri and Takagi, 1984; Takagi and Hizukuri, 1984; Yu and Rollings, 1987).

G. Surface and Internal Lipids

The non-waxy cereal starches appear to be unique, certainly amongst the major food starches, in having monoacyl lipids inside the granules. In the *Triticeae* starches the lipids are the lysophospholipids (LPL), lysophosphatidylcholine, -ethanolamine and -glycerol, while the other cereal starches contain free fatty acids (FFA) and LPL (Morrison, 1988a,b). Starches isolated without due precautions (Section II.A) have variable amounts of surface lipids (artifacts) which are mostly FFA but could include many other monacyl lipids (Morrison, 1981, 1988a). Impure starches may also have more loosely associated non-starch lipids.

Surface lipids (and non-starch lipids) should be extracted first, using any common polar lipid solvent such as chloroform–methanol–water or water-saturated butanol at ambient temperature (Morrison *et al.*, 1975, 1980). However, the efficacy of this extraction is by no means certain and it is much better to prepare clean starch in the first instance. Surface lipids are also removed with surface proteins when starch granules are washed with SDS or sodium laurate (Section III.H), but even after thorough washing with water there will be substantial quantities of SDS or laurate in the starch which will interfere with lipid analysis.

The internal lipids may be extracted under nitrogen at 95–100°C with one of several alcohol–water mixtures in proportions optimised for limited swelling of the granules and solubilisation of the lipids (Morrison and Coventry, 1985).

Total acyl lipids can be determined, as fatty acid methyl esters, by direct acid hydrolysis of starch in the presence of heptadecanoic acid (17:0) internal standard (Morrison *et al.*, 1980). This must be done under strictly anaerobic conditions, otherwise there will be appreciable losses of linoleic and linolenic acids. Total extractable lipids can be quantified similarly, or lipid classes can be separated by thin-layer chromatography for quantification (Morrison *et al.*, 1980). These methods require 200–500 mg starch for each analysis.

Starch phosphorus content can be determined directly (Morrison, 1964; Itaya and Ui, 1966) with greater sensitivity (lower limit *c.* 0.5 μg P) and this requires only 5–10 mg starch. With *Triticeae* starches total starch P should be very close to total lipid P, measured in solvent-extracted lipid, but there is hexose phosphate, mostly in the AP, of maize, rice and root starches (Tables 9.1 and 9.2). Factors to convert lipid-P into LPL are P × 16.4 (wheat, maize) and P × 16.2 (rice). Glycerophosphate (from LPL), hexose phosphate and inorganic phosphate can be determined, after acid hydrolysis of starch, by ion-exchange chromatography (Tabata *et al.*, 1975; Hizukuri and Hisatsuka, 1976) and enzymically (Hizukuri *et al.*, 1970).

H. Surface and Integral Proteins

Most starches have small amounts of protein, and recently attention has focused on the cereal starches which have clearly differentiated surface (<0.2%) and integral (<0.6%) proteins (Lowy *et al.*, 1981; Sano *et al.*, 1985; Greenwell and Schofield, 1986; Villareal and Juliano, 1986; Schofield and Greenwell, 1987; Goldner and Boyer, 1989). The only reliable measure of total granule protein is Kjeldahl nitrogen × 6.25, but for cereal starches a correction is necessary for nitrogen in LPL (molar ratio N:P = 0.9:1, hence lipid N = 40% of lipid P, by wt.). Surface proteins can be recovered quantitatively by shaking the starch with 1–2% SDS (2–5 ml g^{-1} starch) at 20°C for 30 min (Gough *et al.*, 1985; Greenwell and Schofield, 1986) and may be estimated by the Lowry method modified for the presence of SDS (Peterson, 1977). Quantitative extraction of integral proteins has not been achieved yet, but good yields are obtained if the starch is heated in 1–2% SDS (20–30 ml g^{-1}) at 5°C below its gelatinisation temperature when there is sufficient swelling to release much integral protein without undue solubilisation of starch (which interferes in sample preparation for electrophoresis) (Gough *et al.*, 1985; J. B. South and W. R. Morrison, unpubl. res.). Both groups of proteins (typically 200–500 mg starch) may be separated by SDS-polyacrylamide gel electrophoresis (SDS-PAGE) on 10% gels or 12–20% gradients gels with marker proteins to cover the range 14–30 kDa apparent mol. wt. for surface proteins and 50–150 kDa for integral proteins (Greenwell and Schofield, 1986; Villareal and Juliano, 1986; Goldner and Boyer, 1989). Single bands on these gels may be heterogeneous and isoelectric focusing and other systems should also be considered.

I. Damaged Starch Granules

All commercial starches and flours and many laboratory preparations contain damaged starch granules, especially if made by a dry milling process. Methods for measuring damage based on cold leaching of AM, susceptibility to α-amylases, and susceptibility to β-amylases (Evers and Stevens, 1985) may not be appropriate in plant biochemistry.

If granules suspended in 0.1–0.2% Congo Red are viewed under the light microscope the damaged granules which have stained pink can be counted. Hessian Bordeau has been proposed recently as a much more sensitive stain (R. G. Fulcher, pers. commun.).

The best enzymic methods for measuring damage (Williams and LeSeelleur, 1970; Evers and Stevens, 1985) depend on the fact that undamaged granules are totally resistant to pure β-amylase while damaged granules are attacked at a measurable rate. The other starch-degrading enzymes (Table 9.3) all attack undamaged granules to some extent, although their action on damaged granules is much faster, so precise assays based on these enzymes are not possible when there are low levels of damage.

IV. PHYSICAL METHODS OF ANALYSIS

A. Spectroscopic Methods on Intact Granules

Non-invasive spectroscopic methods can be used to study polysaccharide ordering in granules (Blanshard, 1979, 1986, 1987). Small-angle scattering of polarised light gives information on the anisotropic regions of granules. Wide-angle X-ray diffraction has been used extensively (Zobel, 1964, 1988a,b; French and Murphy, 1977; Duprat *et al.*, 1980; Wild and Blanshard, 1986; Imberty *et al.*, 1987a,b) to study packing of parallel double helices, and the majority of starches give either the A-pattern (e.g. cereals) or the B-pattern (tubers, amylomaize). AM-FFA and AM-LPL inclusion complexes give a V-pattern (Takeo *et al.*, 1973) not normally seen in native granules until they have been gelatinised. Cross-polarisation magic-angle spinning nuclear magnetic resonance gives characteristic detailed spectra for the A- and B-crystal polymorphs of AM and the crystallites of native starch granules (Veregin *et al.*, 1986). Wide-angle X-ray scattering may be used also to compare the relative crystallinity of granules (Zobel, 1988b) giving information similar to differential scanning calorimetry (DSC). Small-angle X-ray scattering (SAXS) gives information of longer range periodicity of 9–10 nm (Blanshard et al., 1984; Blanshard, 1987) which may relate to periodicity in AP revealed by biochemical studies (Fig. 9.1). Small-angle neutron scattering is a comparatively new technique in starch research which complements SAXS (Blanshard et al., 1984).

Limited information on elemental composition can be obtained from emission spectra. The technique of ESCA (electron spectroscopy for chemical applications) has been used to study N, P and S in the surface layers of starch granules subjected to various treatments likely to affect granule surface proteins and lipids (Russell et al., 1987). Analysis of γ-radiation emitted by P atoms in thick sections of granules for TEM has also been used to obtain rough measurements of P distribution across the sections (Morrison and Gadan, 1987). Both methods using emission spectra suffer from the fact that the elements of interest are present near their lower limits of detection.

B. Thermal Analysis of Intact Granules

The related techniques of differential thermal analysis, differential scanning calorimetry (DSC) and thermogravimetric analysis have all been used on starches, and provide information of particular value for applied research. However, DSC (the most popular technique) also has relevance to plant biochemistry and some discussion is warranted here.

In DSC small starch samples (3–5 mg) are heated in water, generally at $10°C \, min^{-1}$, from ambient temperature to 90–120°C, according to requirements. The proportion of water is sometimes expressed as the volume fraction (v), i.e. the volume of water (added water + moisture in starch) divided by the volume of anhydrous starch. When $v > 0.7$, nearly all starches give a clear endotherm due to disordering of AP (gelatinisation) and those starches that contain lipid give a second endotherm in the region 90–105°C due to dissociation of the AM–lipid complex (Burt and Russell, 1983; Blanshard, 1987; Morrison, 1988b). On cooling the AM–lipid complex reforms spontaneously, while the AP remains as an amorphous gel which retrogrades slowly. Thus, if excess LPL is added to starch to complex with all available AM, the AM content of the starch can be determined from the enthalpy of the endotherm (Kugimiya and Donovan, 1981). At lower moisture contents ($v < 0.65$) thermograms are complicated by the appearance of new endotherms attributed to glass transitions which are only relevant in certain food-processing situations (Donovan, 1979; Burt and Russell, 1983; Blanshard, 1987). DSC is useful for studying varietal differences and environmental effects on starch gelatinisation since it can be used on starch from single cereal grains, seeds or tubers (see also Chapter 18).

V. INDUSTRIAL UTILISATION OF STARCHES

Starch is the second most abundant carbohydrate (after cellulose) in the biosphere, and annual world production in crops for all food and industrial uses is estimated to be over 10^9 tonnes.

The principal commercial sources of starch are the major cereals (wheat, rice, maize, barley and sorghum), potatoes, cassava, sweet potatoes, and to a lesser extent arrowroot, sago and various legumes. Most starch is consumed as food without complete separation from the surrounding plant tissue, but the technological properties of the starches are nevertheless of prime importance. A thriving modern technology also exists for the production of native granular starches and modified starches from sources such as maize, wheat and potato (depending on local economic considerations). Commercial starches are 98–99% polysaccharide (AM and AP), with small quantities of protein (0.3%), lipid (<1%), and mineral matter (0.1%) which may be slightly altered compared with the pure starches isolated in the laboratory, mainly due to milling, steeping and drying conditions used during processing.

Most native starch is used by the food industry after hydrolysis (e.g. as dextrose, glucose syrups, high-fructose corn syrup), and enzyme technology then matters as much as the properties of the starch. Native starch is also used for its gel properties. The limited range of gel properties in native starches can be extended by chemical modification, but this practice would be superseded if suitable novel starches became available to meet the public demand for natural food.

Starch is also used for the production of industrial alcohol, as a chemical feedstock, and by the paper-making, pharmaceutical and textile industries, among others. For detailed reviews on the industrial utilisation of starches and biotechnological applications (see Radley, 1976b; Galliard, 1987; Munck et al., 1988).

VI. SUMMARY

The plant biochemist now has an unprecedented array of methods for isolating and characterising starch granules, often on a very small scale, and for detailed structural analysis of the major polysaccharides and minor protein and lipid components. These can be used for several important lines of research, for example to study starch biochemistry in vivo using various plant species, mutants and environments, and to develop new types of starch by plant breeding and (eventually) by genetic engineering with structures, compositions and properties tailored to specific industrial requirements. The tools are to hand, and the challenge should be irresistible.

REFERENCES

Abe, J., Takeda, Y. and Hizukuri, S. (1982). *Biochim. Biophys. Acta* **703**, 26–33.
Adkins, G. K. and Greenwood, C. T. (1966). *Staerke* **18**, 213–218.
Adkins, G. K., Greenwood, C. T. and Hourston, D. J. (1970). *Cereal Chem.* **47**, 13–18.
Ahluwalia, B. and Ellis, E. E. (1984). *J. Inst. Brew.* **90**, 254–259.
Ahmed, M. and Lelievre, J. (1978). *Staerke* **30**, 78–79.
Atwell, W. A., Hoseney, R. C. and Lineback, D. R. (1980a). *Cereal Chem.* **57**, 12–16.
Atwell, W. A., Milliken, G. A. and Hoseney, R. C. (1980b). *Staerke* **32**, 362–364.
Baba, T., Uemura, R., Hiroto, M. and Arai, Y. (1987a). *Denpun Kagaku* **34**, 196–202.
Baba, T., Uemura, R., Hiroto, M. and Arai, Y. (1987b). *Denpun Kagaku* **34**, 203–207.
Badenhuizen, N. P. (1969). "The Biogenesis of Starch Granules in Higher Plants". Appleton-Century-Crofts, New York.
Badenhuizen, N. P. (1971). "Struktur und Bildung des Stärkekorns, Handbuch der Stärke in Einzeldarstellung", Vol. VI, Part 2 (M. Ullman, ed.). Paul Parey, Berlin.
Banks, W. and Greenwood, C. T. (1968). *Carbohydr. Res.* **6**, 177–183.

Banks, W. and Greenwood, C. T. (1975). "Starch and its Components". Edinburgh University Press, Edinburgh.

Banks, W. and Muir, D. D. (1980). In "The Biochemistry of Plants", Vol. 3: Carbohydrate Structure and Function (J. Preiss, ed.), pp. 321–369. Academic Press, New York.

Banks, W., Greenwood, C. T. and Khan, K. M. (1971a). Carbohydr. Res. 17, 25–33.

Banks, W., Greenwood, C. T. and Muir, D. D. (1971b). Staerke 23, 118–124.

Banks, W., Greenwood, C. T. and Muir, D. D. (1974). Staerke 26, 73–78.

Banks, W., Greenwood, C. T. and Muir, D. D. (1979). In "Molecular Structure and Function of Food Carbohydrates" (G. G. Birch and L. F. Green, eds), pp. 177–194. Applied Science, London.

Bechtel, D. B. (1983). "New Frontiers in Food Microstructure". American Association of Cereal Chemists, St. Paul, MN, USA.

Blanshard, J. M. V. (1979). In "Polysaccharides in Food" (J. M. V. Blanshard and J. Mitchell, eds), pp. 139–152. Butterworths, London.

Blanshard, J. M. V. (1986). In "Chemistry and Physics of Baking" (J. M. V. Blanshard, P. J. Frazier and T. Galliard, eds), pp. 1–13. Royal Society of Chemistry, London.

Blanshard, J. M. V. (1987). In "Starch: Properties and Potential" (T. Galliard, ed.), pp. 16–54. John Wiley & Sons, Chichester.

Blanshard, J. M. V., Bates, D. R., Muhr, A. H., Worcester, D. L. and Higgins, J. S. (1984). Carbohydr. Res. 4, 427–442.

Bolling, H. and El Baya, A. W. (1975). Chem. Mikrobiol. Technol. Lebensm. 3, 161–163.

Bowler, P., Evers, A. D. and Sargeant, J. (1987). Staerke 39, 46–49.

Brandt, D. A. and Dimpfl, W. L. (1970). Macromolecules 3, 655–664

Briarty, L. G., Hughes, C. E. and Evers, A. D. (1979). Ann. Bot. 44, 641–658.

Burt, D. J. and Russell, P. L. (1983). Staerke 35, 354–360.

Callaghan, P. T. and Lelievre, J. (1987). Carbohydr. Res. 162, 33–40.

Carpita, N. C. and Kanabus, J. (1987). Anal. Biochem. 161, 132–139.

Chabot, J. F., Allen, J. E. and Hood, L. F. (1978). J. Food Sci. 43, 727–730, 734.

Chandrashekar, A., Somashekar, K. and Savitri, A. (1987). Staerke 39, 195–197.

Chrastil, J. (1987). Carbohydr. Res. 159, 154–158.

Cluskey, J. E., Knutson, C. A. and Inglett, G. E. (1980). Staerke 32, 105–109.

Colborne, C. R. and Schoch, T. J. (1964). In "Methods in Carbohydrate Chemistry" (R. L. Whistler, ed.), Vol. 4, pp. 161–165. Academic Press, New York.

Czaja, A. Th. (1971). "Die Mikroskopie der Stärkekörner, Handbuch der Stärke in Einzeldarstellung", Vol. VI, Part 1 (M. Ullman, ed.). Paul Parey, Berlin.

Czaja, A. Th. (1978). Taxonomy 27, 463–470.

Czaja, A. Th. (1980). Staerke 32, 253–257.

Decker, P., and Höller, H. (1962). J. Chromatogr. 7, 392–399.

Dengate, H. N., Baruch, D. W. and Meredith, P. (1978). Staerke 30, 80–84.

DeWilligen, A. H. A. and Ullman, M. (1974). "Die Bestimmung der Stärke. Handbuch der Stärke in Einzeldarstellung". Vol. I, Part 1 (M. Ullman, ed.). Paul Parey, Berlin.

Dische, Z. (1955). In "Methods of Biochemical Analysis", Vol. 2 (D. Glick, ed.), pp. 313–358. Interscience, New York.

Donovan, J. W. (1979). Biopolymers 18, 263–275.

Dubois, M., Gilles, K. A., Hamilton, J. K., Rebers, P. A. and Smith, F. (1956). Anal. Chem. 28, 350–356.

Duffus, C. M. and Murdoch, S. M. (1979). Cereal Chem. 56, 427–429.

Duprat, F., Gallant, D., Guilbot, A., Mercier, C. and Robin, J. P. (1980). In "Les Polymères Végétaux" (B. Monties, ed.), pp. 176–231. Gauthier-Villars.

Echeverria, E., Boyer, C., Liu, K.-C. and Shannon, J. (1985). Plant Physiol. 77, 513–519.

Enevoldsen, B. S. (1985). In "New Approaches to Research on Cereal Carbohydrates" (R. D. Hill and L. Munck, eds), pp. 55–60. Elsevier, Amsterdam.

Entwhistle, G., Tyson, R. H. and apRees, T. (1988). Phytochemistry 27, 993–996.

Evers, A. D. and McDermott, E. E. (1970). Staerke 22, 23–26.

Evers, A. D. and Stevens, D. J. (1985). Adv. Cereal Sci. Technol. 7, 321–349.

Fales, F. W. (1980). Biopolymers 19, 1535–1542.

Fishwick, M. J. and Wright, A. J. (1980). *Phytochemistry* **19**, 55–59.

Fogarty, W. M. (1983). "Microbial Enzymes and Biotechnology" (W. M. Fogarty, ed.), pp. 1–92. Applied Science, London.

Fogarty, W. M. and Kelly, C. T. (1979). *In* "Progress in Industrial Microbiology", Vol. 15 (M. J. Bull, ed.), pp. 87–150. Elsevier, Amsterdam.

Foster, J. F. (1964). *In* "Methods in Carbohydrate Chemistry", Vol. 4: Starch (R. L. Whistler, ed.), pp. 191–202. Academic Press, New York.

French, A. D. and Murphy, V. G. (1977). *Cereal Foods World* **22**, 61–70.

French, D. (1972). *Denpun Kagaku.* **19**, 8–25.

French, D. (1975). *In* "Biochemistry". Series, 1, Vol. 5: Biochemistry of Carbohydrates (W. J. Whelan, ed.), pp. 267–335. Butterworths, London.

French, D. (1984). *In* "Starch: Chemistry and Technology", 2nd edn. (R. L. Whistler, J. N. BeMiller and E. F. Paschall, eds), pp. 183–247. Academic Press, Orlando, FL, USA.

French, D., Pulley, A. O. and Whelan W. J. (1963). *Staerke* **15**, 349–354.

Fuwa, H., Sugimoto, Y. and Takaya, T. (1979). *Carbohydr. Res.* **70**, 233–238.

Fuwa, H., Takaya, T. and Sugimoto, Y. (1980). *In* "Mechanisms of Saccharide Polymerization and Depolymerization" (J. J. Marshall, ed.), pp. 73–118. Academic Press, New York.

Gallant, D. J. and Bouchet, B. (1986). *Food Microstruct.* **5**, 141–155.

Gallant, D. J. and Sterling, C. (1976). *In* "Examination and Analysis of Starch and Starch Products" (J. A. Radley, ed.), pp. 33–59. Applied Science, London.

Gallant, D. J., Mercier, C. and Guilbot, A. (1972). *Cereal Chem.* **49**, 354–365.

Galliard, T. (1987). "Starch: Properties and Potential". John Wiley & Sons, Chichester.

Galliard, T. and Bowler, P. (1987). *In* "Starch; Properties and Potential" (T. Galliard, ed.), pp. 55–78. John Wiley & Sons, Chichester.

Gidley, M. J., Bulpin, P. V. and Kay, S. (1986). *In* "Gums and Stabilisers for the Food Industry". Vol. 3 (G. O. Phillips, D. J. Wedlock and P. A. Williams, eds), pp. 167–176. Elsevier Applied Science, London.

Gilbert, G. A. and Spragg, S. P. (1964). *In* "Methods in Carbohydrate Chemistry", Vol. 4: Starch (R. L. Whistler, ed.), pp. 168–169. Academic Press, New York.

Goldner, W. and Boyer, C. D. (1989). *Staerke* **41**, 250–254.

Gough, B. M., Greenwell, P. and Russell, P. L. (1985). *In* "New Approaches to Research on Cereal Carbohydrates" (R. D. Hill and L. Munck, eds), pp. 99–108. Elsevier, Amsterdam.

Greenwell, P. and Schofield, J. D. (1986). *Cereal Chem.* **63**, 379–380.

Greenwood, C. T. (1960). *Staerke* **12**, 169–174.

Greenwood, C. T. (1964). *In* "Methods in Carbohydrate Chemistry", Vol. 4: Starch (R. L. Whistler, ed.), pp. 179–188. Academic Press, New York.

Gunja-Smith, J., Marshall, J. J., Mercier, C., Smith, E. E. and Whelan, W. J. (1970). *FEBS Lett.* **12**, 101–104.

Gunja-Smith, J., Marshall, J. J. and Smith, E. E. (1971). *FEBS Lett.* **13**, 309–311.

Harris, P. J., Henry, R. J., Blakeney, A. B. and Stone, B. A. (1984). *Carbohydr. Res.* **127**, 59–73.

Hizukuri, S. (1985). *Carbohydr. Res.* **141**, 295–305.

Hizukuri, S. (1986). *Carbohydr. Res.* **147**, 342–347.

Hizukuri, S. and Hisatsuka, T. (1976). *J. Agric. Chem. Soc. Japan* **50**, 489–494.

Hizukuri, S. and Osaki, S. (1978). *Carbohydr. Res.* **63**, 261–264.

Hizukuri, S. and Takagi, T. (1984). *Carbohydr. Res.* **134**, 1–10.

Hizukuri, S., Tabata, S. and Nikuni, Z. (1970). *Staerke* **22**, 338–343.

Hizukuri, S., Takeda, Y., Yasuda, M. and Suzuki, A. (1981). *Carbohydr. Res.* **94**, 205–213.

Hizukuri, S., Kaneka, T. and Takeda, Y. (1983a). *Biochim. Biophys. Acta* **760**, 188–191.

Hizukuri, S., Shirasaka, K. and Juliano, B. O. (1983b). *Staerke* **35**, 348–350.

Hizukuri, S., Takeda, Y., Shitaozono, T., Abe, J., Ohtara, A., Takeda, C. and Suzuki, A. (1988). *Staerke* **40**, 165–171.

Holm, J., Björck, I., Drews, A. and Asp, N.-G. (1986). *Staerke* **38**, 224–226.

Imberty, A., Chanzy, H. and Perez, S. (1987a). *Food Hydrocolloids* **1**, 455–459.

Imberty, A., Chanzy, H., Perez, S., Buleon, A. and Tsan, V. (1987b). *Macromolecules* **20**, 2634–2636.

Itaya, K. and Ui, M. (1966). *Clin. Chim. Acta* **14**, 361–366.

Jorgensen, B. B. and Jorgensen, O. B. (1960). *Acta Chem. Scand.* **14**, 2135–2138.
Kainuma, K., Kobayashi, S. and Harada, T. (1978). *Carbohydr. Res.* **61**, 345–357.
Karkalas, J. (1985). *J. Sci. Food Agric.* **36**, 1019–1027.
Kassenbeck, P. (1978). *Staerke* **30**, 40–46.
Knutson, C. A. (1986). *Cereal Chem.* **63**, 89–92.
Knutson, C. A., Cluskey, J. E. and Dintzis, F. R (1982). *Carbohydr. Res.* **101**, 117–128.
Kobayashi, S., Schwartz, S. J. and Lineback, D. R. (1985). *J. Chromatogr.* **319**, 205–214.
Kodama, M., Noda, H. and Kamata, T. (1978). *Biopolymers* **17**, 985–1002.
Koehler, L. H. (1952). *Anal. Chem.* **24**, 1576–1579.
Krueger, B. R., Knutson, C. A., Inglett, G. E. and Walker, C. E. (1987). *J. Food Sci.* **52**, 715–718.
Kuge, T., Kobayashi, K., Tanahashi, H., Igushi, T. and Kitamura, S. (1984). *Agric. Biol. Chem.* **48**, 2375–2376.
Kugimiya, M. and Donovan, J. W. (1981). *J. Food Sci* **46**, 765–770, 777.
Kunst, A., Draeger, B. and Ziegenhorn, J. (1984). *In* "Methods of Enzymic Analysis", 3rd edn, Vol. VI (H. U. Bergmeyer, ed.), pp. 163–172. Verlag Chemie, Weinheim.
Kurtzman, R. H., Jones, F. T. and Bailey, G. F. (1973). *Cereal Chem.* **50**, 312–322.
Larson, B. L., Gilles, K. A. and Jenness, R. (1953). *Anal. Chem.* **25**, 802–804.
Leach, H. W. and Schoch, T. J. (1962). *Cereal Chem.* **39**, 318–327.
Lelievre, J., Lewis, J. A. and Marsden, K. (1986). *Carbohydr. Res.* **153**, 195–203.
Leonard, R. and Sterling C. (1972). *J. Ultrastruct. Res.* **39**, 85–95.
Libby, R. A. (1970). *Cereal Chem.* **47**, 273–281.
Lii, C. Y. and Lineback, D. R. (1977). *Cereal Chem.* **54**, 138–149.
Lineback, D. R. (1986). *Denpun Kagaku* **33**, 80–88.
Liu, T.-T. Y. and Shannon, J. C. (1981a). *Plant Physiol.* **67**, 518–524.
Liu, T.-T. Y. and Shannon, J. C. (1981b). *Plant Physiol.* **67**, 525–529.
Lorenz, K. and Kulp, K. (1982). *Staerke* **34**, 50–54.
Lowy, G. D. A., Sargeant, J. G. and Schofield, J. D. (1981). *J. Sci. Food Agric.* **32**, 371–377.
MacDonald, F. D. and apRees, T. (1983). *Biochim. Biophys. Acta* **755**, 81–89.
MacGregor, E. A. and Greenwood, C. T. (1980). *In* "Polymers in Nature". pp. 36–73. John Wiley, Chichester.
Macherel, D., Kobayashi, H. and Akazawak, T. (1985). *Biochem. Biophys. Res. Commun.* **133**, 140–146.
Mäkelä, M. J., Korpela, T. and Laakso, S. (1982). *Staerke* **34**, 329–334.
Mäkelä, M. J. and Laasko, S. (1984). *Staerke* **36**, 159–163.
Manners, D. J. (1979). *In* "Polysaccharides in Foods" (J. M. V. Blanshard and J. Mitchell, eds), pp. 75–91. Butterworths, London.
Manners, D. J. (1985a). *In* "Biochemistry of Storage Carbohydrates in Plants" (P. M. Dey and R. A. Dixon, eds), pp. 149–203. Academic Press, London.
Manners, D. J. (1985b). *Cereal Foods World* **30**, 461–467.
Manners, D. J. (1985c). *Cereal Foods World* **30**, 721–727.
Manners, D. J. (1985d). *In* "New Approaches to Research on Cereal Carbohydrates" (R. D. Hill and L. Munck, eds), pp. 45–54, Elsevier, Amsterdam.
Manners, D. J. and Matheson, N. K. (1981). *Carbohydr. Res.* **90**, 99–110.
Marshall, J. J. (1974). *Adv. Carbohydr. Chem. Biochem.* **30**, 257–370.
Marshall, J. J. and Whelan, W. J. (1974). *Arch. Biochem. Biophys.* **161**, 234–238.
Matheson, N. K. and Welsh, L. A. (1988). *Carbohydr. Res.* **180**, 301–313.
McDonald, A. M. L. and Stark, J. R. (1988). *J. Inst. Brew.* **94**, 125–132.
Mercier, C. and Kainuma, K. (1975). *Staerke* **27**, 289–292.
Meredith, P. (1970). *Cereal Chem.* **47**, 492–500.
Meredith, P., Dengate, H. N. and Morrison, W. R. (1978). *Staerke* **30**, 119–125.
Morrison, W. R. (1964). *Anal. Biochem.* **7**, 218–224.
Morrison, W. R. (1981). *Staerke* **33**, 408–410.
Morrison, W. R. (1988a). *J. Cereal Sci.* **8**, 1–15.
Morrison, W. R. (1988b). *In* "Wheat; Chemistry and Technology" 3rd edn, Vol. 1 (Y. Pomeranz, ed.), pp. 373–439. American Association of Cereal Chemists, St. Paul, MN, USA.
Morrison, W. R. and Coventry, A. M. (1985). *Staerke* **37**, 83–87.
Morrison, W. R. and Gadan, H. (1987). *J. Cereal Sci.* **5**, 263–275.

Morrison, W. R. and Laignelet, B. (1983). *J. Cereal Sci.* **1**, 9–20.
Morrison, W. R. and Scott, D. C. (1986). *J. Cereal Sci.* **4**, 13–21.
Morrison, W. R., Mann, D. L., Wong, S. and Coventry, A. M. (1975). *J. Sci. Food Agric.* **26**, 507–521.
Morrison, W. R., Tan, S. L. and Hargin, K. D. (1980). *J. Sci. Food Agric.* **31**, 329–340.
Morrison, W. R., Milligan, T. P. and Azudin, M. N. (1984). *J. Cereal Sci.* **2**, 257–271.
Morrison, W. R., Scott, D. C. and Karkalas, J. (1986). *Staerke* **38**, 374–379.
Moss, G. E. (1976). *In* "Examination and Analysis of Starch and Starch Products" (J. A. Radley, ed.), pp. 1–32. Applied Science, London.
Moulik, S. P. and Gupta, S. (1986). *J. Sci. Industr. Res.* **45**, 173–178.
Munck, L., Rexen, F. and Haastrup Pedersen, L. (1988). *Staerke* **40**, 81–87.
Nelson, N. (1944). *J. Biol. Chem.* **153**, 375–380.
Nelson, O. E. (1980). *Adv. Cereal Sci. Technol.* **3**, 41–71.
Nikuni, Z. (1978). *Staerke* **30**, 105–111.
Ough, L. D. (1964). *In* "Methods in Carbohydrate Chemistry", Vol. 4: Starch (R. L. Whistler, ed.), pp. 91–98. Academic Press, New York.
Park, J. T. and Johnson, M. J. (1949). *J. Biol. Chem.* **181**, 149–151.
Peat, S., Whelan, W. J. and Thomas, G. J. (1952). *J. Chem. Soc.*, 4546–4548.
Peterson, G. L. (1977). *Anal. Biochem.* **83**, 346–356.
Pfannemüller, B. (1986). *Staerke* **38**, 401–407.
Pfannemüller, B., Mayerhofer, H. and Schulz, R. C. (1971). *Biopolymers* **10**, 243–261.
Praznik, W. (1986). *Staerke* **38**, 292–296.
Praznik, W. and Ebermann, R. (1979). *Staerke* **31**, 288–293.
Praznik, W., Burdiceck, G. and Beck, R. H. F. (1986). *J. Chromatogr.* **357**, 216–220.
Praznik, W., Beck, R. H. F. and Eigner, W. D. (1987). *J. Chromatogr.* **387**, 467–472.
Preiss, J. (1982). *Ann. Rev. Plant Physiol.* **33**, 431–451.
Preiss, J. and Levi, C. (1980). *In* "The Biochemistry of Plants", Vol. 3 (J. Preiss, ed.), pp. 371–423. Academic Press, New York.
Radley, J. A. (1976a). "Examination and Analysis of Starch and Starch Products", Applied Science, London.
Radley, J. A. (1976b). "Industrial Uses of Starch and its Derivatives". Applied Science, London.
Raphaelides, S. and Karkalas, J. (1988). *Carbohydr. Res.* **172**, 65–82.
Reichert, E. T. (1913). "The Differentiation and Specificity of Starches in Relation to Genera, Species, etc." Parts I and II. The Carnegie Institute of Washington, Washington D.C.
Rijven, A.-H. G. C. (1984). *Plant Physiol.* **75**, 323–328.
Robin, J. P., Mercier, C., Charbonniere, R. and Guilbot, A. (1974). *Cereal Chem.* **51**, 389–406.
Robyt, J. F. (1984) *In* "Starch: Chemistry and Technology", 2nd edn (R. L. Whistler, J. N. BeMiller and E. F. Paschall, eds), pp. 87–123. Academic Press, Orlando, FL, USA.
Robyt, J. F. and Bemis, S. (1967). *Anal. Biochem.* **19**, 56–60.
Robyt, J. F. and Whelan, W. J. (1968). *In* "Starch and its Derivatives", 4th edn. (J. A. Radley, ed.), pp. 430–476. Chapman and Hall, London.
Russell, P. L., Gough, B. M., Greenwell, P., Fowler, A. and Munro, H. S. (1987). *J. Cereal Sci.* **5**, 83–100.
Saenger, W. (1984). *Naturwissenschaften* **71**, 31–36.
Sano, Y., Maekawa, M. and Kikuchi, H. (1985). *J. Heredity* **76**, 221–222.
Sargeant, J. G. (1982). *Staerke* **34**, 89–92.
Sarko, A. and Wu, H. C. H. (1978). *Staerke* **30**, 73–78.
Shannon, J. C. and Garwood, D. L. (1984). *In* "Starch: Chemistry and Technology", 2nd edn (R. L. Whistler, J. N. BeMiller and E. F. Paschall, eds.), pp. 25–86. Academic Press, Orlando, FL, USA.
Schofield, J. D. and Greenwell, P. (1987). *In* " Cereals in a European Context" (I. D. Morton, ed.), pp. 407–420. Ellis Horwood, Chichester.
Shure, M., Wessler, S. and Federoff, N. (1983). *Cell* **35**, 225–233.
Skerritt, J. H. (1988). *Adv. Cereal Sci. Technol.* **9**, 263–338.
Snyder, E. M. (1984). *In* "Starch: Chemistry and Technology", 2nd edn. (R. L. Whistler, J. N. BeMiller and E. F. Paschall, eds), pp. 661–673. Academic Press, Orlando, FL, USA.
Somogyi, M. (1952). *J. Biol. Chem.* **195**, 19–23.

South, J. B. and Morrison, W. R. (1990). *J. Cereal Sci.* (in press).
Stark, J. R. and Yin, X. S. (1986). *Staerke* **38**, 369–374.
Stute, R. and Konieczny-Janda, G. (1983). *Staerke* **35**, 340–347.
Suzuki, A., Hizukuri, S. and Takeda, Y. (1981). *Cereal Chem.* **58**, 286–290.
Suzuki, A., Takeda, Y. and Hizukuri, S. (1985). *Denpun Kagaku* **32**, 205–212.
Suzuki, A., Kanayama, A., Takeda, Y. and Hizukuri, S. (1986). *Denpun Kagaku* **33**, 191–198.
Tabata, S., Nagata, K. and Hizukuri, S. (1975). *Staerke* **27**, 333–335.
Takagi, T. and Hizukuri, S. (1984). *J. Biochem. (Tokyo)* **95**, 1459–1467.
Takeda, C., Takeda, Y. and Hizukuri, S. (1983). *Cereal Chem.* **60**, 212–216.
Takeda, C., Takeda, Y. and Hizukuri, S. (1987). *Denpun Kagaku* **34**, 31–37.
Takeda, Y. (1987). *Denpun Kagaku* **34**, 225–233.
Takeda, Y. and Hizukuri, S. (1981). *Carbohydr. Res.* **89**, 174–178.
Takeda, Y., Hizikuri, S., Ozono, Y. and Suetake, M. (1983). *Biochim. Biophys. Acta* **749**, 302–311.
Takeda, Y., Shirasaka, K. and Hizukuri, S. (1984). *Carbohydr. Res.* **132**, 83–92.
Takeda, Y., Hizukuri, S. and Juliano, B. O. (1986a). *Carbohydr. Res.* **148**, 299–308.
Takeda, Y., Tokunaga, N., Takeda, C. and Hizukuri, S. (1986b). *Staerke* **38**, 345–350.
Takeda, Y., Hizukuri, S. and Juliano, B. O. (1987a). *Carbohydr. Res.* **168**, 79–88.
Takeda, Y., Hizukuri, S., Takeda, C. and Suzuki, A. (1987b). *Carbohydr. Res.* **165**, 139–145.
Takeda, Y., Shitaozono, T. and Hizukuri, S. (1988a). *Staerke* **40**, 51–54.
Takeda, Y., Suzuki, A. and Hizukuri, S. (1988b). *Staerke* **40**, 132–135.
Takeo, K., Tokumura, A. and Kuge, T. (1973). *Staerke* **25**, 357–362.
Teitelbaum, R. C., Ruby, S. L. and Marks, T. J. (1978). *J. Am. Chem. Soc.* **100**, 3215–3217.
Ullman, M. (1973). "Analytische Kennzeichnung von Amylose und Amylopktin. Handbuch der Stärke in Einzeldarstellung", Vol. VII, Part 2 (M. Ullman, ed.). Paul Parey, Berlin.
Veregin, R. P., Fyfe, C. A., Marchessault, R. H. and Taylor, M. G. (1986). *Macromolecules* **19**, 1030–1034.
Villareal, C. P. and Juliano, B. O. (1986). *Staerke* **38**, 118–119.
Voss, P. and Bergthaller, W. (1982). *Getreide Mehl Brot* **36**, 143–147.
Wharton, D. C. and McCarty, R. E. (1972). "Experiments and Methods in Biochemistry", pp. 197–200. Macmillan, New York.
Whistler, R. L. (1964). "Methods in Carbohydrate Chemistry", Vol. 4: Starch (R. L. Whistler, ed.). Academic Press, New York.
Whistler, R. L. and Daniel, J. R. (1984). *In* "Starch: Chemistry and Technology", 2nd edn. (R. L. Whistler, J. N. BeMiller and E. F. Paschall, eds), pp. 153–182. Academic Press, Orlando, FL, USA.
Wild, D. and Blanshard, J. M. V. (1986). *Carbohydr. Polym.* **6**, 121–143.
Williams, J. M. and Duffus, C. M. (1977). *Plant Physiol.* **59**, 189–192.
Williams, P. C. and LeSeelleur, G. C. (1970). *Cereal Sci. Today* **15**, 4–19.
Williams, P. C., Kuzina, F. D. and Hlynka, I. (1970). *Cereal Chem.* **47**, 411–420.
Wolf, M. J., Melvin, E. H., Garcia, W. J., Dimler, R. J. and Kwolek, W. F. (1970). *Cereal Chem.* **47**, 437–446.
Wu, H.-C. H. and Sarko, A. (1978a). *Carbohydr. Res.* **61**, 7–25.
Wu, H.-C. H. and Sarko, A. (1987b). *Carbohydr. Res.* **61**, 27–40.
Yajima, H. and Nishimura, T. (1987). *Carbohydr. Res.* **163**, 155–167.
Yajima, H., Nishimura, T., Ishii, T. and Handa, T. (1987). *Carbohydr. Res.* **163**, 155–167.
Yamaguchi, M., Kainuma, E. and French, D. (1979). *J. Ultrastruct. Res.* **69**, 249–261.
Yamamoto, M., Sano, T. and Yasunaga, T. (1982). *Bull. Chem. Soc. Japan* **55**, 1886–1889.
Yin, X. S. and Stark, J. R. (1988). *J. Cereal Sci.* **8**, 17–28.
Yu, L.-P. and Rollings, J. E. (1987). *J. Appl. Polymer Sci.* **33**, 1909–1921.
Zobel, H. F. (1964). *In* "Methods in Carbohydrate Chemistry", Vol. 4: Starch (R. L. Whistler, ed.), pp. 109–113. Academic Press, New York.
Zobel, H. (1988a). *Staerke* **40**, 1–7.
Zobel, H. (1988b). *Staerke* **40**, 44–50.

10 Fructans

HORACIO G. PONTIS

Centro de Investigaciones Biológicas, Fundación para Investigaciones Biológicas Aplicadas (FIBA), Casilla de Correo 1348, 7600 Mar del Plata, Argentina

I. GENERAL CONSIDERATIONS

Several reviews on the occurrence, structure and metabolism of fructans have recently appeared (Meier and Reid, 1982: Pontis and Del Campillo, 1985; Pollock, 1986, Nelson and Spollen, 1987; Hendry, 1987). However, no publication has dealt with the experimental aspects of fructan biochemistry since Whelan's chapter in *Handbuch der Pflanzen Physiologie* was published in 1955. The purpose of this review is therefore to

METHODS IN PLANT BIOCHEMISTRY Vol. 2
ISBN 0-12-461012-9

attempt to present the current status of experimental techniques and methods used for studying the different aspects of fructan biochemistry.

A. Historical Background

Fructans are polymers of fructose carrying a D-glucosyl residue at the end of the chain attached via a β-(2→1)-linkage as in sucrose. They constitute a series of homologous oligosaccharides which can be considered as derivatives of sucrose (Hirst, 1957). Fructans were previously called fructosans, a term that probably initiated after Schlubach and Sinh (1940) utilised the word 'polyfructosan' to describe the fructose polymers present in monocotyledons, because they thought they were derivatives of cyclic structures of fructose called fructosan. However, contemporaneously, by similarity with the polymers of D-glucose found in Nature, the name fructan is assigned to describe the polymers of D-fructose, and fructosan is rarely used. There is still a minor point of conflict regarding the use of the plural form of the name: fructans. Some authors feel that only the singular form should be used likening the term to starch, whose plural is never employed. Other authors, however, feel that as there are many polymers of fructose present as a series simultaneously in the same tissue, the plural form is more appropriate. In this chapter the latter will be used.

B. Occurrence

Fructans are widely distributed in the plant kingdom. They are not only present in monocotyledons and dicotyledons, but also in green algae (Meier and Reid, 1982; Pontis and del Campillo, 1985; Hendry, 1987). There is also a report of the presence of a glucofructan in blue-green algae (Tsusué and Fugita, 1964). Nevertheless, it should be pointed out that many early reports on the presence of fructan going back to the 1870s appear to be unreliable when tested using modern analytical techniques. However, even taking a conservative approach regarding reports on fructan presence, Hendry (1987) predicts that the number of species world-wide which can be expected to contain fructans is about 36 000, that is about 12% of the angiospermous flora, Compositae being the dominant family.

Fructans have differences in molecular structure and in molecular weight. They may be classified in three main types: the inulin group, the phlein group, and the branched group (see below). The fructans of the dicotyledons are, as far as we know, all of the inulin group, while most of the fructans of monocotyledons are of the phlein type or of the branched type. However, the first member of the inulin series, isokestose, seems to be ubiquitous in Graminea (Wagner and Wiemken, 1987).

Whatever their source, most fructans are distributed in an unbroken series between a fructosyl-sucrose and the member of the highest molecular weight found in that plant. This fact should be taken into account when considering the many fructans that have been given trivial and highly specialised names (i.e. inulin, asparagosan) on the basis of their origin. It is now clear that these names usually refer to the isolated polymer of highest molecular weight. It should also be borne in mind that the highest member of a series may well be a mixture of neighbouring homologues that are difficult to separate by virtue of their similar physical properties.

In general, fructans are distributed throughout the plants in which they occur, although their concentration in different parts of the same plant varies considerably.

Usually, their amount is very small in leaves and especially large in roots, bulbs, tubers, rhizomes and in some immature fruits (Meier and Reid, 1982). In some grasses, the lower sections of the stem are quite rich in fructans (Smith, 1973). An outstanding characteristic of fructans is that they can be accumulated to very high concentrations and values exceeding 50% dry weight of tissue have been recorded for several members of the Compositae, Liliaceae and Gramineae (Edelman and Jefford, 1968; Smith, 1973; Darbyshire and Henry, 1981; Wagner *et al.*, 1983). Nevertheless, important variations occur during the life-cycle of the plants (Whistler and Smart, 1953). Subcellularly, fructans appear to be located mainly in the vacuole (Frehner *et al.*, 1984).

C. Chemical Nature

Fructans have a unique structural feature within the family of polysaccharides in that no bond of the fructose furanose ring is part of the macromolecular backbone (Marchessault *et al.*, 1980). Besides, they are one of the few natural polymers where the carbohydrate exists in the furanose form (Fig. 10.1). These two structural features play an important role in the final conformation of the molecules in solution. Moreover, the enhanced flexibility of the furanose ring in comparison with the relatively rigid pyranose ring of the majority of reserve polysaccharides brings additional flexibility to the whole fructan molecule (Marchessault *et al.*, 1980).

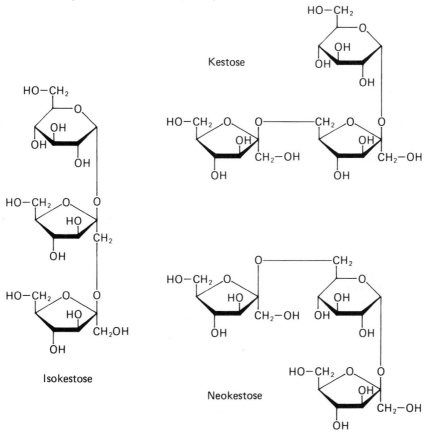

FIG. 10.1. Structure of fructosyl-sucroses.

As indicated before, there are three main groups of fructans: (a) the inulin type or those with β-$(2\rightarrow1)$-D-fructofuranosyl units, (b) the phlein type or those with β-$(6\rightarrow2)$-D-fructofuranosyl units, and (c) the branched type or those with both kinds of glycosidic linkages. The smallest fructose derivative belonging to the inulin type is the trisaccharide isokestose (1-kestose), while the corresponding trisaccharide of the phlein type is 6-kestose (also called kestose; see Fig. 10.1). A minor group is formed by those fructans with β-$(2\rightarrow1)$-D-fructo-furanosyl linkages, but they are derivatives of neokestose or 6-fructofuranosyl-sucrose. By prolonging the chain at either of the two terminal fructoses, fructans with a non-terminal glucose are obtained (Hammer, 1970; Tomoda et al., 1973; Shiomi et al., 1976, 1979; Shiomi, 1981a,b).

The general properties of fructans can be summarised as follows: they are laevorotatory, amorphous or microcrystalline, of varying solubility in cold water, very soluble in hot water, insoluble in absolute alcohol, not precipitated by lead acetate, except in alkaline solutions, but precipitated by baryte, either directly or on addition of alcohol (Colin and Belval, 1940; Whelan, 1955). They are also non-reducing, not hydrolysed by yeast invertase (this may not be true of low molecular-weight fructans), resistant to amylase action, but very susceptible to hydrolysis by acid (Archbold, 1940). They are not stained by iodine, but hydrochloric acid vapour imparts a purple colour that distinguishes them from polysaccharides not containing fructose (Colin and De Cugnac, 1926).

The degree of polymerisation (DP) varies with plant species and with their life-cycle, but all fructans have a low degree of polymerisation compared with starch. The number of fructose units range from ~10 in the fructans of onion (Bacon, 1957) to ~260 in *Phleum pratense* (Groteluschen and Smith, 1968).

D. Role of Fructans

Fructans are reserve polysaccharides, and their role is clearly fulfilled in many fructan-containing plants. Thus, fructans are accumulated in the underground storage organs in the Compositae (Whelan, 1955; Whistler and Smart, 1953), while in Gramineae they are distributed in all parts of the plants. However, the major site of accumulation is the elongated stem internodes, with the highest concentration in the lower ones (Smith, 1967, 1973).

Fructans have been considered to play a role as osmotic regulator, (Pontis, 1971; Darbyshire and Henry, 1981), and in chill and freeze tolerance (Eagles, 1967; Edelman and Jefford, 1968; Gunn and Walton, 1985; Pontis, 1989). These views originated from observations correlating fructan accumulation with cold-hardiness. Nevertheless, Pollock (1984a,b, 1986) has stated a number of reservations. He has noted that the accumulation of fructan occurs as a consequence of growth stoppage at low temperature, and that much of the evidence supporting the cryoprotection theory is hedged with qualifications and exceptions. The issue thus remains a controversial one and further data will be needed to resolve it.

E. Industrial Applications

Owing to its great sweetness and high utilisation in the body, D-fructose has been of special interest in nutrition for many decades. Therefore, cultivation of Jerusalem artichoke (*Helianthus tuberosus* L.) has received widespread interest as a potential new crop because of its reserve carbohydrate inulin. It could be used not only for the production of fructose, high-fructose syrups, ethanol (through microbial fermentation) and other organic solvents, but also for the manufacture of furan-based and other chemicals.

Moreover, a mixture of isokestose, nystose (GF3) and 1-fructofuranosyl nystose (GF4) is commercially produced from sucrose in Japan under the trade name of Neosugar, and it is widely used for humans and animals in health foods and feeds.

II. METHODS

In this section the methods for the isolation, separation and determination of fructans are considered, as well as the assays used for estimating the enzymes associated with their metabolism.

A. Isolation

It is necessary to prepare the plant material in order to prevent enzyme action that may alter the composition and the total amount of fructans present. This is accomplished by freezing the plant material in dry ice, covering the specimen with alcohol or by freeze-drying (Laidlaw and Wylam, 1952). Enzymes are then inactivated by boiling with alcohol, or alternatively the material may be extracted directly in boiling water (see below). Drying at a low oven temperature does not inactivate the enzymes (Archbold, 1938) while drying at 105°C causes a change in the fructan content (Laidlaw and Wylam, 1952; Raguse and Smith, 1965).

The subsequent procedure depends on the presence of low molecular weight fructans. If they are absent, the extraction with boiling 80% ethanol serves both to inactivate enzymes and to extract sugars. When a series of low molecular weight fructans is present, the extraction with boiling alcohol would effect an arbitrary fractionation of the fructans. However, in this case, the alcoholic extract must be combined with the subsequent water extract (Archbold, 1938). Alternatively, sugars and fructans are both extracted with boiling water (Bacon and Loxley, 1952). In each case the problem becomes one of detecting fructans in the presence of sugars. This can be accomplished by analytical procedures or by separation of the accompanying sugar and the fructans.

1. Isolation of fructans from underground organs and leaves

Plant material (tubers, roots, bulbs, rhizomes) is washed, peeled, cut in small pieces, ground in a Waring Blender in 80% ethanol containing a small amount of calcium carbonate and subsequently extracted by boiling over a water bath. The extraction step is repeated three times. The resulting extract is concentrated to near dryness *in vacuo*, and redissolved in water. The residue left after the ethanol extraction is extracted twice further with boiling water. The two water extracts are combined and mixed with the aqueous solution obtained by dissolving the residue remaining after the ethanol has been removed from the ethanol extractions. The aqueous solution is passed through a column of Amberlite MB-3, the carbonate form, and lyophilised. The powder obtained contains glucose, fructose, sucrose and fructans.

Comments. The procedure described has been used with minor variations to isolate fructans from Jerusalem artichoke tubers (Edelman and Jefford, 1964), dandelion, chicory (Rutherford and Weston, 1968), and asparagus roots (Shiomi *et al.*, 1976), onion bulbs (Darbyshire and Henry, 1978), Liliaceae rhizomes (Tomoda *et al.*, 1973), and from *Dactylis glomerata* (Pollock, 1982a), *Lolium temulentum* (Pollock, 1982b), barley (Wagner and Wiemken, 1987), wheat (Calderón and Pontis, 1985), *Festuca* (Schnyder and Nelson, 1987) leaves.

As the ethanol extract contains glucose, fructose, sucrose and fructans of DP < 7, a rough separation is obtained if it is not mixed with the water extracts. The aqueous solutions may need clarification. This is accomplished by centrifugation ($2000 \times g$, 10–20 min) prior to the liophilisation step. The calcium carbonate is added to maintain a slightly alkaline pH during extraction. Care must be taken throughout the procedure in order to prevent fructan hydrolysis in acid pH. The yield of fructan varies according to the stage of development of the plant. From Jerusalem artichoke tubers collected at the beginning of autumn, a yield of up to 36% dry weight of total fructans may be obtained.

When the amount of tissue to be extracted is small (1–10 g), fructans may be also isolated by adding the ground tissue to boiling water (pH 8.0).

Extraction of fructan has also been performed by first drying leaf samples at 70°C for 40 h and then grinding the dried tissue with water in a mortar. Extracts are filtered and the filtrate concentrated *in vacuo* or liophilised. The authors claimed that no difference is found between fructan extracted from fresh or dried tissue (Schnyder and Nelson, 1987).

B. Separation

The difficulty in the resolution of fructan mixtures lies in their extremely similar structural characteristics. The only difference between two consecutive members of any of the homologous fructan series is a single fructosyl residue.

Separation of fructans is based on their differences either in size or in solubility. The first methods for the separation of sugars and fructans, and for the resolution of fructan mixtures, were based on the different solubilities in alcohol solutions. These procedures were long and required many precipitations without obtaining clear-cut separations (Schlubach, 1958).

TABLE 10.1. Solvents used for paper chromatography.

Solvent (v/v)	Comments	Reference
1-Butanol–pyridine–water (6:4:3)	Good resolution up to DP 8 in 24–40 h depending on paper used	Pontis (1966)
Ethylacetate–pyridine–water (12:5:4)	Good resolution between fructosyl-sucrose and higher DP up to 5.	Tognetti *et al.* (1989)
1-Propanol–ethylacetate–water (7:1:2) (6:1:3)	Separation up to DP 14 in 5 days.	Edelman and Dickerson (1966)
1-Propanol–ethylacetate–water (11:2:7)	Separation up to DP 25 in 3 weeks.	Edelman and Dickerson (1966)

Descending paper chromatography (PC) was the method used preferentially during the 1960s, and it is still being used. Table 10.1 presents the usual solvents as well as comments on the resolutions obtained.

Thin layer chromatography (TLC) permits faster separations and even resolves mixtures of the three trisaccharides: kestose, isokestose and neokestose. The systems usually allow clear-cut separation of all oligosaccharides up to a DP of 9 fructosyl units (Table 10.2). The position of fructans on paper chromatograms and TLC plates may be ascertained by developing or spraying with reagents specific for fructose (Table 10.3). These reagents do not distinguish between free and combined fructose; as yet there is no test that can accomplish the differentiation.

TABLE 10.2. Thin layer chromatography of fructans.

Solvent (v/v)	Support	Comments	Reference
1-Propanol–ethylacetate–water (60:10:30)	Cellulose in 33 mM K_2HPO_4 on glass plates	Separates DP 7; three developments	Karlsson (1969)
1-Propanol–ethylacetate–water (6:2:2)	Silica gel on glass plates	DP 2–20, developed twice, 90 min each	Collins and Chandorkar (1971)
1-Propanol–ethylacetate–water (40:50:10)	Silica gel on glass plates	DP 7–8, developed twice, 90 min each	Collins and Chandorkar (1971)
1-Butanol–isopropanol–water (3:12:4)	Silica gel on plastic sheets[a]	Separates up to DP 9[b]; developed twice	Cairns and Pollock (1988)
Acetone–water (87:13)	Silica gel on plastic sheets[a]	Separates up to DP 9[b]; developed twice	Wagner and Wiemken (1987)
1-Butanol–acetic acid–water (2:1:1)	Silica gel on glass plates[c]	Separates up to DP 9; developed twice	Schnyder and Nelson (1987)

[a] Ready-fold, F 1500, Scheiler and Schull, FRG.
[b] Both solvents separate the trisaccharides kestose, isokestose and neokestose.
[c] Ready-plate, Fischer 06-600 A, USA.

TABLE 10.3. Composition of reagents used for detecting fructans on PC and TLC.

Reagent	Composition	Reference
Resorcinol-HCl	A 1:9 (v/v) mixture of 1% ethanolic resorcinol and 2 N HCl. The sprayed paper is heated at 80°C for 10 min. Fructose, red colour; aldohexose no colour.	Forsyth (1950)
Naphthoresorcinol-HCl	Naphthoresorcinol (100 mg) dissolved in a mixture of 2 N HCl (20 ml) and ethanol (80 ml). The sprayed paper is heated at 75°C. Fructose shows with a bright red colour; aldohexose, no colour.	Forsyth (1948)
Urea-HCl	Urea, 5 g dissolved in a mixture of 2 N HCl and 80 ml ethanol or 1-propanol. The sprayed paper is heated at 105°C. Fructose free or combined, dark blue-grey; aldohexose, no colour.	Dedonder (1952)

Adsorption chromatography has been successfully used by Labourel and Péaud-Lenoel (1969) using a silica gel column. Separation was achieved with a gradient solvent system of 1-butanol–ethanol–water exponentially enriched in water and ethanol and gradually impoverished in butanol. This procedure separated fructans present in *Helianthus tuberosus* tubers up to DP 20.

Shiomi *et al.* (1976) have separated fructans using carbon–celite (1:1; w/w) columns eluted first with water and then with increasing concentrations of ethanol. This technique produces bulk separations with usually two or more oligosaccharides in the same fructan peak which must be separated by another method. Nevertheless, it allows fractionation on a preparative scale.

Exclusion chromatography exploits the small difference in size that exists among fructans. It was initially developed by Pontis (1968), who was able to separate fructans from dahlia tubers by gel filtration on a Bio-gel P-2 (200–400 mesh) column, 210 cm long. This method allowed a clear-cut resolution of low molecular weight fructans up to DP 6–8. The procedure was improved by using Bio-gel P-2 (minus 400 mesh) or by running chromatography at 50°C. It has been employed for the separation of fructans present in onion (Darbyshire and Henry, 1978), *Lolium temulentum* (Pollock, 1982b), wheat (Blacklow *et al.*, 1984) and *Helianthus tuberosus* (Cairns and Pollock, 1988).

High performance liquid chromatography (HPLC) using calcium (Cairns and Pollock, 1988) or the silver form (Scobel and Brobst, 1981) of cation-exchange resins permits detection of individual oligosaccharides up to DP 4. Sucrose, glucose and fructose are completely resolved, while the three trisaccharides (kestose, isokestose and neokestose) are eluted as a single peak. The tetrasaccharide is also separated, but oligosaccharides of higher DP are eluted together. On the other hand, by selecting a different column for HPLC (Spherisorb-5-NH2) and running it with a mobile phase of acetonitrile–water (70:30; v/v), a clear-cut separation of oligosaccharides up to DP 7 has been performed by Frehner *et al.* (1984) (see Fig. 10.2). Elution of fructan from the HPLC column is detected by differential refractometry.

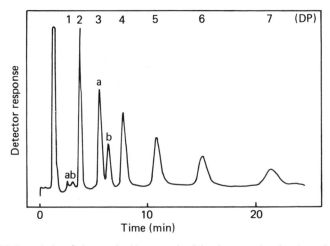

FIG. 10.2. HPLC analysis of the saccharides contained in the vacuolar fraction of resting *Helianthus tuberosus* tubers. The peaks are: 1, monosaccharides: (a) fructose, (b) glucose; 2, sucrose; 3, trisaccharides: (a) isokestose, (b) internal standard (raffinose); 4, 5, 6, 7, tetra-, penta-, hexa- and heptasaccharides of the inulin series, respectively. From Frehner *et al.* (1984).

Recently, the resolution of fructan mixtures has been improved enormously by using strong alkaline conditions (200 mM NaOH) for the elution of anion-exchange columns (AES) and fructans in eluants determined with pulsed amperometric detection (PAD). The technique allows the resolution of the three trisaccharides present in plant tissue (Chatterton *et al.*, 1989a). This methodology has been used to study the fructan composition of 180 selected Gramineae (Chatterton *et al.*, 1988a,b). The speed of the separation (15 min) and the resolution make it the method of choice for assaying the enzymic formation of fructosyl-sucrose.

Gas–liquid chromatography (GLC) also allows the separation of the first members of the series (DP 3-5), but it requires the preparation of fructan trimethylsilyl derivatives (de Miniac, 1970) or fructan methylated alditol acetates (Pollock *et al.*, 1979). The GLC technique has been used to measure the activity of sucrose-sucrose-fructosyl transferase (Frehner *et al.*, 1984) using the procedure of de Miniac (1970).

1. Separation of dahlia and wheat fructans

Fructans present in dahlia tubers collected at the end of summer, and in wheat leaves cut from 15-day-old plants kept at 4°C for the last 7 days, are extracted as indicated in Section II.A.1. A solution of purified fructans is applied to a Bio-gel P 2 (200–400 mesh) column 220 × 1 cm equilibrated with water at pH 8.0. Elution is performed at 10 ml h^{-1} with equilibrating solution. Fractions of 2 ml are collected after 220 ml of eluant has passed through and an aliquot of each fraction is analysed for fructans by the thiobarbituric acid method (see Section II.C; Fig. 10.3). Fractions under a peak are pooled and liophilised. The resulting powder may be kept at $-20°C$ for several months.

Comments. The method described corresponds to the standard procedure used in the author's laboratory by Santoiani and Pontis (unpubl. res.). It is a very simple technique, and permits scaling up without losing resolution. It is the method of choice for obtaining pure separate fructans. However, this is true for homologous series like those present in dahlia or Jerusalem artichoke tubers. For Graminea, where fructans of the β-(2→1) and β-(2→6) series may be found together, this technique does not separate fructans of the same molecular weight belonging to different series. Moreover, the trisaccharide peak may be a mixture of kestoses and may even contain raffinose.

C. Determination

The determination of fructans may be accomplished by measuring the change in reducing power brought about by mild acid hydrolysis. This gives the amount of free plus combined sugars, which would include sucrose, raffinose and stachyose. The initial reducing power gives free sugar content, and hypoiodite oxidation may be used to distinguish free aldoses from free ketoses. Alternatively, glucose present may be estimated with glucose oxidase or by coupling with hexokinase and glucose 6-phosphate dehydrogenase, while fructose may be measured in a similar way using the same enzymes with the addition of phosphoglucoisomerase (see Chapter 1, this volume).

Total fructans, expressed by their fructose content, may be determined by using any of the specific colorimetric methods for ketoses such as the thiobarbituric acid method (Percheron, 1962) or Roe's method (1934). Sucrose and free sugars (glucose and fructose) may be eliminated prior to measuring total fructose content in fructans by hydrolysing sucrose with yeast invertase and destroying the resulting glucose and fructose as well as the existing ones by boiling with NaOH (Pontis, 1966). Invertase may act on the lower fructan members of the inulin series, but Pontis (1966) has shown that it is a rather slow attack and, by selecting a suitable time, it is possible to hydrolyse sucrose completely while barely affecting the fructans.

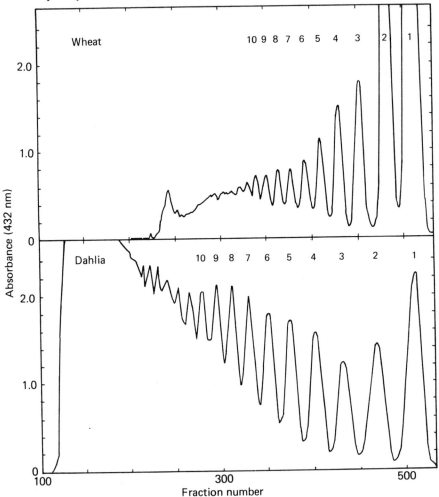

FIG. 10.3. Separation of dahlia (15 mmoles as total fructose) and wheat (3 mmoles as total fructose) fructans by exclusion chromatography. For details, see Section II.B.1. 1, Fructose; 2, sucrose; 3, 4, 5, 6, 7, 8, 9, 10, tri-, tetra-, penta-, hexa-, hepta-, octa-, nona- and decasaccharides, respectively.

Procedures have also been developed for the automated analysis of non-reducing sugars and fructans present in plant tissues (Wolf and Ellmore, 1975; Thome *et al.*, 1986).

A novel approach to the determination of fructans may lie in the anti-fructan activity possessed by some myeloma immunoglobulins (Glaudemans, 1975). They are specific for the inulin type of fructans, i.e. containing β-D-(2→1)-linked D-fructofuranosyl residues (Allen and Kabat, 1957; Streefkerk and Glaudemans, 1977), while others only react with the phlein type, i.e. β-D-(2→6)-linked D-fructofuranosyl residues (Grey et al., 1971; Lundblad et al., 1972; Cisar et al., 1974). Recently, Bona (1989) has published a study on an antibody response to fructans of higher molecular weight. There is a distinct possibility that detection and measurement of higher molecular weight fructans may be possible in a highly specific way through the use of antibodies. However, no attempt has yet been made to use this type of reaction for the quantification of fructans.

Quantitative determination of individual fructans may only be accomplished after their separation by some of the methods previously described.

D. Assays of Fructan-metabolising Enzymes

Although our knowledge of the enzymes involved in the synthesis and degradation of fructans is still limited, three enzymes have been found associated with fructan metabolism.

1. Sucrose-sucrose-fructosyl transferase (SST)

This enzyme catalyses the synthesis of fructosylsucrose according to the following reaction:

$$\text{Sucrose} + \text{sucrose} \underset{}{\overset{\text{SST}}{\rightleftharpoons}} \text{glucose} + \text{fructosyl-sucrose}$$

So far, it seems that there are two SSTs, one catalysing synthesis of isokestose (1F-β-fructosylsucrose) and the other that of kestose (6F-β-fructosylsucrose), as shown by Wagner and Wiemken (1987).

The methods of measuring SST activity are based on the determination of the glucose or the fructosylsucrose formed. Glucose has been estimated using glucose 6-dehydrogenase, hexokinase and phosphoglucoisomerase (Bhatia and Nandra, 1979; Shiomi and Izawa, 1980; Pollock, 1984b). This procedure quantifies changes in glucose and fructose (which appear when invertase is present in the extracts) concentrations rather than measuring the amount of trisaccharide produced. Recently, a procedure has been developed that extends the conventional dehydrogenase-linked assay for fructose and glucose by utilising the intermediary electron carrier, phenazine methosulphate, to couple $NADP^+$ reduction to the production of a formazan dye from the tetrazolium salt, Thiazoyl Blue, in a suitable form for measurement using a microtitre plate (Cairns, 1987). This method offers considerable advantages over the conventional spectrophotometric hexose assay and may be used for screening large numbers of small samples. However, because other reactions may cause changes in the concentration of glucose or fructose, measurements by these procedures provide only empirical estimates of the rates of trisaccharide synthesis. No information is provided by any of these methods regarding the nature of the trisaccharides synthesised in such reactions.

Any methods measuring the trisaccharide directly must separate it from glucose and unreacted sucrose, and from any fructose that may have been produced.

Chromatographic isolation of fructosyl-sucrose may be carried out as indicated in Section II.B by PC (Santoiani and Pontis, unpubl. res.; Tognetti et al., 1989), TLC (Wagner et al., 1983), GLC (Darbyshire and Henry, 1978) and by AES-PAD (Chatterton *et al.*, 1989). The first three methods are slow, each having its own cumbersome step. The position of fructosyl-sucrose must be ascertained in paper chromatograms previous to any determination. Santoiani and Pontis (unpubl. res.) when using [^{14}C]sucrose as substrate for the SST reaction, identified the trisaccharide position by dipping the chromatogram in acetone-HCl and heating at 70°C for 10 min. Fructans appeared as fluorescent spots under UV light, which could be cut for subsequent counting of the labelled fructosyl-sucrose. Similarly, after TLC, spots must be cut from Ready-Foil plates (Wagner *et al.*, 1983), sugars extracted with water and measured colorimetrically. It may be necessary to remove excess sucrose substrate, since high concentrations distort TLC separation and reduce resolution (Cairns and Pollock, 1988). Even taking into account these disadvantages, TLC is able to separate the three trisaccharides although it is difficult to quantify them on account of their proximity. The inconvenience of GLC rests mainly in the need to derivatise samples prior to chromatography. None of these inconveniences is found for the quantification of fructosyl-sucrose by AES-PAD. This method permits the measurement of the three trisaccharides in one sample (see Fig. 10.4), and they may be readily determined from 10 µl samples in concentrations as low as 1 µg ml^{-1}. It will certainly be the method to use in the future, the only disadvantage being the cost of the equipment involved.

(a)   *Assay of SST activity*

Enzyme extract. Primary leaves of barley are homogenised in 50 mM citrate-P-buffer, pH 5.7, 1 mg ml^{-1} BSA at 0°C in a glass tissue grinder or in a chilled mortar (1 ml of buffer per g fr. wt.). The homogenate is centrifuged at $2000 \times g$ for 10 min and the supernatant is desalted by Biogel-P-10, 200–400 mesh, and equilibrated with the same buffer. Alternatively, the supernatant is dialysed against the diluted citrate-P-buffer (5 mM) for 16 h at 4°C.

Incubation mixture. SST is assayed by incubating in a total volume of 0.2 ml, 20 µmol sucrose, 0.2 mg BSA, 4 µmol citrate-P-buffer (pH 5.5) and enzyme (0–0.1 ml) for 1–4 h at 30°C. The reaction is stopped in a boiling water bath for 3 min. After centrifugation for 2 min at $12\,000 \times g$, aliquots of the supernatant are used for determination of glucose or fructosyl-sucrose formed by any of the methods discussed.

2.   *Fructan-fructan-fructosyl transferase (FFT)*

This enzyme in *Helianthus tuberosus* tubers catalyses the transfer of single terminal β-D-fructofuranosyl residues to the same position of another molecule according to:

$$\text{Sucrose-Fru}_n + \text{sucrose-Fru}_m \overset{\text{FFT}}{\rightleftharpoons} \text{sucrose-Fru}_{n-1} + \text{sucrose-Fru}_{m+1}$$

where n may be any member of the fructan series from 1 (trisaccharide) to ~ 35 and m, for the acceptor molecule, is any number from 0 (sucrose) to ~ 35.

Similar transferases have been isolated from asparagus roots (Shiomi *et al.*, 1979) and from onion bulbs (Scott *et al.*, 1966). However, in these cases fructosyl residues are transferred to position 6 of the glucose or fructose moieties of sucrose. The reaction catalysed by fructan-fructosyl transferase has always been awkward to measure because there is a redistribution of fructosyl residues, which in order to be registered, demands separation of the reaction substrates and products that can only be carried out by chromatography.

In Graminea crude extracts, Housley and Pollock (1985) and Housley and Daughtry (1987) estimated the activity of FFT by calculating the increase in the amount of fructose in polymers with DP > 3, after separation on Bio-gel P-2.

Edelman and Dickerson (1966) separated the fructan by PC while Shiomi *et al.* (1979) analysed the reaction products first by carbon–celite column chromatography and then by PC. *Helianthus tuberosus* FFT may also be measured by estimating the formation of fructosyl-sucrose from sucrose and a fructan of DP > 20 (Pontis, 1971), which reduces the separation procedure to that of sucrose and fructosyl-sucrose.

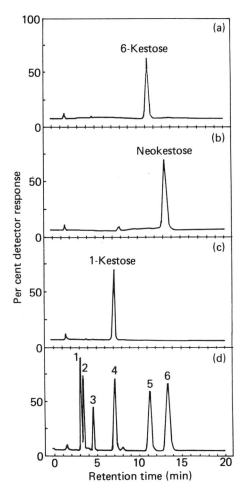

FIG. 10.4. Purified kestoses chromatographed on AES-PAD. (a) 6-Kestose (23 μg ml^{-1}), (b) neokestose (34 μg ml^{-1}) from the action of yeast invertase on sucrose, (c) 1-kestose (26 μg ml^{-1}) from horse chestnut seeds. A composite standard (d) consisting of 8 μg ml^{-1} of each glucose, fructose and sucrose (peaks 1, 2 and 3, respectively) plus 1-kestose (peak 4), 6-kestose (peak 5) and neokestose (peak 6) as in (a), (b) and (c) was also separated. Quantities are expressed as per cent detector response because the individual sugars have different deflections for a given quantity of sugar, as evidenced by three peak weights for 8 μg ml^{-1} each glucose, fructose and sucrose. From Chatterton *et al.* (1989a).

The application of AES-PAD technique may allow the measurement of the reaction in a faster and simpler way. No experimental procedure can be recommended at this stage and the reader is referred to the original contribution already discussed and analysed.

3. Fructan-hydrolases (FH)

This enzyme catalyses liberation of fructose residues from fructans according to the general reaction:

$$F_n-G \xrightarrow{\text{FH}} F_{n-1}-G + F$$

Edelman and Jefford (1964) described the presence of two β-(2→1)-fructan-1-fructan exohydrolases in *Helianthus tuberosus* tubers, and Rutherford and Deacon (1972a,b) in roots of *Taraxacum officinale*. Specific hydrolysis of the β-(2→6)-linked phlein type of fructan occurring mainly in the Graminea was demonstrated with enzymes isolated from tall fescue (Smith, 1976) and *Dactylis glomerata* (Yamamoto and Mino, 1985). Neither of these enzymes attacks sucrose.

The measurement of enzyme activity is carried out by estimating fructose produced. In crude extracts, this is done by TLC (Wagner and Wiemken, 1986) in order to differentiate it from SST and invertase activities. With purified enzyme preparation, activity is estimated by measuring fructose by its reducing power (Yamamoto and Mino, 1985). The enzyme isolated and purified from the stem base of *Dactylis glomerata* (Yamamoto and Mino, 1985) shows high specificity for β-(2→6)-fructosyl linkages, while that from barley leaves presents a wider specificity attacking both β-(2→6)- and β-(2→1)-linkages (Wagner and Wiemken, 1986). The enzyme seems to be confined to the vacuole as shown for barley (Wagner and Wiemken, 1986) and for *Helianthus tuberosus* (Frehner *et al.*, 1984).

(a) *Assay of FH activity.* An enzyme extract is prepared from barley leaves as described for SST.

Incubation mixture. Fructan hydrolase is measured incubating 0.1 ml of enzyme extract, 0.1 ml of 50 mM citrate-P buffer (pH 5.2) and 2 mg of phlein (isolated from stem bases of *Phleum pratense* using the procedure in Section II.A.1) at 30°C for 1–8 h. Reaction is stopped by incubating in a boiling water bath for 3 min. After centrifugation for 2 min at $12\,000 \times g$ (crude extract), aliquots are analysed by TLC or by measuring reducing power by the Somogyi method (1952).

Comments. The procedure described corresponds to that of Wagner and Wiemken (1986) but it is essentially the same as that used by Yamamoto and Mino (1985) for measuring FH in *Dactylis glomerata*. They found that the activity of phleinase was proportional to the molarity of phlein used irrespective of its chain length (mean degree of polymerisation, 30–314). The enzymes isolated from *Taraxacum officinale* and barley are not inhibited by sucrose, while that from *Helianthus tuberosus* is inhibited.

III. CONCLUSIONS—FUTURE PROSPECTS

Research in fructan biochemistry and physiology has increased almost exponentially in the last five years. Recently the First International Symposium on Fructan took place, while the Second Symposium will be held in four year's time. It takes no great insight to predict that techniques of fructan isolation and procedures for measuring enzymes involved in fructan metabolism will improve and expand in the coming years. The arrival of anion-exchange separation and pulsed amperometric detection (AES-PAD) may revolutionise fructan biochemistry, which has been held back for many years for the lack of simpler and faster techniques of analysis. The future will tell if this view is correct.

ACKNOWLEDGEMENTS

The author thanks his colleagues at the Centro de Investigaciones Biológicas for stimulating discussion. HGP is Career Investigator of the Consejo Nacional de Investigaciones Científicas y Técnicas, Argentina.

REFERENCES

Allen, R. Z. and Kabat, E. A. (1957). *J. Exp. Med.* **105**, 383–386.
Archbold, H. K. (1938). *Ann. Bot. (London)* **2**, 183–187.
Archbold, H. K. (1940). *New Phytol.* **39**, 185–219.
Bacon, J. S. D. (1957). *Biochem. J.* **67**, 5p–6p.
Bacon, J. S. D. and Loxley, R. (1952). *Biochem. J.* **51**, 208–213.
Bhatia, I. S. and Nandra, K. S. (1979). *Phytochemistry* **18**, 923–927.
Blacklow, W. M., Darbyshire, B. and Pheloung, P. (1984). *Plant Sci. Lett.* **36**, 213–218.
Bona, C. A. (1989). *J. Plant Physiol.*, in press.
Cairns, A. J. (1987). *Anal. Biochem.* **167**, 270–278.
Cairns, A. J. and Pollock, C. J. (1988). *New Phytol.* **109**, 399–405.
Calderón, P. L. and Pontis, H. G. (1985). *Plant Sci.* **42**, 173–176.
Chatterton, N. J., Harrison, P. A., Thornley, N. R. and Bernett, J. H. (1989a). *Plant Physiol. Biochem.* **27**, 289–295.
Chatterton, N. J., Harrison, P. A., Thornley, N. R. and Bernett, J. H., (1989b). *J. Plant Physiol.* **134**, 169–179.
Cisar, J., Kabat, E. A., Liao, J. and Potter, M. (1974). *J. Exp. Med.* **139**, 159–179.
Colin, H. and Belval, H. (1940). *Bull. Soc. Bot. Fr.* **87**, 341–345.
Colin, H. and De Cugnac, A. (1926). *Bull. Soc. Chim. Biol.* **8**, 621–627.
Collins, F. W. and Chandorkar, K. R. (1971). *J. Chromatogr.* **56**, 163–167.
Darbyshire, B. and Henry, R. J. (1978). *New Phytol.* **81**, 29–35.
Darbyshire, B. and Henry, R. J. (1981). *New Phytol.* **87**, 249–256.
Dedonder, R. (1952). *Bull. Soc. Chim. Biol.* **34**, 144–182.
de Miniác, M. (1970). *C.R. Hebd. Séances Acad. Sci.* **270**, 1583–1587.
Eagles, C. F. (1967). *Ann. Bot. (London). (N.S.)* **31**, 645–651.
Edelman, J. and Dickerson, A. G., (1966). *Biochem. J.* **98**, 787–794.
Edelman, J. and Jefford, T. G. (1964). *Biochem. J.* **93**, 148–161.
Edelman, J. and Jefford, T. G. (1968). *New. Phytol.* **67**, 517–531.
Frehner, M., Keller, F., Wiemken, A. and Matile, Ph. (1984). *J. Plant Physiol.* **116**, 197–208.

Forsyth, W. G. C. (1948). *Nature (London)* **161**, 239–241.
Forsyth, W. G. C. (1950). *Biochem. J.* **46**, 141–145.
Glaudemans, C. P. J. (1975). *Adv. Carbohydr. Chem. Biochem.* **31**, 313–346.
Grey, H. M., Hirst, J. W. and Cohn, M. (1971). *J. Exp. Med.* **133**, 289–304.
Grotelueschen, R. D. and Smith, D. (1968). *Crop. Sci.* **8**, 210–212.
Gunn, T. C. and Walton, D. W. H. (1985). *Polar Biol.* **4**, 237–242.
Hammer, H. (1970). *Acta. Chem. Scand.* **24**, 1294–1300.
Hendry, G. (1987). *New Phytol.* **106**, (Suppl.), 201–216.
Hirst, E. L. (1957). *Proc. Chem. Soc. (London)*, pp. 193–204.
Housley, T. L. and Daughtry, C. S. T. (1987). *Plant Physiol.* **83**, 4–7.
Housley, T. L. and Pollock, C. J. (1985). *New Phytol.* **99**, 499–507.
Karlsson, G. (1969). *J. Chromatogr.* **44**, 413–414.
Labourel, G. and Péaud-Leonel, C. (1969). *Chem. Zvesti* **23**, 765–769.
Laidlaw, R. A. and Wylam, C. B. (1952). *J. Sci. Food. Agric.* **3**, 494–497.
Lundblad, A., Stellar, R., Kabat, E. A., Hirst, J. W., Weigert, H. G. and Cohn, M. (1972). *Immunochemistry* **9**, 535–542.
Marchessault, R. H., Bleha, T., Deslandes, Y. and Revol, J. F. (1980). *Can. J. Chem.* **58**, 2415–2417.
Meier, H. and Reid, J. S. G. (1982). *In* "Encyclopedia of Plant Physiology, New Series" (A. Pirson and M. H. Zimmermann, eds), Vol. 13A, pp. 435–450. Springer, Berlin and New York.
Nelson, C. J. and Spollen, W. G. (1987). *Physiol. Plant.* **71**, 512–516.
Percheron, F. (1962). *C.R. Hebd. Séances Acad. Sci.* **255**, 2521–2522.
Pollock, C. J. (1982a). *New Phytol.* **90**, 645–650.
Pollock, C. J. (1982b). *Phytochemistry* **21**, 2461–2465.
Pollock, C. J. (1984a). *In* "Storage Carbohydrates in Vascular Plants" (D. H. Lewis, ed.), pp. 97–113. Cambridge University Press, Cambridge.
Pollock, C. J. (1984a). *New Phytol.* **96**, 527–534.
Pollock, C. J. (1986). *New Phytol.* **104**, 1–24.
Pollock, C. J., Hall, M. A. and Roberts, D. P. (1979). *J. Chromatogr.* **171**, 411–415.
Pontis, H. G. (1966). *Arch. Biochem. Biophys.* **116**, 416–423.
Pontis, H. G. (1968). *Anal. Biochem.* **23**, 331–333.
Pontis, H. G. (1971). *An. Soc. Cient. Argent. sspl.* 59–63.
Pontis, H. G. (1989). *J. Plant Physiol.* **134**, 148–150.
Pontis, H. G. and del Campillo, E. (1985). *In* "Biochemistry of Storage Carbohydrates in Green Plants" (P. M. Dey and R. A. Dixon, eds), pp. 205–227. Academic Press, London.
Raguse, C. A. and Smith, D. (1965). *J. Agric. Food Chem.* **13**, 306–312.
Roe, J. H. (1934). *J. Biol. Chem.* **107**, 15–18.
Rutherford, P. P. and Deacon, A. C. (1972a). *Biochem. J.* **126**, 569–573.
Rutherford, P. P. and Deacon, A. C. (1972b). *Biochem. J.* **129**, 511–512.
Rutherford, P. P. and Weston, E. W. (1968). *Phytochemistry* **7**, 175–180.
Schlubach, H. H. (1958). *Prog. Chem. Org. Nat. Prod.* **15**, 1–27.
Schlubach, H. H. and Sinh, O. K. (1940). *Justus Liebigs Ann. Chem.* **544**, 111–116.
Schnyder, H. and Nelson, C. J. (1987). *Plant Physiol.* **85**, 548–553.
Scobel, H. D. and Brobst, K. M. (1981). *J. Chromatogr.* **72**, 51–64.
Scott, R. W., Jefford, T. G. and Edelman, J. (1966). *Biochem. J.* **100**, 23p–24p.
Shiomi, N. (1981a). *Phytochemistry* **20**, 2581–2583.
Shiomi, N. (1981b). *Carbohydr. Res.* **106**, 166–169.
Shiomi, N. and Izawa, M. (1980). *Agric. Biol. Chem.* **44**, 603–614.
Shiomi, N., Yamada, J. and Izawa, M. (1976). *Agric. Biol. Chem.* **40**, 567–575.
Shiomi, N., Yamada, J. and Izawa, M. (1979). *Agric. Biol. Chem.* **43**, 2233–2244.
Smith, A. E. (1976). *J. Agric. Food. Chem.* **24**, 476–478.
Smith, D. (1967). *Crop. Sci.* **7**, 62–67.
Smith, D. (1973). *In* "Chemistry and Biochemistry of Herbaces" (G. W. Butler and R. W. Bailey, eds), Vol. 1, pp. 105–155. Academic Press, London.
Somogyi, M. (1952). *J. Biol. Chem.* **195**, 19–25.
Streefkerk, D. G. and Glaudemans, C. P. J. (1977). *Biochemistry* **16**, 3760–3765.

Thome, U., Huber-Reinhard, M. and Kuhbauch, W. (1986).

Tognetti, J. A., Calderón, P. L. and Pontis, H. G. (1988). *J. Plant Physiol.* **134**, 232–236.

Tomoda, M., Satoh, N. and Sugiyama, A. (1973). *Chem. Pharm. Bull.* **21**, 1806–1810.

Tsusué, Y. and Fujita, Y. (1964). *J. Gen. Appl. Microbiol.* **10**, 283–285.

Wagner, W. and Wiemken, A. (1986). *J. Plant Physiol.* **123**, 429–439.

Wagner, W. and Wiemken, A. (1987). *Plant Physiol.* **85**, 706–710.

Wagner, W., Keller, F. and Wiemken, A. (1983). *Z. Pflanzenphysiol.* **112**, 359–372.

Whelan, W. J. (1955). *In* "Hanbuch der Pflanzen Physiologie" (W. Ruhland, ed.), Vol. 6, pp. 184–196. Springer, Berlin and New York.

Whistler, R. L. and Smart, C. L. (1953). "Polysaccharide Chemistry", pp. 276–291. Academic Press, New York.

Wolf, D. D. and Ellmore, T. L. (1975). *Crop. Sci.* **75**, 775–777.

Yamamoto, S. and Mino, Y. (1985). *Plant Physiol.* **78**, 591–595.

11 Mannose-based Polysaccharides

NORMAN K. MATHESON

*Department of Agricultural Chemistry, The University of Sydney,
N.S.W. 2006, Australia*

METHODS IN PLANT BIOCHEMISTRY Vol. 2
ISBN 0-12-461012-9

I. INTRODUCTION

In the 1880s Emil Fischer hydrolysed ivory-nut mannan and identified D-mannose in the hydrolysate. A range of polysaccharides that contain D-mannose has since been found in plant (Whistler and Smart, 1953; Whistler and Richards, 1970; Meier and Reid, 1982; Stephen, 1983) and algal (Percival and McDowell, 1981; Painter, 1983) sources, and these include polymers based on a (1→4)-β-linked mannan backbone. These polysaccharides divide chemically into mannan (Aspinall, 1959), glucomannan (Stepanenko, 1960; Timell, 1964), galactomannan (Smith and Montgomery, 1959; Stepanenko, 1960; Dea and Morrison, 1975; Dey, 1978, 1980) and galactoglucomannan (Timell, 1965a; Dey, 1980). In this introduction the references have been mostly limited to reviews; in later sections selected references containing experimental procedures have also been given.

A. Chemical Structure and Properties

The chemical structures of (1→4)-β-D-mannan, (1→4)-β-D-glucomannan, (1→6)-α-D-galacto-(1→4)-β-D-mannan and (1→6)-α-D-galacto-(1→4)-β-D-glucomannan are represented by formulae 1–4, respectively:

$$- - - 4\text{Man}\beta - 4\text{Man}\beta - 4\text{Man}\beta - 4\text{Man}\beta - - -$$
$$\mathbf{1}$$
$$- - - 4\text{Glc}\beta - 4\text{Man}\beta - 4\text{Man}\beta - 4\text{Glc}\beta - 4\text{Man}\beta - - -$$
$$\mathbf{2}$$

$$
\begin{array}{ccc}
\text{Gal}\alpha & & \text{Gal}\alpha \\
| & & | \\
6 & & 6 \\
\end{array}
$$
$$- - - 4\text{Man}\beta - 4\text{Man}\beta - 4\text{Man}\beta - 4\text{Man}\beta - 4\text{Man}\beta - - -$$
$$\mathbf{3}$$
$$\text{Gal}\alpha$$
$$|$$
$$6$$
$$- - - 4\text{Man}\beta - 4\text{ Glc}\beta - 4\text{Man}\beta - 4\text{Man}\beta - 4\text{Glc}\beta - - -$$
$$\mathbf{4}$$

The division, in terms of sugar content, is not strict. Although the mannans are described as unbranched chains of (1→4)-β-linked D-mannosyl units, small amounts (<5%) of D-glucosyl and D-galactosyl units may be present in the extracted polymer.

Mannan is very insoluble; ivory-nut mannan can be partly extracted by alkali and partly by cuprammonium. The degree of polymerisation (DP) of the extracted material is low (10–80). The *in vivo* value may be higher since alkaline conditions can partly depolymerise. A higher value (90% between 100 and 2500) was determined for a mannan from *Codium fragile*, prepared by removal of non-mannan material, but even here the product was converted into a derivative (nitro-, or permethyl-), allowing the possibility of some depolymerisation. The β-D-mannosyl linkages lead to a negative optical rotation.

In glucomannans a significant proportion of D-mannosyl residues are replaced by D-glucosyl units (20–50%) and there may be partial acetylation. The occurrence of hexosyl residues along the $(1\rightarrow4)$-β-chain appears to involve neighbouring D-mannosyl but not contiguous D-glucosyl residues. The distribution is not regular, and it is probably not statistically random. Small amounts of D-galactose have been detected in some. The glucomannans are more soluble than the mannans and the partial acetylation improves solubility further, such that they become water-soluble. De-acetylation decreases the ease of dissolution. Their solutions are highly viscous and the optical rotation is negative. Estimates of DP have generally been higher than for mannan (100–5000).

In the galactomannans the hydroxymethyl group on D-mannosyl residues (6-OH) is substituted by a single α-linked D-galactosyl group. The percentage of substitution varies with the botanical source—from 30%, e.g. in *Cassia* spp., to almost 100%, e.g. in Trifolieae (Dea and Morrison, 1975; Dey, 1978; Lewis, 1984). The amount of substitution is a mean value but this appears to be reasonably constant for a particular species. The pattern of substitution in those species with less than complete substitution is probably not in blocks nor statistically random but is non-regular, without long relatively unsubstituted regions (Matheson, 1984; McCleary and Matheson, 1986). The galactomannans show high viscosity and, except for those at the lower end of the range of galactosyl contents, are water-soluble. The presence of the α-D-galactosidic bonds confers positive rotation. The molecular size is high but this may reflect the conformation of the polymer as much as molecular weight. Estimates of DP by physico-chemical methods have generally been in the range of about 1000 to 10 000.

Isolated galactoglucomannans from various sources have differed in both the ratio of D-mannosyl to D-glucosyl units in the main $(1\rightarrow4)$-β-chain and in the degree of substitution of this chain by D-galactosyl groups. For example, various gymnosperm preparations have shown Gal:Glc:Man ratios of near 2:2:6, a water-soluble extract from the endosperm of asparagus seed of 1:4:5 and preparations from the endosperm of *Cercis siliquastrum* (Judas tree) of 2:1:11. The oligosaccharide products of partial hydrolysis suggest that, like glucomannan, whereas D-mannosyl residues can be neighbouring, this is much less likely or not so for D-glucosyl residues. The distribution of the two hexose residues in the main chain is probably non-regular. Most of the limited evidence available suggests that only D-mannosyl residues are substituted with D-galactosyl groups and that the pattern is non-regular. Where the extraction procedure has allowed retention of acetyl groups these have commonly been found. Substitution of the glucomannan chain with D-galactosyl, as well as acetyl groups increases the ease of dissolution.

Other structural features have been proposed for the $(1\rightarrow4)$-β-mannan based polysaccharides. e.g. $(1\rightarrow3)$-linkages in the main chain, branching of the main chain, attachment of D-galactosyl groups elsewhere than the *O*-6 of D-mannose, the presence of D-xylose, D-galactosyl side chains longer than one residue and these having $(1\rightarrow2)$-α-linkages. In interpreting these observations when small amounts are involved, consideration should be given to limitations in sensitivity of the method of analysis, e.g. undermethylation, demethylation during hydrolysis; transglycosylation or reversion on partial hydrolysis; incomplete or over-oxidation by periodate and the co-purification of polysaccharides with similar properties, such as galacto-xylo-glucan. Galactomannan has also been found to contain small amounts of protein but any possibility of the presence of covalent linkages is not known.

B. History and Occurrence

Interest in cell-wall storage polysaccharides developed in a descriptive sense in the later part of the nineteenth century and chemical studies started shortly after, but significant advances began in the second and third decades of this century, after recognition of the polymeric nature of these compounds and the introduction of chemical and enzymic techniques for their examination. The recognition of galactoglucomannans as a distinctive class did not emerge until the 1950s.

Mannan has been found as the major endospermic component of palm seeds, where it may constitute up to 60% of the seed (e.g. ivory nut, *Phytelaphas macrocarpa*), and also in seeds of caraway and coffee. It occurs extracellularly as a thickened cell wall. Mannan has also been prepared from the cell-wall of green seaweed (*Codium* spp.) and other Chlorophyceae and has been reported in red algae (Rhodophyceae). A partly sulphated mannan is found in *Codiolum pusillum* (Carlberg and Percival, 1977).

Glucomannan is one of the polysaccharide components of the cell-wall (hemicellulose) of woods. Gymnosperms contain larger amounts (up to 10%) than angiosperms (3–5%). It has also been detected in a number of monocotyledonous seeds; in the endosperm of some Liliaceae, Iridaceae, Amaryllidaceae and Agavaceae. Another source is from vegetative tissues—tubers, bulbs, roots and leaves of members of the Liliaceae, Amaryllidaceae, Dioscoraceae, Orchidaceae and Araceae. Of these, orchid tubers (salep) and *Amorphophallus konjac* (konjac) have been well studied; the former was first prepared in 1924. In these sources glucomannan is not found in specific cell parts but occurs in idioblasts, as large specialised cells containing high amounts of the polysaccharide. In young tuber tissues these can be separated as mucilagenous globules. The amounts of extractable glucomannan can vary from a few per cent to more than 30%, depending on both the source and nature of the tissue.

Galactomannan has a long history of use. The value of carob pods was known in the fourth century BC and in biblical times (St John's bread) and it has been used both as stock and human food. This polymer occurs in the endosperm of legume seeds, as an extracellular deposit, and has also been reported in species of Annonaceae, Convolvulaceae, Ebenaceae, Loganaceae and Palmae. It can be present from a trace, as in soybean, up to nearly 40%, as in carob.

Galactoglucomannan was the last in this group to be described chemically, when, in the 1950s, it was found to be a component of the hemicellulosic fraction of wood, and in some gymnosperms yields of up to 6% have been obtained. Galactoglucomannan has also been extracted from seeds, e.g. asparagus and the legumes *Cercis siliquastrum* and *Bauhinia* spp. In *Cercis siliquastrum* it is present as at least 10% of seed dry weight. Mosses and ferns have also yielded galactoglucomannan. Classification of polysaccharides as glucomannans that contain a small amount of galactose, or as galactoglucomannans, depends on a definition of when the galactose content becomes significant. Where both occur, separation with respect to aldose proportions may reflect the ease of extraction from a polymer distribution (Mills and Timell, 1963).

C. Metabolism and *in vivo* function

The biosynthesis (Dey, 1978; Matheson, 1984; Mackie, 1985; Reid, 1985a,b) in plants of the (1→4)-β-D-mannan chain from D-fructose 6-phosphate (Fru-6-P), derived from sucrose or starch breakdown or photosynthesis, probably proceeds via the sequence:

$$\text{Fru-6-P} \longrightarrow \text{Man-6-P} \longrightarrow \text{Man-1-P} \longrightarrow \text{GDP-Man} \longrightarrow \text{mannan}$$

which is catalysed by the enzymes mannose 6-phosphate isomerase (EC 5.3.1.8) phosphomanno-mutase (EC 5.4.2.8), mannose 1-phosphate guanylyltransferase (GDP-mannose pyrophosphorylase, EC 2.7.7.13) and mannan 4-β-mannosyl transferase (mannan synthase, EC 2.4.1.32). The third reaction requires GTP and releases pyrophosphate and the final reaction releases GDP. Direct interconversion of GDP-Glc and GDP-Man has also been proposed. Synthesis of glucomannan needs an additional activity, a β-glucosyl transferase, and it is believed that the characteristic heteropolymeric structure arises because the glucosyltransferase requires the production of an acceptor molecule with a D-mannosyl group at the non-reducing end, and that, although the mannosyl transferase does not require a D-glucosyl acceptor group and can substitute a D-mannosyl group, it is inhibited by GDP-glucose. Less is known about the enzymes catalysing galactomannan synthesis. The insolubility of (1→4)-β-mannan and the requirement for copolymerisation by a similar type of molecule, xylo-(1→4)-β-glucan, suggests a similar mechanism, possibly involving UDP-D-galactose in combination with GDP-mannose. Mannan synthase activity is associated with a particulate fraction. Microscopy studies have indicated that galactomannan is formed in the intracisternal space of the rough endoplasmic reticulum and expelled to the outside of the plasmalemma.

Little is known about the control of (1→4)-β-mannan synthesis at the enzymic level. One possible effect involves mannose 6-phosphate isomerase. This enzyme, extracted from *Cassia* spp., was very much less inhibited by 6-P-gluconate and erythrose 4-P than was glucose 6-phosphate isomerase. This suggests a possible mechanism whereby an increase in these two metabolites of the pentose phosphate pathway could allow the directing of synthesis from fructose 6-phosphate to mannan.

The catabolism of galactomannan has been studied (Dey, 1978; Ashford and Gubler, 1984; Bewley and Reid, 1985; Reid, 1985a,b,c). The depolymerisation of the β-mannan chain involves *endo*-hydrolysis by β-mannanase (EC 3.2.1.78) to manno-oligosaccharides, mainly mannobiose- and -triose. Further hydrolysis to mannose results from the action of β-mannoside mannohydrolase (EC 3.2.1.25) which, where tested, has been shown to be an *exo*-mannanase. Glucomannan chains presumably also require a β-glucoside glucohydrolase. Where (1→6)-α-galactosyl groups are present these are released by α-D-galactosidase (EC 3.2.1.22). Hydrolysis by β-mannanase can proceed if the extent of galactosyl substitution is limited but, where galactomannan is highly substituted, prior removal of some galactosyl groups is required. All three enzymes are present in the endosperm. Depolymerisation appears to be under little control, apart from limited competitive inhibition by hydrolysis products of low DP. Cotyledons can take up small manno-oligosaccharides as well as monosaccharides and *exo*-mannanase is also present in this tissue. The operation of an *exo*-mannanase, as opposed to a β-mannosidase, could ensure that hydrolysis is limited to (1→4)-β-manno-oligomers, as *exo*-glycanases have a substrate-binding requirement for a sugar residue or residues other than that at the non-reducing end and they are linkage-specific. *Endo*-β-D-mannanase has also been implicated in the hydrolysis of seed mannans and glucomannans in vegetative parts.

The utilisation of the released monosaccharides requires initial phosphorylation. The enzymes of galactose metabolism are galactokinase (EC 2.7.1.6) and either UTP-galactose 1-phosphate uridylyl-transferase (EC 2.7.7.10) or UDP-glucose-galactose-α-1-phosphate uridylyl-transferase (EC 2.7.7.12) and UDP-Glc:UDP-Gal-4-epimerase (EC 5.1.3.2.); of mannose metabolism, hexokinase (EC 2.7.1.1) and mannose 6-phosphate isomerase (EC 5.3.1.8) and of glucose, hexokinase or glucokinase (EC 2.7.1.2) and glucose 6-phosphate isomerase (EC 5.3.1.9).

Mannan-based polysaccharides in seeds are catabolised, and the monosaccharides released are utilised during seedling development. This process occurs at a definite period after imbibition, after substantial depletion of oligosaccharides, which function as the primary reserve carbohydrates for germination. Galactomannan also protects the developing axis from fluctuations in water balance. Further possible roles may be related to an evolutionary advantage of seed size or hardness. The breakdown of galactomannan in legumes can be initiated by imbibition of detached endosperms, leading to liquefaction and the production of mono- and oligosaccharides: thus the process may be independent of hormonal control by the embryo. Glucomannans in vegetative tissues also seem to function as carbohydrate reserve: they are mobilised when bulbs or tubers shoot. *Endo*-β-D-mannanase has been identified; presumably a de-acetylase is also required. The galactoglucomannans and glucomannans found in wood and in stems are extracted in the hemicellulose fraction and appear to be structural. The mannans in algae are also structural.

D.　Commercial Uses

The industrial use of some mannan-based polysaccharides has been reviewed (Whistler and Smart, 1953; Goldstein *et al.*, 1973; Rol, 1973; Dea and Morrison, 1975; Sandford and Baird, 1983). The major polymer in industrial use is galactomannan, in particular from those sources in which the galactosyl substitution is relatively low, such as carob and guar. The properties that are utilised are the very high viscosity, which is produced at low concentrations, and the capacity to form gels in association with other polysaccharides such as agarose, carrageenan and xanthan. It is used in foods as a thickener and stabiliser; for pulp dispersing, sizing and finishing in paper production; in flocculation, filtration and flotation of minerals; in sizing textiles; in oil-drilling muds; for explosive gels; and in stabilising fungicide and herbicide dips. Added to flour it changes the baking properties of a bread. Glucomannan and galactomannan are used for adding fibre to the diet. The presence of glucomannan and galactoglucomannan in pulpwood for paper manufacture is correlated with higher tensile strength and a higher burst point. Historically, an interesting use for ivory-nut mannan was in the production of buttons, when advantage was taken of the hardness and insolubility.

E.　Methods

A number of discussions of methods have appeared, e.g. methodology with plant polysaccharides (Aspinall and Stephen, 1973), isolation and fractionation (Aspinall, 1982a), chemical characterisation and structure determination (Aspinall, 1982b) and spectroscopic methods (Perlin and Casu, 1982). Descriptions of specific experimental procedures are to be found in *Methods in Carbohydrate Chemistry*, Vols 1–8 (R. L. Whistler and J. N. BeMiller, eds, Academic Press, New York) such as the preparation of galactoglucomannan and glucomannan from wood (Timell, 1965b) and galactomannan from guar seed (Whistler and Marx, 1965) and in *Methods of Biochemical Analysis*, Vols 1–33 (D. Glick, ed., John Wiley, New York). Enzymic methods can be found in *Methods in Enzymology* Vols 1, 8, 28, 41, 50, 89, and 90 (S. P. Colowick and N. O. Kaplan, series eds, Academic Press, New York) and in *Methods of Enzymatic Analysis*. 3rd English edition (1983, H. U. Bergmeyer, ed., Verlag Chemie, Weinheim).

II. MICROSCOPY

$(1\rightarrow4)$-β-Mannan based polysaccharides have been studied (Meier and Reid, 1982) *in situ*, using light, electron (Chanzy and Vuong, 1985) and scanning electron microscopy.

A. Light Microscopy

Date and ivory-nut endosperms have been examined (Meier, 1958) using polarised light and phase contrast at magnifications of $\times 140$ and $\times 540$ in tangential and longitudinal sections. Isotropic (middle lamella) and birefringent sections were identified. When the dissolution zone of germinating date seeds was studied (Keusch, 1968), the seed was cut tangentially to the haustorium and viewed with polarised light ($\times 400$). The double refraction, associated with the secondary-wall thickening, disappeared gradually with increasing dissolution. Native walls of the green seaweeds (*Codium* and *Acetabularia* were found to be birefringent not only in section, but also in face view (Frei and Preston, 1968).

Sections of fenugreek seeds (Meier and Reid, 1977) 2 µm thick, prepared at different stages of galactomannan accumulation, were embedded in Araldite and stained with 0.5% Azur II-Methylene Blue–borate, which indicated that deposition started next to the embryo and proceeded outwards. Deposition on the primary wall was irregular but often began at the tangential walls. In outer cells a large lumen, filled with protoplasm, remained at seed maturation.

The galactomannan in germinating fenugreek seeds was observed at various stages (Reid, 1971). Sections 30 µm thick were treated with cold 40% ethanol and then with 0.5% HIO_4 in the same solvent. The aqueous ethanol was exchanged stepwise for water and the section treated with Schiff's reagent and H_2SO_3, embedded in glycerol jelly and examined at $\times 300$–415. The dissolution zone started at the outer surface of the endosperm and spread inwards towards the cotyledons. Carob seed sections (20 µm) were examined (Seiler, 1977) after germination by staining with periodic acid–Schiff's reagent (PAS), which showed that the initiation of galactomannan breakdown was not limited to the outer section of the endosperm but started in all cells, more or less uniformly. Breakdown continued more rapidly near the embryo; it was considered that removal of inhibitors of enzymic activity, the lower molecular weight carbohydrates, allowed degradation to proceed.

Cassia coluteoides seeds (Lee, 1982) were cut and fixed in 2% glutaraldehyde in 0.025 M phosphate at pH 6.8, frozen with CO_2 gas and 10 μm transverse sections cut, and these stained with PAS and counterstained with safranin-fast green. Sections were also treated with fluorescein isothiocyanate (FITC)-lectins from *Griffonia* (formerly *Bandeiraea*) *simplicifolia*, which binds to α-D-galactosyl groups, *Glycine max.* (soybean, α- and β-D-galactosyl groups), *Triticum aestivum* (wheat, *N*-acetyl-D-glucosaminyl groups) and *Canavalia ensiformis* (ConA, α-D-glucosyl and α-D-mannosyl groups) (Baldo *et al.*, 1984). Cryostat sections were incubated for 30 min at room temperature in a humid environment with the FITC-lectin (1 mg ml^{-1}) and gently rinsed twice with physiological saline and then water, air dried and examined in a microscope with an epifluorescence attachment. Filter sets with excitation characteristics (a) exciter filter BP 450–490 nm, chromatic beam splitter FT 510 nm, barrier filter LP 525 nm, and (b) 405 nm, 460 nm and 475 nm, respectively, were used. In developing seeds (25 days post-anthesis) stained with PAS and safranin-fast green, the cotyledons were relatively small and rich in protein. The endosperm contained protein and PAS-staining material was most strongly associated with the section closer to the cotyledons. Staining with FITC-*Griffonia* lectin confirmed this, when the inner third of the endosperm fluoresced. Autofluorescence was shown to be negligible using suitable filter combinations. The results indicated that, similarly to fenugreek (Meier and Reid, 1977), galactomannan is initially deposited in quantity close to the cotyledons and the process spreads towards the seed-coat. In mature seeds, prior to germination, the regions staining with PAS correspond to those fluorescing with FITC-*Griffonia* lectin. Treatment with FITC-ConA and wheat lectins showed few or no affinity sites. FITC-soybean lectin fluoresced in the same parts as *Griffonia*-FITC lectin but less intensely. Points of protein staining were found scattered throughout the whole endosperm and the endosperm was partly separated from the cotyledons. Seeds were then stained 2 days after imbibition. Most of the protein-staining spots showed increased areas surrounding them that did not stain with PAS and this occurred over the whole endosperm. The behaviour of germinating *C. coluteoides* is similar to that of carob (Seiler, 1977). Staining of mature guar with PAS and safranin-fast green gave a pattern of a high level of large protein-staining spots towards the seed coat (one-tenth of the cross-section) surrounded by a small area of polysaccharide, and small protein-staining spots over the remainder of the endosperm surrounded by larger areas of polysaccharide.

The methods and results of detailed studies on storage glucomannan have been reviewed (Meier and Reid, 1982). In vegetative tissues the glucomannan in idioblasts surrounds raphide crystals and it is deposited within the protoplasts. In seeds it is found as endosperm. On germination of asparagus seeds (Goldberg and Roland, 1971) a dissolution zone surrounding the cotyledons developed, into which the latter grew. Polarisation studies on extracted glucomannan from wood of Parana pine (Katz, 1965) indicated that the polymer maintains a randomly oriented structure while acetylated, as it appeared isotropic in polarised light. On deacetylation anisotropy was observed.

B. Electron and Scanning Electron Microscopy

Ultra-thin sections of date palm and ivory-nut seeds (Meier, 1958) were prepared from material embedded in butyl- and methyl-methacrylates (5 : 1) by adding 0.5% dichloro-benzoyl peroxide. The embedding material of microtome sections was dissolved in amyl acetate. Sections were shadowed with chromium at 30° and magnification varied from × 15 000 to × 22 500. Longitudinal sections of fully swollen seeds showed that the wall consisted, at least partly, of fibrillar elements. A sample remaining after extraction, that contained mannans A and B, was disintegrated and examined, when it showed small aggregated grains and microfibrillar material. When a sample from which mannan A had been further extracted was examined, most of the granular material had disappeared. It was concluded that mannan A is built into the cell wall as crystalline grains, 10–20 nm in diameter, whereas mannan B occurs as microfibrils. Isolated mannan A consisted of small crystallites (1–2 nm) that when sectioned showed a somewhat similar structure to native mannan A. Regenerated and purified mannan B had no fibrillar structure. Ground, chlorite-bleached nuts, leached with 1.25 M KOH, were examined by shadow casting with tungsten–tantalum alloy (Chanzy *et al.*, 1984). Two components could be seen: the main fraction (called I) was granular and was extracted and the other (II) was microfibrillar. Cellulose microfibrils were also present. Mannans from the algae *Acetabularia crenulata* and *Codium fragile* (Frei and Preston, 1968; Mackie and Preston, 1968; Chanzy *et al.*, 1984) were also found to exist as alkali-soluble, granular mannan I and alkali-resistant, fibrillar mannan II. The mannan I–II polymorphism could be produced by crystallisation. Mannan microfibrils were detected after washing walls successively with detergent, urea, methanol and water, drying and shadowing with palladium–gold or tungsten–tantalum. Extracted and crystallised ivory-nut mannan has also been examined (Chanzy *et al.*, 1979). The material was transferred to methanol, which was evaporated on carbon-coated mica. This was then floated onto a water interface and mounted. The crystals were shadowed with tungsten–tantalum or were unshadowed. They had a mannan I (A) structure and when the crystallising baths were seeded with cellulose, microfibrils grew in an ordered pattern.

The accumulation of galactomannan in fenugreek seeds was followed (Meier and Reid, 1977) using electron microscopy by (a) fixing in 3% glutaraldehyde in 0.05 M cacodylate buffer, pH 7.2, or (b) fixing in 3% glutaraldehyde, 3% acrolein in 0.05 M collidine buffer, pH 7.1, and post-fixing in 2 or 1.5% OsO_4 respectively, and then dehydrating with ethanol. The fixed, dehydrated tissues were embedded in Araldite and stained with 2% uranyl acetate followed by lead acetate, or with 1% $KMnO_4$ or periodate–thiocarbohydrazide–silver proteinate. The endosperm cells near the embryo filled first. In those cells that had just started to deposit galactomannan, parallel stacks of rough endoplasmic reticulum (ER) formed. Then the intracisternal space of the ER became swollen and vacuolated, with the ER membranes entrapping cytoplasmic pockets. Ribosomes were attached to the inner sides of membranes bordering these pockets. Galactomannan was then discharged to the outside of the plasmalemma with local disruption of the cell membrane. In later stages, galactomannan remained within the protoplast, inside the vacuolated ER, and finally almost the whole lumen of the cell was filled with reserve galactomannan. In the outermost cells of the endosperm, the deposition of galactomannan was restricted; it was deposited mainly where the cells were in contact with the seed coat, with only a small amount being deposited on side walls and the inner walls remaining thin. In germinating fenugreek seeds (Reid and Meier, 1972), sections were examined with similar fixing and staining techniques. Two per cent $KMnO_4$ was also used and magnifications were × 6100 to × 22 400. Galactomannan was inititally depleted in the outermost cells, near the seed coat. Electron microscopy of carob seeds after germination (Seiler, 1977), using similar fixing and staining procedures, showed that in this seed the breakdown of galactomannan started more or less evenly in all cells.

In studies of glucomannan (Meier and Reid, 1982) longitudinal segments from axial root tips of *Vanilla planifolia* and *Monstera deliciosa* (Mollenhauer and Larson, 1966) were fixed in either 2% $KMnO_4$ or 1.5% acrolein and 3% glutaraldehyde in 0.05 M cacodylate buffer, pH 7.4, post-fixed with 1% OsO_4 in this buffer, dehydrated through an ethanol series followed by acetone, and embedded in Araldite. Magnifications ranged from $\times 7500$ to $\times 62\,000$. Accumulation occurred in raphide cells, which developed copious amounts of ER. Material, which was most likely glucomannan, appeared to be produced from the ER. The contents of the vesicles, derived from swollen terminal portions of ER, emptied into the central region of the cell but probably not directly into the vacuole. As product accumulated it remained separated from the cytoplasm by a single membrane, which may have been derived from the ER. The secreted material then appeared to engulf and mix with vacuolar components. Electron micrographs have been published (Chanzy *et al.*, 1982) of two crystalline forms of glucomannans from two tubers (*Tubera salep* and *Amorphallus konjac*) and three woods (*Pinus strobus*, *P. sylvestris* and *Sequoia sempervirens*). Dried specimens were shadow-cast with tungsten–tantalum alloy or negatively stained with 1% phosphotungstate (pH 5). Different forms (I and II) were obtained according to the crystallising conditions. Comparison ($\times 67\,000$) of acetyl and de-acetyl glucomannan from Parana pine wood showed the latter was agglomerated (Katz, 1965): samples were shadowed with palladium. The location of glucomannan in tracheids of spruce wood has been studied using a β-mannanase–gold complex (Ruel and Joseleau, 1984).

Examination (McClendon *et al.*, 1976) of the endosperm of fully imbibed guar seeds and fractured dry endosperm by scanning electron microscopy at magnifications of $\times 206$ to $\times 2500$, indicated that galactomannan is deposited as a secondary wall thickening, inside the primary wall, and that, whereas the outer cells have lumina, these are essentially absent from the inner part of the endosperm, which is a solid mass of galactomannan. For examination, the surfaces were coated with carbon and gold. Freeze-substitution was performed by immersing in methanol in a dry-ice bath and then leaving in methanol at $-20°C$ for 10 days and finally transferring to methanol at room temperature.

III. EXTRACTION AND ESTIMATION

The differing ease of extraction of $(1\rightarrow 4)$-β-mannan-based polymers has required a range of methods, from strong alkali and cuprammonium solutions for mannan to aqueous extraction for highly substituted galactomannans. Also, different types of plant tissue may require different procedures, e.g. glucomannan in woody and vegetative tissues. The presence of acetyl groups alters the ease of extraction. Purification has usually involved precipitation as barium or copper complexes, but methods removing other polysaccharides have also been used. When extraction involves neutral aqueous solution at or near room temperature it is essential that endogenous enzymic activities be destroyed, e.g. by prior boiling in ethanol or inclusion of a denaturing agent in the extracting solution.

A. Extraction, Purification, Fractionation and Examination of Homogeneity

The low solubility of mannan requires the use of strong alkali or cuprammonium solvents or that it be converted to a derivative (acetate or nitrate) and then dissolved. Alternatively, other materials can be selectively removed to leave insoluble mannan. Alkaline extractions of ivory-nut mannan were first performed in the early 1900s. A prior treatment with a delignifying agent assists extraction. Alkali removes mannan A, which gives an X-ray diffraction pattern that indicates crystallinity, but leaves mannan B, which is microfibrillar and extracted with cuprammonium. Procedures for extraction of ivory-nut mannan (Aspinall *et al.*, 1953; Whelan, 1955; Meier, 1958) have used prior extraction of shavings with an organic solvent (benzene, methanol, acetone and ether) to remove wax and colouring materials, followed by a delignifying treatment with sodium chlorite in acetate buffer at pH 4. The residual fibrous material was extracted three times with 7% KOH, the extracts acidified with acetic acid and polysaccharide precipitated with an equal volume of ethanol. To isolate mannan B, the residue, after extraction of mannan A, was extracted with 14% KOH to remove any remaining mannan A, washed with water and shaken with cuprammonium solution. After centrifugation, the supernatant was treated with NaOH until the concentration was 0.2 M. The mixture was again centrifuged, the precipitate dispersed in water and decomposed with acetic acid, and the mannan B precipitated with ethanol and washed. This mannan can be dissolved in anhydrous formic acid and precipitated by ethanol. Mannan has also been extracted from coffee beans with 18% NaOH after a washing with 10% NaOH of chlorite-treated flour (Wolfram *et al.*, 1961), from doum-palm kernels (*Hyphaene thebaica*) with 20% NaOH (Khadem and Sallam, 1967), and from *Erythea edulis* kernels with 7 and 14% KOH (Robic and Percheron, 1973). Only a small proportion of the total mannan in *Carum carvi* seeds (Hopf and Kandler, 1977) could be extracted with alkali, suggesting that most occurred as mannan B.

Mannan has been prepared from algae by removal of other materials. Air-dried samples of the seaweeds *Codium fragile*, *Acetabularia calyculus* and *Halicoryne wrightii* (Iriki and Miwa, 1960) were soaked in 1% HCl for 24 h and then treated successively with 1.25% NaOH and 1.25% H_2SO_4 at 100°C for 30 min and the residue bleached with 1% NaClO. This crude fibre, which was insoluble in NaOH, was purified by dissolving in 50% $ZnCl_2$ and adding acetone. Preparations from *Codium* and *Acetabularia* spp. have been made (Frei and Preston, 1968; Mackie and Preston, 1968; Chanzy *et al.*, 1984) by homogenising briefly and extracting successively with 2% EDTA, 0.3% sodium lauryl sulphate and 50% urea. *Cymopolia barbata* mannan (Nieduszynski and Marchessault, 1972) has been prepared by extraction with 0.1 M HCl, followed by successive aliquots of boiling water. An alternative approach is to convert the mannan to a derivative and then extract. Ivory-nut shavings, previously treated with chlorite and hot (70°C) ammonium oxalate, have been nitrated and the nitrate dissolved in acetone (Timell, 1957). Extracted shavings were kept with HNO_3, H_3PO_4 and P_2O_5 (ratio 64 : 26:10; w/w/w). Acids were removed by filtering and washing, and the dried product extracted with acetone containing 8% water. The mannan from *C. fragile* (Mackie and Sellen, 1969), after blending successively with 10% aqueous *n*-butanol, 0.3% sodium lauryl sulphate and 50% urea solutions, was nitrated with P_2O_5 in 90% fuming HNO_3. Mannan nitrate was recovered by precipitation from acetone with water. The polysaccharide from doum-palm kernel has been isolated as the acetate and then deacetylated (Khadem and Sallam, 1967). Dry kernels were acetylated with a mixture of acetic acid containing Cl_2 and acetic anhydride containing SO_2. The filtrate was poured into cold water, the precipitated acetate washed with water and alcohol, dried and extracted into acetone, and precipitated with water. The acetylated mannan was then de-acetylated by dissolution in boiling acetone, cooling and shaking with 30% aqueous KOH. The aqueous layer was diluted with water, acidified with acetic acid and precipitated in ethanol. These procedures of formation of derivatives require acidic conditions and the products could be expected to be partly depolymerised.

The method of extraction of glucomannan depends on whether the source is vegetative tissue, seeds, wood or fibre. It has been isolated from tubers, bulbs, leaves, seeds and wood. Partial acetylation in the native product increases water solubility and as alkaline conditions deacetylate, the use of alkali changes the solubility. Extraction of glucomannan from vegetative tissue (tubers, bulbs and leaves) (Roboz and Haagen-Smit, 1948; Whelan, 1955; Andrews *et al.*, 1956; McCleary and Matheson, 1983; Koleva and Gioreva, 1986) involves prior maceration and boiling in ethanol to inactivate enzymic activities, and after centrifuging or filtering a flour is obtained. Crude glucomannan can then be extracted with cold aqueous solutions or 50% aqueous ethanol and precipitated in ethanol. Mucilage cells have been prepared from young, fresh tubers of *Orchis morio* (Buchala *et al.*, 1974). These were cut into slices and moistened with 70% ethanol. Scraping the surface with a razor blade released large mucilage globules. These were suspended in 70% ethanol and subjected to mild ultrasonic treatment at 0°C, which released adhering starch granules and cell debris. The debris was removed by decantation and starch granules separated by passing the mixture through a 0.3 mm sieve, which retained the mucilage globules. Seed glucomannan has been prepared by alkaline (Andrews *et al.* 1953; Thompson and Jones, 1964; Goldberg, 1969), or aqueous (Gupta, 1980) extraction. Milled bluebell seeds, after boiling in methanol or ethanol, were extracted first with cold and then boiling water. The residue was stirred with 10% NaOH. After centrifugation the supernatant was neutralised with acetic acid and precipitated in ethanol.

The isolation of wood or fibre glucomannan requires a preliminary delignification. De-acetylated glucomannan can then be extracted with sodium hydroxide (Perila and Bishop, 1961; Timell, 1965b; Gupta *et al.* 1976) or the partly acetylated polymer can be dissolved in boiling water after pre-treatment with dimethyl sulphoxide (Meier, 1961; Katz, 1965). In the former method, ground wood was extracted with an ethanol–benzene mixture and stirred with acetic acid and sodium chlorite at 75°C. The residue was washed and dried. This holocellulose was extracted with 24% NaOH. The residue was washed and shaken with 24% NaOH solution containing 4% boric acid. The extract was neutralised with acetic acid and precipitated. To extract partly acetylated glucomannan from wood, holocellulose was extracted twice overnight with dimethyl sulphoxide. The residue was washed with water and extracted twice with boiling water.

The conditions required for solubilising galactomannan depend on the amount of substitution of the mannan chain with D-galactosyl groups. When the content is above 25% they extract into cold or warm aqueous solution; if 20–25%, then into boiling water; but if less than 20%, alkali is required. If both the seeds and amount of galactomannan are sufficiently large, for example carob, guar and honey locust, then the endosperm can be removed and extracted. As the endosperm contains up to 80–90% of galactomannan this eliminates contamination by polysaccharides from the remainder of the seed. If the galactomannan content is low and the seed large (e.g. soybean) or small (clover or lucerne), then extraction of the whole seed followed by purification is usually more convenient. Seed or endosperm flour, after inactivation of enzymes with boiling alcohol, has been extracted with various solutions, including: cold aqueous salt solution (McCleary and Matheson, 1974); hot salt solution (Whistler and Marx, 1965; Reid and Meier, 1970; Unrau and Choy, 1970a; McCleary et al., 1983); cold dilute acetic acid (Khanna and Gupta, 1967); cold alkali (Srivastava et al., 1968; Jindal and Mukherjee, 1970); and hot alkali (Hirst et al., 1947). Dissolution is assisted by fine dispersion of the endosperm. Polysaccharide has been precipitated in ethanol (after neutralisation if alkali-extracted). Sometimes protein has been denatured with chloroform. A variation of hot-water extraction has been to use ultrasonication (Reid and Meier, 1970). The seed coats broken, the embryos could be removed and then further ultrasonic treatment caused dispersion of the endosperm. Methods have been described (Mallett et al., 1987) for the separation of the endosperm of Gleditsia triacanthos seeds at various stages of development.

Although galactoglucomannans from suspension-cultured cells (Akiyama et al., 1983; Cartier, et al., 1988) and acetylated polymer from gymnosperms (Bouveng and Lindberg, 1965a) have been isolated with neutral solutions, many extractions have used strong alkali or alkaline borate solutions after delignification (if necessary) followed by removal of other polysaccharide material. The filtrate from suspension-cultured tobacco cells was applied to a column of DEAE-cellulose and the wash collected. Ammonium sulphate was added to saturation, the precipitate removed by centrifugation and the supernatant dialysed, concentrated and passed through Sephadex G-75, to give a polymer with a ratio of Gal:Glc:Man of approximately 1:1:1, which contained lesser amounts of arabinose and xylose. Pinewood was delignified with chlorite, extracted twice with dimethyl sulphoxide and then twice with boiling water and polysaccharide precipitated. A solution was applied to a DEAE–cellulose column, when some galactoglucomannan was unbound. Combination of this fraction with a wash of 0.05 M phosphate gave a polysaccharide with a Gal:Glc:Man ratio of 1:4:15 and an O-acetyl content of about 6%. Extraction with alkali and alkali containing borate has been used, e.g. with wood (Hamilton et al., 1960; Schwarz and Timell, 1963; Timell, 1965b), Townsville lucerne (Stylosanthes humilis) stems (Alam and Richards, 1971), red clover (Trifolium pratense) leaves and stems (Buchala and Meier, 1973), seeds of Cercis siliquastrum (Judas tree) (McCleary et al., 1976), cell-walls of suspension-cultured tobacco leaf cells (Eda et al., 1985) and tobacco leaf mid-ribs (Eda et al., 1984). Alkali without borate extracts galactoglucomannan with higher levels of galactose, which can be water-soluble after extraction. Delignified Eastern hemlock wood was extracted with 24% (w/w) aqueous KOH. The precipitated polysaccharides yielded a fraction with a ratio of Gal:Glc:Man of 2:2:6 with small amounts of arabinose and xylose. Further extraction of the residue with 24% NaOH containing 4% boric acid gave a product with a Gal:Glc:Man ratio of 1:5:14.

Galactoglucomannan was prepared from Townsville lucerne stems after prior extraction of pectic substances with ammonium oxalate, delignification with sodium chlorite and removal of hemicellulose with 10% NaOH. The residue was extracted with 24% NaOH–4% borate and further purification gave a product with a Gal:Glc:Man ratio of 1:9:9. Milled leaf and stem tissues of red clover were extracted with EDTA to remove pectic substances, then delignified, and hemicellulose removed. After alkaline borate extraction, galactoglucomannan was purifed and had a Gal:Glc:Man ratio of 2:9:9. An extraction by 5% NaOH of the delignified wood of *Pinus* spp. (Hamilton *et al.*, 1960) was separated into the acetone-insoluble arabinoxylan acetate and acetone-soluble galactoglucomannan acetate. The extract was swollen for 70 h in formamide, pyridine added, followed by three aliquots of acetic anhydride at 0, 1 and 8 h. After 48 h the mixture was poured into dilute aqueous HCl and the acetate recovered by filtration.

Purification of $(1\rightarrow4)$-β-mannan polymers can be achieved by complexing with metal ions, such as copper (Hirst *et al.*, 1947; Andrews *et al.*, 1952; Hamilton *et al.*, 1960; Thompson and Jones, 1964; Jones and Stoodley, 1965) or barium (Meier, 1965; Timell, 1965b; Alam and Richards, 1971; Sieber, 1972; Seth *et al.*, 1984) or with borate (Larson and Smith, 1955; Goldstein *et al.*, 1973). Other procedures involve removing acidic polysaccharides by absorbing these on a cationic ion-exchanger (Meier, 1961; Bouveng and Lindberg, 1965a; Alam and Richards, 1971; Akiyama *et al.*, 1983) precipitating them with quaternary ammonium salts, such as cetyl trimethyl ammonium bromide (cetavlon) (Scott, 1965; Aspinall *et al.*, 1966; Buchala and Meier, 1973) or enzymic removal of starch with hydrolytic enzymes (Hirst *et al.*, 1947) and of xyloglucan with fungal cellulase (Kooiman, 1971), or separating them from other polymers by gel chromatography on cross-linked dextran (Matsuo and Mizuno, 1974a; Manzi *et al.*, 1984; Vilkas and Radjabi-Nassab, 1986) or cross-linked agarose (Alam and Richards, 1971; Eda *et al.*, 1984; Sen *et al.*, 1986).

In copper complexing, the polysaccharide, dissolved in water or alkali, is treated with Fehling's solution or aqueous copper acetate, until precipitation is just complete. The gelatinous precipitate is centrifuged and then dispersed in cold water and the mixture acidified with dilute hydrochloric acid until the precipitate dissolves. Polysaccharide is precipitated in ethanol and washed with aqueous ethanol until free of chloride. Alternatively the copper complex can be destroyed with acidified ethanol. In barium complexing the polymer is dissolved in water or NaOH solution and barium hydroxide (or barium chloride or acetate to an alkaline solution) added dropwise. The precipitate is washed with dilute barium or sodium hydroxide and acidified with acetic acid. Precipitation as metal complexes may also occur with polysaccharides such as xyloglucan or degraded cellulose.

Cellulose ion-exchangers are usually employed as columns, when neutral polymer molecules, such as the $(1\rightarrow4)$-β-mannan based polysaccharides, are unbound. Cetavlon precipitates acidic polysaccharides, allowing neutral polysaccharides to be recovered from the supernatant. Galactoglucomannan has been precipitated as the cetavlon–borate complex after initial addition of cetavlon (which produced no precipitate) and boric acid, followed by the addition of NaOH (Hamilton *et al.*, 1960). Lectin affinity chromatography provides another method of purification. A column of concanavalin A, cross-linked by glutaraldehyde, bound glucomannan, which was then eluted with methyl α-D-mannoside (Koleva and Achtardjieff, 1973). Galactomannan was purified on a column of *Bandeiraea simplicifolia* lectin bound to Sepharose, with elution by D-galactose (Ross *et al.*, 1976).

Mannan-based polysaccharides have been fractionated by sequential extraction, fractional precipitation (both as the parent polysaccharide or as a derivative) or by gel chromatography. The general pattern is that fractions of these polymers show a considerable range of molecular size and monosaccharide composition. Mannan has been sequentially extracted with alkali and cuprammonium (Whelan, 1955); galactomannan, having a relatively low galactose content (about 20%), has been extracted with cold water, hot water and alkali (Hui and Neukom, 1964; McCleary et al., 1976; Mazzini and Cerezo, 1979). Wood glucomannan gave no fractionation on extraction with borate solution of increasing concentration (Hamilton et al., 1956) and galacto-glucomannan was not separated by fractional precipitation with Ba(OH)$_2$ (Hamilton et al., 1960). Galactomannan fractions have differed in their galactose contents, with the cold water extracts having the highest values. Mannan nitrates were fractionally precipitated in acetone–water mixtures (Timell, 1957). Two fractions, one with lower DP values (5–7) and another with higher (300–1200), have been separated and characterised by their viscosities. Glucomannnan acetate has been fractionally precipitated from acetone by the addition of water (Sugiyama et al., 1972). Native glucomannan (Vilkas and Radjabi-Nassab, 1986) has been divided into four fractions by precipitation from water by ethanol, and these fractions differed in the ratios of glucose to mannose and their acetyl contents. Glucomannan acetate has been fractionated between tetrachloro-ethane and alcohol (Hamilton and Kircher, 1958).

Fractional precipitation of galactomannan from water with ethanol has been applied (Heyne and Whistler, 1948; Richards et al., 1968; Manzi et al., 1984; Alam and Gupta, 1986) and the acetate has been fractionated between acetone and water and chloroform and ethanol (Koleske and Kurath, 1964; Leschziner and Cerezo, 1970).

Chromatography on DEAE-Sephadex A-50 gave two glucomannan fractions (Achtardjiev and Koleva, 1973) and glucomannan acetates have been separated on porous glass beads of 37 and 70 nm pore size in methylene chloride–methanol (1:1; v/v) (Villemez, 1974).

Qualitative analysis of component sugars provides a simple way of assessing homogeneity. The presence of monosaccharides other than the appropriate occurrence of mannose, galactose or glucose should signal caution in assuming purity. Xylan, xyloglucan, solubilised cellulose and pectic substances are possible contaminating polysaccharides.

Galactomannans, after reacting with a Procion dye (Dudman and Bishop, 1968), have been examined by electrophoresis on cellulose acetate strips; the original polysaccharides have also been examined on paper in a borate buffer (Tewari et al., 1984). Polysaccharides can be eluted from cut strips and estimated with phenol–sulphuric acid. Glass fibre paper and detection by heating after spraying with H$_2$SO$_4$ can also be used (Akiyama et al., 1983). Glucomannan has been examined on glass fibre paper (Thompson and Jones, 1964) in 2 M KOH and dyed glucomannan on thin-layered Sephadex G-100 in borate buffer (Koleva and Gioreva, 1986). Galactoglucomannans (Alam and Richards, 1971; Gupta and Bose, 1986) have been analysed by free boundary electrophoresis, and galactomannans have been analysed on cellulose acetate as borate complexes and by staining with PAS reagent (Pechanek et al., 1982). There are a number of examples of examination by analytical ultracentrifugation or by gel chromatography (see Section V.A). On immunodiffusion a precipitin reaction was observed between galactomannan and *Ricinus communis* (Manzi et al., 1984) and *B. simplicifolia* lectins (Gupta et al., 1987).

B. Estimation of Polymer Content in Tissues

$(1\rightarrow4)$-β-Mannan-based polysaccharide content can be measured by weighing extracted polymer after purification. This is easier with water-soluble seed galactomannan and tuber or bulb glucomannan than with the polysaccharides in woody tissue. As the precipitated polysaccharides invariably contain firmly bound water, for accurate estimation a portion should be hydrolysed and the reducing sugar content measured, for example by estimating copper-reducing power. Electrophoresis of the polymer, followed by PAS staining, has been combined with densitometric scanning (Pechanek *et al.*, 1982). An enzymic method applied to galactomannan involved extraction of the sample, hydrolysis with a mixture of β-D-mannanase and α-D-galactosidase and estimation of released D-galactose with D-galactose dehydrogenase by spectrophotometric measurement of the reduction of NAD^+ to NADH (McCleary, 1981). The β-mannanase is added to ensure that any mannan, which has D-galactosyl substituents partly removed, does not precipitate. The galactomannan content is then calculated from the known ratio of galactose:mannose in the galactomannan. Using lectins, specific for D-galactosyl groups from *Bandeiraea simplicifolia* and *Artocarpus integrifolia*, in double diffusion and enzyme-linked lectin assays, the galactomannan content in food preparations has been determined (Patel and Hawes, 1987) and the method would appear to be adaptable to plant tissue. In this method the wells in the microtitre plates are first coated with lectin, washed and β-lactoglobulin solution added, incubated and washed again. Solution containing galactomannan is added to some wells and, after incubation, lectin peroxidase is added to all wells. Then after incubation and washing again, peroxidase substrate is added and the level of chromogenic product measured.

IV.   DETERMINATION OF CHEMICAL STRUCTURE

Reviews of methods of determination of chemical structure have appeared (Bouveng and Lindberg, 1960; Aspinall and Stephen, 1973; Aspinall, 1982b; McCleary and Matheson, 1986; Selvendran and O'Neill, 1987) and details of particular methods can be found in the series *Methods in Carbohydrate Chemistry*. Physico-chemical methods are discussed in Section V. Basic aspects of structure involve identification of constituent sugars, including their chirality and proportions, any substituents such as acetyl groups, the positions of glycosidic linkages and the nature of the anomeric linkages. More detailed aspects include the disposition of the sugar residues in the polymeric sequence.

A.   Identification and Estimation of Amounts of Constituent Sugars

The identity and proportions of constituent sugars are now usually estimated, after hydrolysis and conversion to more volatile derivatives, by gas chromatography. Trimethylsilyl compounds (Sweeley *et al.*, 1963; Unrau and Choy, 1970a,b; Achtardjiev and Koleva, 1973; Mazzini and Cerezo, 1979) and their diethyl thioacetal derivatives (Honda *et al.*, 1979; Akiyama *et al.*, 1983), as well as glycitol acetates formed after borohydride reduction (Sloneker, 1972; Eda *et al.*, 1984; Sen *et al.*, 1986; Gupta *et al.*, 1987; Mallett *et al.*, 1987), have been used. Earlier methods of separation employed cellulose column partition or thick paper chromatography, when the sugars could be identified by crystallising or converting to derivatives such as N-phenylglycosylamines (anilides), phenylhydrazones or galactaric (mucic) acid (Hirst *et al.*, 1947; Smith, 1948; Unrau, 1961; Kapoor, 1972). The chirality can also be determined by rotation or enzymic reaction: D-galactose dehydrogenase and D-glucose oxidase react with only one enantiomer.

The optical rotation of galactomannans indicates the proportion of D-galactose and D-mannose (Kooiman, 1972). Formic acid release on periodate oxidation of galactomannan gives the galactose content (see Section IV.D). D-Galactose can be estimated after acidic or enzymic hydrolysis with D-galactose dehydrogenase (EC 1.1.1.48) (McCleary and Matheson, 1975; McCleary, 1981; Mallett et al., 1987) or as galactaric acid (Heyne and Whistler, 1948). D-Glucose in glucomannan can be determined with D-glucose oxidase (EC 1.1.3.4) and total carbohydrate with phenol–sulphuric acid or anthrone reagents (see Chapter 1, this volume). Paper chromatography, after neutralisation of the hydrolysate, separates the component sugars: a variety of methods of estimation have been combined with this procedure. These include estimation of the aromatic imine formed by heating with an amine (Hamilton et al., 1960; Richards et al., 1968; Reid and Meier, 1970); elution of fractions from the paper followed by estimation of periodate uptake (Hirst and Jones, 1949; Andrews et al., 1956; Sømme, 1966); estimation by hypoiodite reduction (Kooiman, 1971); and colorimetric estimation after heating with an amine (Timell, 1957) or with phenol–sulphuric acid (Larson and Smith, 1955; Hamilton et al., 1956).

Acetyl content has been measured by saponification (Gowda et al., 1979; Gowda, 1980) and also by gas chromatography following methanolysis to form methyl acetate (Meier, 1961), using as an external standard methylene chloride and as a calibration standard penta-acetyl galactose.

B. Methylation Analysis

The positions of glycosidic linkages in polysaccharides are established by methylation analysis (Hirst and Percival, 1965; Lindberg, 1972), when unlinked hydroxyl groups are converted to methyl ethers. After hydrolysis, fractionation and estimation of the variously methylated monosaccharide components, the linkages become apparent. Etherification occurs under alkaline conditions: early procedures used dimethyl sulphate with strong (30–45%) aqueous alkali (Smith, 1948; Ahmed and Whistler, 1950; Andrews et al., 1956; Wolfram et al., 1961) (Haworth methylation), alone, or as the first stage in a two-step procedure, when the partly methylated product was reacted with methyl iodide and silver oxide (Purdie methylation) (Andrews et al., 1953; Aspinall et al., 1953; Hamilton et al., 1960; Courtois and Le Dizet, 1963). The insolubility of polysaccharides in methyl iodide requires prior Haworth methylation. Repeated additions are required in both procedures to attain full methylation. Some variations have included reacting the polysaccharide acetate in tetrahydrofuran in the Haworth method, giving simultaneous deacetylation and partial methylation; performing the reaction in acetone with solid NaOH; reacting with methyl iodide and barium oxide in N,N-dimethyl formamide (Kuhn method) (Sømme, 1968) or in dimethyl sulphoxide (Srivastava et al., 1964). Methylation can also be performed using methyl iodide in liquid ammonia (Tipson, 1963) and with thallium derivatives (Hirst et al., 1947; Jones, 1950).

The Hakomori method (Hakomori, 1964; Lindberg, 1972; Gowda et al., 1979; Wankhede et al., 1979) gives substitution much more rapidly, due to the strong basicity of the dimsyl anion. The polysaccharide, dispersed in dimethyl sulphoxide, is mixed with a solution of dimsyl ion (sodium hydride plus dimethyl sulphoxide) and methyl iodide added. Reaction must be performed under strictly anhydrous conditions.

Incomplete methylation can produce wrong conclusions: unmethylated hydroxyl groups in the polymer can be detected by IR spectroscopy or the methoxyl content can be compared with the theoretical value. Methylation occurs readily with $(1\rightarrow4)$-β-mannan-based polysaccharides because they are neutral, but to ensure complete substitution, or to overcome solubility problems, different procedures can be combined. Prior Haworth methylation gives a product that dissolves in dimethyl sulphoxide for Hakomori methylation (Unrau and Choy, 1970b); the Hakomori procedure has been used before Purdie methylation (Alam and Richards, 1971; Robic and Percheron, 1973); and the Haworth and Kuhn methods have been combined (Leschziner and Cerezo, 1970). Haworth methylation has been followed by reaction with barium oxide and methyl iodide in dimethyl sulphoxide (Kapoor, 1972), and these have been combined with Purdie methylation (Srivastava et al., 1968).

Previous reaction of free hydroxyl groups in naturally acetylated glucomannan with phenyl isocyanate to give the carbamate, followed by Kuhn methylation, which hydrolysed the acetyl groups prior to methylation, and then removal of N-methyl carbamoyl groups by reduction with lithium aluminium hydride (Katz, 1965), indicates the positions of acetyl substitution. Reaction of free hydroxyls with methyl vinyl ether to give acetals, followed by ester hydrolysis and methylation analysis, has also shown the positions of substitution (Paulsen et al., 1978).

Hydrolyses of permethylated polysaccharides are not usually performed directly in dilute aqueous acid, as they are insoluble in hot aqueous solution—nevertheless, some have been reported (Larson and Smith, 1955; Jindal and Mukherjee, 1970; Robic and Percheron, 1973). The absence of uronic acid residues allows ready hydrolysis. Depolymerisation by methanolysis with hydrogen chloride in methanol leads to formation of methyl glycosides from which the acetal methyl can be removed with dilute aqueous acid (Andrews et al., 1952, 1953), or the glycosides can be directly examined by gas chromatography (Aspinall, 1963; Buchala et al., 1974). Formic acid (90%) has been used for the initial hydrolytic step: due to the formation of formyl esters this must be followed by hydrolysis with dilute aqueous acid (Bouveng and Lindberg, 1965b). Trifluoracetic acid hydrolyses directly (Manzi et al., 1984): another approach is to treat the polymer with 70–75% sulphuric acid at room temperature to partly depolymerise and then dilute the mixture and heat to fully hydrolyse (Thompson and Jones, 1964; Unrau and Choy, 1970a,b; Alam and Richards, 1971). Demethylation during hydrolysis can occur and, in view of this and the chance of undermethylation, it is unwise to place too much significance on the detection of small amounts of products. Low levels of contaminating polymers such as cellulose or xyloglucan can also lead to possibly erroneous conclusions about structure.

An early method of fractionation used for identification and estimation of the amount of hydrolysate components was chromatography on powdered cellulose (Andrews et al., 1952; Aspinall et al., 1953; Larson and Smith, 1955) or charcoal (Kato et al., 1976), methods that require considerable quantities of material, but which do readily allow the preparation of derivatives and the determination of optical activity. Chromatography on thick paper or thin layers, which uses smaller amounts, has replaced column chromatography (Larson and Smith, 1955; Hamilton et al., 1960; Wolfram et al., 1961; Sømme, 1967; Lal and Gupta, 1972). The methylated monosaccharides have been characterised by direct crystallisation (Andrews *et al.*, 1956; Hamilton *et al.*, 1960; Tewari *et al.*, 1984; Alam and Gupta, 1986) and also by the preparation of crystalline derivatives, such as the N-phenylglycosylamines (anilides) (Andrews *et al.*, 1952, 1956; Aspinall *et al.*, 1953; Unrau, 1961), the nitrobenzoates (Thompson and Jones, 1964; Richards *et al.*, 1968) or the lactones formed after bromine oxidation—or as the phenylhydrazides, or cyclic urethanes derived from these lactones (Andrews *et al.*, 1952, 1956; Aspinall *et al.*, 1953; Hamilton *et al.*, 1960). The amounts of the methylated fractions, after elution from columns, have been estimated by weighing, and from paper by reaction with alkaline hypoiodite (Andrews *et al.*, 1956) or with phenol–sulphuric acid reagent (Larson and Smith, 1955).

Gas chromatography of the glycosides formed by methanolysis (Aspinall, 1963; Richards *et al.*, 1968; Jakimow-Barras, 1973), the trimethylsilylglycosides (Srivastava *et al.*, 1968; Unrau and Choy, 1970a) or the alditol acetates (Alam and Richards, 1971; Lindberg, 1972; Sieber, 1972; Buchala *et al.*, 1974; Manzi *et al.*, 1984), is now commonly used. With the last derivatives the sugars are reduced with sodium borohydride prior to acetylation (Blakeney *et al.*, 1983); this has the advantage that each sugar derivative produces only one peak. The fractions can be identified by their retention times or, more conclusively, by mass spectrometry (Lindberg, 1972; Aspinall, 1982b). In the latter procedure, if the sugar is reduced with borodeuteride, C-1 can be more readily identified in the fragmentation pattern. Gas chromatography is rapid and can be applied on a micro-scale, allowing ready quantitative estimation; the introduction of capillary columns has improved resolution (Geyer *et al.*, 1982).

Most methylation analyses of (1→4)-β-mannan-based polysaccharides have produced as the main products 2,3-di-*O*-methyl and 2,3,6-tri-*O*-methyl D-mannoses, 2,3,4,6-tetra-*O*-methyl-D-galactose and 2,3,6-tri-*O*-methyl-D-glucose, consistent with the generally accepted structures. In the galactoglucomannan from Townsville lucerne stems (Alam and Richards, 1971) the galactosyl residues were linked only to mannosyl units and not to glucosyl units. In some instances (Aspinall *et al.*, 1953; Andrews *et al.*, 1953; Kato *et al.*, 1976) the average chain length of the (1→4)-β-mannan chain has been calculated from the inverse of the proportion of 2,3,4,6-tetra-*O*-methyl mannose released.

There have also been a few reports of unusual linkages from methylation analysis. The galactomannan from *Melilotus indica* seeds has been reported (Gupta and Bose, 1986) to contain (1→2)-linked mannosyl and (1→4)-linked galactosyl residues and the presence of some substitution of galactosyl units was proposed for the galactomannans from *Cassia corymbosa* (Tewari *et al.*, 1984) and *Gleditsia triacanthos* seeds (Leschziner and Cerezo, 1970). Methylation analysis products indicated that the galactoglucomannan from suspension-cultured cells of tobacco and a similar polymer from the midrib contained (1→2)-galactosyl linkages (Akiyama *et al.*, 1983; Eda *et al.*, 1984).

C. Partial Hydrolysis and Determination of Anomeric Linkages

The nature of the anomeric linkages can be established in various ways. These include: (1) optical rotation of the polymer; (2) examination of the oligosaccharides produced by partial hydrolysis with dilute acid or with *endo*-acting enzymes; (3) examination of the monosaccharides released by glycosidases; (4) lectin interaction; (5) oxidation by chromium trioxide; and (6) IR and NMR spectroscopy (Section V.B).

The differences in the optical activity of α- and β-anomeric glycosides (Pigman, 1957) allows the linkages in glycans to be deduced. The optical rotations of the (1→4)-mannan-based polymers are consistent with β-linkage of D-mannosyl and D-glucosyl units and α-linking of D-galactosyl groups (Richards et al., 1968; Kooiman, 1972). The products of partial hydrolysis have been fractionated by chromatography on Bio-Gel P-2, paper, thin layer and charcoal–celite and identified by methylation analysis, glyco-sidase hydrolysis and NMR spectroscopy. The glycose unit at the reducing end can be established by comparison, after hydrolysis, of the original and borohydride-reduced oligosaccharides. The possibility of reversion of monosaccharides to oligosaccharides under acidic conditions and of enzymic transglycosylation to give bonds not present originally in the polysaccharide must be considered: the significance of oligosaccharides isolated in very small amounts should be interpreted cautiously.

Partial hydrolysis with endo-enzymes, provided interfering activities are absent, has the advantage that the specificity of the enzyme defines the type of linkages that have been hydrolysed and produces oligosaccharide products in high yield (Matheson and McCleary, 1985; McCleary and Matheson, 1986). β-Mannanase and other hydrolytic activities have produced oligosaccharides up to a DP of 9 that could be fractionated. The structures have generally been consistent with their derivation from (1→4)-β-D-manno and gluco- and (1→6)-β-D-galacto structures (Whistler and Smith, 1952; Perila and Bishop, 1961; Love and Percival, 1964; Reese and Shibata, 1965; Emi et al., 1972; Shimahara et al., 1975; McCleary et al., 1982, 1983; McCleary and Matheson, 1983; Shimizu and Ishihara, 1983; Takahashi et al., 1984; Kusakabe et al., 1988). The possibility of (1→2)-D-galactosyl linkages has also been proposed (Eda et al., 1984). The residual higher molecular weight fraction from galactomannan hydrolysis has been examined (Courtois and Le Dizet, 1968; McCleary and Matheson, 1983).

In acetolysis (Love and Percival, 1964; Thompson and Jones, 1964; Robic and Percheron, 1973; McCleary et al., 1976) the reaction is catalysed under acidic con-ditions (acetic anhydride and concentrated sulphuric acid) at low temperature, when lysis of glycosidic bonds and acetylation both occur; (1→6)-bonds are split preferen-tially. After extraction into chloroform the products are deacetylated catalytically with sodium methoxide. Oligosaccharides up to a DP of 5 have been detected and have been mostly (1→4)-oligomers of mannose (Aspinall et al., 1958), mixed glucosyl and mannosyl structures (Smith and Srivastava, 1956), and cellobiose. Smaller amounts of 2- and 3-linked structures have also been found.

Partial acidic hydrolysis employs acid of low strength for short periods of time (Sieber, 1972; Jakimow-Barras, 1973; Hopf and Kandler, 1977; Alam and Gupta, 1986; Sen et al., 1986) and has produced similar (1→4)-mannosaccharides, as well as Gal-α-(1→6)-Man (Whistler and Durso, 1952; Meier, 1960), gluco-manno-oligomers (Hamilton and Kircher, 1958) and these with single galactosyl groups attached via (1→6)-linkage (Mills and Timell, 1963). Some other oligomers with unusual linkages have been reported (Eda et al., 1984; Gupta and Bose, 1986).

The linkage of D-galactosyl groups through an α-anomeric linkage has been shown by hydrolysis with α-D-galactosidase (Buchala and Meier, 1973; McCleary et al., 1982, 1983; Gupta et al., 1987). Not all glycosidases will hydrolyse the non-reducing end-groups of polysaccharides, so an enzyme known to react with polysaccharides must be used. Interaction with a lectin of known specificity [such as that of *Griffonia* (*Band-eiraea*) *simplicifolia*, that interacts with α-D- but not β-D-galactosyl] in a gel-diffusion precipitin test indicated the anomeric linkage (Manzi et al., 1984; Gupta et al., 1987).

In chromium trioxide oxidation (Hoffman and Lindberg, 1980) advantage is taken of the much faster rate of oxidation of equatorial anomeric linkages in the chair conformation of pyranose rings (as in β-mannosyl) compared with axial linkages (α-galactosyl) (Angyal and James, 1970). The fully acetylated polysaccharide is stored with chromium trioxide and the aldo-sugars remaining after increasing periods of time, estimated. In galactomannans from the seeds of *Cassia* (Gupta et al., 1987) and *Indigofera* (Sen et al., 1986) the rate of degradation of mannosyl residues was much faster than that of galactosyl residues.

D. Periodate Oxidation and Smith Degradation

Periodate oxidation (Hay et al., 1965a; Aspinall, 1982b) of the structures generally accepted for $(1\rightarrow4)$-β-mannan-based polysaccharides should consume one mole of periodate per mannosyl or glucosyl residue in the main chain and two moles per galactosyl group or mannosyl or glucosyl end-group, releasing one mole of formic acid. Reduction of the oxidised polysaccharide with borohydride, followed by acidic hydrolysis (Smith degradation), produces erythritol (from glucosyl or mannosyl main chain residues) and glycerol (from galactosyl groups) but no hexose. Formic acid release can be estimated by titration, after destruction of excess periodate with ethane 1,2-diol, and periodate consumption measured spectrophotometrically, or by adding excess iodide to the remaining periodate and titrating liberated iodine with thiosulphate or arsenite. Possible over- or under-oxidation can reduce the significance of results from periodate oxidation. Oxidation of mannans and glucomannans (where any acetyl groups had been previously hydrolysed) (Jones, 1950; Andrews et al., 1956; Love and Percival, 1964; Achtardjiev and Koleva, 1973; Maeda et al., 1980) has been in accord with the accepted structure. A comparison of the oxidation of native acetylated and deacetylated glucomannan (Meier, 1961; Buchala et al., 1974) estimated acetyl content, and Smith degradation indicated that substitution occurred on mannosyl units, since mannitol acetate was the main hexitol detected by gas chromatography after reduction and acetylation.

The periodate uptake by galactomannans (Unrau, 1961; Khanna and Gupta, 1967; Leschziner and Cerezo, 1970; Kapoor, 1972; Sen et al., 1986) gives the expected results if substitution by galactosyl groups is low. Where substitution is high, complete oxidation of galactosyl groups occurs, but for mannosyl residues a high concentration and excess of periodate are needed (Andrews et al., 1952; Richards et al., 1968). Formic acid is released from galactosyl groups and (ignoring the small contribution from the ends of the mannan chain) its measurement allows the amount of these to be calculated. Smith degradation produces the predicted amounts of erythritol and glycerol (which can be separated by gas or paper chromatography) and trace amounts of hexose. Underoxidation of galactomannans is apparently due to the formation of inter-residue cyclic hemi-acetals (Cerezo, 1965), leading to anomalous oxidation limits. Hemi-acetals formed by oxidised units substituted on C-6 are much more stable than those formed by unsubstituted units. This behaviour has led to proposals about the distribution of galactosyl substituents along the mannan chain (Hoffman et al., 1975, 1976; Forsberg and Pazur, 1979; Gonzalez and Painter 1980). Products from methylation analysis of the original galactomannan, the polymer oxidised by periodate, and the borohydride reduction product of the latter, have been compared. Analysis of the kinetics of the oxidation reaction has also been applied (Painter et al., 1979).

E. Chemical Estimation of Chain Length and Degree of Polymerisation

Reactions specific for end-groups (Smith and Montgomery, 1956) can estimate chain length or DP of oligomers and polymers, if the end-group is not hindered in reaction or small amounts of contaminants do not interfere. Reactions can occur at the non-reducing or reducing end-groups or at both. Those at non-reducing ends estimate average chain length, which, if there is no branching, will be the DP; those at the reducing end-group estimate DP.

Methylation analysis of $(1\rightarrow 4)$-β-mannan-based polymers gives 1,2,3,6-tetra-O-methyl-D-mannose(-glucose) from a non-reducing end of a chain. The inverse of this, as a mole fraction of the total hydrolysate, estimates average chain length and has been determined for mannans (Aspinall *et al.*, 1953), glucomannans (Andrews *et al.*, 1953; Kato *et al.*, 1976) and galactomannans (Unrau and Choy, 1970a,b; Manzi *et al.*, 1984). On periodate oxidation of a $(1\rightarrow 4)$-linked aldohexopyranosyl polymer, formic acid is released from both types of end-groups and in mannans and glucomannans the average chain length has been estimated (Andrews *et al.*, 1953; Iriki and Miwa, 1960; Thompson and Jones, 1964; Shimahara *et al.*, 1975).

Methods that involve reaction with the single reducing end-group of a polysaccharide should in theory estimate the DP. Radioactive methods measuring the extent of reaction with Na^{14}CN (Isbell, 1965) or reduction with borotritide can be used. Oxidation of end-groups with hypoiodite has been described (Andrews *et al.*, 1953). Another approach is to reduce the end-group with borohydride and then oxidise with periodate. In a $(1\rightarrow 4)$-linked chain the glucitol or mannitol end-group formed will release two molecules of formaldehyde, which can then be estimated with chromotropic acid (Hay *et al.*, 1965b), a procedure that has been applied to galactomannans (Unrau, 1961; Leschziner and Cerezo, 1970). Another approach, which by analogy with glucans and xylans (Sturgeon, 1980) could be used, is to reduce, hydrolyse and then estimate spectro-photometrically (by the reduction of NAD$^+$ to NADH) the hexitol content with D-mannitol dehydrogenase (EC 1.1.1.67) [alone for mannans and galactomannans or admixed with D-glucitol dehydrogenase (EC 1.1.1.14) for glucomannans and galactoglucomannans].

The use of chemical methods nowadays tends to be limited to oligosaccharides. The values obtained for polymers have usually been much lower than those obtained by physical methods.

F. Substituent Distribution and Glucose Interpolation along the $(1\rightarrow 4)$-β-Mannan Chain

Chemical methods of studying the distribution of substituents along the β-mannan chain have included defining the arrangements of galactosyl groups in galactomannans and of acetyl groups in glucomannans. Methods based on periodate oxidation are described in Section IV.D.

In one procedure with galactomannan (Baker and Whistler, 1975), unsubstituted primary hydroxyl groups were converted to C-p-tolylsulphones, which were degraded under alkaline conditions by a β-elimination mechanism that cleaved the ring and methylated concurrently. Splitting of the β-mannan backbone occurred at unsubstituted mannosyl units, giving mannosaccharide derivatives. Acidic hydrolysis, reduction and acetylation produced 1,5-di-O-acetyl-2,3,4,6-tetra-O-methyl-D-mannitol from mono-meric mannose derivatives or the non-reducing end of oligomers and 1,4,5-tri-O-acetyl-2,3,6-tri-O-methyl-D-mannitol from mannosyl residues, linked-$(1 \rightarrow 4)$. The amounts of these, separated by gas chromatography, indicated the degree of isolated or block substitution in the original polymer.

In another procedure (Hoffman and Svensson, 1978), primary hydroxyl groups in the polymer were preferentially substituted as trityl ethers, the polymer methylated by the Hakomori method and the trityl groups selectively removed by dilute acidic hydrolysis. The exposed primary hydroxyl groups were then oxidised with chlorine in dimethyl sulphoxide and the product treated with alkali to induce β-elimination. Further examination of the products (by mild hydrolysis, borodeuteride reduction and methyl-ation) allowed conclusions to be drawn about the arrangement of substitution by galactosyl groups.

The distribution of O-acetyl groups in pine glucomannan was examined (Kenne *et al.*, 1975; Paulsen *et al.*, 1978) by substituting the free hydroxyl groups as an acetal on reaction with methyl vinyl ether. Alkali then hydrolysed the acetyl ester groups and the exposed hydroxyl groups were oxidised to carbonyl groups with chlorine in dimethyl sulphoxide. On alkaline degradation, hexosyl residues containing a keto group at carbon-2 underwent β-elimination, giving a series of oligosaccharides that reflected the original distribution of acetyl groups along the mannan chain. Mild acidic hydrolysis released the acetal substituents and the oligosaccharides were examined by methylation analysis and gel chromatography. These methods require that the reactions proceed in essentially quantitative yield and that interfering side reactions do not occur.

Enzymic methods have also been applied. After hydrolysis with β-mannanase of the fraction from carob galactomannan soluble in hot water, from the structures and amounts of the oligosaccharide products (up to nonasaccharide), as well as consider-ation of the action pattern of the β-mannanase and a possible biosynthetic model, a generalised structure of this polysaccharide was proposed (McCleary and Matheson, 1986). The data on the released oligomers were also used in computer simulation in a chain-extending program, involving a nearest neighbour–second nearest neighbour model, to describe the galactosyl distribution (McCleary *et al.*, 1985).

V. PHYSICO-CHEMICAL METHODS

These divide broadly into methods that determine molecular size, such as analytical ultra-centrifugation, light scattering, osmometry and viscometry; and spectroscopic methods, for example infrared and nuclear magnetic resonance, that yield information about structure. X-ray diffraction also indicates structure. An advantage of these methods is that the polymer can be reclaimed.

A. Determination of Molecular Size

Because polysaccharides consist of distributions of sizes, the average can be a weight average M_w (from ultra-centrifugation, light scattering or viscosity measurements) or a number average M_n (from osmometry and chemical end-group determination). The ratio M_w/M_n indicates the degree of heterogeneity in molecular size (polydispersity).

Polysaccharides, in contrast to globular proteins, give values of this ratio that are significantly higher than 1.

Analytical ultra-centrifugation (Schachman, 1957) has been performed with neutral aqueous solutions of galactomannans (Kubal and Gralen, 1948; Richards et al., 1968; Sharman et al., 1978) and of galactoglucomannan (Akiyama et al., 1983), with alkaline solutions of galactomannan (Hui and Neukom, 1964; Unrau and Choy, 1970a), and with mannan in 90% formic acid (Wolfram et al., 1961). Acetylated glucomannan has been studied in acetone (Kato et al., 1976) and galactomannan acetate in acetonitrile (Koleske and Kurath, 1964). Galactomannans show hyperfine sharpening of the boundary and a non-linear dependence of sedimentation coefficient on concentration (Kubal and Gralen, 1948; Hui and Neukom, 1964; Cerezo, 1965; Unrau and Choy, 1970a; McCleary and Matheson, 1975), making the calculation of molecular weight difficult. The frictional ratio indicates that they are non-spherical.

The advent of gel chromatography has provided a method of estimation of average molecular size and range of sizes (Churms, 1970) that requires simple equipment. As polysaccharides consist of a range of molecular sizes and have solution conformations that differ from the globular shape of most soluble proteins, accurate estimation of molecular weight by gel chromatography is generally not possible. Eluting polysaccharide can be detected by the change in refractive index or with phenol–H_2SO_4. Molecular size can be compared via the K_{av} [(elution volume – void volume)/(total volume – void volume)] and by reference to dextran or pullulan standards. When samples are highly viscous or of high molecular size, slow elution rates should be used.

Mannan, glucomannan, galactomannan and galactoglucomannan have all been examined analytically on a range of column matrices with different exclusion sizes. Mannan has been chromatographed as the nitrate in tetrahydrofuran on polystyrene beads, when a range of DP values of 100–2500 (mean c. 1000) for 90% of the material was indicated (Mackie and Sellen, 1969). Glucomannan, synthesised enzymically, was run on porous glass (37 and 70 nm pore size) in methylene chloride–methanol (1:1; v/v) after conversion to the acetate. The polymer was detected from the incorporated radioactive label (Villemez, 1974). Glucomannan has been chromatographed on porous glass (Matsuo and Mizuno, 1974b), fractions obtained by water–ethanol precipitation have been run on Sephadex G-100 and G-25 (Vilkas and Radjabi-Nassab, 1986), and Procion-dyed glucomannan has been examined on thin layer chromatography (TLC) plates of Sephadex G-100 (Koleva and Gioreva, 1986). Galactomannan has been chromatographed at low pressure on porous glass (Merckogel SI 500 nm) in 1 M NaCl, 5 mM Na_2 ethylene diamine tetra-acetic acid (EDTA) (McCleary and Matheson, 1975; McCleary et al., 1976) and under pressure in the same matrix (Vijayendran and Bone, 1984), as well as on 400 nm pore-sized beads (Barth and Smith, 1981). Sepharose 2B, consistent with its preparation from agarose, absorbed some galactomannan. Galactomannan eluted from Sepharose CL-4B with water as solvent (Sen et al., 1986); a broad elution peak was obtained. Galactomannan has also been run at medium pressure and a low flow rate ($5\,\mathrm{ml\,cm^{-2}\,h^{-1}}$) on Fractogel TSK HW-75(F) (Mallett et al., 1987). Galactomannan, degraded partially by β-mannanase, has been examined in 0.5% NaCl, 0.02% azide by high performance liquid chromatography (HPLC) on Shodex Ionpak K8-806 (Cheetham et al., 1986). It has also been studied on Sephadex G-100 with 7 M urea and 0.1 M NaCl as solvents (Manzi et al., 1984).

Galactoglucomannan has been chromatographed at low pressure on porous glass

(Merckogel SI 500 nm) (McCleary et al., 1976); by HPLC on Toya Soda TSK Gel G 3000 SW (Akiyama et al., 1983) and on Sepharose 4B and CL-6B (Alam and Richards, 1971; Eda et al., 1984, 1985).

Light scattering in solution has been applied to mannan nitrate in ethyl acetate (Mackie and Sellen, 1969), glucomannan (Sugiyama et al., 1972), partially methylated glucomannan (Kishida et al., 1978) and galactomannan (Deb and Mukherjee, 1963; Robinson et al., 1982) in aqueous solution, galactomannan acetate in acetonitrile (Koleske and Kurath, 1964) and permethylated galactomannan in chloroform (Mukherjee et al., 1961). In this procedure of molecular weight determination it is essential that a true solution be obtained, since the presence of aggregated or partly hydrated particles will give erroneously high values.

Membrane osmometry has been used with mannan nitrate in ethyl acetate or butanone (Mackie and Sellen, 1969) and glucomannan nitrate in butyl acetate (Buchala et al., 1974). This method has also been used with the glucomannan acetate in tetrachloroethane–ethanol (Hamilton and Kircher, 1958), and permethylated glucomannan in $CHCl_3$:EtOH, 24:1 (v/v) (Mandal and Das, 1980). Galactomannan (Whistler and Saarnio, 1957) and galactomannan acetate have been studied in acetonitrile (Koleske and Kurath, 1964). Osmometry loses precision as the molecular size increases.

The usefulness of viscometry in determining molecular weight depends on finding a relationship between the specific viscosity and molecular weight (Koleske and Kurath, 1964; Mackie and Sellen, 1969; Kishida et al., 1978; Sharman et al., 1978; Robinson et al., 1982). Viscometers belong generally to two types; those in which the passage of the solution through a uniform, fine tube is timed; or those in which the force required to move one surface, such as a cylinder, relative to a second surface (a concentric cylinder) is measured. Mannan nitrates have been examined in acetone and mannan in cupriethylenediamine (Timell, 1957); partially methylated glucomannan in aqueous solution; and glucomannan and galactomannan in NaOH or aqueous solutions (Hui and Neukom, 1964; McCleary and Matheson, 1975, McCleary et al., 1976, 1981; Elfak et al., 1979; Whitcomb et al., 1980; Mallett et al., 1987). The high viscosity of galactomannan is associated with the mannan chain and this is illustrated by the retention of high values on selective removal of galactosyl groups.

B. Spectroscopic Methods

Infrared spectra (Barker et al., 1956; Marchessault, 1962; Spedding, 1964; Perlin and Casu, 1982) have provided information about the anomeric linkages (in conjunction with the ring size) in mannans (Kato et al., 1973) glucomannans (Thompson and Jones, 1964; Kato et al., 1976) and galactomannans (Cerezo, 1965; Kapoor, 1972; Seth et al., 1984) and the presence of acetyl groups in glucomannans (Katz, 1965; Matsuo and Mizuno, 1974a). Spectra can be obtained from potassium bromide pellets (Seth et al., 1984), polymer films (Leschziner and Cerezo, 1970) or nujol mulls (Alam and Gupta, 1986). Absorption bands at 810–820 cm^{-1} and 870–875 cm^{-1} have been assigned to equatorial and axial C—H bonds, respectively, on a $^4C_1$ conformation of a D-pyranose ring and hence indicate the presence of α-D-galactopyranosyl and β-D-mannopyranosyl units. Absorption bands due to the C=O, C—CH$_3$ and C—O bonds have been detected in partly acetylated glucomannan and galactoglucomannan; these were not detected after ester hydrolysis.

Proton nuclear magnetic resonance (NMR) and carbon-13 NMR spectra (Hall, 1964; Jennings and Smith, 1978; Perlin and Casu, 1982; Casu, 1985) have been recorded for all the types of $(1\rightarrow4)$-β-mannan polysaccharides. These have been used to assign anomeric linkages, to estimate the proportions of constituent monosaccharides, to establish the structures of oligosaccharides derived by partial hydrolysis, and to indicate aspects of the pattern of substitution by α-D-galactosyl groups along the β-mannan backbone of galactomannans. Spectra have been measured in D_2O and carbon-13 NMR spectra have been made in the Fourier-transform mode. When glucomannan was examined (Matsuo and Mizuno, 1974a; Vilkas and Radjabi-Nassab, 1986) by proton NMR spectroscopy, chemical shifts characteristic of acetyl groups and peaks and coupling data characteristic of anomeric protons were identifed, consistent with β-glycosidic linkages of D-mannosyl and D-glucosyl units. Chemical shifts, determined by proton decoupled carbon-13 NMR spectroscopy (Usui et al., 1979), were assigned to the disaccharides $(1\rightarrow4)$-β-mannobiose, Man-β-$(1\rightarrow4)$-Glc and Glc-β($1\rightarrow4$)-Man, and the trisaccharides, $(1\rightarrow4)$-β-mannotriose, Man-β-$(1\rightarrow4)$-Man-β-$(1\rightarrow4)$-Glc and Man-β-$(1\rightarrow4)$-Glc-β-$(1\rightarrow4)$-Man prepared by acidic hydrolysis of glucomannan. Previous data on cellobiose and mannobiose were compared. The peaks in the spectrum of glucomannan were then assigned to the two sugars and also to the acetyl function. Preparations of galactoglucomannan (Akiyama et al., 1983; Eda et al., 1985), studied by carbon-13 NMR showed four anomeric carbon signals that were assigned to α-D-galactosyl, β-D-glucosyl and β-D-mannosyl units, and also β-D-galactosyl, all in the $^4C_1$ conformation. The coupling values $(J_{C-1,H-1})$ were in agreement and similar resonances were found in oligosaccharides produced by enzymic hydrolysis. Proton NMR and proton-decoupled carbon-13 NMR spectra of galactomannan contained signals due to α-D-galactosyl and β-D-mannosyl units in the $^4C_1$ conformation (Gupta et al., 1987), and a comparison of proton NMR spectra, before and after treatment with α-D-galactosidase, demonstrated that the doublet due to α-D-galactopyranosyl groups was not present after enzymic treatment. Galactomannans have been partly depolymerised by acid to an average DP of 20–40 and proton and carbon-13 NMR spectra recorded (Grasdalen and Painter, 1980; Painter, 1982). The proton spectra indicated the presence of α-D-galactopyranosyl and β-D-mannopyranosyl ($^4C_1$ conformation) units and estimates were made of their proportion from the areas of the signals. In the carbon-13 NMR spectrum, the C-4 mannosyl resonance was interpreted to be sensitive to substitution on the neighbouring residue and the relative intensities of three signals were attributed to varying proportions of unsubstituted, mono- and di-substituted pairs of mannosyl residues. Chemical shifts and coupling constants were assigned to substituent positions in mannobiose and mannobiose substituted with an α-galactosyl group on either the reducing mannose residue or the non-reducing mannosyl group (McCleary et al., 1982). The former heterodisaccharide was prepared by β-D-mannanase hydrolysis and the latter by partial acidic hydrolysis. In combination with methylation analysis and enzymic hydrolysis, the structures of oligosaccharides up to a DP of 9, that were released by hydrolysis of galactomannan with β-D-mannanase, were then established (McCleary et al., 1983). Galactomannan–borate complexes have also been studied by NMR (Noble and Taravel, 1988).

X-Ray diffraction provides information about conformation and the packing of the polymer chains in the solid state. The $(1\rightarrow4)$-β-mannan-based polysaccharides show structural similarities to other $(1\rightarrow4)$-diequatorially linked pyranose glycans such as cellulose (Marchessault and Sarko, 1967; Marchessault and Sundararajan, 1983; Atkins *et al.*, 1988) and xylan. The unit cell in the fibre axis is about 1.0 nm, the pyranose rings are $^4C_1$ conformers, there is hydrogen bonding between the 3-OH of one mannosyl unit and the OH of the hydroxymethyl on the next pyranose ring, and hydroxymethyl groups on neighbouring rings lie opposite. The pattern and the fibre repeat unit are similar to those of cellulose (Frei and Preston, 1968; Nieduszynski and Marchessault, 1972; Mackie *et al.*, 1986). Glucomannan (native) acetate from pinewood, when examined (Katz, 1965) as an elongated film, had an unordered structure. Deacetylated glucomannan showed little order, but alkali-extracted polymer (aged for 3 years) showed a slightly ordered structure. Three galactomannans with differing galactosyl contents, examined as elongated fibres (Marchessault *et al.*, 1979), all had similar dimensions for the unit cell but no sign of the diffraction pattern of mannan A. The dimension of the unit cell in one direction was similar to that of β-mannan (Palmer and Ballantyne, 1950), suggesting similar packing forces in this direction. The fibres were formed from thin strips of film that were then elongated in a humid environment. Electron diffraction patterns have been produced from oriented films (Veluraja and Atkins, 1988).

C. Solution Conformation and Interactions

Using known bond lengths and bond angles, and with the mannosyl units in the C_1 conformation, the non-bonded interaction energies of $(1\rightarrow4)$-β-mannobiose and the $(1\rightarrow4)$-β-mannan chain have been computed for various angles of rotation of the linkages involved in the glycosidic bond (Sundararajan and Rao, 1970; Rees and Scott, 1971; Atkins *et al.*, 1973). Theoretical conformations at equilibrium of conformers of methyl $(1\rightarrow4)$-β-mannobioside have been calculated for different solvents (Tvaroska *et al.*, 1987). A consequence of the slightly twisted ribbon-like conformation proposed for the β-mannan chain is that the hydroxymethyl groups and hence galactosyl substituents on neighbouring mannosyl residues project on opposite edges of the ribbon (McCleary *et al.*, 1976). The possible significance of this stereochemistry in biosynthetic processes (Matheson, 1984; McCleary and Matheson, 1986) and in the mechanism of enzymic hydrolysis of galactomannan by the *endo*-enzyme β-mannanase has been discussed (McCleary and Matheson, 1983, 1986).

Galactomannans with lesser amounts of substitution by galactosyl groups and native glucomannans interact in solution with other polysaccharides such as carrageenan, agarose and xanthan to form gels (McCleary *et al.*, 1981, 1984; Rees *et al.*, 1982; Tako and Nakamura, 1986; Brownsey *et al.*, 1988). The process has been studied by chiroptical, viscometric, NMR relaxation, X-ray diffraction and enzymic methods and a model proposed in which unsubstituted regions of the mannan chain interact to form a macromolecular network (Dea and Morrison, 1975; Dea *et al.*, 1977; Morris, 1986). A modification has been proposed that interacting regions need not be completely unsubstituted but can have only one edge of the ribbon-like conformation free of galactosyl groups (McCleary, 1979a; Dea *et al.*, 1986). Removal of sufficient of the galactosyl groups, or deacylation of native glucomannan, leads to precipitation, presumably because the $(1\rightarrow4)$-β chains can then more closely align. It has been suggested that the conformation of the $(1\rightarrow4)$-β-mannan chain has a role in host–pathogen recognition (Morris *et al.*, 1977) and adhesion in cell-wall extension (Sellen, 1980).

VI. METABOLIC ASPECTS

Enzymes metabolising $(1\rightarrow4)\beta$-mannan-based polysaccharides have been purified by ion-exchange, gel and affinity chromatography (Scopes, 1982).

A. Biosynthetic Enzymes

Mannose 6-phosphate isomerase occurs in a number of plant organs that synthesise mannan-based polymers. In konjac tubers (Murata, 1975), after extraction with Tris-HCl (pH 7.8), activity was chromatographed on DEAE-cellulose at pH 7.2 with a KCl gradient to 0.4 M and then at pH 7.8 on Sephadex G-200. The apparent K_m for Man-6-P was 0.73 mM at pH 6.5, with an equilibrium ratio of 1.06. The pH for optimal reaction was 6.5–7.0. Metal-chelating agents inhibited and divalent cations restored activity. Activity was extracted from developing cotyledons of *Cassia coluteoides* (Lee and Matheson, 1984) with 0.1 M Tris-HCl (pH 7.5) using insoluble polyvinylpyrrolidone to bind phenols and purified *Bacillus subtilis* β-mannanase to depolymerise galactomannan, which interfered both with ammonium sulphate precipitation and gel chromatography. Purification involved DEAE-cellulose chromatography (0.01–0.5 M KCl gradient) and gel chromatography on Sephacryl S-200. The K_m for Man-6-P was 1.6 mM at the pH optimum of 7.0. The M_r was estimated as 74 500 by sucrose density gradient centrifugation and 68 000 by gel chromatography on Sephacryl S-200 superfine, values similar to those for the enzymes from developing soybeans, guar and fenugreek. The enzyme can be assayed by estimating the production of ketohexose with resorcinol or by following spectrophotometrically the reduction of $NADP^+$ in the coupled enzyme mixture of glucose 6-phosphate isomerase and glucose 6-phosphate dehydrogenase.

Phosphomannomutase has been studied in both corms accumulating glucomannan (Murata, 1976) and seeds accumulating galactomannan (Small and Matheson, 1979). Activity from konjac was extracted with Tris-HCl (pH 7.8) containing phosphate, chromatographed on DEAE-cellulose (KCl gradient 0–0.43 M) in phosphate buffer (pH 7.2) containing Mg^{2+} EDTA and mercaptoethanol and then on hydroxylapatite in a phosphate gradient (0.01–0.3 M). Phosphoglucomutase overlapped on both columns. The pH optimum was 6.5–7.0 and K_m (Man-1-P) 0.2 mM and K_m (Glcl-6-bisP) 1.8 mM at pH 7.0. Mg^{2+} was required and the equilibrium [Man-6-P]/[Man-1-P] 8.5 at pH 7.0. The M_r, by gel chromatography, was 62 000. The enzyme from developing cotyledons of *Cassia corymbosa* was extracted with acetate buffer (pH 6.2) containing Mg^{2+} and histidine as a metal ion chelator. After chromatography on DEAE-cellulose with a KCl gradient (0–0.3 M) on phosphocellulose, buffer eluted phosphoglucomutase and hexose 6-phosphate isomerase, and 5 mM Fru-1,6-bisP eluted phosphomannomutase. Chromatography on Sephadex G-100 was then applied. K_m (Man-1-P) was 0.15 mM; K_m (Man-6-P), 0.3–0.5 mM; K_m (Glc-1,6-bisP) 0.87 μM and K_m (Man-1,6-bisP), 2.5 μM at pH 6.8, with an optimum pH range of 6.5–7.0 and an equilibrium at pH 7.5 and 30°C of 81:19. Mg^{2+} was optimal at 2.5 mM. Reaction rates can be assayed by estimating the change in copper-reducing activity, or in acid-hydrolysable phosphate.

Evidence for the presence of GDP-Man pyrophosphorylase in *Gleditsia* seeds was obtained by incubating an extract (Tris-HCl, pH 7.9, containing mercaptoethanol) with α-D-^{14}C-Man-1-P and GTP. Mg^{2+} was required (Jiminez de Asua *et al.*, 1966). Enzymic activity has been measured in both directions of the reaction, using radioactively labelled substrates (GDP-Man, pyrophosphate, or Man-1-P), separating the products by chromatography, charcoal absorption (Preiss and Wood, 1964) or electrophoresis, and then counting the label. Activity in mammalian extracts has also been assayed by reacting the nucleoside triphosphate formed with 3-phosphoglycerate in a system containing NADH, 3-phosphoglycerate kinase, glyceraldehyde-3-phosphate dehydrogenase and pyrophosphorylase and measuring NADH oxidation spectrophotometrically (Verachtert *et al.*, 1966).

Mannan synthase has been implicated in glucomannan synthesis by particulate fractions from mung bean and pea seedlings (Elbein, 1969; Heller and Villemez, 1972; Hinman and Villemez, 1975), cambium, xylem and suspension-cultured cells of *Pinus sylvestris* (Dalessandro *et al.*, 1986; Ramsden and Northcote, 1987), sycamore cell cultures (Smith *et al.*, 1976), and orchid tubers (Franz, 1973); in mannan synthesis in *Acetabularia mediterranea* and fenugreek seedlings (Bachmann and Zetsche, 1979; Clermont *et al.*, 1982); and in galactomannan synthesis in developing fenugreek seeds (Campbell and Reid, 1982). Activity has been assayed as incorporation of ^{14}C from GDP[^{14}C]Man into a fraction with the properties of the polysaccharide and K_m GDP-mannose) values of 0.05–0.1 mM recorded.

UDP-glucose 4'-epimerase, which forms UDP-Gal, a probable substrate for galactosyl substitution of galactomannan, has been detected in extracts of mung beans using radioactively labelled substrates (Neufeld *et al.*, 1957), and also in fenugreek seedlings (Clermont and Percheron, 1979), sycamore cambium and xylem, pea seeds (Rubery, 1973) and wheat germ (Fan and Feingold, 1969). In the last tissue K_m (UDP-Glc) was 0.2 mM and K_m (UDP-Gal) was 0.1 mM at the pH optimum of 9.0. NAD^+ was required with a K_a of 0.04 mM. At equilibrium [UDP-Gal]/[UDP-Glc] was 3.1. The M_r of the fenugreek enzyme (gel chromatography) was 70 000. The K_m (UDP-Gal) for the enzyme from pea seeds was 0.01 mM. Activity can be assayed by measuring the formation of UDP-glucuronic acid with UDP-Glc dehydrogenase, the production of which is linked to the reduction of NAD^+.

Models of biosynthesis have been devised to explain the apparently irregular distribution of glucosyl and mannosyl residues in the glucomannan chain (Heller and Villemez, 1972) and of galactosyl substitution along the mannan chain of galactomannan (Matheson, 1984; McCleary and Matheson, 1986). The former is based on the properties of the two transferases involved. The glucosyl- but not the mannosyl-transferase requires the continual production of mannosyl acceptor residues. However, the mannosyl-transferase is inhibited by GDP-Glc and these characteristics would produce a glucomannan with irregular replacement of mannosyl by glucosyl residues. In galactomannan biosynthesis the model proposes that the pattern of substitution is a consequence of the conformation of the mannan chain in conjunction with the transferases and the membrane on which reaction occurs, and a mechanism of synthesis requiring concurrent incorporation of mannosyl and galactosyl residues. Two aspects of the conformation of the mannan chain are its slightly twisted, ribbon-like structure and the positioning of neighbouring hydroxymethyl groups on opposite edges of this ribbon. The degree of hindrance of the galactosyl transferase by previous galactosyl groups then determines the probability of substitution. However, for each species there is a limit to the distance along the chain being synthesised that hindrance to reaction can occur.

B. Depolymerising Enzymes

β-Mannanase has been extracted from seeds and tubers storing mannan-based poly-saccharides (Courtois and Le Dizet, 1964; Hylin and Sawai, 1964); multiple forms have been separated (McCleary and Matheson, 1975; Shimahara et al., 1975; Villarroya et al., 1978). The highly viscous polysaccharide interferes with handling and seeds should be imbibed or extracts stored to reduce its level. The enzyme has been purified and forms separated by chromatography on DEAE-cellulose in acetate buffer (pH 6) with a KCl gradient (McCleary and Matheson, 1975; Matheson et al., 1980); in phosphate (pH 6 or 6.8) with stepwise elution by phosphate (Villarroya and Petek, 1976); on Amberlite CG-50 in acetate (pH 5.5) (Shimahara et al., 1975) and gradient elution with the same buffer; on CM cellulose with an acetate (pH 5) gradient and with a pH gradient (3.5–5.0) in 10 mM acetate (McCleary and Matheson, 1975); on hydroxylapatite in phosphate (pH 6 or 5.3) with stepwise phosphate gradients (Villaroya et al., 1978); by gel chromatography on Sephadex G-100 (Shimahara, 1975; McCleary and Matheson, 1976) or polyacrylamide-agarose (Villarroya et al., 1978); by chromatography on swollen cellulose (Sugiyama et al., 1973); by preparative polyacrylamide gel electro-phoresis (Villaroya et al., 1978); and by substrate affinity chromatography on amino-hexyl Sepharose linked to mannan, with elution by 0.2% mannan solution (McCleary, 1978b). The enzyme can be assayed by determining the amount of reducing activity released on hydrolysis of a mannan chain (Villarroya et al., 1978; Matheson et al., 1980) (reducing activity can be measured by methods such as Cu^{2+} reduction). Relative activities can be established by following the decrease in viscosity (Hylin and Sawai, 1964; Eriksson and Winell, 1968; Shimahara et al., 1975) and the colour released from dyed mannan provides a convenient assay (McCleary, 1978a). Solutions of a galacto-mannan with a low level of galactosyl substitution, such as carob, give a soluble substrate which reacts with β-mannanase—too much substitution blocks reaction and too little causes problems in solubility. The K_m values have varied with both enzyme source and between forms from the same source and with the degree of substitution of the mannan chain, but generally values have been in the range 0.5 to 20 mM mannosyl residues (c. 0.01–0.3% mannan). The pH optima are near to 5 and the M_r has been estimated by gel chromatography (McCleary and Matheson, 1975; Shimahara et al., 1975) and sodium dodecyl sulphate polyacrylamide gel electrophoresis (SDS-PAGE) (Villarroya et al., 1978) as 20 000 to 40 000. Carbohydrases may show adsorption effects on polysaccharide matrices and enzymes with acidic isoelectric points may display abnormal behaviour on SDS-PAGE; lettuce seed enzyme also gave variable estimates (Halmer, 1989). β-Mannanases from microbial sources, with varying extents of purifica-tion, have been used in structural studies (Perila and Bishop, 1961; Reese and Shibata, 1965; Courtois and LeDizet, 1968; McCleary and Matheson, 1983; Shimizu and Ishihara, 1983; Kusakabe et al., 1988).

The action pattern of both plant and microbial β-mannanases has been established as endo from the high rate of reduction in viscosity of a mannan substrate relative to the release of reducing activity (Courtois and Le Dizet, 1968; McCleary and Matheson, 1975), and the production of manno-oligosaccharides (Courtois and Le Dizet, 1964;

Villarroya and Petek, 1976; Clermont-Beaugirard and Percheron, 1968), particularly mannobiose and -triose, as final products. Comparison of relative V_{max}/K_m values for manno-oligosaccharides of increasing DP (McCleary and Matheson, 1983) showed a levelling at DP 5–6, indicating that the enzyme binds to 5 mannosyl residues. The active site, where the glycosidic bond is broken, lies between mannosyl residues 2 and 3 back from the reducing end of the pentaosyl unit. From a paper chromatographic examination of the way reaction products of mannotri-itol change with time, a mechanism involving transglycosylation and re-positioning of the product prior to hydrolysis has been proposed. Differences in the extent of hydrolysis by β-mannanases from different sources, particularly those from plants and microbes, have been detected (Courtois and Le Dizet, 1970; McCleary, 1979b; McCleary and Matheson, 1983). The degree of hydrolysis of galactomannans is dependent on the amount of substitution and there is a linear relationship between hydrolysis and substitution (McCleary and Matheson, 1983, 1986). Hetero-oligosaccharide products have been fractionated and, after separation by gel chromatography on Bio-Rad P-2 and TLC, and structural determination by methylation analysis, proton and carbon-13 NMR and enzymic hydrolysis, the products from carob galactomannan (up to DP 9) have been identified and the manner of binding between enzyme and polysaccharide and mode of hydrolysis proposed. The enzyme binds to the mannan chain in the favoured conformation, in which the hydroxymethyl groups on neighbouring mannosyl residues extend in opposite directions. Where galactosyl substitution occurs, binding can only take place where this group projects away from the enzyme (McCleary et al., 1982, 1983; McCleary and Matheson, 1983, 1986). On reaction with glucomannan (Shimahara et al., 1975; McCleary and Matheson, 1983; Shimizu and Ishihara, 1983; Takahashi et al., 1984; Kusakabe et al., 1988), if a glucosyl residue is found where it presents the edge of the pyranosyl ring resembling the manno-configuration (i.e. containing the ring oxygen and the hydroxymethyl), then binding can occur: thus only a mannose residue can be the reducing-end sugar of released oligosaccharides. Galactoglucomannans produced similar hetero-oligosaccharides.

β-D-Mannoside mannohydrolase activity has been detected in seeds containing glucomannan (Goldberg, 1969); a number of legume seeds storing galactomannan (Reid and Meier, 1973; McCleary and Matheson, 1975; McCleary, 1982, 1983); and the cotyledons of lettuce seeds (Oullette and Bewley, 1986). The enzyme in legume seeds is an exo-mannanase. Extraction requires a relatively high pH (8) and high salt concentrations to avoid co-precipitation with other insoluble protein in the extract. An initial purification (McCleary and Matheson, 1975) from lucerne and carob seeds involved extraction with 0.1 M acetate (pH 5), then DEAE-cellulose chromatography in acetate (pH 6) with elution by a KCl gradient (0.01–0.4 M), followed by carboxymethyl (CM)-cellulose (0.01–0.03 M acetate, pH 5) and gel chromatography on Sephadex G-100. Yield and purity were improved (McCleary, 1982) by using a buffer of higher pH (Tris-(hydroxymethyl) amino methane (Tris), pH 8) and a higher concentration of salt (0.2 M) for extraction, followed by DEAE-cellulose chromatography in this buffer, with elution by a salt gradient (0–0.4 M NaCl), as well as gel chromatography on Ultragel AcA 44 and chromatofocusing on Polybuffer Exchanger PBE118, with elution by Pharmalyte.HCl (pH 8–10.5). Activity can be assayed by the release of nitrophenol from a nitrophenyl β-mannoside, or of reducing activity from β-D-mannopenta-itol. K_m values for p-nitrophenyl-β-D-mannoside of 0.3–0.5 mM and for mannopenta-itol and hexa-itol of

2.8 mM have been recorded. β-D-mannobi-itol is not hydrolysed. The relative V_{max}/K_m values of the penta- and hexa-itols and p-nitrophenyl β-D-mannoside were 345, 345 and 522, with pH maxima for the enzyme from guar of 5–6 and from lettuce of 5.4. M_r values, determined by gel chromatography, were between 50 000 and 60 000. The guar enzyme had little action on mannotri-itol, limited action on the tetra-itol, but hydrolysed the penta-itol readily, indicating that it is an *exo*-mannanase. β-Glucosidase has been detected in asparagus seeds (Goldberg, 1969) that store glucomannan.

α-D-Galactosidase (Dey and Pridham, 1972; Dey, 1978, 1980) has been found in many plant sources that may not contain galactomannan or galactoglucomannan, as it hydrolyses galactosides like raffinose. Methods of isolating and studying α-galacto-sidases from seeds storing galactomannan have been developed (Courtois and Percheron, 1961; Courtois and Petek, 1966; Reid and Meier, 1973; Seiler, 1977) and multiple forms have been separated (McCleary and Matheson, 1974; Williams *et al.*, 1977). α-Galactosidase has been purified by chromatography on DEAE-cellulose with a KCl gradient (0–0.3 M), on CM- or sulphoethyl (SE)-cellulose with an acetate (pH 5) gradient (0.01–0.3 M) (McCleary and Matheson, 1974); on DEAE-cellulose in phosphate (pH 6.2) and on ECTEOLA-cellulose in phosphate (pH 8.1) with stepwise elution by NaCl, on hydroxylapatite (phosphate, pH 6) with stepwise elution by buffer of increasing concentration (Williams *et al.*, 1977); on alumina with stepwise elution by citrate buffer (pH 4.4, 5.1 and 6.0) (Courtois and Petek, 1966); on DEAE-cellulose, in Tris (pH 8) with an NaCl gradient (0–0.4 M), on Sepharose G-100 and by affinity chromatography on N-ε-aminocaproyl α-D-galactopyranosylamine Sepharose 4B (McCleary, 1983). Activity can be assayed with nitrophenyl α-D-galactoside or by estimation of reducing sugar released from galactomannan. The K_m values for nitrophenyl α-D-galactosides have been generally in the range 0.4–1 mM and c. 2–40 mM for galactosyl residues in galactomannan (c. 0.015–0.3%). The optimal pH for reaction has been in the range 4–5 and M_r values of about 25 000–40 000 have been measured by gel chromatography and SDS-PAGE. Enzymes from other plants (Dey and Wallenfals, 1974) and microorganisms (Suzuki *et al.*, 1970) may not hydrolyse galactomannans. Galactosyl oligosaccharides (such as raffinose) are also hydrolysed and transglycosylation can occur.

α-D-Galactosidases from legume seeds that contain galactomannan readily remove galactosyl residues, until the galactosyl content becomes too low to maintain a soluble polymer and precipitation occurs (Hui and Neukom, 1964; Courtois and Le Dizet, 1966; McCleary and Matheson, 1975, 1986; McCleary *et al.*, 1981, 1984). The pattern of hydrolysis appears to involve preferential removal of galactosyl groups at the same edge of the ribbon-like conformation as that at which the initial attack occurs. This is indicated by gel formation with agarose and xanthan of depleted galactomannan at relatively high galactosyl contents. Also, examination of oligosaccharides, produced by hydrolysis of depleted polymers using β-mannanase, showed high levels of 6'-O-α-D-galactosyl-(1→4)-β-D-mannobiose.

C. Enzymes Utilising Hydrolysis Products

Galactokinase has been studied in leguminous seeds that do not store galactomannan (Chan and Hassid, 1975; Dey, 1983)—where its role would be to phosphorylate galactose released from hydrolysis of galactosylsucrose oligosaccharides—as well as in germinating fenugreek (Foglietti and Percheron, 1974, 1976) which contains galacto-mannan. It was located in non-endospermic tissue. Purification has involved: chromatography on DEAE-cellulose and DEAE-Sephadex in phosphate (pH 7.3) containing glycerol and mercaptoethanol, by gradient elution with phosphate; chromatography on hydroxylapatite with 0–0.9 mM NaCl; gel chromatography using Sephacryl S-200 and Sephadex G-100; affinity chromatography on 8-β-aminoethylthio-ATP coupled to Sepharose, with elution by high salt concentrations (1 M NaCl) or 0.5 M D-galactose; or affinity chromatography on galactosamine linked to Sepharose with elution by KCl or galactose. Activity has been assayed by measuring the label incorporated from [^{14}C]-galactose, separating [^{14}C]galactose-α-1-P by paper electrophoresis or absorption onto DEAE cellulose disks, or by estimating ADP from NADH reduction in the enzyme-coupled system of pyruvate kinase–lactate dehydrogenase. K_m values for galactose of 0.5 mM and MgATP^{2-} of 1.5 and 5.0 mM have been found, with a pH optimum of 7.3. The M_r (by gel chromatography) was estimated as 60 000 and SDS-PAGE indicated two identical subunits.

UDP-Galactose pyrophosphorylase was detected in homogenates of mung bean cotyledons (Neufeld *et al.*, 1957) when UDP-Gal formation was followed by paper electrophoretic separation and by conversion to UDP-Glc by UDP-Glc–UDP-Gal-4-epimerase. Activity has been fractionated from fruit peduncles of cucumber (Smart and Pharr, 1981) using gel chromatography on Sephadex G-150 in 10 mM N-(2-hydroxy-ethyl)piperazine-N'-(2-ethanesulphonic acid) (HEPES) buffer (pH 7.2) and chromatography on DEAE cellulose in the same buffer with a 0–0.25 M KCl gradient. The K_m values were 0.14 mM for UTP and 1.2 mM for Gal-α-1-P, with a broad pH optimum of between 6 and 7.5. Activity was assayed by conversion of UDP-Gal to UDP-Glc with UDP-Glc–UDP-Gal-4-epimerase and measurement of the reduction of NAD$^+$ when UDP-Glc was oxidised with UDP-Glc-dehydrogenase.

DEAE-cellulose chromatography of extracts of the cotyledons of honey locust seeds (containing galactomannan in the endosperm) (D. K. Myers and N. K. Matheson, unpubl. res.) gave four fractions that phosphorylated D-glucose, D-mannose and D-fructose. The column pH was 7.0 (Tris-HCl) and protein was eluted with 0.1–0.4 M KCl. From examination of the relative V_{max}/K_m ratios, three could be described as fructo-kinases and one as a hexokinase—no specific mannokinase was detected. Assay involved the spectrophotometric measurement of NADP$^+$ reduction in the coupled enzyme system, mannose 6-phosphate isomerase, glucose 6-phosphate isomerase and glucose 6-phosphate dehydrogenase. K_m values for mannose of 0.08 mM and for glucose of 0.07 mM were obtained. The pH optimum was 8.5.

Conversion of Man-6-P to Glc-6-P reverses a reaction of mannan synthesis. Enzymic activity has been detected in germinating seeds of a number of legumes (McCleary and Matheson, 1976; Lee and Matheson, 1984). Mannose 6-phosphate isomerase from lucerne was purified by DEAE-cellulose chromatography in 10 mM Tris-HCl (pH 7.5) with elution by a 0.01–0.4 M gradient of the same buffer, followed by DEAE-cellulose chromatography with a stepped phosphate gradient (0.005–0.05 then 0.05–0.1 M). The latter separated glucose phosphate isomerase. In germinating *Cassia coluteoides* seeds, the enzyme was separated from glucose phosphate isomerase by gel chromatography on Sephacel S-200 in 0.2 M Tris-HCl (pH 7.5) after DEAE-cellulose fractionation. The K_m for Man-6-P was 0.8 mM for lucerne and 1.9 mM for *Cassia*. The pH maximum was broad (6–11) and the M_r for the enzymes from germinating *Cassia* and lucerne 46 000–

75 000, as measured by gel chromatography and density gradient centrifugation. Levels of activity are very low in seeds that do not accumulate galactomannan and these plants cannot utilise mannose (Herold and Lewis, 1977).

D. Whole-tissue Studies

Dissection of the developing pods of *Cassia* showed that mannose-6-phosphate isomerase was located in the cotyledons and that the level increased rapidly in the later stages of development, coincident with galactomannan accumulation (Lee and Matheson, 1984). Only one form was detected, although two forms of glucose 6-phosphate isomerase were found. The activity of GDP-mannose pyrophosphorylase was high while cells of *Acetabularia mediterranea* (Bachmann and Zetsche, 1979) grew and synthesised mannan, and decreased with the termination of growth. In developing fenugreek seeds (Campbell and Reid, 1982), homogenates, prepared from endosperm tissue, transferred label from GDP-[^{14}C]Man only when prepared at stages after anthesis when galactomannan synthesis occurred. These homogenates had the centrifugation density of endoplasmic reticulum. Mannan synthase also increased as *A. mediterranea* grew and synthesised mannan, and decreased to a low value when synthesis ceased. Levels were lower in cells held in darkness. Both the pyrophosphorylase and the synthase were distributed in an apical–basal gradient, with high activity in the apical region, where mannan synthesis is concentrated. Synthase in a particulate fraction from growing orchid tubers (Franz, 1973) showed seasonal changes in activity, increasing in the spring when glucomannan synthesis occurred and decreasing in the summer. Synthase activity in *Pinus sylvestris* (Dalessandro *et al.*, 1986, 1988), which contains glucomannan in the wood, increased from cambial cells to differentiating xylem cells and then decreased in differentiated xylem cells. When activity in a particulate fraction of suspension-cultured cells was measured, it was higher in culture conditions that favoured tracheid formation (Ramsden and Northcote, 1987). In fenugreek seeds (Reid and Meier, 1970) no change was detected in the galactose-to-mannose ratio of galactomannan during development. The distribution of molecular size of the polymer from *Gleditsia triacanthos* seeds (Mallett *et al.*, 1987) became more disperse during accumulation, and the galactose content, estimated by three methods, combining enzymic hydrolysis, acidic hydrolysis and enzymic estimation and gas chromatography, decreased slightly with age. The degree of hydrolysis with β-mannanase and the chromatographic profile on Bio-Gel P-2 of the oligosaccharides in the hydrolysates supported this and indicated a high component of segments of neighbouring galactosyl substituents.

The variations in the activities of α-D-galactosidase and β-D-mannosidase were estimated (Reid and Meier, 1973) after imbibition, when, following a short lag period, the levels of both in the endosperm increased to a plateau, but in the embryo remained relatively constant. Cycloheximide or abscissic acid completely inhibited the development of both endospermic activities, as well as the depletion of polysaccharide. Actinomycin D and 5-fluorouracil delayed the attainment of full activity. In four legume seeds (carob, guar, lucerne and soybean; McCleary and Matheson, 1974) in which α-D-galactosidase was fractionated into a number of forms, dissection showed that one of these was located in the endosperm. Activity of this form was negligible in the resting seed and increased greatly after imbibition, concurrently with galactomannan depletion, and then decreased. The form(s) in the embryo, whose function would be to hydrolyse galactosylsucrose oligosaccharides, were active prior to germination and, after imbibition, showed smaller changes. Cycloheximide stopped any increase in the endospermic form (McCleary, 1983). In lucerne and guar, the galactose content of galactomannan sampled during degradation initially decreased, allowing hydrolysis by β-mannanase (McCleary and Matheson, 1975); in carob and honey locust, with lower degrees of galactose substitution, no decrease was observed. No change was detected in fenugreek galactomannan (Reid, 1971). Galactomannan depletion does not start immediately after imbibition and is preceded by the utilisation of raffinose series oligosaccharides (Reid, 1971; McCleary and Matheson, 1976). The breakdown of galactomannan in excised endosperm paralleled that of whole seeds. α-D-Galactosidase activity also increased and then decreased in excised, imbibed endosperm and [^{14}C]serine was incorporated (Reid and Meier, 1972; McCleary and Matheson, 1976; Seiler, 1977).

β-Mannanase is found only in the endosperm and its activity increases from a negligible value shortly after imbibition to a high value and then decreases. The peak of activity is near the maximum rate of breakdown of galactomannan (McCleary and Matheson, 1975). The proportion of activities of two fractions from lucerne seed, separated by DEAE-cellulose chromatography, showed little difference after germination (Matheson et al., 1980). In lettuce seeds (Halmer et al., 1976, 1978) activity was found in the endosperm, increased after imbibition and the mannan-based polymer initially present was completely degraded. The effects of light and giberellin on the rate of enzyme production were studied. Production in isolated endosperms was inhibited by abscissic acid and by cycloheximide, cordycepin, actinomycin-D and α-amanitin (Dulson and Bewley, 1989; Halmer, 1989). Glucomannan content in mature orchid tubers decreased in the early spring and this was accompanied by the appearance of manno-oligosaccharides and mannose (Franz and Meier, 1971). Sucrose was transported to growing tubers where glucomannan accumulated. The distribution of glucomannan across lily bulb scales has been studied: differences in molecular size and amount were found on germination (Matsuo and Mizuno, 1974b).

β-Mannoside mannohydrolase (exo-β-mannanase) activity, assayed with a nitro-phenyl β-mannoside substrate, has been found to increase after imbibition of fenugreek (Reid and Meier, 1973), increase and then decrease in carob (Seiler, 1977) and remain relatively constant or decrease in guar, lucerne, honey locust and carob (McCleary and Matheson, 1975; McCleary, 1983). Activity was detected in ungerminated seeds. It can be extracted from both the endosperm and the embryo, and synthesis is not inhibited by cycloheximide. In lettuce seeds activity is located in cotyledons (Oulette and Bewley, 1986).

The monosaccharides released by these hydrolytic enzymes are utilised rapidly by whole seeds. In germinating seeds of fenugreek, honey locust and carob (Reid and Meier, 1972; McCleary and Matheson, 1976; Seiler, 1977) only trace levels of galactose or mannose were detected. However, if the endosperm was separated shortly after imbibition, the concentrations of galactose, mannose, mannobiose and -triose built up. [^{14}C]Galactose, mannose and glucose were readily absorbed by the separated embryonic axis and fractionation (into respired ^{14}CO$_2$, water-soluble extract, lipid, material released by 'pectinase', and 72% H$_2$SO$_4$ hydrolysates) indicated that all fractions contained label and there were no major differences in the distribution in these fractions for each sugar. [^{14}C]Mannobiose was also taken up by the embryo (McCleary, 1983) to be hydrolysed by exo-β-mannanase prior to its utilisation as mannose.

Legume seeds storing galactomannan and galactosylsucrose oligosaccharides, as well as those storing only the latter or the latter and starch, produce new starch in the cotyledons after germination (Hizukuri *et al.*, 1961; Juliano and Varner, 1968; Reid, 1971; Seiler, 1977; Saini and Matheson, 1981). Radiochemical studies were consistent with ready conversion of galactose and mannose to the gluco-configuration. The observations indicated a rapid hydrolysis of galactosylsucrose oligosaccharides to monosaccharide and sucrose, followed by hydrolysis of galactomannan and the conversion of hexose, surplus to immediate metabolic needs, to starch.

The effect of separation of the embryo and endosperm on the development of α-D-galactosidase activity and the breakdown of galactomannan in the latter, as well as water stress on the three hydrolytic enzymes, has been studied (Spyropoulos and Reid, 1985, 1988). Comparison of the growth patterns and drying behaviour of the embryo of fenugreek, with and without the endosperm attached, has led to the proposal that another function of the endosperm is to maintain a moist environment for the germinating seed (Reid and Bewley, 1979).

In germinating *Vicia faba* seeds, which contain galactosylsucrose oligosaccharides but not galactomannan, galactokinase increased after imbibition and then decreased (Dey, 1983). Total mannose phosphorylating activity from the cotyledons of germinating honey locust seeds, after a lag period of 24 h, increased seven-fold and then decreased to near the original level (D. K. Myers and N. K. Matheson, unpubl. res.). Mannose phosphate isomerase in germinating honey locust seeds, which was located in the cotyledons, declined in whole seeds as galactomannan was depleted (McCleary and Matheson, 1976). This decline started earlier in separated cotyledons under conditions of germination (without a supply of mannose). In germinating *Cassia* seeds (Lee and Matheson, 1984) there was an eight-fold increase until galactomannan was depleted, followed by a decrease.

REFERENCES

Achtardjiev, C. Z. and Koleva, M. (1973). *Phytochemistry* **12**, 2897–2900.
Ahmed, Z. F. and Whistler, R. L. (1950). *J. Am. Chem. Soc.* **72**, 2524–2525.
Alam, N. and Gupta, P. C. (1986). *Carbohydr. Res.* **153**, 334–338.
Alam, M. and Richards, G. N. (1971). *Aust. J. Chem.* **24**, 2411–2416.
Akiyama, Y., Eda, S., Mori, M. and Kato, K. (1983). *Phytochemistry* **22**, 1177–1180.
Andrews, P., Hough, L. and Jones, J. K. N. (1952). *J. Chem. Soc.* 2744–2750.
Andrews, P., Hough, L. and Jones, J. K. N. (1953). *J. Chem. Soc.* 1186–1192.
Andrews, P., Hough, L. and Jones, J. K. N. (1956). *J. Chem. Soc.* 181–188.
Angyal, S. J. and James, K. (1970). *Aust. J. Chem.* **23**, 1209–1221.
Ashford, A. E. and Gubler, F. (1984). *In* "Seed Physiology" (D. R. Murray, ed.), Vol. 2, pp.117–162. Academic Press, Sydney.
Aspinall, G. O. (1959). *Adv. Carbohydr. Chem.* **14**, 429–468.
Aspinall, G. O. (1963). *J. Chem. Soc.* 1676–1680.
Aspinall, G. O. (ed.) (1982a). *In* "The Polysaccharides", Vol. 1, pp. 19–34. Academic Press, New York.
Aspinall, G. O. (ed.) (1982b). *In* "The Polysaccharides" Vol. 1, pp. 35–131. Academic Press, New York.
Aspinall, G. O and Stephen, A. M. (1973). *In* "MTP International Review of Science, Carbohydrates". Organic Chemistry, Series One, Vol. 7 (D. H. Hey and G. O. Aspinall, eds), pp. 285–317. Butterworth, University Park Press, Baltimore.

Aspinall, G. O., Hirst, E. L., Percival, E. G. V. and Williamson, I. R. (1953). *J. Chem. Soc.* 3184–3188.

Aspinall, G. O., Rashbrook, R. B. and Kessler, G. (1958). *J. Chem. Soc.* 215–221.

Aspinall, G. O., Hunt, K. and Morrison, I. M. (1966). *J. Chem. Soc.* C1945–1949.

Atkins, E. D. T., Hopper, E. D. A. and Isaac, D. H. (1973). *Carbohydr. Res.* **27**, 29–37.

Atkins, E. D. T., Farnell, S., Burden, C., Mackie, W. and Sheldrick, B. (1988). *Biopolymers* **27**, 1097–1105.

Bachmann, P. and Zetsche, K. (1979). *Planta* **145**, 331–337.

Baker, C. W. and Whistler, R. L. (1975). *Carbohydr. Res.* **45**, 237–243.

Baldo, B. A., Barnett, D. and Lee, J. W. (1984). *Aust. J. Pl. Physiol.* **11**, 179–190.

Barker, S. A., Bourne, E. J. and Whiffen, D. H. (1956). *Methods Biochem. Anal.* **3**, 213–245.

Barth, H. G. and Smith, D. A. (1981). *J. Chromatogr.* **206**, 410–415.

Bewley, J. D. and Reid, J. S. G. (1985). In "Biochemistry of Storage Carbohydrates in Green Plants" (P. M. Dey and R. A. Dixon, eds), pp. 289–304. Academic Press, London.

Blakeney, A. B., Harris, P. J., Henry, R. J. and Stone, B. A. (1983). *Carbohydr. Res.* **113**, 291–299.

Bouveng, H. O. and Lindberg, B. (1960). *Adv. Carbohydr. Chem.* **15**, 53–89.

Bouveng, H. O. and Lindberg, B. (1965a). *Methods Carbohydr. Chem.* **5**, 147–150.

Bouveng, H. O. and Lindberg, B. (1965b). *Methods Carbohydr. Chem.* **5**, 296–357.

Brownsey, G. J., Cairns, P., Miles, M. J. and Morris, V. J. (1988). *Carbohydr. Res.* **176**, 329–334.

Buchala, A. J. and Meier, H. (1973). *Carbohydr. Res.* **31**, 87–92.

Buchala, A. J., Franz, G. and Meier, H. (1974). *Phytochemistry* **13**, 163–166.

Campbell, J. McA. and Reid, J. S. G. (1982). *Planta* **155**, 105–111.

Carlberg, G. E. and Percival, E. (1977). *Carbohydr. Res.* **57**, 223–234.

Cartier, N., Chambat, G. and Joseleau, J.-P. (1988). *Phytochemistry* **27**, 1361–1364.

Casu, B. (1985). In "Polysaccharides. Topics in Structure and Morphology" (E. D. T. Atkins, ed.) pp. 1–40. VCH, Deerfield Beach.

Cerezo, A. S. (1965). *J. Org. Chem.* **30**, 924–927.

Chan, P. H. and Hassid, W. Z. (1975). *Anal. Biochem.* **64**, 372–379.

Chanzy, H. and Vuong, R. (1985). In "Polysaccharides. Topics in Structure and Morphology" (E. D. T. Atkins, ed.), pp. 41–71. VCH, Deerfield Beach.

Chanzy, H., Dubé, M., Marchessault, R. H. and Revol, J. F. (1979). *Biopolymers* **18**, 887–898.

Chanzy, H. D., Grosrenaud, A., Joseleau, J. P., Dubé, M. and Marchessault, R. H. (1982). *Biopolymers* **21**, 301–319.

Chanzy, H. D., Grosrenaud, A., Vuong, R. and Mackie, W. (1984). *Planta* **161**, 320–329.

Cheetham, N. W. H., McCleary, B. V., Teng, G., Lum, F. and Maryanto (1986). *Carbohydr. Polymers* **6**, 257–268.

Churms, S. C. (1970). *Adv. Carbohydr. Chem. Biochem.* **25**, 13–51.

Clermont-Beaugirard, S. and Percheron, F. (1968). *Bull. Soc. Chim. Biol.* **50**, 633–639.

Clermont, S. and Percheron, F. (1979). *Phytochemistry* **18**, 1963–1965.

Clermont, S., Said, R. and Percheron, F. (1982). *Phytochemistry* **21**, 1951–1954.

Courtois, J. E. and Le Dizet, P. (1963). *Bull. Soc. Chim. Biol.* **45**, 731–741.

Courtois, J. E. and Le Dizet, P. (1964). *Bull. Soc. Chim. Biol.* **46**, 535–542.

Courtois, J. E. and Le Dizet, P. (1966). *Carbohydr. Res.* **3**, 141–151.

Courtois, J. E. and Le Dizet, P. (1968). *Bull. Soc. Chim. Biol.* **50**, 1695–1710.

Courtois, J. E. and Le Dizet, P. (1970). *Bull. Soc. Chim. Biol.* **52**, 15–22.

Courtois, J. E. and Percheron, F. (1961). *Bull. Soc. Chim. Biol.* **43**, 167–175.

Courtois. J. E. and Petek, F. (1966). *Methods Enzymol.* **8**, 565–571.

Dalessandro, G., Piro, G. and Northcote, D. H. (1986). *Planta* **169**, 564–574.

Dalessandro, G., Piro, G. and Northcote, D. H. (1988). *Planta* **175**, 60–70.

Dea, I. C. M. and Morrison, A. (1975). *Adv. Carbohydr. Chem. Biochem.* **31**, 241–312.

Dea, I. C. M., Morris, E. R., Rees, D. A., Welsh, E. J., Barnes, H. A., Price, J. (1977). *Carbohydr. Res.* **57**, 249–272.

Dea, I. C. M., Clark, A. H. and McCleary, B. V. (1986). *Carbohydr. Res.* **147**, 275–294.

Deb, S. K. and Mukherjee, S. N. (1963). *Ind. J. Chem.* **1**, 413–414.

Dey, P. M. (1978). *Adv. Carbohydr. Chem. Biochem.* **35**, 341–376.

Dey, P. M. (1980). *Adv. Carbohydr. Chem. Biochem.* **37**, 283–372.

Dey, P. M. (1983). *Eur. J. Biochem.* **136**, 155–159.

Dey, P. M. and Pridham, J. B. (1972). *Adv. Enzymol. Relat. Area. Mol. Biol.* **36**, 91–130.

Dey, P. M. and Wallenfels, K. (1974). *Eur. J. Biochem.* **50**, 107–112.

Dudman, W. F. and Bishop, C. T. (1968). *Can. J. Chem.* **46**, 3079–3084.

Dulson, J. and Bewley, J. D. (1989). *Phytochemistry* **28**, 363–369.

Eda, S., Akiyama, Y., Kato, K., Takahashi, R., Kusabe, I., Ishizu, A. and Nakano, J. (1984). *Carbohydr. Res.* **131**, 105–118.

Eda, S., Akiyama, Y., Kato, K., Atsushi, A. and Nakano, J. (1985). *Carbohydr. Res.* **137**, 173–181.

Elbein, A. D. (1969). *J. Biol. Chem.* **244**, 1608–1616.

Elfak, A. M., Pass, G. and Phillips, G. O. (1979). *J. Sci. Food Agric.* **30**, 439–444.

Emi, S., Fukumoto, J. and Yamamoto, T. (1972). *Agric. Biol. Chem.* **36**, 991–1001.

Eriksson, K.-E. and Winell, M. (1968). *Acta. Chem. Scand.* **22**, 1924–1934.

Fan, D.-F. and Feingold, D. S. (1969). *Plant Physiol.* **44**, 599–604.

Foglietti, M.-J. and Percheron, F. (1974). *Biochemie* **56**, 473–475.

Foglietti, M.-J. and Percheron, F. (1976). *Biochemie* **58**, 499–504.

Forsberg, L. S. and Pazur, J. H. (1979). *Carbohydr. Res.* **75**, 129–140.

Franz, G. (1973). *Phytochemistry* **12**, 2369–2373.

Franz, G. and Meier, H. (1971). *Planta Med.* **19**, 326–332.

Frei, E. and Preston, R. D. (1968). *Proc. R. Soc. Series B. (London)* **169**, 127–145.

Geyer, R., Geyer, H., Kuhnhardt, S., Mink, W. and Stirm, S. (1982). *Anal. Biochem.* **121**, 263–274.

Goldberg, R. (1969). *Phytochemistry* **8**, 1783–1792.

Goldberg, R. and Roland, J. C. (1971). *Révue Gén. Bot.* **78**, 75–102.

Goldstein, A. M., Alter, E. N. and Seaman, J. K. (1973). *In* "Industrial Gums" (R. L. Whistler, ed.), pp. 303–321. Academic Press, New York.

González, J. J. and Painter, T. J. (1980). *Carbohydr. Res.* **91**, 93–96.

Gowda, D. C. (1980). *Carbohdyr. Res.* **83**, 402–405.

Gowda, D. C., Neelisiddaiah, B. and Anjaneyalu, Y. V. (1979). *Carbohydr. Res.* **72**, 201–205.

Grasdalen, H. and Painter, T. (1980). *Carbohydr. Res.* **81**, 59–66.

Gupta, A. K. and Bose, S. (1986). *Carbohdyr. Res.* **153**, 69–77.

Gupta, D. S., Jann, B., Bajpai, K. S. and Sharma, S. C. (1987). *Carbohydr. Res.* **162**, 271–276.

Gupta, O. C. D. (1980). *Carbohydr. Res.* **83**, 85–91.

Gupta, P. C. D., Sen, S. K. and Day, A. (1976). *Carbohydr. Res.* **48**, 73–80.

Hakomori, S. (1964). *J. Biochem. (Tokyo)* **55**, 205–208.

Hall, L. D. (1964). *Adv. Carbohydr. Chem.* **19**, 51–93.

Halmer, P. (1989). *Phytochemistry* **28**, 371–378.

Halmer, P., Bewley, J. D. and Thorpe, T. A. (1976). *Planta* **130**, 189–196.

Halmer, P., Bewley, J. D. and Thorpe, T. A. (1978). *Planta* **139**, 1–8.

Hamilton, J. K. and Kircher, H. W. (1958). *J. Am. Chem. Soc.* **80**, 4703–4709.

Hamilton, J. K., Kircher, H. W. and Thompson, N. S. (1956). *J. Am. Chem. Soc.* **78**, 2508–2514.

Hamilton, J. K., Partlow, E. V. and Thompson, N. S. (1960). *J. Am. Chem. Soc.* **82**, 451–457.

Hay, G. W., Lewis, B. A. and Smith, F. (1965a). *Methods Carbohydr. Chem.* **5**, 357–361; 377–382.

Hay, G. W., Lewis, B. A., Smith, F. and Unrau, A. M. (1965b). *Methods Carbohdyr. Chem.* **5**, 251–253.

Herold, A. and Lewis, D. H. (1977). *New Phytol.* **79**, 1–40.

Heller, J. S. and Villemez, C. L. (1972). *Biochem. J.* **129**, 645–655.

Heyne, E. and Whistler, R. L. (1948). *J. Am. Chem. Soc.* **70**, 2249–2252.

Hinman, M. B. and Villemez, C. L. (1975). *Plant Physiol.* **56**, 608–612.

Hirst, E. L. and Jones, J. K. N. (1949). *J. Chem. Soc.* 1659–1662.

Hirst, E. L., Jones, J. K. N. and Walder, W. O. (1947). *J. Chem. Soc.* 1443–1446.

Hirst, E. L. and Percival, E. (1965). *Methods Carbohydr. Chem.* **5**, 287–296.

Hizukuri, S., Fujii, M. and Nikuni, Z. (1961). *Nature (London)* **192**, 239–240.

Hoffman, J. and Lindberg, B. (1980). *Meth. Carbohydr. Chem.* **8**, 117–122.

Hoffman, J. and Svensson, S. (1978). *Carbohydr. Res.* **65**, 65–71.

Hoffman, J., Lindberg, B. and Painter, T. (1975). *Acta Chem. Scand.* **B29**, 137.

Hoffman, J., Lindberg, B. and Painter, T. (1976). *Acta Chem. Scand.* **B30**, 365–366.

Honda, S., Yamauchi, N. and Kakehi, K. (1979). *J. Chromatogr.* **169**, 287–293.
Hopf, H. and Kandler, O. (1977). *Phytochemistry* **16**, 1715–1717.
Hui, P. A. and Neukom, H. (1964). *Tappi* **47**, 39–42.
Hylin, J. W. and Sawai, K. (1964). *J. Biol. Chem.* **239**, 990–992.
Iriki, Y. and Miwa, T. (1960). *Nature (London)* **185**, 178–179.
Isbell, H. S. (1965). *Methods Carbohydr. Chem.* **5**, 249–250.
Jakimow-Barras, N. (1973). *Phytochemistry* **12**, 1331–1339.
Jennings, H. J. and Smith, I. C. P. (1978). *Methods Enzymol.* **50**, 39–50.
Jimenez de Asua, L., Carminatti, H. and Passeron, S. (1966). *Biochim. Biophys. Acta* **128**, 582–585.
Jindal, V. K. and Mukherjee, S. (1970). *Ind. J. Chem.* **8**, 417–419.
Jones, J. K. N. (1950). *J. Chem. Soc.*, 3292–3295.
Jones, J. K. N. and Stoodley, R. J. (1965). *Methods Carbohydr. Chem.* **5**, 36–38.
Juliano, B. O. and Varner, J. E. (1969). *Plant Physiol.* **44**, 886–892.
Kapoor, V. P. (1972). *Phytochemistry* **11**, 1129–1132.
Kato, K., Nitta, M. and Mizuno, T. (1973). *Agric. Biol. Chem.* **37**, 433–435.
Kato, K., Yamaguchi, Y., Mutoh, K. and Ueno, Y. (1976). *Agric. Biol. Chem.* **40**, 1393–1398.
Katz, G. (1965). *Tappi* **48**, 34–41.
Kenne, L., Rosell, K.-G. and Svensson, S. (1975). *Carbohydr. Res.* **44**, 69–76.
Keusch, L. (1968). *Planta* **78**, 321–350.
Khadem, H. E. and Sallam, M. A. E. (1967). *Carbohydr. Res.* **4**, 387–391.
Khanna, S. N. and Gupta, P. C. (1967). *Phytochemistry* **6**, 605–609.
Kishida, N., Okimasu, S. and Kamata, T. (1978). *Agric. Biol. Chem.* **42**, 1645–1650.
Koleske, J. V. and Kurath, S. F. (1964). *J. Polymer Sci.* **2A**, 4123–4149.
Koleva, M. I. and Achtardjieff, C. (1973). *Carbohydr. Res.* **31**, 142–145.
Koleva, M. and Gioreva, D. (1986). *Carbohydr. Res.* **155**, 290–293.
Kooiman, P. (1971). *Carbohydr. Res.* **20**, 329–337.
Kooiman, P. (1972). *Carbohydr. Res.* **25**, 1–9.
Kubal, J. V. and Gralen, N. (1948). *J. Colloid. Sci.* **3**, 457–471.
Kusakabe, I., Park, G. G., Kumita, N., Yasui, T. and Murakami, K. (1988). *Agric. Biol. Chem.* **52**, 519–524.
Lal, J. and Gupta, P. C. (1972). *Planta Med.* **22**, 71–77.
Larson, E. B. and Smith, F. (1955). *J. Am. Chem. Soc.* **77**, 429–432.
Lee, B. T. (1982). M.Sc thesis, the University of Sydney, pp. 19–20 and 77–84.
Lee, B. T. and Matheson, N. K. (1984). *Phytochemistry* **23**, 983–987.
Leschziner, C. and Cerezo, A. S. (1970). *Carbohydr. Res.* **15**, 291–299.
Lewis, D. H. (1984). In "Storage Carbohydrates in Vascular Plants", *Society Experimental Biology Seminar Series* No. 19 (D. H. Lewis, ed.). pp. 1–52. Cambridge University Press, Cambridge.
Lindberg, B. (1972). *Methods Enzymol.* **28**, 178–195.
Love, J. and Percival, E. (1964). *J. Chem. Soc.* 3345–3350.
Mackie, W. (1985). In "Polysaccharides. Topics in Structure and Morphology" (E. D. T. Atkins, ed.), pp. 73–105 VCH, Deerfield Beach.
Mackie, W. and Preston, R. D. (1968). *Planta* **79**, 249–253.
Mackie, W. and Sellen, D. B. (1969). *Polymer* **10**, 621–632.
Mackie, W., Sheldrick, B., Akrigg, D. and Perez, S. (1986). *Int. J. Biol. Macromol.* **8**, 43–51.
Maeda, M., Shimahara, H. and Sugiyama, N. (1980). *Agric. Biol. Chem.* **44**, 245–252.
Mallett, I., McCleary, B. V. and Matheson, N. K. (1987). *Phytochemistry* **26**, 1889–1894.
Mandal, G. and Das, A. (1980). *Carbohydr. Res.* **87**, 249–256.
Manzi, A. E., Mazzini, M. N. and Cerezo, A. S. (1984). *Carbohydr. Res.* **125**, 127–143.
Marchessault, R. H. (1962). *Pure Appl. Chem.* **5**, 107–129.
Marchessault, R. H. and Sarko, A. (1967). *Adv. Carbohydr. Chem.* **22**, 421–482.
Marchessault, R. H. and Sundararajan, P. R. (1983). In "The Polysaccharides" (G. O. Aspinall, ed.), Vol. 2, pp. 11–95. Academic Press, New York.
Marchessault, R. H., Buleon, A., Deslandes, Y. and Goto, T. (1979). *J. Colloid Interface Sci.* **71**, 375–382.

Matheson, N. K. (1984). *In* "Seed Physiology: Development" (D. R. Murray, ed.), Vol. 1, pp. 167–208. Academic Press, Sydney.

Matheson, N. K. and McCleary, B. V. (1985). *In* "The Polysaccharides" (G. O. Aspinall, ed.), Vol. 3, pp. 1–105. Academic Press, New York.

Matheson, N. K., Small, D. M. and Copeland, L. (1980). *Carbohydr. Res.* **82**, 325–331.

Matsuo, T. and Mizuno, T. (1974a). *Agric. Biol. Chem.* **38**, 465–466.

Matsuo, T. and Mizuno, T. (1974b). *Plant Cell. Physiol.* **15**, 555–558.

Mazzini, M. N. and Cerezo, A. S. (1979). *J. Sci. Food Agric.* **30**, 881–891.

McCleary, B. V. (1978a). *Carbohydr. Res.* **67**, 213–221.

McCleary, B. V. (1978b). *Phytochemistry* **17**, 651–653.

McCleary, B. V. (1979a). *Carbohydr. Res.* **71**, 205–230.

McCleary, B. V. (1979b). *Phytochemistry* **18**, 757–763.

McCleary, B. V. (1981). *Lebensm. Wiss. Technol.* **14**, 188–191.

McCleary, B. V. (1982). *Carbohydr. Res.* **101**, 75–92.

McCleary, B. V. (1983). *Phytochemistry* **22**, 649–658.

McCleary, B. V. and Matheson, N. K. (1974). *Phytochemistry* **13**, 1747–1757.

McCleary, B. V. and Matheson, N. K. (1975). *Phytochemistry* **14**, 1187–1194.

McCleary, B. V. and Matheson, N. K. (1976). *Phytochemistry* **15**, 43–47.

McCleary, B. V. and Matheson, N. K. (1983). *Carbohydr. Res.* **119**, 191–219.

McCleary, B. V. and Matheson, N. K. (1986). *Adv. Carbohydr. Chem. Biochem.* **44**, 147–276.

McCleary, B. V., Matheson, N. K. and Small, D. M. (1976). *Phytochemistry* **15**, 1111–1117.

McCleary, B. V., Amado, R., Waibel, R. and Neukom, H. (1981). *Carbohydr. Res.* **92**, 269–285.

McCleary, B. V., Taravel, F. R. and Cheetham, N. W. H. (1982). *Carbohydr. Res.* **104**, 285–297.

McCleary, B. V., Nurthen, E., Taravel, F. R. and Joseleau, J.-P. (1983). *Carbohydr. Res.* **118**, 91–109.

McCleary, B. V., Dea, I. C. M., Windust, J. and Cooke, D. (1984). *Carbohydr. Polymers* **4**, 253–270.

McCleary, B. V., Clark, A. H., Dea, I. C. M. and Rees, D. A. (1985). *Carbohydr. Res.* **139**, 237–260.

McClendon, J. H., Nolan, W. G. and Wenzler, H. F. (1976). *Am. J. Bot.* **63**, 790–797.

Meier, H. (1958). *Biochim. Biophys. Acta* **28**, 229–240.

Meier, H. (1960). *Acta Chem. Scand.* **14**, 749–756.

Meier, H. (1961). *Acta Chem. Scand.* **15**, 1381–1385.

Meier, H. (1965). *Methods Carbohydr. Chem.* **5**, 45–46.

Meier, H. and Reid, J. S. G. (1977). *Planta* **133**, 243–248.

Meier, A. and Reid, J. S. G. (1982). *In* "Encyclopedia of Plant Physiology, New Series" (F. A. Loewus and W. Tanner, eds), Vol. 13A, pp. 418–471. Springer, New York.

Mills, A. R. and Timell, T. E. (1963). *Can. J. Chem.* **41**, 1389–1395.

Mollenhauer, H. H. and Larson, D. A. (1966). *J. Ultrastruct. Res.* **16**, 55–70.

Morris, V. J. (1986). *In* "Functional Properties of Food Macromolecules" (J. R. Mitchell and D. A. Ledward, eds), pp. 121–170. Elsevier, Amsterdam.

Morris, E. R., Rees, D. A., Young, G., Walkinshaw, M. D. and Darke, A. (1977). *J. Molec. Biol.* **110**, 1–16.

Mukherjee, A. K., Choudrey, D. and Bagchi, P. (1961). *Can. J. Chem.* **39**, 1408–1418.

Murata, T. (1975). *Plant Cell Physiol.* **16**. 953–961.

Murata, T. (1976). *Plant Cell Physiol.* **17**, 1099–1109.

Neufeld, E. F., Ginsberg, V., Putman, E. W., Fanshier, D. and Hassid, W. Z. (1957). *Arch. Biochem. Biophys.* **69**, 602–616.

Nieduszynski, I. and Marchessault, R. H. (1972). *Can. J. Chem.* **50**, 2130–2138.

Noble, O. and Taravel, F. R. (1988). *Carbohydr. Res.* **184**, 236–243.

Ouellette, B. F. F. and Bewley, J. D. (1986). *Planta* **169**, 333–338.

Painter, T. J. (1982). *Lebensm. Wiss. Technol.* **15**, 57–61.

Painter, T. J. (1983). *In* "The Polysaccharides" (G. O. Aspinall, ed.), Vol. 2, pp. 195–285. Academic Press, New York.

Painter, T. J., Gonzalez, J. J. and Hemmer, P. C. (1979). *Carbohydr. Res.* **69**, 217–226.

Palmer, K. J. and Ballantyne, M. (1950). *J. Am. Chem. Soc.* **72**, 736–741.

Patel, P. D. and Hawes, G. B. (1987). Proceedings of the Royal Australian Chemical Institute 8th National Convention: Symposium on Quality Control Issues in the Production of Foods and Chemicals, pp. 81–87.
Paulsen, B. S., Fagerheim, E. and Øverbye, E. (1978). *Carbohydr. Res.* **60**, 345–351.
Pechanek, U., Blaicher, G., Pfannhauser, W. and Woidich, H. (1982). *J. Assoc. Off. Anal. Chem.* **65**, 745–752.
Percival, E. and McDowell, R. H. (1981). *In* "Encyclopedia of Plant Physiology, New Series" (W. Tanner and F. A. Loewus, eds), Vol. 13B, pp. 277–316. Springer, New York.
Perila, O. and Bishop, C. T. (1961). *Can. J. Chem.* **39**, 815–826.
Perlin, A. S. and Casu, B. (1982). *In* "The Polysaccharides" (G. O. Aspinall, ed.), Vol. 1, pp. 133–193. Academic Press, New York.
Pigman, W. (1957). *In* "The Carbohydrates: Chemistry, Biochemistry, Physiology" (W. Pigman, ed.), pp. 70–76. Academic Press, New York.
Preiss, J. and Wood, E. (1964). *J. Biol. Chem.* **239**, 3119–3126.
Ramsden, L. and Northcote, D. H. (1987). *Phytochemistry* **26**, 2679–2683.
Rees, D. A. and Scott, W. E. (1971). *J. Chem. Soc. B*, 469–479.
Rees, D. A., Morris, E. R., Thom, D. and Madden, J. K. (1982). *In* "The Polysaccharides" (G. O. Aspinall, ed.), Vol. 1, pp. 195–290. Academic Press, New York.
Reese, E. T. and Shibata, Y. (1965). *Can. J. Microbiol.* **11**, 167–183.
Reid, J. S. G. (1971). *Planta* **100**, 131–142.
Reid, J. S. G. (1985a). *Adv. Bot. Res.* **11**, 125–155.
Reid, J. S. G. (1985b). *In* "Biochemistry of Storage Carbohydrates in Green Plants" (P. M. Dey and R. A. Dixon, eds), pp. 265–288. Academic Press, New York.
Reid, J. S. G. (1985c). *In* "Biochemistry of Plant Cell Walls", Society for Experimental Biology Seminar Series, No. 28 (C. T. Brett and J. R. Hillman, eds), pp. 259–268. Cambridge University Press, Cambridge.
Reid, J. S. G. and Bewley, J. D. (1979). *Planta* **147**, 145–150.
Reid, J. S. G. and Meier, H. (1970). *Phytochemistry* **9**, 513–520.
Reid, J. S. G. and Meier, H. (1972). *Planta* **106**, 44–60.
Reid, J. S. G. and Meier, H. (1973). *Planta* **112**, 301–308.
Richards, E. L., Beveridge, R. J. and Grimmett, M. R. (1968). *Aust. J. Chem.* **21**, 2107–2113.
Robic, D. and Percheron, F. (1973). *Phytochemistry* **12**, 1369–1372.
Robinson, G., Ross-Murphy, S. B. and Morris, E. R. (1982). *Carbohydr. Res.* **107**, 17–32.
Roboz, E. and Haagan-Smit, A. J. (1948). *J. Am. Chem. Soc.* **70**, 3248–3249.
Rol, F. (1973). *In* "Industrial Gums", 2nd edn (R. L. Whistler and J. N. BeMiller, eds), pp. 323–337. Academic Press, New York.
Ross, T. T., Hayes, C. E. and Goldstein, I. J. (1976). *Carbohydr. Res.* **47**, 91–97.
Rubery, P. H. (1973). *Planta* **111**, 267–269.
Ruel, K. and Joseleau, J.-P. (1984). *Histochemistry* **81**, 573–580.
Saini, H. S. and Matheson, N. K. (1981). *Phytochemistry* **20**, 641–645.
Sandford, P. A. and Baird, J. (1983). *In* "The Polysaccharides" (G. O. Aspinall, ed.), Vol. 2, pp. 411–490. Academic Press, New York.
Schachman, H. K. (1957). *Methods Enzymol.* **4**, 32–103.
Scopes, R. (1982). *In* "Protein Purification: Principles and Practice". Springer, New York.
Scott, J. E. (1965). *Methods Carbohydr. Chem.* **5**, 38–44.
Schwarz, E. C. A. and Timell, T. E. (1963). *Can. J. Chem.* **41**, 1381–1388.
Seiler, A. (1977). *Planta* **134**, 209–221.
Sellen, D. B. (1980). *Symp. Soc. Exp. Biol.* **34**, 315–329.
Selvendran, R. R. and O'Neill, M. A. (1987). *Methods Biochem. Anal.* **32**, 25–153.
Sen, A. K., Banerjee, N. and Farooqui, M. I. H. (1986). *Carbohydr. Res.* **157**, 251–256.
Seth, R. P., Mukherjee, S. and Verma, S. D. (1984). *Carbohydr. Res.* **125**, 336–339.
Sharman, W. R., Richards, E. L. and Malcolm, G. N. (1978). *Biopolymers* **17**, 2817–2833.
Shimahara, H., Suzuki, H., Sugiyama, N. and Nisizawa, K. (1975). *Agric. Biol. Chem.* **39**, 293–299; 301–312.
Shimizu, S. and Ishihara, M. (1983). *Agric. Biol. Chem.* **47**, 949–955.
Sieber, R. (1972). *Phytochemistry* **11**, 1433–1441.

Sloneker, J. H. (1972). *Methods Carbohydr. Chem.* **6**, 20–24.
Small, D. M. and Matheson, N. K. (1979). *Phytochemistry* **18**, 1147–1150.
Smart, E. L. and Pharr, D. M. (1981). *Planta* **153**, 370–375.
Smith, F. (1948). *J. Am. Chem. Soc.* **70**, 3249–3253.
Smith, F. and Montgomery, R. (1956). *Methods Biochem. Anal.* **3**, 153–212.
Smith, F. and Montgomery, R. (1959). *In* "The Chemistry of Plant Gums and Mucilages", pp. 324–393. Reinhold, New York.
Smith, F. and Srivastava, H. C. (1956). *J. Am. Chem. Soc.* **78**, 1404–1408.
Smith, M. M., Axelos, M. and Peaud-Lenoel, C. (1976). *Biochimie* **58**, 1195–1211.
Sømme, R. (1966). *Acta Chem. Scand.* **20**, 589–590.
Sømme, R. (1967). *Acta Chem. Scand.* **21**, 685–690.
Sømme, R. (1968). *Acta Chem. Scand.* **22**, 870–876.
Spedding, H. (1964). *Adv. Carbohydr. Chem.* **19**, 23–49.
Spyropoulos, C. G. and Reid, J. S. G. (1985). *Planta* **166**, 271–275.
Spyropoulos, C. G. and Reid, J. S. G. (1988). *Planta* **174**, 473–478.
Srivastava, H. C., Singh, P. P., Harshe, S. N. and Virk, K. (1964). *Tetrahedron Lett.*, 493–498.
Srivastava, H. C., Singh, P. P. and Subba Rao, P. V. (1968). *Carbohydr. Res.* **6**, 361–366.
Stepanenko, B. N. (1960). *Bull. Soc. Chim. Biol.* **42**, 1519–1536.
Stephen, A. M. (1983). *In* "The Polysaccharides" (G. O. Aspinall, ed.), Vol. 2, pp. 97–193. Academic Press, New York.
Sturgeon, R. J. (1980). *Meth. Carbohydr. Chem.* **8**, 77–80.
Sugiyama, N., Shimahara, H., Andoh, T., Takemoto, M. and Kamata, T. (1972). *Agric. Biol. Chem.* **36**, 1381–1387.
Sugiyama, N., Shimahara, H., Andoh, T. and Takemoto, M. (1973). *Agric. Biol. Chem.* **37**, 9–17.
Sundararajan, P. R. and Rao, V. S. R. (1970). *Biopolymers* **9**, 1239–1247.
Suzuki, H., Li, S.-C. and Li, Y.-T. (1970). *J. Biol. Chem.* **245**, 781–786.
Sweeley, C. C., Bentley, R., Makita, M. and Wells, W. W. (1963). *J. Am. Chem. Soc.* **85**, 2497–2507.
Takahashi, R., Kusakabe, I., Kusama, S., Sakurai, Y., Murakami, K., Maekawa, A. and Suzuki, T. (1984). *Agric. Biol. Chem.* **48**, 2943–2950.
Tako, M. and Nakamura, S. (1986). *FEBS Lett.* **204**, 33–36.
Tewari, K., Khare, N., Singh, V. and Gupta, P. C. (1984). *Carbohydr. Res.* **135**, 141–146.
Thompson, J. L. and Jones, J. K. N. (1964). *Can. J. Chem.* **42**, 1088–1091.
Timell, T. E. (1957). *Can. J. Chem.* **35**, 333–338.
Timell, T. E. (1964). *Adv. Carbohydr. Chem.* **19**, 247–302.
Timell, T. E. (1965a). *Adv. Carbohydr. Chem.* **20**, 409–483.
Timell, T. E. (1965b). *Meth. Carbohydr. Chem.* **5**, 134–138.
Tipson, R. S. (1963). *Meth. Carbohydr. Chem.* **2**, 150–161.
Tvaroska, I., Perez, S., Noble, O. and Taravel, F. (1987). *Biopolymers* **26**, 1499–1508.
Unrau, A. M. (1961). *J. Org. Chem.* **26**, 3097–3101.
Unrau, A. M. and Choy, Y. M. (1970a). *Can. J. Chem.* **48**, 1123–1128.
Unrau, A. M. and Choy, Y. M. (1970b). *Carbohydr. Res.* **14**, 151–158.
Usui, T., Mizuno, T., Kato, K., Tomoda, M. and Miyajima, G. (1979). *Agric. Biol. Chem.* **43**, 863–865.
Veluraja, K. and Atkins, E. D. T. (1988). *Carbohydr. Res.* **183**, 131–134.
Verachtert, H., Rodriguez, P., Bass, S. T. and Hansen, R. G. (1966). *J. Biol. Chem.* **241**, 2007–2013.
Vijayendran, B. R. and Bone, T. (1984). *Carbohydr. Polymers* **4**, 299–313.
Vilkas, E. and Radjabi-Nassab, F. (1986). *Biochimie* **68**, 1123–1127.
Villarroya, H. and Petek, F. (1976). *Biochim. Biophys. Acta* **438**, 200–211.
Villarroya, H., Williams, J., Dey, P., Villarroya, S. and Petek, F. (1978). *Biochem. J.* **175**, 1079–1087.
Villemez, C. L. (1974). *Arch. Biochem. Biophys.* **165**, 407–412.
Wankhede, D. B., Tharanathan, R. N. and Rao, M. R. (1979). *Carbohydr. Res.* **74**, 207–215.
Whelan, W. J. (1955). *In* "Modern Methods of Plant Analysis" (K. Paech and M. V. Tracey, eds), Vol. 2, pp. 181–184. Springer, Berlin.

Whistler, R. L. and Durso, D. F. (1952). *J. Am. Chem. Soc.* **74**, 5140–5141.
Whistler, R. L. and Marx, J. W. (1965). *Methods Carbohydr. Chem.* **5**, 143.
Whistler, R. L. and Richards, E. L. (1970). *In* "The Carbohydrates", 2nd edn (W. Pigman and D. Horton, eds), Vol. IIA, pp. 447–469. Academic Press, New York.
Whistler, R. L. and Saarnio, J. (1957). *J. Am. Chem. Soc.* **79**, 6055–6057.
Whistler, R. L. and Smart, C. L. (1953). *In* "Polysaccharide Chemistry", pp. 152–160; 291–303. Academic Press, New York.
Whistler, R. L. and Smith, C. G. (1952). *J. Am. Chem. Soc.* **74**, 3795–3796.
Whitcomb, P. J., Gutowski, J. and Howland, W. W. (1980). *J. Appl. Polymer. Sci.* **25**, 2815–2827.
Williams, J., Villarroya, H. and Petek, F. (1977). *Biochem. J.* **161**, 509–515.
Wolfram, M. L., Laver, M. L. and Patin, D. L. (1961). *J. Org. Chem.* **26**, 4533–4535.

12 The Pectic Polysaccharides of Primary Cell Walls

MALCOLM O'NEILL, PETER ALBERSHEIM and
ALAN DARVILL

*Complex Carbohydrate Research Center and Department of Biochemistry,
University of Georgia, 220 Riverbend Road, Athens, GA 30602, USA*

I. INTRODUCTION

Pectic substances have historically been considered as those components of the primary cell wall of higher plants extractable with hot water, dilute acid, ammonium oxalate, and other chelating agents (Northcote, 1963). This operationally defined group of

METHODS IN PLANT BIOCHEMISTRY Vol. 2
ISBN 0-12-461012-9

polysaccharides in which D-galactosyluronic acid is a principal constituent contained, depending on the plant source, galacturonans, rhamnogalacturonans, arabans, galactans, and arabinogalactans (Aspinall, 1970). However, since current practice is to classify polysaccharides by their structures rather than the mode of their isolation (Aspinall, 1980, 1983), pectic polysaccharides can be considered as a group of polymers associated with 1,4-linked α-D-galactosyluronic acid residues. Cell-wall pectic polysaccharides, then, are defined as those D-galactosyluronic acid-rich polymers remaining with the insoluble cell wall after removal of cytoplasmic material and non-structural wall components. This definition, though somewhat arbitrary, will serve to distinguish the pectic polysaccharides of the cell wall from the structurally related gums and mucilages exuded by plants (see Chapter 14, this volume).

Pectic polysaccharides are present in the primary cell walls of all seed-bearing plants and are located particularly in the middle lamella (Bacic *et al.*, 1988). These polysaccharides are major components of the primary cell walls of dicotyledons (e.g. sycamore, citrus and legumes) and gymnosperms (e.g. Douglas fir). Though abundant in the primary walls of non-graminaceous monocotyledons (e.g. onion, garlic, Lemna and sisal), pectic polysaccharides account for relatively less of the primary cell wall in graminaceae (e.g. barley, wheat and ryegrass) (Bacic *et al.*, 1988; Jarvis, *et al.*, 1988). Limited data suggest that the primary cell walls of lower plants (e.g. ferns) are rich in pectic polysaccharides (White *et al.*, 1986; A. Koller, unpubl. res., this laboratory).

The distribution within plant cell walls of the component polysaccharides is not known. Cytochemical (Albersheim *et al.*, 1960; Roland and Vian, 1981) and immunochemical (Moore *et al.*, 1986; Moore and Staehelin, 1988) localisation techniques have provided evidence that the middle lamella is enriched with pectic polysaccharides. Monoclonal antibodies with well-defined cell-wall carbohydrate epitopes should lead to identifying more specifically the locations of the various wall polysaccharides, including the pectins.

The general glycosyl residue compositions of pectic polysaccharides have been determined using pectic polysaccharide-containing fractions obtained from commercially important sources such as apple and citrus pulps, sugar beet, lucerne, and alfalfa (Kertesz, 1951). Despite the heterogeneous nature of the fractions and the limitations of the methods available for the early studies, the presence of polymers containing arabinosyl, galactosyl, and galactosyluronic acid residues was demonstrated (Aspinall, 1962). Structural studies of pectic polysaccharides chemically extracted from lucerne leaf and stem (Aspinall and Fanshawe, 1961; Aspinall *et al.*, 1968b), apple (Barrett and Northcote, 1965), soybean cotyledon and hull (Aspinall *et al.*, 1967a,b), lemon peel (Aspinall *et al.*, 1968a), and rape seed (Aspinall and Jiang, 1974) provided evidence for short sequences of some of the glycosyl residues in these partially purified pectins. A significant portion of these extracted polysaccharides was accounted for by regions of homogalacturonan, as oligosaccharides containing only 4-linked α-D-galactosyluronic acid residues were isolated after partial acid hydrolysis, acetolysis, or enzymatic depolymerisation. Acidic oligosaccharides composed of a mixture of 4-linked galactosyluronic acid and 2-linked rhamnosyl residues and neutral oligosaccharides containing only 4-linked β-D-galactosyl residues were also isolated and characterised. Since these materials were obtained from tissues that contained a variety of cell types, it could not be determined whether the polysaccharides were of primary cell-wall origin. However, these studies established that pectic polysaccharides are complex macromolecules and that they are not a mixture of homopolysaccharides, that is, galacturonan, araban, and galactan (Aspinall, 1980).

Three—and to date only three—pectic polysaccharides have been isolated from the primary cell walls of plants and characterised. These are homogalacturonan, rhamnogalacturonan I (RG-I), and a substituted galacturonan referred to as rhamnogalacturonan II (RG-II). Homogalacturonan is a chain of 1,4-linked α-D-galactosyluronic acid residues in which some (perhaps most) of the carboxyl groups are methyl esterified. RG-I is a family of closely related polysaccharides that contain a backbone of the alternating disaccharide →4)-α-D-Gal$p$A-(1→2)-α-L-Rha$p$-(1→. About 50% of the rhamnosyl residues are substituted at C-4 with neutral, and perhaps acidic, oligosaccharides. RG-II is a polysaccharide composed of a 1,4-linked α-D-galactosyluronic acid backbone with aldehydo- and keto-sugar oligosaccharide side chains attached to C-2 and/or C-3. Some of the glycosyl residues of the pectic polysaccharides are O-acetylated, but, in general, the location of the acetyl substituents has not been determined. However, RG-I is known to be O-acetylated on C-3 of some of the backbone galactosyluronic acid residues (Mort et al., 1988).

Recent developments in the study of cell-wall pectic polysaccharides, in addition to elucidating new structural features, have emphasised the importance of their biological regulatory functions (McNeill et al., 1984) and physical properties (Jarvis, 1984; Chapman et al., 1987). The heightened interest in structure–function relationships has led to: (1) greater awareness of the problems associated with isolating these complex macromolecules without altering their structures, (2) improved purification procedures for the preparation of homogeneous polysaccharides, and (3) refinements in established analytical procedures and development of new procedures to facilitate the structural characterisation of the polysaccharides.

II. ISOLATION OF CELL WALLS

Plant tissues are generally composed of a variety of cell types, each of which is enclosed by a primary or secondary cell wall. It is difficult to obtain a homogeneous primary cell-wall preparation from intact tissues of plants unless specific portions of the plant, such as the parenchyma of vegetables (Selvendran, 1975) or the mesophyll of grasses (Chesson et al., 1985), are dissected from an organ. Cell walls of a more uniform type can be obtained from suspension-cultured cells (e.g. sycamore, rose, tobacco, maize, rice and Douglas fir), a convenient source of relatively homogeneous primary cell walls (York et al., 1986).

Primary walls are obtained from plant tissue or suspension-cultured cells by disrupting the cells to remove cytoplasmic components (such as protein, starch, nucleic acid and lipid) by treating the insoluble walls with a combination of aqueous buffers, detergents, and/or enzymes (York et al., 1986; Selvendran and O'Neill, 1987). The chemical treatments are usually performed at low temperature (<4°C) to minimise partial solubilisation or fragmentation of the cell-wall polymers by exo- and endo-glycanases located in the cell wall (York et al., 1986; Selvendran and O'Neill, 1987). Details of the procedures used to isolate primary cell walls from plant tissues (Selvendran and O'Neill, 1987; see Chapter 16, this volume) and from suspension-cultured cells (York et al., 1986) will not be described here.

III. STRUCTURAL ANALYSIS

A. General Considerations

The ultimate goal for the characterisation of pectic polysaccharides is to relate the primary structure of the molecules to their three-dimensional conformations, their physical properties, their biological functions, and their interactions with other polymers in the primary cell wall.

Most of the effort in structurally characterising pectic polysaccharides involves purifying them to homogeneity and determining their primary structures. The pectic polysaccharides in the plant cell wall must be solubilised by using either chemical extractants, attempting to minimise non-specific structural modifications (Selvendran and O'Neill, 1987), or highly purified enzymes, such as α-1,4-*endo*-polygalacturonases (Talmadge *et al.*, 1973; York, *et al.*, 1986). The extracted pectic polysaccharides must then be purified to a high degree of homogeneity prior to attempting to characterise them structurally.

The primary structure of a pectic polysaccharide is known only when all of the following characteristics have been elucidated:

1. the glycosyl-residue composition;
2. the absolute configuration, D- or L-, of each glycosyl residue;
3. the glycosyl-linkage composition;
4. the ring form, furanose or pyranose, of each glycosyl residue;
5. the sequence of glycosyl residues;
6. the anomeric configuration, α- or β-, of each glycosyl linkage;
7. the points of attachment of any non-carbohydrate substituents.

Detailed methods for the accurate determination of points (1) to (7) have been described elsewhere (Lindberg, 1972; Lonngren and Svennson, 1974; Lindberg and Lonngren, 1976, 1978; Leontein *et al.*, 1978; Aspinall, 1982; McNeil *et al.*, 1982b; Waeghe *et al.*, 1983; Vliegenthart *et al.*, 1983; Dell, 1987).

B. Homogalacturonan

Pectic polysaccharides have been released from the walls of suspension-cultured sycamore cells by treatment with a purified *endo*-polygalacturanose isolated from *Colletotrichum lindemuthianum* (Talmadge *et al.*, 1973; McNeill *et al.*, 1980). The analogous enzyme from *Aspergillus niger* is equally effective (unpublished results of this laboratory). Approximately 12% of the wall (~50% of total wall pectin) is solubilised by *endo*-polygalacturonase. Galactosyluronic acid oligomers with a degree of polymerisation (DP) between 1 and 3, homogalacturonan, RG-I and RG-II have been isolated from the *endo*-polygalacturonase-solubilised material by dialysis and size-exclusion and ion-exchange chromatographies (York *et al.*, 1986). The small oligogalacturonides, which do not readily pass through dialysis membranes, are recovered in the partially included volume of the size-exclusion column. The solubilised homogalacturonan was resistant to *endo*-polygalacturonase hydrolysis because the presence of methyl esterified galactosyluronic acid residues prevented enzymatic cleavage (McNeil *et al.*, 1980). De-esterification and re-treatment of the homogalacturonan with *endo*-polygalacturonase resulted in the formation of mono-, di- and trigalacturonides (York *et al.*, 1986). The isolation of oligosaccharides and polysaccharides composed of unbranched, uninterrupted sequences of 1,4-linked α-D-galactosyluronic acid residues established the presence of homogalacturonan in primary cell walls (McNeil *et al.*, 1980).

Oligogalacturonides were also released from isolated cell walls of soybean hypocotyls by partial acid hydrolysis (Hahn *et al.*, 1981). These oligogalacturonides (DP 2–12) were separated and purified to homogeneity by preparative ion-exchange chromatography on QAE-Sephadex (Nothnagel *et al.*, 1983). Recent results from our laboratory (M. O'Neill and R. Lo, unpubl. res.) demonstrated that oligogalacturonides (DP 5–20) can be resolved by high resolution analytical anion-exchange chromatography on a Dionex (CarboPac I) column (Fig. 12.1). This system is more rapid (45 min per analysis) than conventional anion-exchange chromatography (Nothnagel *et al.*, 1983), has higher resolving power, and, with the pulsed amperometric detector, requires relatively small amounts of oligogalacturonides ($< 10 \mu g$ of each oligogalacturonide). This system increases the ability to monitor the production of acidic and neutral oligosaccharides during selective depolymerisation of pectic polysaccharides by chemical and enzymatic treatments.

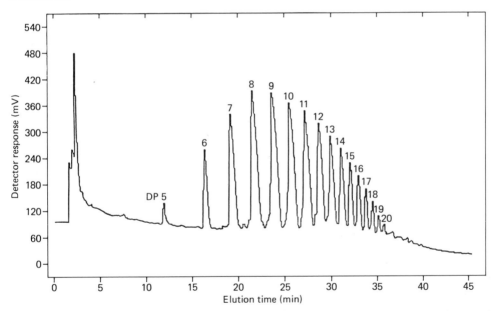

FIG. 12.1. Separation of galactosyluronic acid oligomers, derived from a partial acid hydrolysis of polygalacturonic acid (fraction C, Nothnagel *et al.*, 1983), on a Dionex CarboPac-I column. Approximately 200 μg total uronic acid in 75 μl of deionised water was loaded onto the column and eluted at 1 ml min⁻¹ with 400 mM NaOAc in 150 mM NaOH (0–2 min) followed by a gradient of 400–900 mM NaOAc in 150 mM NaOH (2–45 min). Carbohydrate was detected using a Dionex pulsed amperometric detector with the output range set at 10 μA and a 1 s response time. The applied potentials E_1, E_2 and E_3 were 0.05 V, +0.60 V and −0.60 V, respectively, and the duration of the pulses T_1, T_2 and T_3 set to 480 ms, 180 ms and 60 ms, respectively.

Glycosyl-residue and glycosyl-linkage composition analyses of the oligogalacturonides isolated from the cell walls of soybean hypocotyls established the presence of 4-linked galactosyluronic acid residues. Molecular ions ($[M-H]^-$) were obtained using fast atom bombardment-mass spectrometry (FAB-MS) of the underivatised oligogalacturonides (up to DP 11). Satisfactory FAB mass spectra were obtained for the larger (DP > 11) oligogalacturonides when they were converted to their pentafluorobenzyloxime derivatives (Nothnagel *et al.*, 1983). Recent results in our laboratory (R. Lo and W. York, unpubl. res.) showed that, with negative-ion FAB-MS, [M-H]⁻ ions were obtained from oligogalacturonides (DP 2–14) that had been converted to their ammonium salts.

Oligogalacturonides obtained from the walls of suspension-cultured soybean cells were depolymerised by a purified α-1,4-*endo*-polygalacturonase which, in combination with the results of proton and carbon-13 NMR, established that the galacturonosyl residues were α-1,4-linked (Nothnagel *et al.*, 1983).

Oligogalacturonides with either a saturated or a Δ4,5-unsaturated galactosyluronic acid residue at the non-reducing terminus have been generated from apple pectin by enzymatic digestion and characterised by proton NMR (Tjan *et al.*, 1974). The chemical shift (δ) and coupling constants confirmed that all the D-galactosyluronic acid residues had the C-1 conformation and were α-1,4-linked. Reproducible chemical shift values for the ¹H and ¹³C nuclei of acidic oligoglycosyl residues were only obtained when the sample was placed in a buffered ²H₂O solution of known pH before generating the spectra (Tjan *et al.*, 1974).

Evidence has been obtained that linear polysaccharides composed of unbranched 1,4-linked α-D-galactosyluronic acid residues are present in a range of plant species including *Amabilis* fir (Bhattarcharjee and Timell, 1964), lucerne (Aspinall and Fanshawe, 1961), apple (Barrett and Northcote, 1965), sunflower heads (Zitko and Bishop, 1966), soybean (Aspinall *et al.*, 1967a,b), lemon (Aspinall *et al.*, 1968a), rape seed (Aspinall and Jiang, 1974), cabbage (Stevens and Selvendran, 1984a), sugar beet (Rombouts and Thibault, 1986), and carrot (Massiot *et al.*, 1988).

Most of the carboxyl groups of the galactosyluronic acid residues of the cell-wall pectic polysaccharides are esterified (Aspinall, 1970), but, despite a number of studies, the distribution of the methyl esters is unknown (Taylor, 1982; Tuerena *et al.*, 1982; DeVries *et al.*, 1986). It is possible that homogalacturonans are fully methyl esterified when synthesised and later are chemically or enzymatically partially de-esterified. Pectin methyl esterases have been detected in some plant tissues (Rombouts and Pilnik, 1980).

C. Rhamnogalacturonan I (RG-I)

The major polysaccharide solubilised from suspension-cultured sycamore cell walls after treatment with α-1,4-*endo*-polygalacturonase is RG-I, which accounts for at least 7% and probably twice that amount of the wall material (Talmadge *et al.*, 1973; McNeil *et al.*, 1980). RG-I, as solubilised by the enzyme, had an average molecular weight of between 10⁵ and 10⁶ and eluted as a single, partially included peak when chromatographed on a Bio-Gel Agarose 5 m size-exclusion column. The leading edge of the peak of eluted material was enriched in arabinosyl residues whereas the trailing edge was enriched in rhamnosyl and galacturonosyl residues (McNeil *et al.*, 1980), indicating that RG-I is not uniformly substituted with side chains. Glycosyl residue composition analysis of sycamore RG-I, using gas–liquid chromatography (GLC) analysis of the alditol acetate and trimethylsilyl methyl glycoside derivatives (York *et al.*, 1986), showed that arabinose, galactose, galacturonic acid and rhamnose are the major monosaccharide components of RG-I (Table 12.1). Small amounts of fucose are always present in sycamore RG-I. Glycosyl-linkage analysis (Hakomori, 1964), following carboxyl reduction of the methyl esterified galactosyluronic acid residues (McNeil *et al.*, 1980; York *et al.*, 1986), established that sycamore RG-I is a complex, branched pectic polysaccharide (Table 12.2).

TABLE 12.1. Glycosyl residue composition of RG-I from the walls of suspension-cultured sycamore, Douglas fir, and maize cells.

Glycosyl residue[a]	Sycamore[b]	Sycamore[c]	Douglas fir[d]	Maize[e]
Rha	16	9	9	20
Fuc	2	1	4	0
Ara	32	35	30	32
Xyl	0	2	4	4
Gal	31	43	21	20
Glc	0	0	6	0
GalA	17	10	26	24

[a] Values expressed as mol%.
[b] *Endo*-polygalacturonase-solubilised RG-I (Lau *et al.*, 1985).
[c] Na$_2$CO$_3$-solubilised RG-I (Ishii *et al.*, 1989).
[d] LiCl-solubilised RG-I (Thomas *et al.*, 1987).
[e] *Endo*-polygalacturonase-solubilised RG-I (Thomas *et al.*, 1989a).

The structure of sycamore RG-I has been probed by a number of chemical techniques. A 90% decrease in the content of 2-linked rhamnosyl residues was observed (McNeil *et al.*, 1982a) after methylated sycamore RG-I was subjected to base-catalysed β-elimination (Lindberg and Lonngren, 1976). β-Elimination of the uronosyl residues released those glycosyl residues that were glycosidically attached to C-4 of the uronosyl residues. When the products of β-elimination of methylated RG-I were ethylated and examined by gas–liquid chromatography-mass spectrometry (GLC-MS) of the alkylated alditol acetates, most of the O-ethyl groups were located on the 2- and 2,4-linked rhamnosyl residues in the original polysaccharide. This established that most (or all) of the rhamnosyl residues of RG-I are linked to C-4 of galactosyluronic residues, and that all of the rhamnosyl residues are substituted at C-2 with galactosyluronic acid residues (McNeil, *et al.*, 1982a).

The nature of those glycosyl residues that were directly attached to C-4 of the 2-linked rhamnosyl residues was determined (McNeil *et al.*, 1982a) by subjecting methylated sycamore RG-I to β-elimination (Aspinall and Rosell, 1977) to give 4-linked rhamnosyl residues with free hydroxyls at C-2. The hydroxyl groups were oxidised with a methylsulphoxide–chlorine complex (Corey and Kim, 1973) to yield the 2-keto derivative. A second β-elimination, with the organic base 1,5-diazabicyclo[5.4.0]-undec-5-ene in the presence of acetic anhydride (Aspinall and Chaudhari, 1975) released the substituents on C-4 of the rhamnosyl residue. The free reducing groups of the liberated methylated oligosaccharides were acetylated and thereby protected from further base-catalysed degradation. The released derivatised substituents were analysed as their alkylated alditol acetates. The results of this work established that, in sycamore RG-I, terminal arabinosyl and terminal, 3-, 4-, 6-, 2,6- and 3,6-linked galactosyl residues were attached to C-4 of the 2-linked rhamnosyl residues (McNeil *et al.*, 1982a). These results showed that at least seven different side chains, most commencing with a galactosyl residue, were attached to the RG-I backbone.

TABLE 12.2. Glycosyl-linkage composition of RG-I isolated from the cell walls of sycamore, Douglas fir, and maize.

Glycosyl linkage[a]		Sycamore % Branched rhamnosyl residues			Douglas fir[d]	Maize[e]
		48%[b]	34%[c]	84%[c]		
Rha	T	0	3.0	0.5	0	0.4
	2	7.8	15.2	1.3	1.1	5.7
	4	0	2.8	0.6	0	0
	2,3	0	1.6	0	t	0
	2,4	8.0	8.0	10.6	2.8	11.8
	2,3,4	0.6	1.1	2.0	0.5	2.1
Fuc	T	1.4	0.6	2.4	t	0
	3,4	0	0	0	0.5	0
Ara	T	9.5	2.2	2.8	13.4	14.9
	2	2.2	0	0.4	t	0.8
	3	2.2	0.9	2.6	17.2	2.5
	5	11.2	3.4	8.0	19.8	13.0
	2,5	1.0	1.4	3.2	6.8	1.4
	3,5	3.5	1.1	5.9	7.4	6.1
Xyl	T	2.0	0.7	0.3	t	1.1
	4	0	0	0	t	1.0
	2,4	0	0	0	t	1.1
Gal	T	6.3	8.1	6.6	15.1	14.2
	2	0.6	1.0	2.9	0.8	0.2
	3	2.7	2.2	3.9	4.2	3.5
	4	8.4	2.9	11.3	4.9	4.8
	6	7.5	4.4	8.1	1.0	4.8
	2,4	6.3	0	6.6	15.1	14.2
	2,6	1.2	0.2	2.5	0	1.9
	3,4	0	0.4	1.4	1.2	0.7
	3,6	1.2	0.6	6.5	1.1	2.6
	4,6	2.4	1.5	0.8	1.1	0
Glc	T	0	0	0	1.6	0
GalA	T	1.6	4.4	0.6	0[f]	0[f]
	4	15.2	30.6	12.5	0	0
	2,4	1.0	0.2	0.2	0	0
	3,4	0	1.1	1.9	0	0

[a] Values expressed as mol%.
[b] *Endo*-polygalacturonase-solubilised RG-I (York *et al.*, 1986).
[c] Na$_2$CO$_3$-solubilised RG-I (Ishii *et al.*, 1989).
[d] LiCl-solubilised RG-I (Thomas *et al.*, 1987).
[e] *Endo*-polygalacturonase-solubilised RG-I (Thomas *et al.*, 1989a).
[f] Galactosyluronic acid linkages not determined.
T = non-reducing terminal residue; t = trace.

The distribution of rhamnosyl and galacturonosyl residues in the sycamore RG-I backbone was examined by partial acid hydrolysis of the carboxyl-reduced (Taylor and Conrad, 1972) and methylated polymer (Lau *et al.*, 1985). The resulting fragments were converted to their corresponding oligoglycosyl alditols, deuterioethylated, and separated by reversed phase high performance liquid chromatography (HPLC); the derivatives were detected by liquid chromatography-mass spectrometry (LC-MS) (McNeil *et al.*, 1982b). The anomeric configurations, number and sequence of glycosyl residues, linkage types, and pattern of *O*-ethylation of the alkylated oligoglycosyl alditols were determined by a combination of proton NMR, GLC, chemical ionisation (CI) and electron impact (EI) mass spectrometric and glycosyl-linkage composition analyses. These experiments established that the backbone of sycamore RG-I was entirely composed of repeating disaccharide **1** (Lau *et al.*, 1985). The pattern of labelling with *O*-ethyl groups demonstrated that 50% of the 2-linked rhamnosyl residues were substituted at C-4, but a regular pattern of substitution along the backbone was not discernible (Lau *et al.*, 1985). Indeed, it was established that neighbouring backbone rhamnosyl residues could either be branched, unbranched, or one but not the other could be branched.

$$\rightarrow4)\text{-}\alpha\text{-}D\text{-}GalpA\text{-}(1\rightarrow2)\text{-}\alpha\text{-}L\text{-}Rhap\text{-}(1\rightarrow$$

1

A pectic polysaccharide was isolated from the walls of tobacco mesophyll cells by extraction with ethylenediamine tetra-acetic acid (EDTA) (Eda *et al.*, 1986). The purified polysaccharide was partially acid hydrolysed, and some of the acidic oligosaccharides released were characterised by proton NMR, field desorption-MS, GLC-MS, and glycosyl-linkage analysis. The acidic oligosaccharides were composed of 4-linked α-D-galactosyluronic acid with smaller amounts of α-D-GalpA-(1,2)-α-L-Rhap (**1**) and a tetrasaccharide composed of two copies of repeating unit **1**. Sequences containing up to six repeats (12 glycosyl residues) of repeating disaccharide **1** were tentatively identified. These results for the pectic polysaccharides of tobacco mesophyll cell walls corroborate the repeating disaccharide backbone of RG-I elucidated by studies of the sycamore cell wall polysaccharide (Lau *et al.*, 1985).

The oligoglycosyl side chains attached to C-4 of 2,4-linked rhamnosyl residues of sycamore RG-I were released by lithium degradation of the galactosyluronic acid residues, and many of the side chains were purified and characterised (Lau *et al.*, 1988b). In this procedure, sycamore RG-I in ethylenediamine was treated with lithium metal (Mort and Bauer, 1982; Lau *et al.*, 1988a) and the products reduced to give a mixture of oligoglycosyl alditols. Most of these oligomers had 4-linked rhamnitol at their reducing ends. The presence of small amounts of oligomers with arabinitol and galactitol at their reducing ends could be explained by the ability of the lithium treatment to cleave small amounts of neutral residues (Lau *et al.*, 1988a) or by the presence, in some side chains, of galactosyluronic acid residues. Therefore, it has not yet been possible to establish whether some of the side chains of RG-I contain galactosyluronic acid residues.

The oligoglycosyl alditols formed by lithium treatment of RG-I were methylated and analysed by HPLC and by FAB-MS. These studies confirmed the existence of a large number of side chains (McNeil, *et al.*, 1982a) ranging in size from a single glycosyl residue to more than 15 glycosyl residues. Any chains that contained galactosyluronic acid residues would have been shortened by the lithium treatment. The alkylated oligosaccharide alditols obtained from RG-I were separated by reversed phase HPLC, using direct liquid insertion-mass spectrometry (DLI-MS) detection, and analysed by proton NMR, FAB-MS and GLC-MS (Lau *et al.*, 1988b). The complete structures of eight oligoglycosyl rhamnitols and partial structures of four other oligoglycosyl rhamnitols were determined. Partial structures for oligoglycosyl alditols terminating with arabinitol or galactitol were also obtained (Lau *et al.*, 1988b). Oligoglycosyl rhamnitols **2** to **6** are representative of the RG-I side chains.

β-D-Gal*p*-(1→6)-β-D-Gal*p*-(1→4)-β-D-Gal*p*-(1→4)-Rhamnitol

2

α-L-Fuc*p*-(1→2)-β-D-Gal*p*-(1→4)-β-D-Gal*p*-(1→4)-Rhamnitol

3

L-Ara*f*-(1→5)-α-L-Ara*f*-(1→2)-α-L-Ara*f*-(1→3)-β-D-Gal*p*-(1→4)-Rhamnitol

4

Ara*f*-[Ara*f*]$_{0-3}$-Rhamnitol

5

Gal*p*-[Gal*p*]$_{0-3}$-Rhamnitol

6

A combination of the chemical techniques described above has produced a considerable amount of structural information about sycamore RG-I. We picture RG-I as a long repeating sequence of disaccharide **1** with a variety of arabinosyl-, galactosyl-, and fucosyl- (dicots) containing side chains (**2** to **6**). There is a good possibility that at least some of the side chains contain galactosyluronic acid residues. Work is in progress to determine additional side-chain structures, the distribution of the various side chains along the RG-I backbone, and whether different RG-I backbones, perhaps those in different cells, possess different sets of side chains. Toward these goals, efforts are under way in the authors' laboratory to purify an *endo*-glycanase that cleaves the RG-I backbone, and to generate monoclonal antibodies specific for individual RG-I side chains.

Only about half of the pectic polysaccharides present in the walls of suspension-cultured sycamore cells are solubilised by treatment with α-1,4-*endo*-polygalacturonase (McNeil *et al.*, 1980). About half of the pectic polysaccharides remaining in the *endo*-polygalacturonase-treated wall residue are solubilised by extractions with Na$_2$CO$_3$ at 1°C and 20°C. Most of the polysaccharide extracted with Na$_2$CO$_3$ is similar in glycosyl composition (Table 12.1) to that of RG-I solubilised with α-1,4-*endo*-polygalacturonase (McNeil *et al.*, 1980). However, glycosyl-linkage analysis of various Na$_2$CO$_3$-solubilised fractions showed (Table 12.2) that the degree of side-chain substitution of Na$_2$CO$_3$-solubilised RG-I differs from *endo*-polygalacturonase-solubilised RG-I (Ishi *et al.*, 1989). Some of the Na$_2$CO$_3$-solubilised RG-I had as few as 34% of the rhamnosyl residues branched, whereas other Na$_2$CO$_3$-solubilised RG-I chains had branches attached to as many as 84% of the rhamnosyl residues. By comparison, ∼50% of rhamnosyl residues were branched in *endo*-polygalacturonase-solubilised RG-I.

Only small amounts of RG-I (1% of the wall) can be solubilised by extraction of sycamore cell walls with cyclohexanediamine tetra-acetic acid (CDTA) and Na_2CO_3 prior to treating the walls with *endo*-polygalacturonase (A. Koller, unpubl. res., this laboratory). Therefore, *endo*-polygalacturonase facilitates subsequent chemical extractions of RG-I, perhaps by cleaving covalent linkage of RG-I to homogalacturonan.

Cell walls from suspension-cultured maize, a monocot, were sequentially extracted with hot water, dimethyl sulphoxide, ammonium oxalate, and *endo*-polygalacturonase (Thomas et al., 1989a). The enzymatically solubilised polysaccharide (1% of the wall) had glycosyl-residue and glycosyl-linkage compositions that were characteristic of RG-I (Tables 12.1 and 12.2). Purified maize RG-I was subjected to lithium degradation and the products purified by reversed phase HPLC. Partial or complete structures were obtained for nine oligoglycosyl alditols, and evidence for the presence of eight others was obtained by FAB-MS. All of these structures had previously been found in sycamore RG-I. However, maize RG-I, like maize hemicellulosic xyloglucan (Thomas et al., 1989a), does not contain terminal fucosyl residues (see oligosaccharide **3**) and thus differs from sycamore RG-I in this respect.

Studies with the pectic polysaccharides solubilised from suspension-cultured rice cell walls demonstrated the presence of a polysaccharide with a structure very similar to sycamore RG-I (Thomas et al., 1989b). Also, approximately 1–2% of the cell walls of rice endosperm were solubilised by extraction with hot ammonium oxalate (Shibuya and Nakane, 1984). Two pectic polysaccharides that differed in their content of galactosyluronic acid and arabinosyl residues were purified by ion-exchange and size-exclusion chromatographies. These polysaccharides both possessed glycosyl-residue and glycosyl-linkage compositions similar to those of RG-I isolated from the walls of suspension-cultured rice cells (Thomas et al., 1989b).

Homogalacturonan and RG-I (approximately 8% of the wall) were extracted from the walls of suspension-cultured Douglas fir cells by prolonged treatment with 1 M LiCl; subsequent treatment with α-1,4-*endo*-polygalacturonase only solubilised an additional 1% of the cell wall (Thomas et al., 1987). The glycosyl-linkage composition of Douglas fir RG-I (Table 12.2) was similar to that of sycamore RG-I. However, some differences were detected, especially in the ratios of 2- and 2,4-linked rhamnosyl residues, 3-linked arabinosyl, and 3,6-linked galactosyl residues.

Pectic polysaccharides were released from the walls of suspension-cultured *Rosa glauca* cells by chemical extraction (Chambat et al., 1981, 1984). Sequential extraction with the traditional pectic-solubilising reagents EDTA and ammonium oxalate solubilised 21% of the cell-wall glycosyluronic acid residues; another 7% of the glycosyluronic acid residues were subsequently extracted with cold alkali, leaving ~70% of the glycosyluronic acid residues in the insoluble wall material. The oxalate and EDTA-solubilised pectins were identified as homogalacturonan and RG-I on the basis of glycosyl-composition analyses (Chambat et al., 1984).

Pectic polysaccharides were extracted from cell walls isolated from various tissues (outer pericarp, inner pericarp, locule, and core) of kiwi fruit (Redgwell et al., 1988). A portion (20–30%) of the cell-wall pectic polysaccharides was solubilised by sequential extraction with CDTA and cold Na_2CO_3. Ion-exchange chromatography in combination with glycosyl-residue and glycosyl-linkage composition analyses established that these pectic polysaccharides were composed of homogalacturonan and RG-I. The neutral oligosaccharides attached to C-4 of the 2,4-linked rhamnosyl residues contained predominantly 4-linked galactosyl residues. A small amount ($\sim 7\%$ of the cell wall) of additional pectic polysaccharide was solubilised from kiwi fruit cell walls by treatment with guanidinium thiocyanate and KOH (Redgwell *et al.*, 1988); this contained a higher proportion of branched glycosyl residues than the polysaccharides solubilised by CDTA and Na_2CO_3. The presence of variously branched pectic polysaccharide in kiwi fruit is characteristic of the variously branched RG-Is that have been extracted from sycamore cell walls (Ishi *et al.*, 1989).

Pectic polysaccharides were extracted from apple fruit cell walls with hot water and ammonium oxalate (Stevens and Selvendran, 1984c; Aspinall and Fanous, 1984). Glycosyl-residue and glycosyl-linkage composition analyses suggested that the pectic polysaccharides were homogalacturonan and RG-I (Stevens and Selvendran, 1984c; Aspinall and Fanous, 1984).

The distribution of neutral and acidic glycosyl residues in pectic polysaccharides isolated from sugar beet was examined by a combination of enzymatic and chemical degradations (Rombouts and Thibault, 1986). An α-1,4-linked galacturonan appeared to contain a small amount of 2,4-linked rhamnosyl residues (RG-I) that were substituted through C-4 with oligosaccharides containing 5-linked α-L-arabinosyl and 4-linked β-D-galactosyl residues (Keenan *et al.*, 1985; Rombouts and Thibault, 1986). Thus, these pectic polysaccharides were probably a mixture of homogalacturonan and RG-I.

Pectic polysaccharides released from the cell walls of tomato by treatment with α-1,4-*endo*-polygalacturonase isolated from tomato tissue were partially separated into homogalacturonan- and rhamnogalacturonan-rich fractions (Pressey and Himmelsbach, 1984). Galactosyl and arabinosyl residues were the dominant neutral glycosyl residues of the rhamnogalacturonan-rich fraction. The galactosyl and arabinosyl residues were shown by carbon-13 NMR to be β-1,4-linked and α-1,5-linked, respectively (Pressey and Himmelsbach, 1984). Thus, again, these polysaccharides appear to have been homogalacturonan and RG-I.

Galacturonans and RG-I were chemically extracted from the walls of suspension-cultured carrot cells (Konno *et al.*, 1986) as well as from the alcohol-insoluble residue of carrot root tissue (Stevens and Selvendran, 1984b; Massiot *et al.*, 1988).

Approximately 35% of the cell wall of onion bulbs, a non-graminaceous monocot, can be solubilised by treatment with CDTA and cold alkali (Redgwell and Selvendran, 1986). Most of the solubilised material was shown, by glycosyl-residue and glycosyl-linkage composition analyses, to be homogalacturonan and RG-I. RG-I was primarily substituted through C-4 of the 2,4-linked rhamnosyl residues with oligosaccharides containing 4-linked galactosyl residues. A neutral 4-linked galactan was detected following fractionation of the alkali extract by ion-exchange chromatography; the galactan may have been cleaved from RG-I by the alkali.

Some of the glycosyl residues of RG-I are *O*-acetylated (Bacic *et al.*, 1988), but the location of the *O*-acetyl groups has not been fully established. Structural studies of the acidic oligosaccharides released from the cell walls of cotton (including disaccharide **1**) by treatment with anhydrous hydrogen fluoride indicated that 40% of the galactosyluronic acid residues of cotton cell wall RG-I were *O*-acetylated on C-3 (Mort *et al.*, 1988).

Some of the cell-wall pectic polysaccharides of a limited number of plant families such as the Chenopodiaceae (e.g. spinach and sugar beet) appear to be esterified with phenolics (e.g. ferulic acid) (Fry, 1987; Bacic *et al.*, 1988). Feruloyl residues attached to the pectic polysaccharides of suspension-cultured spinach cells were released by mild alkaline conditions, indicating that the feruloyl groups were attached to the polysaccharides through ester linkages (Fry, 1982). Feruloyl esters of two disaccharides that were released by treatment of spinach cell walls with Driselase, a fungal enzyme that contains a mixture of *exo-* and *endo-*glycanases, have been characterised as 4-*O*-(6-*O*-feruloyl-β-D-Gal*p*)-D-Gal*p* and 3-*O*-(3-feruloyl-α-L-Ara*p*)-Ara*p*. These feruloylated disaccharides accounted for 60% of the total cell wall ferulic acid. Subsequently, larger oligosaccharides (DP 3–5) containing either arabinosyl or galactosyl residues and esterified ferulic acid were isolated from the spinach cell wall pectic polysaccharide and partially characterised (Fry, 1983).

The evidence summarised above established that the primary cell walls of sycamore, tobacco, *Rosa*, carrot, apple, sugar beet, tomato, kiwi fruit, Douglas fir, maize, and rice all contain homogalacturonan and RG-I, and that the RG-Is of these evolutionarily diverse plants have remarkably conserved structures. All of the RG-Is appear to have a backbone of alternating α-4-linked galacturonosyl and α-2- or 2,4-linked rhamnosyl residues with arabinosyl- and galactosyl-rich side chains attached to C-4 of the 2,4-linked rhamnosyl residues. Side chains also appear to be attached to C-3 of some of the rhamnosyl residues (see, for example, Ishii *et al.*, 1989). Moreover, sycamore and maize RG-Is have similar sets of oligoglycosyl side chains (Lau *et al.*, 1988b; Thomas *et al.*, 1989a). The conservation of the structural features of RG-I suggests that RG-I has a crucial function.

D. Rhamnogalacturonan II (RG-II)

The structure of the complex pectic polysaccharide, RG-II, solubilised from the walls of suspension-cultured sycamore and rice cells by α-1,4-*endo*-polygalacturonase, has been extensively studied. The polysaccharide, composed of approximately 30 glycosyl residues, has extremely complex glycosyl-residue (Table 12.3) and glycosyl-linkage compositions (Table 12.4). Several sugars previously not recognised as components of plant polysaccharides have been identified as components of RG-II. For example, an unidentified glycosyl residue was detected during analysis of the alditol acetates derived from sycamore RG-II (Darvill *et al.*, 1978). This component was subsequently identified as the branched chain acidic monosaccharide 3-*C*-carboxy-5-deoxy-L-xylose (aceric acid, **7**) by a combination of chemical derivatisation and degradation methods, and by GLC-MS, FAB-MS, proton and carbon-13 NMR, and X-ray crystallography (Spellman *et al.*, 1983a). Aceric acid is the only branched, acidic, deoxy sugar to have been identified in nature.

7

The positive reaction of sycamore RG-II in the thiobarbituric acid assay (Karkhanis *et al.*, 1978) first suggested the possible presence of 2-keto-3-deoxyoctulosonic acid (York *et al.*, 1985). As these glycosyl residues are subject to acid-catalysed degradation,

TABLE 12.3. Glycosyl-residue composition of RG-II polysaccharides isolated from different plant sources.

Glycosyl residue[a]	Sycamore[b]	Rice[c]	Pectinol[d]
Rha	12.4	18.2	16.1
Fuc	2.8	4.9	3.7
2MeFuc	3.5	5.3	4.1
Ara	10.0	10.0	15.3
2MeXyl	4.8	7.3	4.1
Apiose	12.2	9.0	10.2
Gal	9.0	12.3	14.5
GlcA	3.2	6.3	6.7
GalA	31.2	26.7	29.3
Aceric acid[d]	3.5	+	+
KDO[d]	3.5	+	+
DHA[d]	3.5	+	+

[a] Values expressed as mol%.
[b] Stevenson *et al.* (1988b).
[c] Thomas *et al.* (1989b).
[d] These glycosyl residues are present in rice and Pectinol RG-II but were not quantified.
+ = present but not quantified.

TABLE 12.4. Glycosyl-linkage composition of RG-II polysaccharides from different plant sources.[a]

Glycosyl linkage[b]		Sycamore[c]	Rice[d]	Pectinol[c]
Rha	T[e]	6.6	8.6	5.0
	2	t	t	1.7
	3	5.7	5.9	6.3
	2,3,4	4.5	3.7	3.1
Fuc	3,4	4.5	4.9	2.3
2MeFuc	T	4.8	5.3	5.5
Apiose	3'	10.9	9.0	10.2
Ara	T$_f$	6.1	5.8	6.3
	T$_p$	0	0	4.9
	2$_p$	5.0	4.2	0
2MeXyl	T	4.5	7.3	4.1
Gal	T	4.9	6.8	5.2
	2,4	5.6	5.5	6.5
	3,4	0	0	2.8
GalA	T	10.3	10.2	10.0
	4	8.8	7.9	6.0
	3,4	7.3	3.1	7.6
	2,4	4.6	2.8	4.2
	2,3,4	1.5	2.7	1.5
GlcA	2	6.3	6.3	6.7

[a] 5-linked KDO and 5-linked DHA are also present in these preparations and account for ~5% of the material.
[b] Values expressed as mol%.
[c] Stevenson *et al.* (1988b).
[d] Thomas *et al.* (1989b).
[e] T = non-reducing terminal rhamnosyl, etc., t = trace.

it was not possible to detect the keto sugar using the conventional alditol acetate procedure for sugar analysis (York et al., 1986). To determine whether this acid-labile molecule was present in RG-II, the alditol acetate method was modified by the inclusion of mild acid hydrolysis and reduction steps (York et al., 1985). By these procedures, the 2-keto-3-deoxyoctulosonic acid was converted to 3-deoxyoctitol, which was then acetylated and analysed by GLC-MS. The derivatives of the partially resolved diastereomeric isomers of the acetylated 3-deoxy-octitol co-chromatographed with and gave mass spectral (EI and CI) fragmentation patterns identical to those of the same derivatives of authentic 2-keto-3-deoxy-D-manno-octulosonic acid (KDO, **8**).

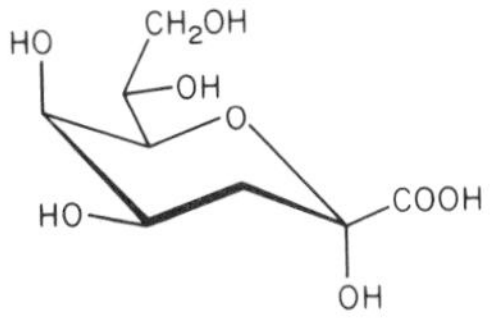

8

The presence of KDO in RG-II was confirmed by analysis of its trimethylsilyl methyl ester (York et al., 1985). This was the first unequivocal demonstration of the presence of KDO in plants and, of course, in a primary cell-wall pectic polysaccharide. By using the modified alditol acetate procedure described above, KDO has been shown to be present in the primary cell walls of a wide range of plants (York et al., 1985; Stevenson et al., 1988a). Since KDO, like aceric acid, is an integral component of RG-II, which has been found in every plant examined for it, it was expected that RG-II would be found in every plant analysed.

The amount of KDO in sycamore RG-II accounts for only 50% of the response in the thiobarbituric acid assay. The remaining 50% is due to the presence of another acid-labile, acidic keto sugar. When the alditol acetates derived from the partial acid hydrolysis products of methylated RG-II were examined by GLC-MS, the mass spectral fragmentation patterns (EI and CI) of two partially resolved diastereomeric components were consistent with the presence in the native polysaccharide of 3-deoxy-2-ulosaric acid. A combination of chemical derivatisation and degradation reactions, GLC-MS and proton NMR established (Stevenson et al., 1988a) that the second acidic keto sugar in RG-II was 3-deoxy-D-lyxo-2-heptulosaric acid (DHA, **9**). It has also been shown, using a modified alditol acetate procedure, that DHA, like KDO and aceric acid, is present in the primary cell walls of each of the diverse group of plants analysed (Stevenson et al., 1988a).

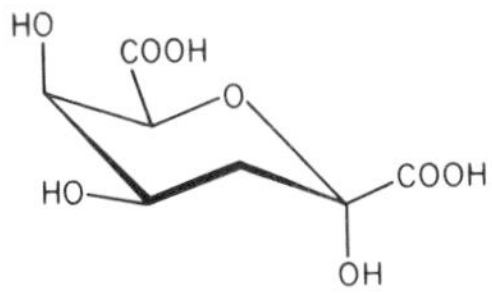

9

Considerable progress has been made toward elucidating the entire sequence of glycosyl residues in sycamore RG-II. The first major step towards elucidating the glycosyl sequence resulted from the selective cleavage of the acid-labile glycosidic linkages of the apiosyl residues of RG-II (Spellman *et al.*, 1983b). The acid-labile keto sugar glycosyl linkages that were later shown to be present in sycamore RG-II were also hydrolysed by the conditions that cleaved the apiosyl residues. However, sequential fractionation of the partial acid hydrolysate by size-exclusion and ion-exchange chromatographies resulted in the purification to homogeneity of an oligosaccharide containing one apiosyl, one arabinosyl, one galactosyl, one aceryl, one 2-*O*-Me fucosyl, and two rhamnosyl residues. FAB-MS analysis established that the oligosaccharide was a heptasaccharide (Spellman *et al.*, 1983b). The heptasaccharide was converted to its hexaglycosyl alditol and deuteriomethylated, which resulted in partial degradation of the aceryl residues (the sugar degraded during this reaction was only later identified as aceric acid) and the release of a number of deuteriomethylated oligoglycosyl glycosides and oligoglycosyl alditols (Spellman *et al.*, 1983b). The deuteriomethylated oligoglycosyl glycosides and oligoglycosyl alditols were separated from the undegraded deuteriomethylated hexaglycosyl alditol by reversed phase HPLC with LC-MS detection. The mixture of deuteriomethylated oligoglycosyl glycosides and oligoglycosyl alditols was then fragmented by partial acid hydrolysis, reduced to partially deuteriomethylated alditols, and deuterioethylated. The partially deuterioethylated and partially deuteriomethylated oligoglycosyl alditols (DP 2–4) were analysed by GLC-MS, and their glycosyl sequences determined. The sequences of these oligoglycosyl alditols allowed six of the glycosyl residues in the heptasaccharide to be defined. Oligoglycosyl alditol derivatives containing aceric acid were not detected (Spellman *et al.*, 1983b).

The sequence of glycosyl residues including the position of the aceryl residue and the anomeric configurations of the glycosyl residues in the intact methylated hexaglycosyl alditol were, with one exception, established by FAB-MS and proton NMR, respectively (Spellman *et al.*, 1983b). Partial acid hydrolysis in combination with methylation analysis of the native heptasaccharide established that the 2-*O*-methyl fucosyl residue was attached to C-2 of the 2,4-linked galactosyl residue, thereby establishing the complete structure of heptasaccharide **10** (Spellman *et al.*, 1983b).

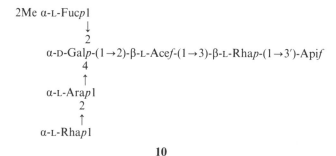

10

Mild acid hydrolysis of RG-II produced, in addition to heptasaccharide II, a disaccharide containing one rhamnosyl residue and one 2-keto-3-deoxy-*manno*-octulosonic acid residue (York *et al.*, 1985). Reduction and methylation of this disaccharide gave a methylated monoglycosyl alditol whose EI-mass spectrum was indistinguishable from that obtained from authentic α-L-Rha*p*-(1,5)-KDO. Proton NMR established the α anomeric configuration of the rhamnosyl residue and, in combination with glycosyl-linkage analysis, established the structure of disaccharide **11** (York *et al.*, 1985).

α-L-Rhap-(1→5)-D-KDOp

11

Alkylated oligoglycosyl alditols were released from methylated and carboxyl-reduced sycamore RG-II by partial acid hydrolysis followed by reduction and acetylation. The presence of Rhap-(1,5)-KDO was confirmed by GLC-MS of the alkylated monoglycosyl alditol (Stevenson *et al.*, 1988a). A second monoglycosyl alditol derivative, composed of a non-reducing terminal arabinosyl and a 5-linked heptulosaryl residue was detected by GLC-MS (EI) and isolated by reversed phase HPLC. The structure of disaccharide **12** was established using a combination of proton NMR and glycosyl-linkage analyses (Stevenson *et al.*, 1988a).

β-L-Araf-(1→5)-D-DHAp

12

Partial acid hydrolysis followed by size-exclusion chromatography was used selectively to remove oligosaccharides **10**, **11** and **12** from sycamore RG-II. The higher molecular weight material remaining in the resulting residue was carboxyl-reduced, methylated, partially acid-hydrolysed, and the oligosaccharide products converted to their partially methylated oligoglycosyl alditols and then re-methylated. The fully methylated oligoglycosyl alditols were separated by reversed phase HPLC using LC-MS detection, and the isolated fractions characterised by proton NMR, FAB-MS, GLC-MS, and glycosyl-linkage analyses (Melton *et al.*, 1986). A total of 23 methylated oligoglycosyl alditols were structurally characterised, and the structures of trisaccharide **13**, pentasaccharide **14** and octasaccharide **15**, all oligosaccharide fragments of RG-II, were established.

L-Araf-(1→3)-D-GalpA-(1→4)-D-GalpA
4
↑

13

α-D-GalpA-(1→4)-[α-D-GalpA]$_{5-7}$-(1→4)-D-GalpA

14

α-D-Galp1   β-D-GalpA1
↓   ↓
2   3
β-D-GlcpA-(1→4)-α-L-Fucp-(1→4)-β-L-Rhap-(1→3')-Apif
3   2
↑   ↑
2Me α-D-Xylp1   α-D-GalpA1

15

An acid-labile 2-*O*-Me-xylosyl residue detected during analysis of RG-II by the formation of its alditol acetate (Darvill *et al.*, 1978) was not present in any of the oligosaccharide fragments of RG-II derived by partial acid hydrolysis. Degradation of

RG-II in ethylenediamine with lithium metal and recovery and structural analysis of the resulting oligosaccharide products demonstrated (Stevenson *et al.*, 1988b) that the 2-*O*-Me-xylosyl residue was attached to C-3 of the 3,4-linked fucosyl residue (see structure **15**).

Experiments have been performed which provided information about the organisation in RG-II of oligosaccharides **10**, **11**, **12**, **13**, **14** and **15**. Partial acid hydrolysis of methylated, carboxyl-reduced, and re-methylated sycamore RG-II, in combination with selective deuteriomethyl labelling of those hydroxyl groups exposed by the partial acid hydrolysis, established the points of attachment of oligosaccharides **10**, **11**, **12** and **15** to the 4-linked galactosyluronic acid residues of **14**, the RG-II backbone (Stevenson *et al.*, 1988b). Disaccharides **11** and **12** were found to be attached to C-3 of the 3,4-linked galactosyluronic acid residues in the homogalacturonan backbone, whereas oligosaccharides **10** and **15** were found to be attached to C-2 of the 2,4-linked galactosyluronic acid residues of the backbone (Stevenson *et al.*, 1988b). Oligosaccharide **13** appears to constitute an arabinosyl side chain to C-3 of the backbone. However, the exact galacturonosyl residues to which oligosaccharides **10**, **11**, **12**, **13** and **15** are attached have not been determined.

A pectic polysaccharide, accounting for approximately 0.1% of the weight of suspension-cultured rice cell walls, has been extracted from the walls with CDTA. The polysaccharide, purified by size-exclusion chromatography, had glycosyl-residue (Table 12.3) and glycosyl-linkage compositions (Table 12.4) indistinguishable from those of sycamore RG-II. Disaccharides **11** and **12**, heptasaccharide **10**, and octasaccharide **15** were identified in the partial acid-hydrolysis products of rice RG-II. The glycosyl sequence of RG-II from sycamore, a dicot, and rice, a monocot, are therefore closely related if not identical.

RG-II has been isolated from the commercial enzyme preparation Pectinol (Spellman *et al.*, 1983a,b). The enzymes in Pectinol were secreted by *Aspergillus niger* when the fungus was grown on the insoluble (wall) residues of sugar beet, apple and citrus pulp following extraction of the water-soluble components of these tissues. The RG-II in Pectinol, which must have arisen from these plant cell walls following digestion of the walls by the fungus, was largely resistant to the *A. niger* enzymes. A comparison of the glycosyl-linkage composition of sycamore RG-II, rice RG-II, and Pectinol RG-II (Table 12.4) shows that, with the exception of the terminal rhamnosyl residue of oligosaccharide **10**, the polysaccharides are indistinguishable. Sycamore and rice RG-II contain 2-linked arabinopyranosyl residues and a non-reducing terminal rhamnosyl residue (see **10**). In Pectinol RG-II the 2-linked arabinopyranosyl residue has been converted into a non-reducing terminal arabinopyranosyl residue by removal of the terminal rhamnosyl residue. This structural difference undoubtedly resulted from the action of an enzyme (α-L-rhamnosidase) known to be present in the Pectinol preparation (D. Gollin, this laboratory, pers. commun.).

The RG-IIs obtained from the primary cell walls of sycamore and rice have been more completely characterised than those from other plants. The presence in the cell walls of bean, apple, tomato, cabbage, potato, Douglas fir, and maize of the RG-II-characteristic monosaccharides 2-*O*-Me xylose, 2-*O*-Me fucose, apiose, aceric acid, KDO and DHA provides strong evidence that RG-II is present in the primary walls of all of these plants and, therefore, of all higher plants (York *et al.*, 1985; Stevenson *et al.*, 1988b; Thomas *et al.*, 1989b). Apiose, 2-*O*-Me xylose, 2-*O*-Me fucose, and aceric acid were also detected in cell walls derived from ferns (White *et al.*, 1986; A. Koller, unpubl. res., this laboratory), and, thus, RG-II is also present in lower plants. RG-II accounts for 4% or less of the primary cell walls of dicots and less than 1% of the cell walls of rice, a monocot. The primary structure of RG-II is closer to being fully elucidated than the structure of any other plant cell-wall polysaccharide except cellulose. Considering the low levels of this polysaccharide in the cell wall, the energy and the large number of enzymes needed to synthesise this polysaccharide, and the precise structure of RG-II, any structural role served by RG-II is unlikely to be its primary function.

A pectic polysaccharide containing galactosyluronic acid and apiosyl residues was extracted by ammonium oxalate from the cell wall of duckweed (*Lemna minor*) (Hart and Kindel, 1970a,b). The ammonium oxalate-soluble polysaccharide accounted for 14% of the wall and contained 20% of the apiose originally present in the cell wall. A considerable amount (76%) of apiose remained in the cell wall residue even after further extraction with hot water and hot ammonium oxalate. Apiogalacturonans were purified from the ammonium oxalate extract by ion-exchange chromatography and fractional precipitation with NaCl, and shown to contain galactosyluronic acid and apiosyl residues in a ratio of 1.0:0.8 (Hart and Kindel, 1970b). Mild acid hydrolysis of the apiogalacturonan gave three products, apiose, apiobiose and homogalacturonan (Hart and Kindel, 1970b). The galacturonan was hydrolysed with a crude pectinase, providing some evidence that the galactosyluronic acid residues were α-1,4-linked. Periodate oxidation of the released apiobiose in combination with proton NMR analysis suggested that the disaccharide had the structure β-D-Api$_f$-(1→3')-D-Api. It was not established whether the apiobiosyl residues were linked to C-2 or C-3 of the galacturonan backbone or if there was a regular distribution of the apiosyl residues along the galacturonan backbone (Hart and Kindel, 1970b). Since no oligosaccharide containing both apiosyl and galacturonosyl residues was isolated and characterised, it cannot be certain the apiosyl residues were covalently attached to the galactosyluronic acid residues. If apiogalacturonans, linear galacturonans with apiobiosyl side chains, are a common constituent of primary cell walls, they would be the second substituted galacturonan in the walls.

An apiogalacturonan was isolated from eel grass as part of a complex pectic polysaccharide containing homogalacturonan and RG-I (Ovodov *et al.*, 1975). In addition to apiosyl residues, this pectic polysaccharide contained xylosyl, arabinosyl, galactosyl, and 2-*O*-Me xylosyl residues (Ovodov *et al.*, 1975). Since 2-*O*-Me xylosyl residues have only been conclusively associated with RG-II (Darvill *et al.*, 1978), these workers may well have isolated RG-II rather than an apiogalacturonan.

A polysaccharide composed of xylosyl (22%), galactosyluronic acid (66%) and arabinosyl (1.5%) residues was extracted from mountain pine pollen with aqueous 12% KOH (Bouveng, 1965). Acidic oligosaccharides (DP 2–4), released by treatment of the polysaccharide with a commercial pectinase enzyme mixture, were structurally characterised by glycosyl-residue and glycosyl-linkage composition analyses, periodate oxidation and enzymatic hyrdolysis. The results of these analyses suggested that the polysaccharide had an α-1,4-linked galacturonan backbone to which non-reducing terminal xylosyl residues were attached to C-3 of every second galactosyluronic acid residue (Bouveng, 1965). It was not established whether this substituted galacturonan was derived from the primary cell wall. The pseudoaldobiouronic acid, D-Xyl*p*-(1,3)-D-Gal*p*A, was isolated from the partially purified pectins of soybean (Aspinall *et al.*,

1967a,b; Kikuchi and Sugimoto, 1976) and lemon peel (Aspinall *et al.*, 1968a). Although about 90% of the galactosyluronic acid residues of the pectins extracted chemically from the alcohol-insoluble residue of apple (DeVries *et al.*, 1982) occurred as chains of homogalacturonan, degradation (β-elimination) of these pectins with hot aqueous base followed by size-exclusion chromatography produced pectic fractions enriched in galactosyluronic acid and xylosyl residues (DeVries *et al.*, 1983). The structures of these pectic fractions were not determined. None of the xylogalacturonans was extracted from isolated primary cell walls, and none of the putative xylogalacturonans was highly purified before fragmentation. We conclude that xylogalacturonans are a third possible substituted galacturonan of primary cell walls, but that their presence in primary walls and their structural characteristics have not been firmly established.

IV. ARABANS, GALACTANS AND ARABINOGALACTANS

Arabans have been isolated from the tissues and cell walls of many dicotyledons (Darvill *et al.*, 1980; Stephen, 1983). It is not known whether arabans exist in primary cell walls as separate homopolysaccharides, or only as covalently linked side chains of RG-I (Stephen, 1983). An araban has not been isolated from primary cell walls without using extraction conditions capable of breaking the covalent connections of the arabans to RG-I.

The structures of arabans have been investigated by methylation analysis, Smith degradation, and proton and carbon-13 NMR spectrometry (Darvill *et al.*, 1980; Stephen, 1983). Arabans are highly branched polysaccharides composed of α-1,5-linked chains of arabinofuranosyl residues that are more often substituted at C-3 than at C-2 (Rees and Richardson, 1966; Aspinall and Cottrell, 1971). The degree of branching of the 5-linked backbone appears to be dependent on the source of the araban. Non-reducing terminal and branched arabinosyl residues accounted for only 8% of the araban isolated from apple juice (Churms *et al.*, 1983). However, apple juice is rich in *endo-* and *exo-*glycanases so that apple juice araban is likely to have been partially deglycosylated. Evidence for the modification of apple juice araban was provided by the hot-water extraction of a highly branched araban from the alcohol-insoluble residue of apple fruit. The highly branched araban was purified by ion-exchange chromatography (Aspinall and Fanous, 1984). The carbon-13 NMR spectrum of the araban suggested a branched structure containing α-L-arabinofuranosyl residues. Glycosyl-linkage analysis showed that non-reducing terminal and 3,5-linked arabinosyl residues accounted for 56% of the polysaccharide.

The distribution in apple araban of the 3-, 3,5- and 5-linked arabinosyl residues has been investigated by Smith degradation (Goldstein *et al.*, 1965), as the 3- and 3,5-linked residues are resistant to periodate oxidation. In this procedure, the araban was subjected to periodate oxidation, sodium borohydride reduction, and then partial acid hydrolysis to yield a mixture of oligoarabinosyl glycerols (from the 5-linked residues). The oligoarabinosyl glycerols were methylated and examined by GLC-MS (EI). Three compounds α-L-Araf-(1→1)-glycerol, α-L-Araf-(1→5)-α-L-Araf-(1→1)-glycerol, and α-L-Araf-(1→3)-α-L-Araf-(1→1)-glycerol, were tentatively identified (Aspinall and Fanous, 1984). The results of these investigations indicated that the dominant structural feature of the araban isolated from apple cell walls was a backbone of 5-linked arabinofuranosyl residues to which terminal or oligoarabinosyl residues were attached at C-3.

Araban extracted by cold water from marsh mallow (*Althaea officinalis*) and purified by ion-exchange and size-exclusion chromatographies contained a high proportion (67%) of non-reducing terminal and branched 3,5-linked arabinosyl residues (Capek *et al.*, 1983). All of the arabinosyl residues were shown by carbon-13 NMR to have α anomeric configuration. Most of the arabinosyl residues in plants are in the furanosyl ring form, but about 1% of the arabinosyl residues of marsh mallow araban were shown, by glycosyl-linkage analysis, to be in the pyranosyl ring form. The presence of arabinopyranosyl residues appears to be dependent on the source of araban (Darvill *et al.*, 1980).

Polysaccharides rich in arabinosyl residues were extracted with hot water from the cell wall of horsebean root (Joseleau *et al.*, 1983). The glycosyl-residue composition analysis and carbon-13 NMR spectra of the arabans indicated that they were composed predominantly of α-L-arabinosyl residues. One of the arabans contained about 15% galactosyluronic acid residues, suggesting that the arabans may have originated from a pectic polysaccharide.

A number of complex pectic polysaccharides isolated from the cell walls of plants have been shown to contain a high proportion of araban (Darvill *et al.*, 1980; Stephen, 1983; Selvendran, 1985). However, direct evidence for the covalent linkage of arabinosyl residues to rhamnogalacturonan has only been obtained with RG-I isolated from the walls of suspension-cultured sycamore cells. Sycamore RG-I contains oligosaccharides (DP 2–5) composed of arabinosyl residues attached to C-4 of the 2,4-linked rhamnosyl residues in the polysaccharide's backbone (Lau *et al.*, 1988b). The glycosyl-linkage pattern of the arabinosyl residues linked to RG-I is similar to that of the arabans. The ratios of the different glycosyl linkages (e.g. non-reducing terminal, 5-, and 3,5-linked arabinosyl residues) appear to vary depending on the plant source (Darvill *et al.*, 1980; Thomas *et al.*, 1989a). The attachment to RG-I of larger oligoarabinosides (DP > 5) or of even larger arabans has not been demonstrated, although larger oligosaccharides of undetermined composition and structure are known to be attached to the rhamnosyl residues of the RG-I backbone (Lau *et al.*, 1988b).

Galactans have been isolated from intact tissue and isolated cell walls of a number of plants (Darvill *et al.*, 1980; Stephen, 1983). As with arabans, homogalactans have not been isolated from primary cell walls under conditions in which cleavage of the galactan from RG-I can be discounted.

The presence of sequences of β-1,4-linked galactosyl residues in primary cell walls was demonstrated by results of glycosyl-linkage analyses and enzymatic cleavage (*endo*-β-1,4-galactanase) of intact cell walls and of polysaccharides extracted from cell walls (Toman *et al.*, 1972; Labavitch *et al.*, 1976; DeVries *et al.*, 1983). Galactosyl-containing oligosaccharides were shown to be attached to C-4 of 2,4-linked rhamnosyl residues of sycamore RG-I (McNeil *et al.*, 1982a; Lau *et al.*, 1988b). The RG-I side chains were variously composed of only galactosyl residues (e.g. **2** and **6**) or of a mixture of galactosyl and other neutral glycosyl residues (**3** and **4**). The attachment to RG-I of larger oligogalactosides or galactans has not been demonstrated. However, partial acid hydrolysis of tobacco RG-I released neutral oligosaccharides (DP 2–12) composed of 4-linked β-D-galactosyl residues (Eda *et al.*, 1986). The attachment of these oligogalactosyl residues to C-4 of 2,4-linked rhamnosyl residues in the rhamnogalacturonan backbone was only inferred from the presence of the branched 2,4-linked rhamnosyl residues in the glycosyl-linkage analysis of tobacco RG-I (Eda *et al.*, 1986). If this possibility is substantiated, it will support the hypothesis that galactans are covalently linked side chains of RG-I.

The glycosyl-linkage patterns of the galactosyl residues in polysaccharides extracted from plants, like those of arabinosyl residues, are dependent on the source. Cell walls prepared from tobacco (Eda *et al.*, 1986), onion (Redgwell and Selvendran, 1986), carrot (Massiot *et al.*, 1988), and kiwi fruit (Redgwell *et al.*, 1988) are rich in 4-linked galactosyl residues. The galactosyl residues of RG-I, isolated from the walls of suspension-cultured sycamore cells, are predominantly non-reducing terminal, 4- and 6-linked (York *et al.*, 1986). The RG-Is isolated from the cell walls of maize (Thomas *et al.*, 1989a) and Douglas fir (Thomas *et al.*, 1987) contain a high proportion of non-reducing terminal galactosyl residues, but also a variety of differently linked galactosyl residues; no particular linear or branched galactosyl residue predominates. The galactosyl residues of RG-I isolated from the walls of suspension-cultured rice cells are predominantly non-reducing terminal, 4- and 3,6-linked (Thomas *et al.*, 1989b). Interestingly, the pectic polysaccharides isolated from the cell walls of rice endosperm contained relatively high proportions of non-reducing terminal and 4-linked galactosyl residues; 3,6-linked galactosyl residues were not detected in the rice endosperm polysaccharides (Shibuya and Nakane, 1984).

Two types of arabinogalactans have been isolated from plants, but their presence in the primary cell wall remains to be established (Darvill *et al.*, 1980). Arabinogalactan I is a cell-wall polysaccharide composed of a β-1,4-linked galactosyl backbone that is substituted through C-3 with arabinosyl and galactosyl side chains (Aspinall, 1980; Stephen, 1983). Arabinogalactan II is a highly branched polysaccharide containing 3-, 6-, and 3,6-linked galactosyl residues with variable amounts of arabinosyl, galactosyl-uronic acid and glucosyluronic acid residues (Aspinall, 1980). Arabinogalactan II is abundant in gymnosperms and angiosperms (Stephen, 1983), but arabinogalactan II is generally not considered a structural component of the primary cell wall. Arabinogalactan II has been found in the lumen of tracheids, cytoplasmic granules, and intercellular vesicles, and at the cytoplasm–cell wall interface of various plants (Clarke *et al.*, 1979). Cultured plant cells are known to secrete arabinogalactan II into the culture medium (Aspinall *et al.*, 1969; Keegstra *et al.*, 1973; Burke *et al.*, 1974; Kato *et al.*, 1977), and this polysaccharide may be covalently attached to hydroxyproline-containing proteins (Pope, 1977; Stephen, 1983).

The presence of arabinogalactans in the primary cell walls of plants has been inferred from the results of glycosyl-linkage analyses of isolated cell walls and of the polysaccharides extracted from isolated cell walls (Talmadge *et al.*, 1973). The pectic polysaccharides released from the walls of suspension-cultured sycamore cells by *endo*-polygalacturonase were relatively rich in 4- and 6-linked galactosyl residues and possessed small amounts of 3-, 6- and 3,6-linked galactosyl residues. These are the same linkages and in the same ratios as the galactosyl linkages found in the oligosaccharide side chains of sycamore RG-I (McNeil *et al.*, 1982a; Lau *et al.*, 1988b). Two pectic fractions extracted from the cell wall of Douglas fir with LiCl contained appreciable quantities of 3- and 3,6-linked galactosyl residues, respectively (Thomas *et al.*, 1987). It was not established whether these galactosyl residues were derived from arabinogalactan II or from oligosaccharides linked to RG-I. Thus, there may be no free arabinogalactan in the wall, but, rather, long arabinogalactan side chains may be attached to the RG-I backbone.

Small amounts of cell-wall pectic polysaccharides have been isolated which, after repeated attempts at purification, were still associated with other polysaccharides (e.g. xyloglucan) and protein (Selvendran, 1985). The possibility that some pectic poly-saccharides are complexed with or covalently bound to other cell-wall macromolecules lends support to the suggestion (Keegstra et al., 1973; Fry, 1987) that covalent cross-linkages between different polysaccharides and structural protein are important in maintaining the structural integrity of the primary cell wall. An oligosaccharide of sufficient size, e.g. >25 residues, that is covalently linked to another polysaccharide, would probably be of separate synthetic origin and cross-linked after synthesis (perhaps in the cell wall). Thus, we would consider that arabans, galactans and arabinogalactans, if they are of sufficient size to suggest that they had an independent synthetic origin, are distinguishable polysaccharides in their own right even if they are covalently linked to RG-I. Evidence for polysaccharide–polysaccharide or polysaccharide–protein cross-linkages in plant cell walls will only be convincing when oligosaccharides (or glyco-peptides) are extracted from the cell wall, purified to apparent homogeneity, and shown to contain fragments from two different, well-defined, cell wall polymers.

V. FUTURE RESEARCH

This chapter describes in some detail the three pectic polysaccharides known to be present in the primary cell walls of plants: homogalacturonan, RG-I, and RG-II. Although much is known about the structures of these polysaccharides, the complete glycosyl sequences of RG-I and RG-II remain to be determined and, despite modern analytical techniques, that remains a considerable challenge.

Some of the questions remaining about RG-I include:

1. What are the structures of all RG-I oligosaccharide side chains, particularly those larger than pentasaccharides? Are there arabans, galactans and/or arabinogalac-tans covalently linked to RG-I through the C-4 of rhamnosyl residues in the backbone?
2. What non-carbohydrate substituents are present on RG-I and on which carbons of which glycosyl residue are they located?
3. How are the many different oligosaccharide side chains distributed along the backbone of a single RG-I molecule and between the backbones of different RG-I molecules? How exact are the side chain structures and side chain distributions of RG-I?
4. Are there families of RG-I which differ in the nature of the oligosaccharide side chains? Do RG-Is from cell to cell and tissue to tissue in the same plant have different arrays of side chains?
5. Where is RG-I located in primary cell walls? Is it only in the middle lamella? Is its distribution different in the walls of cells from different tissues?
6. What are the favoured solution conformations (three-dimensional shapes) of RG-I?
7. What are the structural, regulatory and other functions of RG-I?

Considerably more is known about the structure of the much smaller RG-II. Oligosaccharide fragments that have been isolated from RG-II account for all the glycosyl residues known to be present in this polysaccharide. The exact molecular weight of RG-II must be obtained in order to determine whether RG-II contains any other acid-labile glycosyl or non-carbohydrate constituents. Indeed, at least one isolated RG-II oligosaccharide side chain possesses at least two *O*-acetyl substituents (Spellman *et al.*, 1983b). The exact galactosyluronic acid residues of the backbone to which each of the oligosaccharide side chains is attached remains to be determined.

The keto sugars KDO (**8**) and DHA (**9**), known to be present in RG-II, are structurally related. Do they have common biosynthetic precursors? How are the cell-wall pectic polysaccharides synthesised? Little is known about the biosynthesis of these important biomolecules, and no plant cell-wall polysaccharide has been convincingly synthesised in a cell-free system.

It remains to be determined whether homogalacturonan, RG-I, RG-II, and the arabans, galactans, and arabinogalactans are part of a single pectic polysaccharide primary cell-wall complex or whether some of them exist as individual polysaccharides or at different locations in the cell wall. The pectic polysaccharides that have been chemically extracted from plant cell walls all contain a high proportion of galactosyluronic acid residues. Enzymatically solubilised RG-I and RG-II are closely associated with chains of homogalacturonan, and both are solubilised by *endo*-polygalacturonases. Thus, homogalacturonan, RG-I, and RG-II may be covalently linked to one another, and we discussed above the evidence that araban, galactan, and arabinogalactans may be covalently attached to RG-I. Major advances in our knowledge of the structure of primary cell walls will depend on the availability of an array of cell wall-cleaving enzymes to help dissect the wall structure.

Structural analyses have established that the three pectic polysaccharides isolated from the cell walls of dicots, monocots, and gymnosperms are structurally identical or very closely related, although the primary walls of different plants appear to contain very different amounts of the three pectic polysaccharides. The high degree of structural conservation of these polysaccharides in evolutionary diverse plants points to these molecules having important biological functions.

Portions of pectic polysaccharides, notably oligosaccharides composed of 4-linked galactosyluronic residues, have recently been shown to have biological activity in plants (Darvill and Albersheim, 1984). Other functions of pectic polysaccharides might be more rapidly elucidated if mutants possessing altered synthesis or metabolism of wall polysaccharide were available. Since hundreds of genes are likely to be involved in the synthesis of the primary cell walls, mutations should be frequent. The fact that no mutant of cell-wall synthesis or metabolism has been reported suggests that such mutants are lethal. However, fewer cell-wall mutants are likely to be lethal in suspension-cultured cells grown in an osmotically controlled medium, but the absence of a way to select positively for interesting mutants makes the task difficult. A laboratory could dedicate itself to studying the synthesis of any of the non-cellulosic polysaccharides of primary cell walls. The availability of mutants would likely be of considerable assistance to such a programme. Recently available knowledge of the structures of the homogalacturonan, RG-I, and RG-II should also facilitate studies of their biosynthesis and make it possible to begin studying the three-dimensional conformations of these molecules and how these molecules interact with other cell-wall polymers. Only when the structure, biosynthesis, and functions of primary cell walls are known will a fundamental understanding of plant growth, morphogenesis, and resistance to pests become possible.

ACKNOWLEDGEMENTS

The research reported in this article was supported in part by US Department of Energy grant DE-FG09-87ER13810 as part of the DOE/NSF/USDA Plant Science Centers program; and by Department of Energy grant DE-FG09-85ER13426. We would like to thank D. Gollin, A. Koller, R. Lo, and W. York of this laboratory for making available unpublished results, and R. Nuri for editorial assistance.

REFERENCES

Albersheim, P., Muhlethaler, K. and Frey-Wyssling, A. (1960). *Biochem. Cytol.* **8**, 501–506.
Aspinall, G. O. (1962). *Ann. Rev. Biochem.* **31**, 79–102.
Aspinall, G. O. (1970). "Polysaccharides". Pergamon Press, Oxford.
Aspinall, G. O. (1980). *In* "The Biochemistry of Plants, A Comprehensive Treatise" (P. K. Stumpf and E. E. Conn, eds), Vol. 3, pp. 473–500. Academic Press, New York.
Aspinall, G. O. (1982). *In* "The Polysaccharides" (G. O. Aspinall, ed.), Vol. 1, pp. 35–131. Academic Press, New York.
Aspinall, G. O. (1983). *In* "The Polysaccharides" (G. O. Aspinall, ed.), Vol. 2, pp. 1–9. Academic Press, New York.
Aspinall, G. O. and Chaudhari, A. S. (1975). *Can. J. Chem.* **53**, 2189–2193.
Aspinall, G. O. and Cottrell, I. W. (1971). *Can. J. Chem.* **49**, 1019–1022.
Aspinall, G. O. and Fanous, H. K. (1984). *Carbohydr. Polymers* **4**, 193–214.
Aspinall, G. O. and Fanshawe, R. S. (1961). *J. Chem. Soc.*, 4215–4225.
Aspinall, G. O. and Jiang, K.-S. (1974). *Carbohydr. Res.* **38**, 247–255.
Aspinall, G. O. and Rosell, K. G. (1977). *Carbohydr. Res.* **57**, C23–C26.
Aspinall, G. O., Begbie, R., Hamilton, A. and Whyte, J. N. C. (1967a). *J. Chem. Soc.*, 1065–1070.
Aspinall, G. O., Hunt, K. and Morrison, I. M. (1967b). *J. Chem. Soc.*, 1080–1086.
Aspinall, G. O., Craig, J. W. T. and Whyte, J. L. (1968a). *Carbohydr. Res.* **7**, 442–452.
Aspinall, G. O., Gestetner, B., Molloy, J. A. and Uddin, M. (1968b). *J. Chem. Soc.*, 2554–2559.
Aspinall, G. O., Molloy, J. A. and Craig, J. W. T. (1969). *Can. J. Biochem.* **47**, 1063–1070.
Bacic, A., Harris, P. J. and Stone, B. A. (1988). *In* "The Biochemistry of Plants" (J. Preiss, ed.), Vol. 14, pp. 297–371. Academic Press, San Diego.
Barrett, A. J. and Northcote, D. H. (1965). *Biochem. J.* **94**, 617–627.
Bhattacharjee, S. S. and Timell, T. E. (1964). *Can. J. Chem.* **43**, 758–765.
Bouveng, H. O. (1965). *Acta Chem. Scand.* **19**, 953–963.
Burke, D., Kaufman, P., McNeil, M. and Albersheim, P. (1974). *Plant Physiol.* **54**, 109–115.
Capek, P., Toman, R., Kadosova, A. and Rosik, J. (1983). *Carbohydr. Res.* **117**, 133–140.
Chambat, G. and Joseleau, J.-P. (1980). *Carbohydr. Res.* **85**, C10–C12.
Chambat, G., Joseleau, J.-P. and Barnoud, F. (1981). *Phytochemistry* **20**, 241–246.
Chambat, G., Barnoud, F. and Joseleau, J.-P. (1984). *Plant Physiol.* **74**, 687–693.
Chapman, H. D., Morris, V. J., Selvendran, R. R. and O'Neill, M. A. (1987). *Carbohydr. Res.* **165**, 53–68.
Chesson, A., Gordon, A. H. and Lomax, J. A. (1985). *Carbohydr. Res.* **141**, 137–147.
Churms, S. C., Merrifield, E. H., Stephen, A. M., Walwyn, D. R., Polson, A., van der Merwe, H. C. and Spies, H. S. C. (1983). *Carbohydr. Res.* **113**, 339–344.
Clarke, A. E., Anderson, R. L. and Stone, B. A. (1979). *Phytochemistry* **18**, 521–540.
Corey, E. J. and Kim, C. U. (1973). *Tetrahedron Lett.* **12**, 919–922.
Darvill, A. G. and Albersheim, P. (1984). *Ann. Rev. Plant Physiol.* **35**, 243–275.

Darvill, A. G., McNeil, M. and Albersheim, P. (1978). *Plant Physiol.* **62**, 418–422.
Darvill, A. G., McNeil, M., Albersheim, P. and Delmer, D. (1980). *In* "The Biochemistry of Plants" (N. E. Tolbert, ed.), Vol. 1, pp. 91–162. Academic Press, New York.
Dell, A., (1987). *Adv. Carbohydr. Chem. Biochem.* **45**, 19–72.
DeVries, J. A., Rombouts, F. M., Voragen, A. G. J. and Pilnik, W. (1982). *Carbohydr. Polymers* **2**, 25–34.
DeVries, J. A., Den Uijl, C. H., Voragen, A. G. J., Rombouts, F. M. and Pilnik, W. (1983). *Carbohydr. Polymers* **3**, 193–206.
DeVries, J. A., Hanson, M., Soderberg, J., Glahn, P.-E. and Pederson, J. K. (1986). *Carbohydr. Polymers* **6**, 165–176.
Eda, S., Miyabe, K., Akiyama, Y., Ohnishi, A. and Kato, K. (1986). *Carbohydr. Res.* **158**, 205–216.
Fry, S. C. (1982). *Biochem. J.* **203**, 493–504.
Fry, S. C. (1983). *Planta* **157**, 111–123.
Fry, S. C. (1987). *Ann. Rev. Plant Physiol.* **37**, 165–186.
Goldstein, I. J., Hay, G. W., Lewis, B. A. and Smith, F. (1965). *Methods Carbohydr. Chem.* **5**, 361–370.
Hahn, M. G., Darvill, A. G. and Albersheim, P. (1981). *Plant Physiol.* **68**, 1161–1169.
Hakomori, S. (1964). *J. Biochem. (Tokyo)* **55**, 205–207.
Hart, D. A. and Kindel, P. K. (1970a). *Biochem. J.* **116**, 569–579.
Hart, D. A. and Kindel, P. K. (1970b). *Biochemistry* **9**, 2190–2196.
Ishii, T., Thomas, J. R., Darvill, A. G. and Albersheim, P. (1989). *Plant Physiol.* **89**, 421–428.
Jarvis, M. C. (1984). *Plant Cell Environ.* **7**, 153–164.
Jarvis, M. C., Forsyth, W. and Duncan, H. J. (1988). *Plant Physiol.* **88**, 309–314.
Joseleau, J.-P., Chambat, G. and Lanvers, M. (1983). *Carbohydr. Res.* **122**, 107–113.
Karkhanis, Y. D., Zeltner, J. Y., Jackson, J. L. and Carlo, D. J. (1978). *Anal. Biochem.* **85**, 595–601.
Kato, K., Watanabe, F. and Eda, S. (1977). *Agric. Biol. Chem.* **41**, 533–538.
Keegstra, K., Talmadge, K. W., Bauer, W. D. and Albersheim, P. (1973). *Plant Physiol.* **51**, 188–196.
Keenan, M. H. J., Belton, P. S., Matthew, J. A. and Howson, S. J. (1985). *Carbohydr. Res.* **138**, 168–170.
Kertesz, Z. I. (1951). "The Pectic Substance". Interscience Publishers, London.
Kikuchi, T. and Sugimoto, H. (1976). *Agric. Biol. Chem.* **40**, 87–92.
Konno, H., Yamasaki, Y. and Katoh, K. (1986). *Phytochemistry* **25**, 623–627.
Labavitch, J. M., Freeman, L. E. and Albersheim, P. (1976). *J. Biol. Chem.* **251**, 5904–5910.
Lau, J. M., McNeil, M., Darvill, A. G. and Albersheim, P. (1985). *Carbohydr. Res.* **137**, 111–125.
Lau, J. M., McNeil, M., Darvill, A. G. and Albersheim, P. (1988a). *Carbohydr. Res.* **168**, 219–243.
Lau, J. M., McNeil, M., Darvill, A. G. and Albersheim, P. (1988b). *Carbohydr. Res.* **168**, 245–274.
Leontein, K. B., Lindberg, B. and Lonngren, J. (1978). *Carbohydr. Res.* **62**, 359–362.
Lindberg, B. (1972). *Methods Enzymol.* **28**, 178–195.
Lindberg, B. and Lonngren, J. (1976). *Methods Carbohydr. Chem.* **7**, 142–148.
Lindberg, B. and Lonngren, J. (1978). *Methods Enzymol.* **50**, 3–33.
Lonngren, J. and Svensson, S. (1974). *Adv. Carbohydr. Chem. Biochem.* **29**, 41–106.
Massiot, P., Rouau, X. and Thibault, J.-F. (1988). *Carbohydr. Res.* **172**, 229–242.
McNeil, M., Darvill, A. G. and Albersheim, P. (1980). *Plant Physiol.* **66**, 1128–1134.
McNeil, M., Darvill, A. G. and Albersheim, P. (1982a). *Plant Physiol.* **70**, 1586–1591.
McNeil, M., Darvill, A. G., Aman, P., Franzen, L.-E. and Albersheim, P. (1982b). *Methods Enzymol.* **83**, 3–45.
McNeil, M., Darvill, A. G., Fry, S. C. and Albersheim, P. (1984). *Ann. Rev. Biochem.* **53**, 625–663.
Melton, L. D., McNeil, M., Darvill, A. G. and Albersheim, P. (1986). *Carbohydr. Res.* **146**, 279–305.
Moore, P. J. and Staehelin, L. A. (1988). *Planta* **174**, 433–445.

Moore, P. J., Darvill, A. G., Albersheim, P. and Staehelin, L. A. (1986). *Plant Physiol.* **82**, 787–794.

Mort, A. J. and Bauer, W. D. (1982). *J. Biol. Chem.* **257**, 1870–1875.

Mort, A. J., Komalavilas, P., Maness, N., Ryan, J., Moerschbacher, B. and An, J. (1988). *Abstr. A6, XIV International Carbohyr. Symp.*, Stockholm, Sweden.

Northcote, D. H. (1963). *Int. Rev. Cytol.* **14**, 223–241.

Nothnagel, E., McNeil, M. and Albersheim, P. (1983). *Plant Physiol.* **71**, 916–926.

Ovodov, Y. S., Ovodova, R. G., Shivaeva, V. I. and Micheyskaya, L. V. (1975). *Carbohydr. Res.* **42**, 197–199.

Pope, D. G. (1977). *Plant Physiol.* **59**, 894–900.

Pressey, R. and Himmelsbach, D. S. (1984). *Carbohydr. Res.* **127**, 356–359.

Redgwell, R. J. and Selvendran, R. R. (1986). *Carbohydr. Res.* **157**, 183–199.

Redgwell, R. J., Melton, L. D. and Brasch, D. J. (1988). *Carbohydr. Res.* **182**, 241–258.

Rees, D. A. and Richardson, N. G. (1966). *Biochemistry* **5**, 3099–3115.

Roland, J. C. and Vian, B. (1981). *J. Cell Sci.* **48**, 333–343.

Rombouts, F. M. and Pilnik, W. (1980). *In* "Economic Microbiology" (A. H. Rose, ed.), Vol. 5, pp. 228–283. Academic Press, London.

Rombouts, F. M. and Thibault, J.-F. (1986). *Carbohydr. Res.* **154**, 189–203.

Selvendran, R. R. (1975). *Phytochemistry* **14**, 1011–1017.

Selvendran, R. R. (1985). *J. Cell Sci.* **2**, 51–88.

Selvendran, R. R. and O'Neill, M.A. (1987). *In* "Methods of Biochemical Analysis" (D. Glick, ed.), Vol. 32, pp. 25–153. John Wiley & Sons, London.

Shibuya, N. and Nakane, R. (1984). *Phytochemistry* **23**, 1425–1429.

Spellman, M. W., McNeil, M., Darvill, A. G., Albersheim, P. and Henrick, K. (1983a). *Carbohydr. Res.* **122**, 115–129.

Spellman, M. W., McNeil, M., Darvill, A. G., Albersheim, P. and Dell, A. (1983b). *Carbohydr. Res.* **122**, 131–153.

Stephen, A. M. (1983). *In* "The Polysaccharides" (G. O. Aspinall, ed.), Vol. 2, pp. 97–193. Academic Press, New York.

Stevens, B. J. H. and Selvendran, R. R. (1980). *Phytochemistry* **19**, 559–561.

Stevens, B. J. H. and Selvendran, R. R. (1984a). *Phytochemistry* **23**, 107–115.

Stevens, B. J. H. and Selvendran, R. R. (1984b). *Carbohydr. Res.* **128**, 321–333.

Stevens, B. J. H. and Selvendran, R. R. (1984c). *Carbohydr. Res.* **135**, 155–166.

Stevenson, T. T., Darvill, A. G. and Albersheim, P. (1988a). *Carbohydr. Res.* **179**, 269–288.

Stevenson, T. T., Darvill, A. G. and Albersheim, P. (1988b). *Carbohydr. Res.* **182**, 207–226.

Talmadge, K. W., Keegstra, K., Bauer, W. D. and Albersheim, P. (1973). *Plant Physiol.* **51**, 158–173.

Taylor, A. J. (1982). *Carbohydr. Polymers* **2**, 9–17.

Taylor, R. L. and Conrad, H. E. (1972). *Biochemistry* **11**, 1383–1388.

Thomas, J. T., McNeil, M., Darvill, A. G. and Albersheim, P. (1987). *Plant Physiol.* **83**, 659–671.

Thomas, J. T., Darvill, A. G. and Albersheim, P. (1989a). *Carbohydr. Res.* **185**, 279–305.

Thomas, J. T., Darvill, A. G. and Albersheim, P. (1989b). *Carbohydr. Res.* **185**, 261–277.

Tjan, S. B., Voragen, A. G. J. and Pilnik, W. (1974). *Carbohydr. Res.* **34**, 15–23.

Toman, R., Karacsonyi, S. and Kovacik, V. (1972). *Carbohydr. Res.* **25**, 371–378.

Tuerena, C. E., Taylor, A. J. and Mitchell, J. R. (1982). *Carbohydr. Polymers* **2**, 193–203.

Vliegenthart, J. F. G., Dorland, L. and Van Halbeek, H. (1983). *Adv. Carbohydr. Chem. Biochem.* **41**, 209–374.

Waeghe, T. J., Darvill, A. G., McNeil, M. and Albersheim, P. (1983). *Carbohydr. Res.* **123**, 281–304.

White, A. R., Elmore, H. W. and Watson, M. B. (1986). *Abstr. B 129, XIII Int. Carbohydr. Symp.*, Cornell University, Ithaca, NY, USA.

York, W. S., Darvill, A. G., McNeil, M. and Albersheim, P. (1985). *Carbohydr. Res.* **138**, 109–126.

York, W. S., Darvill, A. G., McNeil, M., Stevenson, T. T. and Albersheim, P. (1986). *Methods Enzymol.* **118**, 3–40.

Zitko, V. and Bishop, C. T. (1966). *Can. J. Chem.* **44**, 1275–1282.

13 Chitin

RAFAEL PONT LEZICA[1] and LUIS QUESADA-ALLUÉ[2]

[1]*Department of Biology, Washington University, St Louis, MO, USA*

[2]*Instituto de Investigaciones Bioquímicas, Fundación Campomar, Buenos Aires, Argentina*

I. INTRODUCTION

Chitin is the most abundant of the polysaccharides containing amino sugars. It consists predominantly, if not entirely, of unbranched chains of β-(1,4)-linked 2-acetamido-2-

METHODS IN PLANT BIOCHEMISTRY Vol. 2
ISBN 0-12-461012-9

deoxy-D-glucose (N-acetyl-D-glucosamine) residues. Chitin has a widespread distribution amongst invertebrates and the lower forms of plant life.

The first report describing an alkali-resistant fraction obtained from higher fungi was presented by Braconnot (1811). In this report the substance was named 'fungine' and a description was given of the lack of reaction of fungine with diluted alkali, the degradation caused by sulphuric acid, the release of acetic acid and several other reactions of chitin. However, the name of chitin (Greek, χιτων: tunic, envelope) was proposed by Odier (1823) for a substance isolated from elytra of May beetles. He did not detect nitrogen in chitin and believed that chitin was identical to the substance that forms the structure of plants (cellulose). Lassaigne (1843) demonstrated the presence of nitrogen in chitin obtained from the exoskeletons of insects. The presence of glucosamine and acetic acid in chitin was clearly demonstrated by Ledderhose (1878) and confirmed by Gilson (1894). Rouget (1859) found that boiling chitin in a concentrated KOH solution rendered it soluble in organic acids. This modified chitin was later named chitosan (Hoppe-Seiler, 1894).

During the first half of this century, research on chitin was mostly directed toward the study of its occurrence in living organisms, its degradation by bacteria and its chemistry. X-Ray analysis became the most reliable method for the differentiation between chitin and cellulose in cell walls of fungi. None of the studies done using X-ray diffraction showed juxtaposition of chitin and cellulose in the same organism, confirming the sharp separation of Phycomycetes into chitin- and cellulose-containing fungi. These results indicate that the occurrence of cellulose or chitin in fungi has a phylogenetic significance (Frey, 1950).

Most of the information now available on chitin has been obtained since 1950. Studies at low resolution have been performed on the structure using X-ray diffraction, infrared and Raman spectroscopy. However, such studies cannot give detailed information in the case of fibres, because they lack the accuracy possible with single crystal studies. High resolution NMR, which is by far the most informative spectroscopic tool, could not previously be applied to solid specimens. Techniques have been improving and the application of NMR at high resolution to polysaccharides is now increasing rapidly.

Most of the present knowledge on chitin biosynthesis has been obtained using fungal systems (Cabib, 1987). However, the use of recombinant DNA techniques has recently shown that the well-known yeast system is not functional in vivo. It is also evident that some differences in the pathway and in the properties of the enzymes involved seem to occur in crustacean chitin synthesis (Horst, 1986).

Chitin and cellulose are the most abundant substances of biological origin found on earth. Both have an important biological role as structural components of cell walls in plants and fungi. In the particular case of chitin, we should include information related to its occurrence, structure and biological role in the exoskeleton of crustaceans, insects, moluscs and other animals, that may exceed the scope of this book, but that are needed in order to have a general view of the subject. The present chapter will not only cover the methods and techniques used in the study of chitin, but will also emphasise those fields in which active research is needed in order to obtain a clear understanding of the area. For detailed information on different aspects of chitin chemistry and biochemistry, the reader may consult several books and review articles (Tracey, 1955;

Foster and Webber, 1960; Jeuniaux, 1963; Muzzarelli, 1977; Berkeley, 1979; Cabib, 1981, 1987; Wright and Retnakaran, 1987; Cohen, 1987).

II. DISTRIBUTION

Chitin is widely distributed, especially among the less evolved taxonomic groups of multicellular organisms. It is also found in certain unicellular Protista, but the most representative groups in which chitin occurs are arthropods, annelids, molluscs, pogono-phores, fungi and some algae. The distribution among living organisms is related to the main function of chitin: it is a supporting material. There are three supporting systems in organisms: cellulose, chitin and collagen. All three form long-chain fibrous molecules that can support the body of multicellular organisms.

In animals the principal skeletal structure is the collagen system, but chitin is a prominent supporting material in invertebrates. The major role of collagen is as an internal connective tissue supporting organs and having a constant turnover. On the other hand, chitin is usually organised as a cuticle at one surface of an epithelium. The types of structures are also different. Chitin–protein systems are formed by a fibrous framework of polysaccharide (chitin) reinforced by a protein matrix. Collagenous connective tissue presents a fibrous framework of protein (collagen) reinforced by a polysaccharide matrix. The presence or abundance of chitin in animal systems seems to depend on the presence or absence of chitin synthase (Jeuniaux, 1963).

In plants each cell is supported by a complete cell wall containing skeletal material (cellulose, chitin or other polysaccharides). Vascular plants use cellulose as supporting polysaccharide, and their inability to synthesise chitin was supposed to be connected with their lack of glucosamine synthesis (LéJohn, 1971; Rudall and Kenchington, 1973). However, it is known that N-acetyl glucosamine is present in higher plant glycoproteins. Thus the pathway is present, but the whole polymerisation system seems to be absent, as in higher animals.

A. Lower Animals

Jeuniaux (1963) has written the most comprehensive review of the distribution of chitin in animal phylla. Chitin is found as a major organic component of skeletal matrix in the segmented worms (Annelida), molluscs and arthropods, as well as in other minor phylla (Table 13.1). It should be pointed out that chitin is only one of several components of chitinous exoskeletons, and its proportion can vary from 1 to 50% of the dry weight.

B. Lower Plants

The occurrence of chitin in the plant kingdom is restricted to green algae (Chlorophy-ceae) and fungi (Table 13.2). All fungi seem to have chitinous cell walls, however the accuracy of many reports is doubtful. Most of the old reports were based on analyses of the cell walls obtained by exhaustive alkali-extraction and used methods known to be non-specific for the detection and quantification of chitin. In yeasts, chitin is located almost exclusively in the primary septum, of which it seems to be the sole component.

TABLE 13.1. Distribution and content of chitinous cuticles or shells in multicellular animals (Eumetazoa).

Phyllum	Class	Genus	Chitin content[a] (%)	Other compounds
Cnidaria	Hydrozoa	*Hydra*	3–30	Prot
		Millepora	NQ	$CaCO_3$
	Anthozoa	*Metridium, Pocillopora*	NQ	$CaCO_3$
	Scyphozoa	*Aurelia*	NQ	—
Rotifera	Monogononta	*Asplanchna*	14	Prot
Nemata	Scernetea	*Ascari*	16	Prot
Priapulida		*Priapulis*	NQ	Prot
Mollusca	Polyplacophora	*Chiton*	12	Prot/$CaCO_3$
	Gastropoda	*Helix*	3–20	Prot/$CaCO_3$
	Cephalopoda	*Nautilus, Sepia*	3–26	Prot/$CaCO_3$
		Loligo, Octopus	18	Prot
		Omnatostrephes	20	Prot
Annelida	Polychaeta	*Amphinoe, Aphrodite*	20–38	Prot
	Oligochaeta	*Lombricus*	NQ	Prot
Pogonophora		*Oligobrachia*	33	Prot
Arthropoda	Aracnida	*Galeodes, Scolopendra*	4–22	Prot
		Buthus, Mygale	32–38	Prot
	Crustacea	*Cancer, Homarus, Palinorus*	58–85	Prot/$CaCO_3$
		Astacus, Carcinus, Eupagurus, Euphausia	48–80	Prot/$CaCO_3$
	Insecta (Blattaria)	*Blatta, Periplaneta*	31–35	Prot
	(Orthoptera)	*Locusta*	31–43	Prot
	(Phasmatoptera)	*Dixippus*	35–45	Prot
	(Hemiptera)	*Rhodnius*	11	Prot
	(Coleoptera)	*Agrianome, Dytiscus, Tenebrio, Tribolium Xylotrupes*	31–47	Prot
	(Diptera)	*Calliphora, Drosophila Glossina, Lucilia, Melophagus, Musca, Phormia, Sarcophagae*	10–60	Prot
	(Lepidoptera)	*Bombyx, Balleria, Sphinx*	38–50	Prot
		Loxostege	1.5–2.5	Prot
Lophophorata	Phoronida		13	Prot
Bryozoa	Phylactolaemata	*Crystatella*	2–6	Prot/$CaCO_3$
Entoprocta		*Pedicellaria*	NQ	Prot
Brachipoda			4–29	Prot/$CaCO_3$

NQ, not quantified; Prot, proteins.
[a] As a percentage of organic components.

TABLE 13.2. Content of chitin in cell walls of protozoan algae and fungi.

Division	Class	Genus	Chitin content (%)	Other compounds
PROTOZOA				
Sarcomastigophora	Mastigophora	*Plagiopyxidae*	NQ	Silica
	Lobosa	*Pelomyra*	NQ	—
	Granuloreticulosa	*Allogramia*	NQ	Prot/Fe
Ciliophora	Kinetofragmino-phora	*Vasicola*	NQ	Prot
Chlorophycota	Chlorophyceae	*Geosiphon, Ulva, Valonia*	NQ	Glucan
Eumycota	Chytridiomycetes	*Blastocladiella*	NQ	Glucan
	Oomycetes	*Apodachyla*	NQ	Glucan
	Zygomycetes	*Mucor, Mortirella, Phycomyces, Rhizopus*	10–15	Chitosan
	Hemiascomycetes	*Eremascus, Nadsonia, Saccharomyces*	1–2	Glucan Mannan
	Pyrenomycetes	*Aspergillus, Neurospora, Penicillum*	20–45	Glucan
	Hymenomycetes	*Agaricus, Armillariella, Boletus, Cantharellus, Psalliota, Schizophyllum*	35	Glucan

NQ, not quantified; Prot, proteins.

III. STRUCTURE

A. Isolation and Properties

Chitin is found in close association with other materials in living organisms. This has led to the use of drastic methods for its isolation. These methods can cause degradation and it is doubtful whether a pure, undegraded product is normally obtained.

1. Demineralisation

The raw chitinous materials most abundantly available are of animal origin: crab and shrimp shells or prawn wastes. These materials have a high content of $CaCO_3$ which is generally removed by digestion with dilute mineral acid. In a typical procedure (Hackman, 1954) 2 N HCl at room temperature for 5 h is used for decalcification. A progressive rise in the concentration of HCl enhances demineralisation. However, concentrations above 1.25 N HCl yield products with a lower degree of polymerisation. When chitin was treated with 2 N HCl at 100°C for 7 h, only 6.6% was recovered as free *N*-acetylglucosamine (Hackman, 1954), but although that treatment produced little complete hydrolysis, considerable partial degradation was obtained. Figure 13.1 shows the effect of different concentrations of HCl on ash content and viscosity (the viscosity of a solution of a macromolecular compound is related to the degree of polymerisation).

A milder method has been used based on the decalcification properties of ethylene-diamine tetra-acetic acid (EDTA) (Foster and Hackman, 1957). Chitin from fungal origin does not contain inorganic salts and demineralisation is not needed.

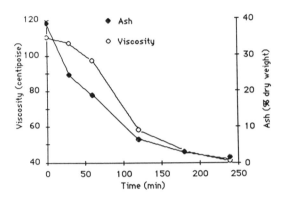

FIG. 13.1. Effect of demineralisation conditions on the viscosity of chitosan (data from Madhavan and Ramachandran, 1974).

2. Removal of proteins

Decalcified shells are subjected to different treatments to remove proteins that are normally present in crustacean shells. One of the best known procedures (Hackman, 1954) involves a 48 h extraction with 1 N NaOH at 100°C. The alkaline solution is replaced with a fresh one every 12 h. Alternative procedures such as the use of concentrated formic acid for an 18 h extraction of chitosan prior to deproteinisation with 2.5 N NaOH have been performed. This can be done at room temperature for 3 days (Whistler and BeMiller, 1962) or at 100°C for 2.5 h (Horowitz et al., 1957). The proteins present in decalcified materials cannot be removed with dimethylformamide or phenol–water mixtures.

The methods that include prolonged digestions in alkaline solutions not only eliminate proteins, but are also known to deacetylate chitin. Proteolytic enzymes have been used as an alternative method for elimination of proteins without deacetylation (Takeda and Abe, 1962; Takeda and Katsuura, 1964; Broussignac, 1968). The protein remaining after the enzymatic digestion (around 5%) can be removed effectively by dimethylformamide or sodium dodecylbenzenesulphonate.

Chitin is associated with other organic compounds in fungal cell walls: β-glucans and pigments. The latter (melanin and carotenoids) can be eliminated with 0.02% potassium permanganate at 60°C (Scholl, 1909; Schmidt, 1936) or sodium metabisulphite (Clark and Smith, 1936), but these reagents probably cause more extensive degradation. On the other hand, fungal β-glucans are alkali-resistant and contaminate chitin preparations that rarely contain more that 4% of nitrogen (pure chitin contains 6.9% nitrogen). The difficulties encountered in isolating pure chitin are probably inherent in methods based on the removal of contaminants rather than extraction of the required material.

3. Isolation from fungal walls

The qualitative or quantitative determination of chitin in fungal cell walls has many technical problems. The first problem associated with cell-wall analysis is to obtain cell walls free from other cellular components. Disruption of fungal cells as well as washing should be performed at low temperature. This will minimise the activity of degradative enzymes from other cell compartments. Prior to fragmentation, cells may be frozen; the use of liquid nitrogen during disruption will keep the tissue in good condition. The optimal conditions for cell fragmentation must be determined for each particular organism. However, the basic principles of tissue fractionation apply in general to cell-wall purification (de Duve, 1964).

The cell walls are washed carefully to remove all traces of cytoplasm and other possible contaminants. Different solutions can be used such as NaCl, 8 M urea, 1 M ammonia and 0.5 M acetic acid (Taylor and Cameron, 1973). Each of these is chosen for its ability to dissolve protein or disrupt hydrogen bonding. Organic solvents or detergents are used for extraction of lipids (Joppien et al., 1972). The removal of contaminant material almost certainly involves the removal of weakly bound cell-well components.

Extraction of cell walls by mild acid or alkali will leave chitin as a residue; β-glucans can contaminate this residue and total neutral sugars and amino-sugars should be determined. Amino-sugar analyses are currently performed using the Elson–Morgan (Elson and Morgan, 1933) method or any of its modifications. However, glucosamine and galactosamine are common fungal cell-wall constituents, and determination of each hexosamine should be achieved.

4. Solubility

Chitin isolated from different sources by any of the above-mentioned procedures is an amorphous solid, insoluble in water, weak acids, dilute or concentrated boiling alkalis, alcohol and other organic solvents. It can be dissolved in concentrated HCl or H_2SO_4, or anhydrous formic acid (Foster and Webber, 1960). Dissolution in mineral acids causes some degradation (hydrolysis of glycosidic bonds and some deacetylation).

Colloidal solutions of chitin can be obtained by dispersion in certain hydrotropic salt solutions such as lithium thiocyanate (von Veimarn, 1927), but these solutions degrade chitin or are inconvenient to use. On the other hand, some fluoroalcohols, e.g. hexafluoroisopropanol or hexafluoroacetone (Capozza, 1975), and chloroalcohols in conjunction with aqueous solutions of mineral or organic acids, are effective for dissolving chitin (Austin, 1975). Dimethylacetamide containing 5% dissolved lithium chloride has been used as a non-degradative solvent giving a syrupy chitin solution. Chitin can be reprecipitated from these salt dispersions by dilution with alcohol or acetone yielding continuous filaments (Austin et al., 1981).

5. Chemical modifications

Chitin can be presented as a new type of polymeric material having greater possibility than cellulose in many respects. Since it is an amino polysaccharide, chitin is capable of undergoing many additional modification reactions.

(a) *Deacetylation.* Alkali treatment of chitin hydrolyses the acetamido groups to give free amino residues. This hydrolysis to chitosan is usually performed with NaOH or

KOH at high temperatures. In addition to deacetylation the polysaccharide is degraded to some extent. The presence of thiophenol, which traps oxygen (preventing degradation) during deacetylation, improves the formation of chitosan with a high degree of polymerisation (Domard and Rinaudo, 1983).

(b)   N-*Acetylation*. N-Acetylation of chitosan leads to fully N-acetylated chitin. Acetic acid is the acetylating reagent and dicyclohexylcarbodiimide is used as condensing agent. The solvent used should provide the penetration of the acetylation agent into the interior of the particle so the amino groups not exposed may be acetylated. When unswollen material is acetylated, only the surface amino groups are derivatised. It was found that maximum reaction rate is achieved using binary mixtures of ethanol and methanol or methanol and formamide (Moore and Roberts, 1981). Complete N-acetylation can be achieved in a few minutes at room temperature using highly swollen chitosan in organic aprotic solvents (Nishi *et al*., 1979).

(c)   *Shiff bases*. The Shiff reaction between chitosan and aldehydes or ketones gives the corresponding alkimines and ketimines which can be reduced to products less susceptible to hydrolysis (Hall and Yalpani, 1980). The attachment of reducing carbohydrates as side chains to the 2-amino functions of chitosan transforms it into branched-chain water-soluble derivatives. Easy conversion can be achieved by reductive alkylation using $NaCNBH_3$ (Yalpani *et al*., 1983).

(d)   O-*Carboxymethylation*. Carboxymethylation is performed with monochloroacetic acid and sodium hydroxide (Zimmermann, 1952). The reaction takes place preferentially at C-6 hydroxyl groups, but sometimes over 50% acetamido groups are hydrolysed because of the strong basic conditions used (Trujillo, 1968).

Other miscellaneous reactions of chitin and chitosans have been described (sulphation, cyanoacetylation, chelation of metal ions, etc). The reader will find detailed reviews on these subjects elsewhere (Muzzarelli, 1977; Kurita, 1985).

B.   Structure of Chitin Saccharides

Isolation of chitin from other materials frequently requires conditions under which some destruction or modification is known to occur. Hydrolysis by chemical means results in variable losses of glucosamine. Enzymic hydrolysis has the advantage that neither deacetylation nor destruction of the products of hydrolysis occurs. It has the advantage of confirming the presence of both acetyl groups and glucosamine units in the material, providing unequivocal evidence of the location of acetyl groups. However, complete enzymic hydrolysis can be performed only under restricted conditions. Nevertheless, the use of chemical and enzymic hydrolysis can provide a reasonable estimation of the chitin content, especially in fungal material.

Original proof of the structure of the chitin disaccharide was provided by studies on its octa-acetate (Bergman *et al*., 1931). It was shown that it contained two N-acetyl and six O-acetyl groups. Oxidation with iodine in alkaline conditions followed by hydrolysis gave 2-amino-2-deoxy-D-glucose and 2-amino-2-deoxy-D-gluconic acid, indicating that C-1 as well as C-2 of the reducing moiety cannot be involved in the glycosidic linkage.

Periodate oxidation of chitin and chitosan has been performed and the results indicate that only 1-4-linkages are present in the polysaccharides (Jeanloz and Forchielli, 1950). However, the oxidation pattern postulated for the reducing unit of the chitin chain does not agree with that observed for the di- and trisaccharide (Barker *et al.*, 1958).

Deamination of chitosan using nitrous acid proceeds at a rate closely similar to that of β-D-glycosides, resulting in the formation of 2,5-anhydro-D-mannose (Foster *et al.*, 1953). However, chitin oligosaccharides and chitin itself showed absorptions in the infrared at 884–890 cm^{-1}, indicative of a β-D-glucopyranosidic linkage (Barker *et al.*, 1957). As a result of the various acidic and enzymic degradative studies on chitin, there can be little doubt that the polysaccharide is a homogeneous polymer composed of 2-acetamido-2-deoxy-D-glucose residues. Moreover, the specificity of chitinases as 2-acetamido-2-deoxy-D-glucosidases confirms the β-D nature of the glycosidic linkages (Fig. 13.2).

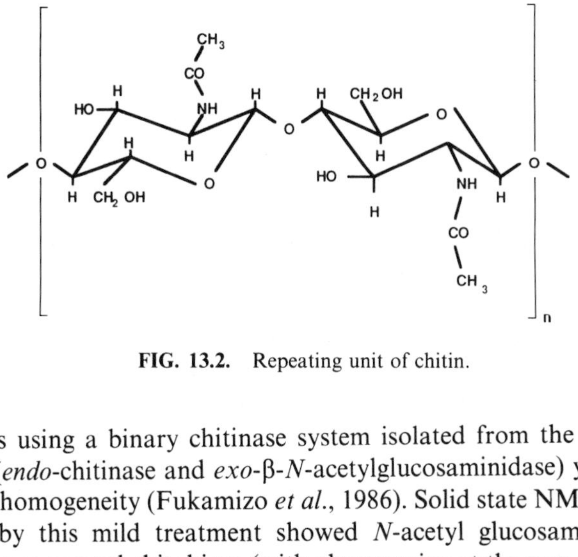

FIG. 13.2.   Repeating unit of chitin.

Recent studies using a binary chitinase system isolated from the moulting fluid of *Manduca sexta* (*endo*-chitinase and *exo*-β-*N*-acetylglucosaminidase) yielded chitin with a high degree of homogeneity (Fukamizo *et al.*, 1986). Solid state NMR spectroscopy of chitin released by this mild treatment showed *N*-acetyl glucosamine as the major product, and *N*-monoacetyl chitobiose (with glucosamine at the non-reducing end) as a minor product. This suggests that intact chitin might not be a homopolymer, but may have some proportion of glucosamine in the chain. These non-acetylated glucosamine residues will provide good reactive groups for cross-linking chitin to other components in the cuticle. The presence of glucosamine in chitin samples has been reported frequently, but deacetylation was attributed to alkali-extraction. It is difficult to evaluate in alkali-treated chitin which glucosamine residues are derived from the treatment and which are naturally present in the intact molecule.

1.   *Qualitative and quantitative methods*

Microchemical and colour tests based on the stability of chitin in alkaline conditions have been used to confirm the occurrence of chitin in many samples, especially among the fungi. Qualitative tests for chitin have mainly been developed in the form of tests based on the properties of chitosan. The iodine–sulphuric acid test, chitosan sulphate test, the Elson–Morgan test and others have been published in a detailed review (Tracey, 1955). However, the absolute specificity of these tests is doubtful and many conflicting reports are found in the literature.

Enzymic methods based on the specificity of chitinase, which releases N-acetyl glucosamine, have been described (Tracey, 1954; Ride and Drysdale, 1971; Bussers and Jeuniaux, 1974). The method is limited by the inability of the enzyme to effect complete digestion of coarsely dispersed or chemically purified chitin. On the other hand, the presence of contaminating glycosidases in enzyme preparations is not a problem since the product of those enzymes is not N-acetyl glucosamine. Methods for the quantitative determination of chitin are of limited accuracy.

The determination of chitin is also feasible by gas chromatography through the determination of acetic acid (Holan et al., 1971) or the trimethylsilyl derivatives (Sweeley et al., 1963). Recently a method that can distinguish between chitin, chitosan and N-acetyl glucosamine has been developed (Davies et al., 1985). It is based on pyrolysis–gas chromatography and seems to be a convenient analytical technique for the chitin family of polymers.

2. Histochemical methods

Histochemical estimation of chitin synthesis and/or deposition was probably the earliest qualitative technique used. As mentioned above, most colorimetric estimations are inaccurate or non-specific (Muzzarelli, 1977; Hackman, 1984). Some histochemical methods for the detection of chitin have been reviewed by Pearse (1985) and Richards (1951). It has been found that classical methods as Schultz chlor-zinc-iodine reagent, alkaline tetrazolium, alcian dyes, periodic acid-Schiff reagent and colloidal iron are not specific, giving positive reactions with a variety of compounds. An interesting approach taking advantage of the specificity of chitinase for its substrate has been described in which the enzyme is labelled with fluorescent conjugates (Benjaminson, 1969). However, this technique has not been developed further, perhaps because of the need for a pure enzyme.

Wheat germ agglutinin (WGA) is a plant lectin having affinity for N-triacetyl chitotriose that also binds chitin. WGA complexes have been used for localisation of chitin; fluorescent derivatives were visualised by fluorescent microscopy and colloidal gold as specific electron-dense markers for electron microscopy (Mauchamp and Schrevel, 1977; Molano et al., 1980; Tronchin et al., 1981; Fristrom and Fristrom, 1982; Fristrom et al., 1982; Arroyo-Bergovich and Carabez-Trejo, 1982; Chamberland et al., 1985; Peters and Latka, 1986). With the WGA–gold complex chitin and chitin precursors were localised and it was possible to differentiate between chitin and N-acetyl glucosamine-containing glycoproteins.

C. Macrostructure of Chitin

Much of the knowledge of the properties and structure of chitin has been derived from studies on 'purified' chitin. Removal of the associated protein increases the degree of orientation of chitin chains, as shown by X-ray diffraction, and the attraction between them is such that they aggregate to give a continuous sheet of chitin. The β-glycosidic bond between hexosamines in the chain causes the amino groups in adjacent residues to be on opposite sides of the chain (Fig. 13.2).

1. X-ray diffraction patterns

Several crystalline forms of the chitin have been described and they differ in the packing and polarity of the adjacent chain (Rudall, 1963). α-Chitin is the tightly compacted, most crystalline polymorphic form where the chains are arranged in anti-parallel orientation. The buckled chain structure has been proposed to be arranged in an orthorhombic cell with dimensions $a = 0.474$, $b = 1.032$, $c = 1.886$ nm with the space group $P2_12_12_1$ (Minke and Blackwell, 1978). Sheets of chains are arranged in stacks along the a-axis, the sheets being linked by $C=O \cdots H-N$ hydrogen bonds approximately parallel to the a-axis. In the plane of the pyranose rings, along the b-axis, there is interchain hydrogen bonding between CH_2OH groups. The strong inter- and intramolecular bonding leads to the formation of long microfibrils. α-Chitin is the most abundant form and it is found in arthropod cuticles and in certain fungi.

The β-chitin is the form where the chains are parallel. The unit cell contains two sugar residues related to the two-fold axis, is monoclinic with dimensions $a = 0.485$, $b = 1.038$, $c = 0.926$ nm and space group $P2_1$ (Gardner and Blackwell, 1975). There are numerous points of analogy between the structures of β-chitin and cellulose I. Both structures contain extended parallel chains and can be visualised as an array of hydrogen-bonded sheets. β-Chitin from Polychaetae, when precipitated from acids, assumes the α-form. It has been found in the cocoons of some beetles.

γ-Chitin is an uncommon form in which the arrangement of chains alternates so that for every three chains, two are parallel. Also γ-chitin can be transformed into α-chitin by treating with lithium thiocyanate (Rudall and Kenchington, 1973).

The three forms of chitin have been found in different parts of the same organisms, suggesting that the three forms are relevant to the different functions and not to animal grouping. However, the interconversion of different forms of 'purified' chitin into the α-form following chemical treatments makes it difficult to assign a physiological role to these crystalline forms. The conclusions drawn from studies on purified chitin may have to be modified when considering chitin in its native form, namely associated with proteins.

2. Chitin associations

Chitin can be found in nature in association with different compounds as carotenoids (insects and crustacean shells), glucans (fungal walls) and proteins (all excepting mould and fungi).

(a) *Carotenoids.* Carotenoids are found as conjugates with chitin and proteins, usually in different species of insects. They give the pink or red colour to crustacean shells, or green colours to many insects.

Decalcification of crustacean shells does not extract pigments, indicating that they are not bound to the calcareous portion. Very little carotenoid can be recovered from chitinous material by treatment with the usual solvents and/or protein denaturants, unless the shell has been decalcified. Pigments from carbonate-free shells can be readily released by ethanol or acetone (Muzzarelli, 1977). More harsh treatments with cold formic acid and mixtures of ammonium sulphate plus sulphuric acid on chitosan also release carotenoids, suggesting that pigments can be combined with chitin amino groups by Schiff's base linkages (Fox, 1973). Removal of carotenoids makes available more free amino groups and introduces a higher degree of crystallinity in chitosans (Muzzarelli, 1973; Muzzarelli and Rochetti, 1974).

(b) *Glucans*. In some fungal cell walls, chitin was found associated with β-glucans and proteins. The alkali-insoluble wall residue of ascomycetes and basidiomycetes consists mainly of β-(1,3)-D/β-(1,6)-D-glucans and chitin (Rosenberg, 1976). A large part of the glucan was solubilised by acetylation. Nitrous acid treatment, which causes oxidative deamination, produced striking changes in solubility. This alkali-insoluble portion of fungal cell walls is not a heteropolymer, but is composed of chitin and glucan which may or may not be bonded (Stagg and Feather, 1973).

It was found that in the walls of different fungi, most of the alkali-insoluble β-glucans of the cell wall can be extracted with dimethyl sulphoxide followed by treatment with 40% NaOH at 100°C. The remaining β-glucan fraction can only be extracted after selective depolymerisation of deacetylated chitin by HNO_2. This has been used as the criterion for linkage of β-glucans to chitin. Species as *Schizophyllum commune* and *Agaricus biporus* had nearly all the β-glucans linked to chitin. On the other hand, *Saccharomyces cerevisiae*, *Neurospora crassa*, *Aspergillus nidulans* and *Coprinus cinereus* seem to have the β-glucan not covalently linked to chitin (Sietsma and Wessel, 1977, 1981). Figure 13.3 shows a proposed model for the structure of the glucan/chitin complexes (Sietsma and Wessels, 1979), it involves amino acids, especially lysine and citrulline, in the linkages between glucan and chitin. The sequence and mode of linkage between these monomers remain largely unknown.

(c) *Proteins*. Mollusc shells and arthropod cuticle are composed mainly of chitin and proteins. Proteins are the major component (50–80% of the organic dry weight in mollusc shells) according to Jeuniaux (1963) and Gerhard and Jeuniaux (1979). Proteins are bound non-covalently and possibly covalently to chitin, and the link may be through aspartic acid (Hackman and Goldberg, 1958, 1978; Rudall and Kenchington, 1973; Brine, 1982). However, the nature of the association between chitin and cuticular proteins has not yet been firmly established (Hackman, 1987).

Arthropodins are characterised by a low proportion of glycine, absence of sulphur amino acids, and a high content of tyrosine (Jeuniaux, 1971). They are soluble in hot water. Sclerotins are water-insoluble. Sclerotisation of the insect cuticle results from the incorporation of diphenolic compounds, but very little is known of the molecular mechanism of sclerotisation. It has been suggested that chitin may be involved in the O-quinone cross-link (Lipke *et al.*, 1983) and a model for chitin–protein linkages in dipteran cuticle has been proposed (Lipke and Geoghegan, 1971).

In a recent study Schaefer *et al.* (1987) labelled the cuticle of *Manduca sexta* with ^{13}C and [15]N. The intact cuticle was analysed by solid-state NMR spectroscopy showing direct covalent linkages between the imidazole nitrogen from histidine and ring carbons derived from catechol. This histidyl–catechol complex is probably bonded to chitin and a model for this aromatic cross-link in insect cuticle has been proposed (Fig. 13.4). On the other hand, *in situ* studies using carbon-13 NMR spectroscopy show that N-acetyldopamine oxidation products bind to chitin by non-covalent interactions (Peter *et al.,* 1986). Non-covalent interactions may contribute significantly, if not predominantly, to the stability of sclerotised insect cuticle (Hackman and Goldberg, 1977; Vincent and Hillerton, 1979; Peter, 1980).

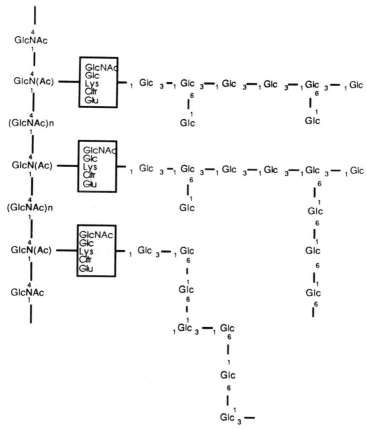

FIG. 13.3. Proposed model for the structure of glucan–chitin complexes of *Schizophyllum commune.* Glc, glucose; Lys, lysine; Citr, citrulline; Glu, glutamic acid (from Sietsma and Wessels, 1979).

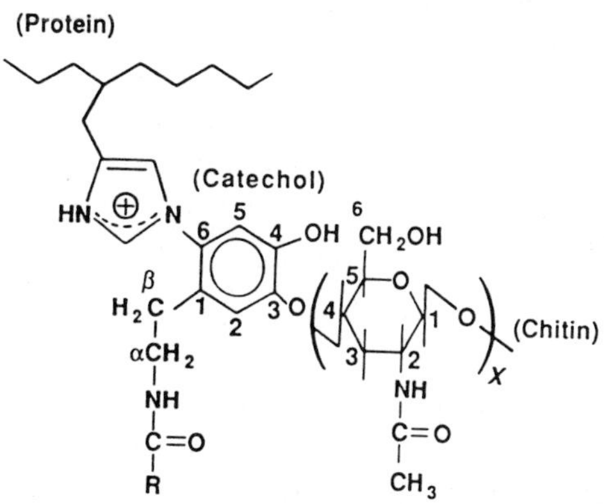

FIG. 13.4. Proposed model for the aromatic cross-links of proteins, catechols and chitin in insect cuticles (from Schaefer *et al.,* 1987: Copyright 1987 by the AAAS).

The chitin–protein matrix of arthropod cuticles shows a supramolecular organisation geometrically comparable to twisted plywood. This arrangement is also similar to that found in liquid crystals. Liquid crystals were first described in cholesterol esters and the term cholesteric was given to this particular state of matter. Ultrastructural observations of the chitin–protein molecular organisation have been performed and models have been proposed in which rods of chitin crystallites are cemented by a protein matrix (Rudall, 1965; Neville, 1975; Blackwell and Weih, 1980) as shown in Fig. 13.5a and b. A different interpretation of the X-ray diffraction studies led to a new model proposed by Giraud-Guille and Bouligand (1986) and shown in Fig. 13.5c and d.

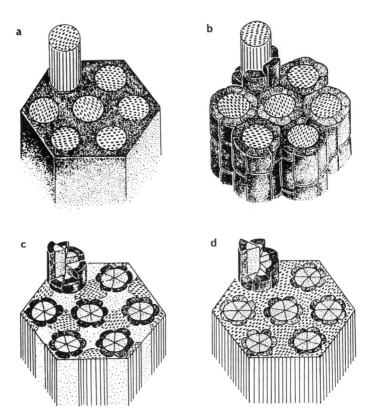

FIG. 13.5. Different cuticle models. (a) Parallel chitin crystallites cemented with proteins. (b) Parallel chitin crystallites surrounded by helically arranged proteins. (c) Polar and non-polar parts of proteins segregate; clear rods are thought to be formed by apolar amino acids and electron-dense alveolar walls are supposed to contain polar amino acids and separated chitin crystallites. (d) Model slightly different from that proposed in (c). The chitin lattice is connected and more or less regularly ordered (from Giraud-Guille and Bouligand, 1986).

IV. CHITIN BIOSYNTHESIS

Chitin is an idealised polymer operationally defined as the cell wall or cuticular material, insoluble in hot, dilute alkali, formed by linear chains of poly (N-acetylglucosamine). However, traces of covalently bound amino acids are usually found in chitin analysis. In addition, particularly in arthropods, some of the monomers appear deacetylated, raising doubts about the fully acetylated state of the native polymer (Hackman and Goldberg, 1965; Jacobs, 1978, 1985; Kramer and Koga, 1986).

Most of the recent reviews (Cabib, 1981; Kramer *et al.*, 1985; Kramer and Koga, 1986; Cohen, 1987) show a clear pathway leading to the biosynthesis of UDP-*N*-acetylglucosamine (UDP-GlcNAc), the glycosyl donor for the polymer (Fig. 13.6). The polymerisation step has been studied extensively in fungi and the enzyme responsible, chitin synthase, is well known.

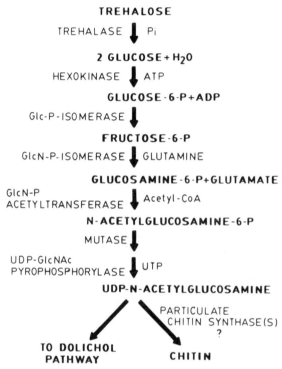

FIG. 13.6. Pathway leading to the synthesis of UDP-GlcNAc as precursor of chitin and Asn-glycoproteins which are glycosylated by the dolichol cycle.

For many years it has been accepted that a unique fungal chitin synthase (chitin: UDP-acetyl glucosaminyl transferase, EC 2.4.1.16) existed (Muzzarelli, 1977; Cabib, 1987). However, recent results (Bulawa *et al.*, 1986; Sburlati and Cabib, 1987; Orlean, 1987) have demonstrated that, at least in yeast, the well known fungal enzyme now called chitin synthase I (ChS I) is non-essential since mutants with a disrupted gene ChS I still synthesise normal levels of chitin and are able to bud (Bulawa *et al.*, 1986). The main characteristic of *Saccharomyces* chitin synthase I as well as of many similar enzymes from other fungi is that most of the enzymatic activity appears only after treatment with proteases (Muzzarelli, 1977; Cabib, 1987). For the purposes of this review we will assume that most of the putative fungal chitin synthases described up to now are involved at least partly in chitin synthesis or that—alternatively—even if non-physiological in certain species they are still related, in terms of amino acid sequences, to active enzymes in other organisms.

The present status of the problem is that biosynthesis of chitin is not well understood. Distinct forms of chitin exist in nature (usually associated with proteins, glucans, lipids, catechols, etc.) and it is difficult to assess which are the first enzymatic steps leading to the synthesis of a primary linear chain of 1,4-β-linked 2-acetamide-2-deoxy-D-glucose (poly-N-acetylglucosamine). The existence, at least in invertebrates, of chitin–protein covalent and non-covalent linkages has been demonstrated in a number of cases (Attwood and Zola, 1967; Hackman and Goldberg, 1971, 1978; Lipke, 1971; Lipke and Geoghegan, 1971; Lipke and Strout, 1972; Hillerton, 1980; Lipke et al., 1981; Hackman, 1982). Different chitin 'domains' have also been inferred as judged by chitinase treatment experiments (Molano et al., 1979; Stirling et al., 1979; Quesada Allué, 1982; Cabib, 1987). Since both linkage to protein and deacetylation of some monomers can be explained by post-synthetic modifications, the whole chitin metabolism remains obscure. Inhibition of protein synthesis immediately affects fungal chitin synthesis (Ryder and Peberdy, 1979; Farkas and Svoboda, 1980). Since most synthases are accumulated in zymogen form, a possible explanation for such sensitivity to protein synthesis inhibitors could be the requirement of an activator or a cofactor synthesis. Such hypothetical molecules might be locally synthesised and their synthesis would be dependent on unknown regulatory signals related to the life-cycle. In insects, ecdysterone levels trigger or inhibit chitin synthesis (Silvert and Fristrom, 1980; Fristrom et al., 1982), whereas mating pheromone in yeast raises chitin synthesis activity.

As with other homopolysaccharides, the main question is whether or not there is an *in vivo* requirement for a primer acting as receptor substrate for the elongation of the chain by a transferase (Krisman, 1972; Barengo and Krisman, 1978; Stoddart, 1984; Moreno et al., 1986, 1987; Rivas and Pont Lezica, 1987a,b). The latter, in turn, could or could not be the polymerising enzyme. As judged from cell-free experiments, fungal chitin synthases seemed not to require any primer (Cabib, 1981).

The picture in insects and other invertebrates has never seemed to be as clear as in filamentous fungi and yeasts. The postulated putative insect chitin synthase has never been purified and chitin synthesis is difficult to obtain *in vitro* (Quesada-Allué et al., 1976b; Muzzarelli, 1977; Kramer and Koga, 1986). Most of the early doubts about the existence of a single enzyme involved in chitin synthesis arise from work done with invertebrates, mostly with insect tissues (Neville, 1975; Hepburn, 1976; Glenn Richards, 1978; Quesada-Allué, 1978; Andersen, 1979; Hackman, 1984; Kramer and Koga, 1986). A search for postulated new enzymatic activities involved in chitin synthesis might be a way to discover the real sequence of biochemical events. A promising approach might be to obtain chitin synthesis mutants in suitable organisms like the already isolated *Saccharomyces* mutants or *Drosophila*.

A. Measurement of Chitin Synthesis

Biosynthesis of chitin during the life-cycle as well as deposition in different structures and organs can be estimated by physico-chemical or chemical analysis and/or by histochemistry. In the past, most assays usually lacked specificity and/or accuracy (Kramer and Koga, 1986). Radioactive precursors are routinely used to follow chitin synthesis both *in vivo* and in cell-free systems. Any of the intermediates of the UDP-GlcNAc biosynthetic pathway can be used (Fig 13.6). ^{3}H or ^{14}C-labelled monosaccharides (mainly N-acetyl glucosamine, glucosamine and glucose) have been used as chitin precursors in fungal and insect systems (Agui *et al.*, 1969; Oberlander and Leach, 1975; Sowa and Marks, 1975; Quesada-Allué *et al.*, 1976a; Vardanis, 1976; Isaac *et al.*, 1978; Quesada-Allué, 1982). Radioactive UDP-GlcNAc or GlcNAc-1-P have been used in cell-free systems (Bartniki-García, 1968; Cabib, 1972, 1987; Muzzarelli, 1977; Hajjar, 1985; Kramer and Koga, 1986), and [^{14}C]trehalose, the physiological haemolymph sugar, is a good precursor for insect and crustacean systems (Quesada-Allué *et al.*, 1976a; Quesada-Allué, 1978).

In most of the published work, radiolabelled chitin identification is based on retention on filters as proof of relatively high molecular weight and insolubility, usually after acid precipitation (Brillinger, 1979; Leighton *et al.*, 1981; Kang *et al.*, 1984). Some authors (Doctor, 1984) previously treated the sample with dilute alkali to define chitin as a 'base non-hydrolysable, acid precipitable' material. In our experience, putative nascent chitin chains (i.e. 'incorporated radiolabelled GlcNAc') could be very difficult to distinguish from N,N'-diacetyl chitobiose-containing glycoconjugates (glycoprotein or glycolipids) (Quesada-Allué *et al.*, 1976a,b; Parodi and Leloir, 1979; Quesada-Allué, 1982). In fact, some GlcNAc-labelled lipid-bound oligosaccharides which are water-insoluble may remain trapped within protein aggregates and are neither extracted nor degraded by cold 10% trichloroacetic acid (Quesada-Allué, 1978, 1980, 1981, 1982; Quesada-Allué and Belocopitow, 1978). Heating the acid mixture for 7 min at 98°C is required to destroy dolichyl-P-P-GlcNAc$_2$-Man$_9$-Glc$_{1-3}$ (Quesada-Allué, 1980).

A more accurate way of identifying chitin seems to be the digestion of samples with chitinase. However, as already mentioned, different chitin domains seem to exist (Molano *et al.*, 1979, 1980; Quesada-Allué, 1982; Zarain-Herzberg and Arroyo-Bergovich, 1983) with different sensitivity to a given chitinase. Since digestion of newly synthesised chitin with chitinase is apparently dependent on both the particular synthesising enzymatic complex and on the origin of the degrading enzyme, a variety of contradictory results have been reported (Kramer *et al.*, 1985; Kramer and Koga, 1986; Cabib, 1987). Cabib (1987) interprets differences in rate of chitin digestion as due to different degrees of organisation of chains. The sensitivity of nascent chitin to chitinase would be greater than that of pre-formed chitin in which more of the chains would be highly organised in crystallites. When measuring synthesised chitin, a possible endogenous chitinase activity should be taken into account. Final products of chitinase degradation are GlaNAc$_3$ and GlcNAc$_2$. If chitobiases are present, the final product is GlcNAc.

B. Chitin Synthases

Leloir and Cardini suggested in 1953 that UDP-GlcNAc was involved in chitin synthesis. The pathway leading to UDP-GlcNAc synthesis is shown in Fig. 13.6. Glaser and Brown (1957) reported later that a particulate enzyme from *Neurospora crassa* catalysed the synthesis of poly-*N*-acetyl glucosamine from UDP-[^{14}C]GlcNAc, confirming the hypothesis. This was the first report showing that a sugar nucleotide was the monosaccharide donor in the polymerisation reaction leading to the formation of a polysaccharide. It was assumed that the poly-GlcNAc polysaccharide synthesised *in vitro* was identical to chitin, that is, the insoluble material obtained after hot alkali treatment. As mentioned before, it is difficult to know whether radiolabelled material synthesised *de novo* in this kind of experiment is bound to (or grows upon) protein or another glycoconjugate, or if all the monosaccharides remain acetylated. Most of the general properties described for *Neurospora* chitin synthase (i.e. a membrane-bound enzyme activated by proteinase treatment and by magnesium salts, inhibited by UDP) were later found to be similar for many fungal chitin synthases.

1. Algal chitin synthesis

Very little is known about algal chitin synthesis. Algal chitin is apparently synthesised at plasma membrane level (Schnepf *et al.*, 1975). Chitin synthesis and deposition have been reported in the chrysoflagellate unicellular alga *Ochromonas malhamensis* (Herth *et al.*, 1977; Herth and Zugenmaier, 1977; Herth and Barthlott, 1979; Herth, 1979, 1980). Chitin was characterised using microscopic techniques. Mulisch *et al.* (1983) described the synthesis of lorica chitin in *Eufollicularia uhligi*. As far as we know, no studies have been carried out with algal cell-free systems.

2. Fungal chitin synthases

One puzzling fact emerging from the literature is that most of the chitin synthases described resemble the *Saccharomyces cerevisiae* proteinase-activated chitin synthase, now called 'chitin synthase I' (Bulawa *et al.*, 1986, Orlean, 1987), which apparently is not involved in yeast cell-wall chitin synthesis (Cabib, 1987). The logical conclusion is that all the fungal enzymes polymerising GlcNAc in cell-free systems are related. Another assumption is that most of these fungal type I enzymes should be involved in some way in GlcNAc transfer, contributing to the building, at least in part, of chitin molecules. In fungi, unlike yeast, these enzymes might be the main chitin synthases involved in chitin cell-wall synthesis. It is probable that inhibition of these enzymes by aminoglycosidic and other antibiotics (see Section IV.C and Müller *et al.*, 1981) reflects a well conserved evolutionary similarity at the substrate binding site. The whole protein itself may have been conserved through evolution. When more chitin synthase genes have been cloned these questions will be answered (Bartniki-García, 1970; Hänseler *et al.*, 1983). However, Bulawa *et al.* (1986) showed that low-stringency Southern analysis of digested yeast DNA, using sequences of DNA coding for chitin synthase I as probes, failed to detect additional putative isozymes.

(a) *Yeast chitin synthase.* Table 13.3 shows the differences and similarities between yeast chitin synthases I and II. Both are zymogenic according to Sburlati and Cabib (1986), but only ChS I is found to be activated by trypsin (Orlean, 1987). It has been proposed that ChS II is the physiological enzyme responsible for chitin synthesis, whereas the non-essential ChS I is not normally active *in vivo* (Orlean, 1987). Previous studies on *S. cerevisiae* and other species of yeast should now be re-evaluated and data interpreted in the light of current knowledge. A role should be found for type I enzymes since it is difficult to explain ChS I zymogen accumulation without some physiological output. Since these enzymes seem to be located in the plasma membrane, some cell-wall repair function might be involved and this could explain the necessity of an *in situ* protease activation. This putative endogenous activation could be the one detected at 45°C in yeast wild-type membranes (Orlean, 1987). Apparently, chitin synthase II seems responsible for the increase in chitin synthesis in MATa haploids, induced by the α-factor from compatible strain. Therefore, the enzyme required for mating, which implies cell to cell contact and protease participation, might be different from the enzyme involved in wound healing.

Whether both yeast-like chitin synthases are present in multicellular fungi remains an open question. Dimorphic species should be the systems of choice to study different chitin synthases (Sburlati and Cabib, 1987; Orlean, 1987).

TABLE 13.3. Properties of yeast chitin synthases.[a]

	ChS I	ChS II
No trypsin treatment	Inactive	Active
Trypsin treatment	Active (90%)	Active (10%)
Digitonin	Stimulates	Inhibition (80%)
Substrate	UDP-GlcNAc	UDP-GlcNAc
Product	poly-GlcNAc	poly-GlcNAc
Polyoxin D (K_i) competitive inhibition	Sensitive (1 μM)	Sensitive (1 μM)
NaCl inhibition (0.15 M)	Inhibition (86%)	Inhibition (29%)[b]
Optimum pH	7.0	8.0
Optimum temperature	40°C	25°C
Activity at 45–50°C	Active	Inactive
GlcNAc stimulation (80 mM)	3-fold	1.4-fold
Divalent cations		
$\quad$Mg^{2+}	Best[a,b]	Best[a,b]
$\quad$Co^{2+}	Inhibit[a,b]	Stimulate[a,b]
$\quad$Ca^{2+}	Inhibit (30%)	Inhibit (30%)
$\quad$Zn^{2+}	Inhibit (90%)	Inhibit (90%)
Kinetics		
$\quad K_m$ UDP-GlcNAc	0.5 mM[a]	0.6–0.8 mM[a]
$\quad V_{max}$	6 nmol min^{-1} mg^{-1}	0.55 nmol min^{-1} mg^{-1}

[a] Most data on the table are taken from Orlean (1987). ChS II activity corresponds to chs 1:URA 3 (disrupted gene) mutant and also to wild type, ChS I strain (no trypsin activation).
[b] Data from Sburlati and Cabib (1986).

(b) *Filamentous fungal chitin synthases.* In filamentous fungi, chitin is deposited at the tip of growing hyphae (Bartnicki-García and Lippmann, 1969; Grove and Bracher, 1970; Gooday, 1971, 1978; Muzzarelli, 1977; Cabib and Shematek, 1981; Wessels *et al.*, 1983). Most of the described particulate protease-activated chitin synthases from multicellular mycelial fungi are similar to the *Neurospora* enzyme (Glaser and Brown, 1957). It has been described in several systems as *Allomyces macrogynus* (Porter and Jaworski, 1966), *Blastocladiella emersonii* (Plessman-Camargo *et al.*, 1967), *Mucor rouxii* (McMurrough and Bartnicki-García, 1971; McMurrough *et al.*, 1971; Ruiz-Herrera *et al.*, 1977), *Aspergillus flavus* (Moore and Peberdy, 1976; López-Romero and Ruiz-Herrera, 1976), *Phycomyces blakesleanus* (Jan, 1974), etc.

(c) *Cell-free systems.* Most of the particulate enzymes showed a general requirement for divalent cations (Mg^{2+} is usually the best activator), maximum activity at neutral pH and were inhibited by Polyoxin D (Muzzarelli, 1977; Cabib, 1987). K_m values for UDP-GlcNAc usually range from 0.5 to 3.0 mM. GlcNAc has been described as an allosteric activator of several chitin synthases (McMurrough and Bartnicki-García, 1971; Gooday and Rousset-Hall, 1975; Rousset-Hall and Gooday, 1975; Ruiz-Herrera *et al.*, 1975, 1977; Gooday, 1978; Hänseler *et al.*, 1983); however, it does not stimulate other fungal systems such as *Martierella vinacea* (Peberdy and Moore, 1975). Soluble chitodextrins have also been described as activators (Keller and Cabib, 1971). Diacetyl chitobiose was proposed as an intermediate in chitin synthesis in *Blastocladiella emersonii* (Plesman-Camargo *et al.*, 1967) but probably was an activator in that system.

Only a few chitin synthases, like the *Coprinus cinereus* enzyme, have been isolated in an active form (Gooday and Rousset-Hall, 1975) but, even in that case, activation by endogenous proteinases prior to or during the extraction, could not be discounted. As mentioned above concerning the assay of yeast chitin synthase, most fungal chitin synthesis systems *in vitro* should be activated (up to 40-fold) by proteases. Several reported physiological regulatory molecules were found to be proteases (Stoddart, 1984). Trypsin is the most frequently used protease and it is active in *Neurospora* (Bartnicki-García *et al.*, 1978b; Arroyo-Bergovich and Ruiz-Herrera, 1979), *Phycomyces* (Van Laere and Carlier, 1978), *Aspergillus* (Ryder and Peberdy, 1971; Archer, 1977) and many other systems (Bartnicki-García *et al.*, 1978a; Braun and Calderone, 1978; Vermeulen *et al.*, 1979). Enzyme from *Mucor rouxii*, *Agaricus bisporus* or *Aspergillus flavus* could be activated by acid proteases (López-Romero and Ruiz-Herrera, 1976; Ruiz-Herrera *et al.*, 1977; Hänseler *et al.*, 1983), but yeast enzyme can also be activated by neutral proteases (Hasilik and Holzer, 1973; Ulane and Cabib, 1976). Duran and Cabib (1978) demonstrated that the yeast enzyme exists in a real zymogen form since after solubilisation the enzyme still remained inactive and was further activated by trypsin treatment.

Farkas (1979) suggested that inhibition of chitin synthases occurs through the binding of a protein inhibitor. In that case, zymogen would be formed by both the active enzyme and its inhibitor, and would be activated by local proteases. This model is not supported by clear experimental evidence, but cannot be ruled out. Inactivation of chitin synthase preparations by proteolytic enzymes was discussed (Hasilik, 1974).

At low concentrations (about 10 µM) amphotericin B methyl ester slightly stimulated (15%) *A. bisporus* chitin synthase activity (Hänseler *et al.*, 1983). Substances showing antimycotic activity like digitonin (Wulff, 1968) and other saponins had a stimulatory effect on the same enzymatic system (Ruiz-Herrera *et al.*, 1980; Hänseler *et al.*, 1983). Interactions of these saponin molecules with components of chitin synthesis are poorly understood but probably a detergent effect is involved. Dissociation of chitin synthase into subunits and also interactions with sterols have been described (Ruiz-Herrera *et al.*, 1980; Hernández *et al.*, 1981).

From activation experiments of different zymogenic chitin synthases with different proteases and other reagents it can be concluded that the synthases themselves or the accompanying lipids and/or glycoproteins are different, (Ruiz-Herrera and Bartnicki-García 1976; Bartnicki-García *et al.*, 1978a,b, 1979: Ruiz-Herrera *et al.*, 1980).

Synthesis of fungal chitin can be achieved with a soluble enzyme preparation (Cabib, 1972, 1981). As it occurs with other membrane-bound enzymes, detergents, butanol and digitonin were found to be useful in partial or total solubilisation of chitin synthases (Cabib, 1972, 1981, 1987; Muzzarelli, 1977; Stoddart, 1984). Depending on the procedure of solubilisation and purification, multicomponent vesicles, aggregates or almost pure soluble enzymes can be obtained. The apparent specific activity of the solubilised enzymes from a given mycelium can be very different from the corresponding crude preparations. Both activation and loss of activity can occur. Digitonin is a powerful tool for chitin synthase solubilisation. It was first used by Gooday and Rousset-Hall (1975) to isolate a *Coprinus cinereus* enzyme from a particulate crude fraction. A number of successful preparations from different fungi have been obtained with a similar solubilisation method (Duran and Cabib, 1978; Braun and Calderone, 1979; Ruiz-Herrera *et al.*, 1980; Hänseler *et al.*, 1983; Vermeulen and Wessels, 1983). In some cases, as in the yeast form of *Candida albicans*, a combination of detergent treatment (2.5% Na-deoxycholate) followed by digitonin extraction (1%) was found to be the best solubilising procedure (Braun and Calderone, 1979). Recently the yeast chitin synthase I has been purified to near homogeneity (Kang *et al.*, 1984). The ability of the *in vitro* synthesised nascent chitin-like chains to bind the enzyme was used as the main purification step. An apparent molecular weight of 570 000 was found for the native enzyme, which is similar to that reported for *Agaricus bisporus* and *Mucor rouxii* enzymes (Ruiz-Herrera *et al.*, 1980; Hänseler *et al.*, 1983). The *C. cinereus* chitin synthase has been purified by Montgomery *et al.* (1984). The enzyme does not require activation by protease. When denatured and subjected to electrophoresis in polyacrylamide gels a major band of 65 000 Da was found together with other minor bands.

Chitin synthases from a variety of fungi have been isolated as components of vesicles. Myers and Cantino (1974) reported that γ-particles involved in the genesis of the cyst wall of *Blastocladiella emersonii* contain chitin synthase bound to (glyco)lipids. These particles were sedimented at 20 000 × g and could be separated in sucrose gradients (Mills and Cantino, 1978). It is not clear if such γ-particles are related to the vesicles called 'chitosomes' (30–100 nm) which are assumed to be a kind of microorganelle by a number of authors (Bracker *et al.*, 1976; Ruiz-Herrera *et al.*, 1977; Bartnicki-García *et al.*, 1978a,b; Hänseler *et al.*, 1983). These chitosomes remain suspended in a 54 000 × g supernatant and have been observed in electron microscopy images (Ruiz-Herrera *et al.*, 1975; Hänseler *et al.*, 1983). The standard system for the isolation of chitosomes has been described by Bartnicki-García *et al.* (1978a,b) and Mills and Cantino (1981). As noted by Kang *et al.* (1985), the method of enzyme preparation seems to be crucial in trying to understand the real nature of chitosomes. The method described (Bartnicki-García *et al.*, 1978a) requires the rupture of the mycelium with glass beads, although a milder method has also been used (Scarborough, 1975; Bartnicki-García *et al.*, 1984).

Most of the chitosomal chitin synthases show substantial activity only when they are treated with proteases (Gooday and Rousset-Hall, 1975; Ruiz-Herrera and Bartnicki-García, 1976; Duran and Cabib, 1978; Gooday, 1978; Hänseler *et al.*, 1983). Artificial or not, when incubated with the sugar donor UDP-GlcNAc and a divalent cation these protease-activated particles give rise to chitin microfibrils which appear to be insoluble and are visualised by electron microscopy (Ruiz-Herrera and Bartnicki-García, 1974; Jan, 1974; Hänseler *et al.*, 1983).

(d) *Requirement of membrane cofactors*. From the data in the literature it can be deduced that, as for many membrane-bound enzymes, chitin synthases seem to require other membrane components to reach maximum stimulation, probably through conformational changes. These putative allosteric changes can explain a number of contradictory data concerning the *in vitro* action by lipids, surfactants and detergent agents (Stoddart, 1984). Experiments with particulate preparations containing membrane-bound glycosyl transferases have shown that detergents interact with both the proteins, including the enzyme itself, and the accompanying lipids, in a coordinated way (Belocopitow *et al.*, 1977; Ochoa, A. pers. commun.).

The *Neurospora* enzymatic complex (Glaser and Brown, 1957) can be solubilised with *n*-butanol. The resulting extracted preparation is still active in synthesis but loses its apparently allosteric stimulation by GlcNAc. The same effect has been observed in other purified systems (Kang *et al.*, 1984). Chitin synthase-containing particles from *Blastocladiella emersonii* were very rich in lipids (Mills and Cantino, 1974). Glycolipids accounted for around 15% of the total lipids. The same authors obtained direct evidence for the role of a palmitate-rich glucosyl diacylglycerol as a component of the chitin synthase milieu. They apparently demonstrated that this glycolipid acted as receptor for GlaNAc transferred from UDP-GlcNAc. It was suggested that the glycolipid was acting as a true primer (Mills and Cantino, 1980). Involvement of lipid-bound intermediates in chitin synthesis has been investigated in fungal systems (Endo *et al.*, 1970). Some lipidic extracts of *Mucor rouxii* were highly stimulatory for chitin synthesis (McMurrough and Bartnicki-García, 1971). In addition to the above described glycolipid requirement of the *Blastocladiella* enzyme, some other chitin synthase preparations have been found to require phospholipids, which are natural detergents (Duran and Cabib, 1978; Vermeulen and Wessels, 1983; Kang *et al.*, 1984). *In vitro* chitin synthesis is frequently stimulated at low concentrations of neutral detergent and inhibited at higher concentrations (Lyr and Seyd, 1978, 1979).

(e)   *Formation of chitin microfibrils.* The deposition of chitin microfibrils in both fungi cell walls and insect cuticles is functionally linked to their biosynthesis. In *S. cerevisiae* chitin seems to be deposited on the external face of isolated plasma membranes (Cabib *et al.*, 1983). The enzyme itself seems to be attached to the plasma membrane. Duran *et al.* (1975) stabilised protoplast using the Scarborough technique (1975), coating the external membrane with bound concanavalin A. The experiments showed that after osmotic disruption most of chitin synthase activity remained associated with the plasma membrane ghosts. These results have apparently been confirmed in yeast and other fungi (Braun and Calderone, 1978; Duran *et al.*, 1979; Dalley and Sonneborg, 1982; Kang *et al.*, 1985). However, using similar methods, Vermeulen *et al.* (1979) found only half of the *Schizophyllum commune* chitin synthase associated with plasma membrane. The same result was obtained with yeast plasma membranes isolated from homogenates (Schekman and Brawley, 1979). Using digitonin, mycelia from *A. bisporus* were extracted obtaining two different fractions showing chitin synthase activity. It was concluded that one of the enzymes was probably microsomal whereas the other might be located in the vicinity of the cell wall (Hänseler *et al.*, 1983).

Unfortunately it is difficult to correlate this kind of result with electron microscopy images of nascent chitin fibrils (1–10 nm thick) extruding from the chitosomes (30–50 nm diameter), since preparation of these vesicles involves fragmentation and, possibly, reconstitution of mixed membranes. The main question remains to be answered whether chitosomes are true microorganelles, as postulated by Bartnicki-García's group, or only 'artificially generated by disruption of other organelles or membranes' as postulated by Cabib (1987).

Sphaeroplasts of yeast are able to regenerate cell wall when incubated in a suitable medium (Phaff, 1971). However, the composition and organisation of the newly synthesised cell wall is not usually similar to that in the intact cells (Hunter and Rose, 1971). When protoplasts of *C. utilis* synthesised and secreted cell-wall components, chitin was overproduced (García Mendoza and Navaes Ledieu, 1968). This seems to indicate that the enzyme regulation was deranged and suggests that disturbance of the surface cell and/or temporary disappearance of the compartment between the plasma membrane and cell wall affects some regulatory component of the chitin synthesis machinery.

From several different approaches it can be inferred that *in vivo* deposition of microfibrils probably occurs at the external face level of the plasma membrane. Whether or not the chitin synthase(s) travel to the plasma membrane associated to special vesicles (chitosomes) or via the normal traffic of vesicles (together with other membrane-bound glycoproteins) remains to be studied. Chitosomes seem too large and too regular in size to be interpreted as an artifact originating from plasma membrane disruption. The so-called 'vectorial synthesis' model for chitin proposed by Cabib (1987) seems logical but it has been proposed taking into account properties of an enzyme not synthesising chitin *in vivo* (Duran *et al.*, 1975; Cabib *et al.*, 1983; Cabib, 1987).

(f) *Genetic and regulatory aspects.* A brute-force screening technique was used by Bulawa *et al.* (1986) to isolate yeast mutants defective in chitin synthase obtained by nitroso guanidine treatment (Hartwell, 1967). These mutant ChS I lacking *in vitro* chitin synthesis activity defined the gene ChS I responsible for the above described trypsin-activated chitin synthase I activity (Cabib and Farkas, 1971; Duran and Cabib, 1978). The heterozygous diploids obtained through mating to wild-type exhibited half of the chitin synthase activity, thus showing that gene dosage compensation does not occur. ChS I was found to be a single recessive chromosomal mutation by tetrad analysis.

The ChS I gene was cloned by transformation of a ChS I Ura 3 yeast strain with a yeast genomic library in the plasmid vector YEp24, followed by selection of uracil prototrophs and monitoring of chitin synthase activity (Bulawa *et al.*, 1986). The recovered ChS I transformants were isolated and the DNA sequence of the gene was obtained. The DNA sequence contained a single open reading frame encoding a protein of 1131 amino acids with a putative molecular weight of 130 000 Da. The protein seems to contain a hydrophobic carboxyl terminus, a neutral central sequence and a hydro-phillic amino terminus. The sequence contained eleven Asn—X—Ser/Thr sequences as possible sites for Asn-glycosylation.

After transformation of *Saccharomyces pombe*, a fission yeast lacking chitin and chitin synthase activity (Phaff, 1977), with the cloned ChS I fused to β-galactosidase, both enzymatic activities were demonstrated in the transformants. However, as men-tioned before, the ChS I mutants lacking chitin synthase I were able *in vivo* to synthesise normal levels of chitin (Bulawa *et al.*, 1986). The ChS I gene in a wild-type strain was disrupted with the Ura 3 gene and the new mutants were able to bud and contained normal levels of chitin, thus showing that the cloned gene is not essential for chitin synthesis nor for normal growth (Bulawa *et al.*, 1986).

A second enzyme, chitin synthase II, was detected in the ChS I :: Ura 3 strain by Sburlati and Cabib (1987) and Orlean (1987). Although this enzyme represents only 5% of the total *in vitro* chitin synthesis activity, it seems to be the physiological enzyme. This new enzyme seems responsible for the α-mating pheromone-induced chitin deposition in MATa cells described by Schekman and Brawley (1979). However, this hormone also stimulates twice the levels of chitin synthase I zymogen (Orlean, 1987). As far as we know, the gene responsible for this physiological chitin synthase II has not yet been cloned.

3. Arthropod chitin synthesis

As already mentioned, chitin synthesis in animals is difficult to study *in vitro*. Except for the case of certain molluscs, chitin fibrils from metazoa are always bound to proteins and/or glucans. Such association could be covalent or non-covalent.

Chitin from arthropod cuticular structures is partially digested by chitinases from the moulting fluid at the beginning of each instar (Powning and Irzykiewicz, 1963; Bade, 1975, 1978; Spindler, 1976; Mommsen, 1978, 1980). The monosaccharides produced (deacetylated or not) enter the sugar nucleotide pathway and are reutilised, at least in part, for new chitin synthesis. This was demonstrated by Gwinn and Stevenson (1973a,b) in the moulting crayfish and by Surholt (1975a) in larvae of locust (*Locusta migratoria*). UDP-GlcNAc was detected in crustacean epidermis (Kent and Lundt, 1958) and in the haemolymph of *Cecropia* adults (Carey and Wyatt, 1960). The *in vivo* synthesis of chitin during the first few days after moulting was demonstrated in the desert locust *Schistocerca gregaria* (Candy and Kilby, 1962). *In vivo* studies using the radiolabelled sugar tracer technique have been carried out in a number of insects and crustacea (Kramer and Koga, 1986). With pure preparations of epidermis (wing extracts) it was confirmed that the pathway leading to the synthesis of UDP-GlcNAc in insects was operative, but the direct transfer of *N*-acetyl glucosamine to the polysaccharide *in vitro* was not detected (Candy and Kilby, 1962). The UDP-GlcNAc synthetic pathway was also found in crustacean systems (Hohnke, 1971; Gwinn and Stevenson, 1973b).

Jaworski *et al.* (1963), working with cell-free extracts from southern armyworm (*Prodenia eridania*) larvae incubated with UDP-[^{14}C]GlcNAc, obtained incorporation of radioactivity into alkali-insoluble material which was shown to be chitin-like (Porter and Jaworski, 1966). The latter identification is, unfortunately, not always performed and, as mentioned above, wrong interpretation of data may occur.

The first UDP-GlcNAc-dependent chitin synthesis activity in crustacean preparations was described by Carey (1965). The amount of radiolabelled chitin obtained from injection experiments is usually much higher than that synthesised in culture by similar amounts of tissue (Quesada-Allué *et al.*, 1976a). These difficulties arise from the fact that the performance of commonly used insect tissue culture media depends on the species and usually requires haemolyph factors. For example, when imaginal disks of *Drosophila* are cultured *in vitro* (Fristrom, 1981), different kinds of glycosyl transferases, including those belonging to the dolichol pathway as well as those involved in chitin synthesis, are much more active in Robb's medium (Robb, 1969) prepared according to Eugene *et al.* (1979) than in Grace's or other media (Grace, 1962). Labelled glucosamine or *N*-acetyl glucosamine are the best and most used tracers (Quesada-Allué, 1973, 1978, 1982; Oberlander and Leach, 1975; Sowa and Marks, 1975; Surholt, 1975a; Fristrom *et al.*, 1982; Turnbull and Howells, 1982). Glucose has been successfully used as precursor (Hornung and Stevenson, 1971; Surholt, 1975a,b; Vardanis, 1976; Quesada-Allué *et al.*, 1976b; Ratnakaran and Hackman, 1985), although in long *in vivo* experiments (more than 4 h) first glycerolipids and then amino acids became labelled (L. A. Quesada-Allué, unpubl. res.). The use of the physiological precursor trehalose, which is the main circulating saccharide in most arthropods, is very useful in detecting preferential biosynthetic pathways in a given tissue (Quesada-Allué *et al.*, 1976b). Fructose has been used as chitin precursor in *Stomoxys* (Mayer *et al.*, 1979, 1980a).

(a) *Insect chitin synthases*. As mentioned above, particulate preparations showing chitin synthase activity are difficult to obtain and to characterise. *In vitro* cell-free synthesis of chitin always seems to require a primer to initiate polymerisation. This putative primer could be pre-existing short chitin molecules, or unknown (glyco)proteins bound to the enzyme or some domain of the enzyme itself.

Most of the cell-free experiments using insects were performed using microsome-enriched preparations (Quesada-Allué *et al.*, 1976b; Cohen and Casida, 1980a,b, 1982a). Similar particulate preparations have been obtained from crustacean or aracnid tissues (Carey, 1965; Hohnke, 1971; Mayer *et al.*, 1980a, 1981; Horst, 1981, 1983). As in the case of fungal synthases, arthropod enzymes are usually dependent on Mg^{2+} ions for significant activity. Optimum pH is usually neutral, being 7.2 in insects (Quesada-Allué *et al.*, 1976b; Cohen and Casida, 1982b), or slightly basic (7.0–8.0 in crustaceans: Carey, 1965; Hohnke, 1971; Horst, 1981). In a few cases it was reported that insect chitin synthesis systems were associated with 'chitosomal-like vesicles' (Cohen and Casida, 1982a). Mayer *et al.* (1980a) reported the activation of stable fly microsome chitin synthase by incubation with trypsin.

No significant purification of any arthropod enzyme has been achieved (Sowa and Marks, 1975; Muzzarelli, 1977; Andersen, 1979; Mayer *et al.*, 1980a,b) and the enzyme complex seems to be quite unstable. In our experience, both sucrose and glycerol are required to stabilise the particulate enzyme. In hemimetabolous insects like *Triatoma* or *Tribolium* (Cohen and Casida, 1980a,b; Quesada-Allué, 1982) and in holometabolous insects, chitin synthesis is triggered simultaneously with cuticular (glyco)protein synthesis (Silvert and Fristrom, 1980; Silvert *et al.*, 1984; Doctor, 1984). As mentioned above, nascent radiolabelled chitin is very difficult to distinguish from radioactive N,N'-diacetylchitobiose-containing glycoproteins from the cuticle. These cuticular glycoproteins are usually very insoluble and this fact, together with the usual difficulties inherent in analysing chitin, creates doubts about the interpretation of certain published data on insect chitin synthase activities. In some cases, certain particulate preparations seem to synthesise GlcNAc-containing material other than (or in addition to) chitin (Mayer *et al.*, 1980a,b). From a number of indirect data it can be postulated as a working hypothesis that chitin synthesis in arthropods probably occurs in two steps: first a proximal, probably very short chain is synthesised upon a (glyco)protein (Quesada-Allué *et al.*, 1976b; Quesada-Allué, 1978, 1982; Kramer and Koga, 1986) or a glycolipid (Horst, 1981, 1983) primer. Secondly, a classic polymerase activity, probably similar to yeast chitin synthase I, would eventually elongate the nascent chitin at a very fast rate. This would explain kinetic data (Quesada-Allué, 1982) which can be interpreted by either conformation-dependent changes of a unique allosteric enzyme or by participation of two different enzymes. Inhibition studies also point to the possibility of there being more than one step in arthropod chitin synthesis.

The putative role of lipid-bound intermediates of the dolichol type in insect chitin synthesis cannot be ruled out (Quesada-Allué *et al.*, 1976b; Quesada-Allué and Belocopitow, 1978; Parodi and Leloir, 1979; Quesada-Allué, 1982). Many indirect data support this idea but no conclusive results have been obtained as yet. However, lipid intermediates seem to be involved in *Artemia* chitin synthesis (Horst, 1981, 1983). Lipid-bound GlcNAc-labelled oligosaccharides (2–8 monosaccharide units) were synthesised and are postulated to be putative chitin nascent chains (Horst, 1983), but these results can have alternative interpretations.

Tunicamycin was found to inhibit both fungal (Selitrenikoff, 1979) and insect chitin synthesis (Marks *et al.*, 1982; Quesada-Allué, 1982), probably through a competitive effect (Mayer and Chen, 1985). Inhibition did not occur in other systems (Mayer *et al.*, 1981; Bade, 1983: Doctor, 1984). Kramer and Koga (1986) suggested that there may be two different pathways for chitin synthesis, one involving dolichyl derivatives as intermediates and another using only UDP-GlcNAc as direct donor. These ideas are compatible with the possibility of a two-step chitin synthesis.

Insect and probably crustacean chitin synthesis is dependent on ecdysteroid circulating levels (Marks, 1973, 1980; Silvert and Fristrom, 1980; Fristrom *et al.*, 1982; Kramer and Koga, 1986). On the other hand, some data suggested a regulated reciprocal relationship between glycogen catabolism and chitin syntheses.

As far as we know, mutations clearly affecting chitin synthesis have not been reported in insects. A *Drosophila melanogaster* mutant, lethal cryptocephal (CRC), was proposed as possibly having affected chitin synthesis since an excess of chitin in the cephalic region would prevent the eversion of the head during pupal development (Fristrom, 1965). Sparrow and Chadfield (1982) were unable to find differences in chitin synthesis activity in these mutants, suggesting that only local changes in the amount of cuticle synthesis may occur.

4.   Chitosan synthesis

Chitosan was detected in the mycelium and sporangia of *Phycomyces blakesleanus* (Kreger, 1954). It was later found in the cell wall of some fungi together with chitin and other glucans (Bartnicki-García and Nickerson, 1962; Bartnicki-García, 1968; Muzzarelli, 1977). Like chitin, chitosan is not a uniquely defined compound, since the amount of deacetylation fluctuates, depending on the organism and on the isolation procedure. Sometimes the fully deacetylated product is called chitan (Muzzarelli, 1977). Changes in the chitosan content of different structures have been recorded during the life-cycle of *Mucor rouxii* (Bartnicki-García, 1968). Hyphae cell walls contain 33% chitosan whereas yeast cell walls contain 28% and spores only 10%. The ratio of chitosan to chitin was roughly maintained (3.4, 3.3, and 4.5, respectively), suggesting some coordinated balance between synthesis and degradation.

Araki and Sito (1974, 1975) demonstrated that chitosan is a chitin degradation product since it results from chitin deacetylation. These authors purified a soluble chitin deacetylase from *Mucor rouxii*. To study the enzyme they used a soluble patially *O*-hydroxyethylated chitin tritiated in the *N*-acetyl groups. The K_m for the monosaccharide residues was 2.6 mM and the pH optimum was 5.5, and no divalent cations were required. Therefore the 'chitosan synthase' could be easily measured in the absence of significant chitin synthesis. Chitosan seems to originate preferentially from nascent chitin chains (Davis and Bartnicki-García, 1984a,b). However, since chitin deacetylase is soluble, deacetylation probably occurs in a hydrophillic environment and not inside or in the vicinity of the membranes where chitin synthesis initiates. The coupled synthesis of both polymers was followed by Davis and Bartnicki-García (1984a). There are no studies on invertebrate putative chitin deacetylase.

V. INHIBITION OF CHITIN SYNTHESIS

Since no chitin is found in the evolutionary branch of deterostomiates leading to vertebrates nor in higher plants, hundreds of potential specific chitin synthesis inhibitors have been tested as fungicides and/or insecticides (Leighton *et al.*, 1981). Many advances in the study of chitin synthases were made possible by using *in vitro* chitin inhibitors (Brillinger, 1979; Suhadolnik, 1983; Hajjar, 1985; Cabib 1987). Although not directly related to plant metabolism, the eventual undesirable effects of chitin synthesis inhibitors, both at the physiological and ecological levels, should be taken into account. A complete compilation of chitin synthase inhibitors is beyond the scope of this review, but we will briefly discuss the most important ones.

A. Polyoxins and Neopolyoxins

The polyoxins are a group of peptidyl-pyrimidine nucleoside antibiotics from *Streptomyces cacaoi* strain called *asoensis* (Isono *et al.*, 1965, 1967, 1969, 1971a,b; Isono and Suzuki, 1966; Suhadolnik, 1970, 1983; Endo *et al.*, 1970). Polyoxin D is the major and most potent chitin synthesis inhibitor among 12 or more active polyoxin complex components (Endo *et al.*, 1970). As shown in Fig. 13.7, the polyoxins are structurally similar to UDP-GlcNAc, thus acting as competitive inhibitors of chitin synthesis. This was first suggested by Isono *et al.* (1969) and then demonstrated by Endo and Misato (1969) working with filamentous fungi. Chemical data on polyoxins have been reviewed by Suhadolnik (1983).

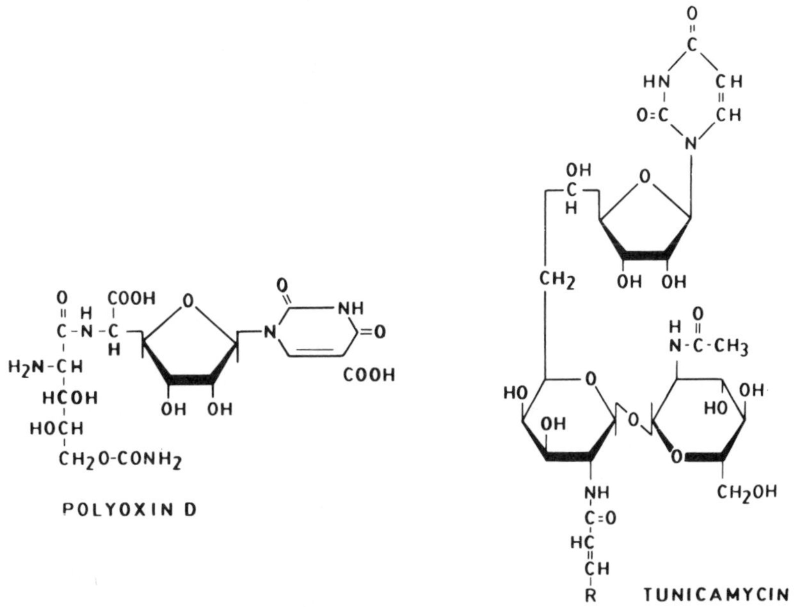

FIG. 13.7. Structures of polyoxin D and tunicamycin, inhibitors of chitin synthase and dolichyl-P-P-GlcNAc synthase, respectively.

The active groups of polyoxin D contributing to the competitive inhibition are the free carboxyl group attached to C-5 and its proximal α-L amino group (Isono *et al.*, 1971a,b; Hori *et al.*, 1974a,b). The possible mechanism of interaction between the chitin synthase binding and catalytic sites and polyoxins has been discussed by Hori *et al.* (1974b).

1. Cell-free studies

Cell-free studies showed that polyoxin D inhibition constants of fungal chitin synthases for the substrate UDP-GlcNAc were in the micromolar range (K_i = 1.4–6.5 µM for polyoxin D; K_i = 7.0–8.0 µM for polyoxins A, B and L) (Ohta *et al.*, 1970; Keller and Cabib, 1971; Isono *et al.*, 1973; Gooday *et al.*, 1976). This compares well with an UDP-GlcNAc K_m of 18 µM for *Mucor* chitin synthase (López-Romero and Ruiz-Herrera, 1976). In *Mucor rouxii* polyoxin D strongly inhibits both mycelial growth and spore germination showing a K_i of 0.6 µM for chitin synthase (Bartnicki-García and Lippmann, 1972).

Polyoxin D is a strong inhibitor of insect chitin synthesis systems in tissue culture and cell-free assays (Nishioka *et al.*, 1979; Mayer *et al.*, 1980a,b; Fristrom *et al.*, 1982; Cohen and Casida, 1982a; Turnbull and Howells, 1983; Cohen, 1985). Since insect chitin synthase preparations are usually crude extracts, inhibition constants for polyoxin D are difficult to compare with those obtained with fungal tissues. A crude enzyme of the fly *Lucilia cuprina* showed an apparent K_i of 40 nM (Turnbull and Howells, 1983). *Tribolium* chitin synthases were 50% inhibited by 4 µM polyoxin D (Cohen and Casida, 1980a,b) or by 1.1 µM in another *Tribolium* species (Cohen and Casida, 1982a,b). Polyoxin D inhibited 50% of chitin synthesis when epidermal tissue of *Triatoma infestans* was cultured together with 2 µg ml^{-1} of the antibiotic (Quesada-Allué, 1982). Significantly, the remaining 'non-inhibited' chitin was found to be more resistant to chitinase digestion than chitin in controls. This reinforces the idea of more than one chitin domain and perhaps reflects different synthesis steps (Quesada-Allué, 1982; Cabib, 1987). *Artemia salina* chitin synthase was not affected by polyoxin D *in vitro* (Horst, 1981, 1983), but 70% inhibition was found *in vivo* (Calcott and Fatig, 1984).

2. Uses as pesticides

The polyoxins are used as field fungicides in Japan, but their effects *in vivo* vary, probably because of problems of permeability since some peptides can reverse the polyoxin fungicidal effect through competition for the same incorporation channel (Mitani and Inone, 1968; Suhadolnik, 1983). Polyoxin D is only slightly inhibitory of *Alternaria kikuchiana* growth (Suhadolnik, 1983) but very effective against *Pellicularia* from infected rice (Shibuya *et al.*, 1972). On the other hand, polyoxin L is active against *Alternaria* from trees (Isono *et al.*, 1967, 1972). *Trichoderma viride* spores were sterile when treated with polyoxin D (Benítez *et al.*, 1976) and the effect was shown to be concomitant with an 83% inhibition of chitin synthesis. Eguchi *et al.* (1968) reported that polyoxins caused swelling of the cell wall. Polyoxins have not been used as field insecticides, probably because of penetration problems. Topication in larval stages has been used in trials with little success.

3. Neopolyoxins

Dähn et al. (1976) isolated from the culture medium of Streptomyces tendai several pyrimidine nucleosides related to polyoxins, called nikkomycins. These antibiotics are also structural analogues of UDP-GlcNAc and were shown to be powerful competitive inhibitors of chitin synthases from different fungal tissues (Dähn et al., 1976; Brillinger, 1979; Leighton et al., 1981). Apparently nikkomycins penetrate better than polyoxins in certain systems (Furter and Rast, 1985; Cabib, 1987). Inhibition constants for nikkomycins X and Z against Agaricus bisporus chitin synthase were found to be 1.5 and 8.5 µM, respectively, compared with amphotericin ($K_i = 0.1$ mM) and digitonin ($K_i = 4$ mM) (Hänseler et al., 1983).

Nikkomycins also inhibit in vitro synthesis of insect chitin (Cohen and Casida, 1980a,b, 1982a,b). In Tribolium gut chitin synthase, these antibiotics were more potent than polyoxin D, Captan or even UDP (Cohen and Casida, 1980a,b).

B. Benzophenylureas

Benzophenylureas were found apparently to block chitin synthesis in insects and other arthropods (Hajjar, 1985) but to be totally ineffective in the case of fungal tissues (Verloop and Ferrell, 1977; Misato and Kakiki, 1977; Lyr and Seyd, 1978; Van Eck, 1979; Marks et al., 1982). It is now clear that these substances seem to act as inhibitors of a metabolic step different from chitin synthase itself (Kramer and Koga, 1986), while strongly disturbing cuticle component deposition, including chitin (Soltani et al., 1983, 1984; Doctor, 1984).

Diflubenzuron [1-(2,6-difluorobenzoyl)-3-(4-chlorophenyl)urea] has been widely tested as an insecticide (Hajjar, 1985) and several analogues have been synthesised with the same purpose. Leighton et al. (1981) suggested that benzophenylurea derivatives may interact with some unknown regulator of the metabolic pathway leading to chitin synthesis, perhaps a proteolytic enzyme. In any case, these insecticides are very useful tools with which to study the different enzymatic activities required for cuticle building in insects. Accumulation of UDP-GlcNAc was observed in a number of insect tissue culture systems when treated with diflubenzuron (Marks and Sowa, 1976; Van Eck, 1979). In vivo experiments also give similar results (Mitsui et al., 1981). However, such apparent blockage of GlcNAc transfer was not observed in other insect systems (Turnbull and Howells, 1983). Direct inhibition of chitin synthase by benzophenylurea was reported for microsomal preparation of the brine shrimp Artemia salina (Horst, 1981). This seems to be a peculiar enzyme since it was found to be resistant to polyoxin D inhibition.

C. Other Pesticides Inhibiting Chitin Synthesis

Captan has been found to be a non-competitive inhibitor of chitin synthase activity in Phycomyces blakesleanus (Leighton et al., 1981) and in S. cerevisiae (Cohen and Casida, 1980b). No inhibition was found in Coprinus cinereus (Brillinger, 1979). Significant inhibitory activity has been described in a number of insect chitin synthesis systems like Leucophaera, Tribolium or Calliphora erythrocephala (Marks and Sowa, 1976; Becker, 1978; Cohen and Casida, 1980b, 1982a). The effect of captan could be due to binding and/or reaction of its breakdown product, thiophene, to any of the proteins involved in the chitin synthesis system.

Congo red (Serva) and calcofluor white ST (American Cyanamid), two substances having a strong affinity to polysaccharides, were found to inhibit chitin synthesis in the alga *Ochromonas malhamensis* (Herth, 1980). Calcofluor white H2R also inhibits fungal chitin synthase *in vitro* (Selitrenikoff, 1984; Roncero and Duran, 1985). Surprisingly, these dyes have also been found to stimulate *in vivo* chitin synthesis in *Saccharomyces* and other fungi (Elorza *et al.*, 1983; Vannini *et al.*, 1983).

Plumbagin, a substance isolated from a tropical plant *Plumbago*, inhibited insect chitin synthesis at the same level as polyoxins. No data are available on the mechanism (Kubo *et al.*, 1983). Several other herbicides were inhibitors of insect chitin synthesis whereas DDT seems to inhibit the *Phycomyces blakesleanus* enzyme (Leighton *et al.*, 1981; Miller *et al.*, 1981; Marks *et al.*, 1982). Non-specific effects of phenylcarbamates were studied by Cohen and Casida (1980b). Amphotericin B, nystatin and other polyene antibiotics have been proposed as chitin synthase inhibitors (Rast and Bartnicki-García, 1981). Avermectins, a family of potent antihelminthic drugs (Burg *et al.*, 1979) inhibit *Artemia* chitin synthesis (Calcott and Fatig, 1984) but recent results by Gordnier *et al.* (1987) showed no inhibition of chitin synthase from *Streptomyces*, *Tenebrio*, *Acheta* and *Galleria*. The same was found for the similar acaricidal antibiotics, milbemycins (Gordnier *et al.*, 1987). Avermectins and similar toxicants probably act as neurotoxins (Hajjar, 1985). Some isoprenoid derivatives (1-geranyl-2-methyl-, 1-geranyl- and 1-neryl-benzimidazole) slightly inhibited chitin synthase from *Tribolium* (Cohen *et al.*, 1984; Cohen, 1985).

D. Tunicamycin

Tunicamycin (Fig. 13.7) was isolated by Takatsuki *et al.* (1971) from both culture medium and mycelium of *Streptomyces lysoperficus* as a complex of several antibiotics. They have structural analogy with both UDP-GlcNAc and the isoprenoid dolichyl phosphate. Takatsuki and Tamura (1971a,b) first demonstrated that tunicamycin inhibited sugar incorporation, particularly GlcNAc, into glycoproteins. Several authors further demonstrated that tunicamycin specifically inhibits the transfer of GlcNAc-1-P from UDP-GlcNAc to dolichyl-P, thus blocking the synthesis of dolichyl-P-P-GlcNAc (Tkacz and Lampen, 1975; Takatsuki *et al.*, 1975; Lehle and Tanner, 1976; Schwarz and Datema, 1982). This is the first lipid intermediate in the Asn-glycosylation of proteins from which the glycosyl chain grows to give dolichyl-P-P-GlcNAc$_2$-Man$_9$-Glc$_3$, except in certain lower eukaryotes (Parodi *et al.*, 1981; Parodi and Quesada-Allué, 1982; Quesada-Allué and Parodi, 1983). This lipid–oligosaccharide is the donor of the glycosyl moiety which is transferred *en bloc* to proteins (Parodi and Leloir, 1979). The tunicamycin complex also inhibits protein synthesis through an unknown mechanism (Mahoney and Duskin, 1979; Miller and Silhacek, 1982).

1. Effects of tunicamycin on chitin synthesis

From kinetic data we first suspected some relationship between insect chitin synthesis and the lipid-bound intermediates involved in Asn-glycosylation (Quesada-Allué, 1973,

1978, 1979, 1980, 1981, 1982; Quesada-Allué *et al.*, 1975, 1976a,b; Belocopitow *et al.*, 1977; Quesada-Allué and Belocopitow, 1978; Mayer *et al.*, 1983). The hypothesis is difficult to test because of the similar subcellular locations of both membrane-bound systems and the low tracer incorporation. However, the idea was reinforced by data on tunicamycin inhibition.

Tunicamycin was reported as a linear competitive inhibitor of the *Neurospora crassa* chitin synthesis system with an apparent K_i of around 0.5 mM (Selitrenikof, 1979). Surprisingly, Marks *et al.* (1982) found a stimulation of chitin synthesis in *Phycomyces blakesleanus* produced by the antibiotic. The effect was attributed to the enhancement of substrate availability (UDP-GlcNAc) due to the blockage of the dolichol pathway.

Results in insect systems are also contradictory. Insect chitin synthesis was inhibited by tunicamycin in the cockroach leg regeneration system (Marks *et al.*, 1982). Quesada-Allué (1982) found that the antibiotic consistently inhibited the synthesis of radioactive chitin *in vitro* using the blood-sucking bug system. As previously mentioned, it was found that the 'tunicamycin sensitive' portion of chitin was more susceptible to chitinase degradation than the 'resistant' one. Similar results were obtained with *Galleria* larval epidermis (Quesada-Allué, 1982). From results obtained with the stable fly Mayer and Chen (1985) have suggested that tunicamycin may competitively inhibit the insect chitin synthases. If dolichyl derivatives are in fact crustacean chitin precursors (Horst, 1981, 1983), tunicamycin should block chitin synthesis. Tunicamycin did not inhibit chitin synthesis in *Drosophila* imaginal disks (Fristrom *et al.*, 1982) nor in epidermis from tobacco hornworm larvae (Bade, 1983). Chitin synthase was not inhibited in cell-free assays using tissues from *Tribolium castaneum* (Cohen and Casida, 1980b) and *Stomoxis calcitrans* (Mayer *et al.*, 1981). A summary of the effects of tunicamycin on chitin synthesis in different invertebrate and fungal systems is shown in Table 13.4.

TABLE 13.4. Effect of tunicamycin on chitin synthesis.

System	Inhibition	Source
Tissue culture		
Cockroach leg regeneration	+	Marks *et al.* (1982)
Triatoma abdominal epidermis	+	Quesada-Allué (1982)
Galleria larval epidermis	+	Quesada-Allué (1982)
Drosophila imaginal disks	−	Doctor (1984)
Particulate preparations		
Phycomyces blakesleanus	−[a]	Marks *et al.* (1982)
Neurospora crassa	−	Selitrenikoff (1979)
Ptribolium spp.	−	Cohen and Casida (1980b)
Stomoxis calcitrans	−	Mayer *et al.* (1980a, 1981)

[a] In this particular case stimulation was observed.

Kramer and Koga (1986) suggested that the variable effects of tunicamycin and other inhibitors could be due to both multiple types of insect chitin synthases and to inconsistencies in the experimental conditions of the different assays.

VI. CONCLUSIONS

The present knowledge of chitin chemistry indicates that chitin is not a defined substance. It is present in nature as a family of compounds, having in common the basic backbone of a linear chain of poly-N-acetylglucosamine. However, the degree of polymerisation and deacetylation is variable from one organism to another. On the other hand, chitin is found in close association with different molecules (glycans, proteins, pigments). The knowledge on the type of interaction between chitin chains and these molecules is just emerging. This is a consequence of the application of non-destructive techniques to the study of naturally occurring chitinous materials.

The biosynthetic pathway of chitin seems clear. The only step that still remains unclear is the polymerisation of GlcNAc. A few years ago this step was thought to be well understood, but application of the new recombinant DNA techniques surprisingly showed that the enzyme so far studied was not essential for the synthesis of chitin *in vivo*. The lesson to be learned from this is that we should be very cautious about the interpretation of data obtained from the cell-free systems, since they may have little or no physiological significance. The development of new techniques for the *in vivo* study of biochemical effects is becoming a necessity for achieving better understanding of nature.

ACKNOWLEDGEMENTS

The authors are indebted to Professor J. E. Varner from Washington University for his advice and criticism of the manuscript. Work in our laboratories was supported by grants from the Consejo Nacional de Investigaciones Científicas y Técnicas (CONICET) and the Comisión de Investigaciones Científicas de la Provincia de Buenos Aires (CIC) Argentina. RPL is on sabbatical leave from the Instituto de Investigaciones Biológicas, FCEyN, Universidad Nacional de Mar del Plata. RPL and LQA are established researchers from CIC and CONICET, respectively.

REFERENCES

Agui, N., Yagui, S. and Fukaya, M. (1969). *Appl. Ent. Zool.* **4**, 156–157.
Andersen, S. O. (1979). *Ann. Rev. Entomol.* **24**, 29–61.
Araki, Y. and Ito, E. (1974). *Biochem. Biophys. Res. Commun.* **56**, 669–675.
Araki, Y. and Ito, E. (1975). *Eur. J. Biochem.* **55**, 71–78.
Archer, D. B. (1977). *Biochem. J.* **164**, 654–658.
Arroyo-Bergovich, A. and Carabez-Trejo, A. (1982). *J. Parasitol.* **68**, 253–258.
Arroyo-Bergovich, A. and Ruiz-Herrera, J. (1979). *J. Gen. Microbiol.* **113**, 339–345.
Attwood, M. M. and Zola, H. (1967). *Comp. Biochem. Physiol.* **20**, 993–998.
Austin, P. R. (1975). US Patent 3,731,892.
Austin, P. R., Brine, C. J., Castle, J. E. and Zikakis, J. P. (1981). *Nature* **212**, 749–753.
Bade, M. L. (1975). *FEBS Lett.* **51**, 161–163.
Bade, M. L. (1978). *In* "Proceedings of the 1st International Conference on Chitin/Chitosan" (R. A. Muzzarelli and E. R. Pariser, eds), pp. 472–479, M.I.T. Press, Cambridge, MA.
Bade, M. L. (1983). *J. Appl. Polymer Sci.* **37**, 165–178.
Barengo, R. and Krisman, C. L. (1978). *Biochem Biophys. Acta* **540**, 190–196.
Barker, S. A., Foster, A. B., Stacey, M. and Webber, J. M. (1957). *Chem. and Ind. (London)*, 208–209.

Barker, S. A., Foster, A. B., Stacey, M. and Webber, J. M. (1958). *J. Chem. Soc.*, 2218–2227.
Bartnicki-García, S. (1968). *Ann. Rev. Microbiol.* **22**, 87–108.
Bartnicki-García, S. (1970). *In* "Phytochemical Phylogeny" (J. B. Harborne, ed.), pp. 81–103. Academic Press, London.
Bartnicki-García, S. and Lippmann, E. (1969). *Science* **165**, 302–304.
Bartnicki-García, S. and Lippmann, E. (1972). *J. Gen. Microbiol.* **71**, 301–308.
Bartnicki-García, S. and Nickerson, W. Y. (1962). *Biochem. Biophys. Acta* **58**, 102–119.
Bartnicki-García, S., Bracker, C. E. and Ruiz-Herrera, J. (1978a). *In* "Proceedings of the First International Conference on Chitin/Chitosan" (R. A. Muzzarelli and E. R. Parisier, eds), pp. 450–463. M.I.T. Press, Cambridge, MA.
Bartnicki-García, S., Bracker, C. E., Reyes, E. and Ruiz-Herrera, J. (1978b). *Exp. Mycol.* **2**, 173–192.
Bartnicki-García, S., Ruiz-Herrera, J. and Bracker, C. E. (1979). *In* "Fungal Walls and Hyphal Growth" (J. H. Burnett and A. P. J. Trinci, eds), pp. 149–169. Cambridge University Press, Cambridge.
Bartnicki-García, S., Bracker, C. E., Lippmann, E. and Ruiz-Herrera, J. (1984). *Arch. Microbiol.* **139**, 105–112.
Becker, P. (1978). *J. Insect Physiol.* **24**, 699–795.
Belocopitow, E., Maréchal, L. R. and Quesada-Allué, L. A. (1977). *Mol. Cell. Biochem.* **16**, 127–134.
Benítez, T., Villa, T. G. and García Acha, I. (1976). *Arch. Microbiol.* **108**, 183–189.
Benjaminson, A. A. (1969). *Stain Technol.* **44**, 27–31.
Bergman, M., Zervas, L. and Silberkweit, E. (1983). *Naturwissenschaften* **19**, 20.
Berkeley, R. C. W. (1979). *In* "Microbial Polysaccharides and Polysaccharases" R. C. W. Berkeley and G. W. Gooday, eds), pp. 205–236. Academic Press, London.
Blackwell, J. and Weih, M. A. (1980). *J. Mol. Biol.* **137**, 49–60.
Bracker, C. E., Ruiz-Herrera, J. and Bartnicki-García, S. (1976). *Proc. Natl. Acad. Sci. USA* **73**, 4570–4574.
Braconnot, H. (1811). *Ann. Chim. Phys.* **79**, 265–304.
Braun, P. C. and Calderone, R. A. (1978). *J. Bacteriol.* **133**, 1472–1477.
Braun, P. C. and Calderone, R. A. (1979). *J. Bacteriol.* **140**, 666–670.
Brillinger, G. U. (1979). *Arch. Microbiol.* **121**, 71–74.
Brine, C. J. (1982). *In* "Proceedings of 2nd International Conference on Chitin and Chitosan" (S. Hirano and S. Tokura, eds), pp. 105–110. Sapporo, Japan.
Broussignac. P. (1968). *Chim. Ind. Génie Chim.* **99**, 1241–1247.
Bulawa, C. E., Slater, M., Cabib, E., An-Young, J., Sburlati, A., Adair, W. L. and Robbins, P. W. (1986). *Cell* **46**, 213–225.
Burg, R. W., Miller, B. M., Baker, E. E., Birnbaum, J., Currie, S. L., Hartman, R., Kong, Y. L., Managhan, R. L., Olson, G., Putter, I., Tunac, J. B., Wallick, H., Stapley, E. O., Oiwa, R. and Ounira, S. (1979). *Antimicrob. Agents Chemother.* **15**, 361–367.
Bussers, C. J. and Jeuniaux, Ch. (1974). *Protistologica* **10**, 43–46.
Cabib, E. (1972). *Methods Enzymol.* **28**, 572–580.
Cabib, E. (1981). *In* "Encyclopedia of Plant Physiology" (W. Tanner and F. W. Loewus, eds), pp. 395–415. Springer, Heidelberg.
Cabib, E. (1987). *Adv. Enzymol.* **59**, 59–101.
Cabib, E. and Farkas, V. (1971). *Proc. Natl. Acad. Sci. USA* **68**, 2052–2056.
Cabib, E. and Shematek, E. M. (1981). *In* "Biology of Carbohydrates" (V. Ginsburg and P. W. Robbins, eds), Vol. 1, pp. 51–90. John Wiley & Sons, New York.
Cabib, E., Bowers, B. and Roberts, R. L. (1983). *Proc. Natl. Acad. Sci. USA* **80**, 3318–3321.
Calcott, P. H. and Fatig, R. O. (1984). *J. Antibiotic.* **37**, 253–259.
Candy, D. J. and Kilby, B. A. (1962). *J. Exp. Biol.* **39**, 129–140.
Capozza, R. C. (1975). Ger. Patent 2,255,305.
Carey, F. G. (1965). *Comp. Biochem. Physiol.* **16**, 155–158.
Carey, F. G. and Wyatt, G. R. (1960). *Biochem. Biophys. Acta* **41**, 178–179.
Chamberland, H., Charest, P. M., Ollette, G. and Pallze, F. J. (1985). *Histochem. J.* **17**, 313–321.
Clark, G. L. and Smith, A. F. (1936). *J. Phys. Chem.* **40**, 863–879.

Cohen, E. (1985). *Experientia* **41**, 470–472.

Cohen, E. (1987). *Ann. Rev. Entomol.* **32**, 71–93.

Cohen, E. and Casida, J. E. (1980a). *Pestic. Biochem. Physiol.* **13**, 121–128.

Cohen, E. and Casida, J. E. (1980b). *Pestic. Biochem. Physiol.* **13**, 129–136.

Cohen, E. and Casida, J. E. (1982a). *Pestic. Biochem. Physiol.* **17**, 301–306.

Cohen, E. and Casida, J. E. (1982b). *In* "Pesticide Chemistry, Human Welfare and the Environment" (S. Matsunaka, D. H. Hutson and S. D. Murphy, eds), Vol. 3, pp. 25–32. Pergamon Press, New York.

Cohen, E., Kuwano, E. and Eto, M. (1984). *Agric. Biol. Chem.* **48**, 1617–1620.

Dähn, U., Hagenmaier, J. H., Höhne, H., König, W. A., Wolf, G. and Zähner, J. (1976). *Arch. Microbiol.* **107**, 143–149.

Dalley, N. E. and Sonneborg, D. R. (1982). *Biochim. Biophys. Acta* **686**, 65–75.

Davies, D. H., Hayes, E. R. and Lal, G. S. (1985). *In* "Chitin in Nature and Technology" (R. Muzzarelli, Ch. Jeuniaux and G. W. Gooday, eds), pp. 365–370. Plenum Press, New York.

Davis, L. L. and Bartnicki-Garcia, S. (1984a). *J. Gen. Microbiol.* **130**, 2095–2102.

Davis, L. L. and Bartnicki-Garcia, S. (1984b). *Biochemistry* **23**, 1065–1073.

de Duve, C. (1964). *J. Theoret. Biol.* **6**, 33–59.

Domard, A. and Rinaudo, M. (1983). *Int. J. Biol. Macromol.* **5**, 49–52.

Doctor, J. S. (1984). Ph.D. Thesis, University of California, Berkeley.

Duran, A. and Cabib, E. (1978). *J. Biol. Chem.* **253**, 4419–4425.

Duran, A., Bowers, B. and Cabib, E. (1975). *Proc. Natl. Acad. Sci. USA* **72**, 3952–3955.

Duran, A., Cabib, E. and Bowers, B. (1979). *Science* **203**, 363–365.

Eguchi, J., Sasaki, S., Ohta, N., Akashiba, T., Tsuchiyama, T. and Suzuki, S. (1968). *Am. Phytopathol. Soc.* **34**, 280–283.

Elorza, M. V., Rico, H. and Sentandreu, R. (1983). *J. Gen. Microbiol.* **129**, 1577–1582.

Elson, L. A. and Morgan, W. T. J. (1933). *Biochem. J.* **27**, 1824–1828.

Endo, A. and Misato, T. (1969). *Biochem. Biophys. Res. Commun.* **37**, 718–722.

Endo, A., Kakiki, K. and Misato, T. (1970). *J. Bacteriol.* **104**, 189–196.

Eugene, O., Yund, M. A. and Fristrom, J. W. (1979). *Tissue Culture Assoc. Man.* **5**, 1055–1062.

Farkas, V. (1979). *Microbiol. Rev.* **43**, 117–144.

Farkas, V. and Svoboda, A. (1980). *Curr. Microbiol.* **4**, 99–103.

Foster, A. B. and Hackman, R. H. (1957). *Nature* **180**, 40–41.

Foster, A. B. and Webber, J. M. (1960). *Adv. Carbohydr. Chem. Biochem.* **15**, 371–393.

Foster, A. B., Martlew, E. F. and Stacey, M. (1953). *Chem. and Ind.* (*London*), 825–826.

Fox, D. L. (1973). *Comp. Bioch. Physiol.* **B44**, 953–962.

Frey, R. (1950). *Ber. Schweitz. Botan. Ges.* **60**, 199–220.

Fristrom, D. K. and Fristrom, J. W. (1982). *Develop. Biol.* **92**, 418–427.

Fristrom, J. W. (1965). *Genetics* **52**, 297–318.

Fristrom, J. W. (1981). *In* "Metamorphosis, a Problem in Developmental Biology" (L. Gilbert and E. Frieden, eds), pp. 217–240. Plenum Press, New York.

Fristrom, J. W., Doctor, J., Fristrom, D. K., Logan, W. R. and Silvert, D. J. (1982). *Develop. Biol.* **91**, 337–350.

Fukamizo, T., Kramer, K. J., Mueller, D. D., Schaefer, J., Garbow, J. and Jacob, G. S. (1986). *Arch. Biochem. Biophys.* **249**, 15–26.

Furter, R. and Rast, D. M. (1985). *FEBS Microbiol. Lett.* **28**, 208–211.

García Mendoza, C. and Navaes Ledieu, M. (1968). *Nature* **220**, 1035.

Gardner, K. H. and Blackwell, J. (1975). *Biopolymers* **14**, 1481–1495.

Gerhard, G. and Jeuniaux, Ch. (1979). *Cah. Biol. Mar.* **20**, 341–349.

Gilson, E. (1894). *Bull. Soc. Chim. Paris* **3**, 1099.

Giraud-Guille, M. M. and Bouligand, Y. (1986). *In* "Chitin in Nature and Technology" (R. Muzzarelli, Ch. Jeuniaux and G. W. Gooday, eds), pp. 29–35. Plenum Press, New York.

Glaser, L. and Brown, D. H. (1957). *J. Biol. Chem.* **228**, 729–742.

Glenn Richards, A. (1978). *In* "The Chemistry of Insect Cuticle" (M. Rockstein, ed.), pp. 205–232. Academic Press, New York.

Gooday, G. W. (1971). *J. Gen. Microbiol.* **67**, 125–133.

Gooday, G. W. (1978). *In* "Development Mycology" (J. E. Smith and D. R. Berry, eds), Vol. 3, pp. 51–77. Edward Arnold, London.

Gooday, G. W. and Rousset-Hall, A. (1975). *J. Gen. Microbiol.* **89**, 137–145.

Gooday, G. W., Rousset-Hall, A. and Hunsley, D. (1976). *Trans. Brit. Mycol. Soc.* **67**, 193–198.

Gordnier, P. M., Brezner, J. and Tanenbaum, S. W. (1987). *J. Antibiot.* **60**, 110–112.

Grace, T. C. C. (1962). *Nature* **195**, 788–789.

Grove, S. N. and Bracher, C. E. (1970). *J. Bacteriol.* **104**, 989–1009.

Gwinn, J. and Stevenson, J. R. (1973a). *Comp. Biochem. Physiol.* **45B**, 769–776.

Gwinn, J. and Stevenson, J. R. (1973b). *Comp. Biochem. Physiol.* **45B**, 777–785.

Hackman, R. H. (1954). *Aust. J. Biol. Sci.* **7**, 168–178.

Hackman, R. H. (1982). *In* "Proceedings of the 2nd Conference on Chitin/Chitosan" (S. Hirano and S. Tokura, eds), pp. 5–9. JSCC Japan.

Hackman, R. H. (1984). *In* "Biology of the Integument" (J. Bereiter-Hahn, A. G. Matoltsy and K. S. Richards, eds), Vol. 1, pp. 583–610. Springer, Berlin.

Hackman, R. H. (1987). *In* "Chitin and Benzoylphenyl Ureas" (J. E. Wright and A. Retnakaran, eds), pp. 1–32. W. Junk, Boston.

Hackman, R. H. and Goldberg, M. (1958). *J. Insect Physiol.* **2**, 221–231.

Hackman, R. H. and Goldberg, M. (1965). *Aust. J. Biol. Sci.* **18**, 935–946.

Hackman, R. H. and Goldberg, M. (1971). *J. Insect Physiol.* **17**, 335–347.

Hackman, R. H. and Goldberg, M. (1977). *Insect Biochem.* **7**, 175–184.

Hackman, R. H. and Goldberg, M. (1978). *Insect Biochem.* **8**, 353–357.

Hajjar, N. P. (1985). *In* "Insecticides" (D. H. Hutson and T. H. Roberts, eds), pp. 275–310. J. Wiley and Sons, New York.

Hall, L. D. and Yalpani, M. (1980). *J. Chem. Soc., Chem. Commun.*, 1153–1154.

Hänseler, E., Nyhlén, L. E. and Rast, D. M. (1983). *Exp. Mycol.* **7**, 17–30.

Hartwell, L. H. (1967). *J. Bacteriol.* **93**, 1662–1670.

Hasilik, A. (1974). *Arch. Microbiol.* **101**, 295–301.

Hasilik, A. and Holzer, H. (1973). *Biochem. Biophys. Res. Commun.* **53**, 552–559.

Hepburn, H. R. (1976). "The Insect Integument". Elsevier, Amsterdam.

Hernández, J., López-Romero, E., Cerben, J. and Ruiz-Herrera, J. (1981). *Exp. Mycol.* **5**, 349–356.

Herth, W. (1979). *J. Ultrastruct, Res.* **68**, 16–27.

Herth, W. (1980). *J. Cell. Biol.* **87**, 442–450.

Herth, W. and Bartholtt, W. (1979). *J. Ultrastruct, Res.* **68**, 6–15.

Herth, W. and Zugenmaier, P. (1977). *J. Ultrastruct. Res.* **61**, 230–239.

Herth, W., Kupel, A. and Schnepf, E. (1977). *J. Cell Biol.* **73**, 311–312.

Hillerton, J. E. (1980). *J. Mater. Sci.* **15**, 3109–3112.

Hohnke, L. A. (1971). *Comp. Biochem. Physiol.* **40B**, 757–779.

Holan, Z., Beran, K., Prochazcova, V. and Baldrian, J. (1971). *Sci. Tech. Proc. Spec. Int. Symp.* **71**, 239–260.

Hoppe-Seiler, F. (1894). *Ber. Deut. Chem. Gesell.* **27**, 3329–3331.

Hori, M., Kakiki, K. and Misato, T. (1974a). *Agric. Biol. Chem.* **38**, 691–698.

Hori, M., Kakiki, K. and Misato, T. (1974b). *Agric. Biol. Chem.* **38**, 699–705.

Hornung, D. E. and Stevenson, J. R. (1971). *Comp. Biochem. Physiol.* **40B**, 341–346.

Horowitz, S. T., Roseman, S. and Blumenthal, H. J. (1957). *J. Am. Chem. Soc.* **79**, 5046–5049.

Horst, M. N. (1981). *J. Biol. Chem.* **256**, 14112–14119.

Horst, M. N. (1983). *Arch. Biochem. Biophys.* **223**, 254–263.

Horst, M. N. (1986). *In* "Chitin in Nature and Technology (R. Muzzarelli, Ch. Jeuniaux and G. W. Gooday, eds), pp. 45–52. Plenum Press, New York.

Hunter, K. and Rose, A. (1971). *In* "The Yeast" (A. Rose and J. Harrison, eds), Vol. 2, pp. 211–270. Academic Press, London.

Isaac, S., Ryder, N. S. and Peberdy, J. F. (1978). *J. Gen. Microbiol.* **105**, 45–50.

Isono, K. and Suzuki, S. (1966). *Agric. Biol. Chem.* **30**, 813–816.

Isono, K., Nagatsu, J., Kawashima, Y. and Suzuki, S. (1965). *Agric. Biol. Chem.* **29**, 848–853.

Isono, K., Nagatsu, J., Kobinata, K., Sasaki, K. and Suzuki, S. (1967). *Agric. Biol. Chem.* **31**, 190–193.

Isono, K., Asahi, K. and Suzuki, S. (1969). *J. Am. Chem. Soc.* **91**, 7490–7496.
Isono, K., Azuma, T. and Suzuki, S. (1971a). *Chem. Pharmacol. Bull.* **19**, 505–511.
Isono, K., Suzuki, S. and Azuma. T. (1971b). *Agric. Biol. Chem.* **35**, 1986–1989.
Isono, K., Suzuki, S., Tanaka, M., Nambata, T. and Shibuya, K. (1972). *Agric. Biol. Chem.* **36**, 1571–1576.
Isono, K., Crain, T., Odiorne, J. J., MacCloskey, J. A. and Suhadolnik, R. J. (1973). *J. Am. Chem. Soc.* **95**, 5788–5794.
Jan, Y. N. (1974). *J. Biol. Chem.* **249**, 1973–1979.
Jacobs, M. E. (1978). *Insect Biochem.* **8**, 37–41.
Jacobs, M. E. (1985). *J. Insect Physiol.* **31**, 509–515.
Jaworski, E. L., Wang, L. and Marco, G. (1963). *Nature* **193**, 790–791.
Jeanloz, R. W. and Forchielli, E. (1950). *Helv. Chim. Acta* **33**, 1690–1697.
Jeuniaux, Ch. (1963). "Chitine et Chitinolyse". Masson, Paris.
Jeuniaux, Ch. (1971). *In* "Comprehensive Biochemistry" (M. Florkin and E. H. Stotz, eds), Vol. 26-C, pp. 450–632. Elsevier, Amsterdam.
Joppien, S., Burger, A. and Reisener, H. J. (1972). *Arch. Microbiol.* **82**, 337–352.
Kang, M. S., Au-Young, J. and Cabib, E. (1985). *J. Biol. Chem.* **260**, 12680–12684.
Kang, M. S., Elango, N., Mattia, E., Au-Young, J., Robbins, P. W. and Cabib, E. (1984). *J. Biol. Chem.* **259**, 14966–14972.
Keller, F. A. and Cabib, E. (1971). *J. Biol. Chem.* **246**, 160–166.
Kent, P. W. and Lundt, M. P. (1958). *Biochem. Biophys. Acta* **28**, 657–658.
Kramer, K. J., Dziadk-Turner, C. and Koga, D. (1985). *In* "Comprehensive Insect Physiology, Biochemistry and Pharmacology" (G. Kerburt and L. Gilbert, eds), Vol. 3, pp. 75–115, Pergamon Press, New York.
Kramer, K. J. and Koga, D. (1986). *Insect Biochem.* **16**, 851–877.
Kreger, D. R. (1954). *Biochem. Biophys. Acta* **13**, 1–9.
Krisman, C. R. (1972). *Biochem. Biophys. Res. Commun.* **46**, 1206–1212.
Kubo, I., Uchida, M. and Klocke, J. A., (1983). *Agric. Biol. Chem.* **47**, 911–913.
Kurita, K. (1985). *In* "Chitin in Nature and Technology" (R. Muzzarelli, Ch. Jeuniaux and G. W. Gooday, eds), pp. 287–293. Plenum Press, New York.
Lassaigne, J. L. (1843). *Comp. Rend.* **16**, 1087–1089.
Ledderhose, G. (1878). *Z. Physiol. Chem.* **2**, 213–227.
Lehle, L. and Tanner, W. (1976). *FEBS Lett.* **71**, 167–170.
Leighton, T., Marks, E. and Leighton, F. (1981). *Science* **213**, 905–907.
LéJohn, H. B. (1971). *Nature* **231**, 164–168.
Leloir, L. F. and Cardini, C. E. (1953). *Biochem. Biophys. Acta* **12**, 15–22.
Lipke, H. (1971). *Insect Biochem.* **1**, 189–198.
Lipke, H. and Geoghegan, T. (1971). *Biochem. J.* **125**, 703–715.
Lipke, H. and Strout, V. (1972). *Israel J. Entomol.* **7**, 117–128.
Lipke, H., Strout, V., Henzel, W. and Sugumaran, M. (1981). *J. Biol. Chem.* **256**, 4241–4246.
Lipke, H., Sugumaran, M. and Henzel, W. (1983). *Adv. Insect Physiol.* **17**, 1–84.
López-Romero, E. and Ruiz-Herrera, J. (1976). *J. Microbiol. Serol.* **42**, 261–276.
Lyr, H. and Seyd, W. (1978). *Z. Allerg. Mikrobiol.* **18**, 721–729.
Lyr, H. and Seyd, W. (1979). "Effects of some Fungicides and other Compounds on Chitin Synthesis". Akad. Wiss., Berlin, D.D.R.
Madhevan, P. and Ramachandran, N. K. G. (1974). *Fish. Technol.* **11**, 50–53.
Mahoney, W. C. and Duskin, D. (1979). *J. Biol. Chem.* **254**, 6572–6576.
Marks, E. P. (1973). *Gen. Comp. Endocrinol.* **21**, 472–477.
Marks, E. P. (1980). *Ann. Rev. Entomol.* **25**, 73–101.
Marks, E. P. and Sowa, B. A. (1976). *In* "The Insect Integument" (H. R. Hepburn, ed.), pp. 339–357. Elsevier, Amsterdam.
Marks, E. P., Leighton, T. and Leighton, F. (1982). *In* "Insecticide Mode of Action" (J. Coats, ed.), pp. 281–313. Academic Press, New York.
Mauchamp, B. and Schrevel, J. (1977). *C.R. Acad. Sci. Paris Ser. D* **285**, 1107–1110.
Mayer, R. T. and Chen, A. C. (1985). *Arch. Insect Biochem. Physiol.* **2**, 161–179.
Mayer, R. T., Meola, J. M., Coppage, D. L. and DeLoach, J. R. (1979). *J. Insect Physiol.* **25**, 677–683.

Mayer, R. T., Chen, A. C. and DeLoach, J. R. (1980a). *Insect Biochem.* **10**, 549–556.
Mayer, R. T., Coppage, D. L., Meolas, S. M. and DeLoach, J. R. (1980b). *J. Econ. Ent.* **73**, 76–80.
Mayer, R. T., Chen, A. C. and DeLoach, J. R. (1981). *Experientia* **37**, 2–3.
Mayer, R. T., Chen, A. C. and DeLoach, J. R. (1983). *Arch. Insect Biochem. Physiol.* **1**, 1–15.
McMurrough, I. and Bartnicki-García, S. (1971). *J. Biol. Chem.* **246**, 4008–4016.
McMurrough, I., Flores-Carreon, A. and Bartnicki-García, S. (1971). *J. Biol. Chem.* **246**, 3999–4007.
Miller, R. W., Corley, C., Cohen, C. F., Robbins, W. and Marks, E. P. (1981). *Southwestern Entomol.* **6**, 272–278.
Miller, S. G. and Silhacek, D. L. (1982). *Insect Biochem.* **12**, 301–309.
Mills, G. L. and Cantino, E. C. (1974). *J. Bacteriol.* **118**, 192–201.
Mills, G. L. and Cantino, E. C. (1978). *Exp. Mycol.* **2**, 99–109.
Mills, G. L. and Cantino, E. C. (1980). *Exp. Mycol.* **4**, 175–180.
Mills, G. L. and Cantino, E. C. (1981). *Arch. Microbiol.* **130**, 72–77.
Minke, R. and Blackwell, J. (1978). *J. Mol. Biol.* **120**, 167–181.
Misato, T. and Kakiki, T. (1977). "Antifungal Compounds", Vol. 2. Marcel Dekker, New York.
Mitani, M. and Inone, Y. (1968). *J. Antibiot.* **21**, 492–496.
Mitsui, T., Nobusawa, C. and Fukami, J. (1981). *J. Pesticide Sci.* **6**, 155–161.
Molano, J., Polacheck, I., Duran, A. and Cabib, E. (1979). *J. Biol. Chem.* **254**, 4901–4907.
Molano, J., Bowers, B. and Cabib, E. (1980). *J. Cell. Biol.* **85**, 199–212.
Mommsen, T. P. (1978). *Comp. Biochem. Physiol.* **60A**, 371–375.
Mommsen, T. P. (1980). *Biochem. Biophys. Acta* **612**, 361–372.
Montgomery, G. W., Adams, D. J. and Gooday, G. W. (1984). *J. Gen. Microbiol.* **130**, 291–297.
Moore, G. K. and Roberts, G. A. F. (1981). *Int. J. Biol. Macromol.* **3**, 292–296.
Moore, R. M. and Peberdy, J. F. (1976). *Can. J. Microbiol.* **22**, 915–921.
Moreno, S., Cardini, C. E. and Tandecarz, J. S. (1986). *Eur. J. Biochem.* **157**, 539–545.
Moreno, S., Cardini, C. E. and Tandecarz, J. S. (1987). *Eur. J. Biochem.* **162**, 609–614.
Mulisch, M., Herth, W., Zügenmaier, P. and Hausmann, K. (1983). *Biol. Cell* **49**, 169–178.
Müller, H., Furter, R., Zähner, H. and Rast, D. (1981). *Arch. Microbiol.* **130**, 195–197.
Muzzarelli, R. A. A. (1973). "Natural Chelating Polymers". Pergamon Press, Oxford.
Muzzarelli, R. A. A. and Rochetti, R. (1974). *Anal. Chim. Acta* **70**, 283–289.
Muzzarelli, R. A. A. (1977). "Chitin". Pergamon Press, Oxford.
Myers, R. B. and Cantino, E. C. (1974). *In* "Monograph in Developmental Biology" (A. Wolsky, ed.), Vol. 8, pp. 117–128. S. Karger, Basel.
Neville, A. C. (1975). "Biology of the Arthropod Cuticle". Springer, Berlin.
Nishi, N., Noguchi, J., Tokura, S. and Shiota, H. (1979). *Polymer J.* **11**, 27–32.
Nishioka, T., Fryita, T. and Nakajuma, H. (1979). *J. Pesticide Sci.* **4**, 367–374.
Oberlander, H. and Leach, C. (1975). "Proceedings of the First International Congress Stored Products/Entomology", pp. 651–655. Savanah, Georgia.
Odier, A. (1823). *Mém. Soc. Hist. Nat. Paris* **1**, 29–42.
Ohta, N., Kakiki, K. and Misato, T. (1970). *Agric. Biol. Chem.* **34**, 1224–1230.
Orlean, P. (1987). *J. Biol. Chem.* **262**, 5732–5739.
Parodi, A. J. and Leloir, L. F. (1979). *Biochem. Biophys. Acta* **559**, 1–37.
Parodi, A. J. and Quesada-Allué, L. A. (1982). *J. Biol. Chem.* **257**, 7637–7640.
Parodi, A. J., Quesada-Allué, L. A. and Cazzulo, J. J. (1981). *Proc. Natl. Acad. Sci. USA* **78**, 6201–6205.
Pearse, A. G. E. (1985). "Histochemistry Theoretical and Applied", 4th edn, Vol. 2. Churchill Livingstone, Edinburgh.
Peberdy, J. F. and Moore, P. M. (1975). *J. Gen. Microbiol.* **90**, 228–236.
Peter, M. G. (1980). *Insect Biochem.* **10**, 221–228.
Peter, M. G., Kegel, G. and Keller, R. (1986). *In* "Chitin in Nature and Technology" (R. Muzzarelli, Ch. Jeuniaux and G. W. Gooday, eds), pp. 21–28. Plenum Press, New York.
Peters, W. and Latka, I. (1986). *Histochemistry* **84**, 155–160.
Phaff, H. J. (1971). *In* "The Yeast" (A. Rose and J. Harrison, eds), Vol. 2, pp. 135–209. Academic Press, London.

Phaff, H. J. (1977). *Adv. Chem. Ser.* **160**, 244–282.
Plessman-Camargo, E., Dietrich, C. P., Sonneborn, D. and Strominger, J. L. (1967). *J. Biol. Chem.* **242**, 3121–3128.
Porter, C. A. and Jaworski, E. (1966). *Biochemistry* **5**, 1140–1154.
Powning, R. F. and Irzykiewicz, H. (1963). *Nature* **200**, 1128.
Quesada-Allué, L. A. (1973). MS Dissertation, University of Buenos Aires, FCEyN.
Quesada-Allué, L. A. (1978). Ph.D. Thesis, University of Buenos Aires, FCEyN.
Quesada-Allué, L. A. (1979). *FEBS Lett.* **97**, 225–229.
Quesada-Allué, L. A. (1980). *Mol. Cell. Biochem.* **33**, 149–155.
Quesada-Allué, L. A. (1981). *Biochem. J.* **198**, 420–432.
Quesada-Allué, L. A. (1982). *Biochem. Biophys. Res. Commun.* **105**, 312–319.
Quesada-Allué, L. A. and Belocopitow, E. (1978). *Eur. J. Biochem.* **88**, 529–541.
Quesada-Allué, L. A. and Parodi, A. J. (1983). *Biochem. J.* **212**, 123–128.
Quesada-Allué, L. A., Belocopitow, E. and Maréchal, L. R. (1975). *Biochem. Biophys. Res. Commun.* **66**, 1201–1208.
Quesada-Allué, L. A., Maréchal, L. R. and Belocopitow, E. (1976a). *Acta Physiol. Latinoam.* **26**, 349–363.
Quesada-Allué, L. A., Maréchal, L. R. and Belocopitow, E. (1976b). *FEBS Lett.* **67**, 243–249.
Rast. D. M. and Bartnicki-García, S. (1981). *Proc. Natl. Acad. Sci. USA* **78**, 1233–1236.
Ratnakaran, A. and Hackman, R. H. (1985). *Arch. Insect Biochem. Physiol.* **2**, 251–263.
Richards, A. G. (1951). "The Integument of Arthropods". University of Minnesota, Press, Minneapolis, MN.
Ride, J. P. and Drysdale, R. B. (1971). *Physiol. Plant Pathol.* **1**, 409–420.
Rivas, L. A. and Pont Lezica, R. (1987a). *Eur. J. Biochem.* **163**, 129–134.
Rivas, L. A. and Pont Lezica, R. (1987b). *Eur. J. Biochem.* **163**, 135–140.
Robb, J. (1969). *J. Cell. Biol.* **41**, 876–884.
Roncero, C. and Duran, A. (1985). *J. Bacteriol.* **163**, 1180–1185.
Rosenberg, R. F. (1976). In "The Filamentous Fungi" (J. E. Smith and D. R. Berry, eds), Vol. 2, pp. 328–344. Edward Arnold, London.
Rouget, C. (1859). *Comp. Rend.* **48**, 792–795.
Rousset-Hall, A. and Gooday, G. (1975). *J. Gen. Microbiol.* **89**, 146–154.
Rudall, K. M. (1963). *Adv. Insect Physiol.* **1**, 257–313.
Rudall, K. M. (1965). *Biochem. Soc. Symp.* **25**, 83–92.
Rudall, K. M. and Kenchington, W. (1973). *Biol. Rev.* **48**, 597–636.
Ruiz-Herrera, J. and Bartnicki-García, S. (1974). *Science* **186**, 357–359.
Ruiz-Herrera, J. and Bartnicki-García, S. (1976). *J. Gen. Microbiol.* **97**, 241–249.
Ruiz-Herrera, J., Sing, V. O., Van der Woude, W. J. and Bartnicki-García, S. (1975). *Proc. Natl. Acad. Sci. USA* **72**, 2706–2710.
Ruiz-Herrera, J., López-Romero, E. and Bartnicki-García, S. (1977). *J. Biol. Chem.* **252**, 3338–3343.
Ruiz-Herrera, J., Bartnicki-García, S. and Bracker, C. E. (1980). *Biochim. Biophys. Acta* **629**, 201–216.
Ryder, N. S. and Peberdy, J. F. (1971). *J. Gen. Microbiol.* **99**, 69–76.
Ryder, N. S. and Peberdy, J. F. (1979). *Exp. Mycol.* **3**, 259–269.
Sburlati, A. and Cabib, E. (1987). *J. Biol. Chem.* **261**, 15142–15147.
Scarborough, G. A. (1975). *J. Biol. Chem.* **250**, 1106–1111.
Schaefer, J., Kramer, K. J., Garbow, J. R., Jacob, G. S., Stejskal, E. O., Hopkins, T. L. and Speirs, R. D. (1987). *Science* **235**, 1200–1204.
Schekman, R. and Brawley, V. (1979). *Proc. Natl. Acad. Sci. USA* **76**, 645–649.
Schmidt, M. (1936). *Arch. Mikrobiol.* **7**, 241–260.
Schnepf, E., Röderer, G. and Herth, W. (1975). *Planta* **125**, 45–62.
Scholl, E. (1909). *Monatsh* **29**, 1023–1036.
Schwarz, R. T. and Datema, R. (1982). *Adv. Carbohydr. Chem. Biochem.* **40**, 287–379.
Selitrenikoff, C. P. (1979). *Arch. Biochem. Biophys.* **195**, 243–244.
Selitrenikoff, C. P. (1984). *Exp. Mycol.* **8**, 269–272.
Shibuya, K., Tanaka, M., Nanbata, T., Isono, K. and Suzuki, S. (1972). *Agric. Biol. Chem.* **36** 1229–1233.

Sietsma, J. H. and Wessels, J. G. H. (1977). *Arch. Microbiol.* **113**, 79–82.
Sietsma, J. H. and Wessels, J. G. H. (1979). *J. Gen. Microbiol.* **114**, 99–108.
Sietsma, J. H. and Wessels, J. G. H. (1981). *J. Gen. Microbiol.* **125**, 209–212.
Silvert, D. J. and Fristrom, J. W. (1980). *Insect Biochem.* **10**, 341–353.
Silvert, D. J., Doctor, J., Quesada, L. and Fristrom, J. W. (1984). *Biochemistry* **23**, 5767–5774.
Soltani, N., Benson, M. T. and Delachambre, J. (1984). *Pestic. Biochem. Physiol.* **21**, 256–264.
Soltani, N., Delbecque, J. P. and Delachambre, J. (1983). *Pestic. Sci.* **14**, 615–622.
Sowa, B. A. and Marks, E. P. (1975). *Insect Biochem.* **5**, 855–859.
Sparrow, J. C. and Chadfield, C. G. (1982). *Devel. Genet.* **3**, 235–245.
Spindler, K. (1976). *Insect Biochem.* **6**, 663–667.
Stagg, C. M. and Feather, M. S. (1973). *Biochim. Biophys. Acta* **320**, 64–72.
Stirling, J. O., Cook, G. A. and Pope, A. M. (1979). In "Fungal Walls and Hyphal Growth", (J. Burnett and A. P. Trinci, eds), pp. 169–188. Cambridge University Press, Cambridge.
Stoddart, R. W. (1984). "The Biosynthesis of Polysaccharides". Macmillan, New York.
Suhadolnik, R. J. (1970). "Nuceloside Antibiotics". John Wiley & Sons, New York.
Suhadolnik, R. J. (1983). "Nucleosides as Biological Probes". John Wiley & Sons, New York.
Surholt, B. (1975a). *J. Comp. Physiol.* **102**, 135–147.
Surholt, B. (1975b). *Insect Biochem.* **5**, 585–593.
Sweeley, C. C., Bentley, R., Makita, M. and Wells, W. W. (1963). *J. Am. Chem. Soc.* **85**, 2497–2507.
Takatsuki, A. K. and Tamura, G. (1971a). *J. Antibiot.* **24**, 224–231.
Takatsuki, A. K. and Tamura, G. (1971b). *J. Antibiot.* **24**, 785–790.
Takatsuki, A. K., Arima, K. and Tamura, G. (1971). *J. Antibiot.* **24**, 215–223.
Takatsuki, A. K., Arima, K. and Tamura, G. (1975). *Agric. Biol. Chem.* **39**, 2089–2091.
Takeda, M. and Abe, E. (1962). *Norisho Soisan Koshusho Kenkyu Hokoku* **11**, 339–406.
Takeda, M. and Katsuura, H. (1964). *Suisan Daigaku Kenkyu Hokoku* **13**, 109–116.
Taylor, I. E. P. and Cameron, D. S. (1973). *Ann. Rev. Microbiol.* **27**, 243–259.
Tkacz, J. S. and Lampen, J. D. (1975). *Biochem. Biophys. Res. Commun.* **65**, 248–257.
Tracey, M. V. (1954). *Biochem. J.* **61**, 579–685.
Tracey, M. V. (1955). In "Modern Methods of Plant Analysis" (K. Paech and M. V. Tracey, eds), Vol. 2, pp. 264–274. Springer, Berlin.
Tronchin, G., Pouloin, D., Herbaut, J. and Biguet, J. (1981). *Eur. J. Cell. Biol.* **26**, 121–128.
Trujillo, R. (1968). *Carbohydr. Res.* **7**, 483–485.
Turnbull, I. F. and Howells, A. J. (1982). *Austr. J. Biol. Sci.* **35**, 491–503.
Turnbull, I. F. and Howells, A. J. (1983). *Austr. J. Biol. Sci.* **36**, 251–262.
Ulane, E. R. and Cabib, E. (1976). *J. Biol. Chem.* **251**, 3367–3374.
Van Eck, W. H. (1979). *Insect Biochem.* **9**, 295–300.
Van Laere, A. J. and Carlier, A. R. (1978). *Arch. Microbiol.* **116**, 181–184.
Vannini, G. L., Poli, F., Donini, A. and Pancaldi, S. (1983). *Plant Sci. Lett.* **31**, 9–17.
Vardanis, A. (1976). *Life Sci.* **15**, 1949–1956.
Verloop, A. and Ferrell, C. D. (1977). *ACS Symposium Series* **37**, 237–270.
Vermeulen, C. A., Raeven, M. B. and Wessels, J. G. (1979). *J. Gen. Microbiol.* **114**, 87–97.
Vermeulen, C. A. and Wessels, J. G. (1983). *Curr. Microbiol.* **8**, 67–71.
Vincent, J. F. V. and Hillerton, J. E. (1979). *J. Insect Physiol.* **25**, 653–659.
von Veimarn, P. P. (1927). *Kolloid Z.* **42**, 134–140.
Wessels, J. G., Sietsma, J. and Sonneborg, A. S. (1983). *J. Gen. Microbiol.* **129**, 1607–1616.
Whistler, R. S. and BeMiller, J. N. (1982). *J. Org. Chem.* **27**, 1161–1163.
Wright, J. E. and Retnakaran, A. (1987). "Chitin and Benzoylphenyl Ureas". W. Junk, Boston, MA.
Wulff, G. (1968). *Dtsch. Apoth. Ztg.* **23**, 797–808.
Yalpani, M., Hall, L. D., Tung, M. A. and Brooks, D. E. (1983). *Nature* **302**, 812–814.
Zarain-Herzberg, A. and Arroyo-Bergovich, A. (1983). *J. Gen. Microbiol.* **129**, 3319–3326.
Zimmermann, A. (1952). Swed. Patent 136,717.

14 Exudate Gums

A. M. STEPHEN, S. C. CHURMS and D. C. VOGT

Department of Organic Chemistry, University of Cape Town, Rondebosch 7700, South Africa

METHODS IN PLANT BIOCHEMISTRY Vol. 2
ISBN 0-12-461012-9

I. GENERAL CONSIDERATIONS

A. Introduction and Background

Many uses have been developed over the years for the exudate gums which are formed, usually as a result of trauma of some kind, on the exterior surfaces of plants (Smith and Montgomery, 1959a; Aspinall, 1970; Whistler and BeMiller, 1973; Sandford and Baird, 1983; Boothby, 1983). The exudates are complex, uronic acid-containing polysaccharides of molecular weight ranging from 2×10^4 to well over 10^6, and some are associated with protein or terpenoid material. Molecular structures depend on taxa of origin (Cottrell and Baird, 1980; Stephen, 1980), and have been grouped according to their chemical relationship to cell-wall components (Aspinall, 1967, 1969, 1973) such as arabinogalactan-protein (AG-P), pectins and xylans (Stephen, 1983). Proximate analyses (Anderson and Stoddart, 1966; Anderson and Munro, 1969) and the standard methodology of polysaccharide structure determination (Aspinall, 1982) have been applied to exudates isolated from natural sources or as found in food, pharmaceutical and other commercial preparations (Smith and Montgomery, 1959a; Glicksman, 1969a, 1983; Whistler and BeMiller, 1973). The quantities of material available for analysis are often quite large, so that sophisticated micro-methods are not always necessary, but for forensic purposes and quality control the techniques that have been developed (Valent *et al.*, 1980; Waeghe *et al.*, 1983) for similar compounds from animal (Fransson, 1985) and microbiological (Åman *et al.*, 1982; Gorin and Barreto-Bergter, 1983; Kenne and Lindberg, 1983; O'Neill *et al.*, 1986) sources come into their own. Mass spectrometry, following hydrolysis, and immunological procedures (Bishop and Jennings, 1982; Strobel *et al.*, 1982) are of paramount importance.

Methods of isolation applicable to polysaccharides from plant material generally can be applied to gum exudates, though by contrast the raw product is as a rule sufficiently free of contaminants that mere dissolution in water, filtration through Millipore apparatus or centrifugation (Anderson and Munro, 1969), and freeze-drying is adequate (Vandevelde and Fenyo, 1985). Gums are precipitated from aqueous solution by addition of miscible organic solvents or salts, but as in the freeze-drying operation there may be attendant changes in conformation or structure (Anderson *et al.*, 1969).

Certain gums outweigh all others in commercial value (Glicksman, 1983) and for this reason the analysis and molecular structural determination of gum arabic, gum tragacanth, gum karaya, gum ghatti, and larch arabinogalactan are emphasised in this chapter. On the other hand, the search for substitutes and the desirability of measuring analytical parameters of exudates from as wide a range of plant sources as possible in order to develop a better understanding of their biological significance (James *et al.*, 1985) necessitates a wider coverage.

B. Occurrence, Taxonomy and Molecular Structure

The relationship between the molecular structures of gum exudates and their taxa of origin has been reviewed (Stephen, 1980). Angiosperms from at least 25 orders (out of 92) and several gymnosperms yield gums that have been studied sufficiently for them to be classified according to structural type (A–D) (Stephen, 1983). The largest group (A) is based on a 3,6-linked D-galactopyranan core glycosylated by L-arabinofuranosyl units; the branched arabinogalactan (Aspinall's type II) is invariably enclosed by ramified sequences of other sugar (L-rhamnopyranosyl and D-xylopyranosyl) and uronic acid (D-glucopyranuronosyl) units. Variable amounts of protein, up to ~65%, may be associated with the polysaccharide, probably through covalent linkage (hence the terms proteoglycan and glycoprotein; Clarke et al., 1979; Fincher et al., 1983). A measure of heteropolymolecularity (molecular weight variation and variability in composition) is observed (Anderson and Stoddart, 1966), which may be intrinsic or due to the isolation of mixtures of discrete molecular types (Banks and Greenwood, 1963). The *Acacia* gums are a well-studied set of similarly-constituted substances of molecular type A (Anderson and Dea, 1969b, 1971; Anderson, 1978).

Type B comprises gums related to the pectins (Chapter 12), chains of 4-linked α-D-GalA units forming a basis for attachment of short sugar sequences that include L-Araf, D-Galp and D-GlcpA. An important variant is the interpolation of 2-linked L-Rhap residues in the main chain, sometimes as many as one for every GalpA unit (Aspinall and Bhattacharjee, 1970). The *Sterculia* gums are good examples (Aspinall and Fraser, 1965; Aspinall et al., 1965b, 1981, 1987). Sometimes, as for *Khaya* gum (Aspinall et al., 1960), an exudate may include molecules of both types A and B.

A less common structural type (C) is a heavily-glycosylated, 4-linked β-D-xylan, the branches not being long but containing a diversity of units (L-Araf, D-Galp, and D-GlcpA in particular; Aspinall, 1967). The solubility of these gums in water follows from there being substituents at both O-2 and O-3 of the xylan, their removal by chemical means leading to precipitation.

In type D the core structure may be a mannoglucuronoglycan, consisting of alternating units of 4-linked β-D-GlcA and →2)-α-D-Manp (Aspinall, 1967; di Fabio et al., 1982; Stephen, 1983; Redgwell et al., 1986a). Sugar sequences resembling those in gums of type A are attached at O-3 of most Man units, and sometimes to GlcA. Gum ghatti (Aspinall et al., 1965a) and the related exudate from *Anogeissus leiocarpus* (Aspinall and Chaudhari, 1975) are examples.

Some degree of methylation (at O-4 of GlcA or O-3 of Rha), and acetylation, is encountered in most gum structures. Gums of various structural types are found throughout the plant kingdom, but are concentrated in certain orders and families. Within a genus there is a notable constancy with regard to constituent sugar units and their modes of linkage, while for any given species the gum exudate is more or less typical, being subject to minor changes according to conditions and locality of growth and the part of the plant involved.

No particular climatic condition favours production of gums, and although the bulk of commercial supplies come from arid lands, every type of region from rain-forest to the temperate and arctic zones may be a source. Rapid translocation of carbohydrate constituents, following heavy rain, and the wounding of external tissues, are contributing causes, and microorganisms no doubt have a part to play. Incisions are made in bark or roots, following which gum is produced, often erratically. Mechanical damage is thus sealed by a covering of exudate.

Mechanisms of polysaccharide formation are generally understood to follow the transformation, as nucleotide phosphate derivatives, of D-glucose to the sugar units referred to above (James et al., 1985); these are assembled under enzymatic control to form a narrow range of inter-sugar linkages. In view of the limited number of glycosidic bond types found amongst the exudate gums there can be no reason to suppose that any different mechanism operates, though it is as yet uncertain whether biosynthetic capability is always fully exploited, and whether the limitation of molecular size is strictly controlled. The microheterogeneity referred to earlier relates to samples as isolated, and may be a consequence of the latitude permitted by conditions prevailing during biosynthesis. Of possible significance in considering mechanisms of assembly is the occurrence in an *Acacia* gum (from the injured stems of *A. erioloba*) of glycosylation of contiguous hydroxyproline residues in a protein structure by the very sugar and uronic acid units that typically surround the arabinogalactan core of exudates from other taxonomically-related species (Gammon and Stephen, 1986; Gammon *et al.*, 1986b).

C. Industrial Applications

These have been well reviewed (Smith and Montgomery, 1959a; Whistler and BeMiller, 1973; Lawrence, 1976; Davidson, 1980; Sandford and Baird, 1983). Major uses of gums are in the food and pharmaceutical industries, where their emulsifying, stabilising, thickening and gel-forming properties are the main physical requirements (Glicksman, 1982; Walker, 1984; Szczesniak, 1986) and purity is essential. Bulk quantities are employed in the mining industry and in the manufacture of paper and textiles. In all cases some measure of analytical control is necessary. What has yet to be determined in greater detail is the influence of different aspects of gum exudates upon their physical properties, such as the extent of association of the polysaccharide with protein or polyphenolic substances, the degree of ramification of the polysaccharide, and the distributions of molecular sizes. Table 14.1 illustrates uses of different types of gum, and the properties that are required of them.

D. Detection and Testing of Gums

Most gums are soluble in water, some only partially. If the exudate is associated with terpenoid compounds, the water-insoluble non-carbohydrate material is readily removed by extraction with acetone, leaving the polysaccharide component (possibly associated with protein) as a white powder. *Araucaria* gums (Aspinall and Fairweather, 1965; Anderson and Munro, 1969), and those from the Burseraceae such as *Boswellia papyrifera* gum (Anderson *et al.*, 1965a), myrrh and frankincense [from *Commiphora myrrha* (Jones and Nunn, 1955a) and *Boswellia carteri* (Jones and Nunn, 1955b)] are examples. Part only of a gum may dissolve, the remainder being a gel. Insolubility in water may develop on storage or upon drying (Anderson and Smith, 1967; Anderson *et al.*, 1969). Most gum exudates have a high viscosity, in keeping with their macromolecular character, though extensive molecular branching, as for gum arabic and other *Acacia* gums, offsets the effect of size (Anderson *et al.*, 1966). If the gum is part of a food preparation, all but the polysaccharide portion may be removed by standard methods (Jacobs, 1958; Thier, 1984).

Addition of EtOH or Me_2CO to a gum solution, at 1% concentration or even less, yields a stringy or flocculent precipitate, which on recovery responds to the Molisch (1-naphthol and H_2SO_4) reagent, a purple ring at the water–acid interface being readily observed within one minute. The concentration of carbohydrate in very dilute aqueous solution is best determined by the $PhOH$–H_2SO_4 method of F. Smith (Dubois *et al.*,

1956), though the quantification of the colour density produced (at 490 nm) depends on identification of the gum or at least an analysis of its monosaccharide composition. Elemental combustion analysis (C, H, N) yields figures characteristic of polysaccharides, and indicates a maximum value for protein content; nitrogen as determined by Kjeldahl is confirmatory, an appropriate factor being applied (Anderson, 1986a; Anderson and Douglas, 1988). The neutralisation equivalent (g per mol of carboxyl) may be found by titration using an indicator (phenolphthalein) provided all cations are rigorously removed; potentiometric titration does not demand this. In the natural state the exudate gums are neutral, and analysis by flame photometry or atomic absorption should be carried out, particularly as divalent cations might have an adverse effect on water solubility (Jefferies et al., 1978). Colour reactions using carbazole (Dische, 1962b) or m-hydroxybiphenyl (Blumenkrantz and Asboe-Hansen, 1973) may be made quantitative if the uronic acid responsible (GlcA or GalA) is known (Knutson and Jeanes, 1968). Uronic acid content can also be determined by decarboxylation methods (Section IV.A), the results of which have been compared favourably with those obtained by colorimetric procedures (Anderson and Garbutt, 1963).

TABLE 14.1. Gums and their uses.[a]

Gum source	Structural type	Uses	Function
Acacia senegal (gum arabic)	Substituted, acidic arabinogalactan (protein)	Bakery and meat products, beverages, confectionery, sauces, flotation, adhesives, cosmetics, pharmaceutical, printing, textiles	Emulsifier, binder, stabiliser, adhesive, protective colloid, encapsulating agent, crystallisation inhibitor
Larix occidentalis (larch gum)	Arabinogalactan	Beverages, essential oils, confectionery, pudding mixes, dietetic sweeteners, cosmetics, pharmaceutical, flotation, printing	Emulsifier, binder, stabiliser
Astragalus gummifer (gum tragacanth)	Arabinogalactan and glycanogalacturonan	Bakery and dairy products, salad dressings, paper making, textiles, pharmaceutical, cosmetics, polishes	Emulsifier, binder, stabiliser, thickening agent, crystallisation inhibitor
Sterculia urens (gum karaya)	Glycanorhamno-galacturonan	Bakery, meat and dairy products, paper making, printing, cosmetics, pharmaceutical, dental adhesive, textiles	Adhesive, binder, stabiliser, thickener, bulking agent, crystallisation inhibitor
Anogeissus latifolia (gum ghatti)	Glycanomanno-glucuronan	Bakery products, syrups, pharmaceutical, textiles, oil-well drilling fluids	Emulsifier, stabiliser, thickener, adhesive

[a] Compiled from Smith and Montgomery (1959a), Glicksman (1969a, 1983), Whistler and BeMiller (1973), Lawrence (1976), Cottrell and Baird (1980), Davidson (1980).

Numerous spot tests have been developed for specific, industrially important gums (Jacobs, 1958; Proszynski *et al.*, 1965; Glicksman, 1969b; British Pharmacopoeia, 1980). In addition, the optical activity expressed as $[\alpha]_D$ for dilute aqueous solutions of stated concentration (w/v) is an important diagnostic parameter (Snyder *et al.*, 1962); e.g. a fractionated sample of gum arabic (from *A. senegal*) gave values from $-26°$ to $-34°$, limits which embraced the mean values found for several different samples (Vandevelde and Fenyo, 1985).

A more fundamental approach is to carry out small-scale hydrolyses of the gum sample (2–10 mg) in acid, and to identify qualitatively the monosaccharide (neutral and acidic) components released, using paper chromatography (Hough and Jones, 1962: cheap and efficient) and thin layer chromatography (Wing and BeMiller, 1972; Ghebregzabher *et al.*, 1976: usually on silica gel plates, a more rapid procedure), with authentic carbohydrates as markers. Use of specific spray reagents from the wide range available (Merck, 1980; Churms, 1982a) permits a reasonably reliable conclusion to be drawn as to the type of gum polysaccharide, e.g. predominance of galactose and arabinose clearly reveals type A, while large amounts of galacturonic acid and xylose indicate types B and C gums, respectively.

The next step is a qualitative and quantitative analysis of the sugars released on acidic hydrolysis, gas–liquid chromatography (GLC) or derivatives prepared by standard methods (Section IV.A) being the usual approach. It is important to correct the values obtained by using a calibration based on weighed quantities of known sugars or by calculation of molar response factors (Sweet *et al.*, 1975), by using, in addition, factors which represent the degree of decomposition of sugar while in contact with acid (Hough *et al.*, 1972; Churms, 1982b; Churms and Stephen, 1984; see Section IV.A). Where significant amounts of protein are found in the gum sample, acid hydrolysis under more drastic conditions and quantitative analysis of the mixture of amino acids released [using ion-exchange chromatography, after paper chromatography (PC) or thin layer chromatography (TLC) with a ninhydrin spray] may be advantageous (Anderson *et al.*, 1972, 1985a,b,c, 1986; Anderson, 1986b).

The question as to enantiomeric configuration of the sugar units does not often arise, the assumption being that the common forms of the sugars are incorporated in gum structures. However, both D- and L-Gal may occur in a mucilage (Hunt and Jones, 1962), and for certainty of identification the isolated sugar is examined by optical rotation measurement (at chosen wavelengths), or if the quantity available is small (1–2 mg) by circular dichroism (Bebault *et al.*, 1973) of the derived alditol acetate (for Gal this is zero). In the particular case of galactose, absence of the L form in gum arabic and its presence in flax seed mucilage was elegantly demonstrated by Roberts and Harrer (1973) using D-galactose oxidase and assaying the residual galactose; Little (1982) used a GLC technique. Conversion of the sugar into its glycosides with an optically pure form of a chiral alcohol, followed by derivatisation and measurement of the retention time (different from that of the derivatives of the enantiomer of the sugar) on capillary GLC, is applicable to small amounts (Gerwig *et al.*, 1978, 1979; Leontein *et al.*, 1978). Mixtures of sugars can be investigated by these and other variants of the method (Oshima *et al.*, 1983; Schweer, 1983), and a related approach separates sugars as derivatives using columns coated with chiral stationary phases (Leavitt and Sherman, 1982; König *et al.*, 1988a,b). Instead of capillary GLC, high performance liquid chromatography (HPLC) has been employed to separate diastereomeric 1-[*N*-acetyl-α-methylbenzylamino]-1-deoxyalditol acetates derived from enantiomeric sugars (Oshima and Kumanotani, 1984).

Methyl ethers of sugar or uronic acid units are detected by the chromatographic methods indicated, a quantitative measure of methoxyl content using the Zeisel (Horwitz, 1980) or other methods (Anderson et al., 1963, 1964) being important. Acetate and other ester groupings are determined by saponification and titration (Jefferies et al., 1977), though proton, carbon-13, and correlation nuclear magnetic resonance (NMR) spectroscopy is an established method for characterising sugar and uronic acid units (particularly through their anomeric H and C resonances); C_5-Me groups in 6-deoxyhexose units, carboxyl, and ester groupings and their points of attachment are also readily discerned (Section III.D.3).

II. ISOLATION AND FRACTIONATION

Although in general exudate gums are extracted by stirring in water, much depends on the age of the exudate, whether on the tree or after storage. It is noticeable that a relatively fluid, recent exudate on the bark becomes jelly-like or even intensely hard with time. In comparison with other plant polysaccharides, the isolation and purification of exudates may be of such a mild nature that inadvertent cleavage of labile (furanosidic) sugar units or ester linkages (such as acetate), or detachment of bound protein or polyphenols (by oxidative delignification), is avoided. Extraction is usually carried out at pH ~ 7, though for the *Sterculia* and related gums (type B) solubilisation is dependent on treatment with warm alkali to release acetate (Aspinall and Nasir-ud-din, 1965). Grinding and milling of commercial gum samples assists their subsequent dissolution, and mild sonication in water, though attended by the risk of depolymerisation (Glicksman and Sand, 1973), may be applied. In structural studies, addition of traces of borohydride to the exudate in the process of extraction stabilises the molecules against 'peeling' from the reducing end (Aspinall, 1977, 1982); on the other hand, ester linkages are then at risk. The process of aqueous extraction frequently divides a gum exudate into fractions of greater or lesser solubility (Aslam et al., 1978; Jefferies et al., 1978), the soluble portion being required for most practical purposes. Sometimes the gel-like residue is similar in sugar composition to the soluble gum, in which case addition of sodium carbonate, oxalate or EDTA (ethylenediaminetetra-acetic acid) may improve the yield. Gum exudates may contain two distinctly different types of polysaccharide (Aspinall et al., 1960), or may be fractionated, usually on a size-exclusion basis (Vandevelde and Fenyo, 1985), into their protein-rich and pure polysaccharide components. In extreme cases, 4-methylmorpholine-*N*-oxide can be employed for extraction (Joseleau et al., 1981).

Where emphasis is on yield, and loss of a small proportion of peripheral sugar units is not detrimental, partial acid hydrolysis of the gum exudate (e.g. by heating at 100°C for 30 min at pH 2) is an effective process for bringing the polysaccharide into solution. A marked lowering in viscosity is observed, and subsequent neutralisation of the acid is necessary.

Any fractionation procedure should be planned with its purpose in mind, and may not be necessary if bulk quantities are employed.

1. The native air-dried gum, milled fine, could be solvent-extracted with Me_2CO or EtOH in order to remove less polar contaminants, mono- and oligosaccharides, and colouring material.

2. A time-honoured method of fractionation (Smidsrød and Haug, 1967) is to add, with stirring, a miscible organic liquid (EtOH, MeOH, Me_2CO) or a soluble salt such as $(NH_4)_2SO_4$ to the aqueous extract of gum after clarification (e.g. by passing through muslin, Millipore filtration, or centrifugation). Polyethylene glycol (PEG) may also be used as precipitant (Polson, 1977). Solvent fractionation can be performed in stages, with removal of precipitates after successive additions of the precipitating agent. Typically, the bulk of the gum is obtained at an ethanol concentration of ~70%. In structural studies all the material submitted to partial precipitation should be recovered, lest discarding the more soluble fractions should result in loss of low molecular weight analogues of the major polysaccharide (Adam *et al.*, 1977; Churms *et al.*, 1978a) or a second, dissimilar polysaccharide component. Some gums may even separate from concentrated aqueous solutions on near-freezing. Repetition of the fractional precipitation process (to minimise contamination by co-precipitants) should yield material adequate for most industrial purposes, a decision being taken on the basis of uniformity of composition and physical properties of the fractions. Freeze-drying of aqueous solutions is the standard method of obtaining a satisfactory product, though there is a risk of insolubilising the gum (Anderson and Smith, 1967; Anderson *et al.*, 1969); spray-drying is a less satisfactory alternative suited to bulk handling (Glicksman, 1969a; Meer, 1980).

3. An efficient means of fractionating gum solutions to yield components differing in acidity is to add increasing quantities of a long-chain quaternary salt (Scott, 1965), such as cetavlon (cetyltrimethylammonium bromide, CTAB), the pyridinium analogue, dodecyloxycarbonylmethyltrimethylammonium chloride (Fischer *et al.*, 1986), or others. 20-Gram quantities may be handled, though large volumes of liquid will become involved as the precipitated fractions are redissolved in aqueous NaCl and reprecipitated in EtOH.

4. Both the cetavlon approach and the smaller scale separation of acidic polysaccharide components on suitable ion-exchange columns are described for gum tragacanth (Aspinall and Baillie, 1963) and the well-studied exudate of *Anogeissus leiocarpus* (Aspinall *et al.*, 1969). A variety of materials such as diethylaminoethyl-cellulose (Neukom and Kuendig, 1965; Siddiqui and Wood, 1971), Duolite A4 (which has good capacity) and diethylaminoethyl-Sepharose (which also has steric-exclusion capability: Barsett and Paulsen, 1985) have been used for purification by column chromatography, elution proceeding with water through phosphate buffers to NaOH. Polyvinylpolypyrrolidone adsorbs preferentially carbohydrate attached to polyphenols, $HCONH_2$ and 8 M $CO(NH_2)_2$ being required for elution (Stephen and Churms, 1988).

5. Where small samples of a gum are needed for analytical purposes, the standard techniques of dialysis, electrodialysis and Millipore filtration are applied, bearing in mind the high viscosity of many gums even in dilute solution.

6. Complexing with borate may facilitate the fractionation of gum solutions on ion-exchange columns (Hough *et al.*, 1972), as in this way anionic character may be artificially conferred on neutral or high equivalent-weight components; non-complexed material is washed through the column with water.

7. Affinity columns, e.g. those prepared for the binding of terminal α-D-Galp units, as found in *Acacia* gums, may be employed (Goldstein and Hayes, 1978; Blake and Goldstein, 1980). Extensive use has been made of immobilised lectin from the giant clam (*Tridachna maxima*), which binds terminal β-D-Galp in arabinogalactan-proteins (Baldo *et al.*, 1978; Gleeson *et al.*, 1979).

In view of the heteropolymolecularity that is probably inherent in plant gums, a satisfactory criterion by which a sample might be considered pure for practical purposes would be to ensure that such properties as molecular size, optical activity, ratio between sugar components and degree of branching of the polysaccharide are not altered on further attempted fractionation. In addition to using the separation techniques 1–7 above, the isolated material may be analysed by other means, for example electrophoresis (see Section III.C) and techniques involving antigen–antibody reactions (Pazur *et al.*, 1986; Misaki *et al.*, 1988).

Regarding certain industrial operations, such as flotation in the mining industry, it is possible that contaminants could obstruct the efficacy of the gum used; refinement would then be carried out, until no longer cost effective. In any event the fewer and less drastic the purification steps the less the risk of structural modification and attendant changes in chemical and physical behaviour.

III. MEASUREMENT OF PHYSICAL PROPERTIES

A. Ultracentrifugation; Measurement of Viscosity and Diffusion; Light Scattering

The determination of physical properties that depend on molecular size distribution, molecular shape and the distribution of ionisable, acidic residues is clearly vital to the proper assessment of industrial gums. Solubility relationships, viscosity, gel-forming ability, rheological behaviour and emulsifying power may in some degree be correlated with molecular size and branching of the polysaccharide component, but protein may exert a disproportionate influence (Randall *et al.*, 1988). Measurements made on solutions (Banks and Greenwood, 1963) can be supplemented by electron microscopy (Fowle *et al.*, 1964), and sometimes by X-ray diffraction of specimens prepared by stretching (Lelliott *et al.*, 1978; cf. Chanzy *et al.*, 1978). Techniques of molecular weight measurement are standard for macromolecules in solution, with the caveat that loose aggregation, through hydrogen bonding for example, may produce inflated values. Some methods are absolute, while others depend on suitable standards (as similar as possible in molecular properties and shape to the gum under examination: Harding, 1988).

1. Sedimentation and diffusion

Molecular weights have been recorded for industrial gums using the Svedberg formula $M = (R T S)/[D(1 - v\rho)]$ (cf. Gralén and Karrholm, 1950), the essential measurements being those of sedimentation (S, determined by ultracentrifugation) and diffusion (D) constants, taken with the density (ρ) of the solution and the partial specific volume (v). Typical values for gum arabic (Säverborn, 1945; Churms *et al.*, 1983) and tragacanth

(Stauffer, 1980) average 3×10^5 and 8.5×10^5, though these substances are well known to be heterogeneous.

2. *Viscosity*

Measurement of intrinsic viscosity, important for all plant gums, leads to a molecular weight value from an expression (Mark-Houwink) of the form $[\eta] = K'.\bar{M}_w{}^a$, K' being 1.3×10^{-2} and the power term 0.54 for gum arabic (Anderson and Rahman, 1967). The Staudinger constants K' and a will vary according to chemical differences in the polysaccharides (the presence of methoxyl groups and associated protein for example) and the intrinsic viscosity $[\eta]$ with temperature, pH, and ionic strength of the solvent medium.

3. *Light scattering*

Development of the standard light scattering method for determining $\bar{M}_w$, which is affected by any degree of turbidity or lack of optical clarity, has led to precise measurements when a laser source is used. Analyses of fractionated samples of gum arabic were reported by Anderson *et al.* (1967), a higher figure of $\bar{M}_w$ 5.8×10^5 being found, and values for some 22 other *Acacia* gums appeared in subsequent papers (Anderson and Dea, 1969a, 1971).

Laser light scattering (Vandevelde and Fenyo, 1985) gave $\bar{M}_w$ values from 4.4×10^5 to 2.2×10^6 for gum arabic samples. The technique affords information as to the general shape and dimensions of gum molecules in solution (Banks and Greenwood, 1963).

4. *Osmometry*

Measurement of osmotic pressure lowering by gum exudates gives a value for $\bar{M}_n$, the number of particles being significant (Towle, 1972); values of $\sim 2 \times 10^5$ are recorded (Oakley, 1936) for gum arabic.

B. Molecular Weight Distribution

Disparity in $\bar{M}_n$ and $\bar{M}_w$ for gum samples gives an indication of the irregularity of molecular weights, the figures quoted above showing that a range of components is to be found in gum arabic. A simple and effective method of determining the proportions of molecules of different sizes is by the use of steric-exclusion chromatography (SEC; synonyms: gel filtration; gel permeation, molecular sieve, size-exclusion chromatography) (Granath, 1965; Churms, 1970; Whistler and Anisuzzaman, 1980). This process depends on the availability of standard polysaccharides of known molecular weights, lying within a narrow range (Andersson *et al.*, 1985), and which ideally should resemble the material under test in molecular shape and form. Clearly this is impossible to realise in practice so that the chromatographic profiles for the gum sample will give at best a series of relative values for the molecular weights of its components. Calibration of columns, using chosen standard substances, gives a negative linear relationship between $\log \bar{M}$ and peak elution volume, V_e. A better method, which takes some account of the disparity in shape of sample and standard, used the correlation of V_e with $\log[\eta].\bar{M}$, where $[\eta]$ is the intrinsic viscosity of the polymer, a so-called 'universal calibration' being obtained (Kuge *et al.*, 1984; Vandevelde and Fenyo, 1985; Churms and Stephen, 1988). The function $[\eta].\overline{M}$ is related to the hydrodynamic volume of the polymer (Grubisic *et al.*, 1967). The concentration of eluted carbohydrate is determined by differential refractometry or colorimetrically by the $PhOH-H_2SO_4$ method (Dubois *et al.*, 1956), or the anthrone method (Granath and Kvist, 1967). The heteropolymolecularity of gums and their derived products makes imperative the use of calibration curves that show the differing colour responses of components eluted during the course of the run. Fractions should be assayed for other parameters in addition, such as uronic acid content or ultraviolet absorption, in order to derive maximum advantage from the chromatographic separation.

The gels used are preferably non-carbohydrate in character, for example, cross-linked polyacrylamides such as the Bio-Gel P series (Anderson *et al.*, 1965b; Churms, 1970), though the Sephacryl and Superose gels have superior flow characteristics (Praznik *et al.*, 1986). The porosity of the gel is chosen according to the molecular weight range of the gum or derived carbohydrate products under test. Sodium chloride in water (1 M) is a preferred eluent, though water or volatile buffers such as pyridinium acetate solutions can be used if quantitative separation and recovery of carbohydrate fractions is required.

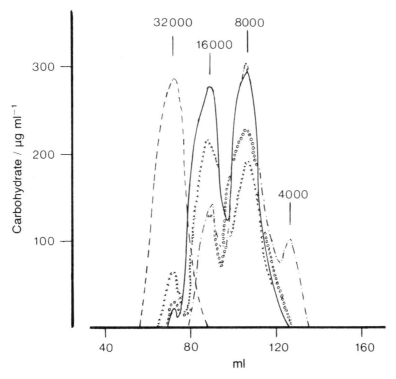

FIG 14.1. Chromatography on Bio-Gel P-300, with 1 M NaCl eluant, of five samples (collected from different localities and parts of single trees) of untreated *Acacia podalyriaefolia* gum; the molecular weights are based on a dextran calibration (reprinted by permission, from *S. Afr. J. Chem.*, 1987, **40**, 89–99).

Anderson and Stoddart (1966) studied the molecular weight distribution analyses of components of gum arabic using Bio-Gel columns, aqueous NaCl, and sample sizes of

~5 mg. The calibrants were commercial fractionated dextrans, characterised by independent, absolute measurements of molecular weight (see above). The chromatograms, which were plots of carbohydrate content against elution volume V_e, provided the prototype for numerous subsequent investigations of the molecular weight distribution of gums, with mean values ranging from ~2×10^4 to >10^7. One such chromatogram is shown as Fig. 14.1.

The technique has also been applied widely in studying the degradation of gums by chemical means, hydrolyses in particular (Churms and Stephen, 1972). The Smith degradation procedure, whereby certain inter-sugar linkages are rendered sensitive to acid hydrolysis, has been monitored to good effect by steric-exclusion chromatographic analysis (Churms *et al.*, 1977; see Section IV.E). Current interest in the role of the proteinaceous components of gums, stimulated by the finding that these may play a predominant role in such processes as emulsification (Randall *et al.*, 1988), has led to their preparative scale isolation by SEC (Vandevelde and Fenyo, 1985); further examination of pronase-digested gum specimens (Connolly *et al.*, 1988) has assisted in developing a clearer picture of the whole gum arabic structure. There is in fact no feasible substitute for the method, as the alternative approach of fractionating the material under study and carrying out independent molecular weight measurements is usually impracticable because of time and very often of sample limitation.

Arabinogalactan-proteins may be analysed for polydispersity, but meaningful molecular weights are in jeopardy when the protein content rises above ~20%. Similarly, if polyphenolic substances are associated with polysaccharide in the test sample, retardation of flow through the gel column beyond the 'total volume' V_t may result from adsorptive interaction, and calibration would be almost impossible. A methylated mucilage (from the seed husk of *Plantago ovata*) was successfully fractionated into a number of components of varying size by SEC on Sephadex LH-60 with 1,4-dioxane as eluent (Sandhu *et al.*, 1981).

C. Electrophoresis and Electrodialysis

Electrodialysis constitutes an important technique for the purification of sample material for analysis (Anderson and Stoddart, 1966), some knowledge of molecular size being required in order to select the correct membranes. Early work on free-boundary electrophoresis of *Acacia* gums (Joubert, 1954), and the separation of gum components on glass-fibre sheets (Lewis and Smith, 1957) or cellulose acetate strips (Dudman and Bishop, 1968) in alkaline borate, has been followed by disc electrophoresis (Pavlenko and Ovodov, 1970), an analytical tool of great importance in the examination of protein-containing gums (Gammon *et al.*, 1986b).

Thin layer electrophoresis is a useful technique for testing and conducting preliminary investigations of exudates (Anderson and Munro, 1969) and gums incorporated in foodstuffs (Scherz, 1985).

D. Spectroscopic Properties

1. Ultraviolet

Application of ultraviolet (UV) spectrometry to gum exudates takes on a special significance when the gums, or derived products obtained by chemical means, are associated with protein. Eluates from columns using steric-exclusion or ion-exchange (or both) are readily monitored by UV spectrometry at appropriate wavelengths (e.g. 220 or 280 nm), and the coincidence of the resulting plots with those for total carbohydrate, uronic acid and specific amino acids, notably hydroxyproline (determined independently in the eluted fractions by colour reactions) is a significant aid in establishing the nature of the carbohydrate–protein linkage (Churms and Stephen, 1984; Gammon and Stephen, 1986; Gammon et al., 1986a,b).

2. Infrared

Attempts have been made to characterise gums by their infrared (IR) spectra using KBr or KCl disks, Nujol mulls or cast-films (Spedding, 1964; Proszynski et al., 1965; Glicksman, 1969b; Rosík et al., 1971; Perlin and Casu, 1982; Mathlouthi and Koenig, 1986). IR spectra in $CHCl_3$ of methylated polysaccharide gums are routinely employed to assess extent of methylation by noting the presence or absence of residual HO absorptions at $3700–3000\ cm^{-1}$. By measuring the ester carbonyl absorption at $1740\ cm^{-1}$, and relating this to the absorption at $1100\ cm^{-1}$ which reflects the total methoxyl content of the polysaccharide, an estimate of uronic acid content may be obtained. For a series of methylated gums of similar structural type and known sugar composition, this absorption ratio may be plotted against carboxylic acid content and the graph used to estimate the proportion of uronic acid in an unknown specimen. Biouronic acids derived from gums have been characterised by IR spectroscopy (Rosík et al., 1973).

3. Nuclear magnetic resonance

Advances in NMR spectroscopic techniques are stimulating renewed interest in their use for characterising structural features in gum exudates (e.g. gum arabic) or their degradation products at oligosaccharide level (Defaye and Wong, 1986). Sequences of sugar units in oligosaccharides, even of DP ~ 6, may be determined by 2D methods, an approach which has been found useful in the study of related bacterial polysaccharides. As for IR spectroscopy, NMR methods (Hall, 1964; Gorin, 1981; Barker et al., 1982; Bock et al., 1984; Bradbury and Jenkins, 1984; João et al., 1988) are now routine not only for the characterisation of anomeric and other easily recognised signals (di Fabio et al., 1982), but also for the complete elucidation of neutral and acidic oligosaccharides.

4. Chiroptical

Although not extensively applied to plant gums, optical rotatory dispersion and circular dichroism afford a general approach to the constituents and conformations of macromolecules in solution (Stone, 1976; Woody, 1977; Johnson, 1987), the carboxylate chromophore being of particular interest. GlcA and GalA at various degrees of ionisation are readily detected. Procedures that have been applied to mannose-based polysaccharides are to be found elsewhere in this volume (Chapter 11).

IV. STRUCTURAL ANALYSIS

A. Acid Hydrolysis and Methanolysis

Hydrolysis of glycosidic bonds to show the identity and proportions of sugar and uronic acid units is an essential prerequisite in identifying the gum exudate and in providing a foundation for molecular structure determination. While results may be obtained rapidly and with a minimum of expense, there are problems:

(a) As the bonds between units vary greatly in susceptibility to hydrolysis in acid, reducing monosaccharides are released in the order of ease of bond fission: furanosidic > pyranosidic, 6-deoxyhexosidic > hexosidic, and neutral hexosidic > uronosidic (Adams, 1965; Albersheim *et al.*, 1967; Dutton, 1973; Selvendran *et al.*, 1979; Morrison, 1988). Progressive decomposition of the free sugars formed takes place, as illustrated in Table 14.2.

TABLE 14.2. Losses of sugars under various hydrolytic conditions.

Sugar	Loss (% by weight)[a]			
	2 M TFA (8 h)	2 M TFA (18 h)	M H_2SO_4[b] (6 h)	M HCl[b] (6 h)
Ara	10	25		
Xyl	20	38	31	54
Rib			48	70
Rha	6	8	7	19
Fuc			7	12
Glc	2	2	3	2
Gal	2	2	10	11
Man	5	16	14	32

[a] Hydrolysis at 100°C, in sealed tubes under N_2.
[b] Hough *et al.* (1972).
See also the review by Bierman (1988).

(b) The hydrolysis process is further complicated if proteins or polyphenols are present in the gum sample, interaction with the reducing sugars then being possible (cf. Rahman and Richards, 1988); hence the decision as to what conditions to use (molarity of acid, temperature, duration, which acid) is important. The polysaccharide might have limited solubility, so aggravating the problem.

(c) The assay of sugars in hydrolysates has been the subject of much research, discussed elsewhere in this book, but clearly a flexible approach is needed where different gums are to be analysed (Churms, 1982c; Blaschek, 1983; Honda, 1984; Lawrence and Iyengar, 1985; Al-Hazmi and Stauffer, 1986; Hicks, 1988). On the other hand, standardised conditions must be followed rigidly for purposes of comparison among samples of a particular gum (Anderson and Stoddart, 1966; Anderson *et al.*, 1971; Anderson, 1986b).

(d) Some difficulties, such as the slow release of sugar units to which uronic acid residues are joined, and decarboxylation of the acids themselves, are obviated if the polysaccharide is pre-reduced to the maximum practicable extent before hydrolysis. If borodeuteride is used, the hexoses derived from hexuronic acids are readily identifiable.

(e) Because of the differing labilities of sugars once liberated, methanolysis is a safer procedure, provided the gum is sufficiently soluble (Chambers and Clamp, 1971; Cheetham and Sirimanne, 1983; Roberts *et al.*, 1987). If sugars, not the methyl glycosides, are required for derivatisation, however, the problem reappears, although to a lesser extent. MeOH–HCl rapidly loses its effectiveness through conversion to MeCl and H_2O at elevated temperatures (Thier, 1984), and it has been suggested that the reaction time be limited accordingly. If the preferred technique for analysis is silylation and GLC separation of the methyl glycosides of sugar and methyl hexuronate, the capillary columns used must be highly efficient.

Acceptable procedures for glycosidic cleavage are as follows:

(i) A 10 mg sample of gum is heated in a sealed glass tube, under nitrogen or preferably the denser argon, with 2 M aqueous trifluoroacetic acid (TFA) at 100°C for 8 h; one-half of the solution is removed, and the remainder heated as before for a total of 18 h. An accurately weighed sample of *meso*-inositol should be included with the gum, as an internal standard and to check the total yield of sugars released. Biouronic acid remaining after 18 h should be estimated independently, e.g. by steric-exclusion chromatography, or by HPLC (Honda *et al.*, 1983). The hydrolysates are diluted and freeze-dried. Qualitative identification is carried out by PC or TLC, and quantitative assay of the sugars by GLC after conversion to alditol acetates (Sawardeker *et al.*, 1965; Blakeney *et al.*, 1983) or trifluoroacetates (Haga and Nakajima, 1988), peracetylated aldononitriles (Dmitriev *et al.*, 1971; Morrison, 1975; Seymour *et al.*, 1979) or TMS ethers (Sweeley *et al.*, 1963), or by HPLC (McGinnis and Fang, 1980; Honda, 1984; Hicks, 1988), ion-exchange chromatography of borate complexes (Kennedy and Fox, 1980a; Honda *et al.*, 1981), ion chromatography in alkali (Edwards *et al.*, 1987), derivatisation into ninhydrin-positive amines (Hara *et al.*, 1979; Perini and Peters, 1982; Shinomiya *et al.*, 1987), or any other appropriate means. Partial conversion of GlcA to Glc takes place during the reduction step in AA (alditol acetate) formation but not in the conversion to PAAN (peracetylated aldononitrile) derivatives (but see also Furneaux, 1983). Percentage losses of sugars are determined independently on weighed samples under practical conditions, as are the response factors. Colour reactions may be used to supplement the assay of neutral sugars and deoxy sugars (Dische, 1962a). Amino acid assays are made on hydrolysates obtained under the more drastic acid hydrolysis conditions applicable to proteins. Uronic acids are determined independently by colorimetric methods, decarboxylation by heating in HCl (Anderson *et al.*, 1963; Whyte and Englar, 1974; Kosheleva *et al.*, 1976), and from the equivalent weight of the gum. Uronic acids in hydrolysates may be separated by electrophoresis on paper (Haug and Larsen, 1961), or by HPLC on an anion-exchange resin (Honda *et al.*, 1983), using borate buffers in each case.

(ii) Pre-hydrolysis in 72% aqueous H_2SO_4 at room temperature for 4 h, dilution with water to 1 M, and heating at 100°C for 6–18 h is an alternative for samples that are difficult to dissolve. The hydrolysate is neutralised with $BaCO_3$, filtered through sintered glass or centrifuged, and a mixture of sugars and barium salts is obtained by freeze-drying.

(iii) Methanolysis as described by Thier (1984) in which 10 mg samples of gums are heated at 100°C for 4 h in 2 M HCl in MeOH, may be advantageous. A mixture of methyl glycosides of the constituent sugars of the gum results. Biouronic esters resist glycosidic fission. An advantage is that the methyl glycosides of methyl uronates are included in the GLC assay, as TMS ethers (Ha and Thomas, 1988) or acetates.

Other procedures, e.g. mercaptolysis (Wolfrom and Thompson, 1963) should be consulted and tested. Anhydrous HF or TfOH (triflic acid = trifluoromethane sulphonic acid) are effective reagents for the removal of intact carbohydrate chains from glycoprotein (Mort and Lamport, 1977; Edge *et al.*, 1981). Glycosylated hydroxy-prolines formed by alkaline hydrolysis may be separated on cation-exchange columns (Lamport and Miller, 1971).

If the scale of operation permits, as will be the case for commercial gums, the sugars and uronic acids in hydrolysates can be isolated in pure form by liquid chromatography on a cellulose column, aqueous BuOH or EtCOMe being typical eluting agents (Whistler and BeMiller, 1962). Hydrolysis (preferably in dilute H_2SO_4) is monitored by changes in optical rotation, or by measurement of reducing power. Individual mono-saccharide components are then identified unambiguously by $[\alpha]_D$, melting point if crystalline, and by conversion into standard derivatives. Uronic acids and accompany-ing biouronic acids, obtained initially as barium salts after neutralisation of the acid used for hydrolysis, may be converted into methyl ester glycosides and derivatised as amides (Smith and Montgomery, 1959c), or characterised spectroscopically. As well as providing convincing evidence for the identity of the sugar units, large-scale hydrolysis is at least semi-quantitative. An added advantage is the production of the constituent sugars, e.g. L-Ara and L-Rha, which otherwise are not readily available.

B. Partial Hydrolysis and Acetolysis

(a) A first and important step towards determining sequences of sugar units is to arrest glycosidic cleavage of the gum at intermediate stages (BeMiller, 1967). If hydrolysis is controlled so as to permit release only of labile L-Ara*f* linkages, or continued until L-Rha*p* and L-Fuc*p* are liberated, the residual, partially degraded polysaccharide serves as a useful starting point for characterisation by the standard methods applied to intact gum. Points of removal of sugar units are revealed by methylation analysis (Section IV.D), and newly-exposed hydroxyl groups may well generate diol systems that are vulnerable to periodate (Section IV.E). To achieve this, gums in the acid form, generated by passage through a column of cation-exchanger such as Amberlite IR-120 (H^+) and freeze-drying, are boiled in water (pH ∼2). Hydrolysis is monitored by $[\alpha]_D$ and reducing power (Baker and Hulton, 1920; Dische, 1962c). Certain *Acacia* gums yield 'autohydro-lysed' products for which there is a correlation between the percentages of L-Ara and L-Rha removed from the periphery of the branched molecules and the fall in molecular weight of the residual core (Churms and Stephen, 1972). For others, there is a marked drop in molecular weight, suggesting frequent points of cleavage in the interior (Stephen, 1983). When the L-Ara content is low, the stoppage-point for autohydrolysis is usually less clearly defined.

Controlled acid hydrolysis frequently provides numerous oligosaccharides, neutral and acidic, the approximate quantities of which may be gauged by PC analysis of samples taken at intervals. The products are recovered by aqueous EtOH extraction of freeze-dried hydrolysates, and separated by ion-exchange or other chromatographic means based on partition, adsorption (Whistler and Durso, 1950) and steric exclusion (Kennedy and Fox, 1980b). Useful structural information was obtained, for example, by partial hydrolysis (Aspinall and Whitehead, 1970) of mesquite gum (type A, from *Prosopis juliflora*), 20 oligosaccharides being identified, many of them neutral. These included a linear L-Ara-containing chain comprising seven units and a shorter sequence interrupted by 6-linked α-D-Gal, together with the biouronic acids 4-Me-β-D-GlcA-(1→6)-D-Gal, 4-Me-α-D-GlcA-(1→4)-D-Gal, and some of β-D-GlcA-(1→6)-D-Gal. These biouronic acids, qualitatively identified by PC, are useful markers for the gum; others, found as units in types A, B and D gums, are α-D-GlcA-(1→4)-D-Gal, α-D-GalA-(1→2)-L-Rha and β-D-GlcA-(1→2)-D-Man, respectively. Triouronic acids, which reveal the identity of the next interior sugar unit, may be isolated. As for total hydrolyses of gums, partial hydrolysis is carried out on a scale commensurate with the availability of starting material and the purpose in mind; rapidity of analysis of small samples using HPLC, NMR and GLC-MS methods (Valent *et al.*, 1980) is counterbalanced by the benefit of obtaining larger quantities of oligosaccharides, neutral and acidic, that are not easily obtained by synthesis. In particular, NMR spectroscopy provides evidence for anomeric configurations and modes of linkage as long as adequate specimens of the oligosaccharide are available.

(b) Acetolysis of acetylated gums or better, acetylated, carboxyl-reduced gums, has advantages in that certain linkage modes, otherwise cleaved by hydrolysis, may be preserved (Guthrie and McCarthy, 1967; Aspinall and McKenna, 1968; Aspinall and McNab, 1969; Aspinall and Bhattacharjee, 1970; Aspinall and Sanderson, 1970; Lindberg *et al.*, 1975; Aspinall, 1976a; Narui *et al.*, 1987); the acetylated oligosaccharides are separated by column chromatography (Thompson, 1962; Wells and Lester, 1979), saponified in the cold using sodium or barium methoxides, and characterised. Rhamnosyl and fucosyl linkages tend to resist acetolysis; thus terminal Rha units have been proved to be linked to O-4 of GlcA in carboxyl-reduced gum arabic through the isolation of α-L-Rha-(1→4)-D-Glc after acetolysis (Aspinall *et al.*, 1963). Important evidence for the presence of (D-Man→D-GlcA)_n sequences in *Anogeissus leiocarpus* gum was obtained in the same way (Aspinall and McNab, 1965, 1969).

(c) Enzymatic hydrolysis of peach gum (from *Prunus persica*) using extracellular enzymes from *Aspergillus flavus* on the gum surface yielded a number of oligosaccharides that demonstrate the presence of (1→3)- and of (1→6)-linkages within chains of β-D-Gal units, and a (1→3)-linkage between α-D-Man*p* and D-Gal*p* (Kardošová *et al.*, 1979). The latter has a bearing on the general question whether Man or Gal residues interrupt sequences of alternating D-Man and D-GlcA in the core. Apart from this successful result, there are few reported indications of the enzymatic breakdown of gums. Gum karaya was degraded by a fungal isolate from *Cephalosporum* (Raymond and Nagel, 1973), and partially-hydrolysed gum tragacanth by α-D-galacturonanase (Aspinall and Baillie, 1963; Matheson and McCleary, 1985). Arabinan-degrading enzymes have been isolated from *Aspergillus niger* extracts (Voragen *et al.*, 1986, 1987) and an α-L-arabino-furanosidase from *A. niger* promotes the release of terminal L-Ara from the periphery of complex gum structures (Kaji, 1984). Modification of gum arabic has been achieved (Aspinall and Knebl, 1986a) by removal of terminal D-Gal groups under the action of an α-D-galactosidase (Dey, 1980).

C. Derivatisation

The three most important forms of derivatisation of plant gum exudates are etherifica-tion (methylation), esterification (acetylation) and reduction (of carboxyl groups, and of the reducing end group where for any reason the molecular weight is low).

1. Etherification

Methylation analysis of gums, which provides quantitative information as to the modes of linkage of each type of sugar unit, is dependent on the successful conversion of all hydroxyl groups to methoxyl, with a minimum of alteration of the molecular structure during the processes of dissolving the gum, methylating, and purifying the permethy-lated product. Each exudate gum poses its own problems in these respects. Very often complete methylation is achieved only after several treatments, preferably using different methylating systems.

Dissolution in the usual solvents, H_2O (strongly alkaline), DMF and DMSO may take many hours and only be partial. Sonication is often helpful, but could be disruptive. $CO(NMe_2)_2$ used as an H-bonding antagonist in non-aqueous solvents (Narui *et al.*, 1982) may well be invaluable in reducing interaction between gum molecules and so facilitating approach of the methylating agent to hydroxylic sites; clearly the complex nature of the gums makes assistance of this kind particularly important.

The classical Haworth methylation process (Haworth, 1915), in which the gum (preferably pre-treated, in order to protect reducing end-groups, with a small quantity of $NaBH_4$ in water) is stirred vigorously, in the cold at first and under nitrogen, with simultaneous, dropwise addition of concentrated aqueous NaOH, or KOH, and Me_2SO_4, is ideal for large-scale work (gram quantities) and is suitable for gums of high uronic acid content. Equally valuable, and indeed indispensable if milligram amounts are to be methylated, is the Hakomori (1964) procedure as adapted and modified by others (Anderson and Cree, 1966; Sandford and Conrad, 1966; Anderson and Stefani, 1979; Phillips and Fraser, 1981; Waeghe *et al.*, 1983; Harris *et al.*, 1984; Blakeney and Stone, 1985; Paz Parente *et al.*, 1985; Kvernheim, 1987). The dried polysaccharide (protonated form) is dissolved in dry DMSO, under N_2 or Ar. Mild sonication may be needed, although there is a danger of septa becoming disinte-grated. Alkoxide formation, using a moderate excess of Na, K or Li methylsulphinyl-methanide (dimsyl) in DMSO, should be carried out for no longer than is considered essential, to guard against molecular degradation. Excess of the reagent must be tested for, using Ph_3CH as internal or preferably external indicator (sample withdrawn by syringe and tested in an inert atmosphere). The methylation step with CH_3I or CD_3I follows, an excess of the reagent being added, cautiously, with stirring, to the chilled alkoxide solution. Success in methylation depends critically on these procedures being carried out with great care. A recent modification, using NaOH as base (Ciucanu and Kerek, 1984), is easier and may prove satisfactory. The Kuhn method (Ag_2O or BaO, in DMF) is effective once partial methylation, by the Haworth procedure, has given an intermediate product soluble in DMF. The work-up involves aqueous cyanide (Kuhn et al., 1955).

Reaction is terminated by neutralisation (Haworth method) or by quenching in ice-water (Hakomori), inorganic and low molecular-weight material being then removed by dialysis, if the molecular size of the product permits. The methylated polysaccharide is normally extracted into $CHCl_3$ or CH_2Cl_2, but it is prudent to check for carbohydrate in the washings, after concentration, as substantial quantities of the gum, possibly under-methylated, may not be extracted into the organic phase.

The extent of methylation may be assessed by acid hydrolysis of a small sample, and PC or TLC of the hydrolysate. Unmethylated sugars or a preponderance of those singly-methylated indicate that further methylation is necessary.

Successive treatments with CH_3I and dry Ag_2O (in presence of $CaSO_4$) under reflux (Purdie and Irvine, 1903) are applied until AgI is no longer observed on the surface of the black Ag_2O. Satisfactory analysis of the product demands further purification by fractional precipitation with light petroleum–$CHCl_3$ mixtures, by use of lipophilic gel columns (solvents $CHCl_3$ and EtOH, the effluent being monitored with the anthrone–H_2SO_4 reagent), or by HPLC on Sep-Pak C_{18} cartridges (Mort et al., 1983; Waeghe et al., 1983). The methylated product should show no HO absorption (IR in $CHCl_3$), and is normally characterised by its specific rotation (in $CHCl_3$), and methoxyl content (Zeisel method).

Special techniques are needed if ester groups (notably acetate) are to be preserved during methylation; to this end the TfOMe–2,6-di-(tert-butyl)pyridine combination of Prehm (1980) may be used. During Hakomori methylation any methyl uronate groups in contact with strong base (dimsyl ion) confer lability to β-elimination on the glycosyl or alkyl groups at O-4 of the ester, leading to destruction of the uronate group and severing of the carbohydrate chain interior to the newly unsaturated ester (Section IV.F.1). This possibility must be taken into account and considered when the structural significance of the results of methylation analyses is assessed.

Finally, methylation has been scaled down to submilligram levels using a rigid protocol (Waeghe et al., 1983).

2. Esterification

Acetylation (or propionylation) of gums may be regarded as a prelude to methylation (in Me_2CO using NaOH and Me_2SO_4), or to checking anomeric configurations (CrO_3 oxidation; Lindberg et al., 1975) or to carboxyl reduction (B_2H_6). The method of Carson and Maclay (1946) requires dissolution of the dried gum in anhydrous, distilled $HCONH_2$, acetylation by Ac_2O at room temperature, and work up by adding to ice-

water, dialysis and extraction with CHCl₃. Acetate groups are determined by saponification (aqueous alkali) and titration, or trans-esterification (EtOH–HCl, followed by
distillation and assay of the EtOAc formed; Matchett and Levine, 1941).

3. *Reduction*

(a) Further investigation of the molecular structure of gums, all of which contain
    carboxyl, may well involve reduction of these groups in aqueous solution. The
    method of Conrad (Taylor *et al.*, 1976; M. A. Anderson and Stone, 1985)
    employs NaBH₄ (NaBD₄) reduction, at controlled pH, of the derivative formed
    from the gum and a water-soluble carbodiimide, and is followed by estimation of
    carboxyl groups to assess the extent of reaction. The reduction may need to be
    repeated. Only 70% of the carboxyl groups in the mannoglucuronoglycan from
    Kiwi fruit stem mucilage were reduced after seven treatments, but 90% of the
    carboxyl groups in the partially-hydrolysed mucilage underwent reduction after
    only two treatments (Redgwell *et al.*, 1986a). Physical properties of gums are
    modified by removal of carboxylate groups; insolubility may result, though the
    molecular size is not necessarily affected.
(b) Suitable on a large scale is the use of borane (generated *in situ* from NaBH₄ and
    BF₃.Et₂O, or led in from an external generator) on previously hydroxyl-esterified
    gum dispersed in diglyme (Smith and Stephen, 1960; Aspinall and Fanshawe,
    1961).

D.   Methylation Analysis

Complete methylation of purified gum is a prerequisite before glycosidic bond cleavage
and analysis of the relative molar proportions of the resulting methylated sugars and
uronic acids is carried out (Jansson *et al.*, 1976; Aspinall, 1982). Correspondence of end-
groups and branch-points is one result that can be checked and the total proportions of
methyl ethers of each sugar and uronic acid, found by more than one method, must
agree with those determined by hydrolysis of the gum or other means in order to
establish the validity of the methylation analysis.

   Hydrolysis or methanolysis of the permethylated gum releases methylated monosaccharides or their glycosides with varying ease (Section IV.A), some only incompletely
under practical conditions. For instance, 10% and 30% of the glycosidic bonds of fully
methylated biouronic acids β-D-GlcA-(1→6)-D-Gal and β-D-GlcA-(1→2)-D-Man were
found to have remained unchanged after prolonged (15 h) methanolysis (cf. Stephen *et
al.*, 1966). Qualitative and quantitative analysis of the methylated sugars and uronic
acids that are released involves procedures similar to those adopted for hydrolysates of
the original gum. Most gums being acidic, a standard procedure is to reduce methyl
carboxylate groups in methylated gum with LiAlH₄ or LiAlD₄ (Åman *et al.*, 1982).
Hydrolysis or methanolysis of the reduced methylated gum, followed by analysis, shows
directly the modes of linkage of uronic acid units in the gum; there is usually an increase
in the proportions of the methylated sugar interior to uronic acid units.

   The simplest qualitative approach to the analysis of mixtures of methylated sugars
and uronic acids is by PC or TLC (on cellulose or silica gel) using several solvent
systems, and spray reagents that indicate the pattern of methylation and the class of
monosaccharide. Hydrolysates of methylated gums already characterised form the best

standards for comparison, and remarkably good correlations may be achieved. Spraying of chromatograms with *p*-anisidine hydrochloride in H_2O–EtOH–BuOH and heating at 105°C produces a series of spots which are easily recognisable from their colour and fluorescence under UV (Smith and Montgomery, 1959b).

Derivatisation of the hydrolysates from methylated gums takes the form of one of the following:

(i) Reduction with aqueous $NaBH_4$ at room temperature during 2 days, acidification with HOAc (which should release H_2, indicating an excess of the reductant), volatilisation of borate as methyl ester, and acetylation. Work up of the mixture of partially-methylated alditol acetates is followed by injection of samples (in $CHCl_3$ or CH_2Cl_2) into a gas chromatograph (packed or capillary columns), coupled for preference to a mass spectrometer with capability of storing, collating and manipulating the data (Björndal *et al.*, 1970; Dutton, 1973, 1974; Parolis and McGarvie, 1978; Gorin *et al.*, 1982; Klok *et al.*, 1982; Bacic *et al.*, 1984). Identification of the methylated sugar derivatives is achieved with near certainty by a comparison of retention times and mass spectra (from selected combinations of ions as well as total ion currents) with those of known standard compounds. Hexuronic acid derivatives are identified in hydrolysates of reduced methylated gums (Åman *et al.*, 1982; cf. Valent *et al.*, 1980), ester functions having been converted to CH_2OH (CD_2OH) with $LiAlH_4$ ($LiAlD_4$) in THF (tetrahydrofuran) solution. This reaction is terminated by addition of moist EtOAc, and the reduced product is extracted into $CHCl_3$ or CH_2Cl_2 with concomitant addition of aqueous H_2SO_4 or tartaric acid. The neutralised aqueous phase should be freeze-dried and checked for absence of carbohydrate before being discarded. If the test proves to be positive, the carbohydrate material needs to be recovered and re-treated. The newly-formed hydroxyl groups may be methylated before hydrolysis and GLC if the nature of the GLC trace warrants this. Response factors for methylated, as for ethylated, alditol acetates have been calculated (Sweet *et al.*, 1975). In all quantitative work involving methyl sugars, their differing rates of release and sensitivity to acid must be taken into account (cf. Section IV.A(a)).

(ii) The methylated sugar mixture is converted to corresponding oximes by heating with $HONH_3Cl$ in pyridine, submitting to acetylation conditions, and analysing by GLC-MS the partially methylated, acetylated aldononitriles produced (Seymour *et al.*, 1975). Uronic acids are lost on GLC when this derivatisation method is used.

(iii) The trimethylsilylated derivatives of methylated sugars (Sweeley *et al.*, 1963) or alditols (Freeman *et al.*, 1972) may be used for GLC identification.

In methylation analysis it is possible to determine uronic acids and their modes of linkage together with the neutral sugar components.

The neutralised (Ag_2CO_3) methanolysate of a methylated gum and an internal standard (e.g. methyl 2,3,4,6-tetra-*O*-methyl-β-D-glucoside) are injected on to a suitable GLC column, without further derivatisation of exposed hydroxyl groups (Stephen *et al.*, 1966). Packed or capillary columns of ethylene glycol succinate and of OV-225 have been used successfully, preferably with a temperature programme rising to 250°C. Standard mixtures of methyl glycosides prepared from known methylated sugars are required (Aspinall, 1963; Bishop, 1964), but the method is an excellent one for comparing a series of gum samples that are known to contain similar ranges of sugar units, e.g. gums from different *Acacia* species (Kaplan and Stephen, 1967). The methanolysates may be trimethylsilylated, acetylated, trifluoroacetylated or otherwise derivatised prior to GLC analysis. Quantitative results are dependent on response factors obtained for pure methylated sugars, from methylated disaccharides and from methylated aldobiouronic acids.

Figure 14.2 compares GLC traces obtained for derivatives from an hydrolysate and a methanolysate of methylated *Encephalartos longifolius* gum. It is usual to record methylation analyses both in the form of proportions of methyl sugars found and of the linkages of the sugar units in the gum from which they are derived.

Vital as these techniques are, they are common to the analysis of most types of plant polysaccharides and are not discussed in any greater detail at this point. Methylation of a gum on a large scale, followed by hydrolysis, enables individual methyl sugars to be isolated using column chromatography (Aspinall *et al.*, 1958a; Stephen, 1962; Stephen and Vogt, 1967; Aspinall and McNab, 1969), yielding valuable samples, and permitting their identification with maximum certainty (including D or L configuration); de-*O*-methylation gives proof of the identity of the parent sugar (Hough *et al.*, 1950), while PC of IO_4^--oxidised samples characterises the methylation pattern (Lemieux and Bauer, 1953). This approach is semi-quantitative. Partial hydrolysis of methylated gum is of potential value in locating the types of sugar unit interior to uronic acid residues (Aspinall *et al.*, 1976), a structural problem which is described later (Section IV.F).

E. Smith Degradation

Inspection of the results of methylation analysis shows the proportions of sugar units in a gum that contain 1,2-diols and are consequently oxidised, in aqueous solution, by periodate ion. The Smith degradation procedure (Goldstein *et al.*, 1965; Hay *et al.*, 1965a) is carried out typically as follows:

> The gum is dissolved in water at a concentration of $\sim 1\%$, and a solution of $NaIO_4$, in approximately 100% excess of the theoretical, is added. The mixture is allowed to stand in the dark (at room temperature, or as low as 5°C). A blank experiment omitting the gum is carried out simultaneously. The scale of the oxidation depends on whether analytical determination of periodate uptake and release of HCO_2H or HCHO (Hay *et al.*, 1965a,b) is what is required, or the isolation of workable quantities of the non-oxidised carbohydrate. Aliquot portions are assayed for the uptake of periodate (stoichiometrically, one mole per vicinal diol), at time intervals, by a chemical method (Fleury and Lange, 1933) which can be scaled down (Churms *et al.*, 1981a,b), or spectrophotometrically (Aspinall and Ferrier, 1957). Owing to the relative inaccessibility of some sugar residues, such as those at branch points, the oxidation is allowed to continue for up to 7 days, by which time periodate consumption has usually levelled off. If work is being conducted on a preparative scale, excess of periodate is destroyed by addition of ethan-1,2-diol, and the bulk of the iodate is precipitated by cautious addition of aqueous $Ba(OAc)_2$ and centrifugation. Alternatively, inorganic ions may be removed by dialysis, and the oxidised

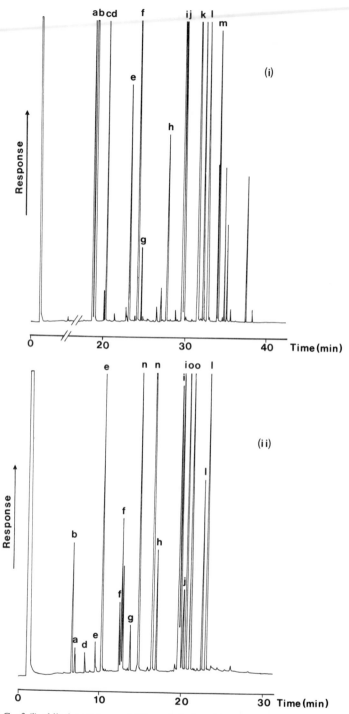

FIG 14.2. GLC of (i) alditol acetates and (ii) methyl glycosides derived from methylated *Encephalartos longifolius* gum. OV-225 OH-terminated bonded glass column, carrier gas He, in a Carlo Erba 6000 Vega Series 2 instrument with cold on-column injection; programme from 100° to 250°C at 4°C min^{-1}. Key: a, Ara*f*; b, Rha*p*; c, Ara*p*; d, Xyl*p*; e, →4-Rha; f, →3-Ara*p*; g, Gal*p*; h, →3-Gal*p*; i, →6-Gal*p*; j, →2,3-Man*p*; k, 2,3,4-Man; l, →3,6-Gal*p*; m, 3,4,6-Gal; n, GlcA; o, →4-GlcA (Stephens, D. C., Stephen, A. M. and Burger, J. A., unpubl. res.)

polysaccharide solution concentrated *in vacuo*. The sensitive aldehyde groups in the oxidised polysaccharide are then reduced at room temperature by the addition of an excess of $NaBH_4$; this step is monitored by testing the reducing power (e.g. by Fehling's reagent). After some days the solution is decationised by passage through a column of ion-exchanger, e.g. Amberlite IR-120 (H^+), in a cold room, and the eluate is freeze-dried to yield the reduced, oxidised polysaccharide. Some hydrolysis of acid-sensitive acetals may occur. A brown discoloration (due to release of iodine) may be apparent; careful addition of $Na_2S_2O_3$ improves the appearance after the commencement of the crucial, controlled hydrolysis step which follows.

The success of the Smith degradation depends on prolonging hydrolysis, at room temperature in 1 M TFA (alternatively 0.5 M H_2SO_4 or 1 M HCl) until all glycolaldehyde–acetal linkages are cleaved (Erbing *et al.*, 1973). This step liberates the sugar or degraded sugar residue interior to each sugar unit oxidised and reduced. In order to decide when to terminate acid hydrolysis, samples containing 2–5 mg of carbohydrate should be removed at intervals for molecular weight distribution analysis by SEC (Section III.B), as most of the plant gums retain blocks of contiguous, periodate-resistant sugar units, the molecular weight of which can be estimated thereby (Stephen, 1987). Alternatively, a sample of the oligomeric product is recovered, and any bound glycolaldehyde present determined by brief hydrolysis in hot 1 M TFA and colorimetric assay of the aldehyde (Dische and Borenfreund, 1949). Combined glycerol can be determined by chromotropic acid assay (Hay *et al.*, 1965b) of formaldehyde released on periodate oxidation of an aliquot (Churms and Stephen, 1971; Bekker *et al.*, 1972). Appearance of free monosaccharide, arabinose in particular, during the cold acid hydrolysis step is often a first indication of the need to terminate the degradation, as the principle of the Smith degradation is that oxidised sugars only should provide positions at which acetal (and not glycosidic) linkages are severed. The point is well made by Dutton and Gibney (1972).

If a uronic acid is modified by the oxidation and reduction sequence (Aspinall *et al.*, 1965a), acetal cleavage is particularly slow and proceeds in two stages, yielding products of the first and second limits of degradation. Whereas first-limit products are obtained after 2–5 days of acid treatment, prolongation of up to several weeks may be necessary to approach the second limit of fission. Clearly, if the slow release of glycolaldehyde (monitored by colorimetric assay) is to be avoided, the Smith degradation can be performed on carboxyl-reduced gum (Section IV.C.3).

The acid hydrolysis is terminated by freeze-drying, and glycerol, other polyols and low molecular-weight glycosides are extracted into Me_2CO–MeOH mixtures, repeating extraction of the syrupy, soluble products with 2-propanol. The products are analysed by PC and GLC (Dutton *et al.*, 1968).

The main objective in structural analysis of the gum is to obtain what has been called the first Smith-degradation product (termed SD1, as opposed to first-limit product of Smith degradation) as relatively high molecular-weight material, from the interior of the molecule, uncontaminated by fragments from the periphery. Analysis of the products of Smith degradation constitutes an invaluable means of locating the variously-linked sugar units in the gum structure (cf. Sections V.A,C).

Polyols and aldonic acids (the latter arising from modification of uronic acid units in the molecule, see below) can be detected and possibly quantified by GLC as TMS derivatives (Petersson, 1974, 1977; Bradbury *et al.*, 1981; Niemelä, 1987) or HPLC (Pecina *et al.*, 1984; Hicks *et al.*, 1985).

Entry into the molecular core by oxidation is facilitated by the prior removal of furanosidically-linked sugars (Section IV.B). Autohydrolysed gum is then submitted to Smith degradation, and products of much lower molecular weight are obtained, e.g. from modified gum arabic one of mol. wt. 2000 results, whereas the first Smith degradation of the gum itself gives two products, of mol. wt. $\sim 6.7 \times 10^4$ and $\sim 3.2 \times 10^4$ (Churms *et al.*, 1983). A variation in the sequence of reactions is to methylate the reduced, oxidised polysaccharide, prior to the acid hydrolysis step (Nánási and Lipták, 1973). If analysis of the product indicates methylated sugars with methoxyl groups on contiguous carbon atoms, this is a sign that periodate oxidation of the gum polysaccharide had not been carried to completion. The pattern of methylated sugars and polyols is informative, though the total hydrolysis of methylated oxidised-and-reduced gum does not strictly constitute a Smith degradation.

Smith degradation was developed in 1955 using gum arabic, the presence of a 3-linked β-D-galactan core being established thereby (Smith and Spriestersbach, 1955). D. M. W. Anderson and his co-workers conducted extensive experiments in which *Acacia* gums, including gum arabic and other arabinogalactans (of type A), were examined by a series of sequential Smith degradations, each polymeric product serving as the starting material for the complete series of reaction steps (Anderson *et al.*, 1966; Anderson and Cree, 1968; Anderson and Dea, 1969b). Very often the yields of the ultimate Smith degradation product were low, but a great deal of information was obtained about the structures and molecular sizes of chains of (1→3)-linked (and consequently protected) β-D-Gal units within the gum structures. The analysis of products by steric-exclusion chromatography was an integral part of the experimental approach. Analysis of the amino acid compositions of Smith degradation products from arabinogalactan proteins (AG-P) is a recent advance, which has a bearing on the composition of gums that are covalently associated with protein (Anderson and McDougall, 1987).

Detailed inspection of the molecular-weight distributions of Smith degradation products has led to the conclusion that for all arabinogalactan and AG-P exudates from *Acacia* species, and many other gums of similar type, e.g. mesquite (from *Prosopis* spp.; Dutton and Unrau, 1963; Churms *et al.*, 1981a) and larch (*Larix*) (Churms *et al.*, 1978b), the core structure comprises blocks of 3-linked β-D-Galp residues which are of practically uniform size. These blocks are evidently joined through sugar units, probably 6-linked β-D-Gal, which provide a means for disengaging the periodate-resistant blocks upon Smith degradation (Stephen, 1987). This has been observed elsewhere in the plant kingdom, e.g. *Brassica* seed arabinogalactan (Churms *et al.* 1981b) and cell-cultured *Lolium multiflorum* AG-P (Bacic *et al.*, 1987). As these examples show, the Smith degradation process causes removal of about one-half of the peripheral sugar and uronic acid units in the highly-ramified substituted arabinogalactans. The resulting polyols and other low molecular-weight products are recovered from the organic solvent extracts, and their identification provides additional insight into the modes of combinations of peripheral sugars. Xylan-based (type C) gums may be examined similarly (Dutton and Unrau, 1962), three types of structural domain being found for one example (Mabusela and Stephen, 1987). Type D gums and other polysaccharides in which the unit β-D-GlcA-(1→2)-D-Manp is prominent undergo acid-catalysed cleavage in two definite stages. The first-limit product contains not only galactose-rich material which resembles the degradation products from type A gums but also Man, erythronic acid and glycolaldehyde residues indicative of GlcA to Man sequences. The second-limit products are formed only after the glycolacetal linkages to interior sugars are broken (Stephen and Churms, 1986; Churms and Stephen, 1987; Stephens and Stephen, 1988).

SCHEME 14.1. A simplified outline of the Smith degradation procedure and its consequences (reprinted, by permission, from *S. Afr. J. Chem.*, 1987, **40**, 88–99).

(i)   1,2-Diols oxidised by IO_4^-
(ii)   Aldehydes reduced by BH_4^-
(iii)   Cold-acid hydrolysis
●   Sugars so attacked and modified are released from next in chain
●   Unattacked sugars remain joined
●   Process repeated on coherent blocks of sugar units remaining

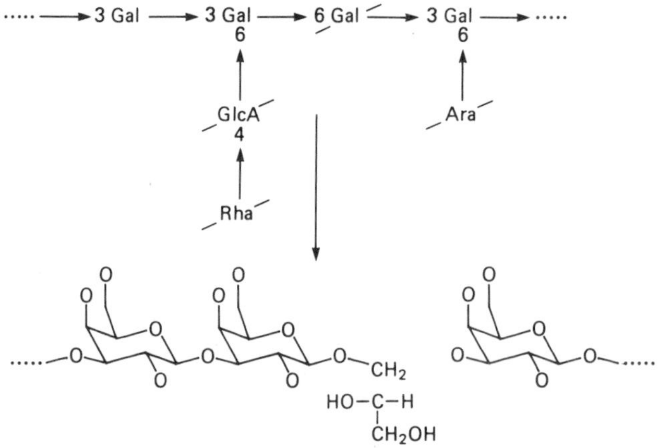

The Smith degradation procedure, outlined in Scheme 14.1, causes cleavages as shown for a fragment of a hypothetical, substituted arabinogalactan of type II. The consequence of there being uronic acids in the main chain, even though these may be oxidised and reduced, is illustrated in Fig. 14.3. Hydrolysis of the glycolacetal is retarded, possibly on account of lactone formation. Erythronic acid, if terminal, may bind glycolaldehyde in the form of a six-membered cyclic acetal (C_1—O—C_3 bond, glucuronic acid numbering); this may persist in the hydrolysis step.

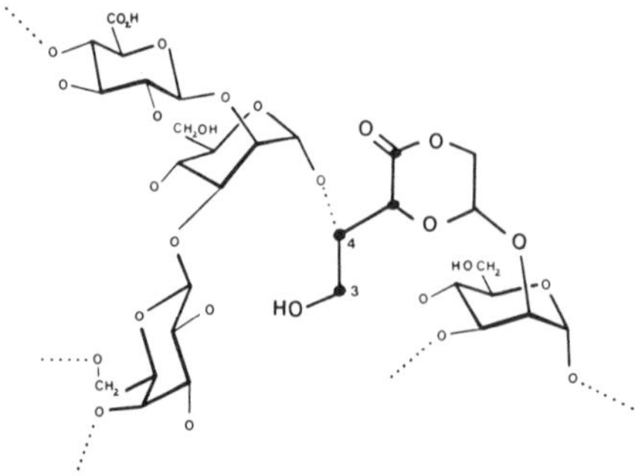

FIG 14.3. Sequence of sugar residues, unchanged and modified, in the Smith degradation product (first-limit) of autohydrolysed *Grevillea robusta* gum (reprinted, by permission, from *S. Afr. J. Chem.*, 1987, **40**, 88–99).

F. Specific Degradation Procedures

1. Base-catalysed elimination reactions

When the carboxyl function of a uronic acid unit is esterified, as in all permethylated gums and in pectins where methyl ester groups are included in the galacturonan chains, application of a strong base results in β-elimination, consequent upon abstraction of H-5, of the glycosyl substituent at O-4 (Lindberg et al., 1973, 1975; Lindberg and Lönngren, 1976; Aspinall, 1976b, 1977, 1982; Aspinall and Rosell, 1977). In the field of exudate chemistry, the reaction is typically carried out under strictly anhydrous conditions in DMSO solution, at room temperature, with dimsyl ion (generated from a metal hydride) as base. After a chosen period (from minutes to days) the reaction is terminated by cooling and the addition of CD_3I or EtI. In the example shown (1), the

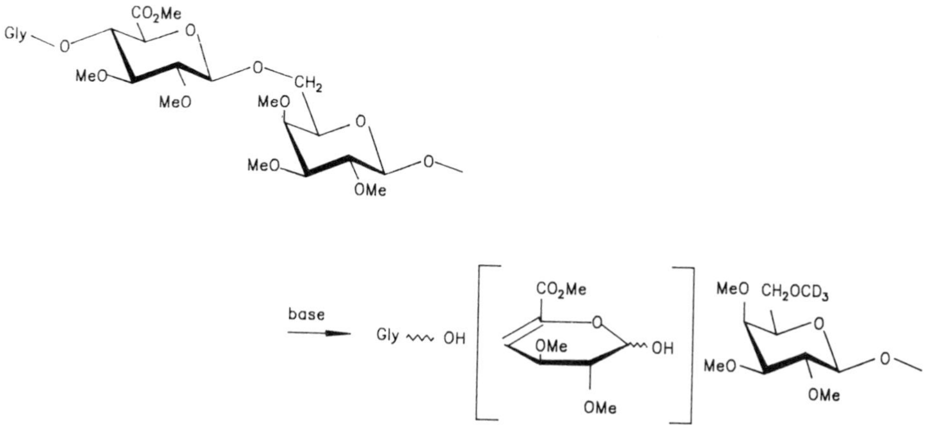

1

point of attachment and, by implication, also the mode of linkage of the interior sugar unit are determined by hydrolysis and analysis by GLC as described above, the sugar to which the uronic acid was attached being a 6-linked D-Gal chain unit. The uronic acid is no longer found on methylation analysis, being converted to the unsaturated compound and further breakdown products, and decomposition of the exterior unit (Gly) occurs. This is illustrated by structure 2, in which Gly is represented by $\rightarrow$3)-D-Gal, H-3 and MeO-2 being eliminated to give an unsaturated sugar derivative.

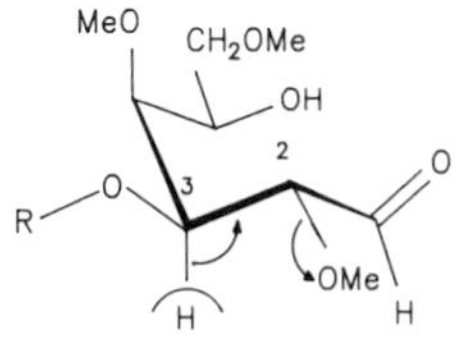

2

An improvement in the base degradation procedure was introduced by G. O. Aspinall and co-workers (Aspinall and Chaudhari, 1975; Aspinall et al., 1975), the exterior sugar residue being protected by acetylation upon its release from the methylated uronic ester.

The chosen base was DBU (1,5-diazabicyclo[5.4.0]undec-5-ene), or an equivalent non-nucleophilic base (Hünig's, for example) may be used, and Ac_2O was utilised *in situ* to protect the potential aldehydic function. Subsequently mild saponification and reduction with $NaBH_4$ ($NaBD_4$) converted the exterior sugar unit to a substituted galactitol (**3**) with hydroxyls free at *C*-1 and *C*-5 which were then deuteriomethylated. Clearly, if borohydride is used to form the alditols after hydrolysis of the degraded polysaccharide and prior to acetylation, it is preferable to use $NaBD_4$ in obtaining **3**. GLC-MS then characterises the exterior sugar unit, as well as the interior.

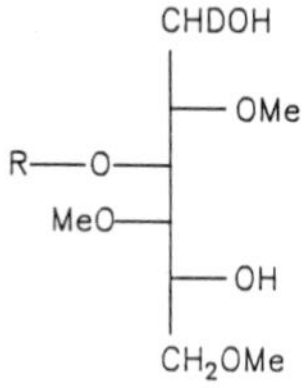

CHDOH
— OMe
R—O—
MeO—
— OH
CH₂OMe

3

Base-catalysed β-elimination has been used to demonstrate the attachment of Rha to GlcA at *O*-4 in *Acacia* gums (Aspinall and Rosell, 1977; Churms *et al.*, 1980), and the DBU method has identified the linkage of Man to GlcA in other cases, together with detail concerning the modes of bonding of the exterior and interior sugars (Aspinall and Chaudhari, 1975; Stephen and Churms, 1986; Churms and Stephen, 1987).

2. Other specific degradation methods

A method that has proved successful in degradation of GlcA-containing gums is that based on a procedure for $Pb(OAc)_4$ decarboxylation–acetoxylation (Kitagawa *et al.*, 1978), which is applied to the free carboxylic acid form of an otherwise fully methylated gum (Aspinall, 1982, 1987). Glycosyl substituents exterior to the degraded acid unit remain unchanged and attached to the xylitol stub produced on reductive work-up; as with base-catalysed degradation, the interior unit can be specifically labelled and so identified. From leiocarpan A (from the gum of *Anogeissus leiocarpus*), one structure produced (Aspinall and Puvanesarajah, 1983) is:

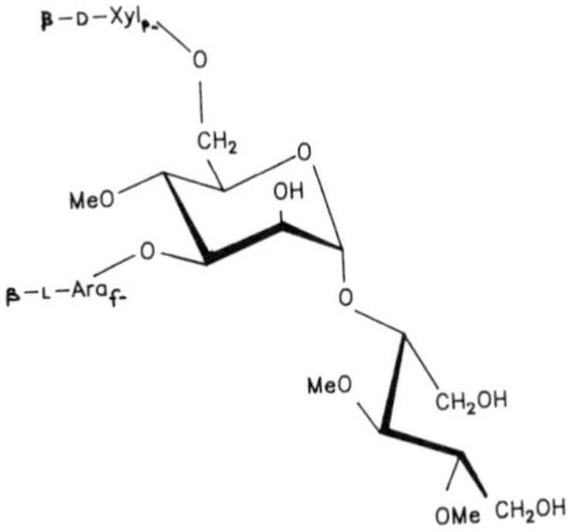

β—D—Xyl*p*
O
CH₂
O
MeO
OH
β—L—Ara*f*
O
O
MeO
CH₂OH
OMe CH₂OH

4

β-D-Xyl*p*

1

↓

6

obtained from → →2)-α-D-Man-(→4)-D-GlcA-

3

↑

1

β-L-Ara*f*

There are other examples in the structural analysis of gums that support the validity of this approach, which has afforded useful information on gum arabic and *Sterculia urens* gum (Aspinall *et al.*, 1981), as well as the mannoglucuronoglycan component of mucin from *Drosera* spp. (Aspinall *et al.*, 1984b).

A procedure applicable to pectic (type B) gums enabled the sequences of neutral sugars in side chains attached to GalA to be determined in the case of tragacanthic acid. In the hex-5-enose degradation (Aspinall *et al.*, 1984a), the fully methylated gum is carboxyl-reduced with LiAlH$_4$, triflated and converted to the corresponding CH$_2$I derivative, e.g. with Bu$_4$NI in benzene. Solubility problems may be anticipated. Zn dust in PrOH effects the elimination of *H*-5 and I, and NaBH$_4$ treatment liberates the interior sugar residue. Thus tragacanthic acid yields **5** and the corresponding fucose-free derivative after acetylation (Aspinall and Puvanesarajah, 1984; Aspinall, 1987). This procedure has also been applied to gums from *Sterculia* spp. (Aspinall, 1987; Aspinall *et al.*, 1987) and from *Khaya ivorensis* (Aspinall *et al.*, 1988).

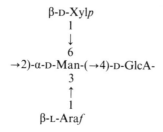

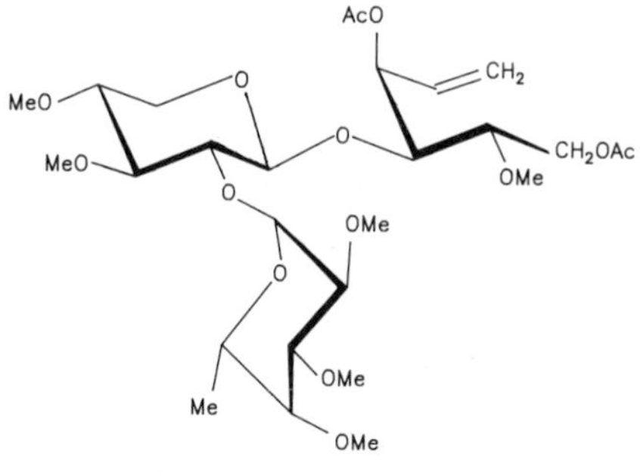

5

There are other specific degradations which might afford new structural information about the environment of the uronic acid residues in gums, such as application of the Hofmann–Weerman degradation (Aspinall and Rosell, 1978; Kochetkov *et al.*, 1980) or Curtius rearrangement (Aspinall and Puvanesarajah, 1983; Aspinall *et al.*, 1983; Aspinall and Knebl, 1986b) to the glycosiduronamide derived from the fully methylated acidic polysaccharide. Without doubt there is scope for the novel approach of G. R. Gray and his co-workers (Rolf and Gray, 1982; D'Ambra *et al.*, 1988; Lee and Gray, 1988; Vodonik and Gray, 1988); the procedure is to degrade the methylated polysaccharide by reductive cleavage using Et$_3$SiH with an appropriate catalyst (BF$_3$.Et$_2$O or TMSOTf). The process is selective, following more or less the ease of acid hydrolysis of the various glycosidic bonds in complex polysaccharides. The necessary data base for

GLC and MS of the derived anhydroalditols (1,5 from the pyranosides, 1,4 from the furanosides) is being established (Van Langenhove and Reinhold, 1985), and it remains to test the applicability of the method to structural analysis of gums. Sequencing is possible if HPLC is combined with spectroscopic measurements.

3.   Pyrolysis—GLC—mass spectrometry

Increasing use is being made of pyrolysis and GLC in the identification of gums, mucilages and other heteroglycans. Further characterisation of the products of decomposition, which are indicative of the types of sugar unit present in the polysaccharide, is achieved by mass spectrometry (Schulten *et al.*, 1982; Budgell *et al.*, 1987; Helleur *et al.*, 1987; Sugiyama *et al.*, 1987).

4.   Sequencing by mass spectrometry

There are established techniques for sequencing methylated oligosaccharides, their alditols, and methylated polysaccharides of mass as high as 4000, using electron-impact, chemical-ionisation, and fast atom bombardment mass spectrometry (Kochetkov and Chizhov, 1966, 1972; Lönngren and Svensson, 1974; Chizhov *et al.*, 1976; McNeil *et al.*, 1982; Dell, 1987). Applied to hydrolysates of Kiwi fruit stem mucilage, the products of reverse phase HPLC separation of derived oligosaccharide alditols were identified by a combination of mass spectrometric methods (Redgwell *et al.*, 1986b).

V. GUM STRUCTURES

The evidence accumulated from using the techniques outlined above enables formulae to be constructed which give a visual impression of possible modes of inter-sugar linkages, and of those limited regions in which sequences have been established. Part-formulae presented in Sections V.A, B and C indicate some of the structural elements defined by the processes named. The proportions of sugar residues, determined by analysis of hydrolysates of the polysaccharides, and their positions of linkage which follow from methylation analysis, have been determined for all three gums; what are shown are approaches to the sequencing of component monosaccharide units.

A. Gum Arabic

Structural features of *Acacia senegal* gum (gum arabic):

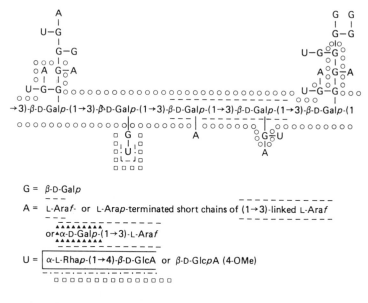

G = β-D-Gal*p*

A = L-Ara*f*- or L-Ara*p*-terminated short chains of (1→3)-linked L-Ara*f*

or α-D-Gal*p*-(1→3)-L-Ara*f*

U = α-L-Rha*p*-(1→4)-β-D-GlcA or β-D-Glc*p*A (4-OMe)

———————— Acetolysis of acetylated, reduced polysaccharide.
— — — Partial, acid hydrolysis.
○ ○ ○ ○ Smith degradation (two stages).
— · — · Decarboxylation–acetoxylation, and by Hofmann–Curtius rearrangement.
▲ ▲ ▲ ▲ α-D-Galactosidase.
□ □ □ □ Base degradation, Hofmann degradation.

B. Gum Karaya

Structural features of *Sterculia urens* gum (karaya):

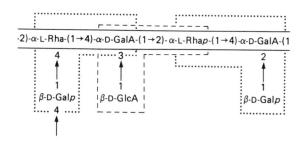

———————— Acetolysis of acetylated, reduced polysaccharide or Smith-degraded polysaccharide.
— — — Partial, acid hydrolysis.
········· Hex-5-enose degradation.

C. Gum Ghatti

Structural features of *Anogeissus* gums (*A. latifolius* ≡ ghatti):

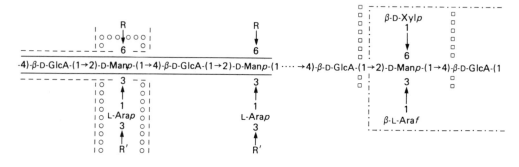

R = L-Araf-(1 or L-Araf-(1→2,3 or 5)-L-Araf-(1

R' = β-D-GlcpA- or β-D-Galp- [(1→6)-D-Galp 3]ₙ -(1
R

———— Acetolysis of acetylated, reduced polysaccharide.
— — — Partial, acid hydrolysis.
○ ○ ○ ○ Smith degradation.
□ □ □ □ Base degradation (*A. leiocarpus*).
— · — · Decarboxylation–acetoxylation, and by Hofmann–Curtius rearrangement (*A. leiocarpus*).

ACKNOWLEDGEMENTS

The authors are grateful for the financial support of the Foundation for Research Development of the Council for Scientific and Industrial Research, and of the Council of the University of Cape Town. We thank Professor G. O. Phillips and Dr P. A. Williams (N.E. Wales Institute of Higher Education) for helpful discussions and Mrs P. Alexander for typing the manuscript.

REFERENCES

Adam, J. W. H., Churms, S. C., Stephen, A. M., Streefkerk, D. G. and Williams, E. H. (1977). *Carbohydr. Res.* **54**, 304–307.
Adams, G. A. (1965). *Methods Carbohydr. Chem.* **5**, 269–276.
Albersheim, P., Nevins, D. J., English, P. D. and Karr, A. (1967). *Carbohydr. Res.* **5**, 340–345.
Al-Hazmi, M. I. and Stauffer, K. R. (1986). *J. Food Sci.* **51**, 1091–1092, 1097.
Åman, P., Franzén, L.-E., Darvill, J. E., McNeil, M., Darvill, A. G. and Albersheim, P. (1982). *Carbohydr. Res.* **103**, 77–100.
Anderson, D. M. W. (1978). *Kew Bull.* **32**, 529–536.
Anderson, D. M. W. (1986a). *Food Additives and Contaminants* **3**, 231–234.

Anderson, D. M. W. (1986b). *In* "Gums and Stabilisers for the Food Industry" (G. O. Phillips, D. J. Wedlock and P. A. Williams, eds), Vol. 3, pp. 79–86. Elsevier Applied Science, Barking, UK.

Anderson, D. M. W. and Cree, G. M. (1966). *Carbohydr. Res.* **2**, 162–166.

Anderson, D. M. W. and Cree, G. M. (1968). *Carbohydr. Res.* **6**, 385–403.

Anderson, D. M. W. and Dea, I. C. M. (1969a). *Carbohydr. Res.* **10**, 161–164.

Anderson, D. M. W. and Dea, I. C. M. (1969b). *Phytochemistry* **8**, 167–176.

Anderson, D. M. W. and Dea, I. C. M. (1971). *J. Soc. Cosmet. Chem.* **22**, 61–76.

Anderson, D. M. W. and Douglas, D. M. B. (1988). *Food Hydrocolloids* **2**, 247–253.

Anderson, D. M. W. and Garbutt, S. (1963). *Anal. Chim. Acta* **29**, 31–38.

Anderson, D. M. W. and McDougall, F. J. (1987). *Food Additives and Contaminants* **4**, 125–132.

Anderson, D. M. W. and Munro, A. C. (1969). *Carbohydr. Res.* **11**, 43–51.

Anderson, D. M. W. and Rahman, S. (1967). *Carbohydr. Res.* **4**, 298–304.

Anderson, D. M. W. and Smith, R. N. (1967). *Carbohydr. Res.* **4**, 55–62.

Anderson, D. M. W. and Stefani, A. (1979). *Anal. Chim. Acta* **105**, 147–152.

Anderson, D. M. W. and Stoddart, J. F. (1966). *Carbohydr. Res.* **2**, 104–114.

Anderson, D. M. W., Garbutt, S. and Zaidi, S. S. H. (1963). *Anal. Chim. Acta* **29**, 39–45.

Anderson, D. M. W., Cree, G. M., Herbich, M. A., Karamalla, K. A. and Stoddart, J. F. (1964). *Talanta* **11**, 1559–1560.

Anderson, D. M. W., Cree, G. M., Marshall, J. J. and Rahman, S. (1965a). *Carbohydr. Res* **1**, 320–323.

Anderson, D. M. W., Dea, I. C. M., Rahman, S. and Stoddart, J. F. (1965b). *Chem. Commun.*, p. 145.

Anderson, D. M. W., Hirst, Sir Edmund and Stoddart, J. F. (1966). *J. Chem. Soc. C*, pp. 1959–1966.

Anderson, D. M. W., Hirst, Sir Edmund, Rahman, S. and Stainsby, G. (1967). *Carbohydr. Res.* **3**, 308–317.

Anderson, D. M. W., Dea, I. C. M. and Munro, A. C. (1969). *Carbohydr. Res.* **9**, 363–365.

Anderson, D. M. W., Bell, P. C. and McNab, C. G. A. (1971). *Carbohydr. Res.* **20**, 269–274.

Anderson, D. M. W., Hendrie, A. and Munro, A. C. (1972). *Phytochemistry* **11**, 733–736.

Anderson, D. M. W., Howlett, J. F. and McNab, C. G. A. (1985a). *Food Additives and Contaminants* **2**, 153–157.

Anderson, D. M. W., Howlett, J. F. and McNab, C. G. A. (1985b). *Food Additives and Contaminants* **2**, 159–164.

Anderson, D. M. W., Howlett, J. F. and McNab, C. G. A. (1985c). *Food Additives and Contaminants* **2**, 231–235.

Anderson, D. M. W., Bell, P. C., Gill, M. C. L., McDougall, F. J. and McNab, C. G. A. (1986). *Phytochemistry* **25**, 247–249.

Anderson, M. A. and Stone, B. A. (1985). *Carbohydr. Polymers* **5**, 115–129.

Andersson, T., Carlsson, M., Hagel, L., Pernemalm, P.-Å. and Janson, J.-C. (1985). *J. Chromatogr.* **326**, 33–44.

Aslam, M., Pass, G. and Phillips, G. O. (1978). *J. Sci. Food Agric.* **29**, 563–568.

Aspinall, G. O. (1963). *J. Chem. Soc.*, pp. 1676–1680.

Aspinall, G. O. (1967). *Pure Appl. Chem.* **14**, 43–55.

Aspinall, G. O. (1969). *Adv. Carbohydr. Chem. Biochem.* **24**, 333–379.

Aspinall, G. O. (1970). *In* "The Carbohydrates" (W. Pigman and D. Horton, eds), Vol. IIB, pp. 515–536. Academic Press, New York.

Aspinall, G. O. (1973). *In* "Biogenesis of Plant Cell-Wall Polysaccharides" (F. Loewus, ed.), pp. 95–115. Academic Press, New York.

Aspinall, G. O. (1976a). *In* "MTP International Review of Science" (G. O. Aspinall, ed.), Org. Chem. Ser. 2, Vol. 7, pp. 205–206. Butterworths, London.

Aspinall, G. O. (1976b). *In* "MTP International Review of Science" (G. O. Aspinall, ed.), Org. Chem. Ser. 2, Vol. 7, pp. 208–209. Butterworths, London.

Aspinall, G. O. (1977). *Pure Appl. Chem.* **49**, 1105–1134.

Aspinall, G. O. (1982). *In* " The Polysaccharides" (G. O. Aspinall, ed.), Vol. 1, pp. 35–131. Academic Press, New York.

Aspinall, G. O. (1987). *Acc. Chem. Res.* **20**, 114–120.
Aspinall, G. O. and Baillie, J. (1963). *J. Chem. Soc.*, pp. 1702–1714.
Aspinall, G. O. and Bhattacharjee, A. K. (1970). *J. Chem. Soc. C*, pp. 365–369.
Aspinall, G. O. and Chaudhari, A. S. (1975). *Can. J. Chem.* **53**, 2189–2193.
Aspinall, G. O. and Christensen, T. B. (1961). *J. Chem. Soc.*, pp. 3461–3467.
Aspinall, G. O. and Fairweather, R. M. (1965). *Carbohydr. Res.* **1**, 83–92.
Aspinall, G. O. and Fanshawe, R. S. (1961). *J. Chem. Soc.*, pp. 4215–4225.
Aspinall, G. O. and Ferrier, R. J. (1957). *Chem. and Ind. (London)*, p. 1216.
Aspinall, G. O. and Fraser, R. N. (1965). *J. Chem. Soc.*, pp. 4318–4325.
Aspinall, G. O. and Knebl, M. C. (1986a). *Carbohydr. Res.* **157**, 257–260.
Aspinall, G. O. and Knebl, M. C. (1986b). *Carbohydr. Res.* **157**, 261–268.
Aspinall, G. O. and McKenna, J. P. (1968). *Carbohydr. Res.* **7**, 244–254.
Aspinall, G. O. and McNab, J. M. (1965). *Chem. Commun.* (No. 22), 565–566.
Aspinall, G. O. and McNab, J. M. (1969). *J. Chem. Soc. C*, pp. 845–851.
Aspinall, G. O. and Nasir-ud-din, (1965). *J. Chem. Soc.*, pp. 2710–2720.
Aspinall, G. O. and Puvanesarajah, V. (1983). *Can. J. Chem.* **61**, 1864–1868.
Aspinall, G. O. and Puvanesarajah, V. (1984). *Can. J. Chem.* **62**, 2736–2739.
Aspinall, G. O. and Rosell, K.-G. (1977). *Carbohydr. Res.* **57**, C23–C26.
Aspinall, G. O. and Rosell, K.-G. (1978). *Can. J. Chem.* **56**, 685–690.
Aspinall, G. O. and Sanderson, G. R. (1970). *J. Chem. Soc. C*, pp. 2259–2264.
Aspinall, G. O. and Whitehead, C. C. (1970). *Can. J. Chem.* **48**, 3850–3855.
Aspinall, G. O., Auret, B. J. and Hirst, E. L. (1958a). *J. Chem. Soc.*, pp. 221–230.
Aspinall, G. O., Auret, B. J. and Hirst, E. L. (1958b). *J. Chem. Soc.*, pp. 4408–4414.
Aspinall, G. O., Johnston, M. J. and Stephen, A. M. (1960). *J. Chem. Soc.*, pp. 4918–4927.
Aspinall, G. O., Charlson, A. J., Hirst, E. L. and Young, R. (1963). *J. Chem. Soc.*, pp. 1696–1702.
Aspinall, G. O., Bhavanandan, V. P. and Christensen, T. B. (1965a). *J. Chem. Soc.*, pp. 2677–2684.
Aspinall, G. O., Fraser, R. N. and Sanderson, G. R. (1965b). *J. Chem. Soc.*, pp. 4325–4329.
Aspinall, G. O., Carlyle, J. J., McNab, J. M. and Rudowski, A. (1969). *J. Chem. Soc. C*, pp. 840–845.
Aspinall, G. O., Krishnamurthy, T. N., Mitura, W. and Funabashi, M. (1975). *Can J. Chem.* **53**, 2182–2188.
Aspinall, G. O., Chaudhari, A. S. and Whitehead, C. C. (1976). *Carbohydr. Res.* **47**, 119–127.
Aspinall, G. O., Fanous, H. K., Kumar, N. S. and Puvanesarajah, V. (1981). *Can. J. Chem.* **59**, 935–940.
Aspinall, G. O., Fanous, H. K., Kumar, N. S. and Puvanesarajah, V. (1983). *Can. J. Chem.* **61**, 1858–1863.
Aspinall, G. O., Chatterjee, D. and Khondo, L. (1984a). *Can. J. Chem.* **62**, 2728–2735.
Aspinall, G. O., Puvanesarajah, V., Reuter, G. and Schauer, R. (1984b). *Carbohydr. Res.* **131**, 53–60.
Aspinall, G. O., Khondo, L. and Williams, B. A. (1987). *Can. J. Chem.* **65**, 2069–2076.
Aspinall, G. O., Khondo, L. and Kinnear, J. A. (1988). *Carbohydr. Res.* **179**, 211–221.
Bacic, A., Harris, P. J., Hak, E. W. and Clarke, A. E. (1984). *J. Chromatogr.* **315**, 373–377.
Bacic, A., Churms, S. C., Stephen, A. M., Cohen, P. B. and Fincher, G. B. (1987). *Carbohydr. Res.* **162**, 85–93.
Baker, J. L. and Hulton, H. F. E. (1920). *Biochem. J.* **14**, 754–756.
Baldo, B. A., Sawyer, W. H., Stick, R. V. and Uhlenbruck, G. (1978). *Biochem. J.* **175**, 467–477.
Banks, W. and Greenwood, C. T. (1963). *Adv. Carbohydr. Chem.* **18**, 357–398.
Barker, R., Nunes, H. A., Rosevear, P. and Serianni, A. S. (1982). *Methods Enzymol.* **83**, 58–69.
Barsett, H. and Paulsen, B. S. (1985). *J. Chromatogr.* **329**, 315–320.
Bebault, G. M., Berry, J. M., Choy, Y. M., Dutton, G. G. S., Funnell, N., Hayward, L. D. and Stephen, A. M. (1973). *Can. J. Chem.* **51**, 324–326.
Bekker, P. I., Churms, S. C., Stephen, A. M. and Woolard, G. R. (1972). *J. S. Afr. Chem. Inst.* **25**, 115–130.
BeMiller, J. N. (1967). *Adv. Carbohydr. Chem.* **22**, 25–108.
Biermann, C. J. (1988). *Adv. Carbohydr. Chem. Biochem.* **46**, 251–271.

Bishop, C. T. (1964). *Adv. Carbohydr. Chem.* **19**, 95–147.
Bishop, C. T. and Jennings, H. J. (1982). *In* "The Polysaccharides" (G. O. Aspinall, ed.), Vol. 1, pp. 291–330. Academic Press, New York.
Björndal, H., Hellerqvist, C. G., Lindberg, B. and Svensson, S. (1970). *Angew. Chem. Int. Ed. Engl.* **9**, 610–619.
Blake, D. A. and Goldstein, I. J. (1980). *Anal. Biochem.* **102**, 103–109.
Blakeney, A. B. and Stone, B. A. (1985). *Carbohydr. Res.* **140**, 319–324.
Blakeney, A. B., Harris, P. J., Henry, R. J. and Stone, B. A. (1983). *Carbohydr. Res.* **113**, 291–299.
Blaschek, W. (1983). *J. Chromatogr.* **256**, 157–163.
Blumenkrantz, N. and Asboe-Hansen, G. (1973). *Anal. Biochem.* **54**, 484–489.
Bock, K., Pedersen, C. and Pedersen, H. (1984). *Adv. Carbohydr. Chem. Biochem.* **42**, 193–225.
Boothby, D. (1983). *J. Sci. Food Agric.* **34**, 1–7.
Bradbury, A. G. W., Halliday, D. J. and Medcalf, D. G. (1981). *J. Chromatogr.* **213**, 146–150.
Bradbury, J. H. and Jenkins, G. A. (1984). *Carbohydr. Res.* **126**, 125–156.
British Pharmacopoeia (1980). Vol. 1, p. 13.
Budgell, D. R., Hayes, E. R. and Helleur, R. J. (1987). *Anal. Chim. Acta* **192**, 243–253.
Carson, J. F. and Maclay, W. D. (1946). *J. Am. Chem. Soc.* **68**, 1015–1017.
Chambers, R. E. and Clamp, J. R. (1971). *Biochem. J.* **125**, 1009–1018.
Chanzy, H., Booy, F. P. and Atkins, E. D. T. (1978) *Polymer* **19**, 368–369.
Cheetham, N. W. H. and Sirimanne, P. (1983). *Carbohydr. Res.* **112**, 1–10.
Chizhov, O. S., Kadentsev, V. I., Solovýov, A. A., Levonowich, P. F. and Dougherty, R. C. (1976). *J. Org. Chem.* **41**, 3425–3428.
Churms, S. C. (1970). *Adv. Carbohydr. Chem. Biochem.* **25**, 13–51.
Churms, S. C. (1982a). "CRC Handbook of Chromatography Series: Carbohydrates", Vol. I, pp. 187–207. CRC Press, Boca Raton, FL.
Churms, S. C. (1982b). "CRC Handbook of Chromatography Series: Carbohydrates", Vol. I, p. 215. CRC Press, Boca Raton, FL.
Churms, S. C. (1982c). "CRC Handbook of Chromatography Series: Carbohydrates", Vol. I, pp. 219–230. CRC Press, Boca Raton, FL.
Churms, S. C. and Stephen, A. M. (1971). *Carbohydr. Res.* **19**, 211–221.
Churms, S. C. and Stephen, A. M. (1972). *Carbohydr. Res.* **21**, 91–98.
Churms, S. C. and Stephen, A. M. (1984). *Carbohydr. Res.* **133**, 105–123.
Churms, S. C. and Stephen, A. M. (1987). *Carbohydr. Res.* **167**, 239–255.
Churms, S. C. and Stephen, A. M. (1988). *S. Afr. J. Sci.* **84**, 855.
Churms, S. C., Merrifield, E. H. and Stephen, A. M. (1977). *Carbohydr. Res.* **55**, 3–10.
Churms, S. C., Merrifield, E. H. and Stephen, A. M. (1978a). *Carbohydr. Res.* **63**, 337–341.
Churms, S. C., Merrifield, E. H. and Stephen, A. M. (1978b). *Carbohydr. Res.* **64**, C1–C2.
Churms, S. C., Merrifield, E. H. and Stephen, A. M. (1980). *S. Afr. J. Chem.* **33**, 39–40.
Churms, S. C., Merrifield, E. H. and Stephen, A. M. (1981a). *Carbohydr. Res.* **90**, 261–267.
Churms, S. C., Stephen, A. M. and Siddiqui, I. R. (1981b). *Carbohydr. Res.* **94**, 119–122.
Churms, S. C., Merrifield, E. H. and Stephen, A. M. (1983). *Carbohydr. Res.* **123**, 267–279.
Ciucanu, I. and Kerek, F. (1984). *Carbohydr. Res.* **131**, 209–217.
Clarke, A. E., Anderson, R. L. and Stone, B. A. (1979). *Phytochemistry* **18**, 521–540.
Connolly, S., Fenyo, J.-C. and Vandevelde, M. C. (1988). *Carbohydr. Polymers* **8**, 23–32.
Cottrell, I. W. and Baird, J. K. (1980). *In* "Kirk-Othmer Encyclopedia of Chemical Technology". 3rd edn, Vol. 12, pp. 45–66. John Wiley, New York.
D'Ambra, A. J., Rice, M. J., Zeller, S. G., Gruber, P. R. and Gray, G. R. (1988). *Carbohydr. Res.* **177**, 111–116.
Davidson, R. L. (ed.) (1980). "Handbook of Water-Soluble Gums and Resins". McGraw-Hill, New York.
Defaye, J. and Wong, E. (1986). *Carbohydr. Res.* **150**, 221–231.
Dell, A. (1987). *Adv. Carbohydr. Chem. Biochem.* **45**, 19–72.
Dey, P. M. (1980). *Adv. Carbohydr. Chem. Biochem.* **37**, 283–372.
Di Fabio, J. L., Dutton, G. G. S. and Moyna, P. (1982). *Carbohydr. Res.* **99**, 41–50.
Dische, Z. (1962a). *Methods Carbohydr. Chem.* **1**, 478–481, 484–494, 501–503, 512–514.
Dische, Z. (1962b). *Methods Carbohydr. Chem.* **1**, 497–501.

Dische, Z. (1962c). *Methods Carbohydr. Chem.* **1**, 512–514.
Dische, Z. and Borenfreund, E. (1949). *J. Biol. Chem.* **180**, 1297–1300.
Dmitriev, B. A., Backinowsky, L. V., Chizhov, O. S., Zolotarev, B. M. and Kochetkov, N. K. (1971). *Carbohydr. Res.* **19**, 432–435.
Dubois, M., Gilles, K. A., Hamilton, J. K., Rebers, P. A. and Smith, F. (1956). *Anal. Chem.* **28**, 350–356.
Dudman, W. F. and Bishop, C. T. (1968). *Can. J. Chem.* **46**, 3079–3084.
Dutton, G. G. S. (1973). *Adv. Carbohydr. Chem. Biochem.* **28**, 11–160.
Dutton, G. G. S. (1974). *Adv. Carbohydr. Chem. Biochem.* **30**, 9–110.
Dutton, G. G. S. and Gibney, K. B. (1972). *Carbohydr. Res.* **25**, 99–105.
Dutton, G. G. S. and Unrau, A. M. (1962). *Can. J. Chem.* **40**, 348–352.
Dutton, G. G. S. and Unrau, A. M. (1963). *Can. J. Chem.* **41**, 1417–1423.
Dutton, G. G. S., Gibney, K. B., Jensen, G. D. and Reid, P. E. (1968). *J. Chromatogr.* **36**, 152–162.
Edge, A. S. B., Faltynek, C. R., Hof, L., Reichert, L. E., Jr and Weber, P. (1981). *Anal. Biochem.* **118**, 131–137.
Edwards, W. T., Pohl, C. A. and Rubin, R. (1987). *Tappi* **70**, 138–140.
Erbing, B., Larm, O., Lindberg, B. and Svensson, S. (1973). *Acta Chem. Scand.* **27**, 1094–1096.
Fincher, G. B., Stone, B. A. and Clarke, A. E. (1983). *Ann. Rev. Plant Physiol.* **34**, 47–70.
Fischer, E., Schlingmann, M., Duersch, W. and von Halasz, S. P. (1986). *Chem. Abstr.* **104**, 90796g.
Fleury, P. F. and Lange, J. (1933). *J. Pharm. Chim.* **17**, 107–113.
Fowle, L. G., Blair, J. McD. and Stephen, A. M. (1964). *S. Afr. Med. J.* **88**, 155.
Fransson, L.-Å. (1985). *In* "The Polysaccharides" (G. O. Aspinall, ed.), Vol. 3, pp. 337–415. Academic Press, New York.
Freeman, B. H., Stephen, A. M. and van der Bijl, P. (1972). *J. Chromatogr.* **73**, 29–33.
Furneaux, R. H. (1983). *Carbohydr. Res.* **113**, 241–255.
Gammon, D. W. and Stephen, A. M. (1986). *Carbohydr. Res.* **154**, 289–295.
Gammon, D. W., Churms, S. C. and Stephen, A. M. (1986a). *Carbohydr. Res.* **151**, 135–146.
Gammon, D. W., Stephen, A. M. and Churms, S. C. (1986b). *Carbohydr. Res.* **158**, 157–171.
Gerwig, G. J., Kamerling, J. P. and Vliegenthart, J. F. G. (1978). *Carbohydr. Res.* **62**, 349–357.
Gerwig, G. J., Kamerling, J. P. and Vliegenthart, J. F. G. (1979). *Carbohydr. Res.* **77**, 1–7.
Ghebregzabher, M., Rufini, S., Monaldi, B. and Lato, M. (1976). *J. Chromatogr.* **127**, 133–162.
Gleeson, P. A., Jermyn, M. A. and Clarke, A. E. (1979). *Anal. Biochem.* **92**, 41–45.
Glicksman, M. (1969a). "Gum Technology for the Food Industry", pp. 94–129. Academic Press, New York.
Glicksman, M. (1969b). "Gum Technology for the Food Industry", pp. 509–554. Academic Press, New York.
Glicksman, M. (1982). *In* "Food Carbohydrates" (D. R. Lineback and G. E. Inglett, eds), pp. 270–295. AVI Publishing Company, Westport, CT.
Glicksman, M. (ed.) (1983). *In* "Food Hydrocolloids" Vol. II, pp. 7–60. CRC Press, Boca Raton, FL.
Glicksman, M. and Sand, R. E. (1973). *In* "Industrial Gums" (R. L. Whistler and J. N. BeMiller, eds), 2nd edn, p. 227. Academic Press, New York.
Goldstein, I. J. and Hayes, C. E. (1978). *Adv. Carbohydr. Chem. Biochem.* **35**, 127–340.
Goldstein, I. J., Hay, G. W., Lewis, B. A. and Smith, F. (1965). *Methods Carbohydr. Chem.* **5**, 361–370.
Gorin, P. A. J. (1981). *Adv. Carbohydr. Chem. Biochem.* **38**, 13–104.
Gorin, P. A. J. and Barreto-Bergter, E. (1983). *In* "The Polysaccharides" (G. O. Aspinall, ed.), Vol. 2, pp. 365–409. Academic Press, New York.
Gorin, P. A. J., Giblin, E. M., Slater, G. P. and Hogge, L. (1982). *Carbohydr. Res.* **106**, 235–238.
Gralén, N. and Karrholm, M. (1950). *J. Colloid Sci.* **5**, 21–36.
Granath, K. A. (1965). *Methods Carbohydr. Chem.* **5**, 20–28.
Granath, K. A. and Kvist, B. E. (1967). *J. Chromatogr.* **28**, 69–81.
Grubisic, Z., Rempp, P. and Benoit, H. (1967). *J. Polymer. Sci. B.* **5**, 753–759.
Guthrie, R. D. and McCarthy, J. F. (1967). *Adv. Carbohydr. Chem.* **22**, 11–23.

Ha, Y. W. and Thomas, R. L. (1988). *J. Food Sci.* **53**, 574–577.

Haga, H. and Nakajima, T. (1988). *Chem. Pharm. Bull.* **36**, 1562–1564.

Hakomori, S. (1964). *J. Biochem. (Tokyo)* **55**, 205–208.

Hall, L. D. (1964). *Adv. Carbohydr. Chem.* **19**, 51–93.

Hara, S., Ikegami, H., Shono, A., Mega, T., Ikenaka, T. and Matsushima, Y. (1979). *Anal. Biochem.* **97**, 166–172.

Harding, S. E. (1988). In "Gums and Stabilisers for the Food Industry" (G. O. Phillips, P. A. Williams and D. J. Wedlock, eds), Vol. 4, pp. 15–23. IRL Press, Oxford.

Harris, P. J., Henry, R. J., Blakeney, A. B. and Stone, B. A. (1984). *Carbohydr. Res.* **127**, 59–73.

Haug, A. and Larsen, B. (1961). *Acta Chem. Scand.* **15**, 1395–1396.

Haworth, W. N. (1915). *J. Chem. Soc.* **CVII**, pp. 8–16.

Hay, G. W., Lewis, B. A. and Smith, F. (1965a). *Methods Carbohydr. Chem.* **5**, 357–361.

Hay, G. W., Lewis, B. A. and Smith, F. (1965b). *Methods Carbohydr. Chem.* **5**, 377–380.

Helleur, R. J., Budgell, D. R. and Hayes, E. R. (1987). *Anal. Chim. Acta* **192**, 367–372.

Hicks, K. B. (1988). *Adv. Carbohydr. Chem. Biochem.* **46**, 17–72.

Hicks, K. B., Lim, P. C. and Haas, M. J. (1985). *J. Chromatogr.* **319**, 159–171.

Honda, S. (1984). *Anal. Biochem.* **140**, 1–47.

Honda, S., Takahashi, M., Nishimura, Y., Kakehi, K. and Ganno, S. (1981). *Anal. Biochem.* **118**, 162–167.

Honda, S., Suzuki, S., Takahashi, M., Kakehi, K. and Ganno, S. (1983). *Anal. Biochem.* **134**, 34–39.

Horwitz, W. (1980). In "Official Methods of Analysis of the A.O.A.C." (W. Horwitz, ed.), 13th edn, pp. 862–863. A.O.A.C., Washington D.C.

Hough, L. and Jones, J. K. N. (1962). *Methods Carbohydr. Chem.* **1**, 21–31.

Hough, L., Jones, J. K. N. and Wadman, W. H. (1950). *J. Chem. Soc.*, pp. 1702–1706.

Hough, L., Jones, J. V. S. and Wusteman, P. (1972). *Carbohydr. Res.* **21**, 9–17.

Hunt, K. and Jones, J. K. N. (1962). *Can. J. Chem.* **40**, 1266–1279.

Jacobs, M. B. (1958). "The Chemical Analysis of Foods and Food Products", 3rd edn, pp. 476–508. Van Nostrand, New York.

James, D. W., Jr, Preiss, J. and Elbein, A. D. (1985). In "The Polysaccharides" (G. O. Aspinall, ed.), Vol. 3, pp. 107–207. Academic Press, New York.

Jansson, P.-E., Kenne, L., Liedgren, H., Lindberg, B. and Lönngren, J. (1976). *Chem. Commun. (Univ. of Stockholm)*, No. 8, 1–75.

Jefferies, M., Pass, G. and Phillips, G. O. (1977). *J. Appl. Chem. Biotechnol.* **27**, 625–630.

Jefferies, M., Pass, G., Phillips, G. O. and Zakaria, M. B. (1978). *J. Sci. Food Agric.* **29**, 193–200.

João, H. I., Jackson, G. E., Ravenscroft, N. and Stephen, A. M. (1988). *Carbohydr. Res.* **176**, 300–305.

Johnson, W. C., Jr (1987). *Adv. Carbohydr. Chem. Biochem.* **45**, 73–124.

Jones, J. K. N. and Nunn, J. R. (1955a). *J. Chem. Soc.*, pp. 3001–3004.

Jones, J. K. N. and Nunn, J. R. (1955b). *J. Am. Chem. Soc.*, **77**, 5745–5746.

Joseleau, J.-P., Chambat, G. and Chumpitazi-Hermoza, B. (1981). *Carbohydr. Res.* **90**, 339–344.

Joubert, F. J. (1954). *J. S. Afr. Chem. Inst.* **7**, 107–113.

Kaji, A. (1984). *Adv. Carbohydr. Chem. Biochem.* **42**, 383–394.

Kaplan, M. and Stephen, A. M. (1967). *Tetrahedron* **23**, 193–198.

Kardošová, A., Rosík, J., Kubala, J. and Kováčik, V. (1979). *Collect. Czech. Chem. Commun.* **44**, 2250–2254.

Kenne, L. and Lindberg, B. (1983). In "The Polysaccharides" (G. O. Aspinall, ed.), Vol. 2, pp. 287–363. Academic Press, New York.

Kennedy, J. F. and Fox, J. E. (1980a). *Methods Carbohydr. Chem.* **8**, 3–12.

Kennedy, J. F. and Fox, J. E. (1980b). *Methods Carbohydr. Chem.* **8**, 13–19.

Kitagawa, I., Yoshikawa, M. and Kadota, A. (1978). *Chem. Pharm. Bull.* **26**, 484–496.

Klok, J., Cox, H. C., de Leeuw, J. W. and Schenck, P. A. (1982). *J. Chromatogr.* **253**, 55–64.

Knutson, C. A. and Jeanes, A. (1968). *Anal. Biochem.* **24**, 470–481.

Kochetkov, N. K. and Chizhov, O. S. (1966). *Adv. Carbohydr. Chem.* **21**, 39–93.

Kochetkov, N. K. and Chizhov, O. S. (1972). *Methods Carbohydr. Chem.* **6**, 540–554.

Kochetkov, N. K., Chizhov, O. S. and Sviridov, A. F. (1980). *Methods Carbohydr. Chem.* **8**, 123–125.

520 A. M. STEPHEN *ET AL.*

König, W. A., Lutz, S., Mischnick-Lübbecke, P., Brassat, B. and Wenz, G. (1988a). *J. Chromatogr.* **447**, 193–197.

König, W. A., Mischnick-Lübbecke, P., Brassat, B., Lutz, S. and Wenz, G. (1988b). *Carbohydr. Res.* **183**, 11–17.

Kosheleva, L. P., Ilchenko, G. Y. and Glebko, L. I. (1976). *Anal. Biochem.* **73**, 115–119.

Kvernheim, A. L. (1987). *Acta Chem. Scand.* **B41**, 150–152.

Kuge, T., Kobayashi, K., Tanahashi, H., Igushi, T. and Kitamura, S. (1984). *Agric. Biol. Chem.* **48**, 2375–2376.

Kuhn, R., Trischmann, H. and Löw, I. (1955). *Angew. Chem.* **67**, 32.

Lamport, D. T. A. and Miller, D. H. (1971). *Plant Physiol.* **48**, 454–456.

Lawrence, A. A. (1976). "Natural Gums for Edible Purposes". Noyes Data Corporation, Park Ridge, NJ.

Lawrence, J. F. and Iyengar, J. R. (1985). *J. Chromatogr.* **350**, 237–244.

Leavitt, A. L. and Sherman, W. R. (1982). *Carbohydr. Res.* **103**, 203–212.

Lee, C. K. and Gray, G. R. (1988). *J. Am. Chem. Soc.* **110**, 1292–1293.

Lelliott, C., Atkins, E. D. T., Juritz, J. W. F. and Stephen, A. M. (1978). *Polymer* **19**, 363–367.

Lemieux, R. U. and Bauer, H. F. (1953). *Can. J. Chem.* **31**, 814–820.

Leontein, K., Lindberg, B. and Lönngren, J. (1978). *Carbohydr. Res.* **62**, 359–362.

Lewis, B. A. and Smith, F. (1957). *J. Am. Chem. Soc.* **79**, 3929–3931.

Lindberg, B. and Lönngren, J. (1976). *Methods Carbohydr. Chem.* **7**, 142–148.

Lindberg, B., Lönngren, J. and Thompson, J. L. (1973). *Carbohydr. Res.* **28**, 351–357.

Lindberg, B., Lönngren, J. and Svensson, S. (1975). *Adv. Carbohydr. Chem. Biochem.* **31**, 185–240.

Little, M. R. (1982). *Carbohydr. Res.* **105**, 1–8.

Lönngren, J. and Svensson, S. (1974). *Adv. Carbohydr. Chem. Biochem.* **29**, 41–106.

Mabusela, W. T. and Stephen, A. M. (1987). *S. Afr. J. Chem.* **40**, 7–11.

Matchett, J. R. and Levine, J. (1941). *Ind. Eng. Chem. Anal. Ed.* **13**, 98–99.

Matheson, N. K. and McCleary, B. V. (1985). *In* "The Polysaccharides" (G. O. Aspinall, ed.), Vol. 3, pp. 1–105. Academic Press, New York.

Mathlouthi, M. and Koenig, J. L. (1986). *Adv. Carbohydr. Chem. Biochem.* **44**, 7–89.

McGinnis, G. D. and Fang, P. (1980). *Methods Carbohydr. Chem.* **8**, 33–43.

McNeil, M., Darvill, A. G., Åman, P., Franzén, L.-E. and Albersheim, P. (1982). *Methods Enzymol.* **83**, 3–45.

Meer, W. (1980). *In* "Handbook of Water-Soluble Gums and Resins" (R. L. Davidson, ed.), pp. 8.1–8.24. McGraw-Hill, New York.

Merck (1980). "Dyeing Reagents for Thin Layer and Paper Chromatography", Merck, Darmstadt, F.R.G.

Misaki, A., Kaku, H., Sone, Y. and Shibata, S. (1988). *Carbohydr. Res.* **173**, 133–144.

Morrison, I. M. (1975). *J. Chromatogr.* **108**, 361–364.

Morrison, I. M. (1988). *Phytochemistry* **27**, 1097–1100.

Mort, A. J. and Lamport, D. T. A. (1977). *Anal. Biochem.* **82**, 289–309.

Mort, A. J., Parker, S. and Kuo, M.-S. (1983). *Anal. Biochem.* **133**, 380–384.

Nánási, P. and Lipták, A. (1973). *Carbohydr. Res.* **29**, 193–199.

Narui, T., Takahashi, K., Kobayashi, M. and Shibata, S. (1982). *Carbohydr. Res.* **103**, 293–295.

Narui, T., Takahashi, K. and Shibata, S. (1987). *Carbohydr. Res.* **168**, 151–155.

Neukom, H. and Kuendig, W. (1965). *Methods Carbohydr. Chem.* **5**, 14–17.

Niemelä, K. (1987). *J. Chromatogr.* **399**, 235–243.

Oakley, H. B. (1936). *Trans. Faraday Soc.* **32**, 1360–1364.

O'Neill, M. A., Selvendran, R. R., Morris, V. J. and Eagles, J. (1986). *Carbohydr. Res.* **147**, 295–313.

Oshima, R. and Kumanotani, J. (1984). *Carbohydr. Res.* **127**, 43–57.

Oshima, R., Kumanotani, J. and Watanabe, C. (1983). *J. Chromatogr.* **259**, 159–163.

Paz Parente, J., Cardon, P., Leroy, Y., Montreuil, J., Fournet, B. and Ricart, G. (1985). *Carbohydr. Res.* **141**, 41–47.

Parolis, H. and McGarvie, D. (1978). *Carbohydr. Res.* **62**, 363–367.

Pavlenko, A. F. and Ovodov, Yu. S. (1970). *J. Chromatogr.* **52**, 165–168.

Pazur, J. H., Kelly-Delcourt, S. A., Miskiel, F. J., Burdett, L. and Docherty, J. J. (1986). *J. Immunol. Methods* **89**, 19–25.

Pecina, R., Bonn, G., Burtscher, E. and Bobleter, O. (1984). *J. Chromatogr.* **287**, 245–258.

Perini, F. and Peters, B. P. (1982). *Anal. Biochem.* **123**, 357–363.

Perlin, A. S. and Casu, B. (1982). *In* "The Polysaccharides" (G. O. Aspinall, ed.), Vol. 1, pp. 133–193. Academic Press, New York.

Petersson, G. (1974). *Carbohydr. Res.* **33**, 47–61.

Petersson, G. (1977). *J. Chromatogr. Sci.* **15**, 245–255.

Phillips, L. R. and Fraser, B. A. (1981). *Carbohydr. Res.* **90**, 149–152.

Polson, A. (1977). *Prep. Biochem.* **7**, 129–154.

Praznik, W., Burdicek, G. and Beck, R. H. F. (1986). *J. Chromatogr.* **357**, 216–220.

Praznik, W., Beck, R. H. F. and Eigner, W. D. (1987). *J. Chromatogr.* **387**, 467–472.

Prehm, P. (1980). *Carbohydr. Res.* **78**, 372–374.

Proszynski, A. T., Michell, A. J. and Stewart, C. M. (1965). Division of Forest Products Technological Paper No. 38. C.S.I.R.O., Melbourne, Australia.

Purdie, T. and Irvine, J. C. (1903). *J. Chem. Soc.* **83**, 1021–1037.

Rahman, M. D. and Richards, G. N. (1988). *J. Wood Chem. Technol.* **8**, 111–120.

Randall, R. C., Phillips, G. O. and Williams, P. A. (1988). *Food Hydrocolloids* **2**, 131–140.

Raymond, W. R. and Nagel, C. W. (1973). *Carbohydr. Res.* **30**, 293–312.

Redgwell, R. J., O'Neill, M. A., Selvendran, R. R. and Parsley, K. J. (1986a). *Carbohydr. Res.* **153**, 97–106.

Redgwell, R. J., O'Neill, M. A., Selvendran, R. R. and Parsley, K. J. (1986b). *Carbohydr. Res.* **153**, 107–118.

Roberts, E. J., Godshall, M. A., Clarke, M. A., Tsang, W. S. C. and Parrish, F. W. (1987). *Carbohydr. Res.* **168**, 103–109.

Roberts, E. M. and Harrer, E. (1973). *Phytochemistry* **12**, 2679–2682.

Rolf, D. and Gray, G. R. (1982). *J. Am. Chem. Soc.* **104**, 3539–3541.

Rosík, J., Kardošová, A. and Kubala, J. (1971). *Carbohydr. Res.* **18**, 151–156.

Rosík, J., Kardošová, A. and Kubala, J. (1973). *Chem. Zvesti* **27**, 551–553.

Sandford, P. A. and Baird, J. (1983). *In* "The Polysaccharides" (G. O. Aspinall, ed.), Vol. 2, pp. 411–490. Academic Press, New York.

Sandford, P. A. and Conrad, H. E. (1966). *Biochemistry* **5**, 1508–1516.

Sandhu, J. S., Hudson, G. J. and Kennedy, J. F. (1981). *Carbohydr. Res.* **93**, 247–259.

Säverborn, S. (1945). "Contribution to the Knowledge of the Acid Polyuronides". Almqvist and Wiksells Boktrycken, Uppsala.

Sawardeker, J. S., Sloneker, J. H. and Jeanes, A. (1965). *Anal. Chem.* **37**, 1602–1604.

Scherz, H. (1985). *Z. Lebensm.-Unters. Forsch.* **181**, 40–44 (see *Chem. Abstr.* **103**, 103550d).

Schulten, H. R., Bahr, U., Wagner, H. and Hermann, H. (1982). *Biomed. Mass Spectrom.* **9**, 115–118.

Schweer, H. (1983). *J. Chromatogr.* **259**, 164–168.

Scott, J. E. (1965). *Methods Carbohydr. Chem.* **5**, 38–44.

Selvendran, R. R., March, J. F. and Ring, S. G. (1979). *Anal. Biochem.* **96**, 282–292.

Seymour, F. R., Plattner, R. D. and Slodki, M. E. (1975). *Carbohydr. Res.* **44**, 181–198.

Seymour, F. R., Chen, E. C. M. and Bishop, S. H. (1979). *Carbohydr. Res.* **73**, 19–45.

Shinomiya, K., Toyoda, H., Akahoshi, A., Ochiai, H. and Imanari, T. (1987). *J. Chromatogr.* **387**, 481–484.

Siddiqui, I. R. and Wood, P. J. (1971). *Carbohydr. Res.* **16**, 452–454.

Smidsrød, O. and Haug, A. (1967). *J. Polymer Sci., Part C*, No. 16, 1587–1598.

Smith, F. and Montgomery, R. (1959a). "The Chemistry of Plant Gums and Mucilages", pp. 1–39, 501–513. Reinhold, New York.

Smith, F. and Montgomery, R. (1959b). "The Chemistry of Plant Gums and Mucilages", pp. 77–132. Reinhold, New York.

Smith, F. and Montgomery, R. (1959c). "The Chemistry of Plant Gums and Mucilages", pp. 514–553. Reinhold, New York.

Smith, F. and Spriestersbach, D. R. (1955). Abstr. 128th Am. Chem. Soc. Meeting, Minneapolis, MN, p. 15D.

Smith, F. and Stephen, A. M. (1960). *Tetrahedron Lett.* (No. 7), 17–23.

Snyder, C. F., Frush, H. L., Isbell, H. S., Thompson, A. and Wolfrom, M. L. (1962). *Methods Carbohydr. Chem.* **1**, 524–534.

Spedding, H. (1964). *Adv. Carbohydr. Chem.* **19**, 23–49.

Stauffer, K. R. (1980). *In* "Handbook of Water-Soluble Gums and Resins" (R. L. Davidson, ed.), pp. 11.1–11.31. McGraw-Hill, New York.

Stephen, A. M. (1962). *J. Chem. Soc.*, pp. 2030–2036.

Stephen, A. M. (1980). *Encycl. Plant Physiol., New Ser.* **8**, 555–584.

Stephen, A. M. (1983). *In* "The Polysaccharides" (G. O. Aspinall, ed.), Vol. 2, pp. 97–193. Academic Press, New York.

Stephen, A. M. (1987). *S. Afr. J. Chem.* **40**, 89–99.

Stephen, A. M. and Churms, S. C. (1986). *S. Afr. J. Chem.* **39**, 7–14.

Stephen, A. M. and Churms, S. C. (1988). Abstracts of the XIVth International Carbohydrate Symposium, Aug 14–19, Stockholm, p. A118.

Stephen, A. M. and Vogt, D. C. (1967). *Tetrahedron* **23**, 1473–1478.

Stephen, A. M., Kaplan, M., Taylor, G. L. and Leisegang, E. C. (1966). *Tetrahedron Suppl.* **7**, 233–240.

Stephens, D. C. and Stephen, A. M. (1988). *S. Afr. J. Sci.* **84**, 263–266.

Stone, A. L. (1976). *Methods Carbohydr. Chem.* **7**, 120–138.

Street, C. A. and Anderson, D. M. W. (1983). *Talanta* **30**, 887–893.

Strobel, S., Ferguson, A. and Anderson, D. M. W. (1982). *Toxicol. Lett.* **14**, 247–252.

Sugiyama, N., Saito, K. and Sato, H. (1987). *Nippon Shokuhin Kogyo Gakkaishi* **34**, 370–375 (see *Chem. Abstr.* **107**, 196580j).

Sweeley, C. C., Bentley, R., Makita, M. and Wells, W. W. (1963). *J. Am. Chem. Soc.* **85**, 2497–2507.

Sweet, D. P., Shapiro, R. H. and Albersheim, P. (1975). *Carbohydr. Res.* **40**, 217–225.

Szczesniak, A. S. (1986). *In* "Gums and Stabilisers for the Food Industry" (G. O. Phillips, D. J. Wedlock and P. A. Williams, eds), Vol. 3, pp. 311–323. Elsevier Applied Science Publishers, Barking, UK.

Taylor, R. L., Shively, J. E. and Conrad, H. E. (1976). *Methods Carbohydr. Chem.* **7**, 149–151.

Thier, H. P. (1984). *In* "Gums and Stabilisers for the Food Industry" (G. O. Phillips, D. J. Wedlock and P. A. Williams, eds), Vol. 2, pp. 13–19. Pergamon Press, Oxford.

Thompson, A. (1962). *Methods Carbohydr. Chem.* **1**, 36–42.

Towle, G. A. (1972). *Methods Carbohydr. Chem.* **6**, 510–512.

Valent, B. S., Darvill, A. G., McNeil, M., Robertsen, B. K. and Albersheim, P. (1980). *Carbohydr. Res.* **79**, 165–192.

Vandevelde, M.-C. and Fenyo, J.-C. (1985). *Carbohydr. Polymers* **5**, 251–273.

Van Langenhove, A. and Reinhold, V. N. (1985). *Carbohydr. Res.* **143**, 1–20.

Vodonik, S. A. and Gray, G. R. (1988). *Carbohydr. Res.* **175**, 93–102.

Voragen, A. G. J., Schols, H. A., Searle-van Leeuwen, M. F., Beldman, G. and Rombouts, F. M. (1986). *J. Chromatogr.* **370**, 113–120.

Voragen, A. G. J., Rombouts, F. M., Searle-van Leeuwen, M. F., Schols, H. A. and Pilnik, W. (1987). Abstracts of the 2nd International Workshop on Plant Polysaccharides, Structure and Function, July 8–10, Grenoble, France, p. 14.

Waeghe, T. J., Darvill, A. G., McNeil, M. and Albersheim, P. (1983). *Carbohydr. Res.* **123**, 281–304.

Walker, B. (1984). *In* "Gums and Stabilisers for the Food Industry" (G. O. Phillips, D. J. Wedlock and P. A. Williams, eds), Vol. 2, pp. 137–161. Pergamon Press, Oxford.

Wells, G. B. and Lester, R. L. (1979). *Anal. Biochem.* **97**, 184–190.

Whistler, R. L. and Anisuzzaman, K. M. (1980). *Methods Carbohydr. Chem.* **8**, 45–53.

Whistler, R. L. and BeMiller, J. N. (1962). *Methods Carbohydr. Chem.* **1**, 47–50.

Whistler, R. L. and BeMiller, J. N. (1973). "Industrial Gums". 2nd edn. Academic Press, New York.

Whistler, R. L. and Durso, F. (1950). *J. Am. Chem. Soc.* **72**, 677–679.

Whyte, J. N. C. and Englar, J. R. (1974). *Anal. Biochem.* **59**, 426–435.

Wing, R. E. and BeMiller, J. N. (1972). *Methods Carbohydr. Chem.* **6**, 42–53.

Wolfrom, M. and Thompson, A. (1963). *Methods Carbohydr. Chem.* **3**, 150–153.

Woody, R. W. (1977). *J. Polymer. Sci., Macromol. Rev.* **12**, 181–320.

15 Algal Polysaccharides

ELIZABETH PERCIVAL and RICHARD H. McDOWELL

Chemistry Department, Royal Holloway and Bedford New College (University of London), Egham, Surrey, TW20 0EX, UK

METHODS IN PLANT BIOCHEMISTRY Vol. 2
ISBN 0-12-461012-9

I. INTRODUCTION

Although the early stages of carbohydrate biosyntheses are the same in the algae as in land plants, many of the polysaccharides which are finally elaborated are found only in algae. They are present mainly in the cell walls, giving mechanical strength to the plants. The requirements in water are for flexibility and smoothness to withstand currents rather than rigidity to hold them upright against gravity.

The polysaccharides which predominate in algae are therefore those which have the properties of gels or mucilages, and fibriller polysaccharides such as cellulose, though present, are found in much smaller amounts than in land plants. A notable difference in chemical composition is the presence in all multicellular algae of at least one polysaccharide substituted by sulphate ester groups. Such compounds are not found in any land plants but are present in animal polysaccharides.

The three main groups of algae have their characteristic polysaccharides. Brown seaweeds contain alginic acid, a polymer of mannuronic and guluronic acids, together with complex fucans containing fucose with other sugar residues and sulphate esters. The main polysaccharides of the red algae are sulphated galactans comprising sulphated galactoses and 3,6-anhydrogalactose units. Green seaweeds contain complex polysaccharides, many of them sulphated. In addition, all the algae synthesise glucans as storage polysaccharides—starch in the red and green algae and laminaran, a $\beta(1 \rightarrow 3)$-glucan, in the brown algae. As the methods used for studying these polysaccharides are the same as for studying the glucans present in land plants, no further mention is made of them in this chapter; for details of occurrence and structure see Painter (1983) and Percival and McDowell (1985).

Only alginates and the galactans—agar and carrageenan—have become commercially important, and for this reason have been the subjects of the most intensive study. Much effort has gone into discovering methods of extraction, purification and structure determination and elucidatinag the relationship between structure and physical properties. Much of this chapter is therefore devoted to these substances.

II. GENERAL ANALYTICAL METHODS

The methods included in this section, although of general application, are particularly useful in the study of algal polysaccharides. The presence of uronic acid units and sulphate esters in many of them calls for special techniques.

A. Determination of Carbohydrate Content

The carbohydrate content is assayed by the phenol sulphuric acid method (Dubois *et al.*, 1956). Water (1 ml) containing 10–100 µg sugar is added to 4% phenol (1 ml) in a colorimetric tube, and concentrated sulphuric acid (5 ml), reagent grade 95.5%, sp. gr. 1.84, is added rapidly. (Fast-delivery 5 ml pipettes to deliver 5 ml of concentrated sulphuric acid in 10–20 s.) The stream of acid is directed against the liquid surface in order to obtain good mixing. The tubes are allowed to stand for 10 min, then they are shaken and placed for 10–20 min in a water bath at 25–30°C before readings are taken. The colour developed is read at 487 nm. Standard graphs must be prepared for each sugar and mixture of sugars in the ratios corresponding to those of the particular polysaccharide.

B. Methods for Uronic Acids

1. Characterisation

(a) *Paper chromatography.* The uronic acids found in the hydrolysates of algal polysaccharides are well separated on paper using the solvent system pyridine–ethyl acetate–water (11:40:6; v/v) (Fischer and Dorfel, 1955).

(b) *Paper chromatography after reduction to the parent aldoses*
  (i) *With carbodiimide* (Taylor and Conrad 1972). To the polysaccharide (25 μ equivalents of carboxyl groups) in water (50 ml) 1-ethyl-3-(3-dimethylaminopropyl)-carbodiimide hydrochloride (25 mg, 0.25 mmol) is aded after the pH has been adjusted to 4.75 with 0.1 M HCl. During the reaction this pH is maintained by adding 0.1 M HCl. When hydrogen ion uptake ceases sodium borohydride solution (2 M, 10 ml) is added and the mixture stirred for 4 h. It is then dialysed against distilled water for 3 days and then freeze-dried. The resulting glycan is then hydrolysed with acid and the liberated sugars separated on paper using the solvent system 1-butanol–pyridine–water (6:4:3; v/v).
  D-glucose derived from D-glucuronic acid in the original polysaccharide can be characterised with D-glucose oxidase spray (Jabbar Mian and Percival, 1973a).
  (ii) *After esterification.* The uronic acids in the polysaccharides are first esterified in dry conditions with 3% HCl-MeOH and the product reduced with sodium borohydride, followed by treatment as under (i) above.

(c) *Gas chromatography–mass spectroscopy of alditol acetates (GC-MS).* The esterified polysaccharide (see (ii) above) is reduced with sodium borodeuteride (so that aldoses derived from uronic acids have different weight fragments from those originally present in the polysaccharide). After hydrolysis the derived aldoses are reduced to alditols with sodium borohydride and after drying these are acetylated in a pyridine–acetic anhydride (1:1; v/v) mixture. The resulting alditol acetates are characterised on GC-MS.

2. Determination

(a) *Titration with standard alkali.* This is the simplest method if the polysaccharide has been converted to the free acid form (see for example Section III.E.2). However, it cannot be used if sulphate ester groups are present (see also Section II.C.1).

(b) *Carbon dioxide evolution.* Hexuronic acids are quantitatively decarboxylated by heating with hot 12–19% hydrochloric acid, and measurement of the carbon dioxide evolved has been used for many years as a standard method for determining uronic acids. Many modifications of equipment and details of operations have been made. A method suitable for determination of uronic acids in seaweeds and the polysaccharides extracted from them has been described by Whyte and Englar (1974). The sample containing approximately 50 mg of uronide is heated for 10 h with 19% hydrochloric acid at 115°C in a sealed tube which is subsequently broken in an absorption apparatus and the carbon dioxide carried into barium hydroxide in a stream of nitrogen.

(c) *Colorimetric methods.*

  (i) Metahydroxydiphenyl (Blumenkranz and Asboe-Hansen, 1973). An acid–tetra-borate solution (5 ml) (0.0125 M sodium tetraborate decahydrate in concentrated Analar sulphuric acid, sp. gr. 1.84) is placed in Quickfit tubes cooled to 4°C. The sample (1 ml) containing 10–60 µg of uronic acid in water is then added. After mixing, and allowing the solution to reach room temperature, the tubes are heated at 100°C for 5 min. After cooling to room temperature the metahydroxydiphenyl reagent (0.1 ml, 0.15% in 0.5% NaOH) is added. The colour is developed by vigorous mixing and the absorbance read within 5 min at 520 nm. If kept at 4°C in the dark the reagent is stable for 5 weeks.

This method is simpler and quicker than other methods. Standard graphs must be prepared for the different uronic acids,

  (ii) Carbazole. The colour developed when carbazole in alcoholic solution is mixed with concentrated sulphuric acid in which uronic acids have been heated has long been used as a test for the presence of these acids, but the original method was unsuitable for the determination of mixtures of uronic acids such as are found in algal polysaccharides, on account of the wide differences in response from the various acids. In particular mannuronic acid gives a very low response compared with glucuronic and galacturonic acids.

Bitter and Muir (1962) standardised a method in which borate was added to the sulphuric acid to make it 0.1 M in H_3BO_3. This increased the response of mannuronic acid, after heating in the acid at 100°C for 15 min, approximately ten-fold, with much smaller increases for the other acids.

Knutson and Jeanes (1968) made a comprehensive study of the effects of borate and the times and temperatures of heating, and described standard methods for carrying out the determinations on mixtures of the acids with and without borate at both 55 and 100°C. In all cases the colour is measured by the absorbance at 530 nm. Strict adherence to operational procedures at all stages is essential, and standard graphs for the acids likely to be encountered must be prepared. A range of 0 to 100 µg uronic acid in the test quantities (1 ml sample solution, 6 ml acid) gives straight lines. The original paper should be consulted for the exact practical details and the method of calculating amounts of each acid in mixtures.

(d) *IR absorbance.* Bociek and Welti (1975) determined the uronate content of alginate and pectate samples by measuring the absorbance due to the stretching band of the COO^- ion at 1607 cm^{-1}. Solution in D_2O is preferred to the use of KBr disks as this avoids the need for rigorous drying. A buffer to ensure that the acid is completely neutralised can also be included. The samples (30–40 mg) are dissolved in 0.28% NaH_2PO_4–0.86% Na_2HPO_4–D_2O buffer making up to 2.0 ml. Bociek and Welti standardised their results using sodium alginate samples which were ashed and the sodium determined by atomic absorption, making allowance for any free acid in the samples. The method is accurate to within ± 2–4%.

(e) *UV absorbance*. In the separation of uronic acids in an alginic acid hydrolysate (see Section III.F.1) by anion-exchange chromatography, Gacesa *et al.* (1983) measured the amount of each acid by UV absorbance at 210 nm. The sample solution contains 1% triethylamine to convert any lactones to the free acid. A suitable amount of sample contains 20–50 μg of each acid. Using a Whatman Partisil 10-SAX anion-exchange column the sample is eluted with 0.02 M KH_2PO_4 containing 5% methanol at a flow rate of 1.0 ml min^{-1}. The publication gives retention times at different flow rates and the response curves for the four acids: D-glucuronic, D-mannuronic, L-guluronic and D-galacturonic acids. All give straight lines but with different slopes.

C. Methods for Sulphate Determination

1. *Precipitation with cetyl pyridinium chloride*

This is a simple titration procedure (Scott, 1960) which, while not exact, is very useful for routine analyses. Polysaccharide (0.5–2.0 mg) in deionised water (1–2 ml) is titrated dropwise from a microburette with 0.1% aqueous cetyl pyridinium chloride (cpc) solution. The solution is viewed against a dark background. The opacity of the solution increases until precipitation suddenly occurs.

Since many of these polysaccharides contain carboxyl groups as well as half ester sulphate, two titrations can be carried out in parallel: (1) with aqueous polysaccharide solution (titration A); and (2) with polysaccharide dissolved in 0.025 M sulphuric acid (titration B). The sulphate content is given by titration B and the uronic acid content is that of B subtracted from A.

2. *Precipitation with 4-chloro-4′amino-diphenyl*

After digestion of the polysaccharide the resulting material is treated with 4-chloro-4′amino diphenyl (Jones and Letham, 1954).

The polysaccharide (10 mg) is digested in a sealed tube with Analar nitric acid (1 ml + a few mg of sodium chloride) for 12 h at 100°C. The solution in the opened tube is evaporated to dryness, treated with concentrated hydrochloric acid (1 ml) and again evaporated to dryness. After a further treatment with water (1 ml) and evaporation the sample is dried at 110°C for 2 h. It is then ready for sulphate estimation.

To the sulphate solutions in deionised water (1 ml) containing 20–100 μg of sulphate in microcentrifuge tubes, the reagent 4-chloro-4′-amino diphenyl (1 ml of 0.19% in 0.1 M hydrochloric acid) and a trace of solid hexadecyltrimethylammonium bromide are added. After mixing, the solutions, including a blank, are allowed to stand for 2 h and are then centrifuged. Aliquots of the supernatants (0.2 ml) are removed and diluted to 25 ml with 0.1 M hydrochloric acid. The optical densities of the resulting solutions are measured at 254 nm against a blank. The reduction in absorbance from the blank reading gives the sulphate content by comparison with a standard graph constructed with anhydrous potassium sulphate. This in the authors' opinion is the most accurate determination of sulphate content.

3.   As barium sulphate

Sulphate is estimated turbimetrically as barium sulphate, with gelatin being used as a cloud stabiliser (Dodgson and Price, 1962). The sulphate contents of porphyran, fucoidan and carrageenan were determined. The results obtained agreed closely with those obtained by large-scale gravimetric procedures.

The polysaccharide sulphate accurately weighed (2–4 mg) is dissolved in sufficient hydrochloric acid to give a final concentration of SO_4^{2-} ion of between 40 and 90 µg per 0.2 ml. A portion (0.5 ml) is sealed in a glass tube (0.5 cm × 10 cm internal diam.) and kept in an oven at 105–110°C for 5 h. After being cooled the contents are mixed before opening the tube. A portion (0.2 ml) is transferred to a 10 ml tube containing 3.8 ml of 3% (w/v) trichloroacetic acid barium chloride–gelatin reagent (Dodgson, 1961) and, after mixing, the whole is kept at room temperature for 15–20 min. The extinction of the solution is measured at 360 nm against a reagent blank prepared as above except that 0.2 ml of water is substituted for the sulphate-containing solution.

D. Methylation

It is often difficult to methylate these charged polysaccharides completely. The Hako-mori method (1964), modified by Bjorndal and Lindberg (1969), has proved to be the most satisfactory. The polysaccharide (5–20 mg) is dissolved or swelled in dry dimethyl sulphoxide (2 ml) in a serum bottle under nitrogen. Dimethyl sulphinyl carbanion (1 ml) is injected into the bottle and the mixture shaken for 8 h. For a single methylation freshly distilled methyl iodide is added (1 ml) with cooling and the mixture again shaken for 8 h. A second methylation is usually necessary, in which case methyl iodide is added (0.1 ml) with cooling and after 8 h shaking additional carbanion (1 ml) is added and after another 8 h shaking methyl iodide (1 ml) is added with cooling. After a final period of shaking the mixture is poured into water (25 ml) and dialysed when the oily layer becomes solid. The mixture is evaporated to dryness. The methylated polysaccharide is analysed by thin layer chromatography with solvent benzene–ethanol (20:3; v/v) (Hamer, 1970). A single spot indicates complete methylation.

E. Location of Polysaccharides in Algal Tissues

The well established methods of examining stained sections of plant materials under the microscope have been used extensively in the study of algae (Percival and McDowell, 1967). These methods, however, have not been able to distinguish between alginates and fucans, which are both present in brown seaweeds, and the different sulphated galactans present in red algae.

Specific antibodies to the various polysaccharides have been developed and when they are coupled with fluorescent dyes they can be used for detection and differential marking of polysaccharides in the tissues (Vreeland, 1972; Gordon Mills and McCandless, 1975). Preparation of these antibodies involves injection of the appropriate antigens into rabbits or goats, and the antibodies isolated from the animals' serums are not completely specific.

More recently, monoclonal antibodies, which can be produced in large quantities and are specific to sequences up to seven residues in a polysaccharide, have been developed

(Vreeland *et al.*, 1984, 1987). A large number of different antibodies have been prepared and their specificity examined. One strain has been shown to be specific for G blocks in alginates (see Section III.B) (Larsen *et al.*, 1985). Others which react with MG blocks in alginates and some which react with different red algal polysaccharides are also being studied.

Another approach to specificity is the use of lectins. Quatrano *et al.* (1978) coupled ricin, which is specific to galactose, with fluorescein to locate fucans (see Section V) in developing zygotes of *Fucus*.

Gel formation with alginates (see Section III.B) involves the association of G blocks in the polymer in the presence of calcium. Vreeland *et al.* (1987) have combined isolated G blocks with fluorescein and found that in the presence of calcium they will label tissues rich in alginate G blocks. They also obtained promising results with a labelled κ-carrageen.

The technique of X-ray electron microscope micro analysis has been applied by Callow and Evans (1976) to locate sulphur in *Fucus* tissues.

F. Sequential Extraction

Some separation of the polysaccharides in an alga is often possible by sequential extraction, and is useful to provide information on the range of polysaccharides present before using specific extraction methods for the different materials. This procedure has been widely used in the examination of brown and green algae. A convenient sequence is extraction with cold water, followed by hot water, cold dilute acid, hot dilute acid, cold dilute alkali, hot dilute alkali, then after chlorite treatment further extraction with stronger alkali. All the extracts contain a mixture of polysaccharides and further fractionation is necessary before structural studies can be made. Solvents can be modified to reduce the range of compounds extracted, for example in extracting brown algae dilute calcium chloride solution can be used instead of water to minimise the extraction of low molecular-weight alginate. The full sequence is likely to degrade or modify polysaccharides. Hot acid will degrade most of them, and alkali will remove sulphate groups from some positions. Preliminary treatment with borohydride has been used to limit degradation with alkali.

G. Viscosity Measurements

As with other polymers, the molecular weights of seaweed polysaccharides can be determined by measurements of their intrinsic viscosities $[\eta]$. Precautions to be observed in obtaining these values are discussed by Morris *et al.* (1981). These authors also showed that for the polymers which they examined, including sodium alginate and λ-carrageenan, a plot of $\log(\eta_{sp})$ against $\log c \times [\eta]$ produces two straight lines, one for η_{sp} 0–10 with a slope of about 1.4 and another for $\eta_{sp} > 10$ with a slope of about 3.3 (η_{sp} is the specific viscosity, which for solutions in water at 25°C is approximately equal to the viscosity in centipoises, and c is the concentration in grams per 100 ml). Using this relationship it is possible to obtain an approximate value for $[\eta]$ by measurement of the viscosity at any concentration. It must be noted however that the relationship applies to viscosities extrapolated to zero shear, and measured viscosities decrease as the rate of shear is increased.

Many applications of sodium alginate and λ-carrageenan make use of the high viscosity of their solutions. Solutions (1%) of these polysaccharides are commonly measured at 25°C using a Brookfield rotational viscosimeter. The rate of shear is a function of the spindle and speed of rotation, and these parameters are included in viscosity specifications for commercial products.

III. ALGINIC ACID

A. Occurrence

Alginic acid is a linear polymer of (1→4)-linked D-mannuronic acid and L-guluronic acid which occurs combined with calcium and other bases in the cell walls and intercellular matrix of brown seaweeds. The amount present varies widely from one species to another, for example 50% of dry weight in some samples of *Durvillea antarctica* (South, 1978) to 13% in *Padina pavonia* (Jabbar Mian and Percival, 1973a). Both alginic acid and its calcium salt are insoluble in water, but the alkali metal salts are soluble so that base exchange affords a simple method of extraction. Careful extraction can give sodium alginate with chain lengths of up to 10 000 units, so presumably it is present in larger molecules in the plants.

B. Composition and Structure

The residues are present in blocks of each monomer, separated by regions in which they are randomly arranged or are alternating. The proportions of M (mannuronic) and G (guluronic) residues, and the lengths of the blocks, can vary considerably, depending on the source of the alginate. These variations are largely responsible for the differences in properties observed from one alginate sample to another. It is the size and proportion of the G blocks in particular which determines the formation and strength of gels formed with calcium.

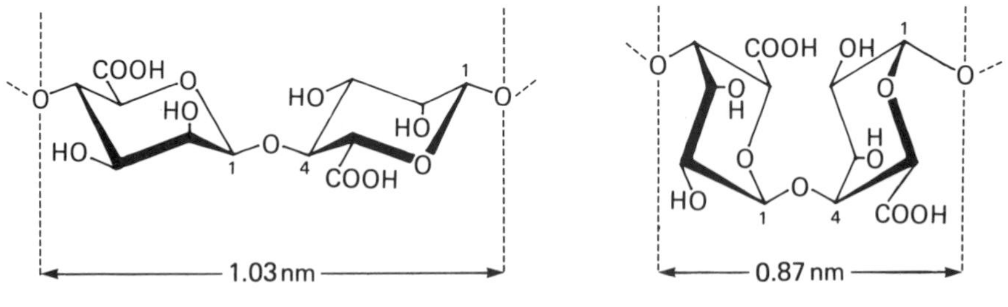

FIG. 15.1. Parts of the alginic acid molecule: (a) mannuronic and (b) guluronic acid residues.

It was shown by X-ray diffraction (Atkins *et al.*, 1971) that the M and G blocks have different conformations (see Fig. 15.1). The G blocks have a strong affinity for calcium ions which leads to aggregation of alginate chains (Smidsrød and Haug, 1972; Morris *et al.*, 1973, 1978) and gel formation (Grant *et al.*, 1973). D-Mannuronic acid residues can be epimerised to L-guluronic residues by an enzyme produced by the bacterium *Azotobacter vinlandii* (Haug and Larsen, 1971), and in experiments designed to increase the strength of calcium alginate gels by this epimerisation reaction Skjåk-Bræk *et al.* (1986) found that the gel strength was determined as much by the length of the G blocks as by the proportion of G units. The mean lengths of the G blocks (sequences of two or more G units) were determined by nuclear magnetic resonance spectroscopy (see Section III.F.3) and ranged from 4 to 17.5 in the samples studied.

As epimerases have been isolated from algae (Madgwick *et al.*, 1973b, 1978; Ishikawa and Nisizawa, 1981) it is likely that this epimerisation is involved in the biosynthesis of L-guluronic acid units in alginates, although it is possible they may also be produced by an independent route from sorbitol (Quillet and Lestang Bremond, 1985). Sodium alginate retains one molecule of water per residue on drying at 110°C and in the Karl Fischer method of moisture determination (Bociek and Welti, 1975). A figure of 216 should therefore be used in calculating equivalents.

Alginates are highly polydisperse so that there is a great difference between number average and weight average molecular weights. For a discussion of the application of viscosity measurements to molecular weight, see Section II.G and Smidsrød and Haug (1968).

C. History and Uses

Sodium alginate (algin) was first extracted from seaweed and recognised as a distinct compound by Stanford in the early 1880s, but it was not until 1930 that alginic acid was shown to be a polyuronic acid and in 1955 a copolymer of D-mannuronic acid and L-guluronic acids (Percival and McDowell, 1967).

Stanford proposed a number of uses for algin but there was no successful commercial production of alginate until the 1930s in the USA. Large-scale manufacture is now carried on in the USA, Scotland, Norway, France, China and Japan, with some smaller-scale production in other countries.

The main uses of alginates arise from the high viscosity and polyelectrolyte character of alginate solutions (e.g. textile printing, food stabilising), and their capacity for gel formation (e.g. food products) and film formation (e.g. paper sizing). Recent accounts are given by Sandford (1985) and Glicksman (1987).

D. Function in the Plant

The alginate is present in seaweeds in most cases in an insoluble form and appears to be the main structural component. The strength of alginate gels depends on the proportion and size of the G blocks and it has been found in some seaweeds that the alginate composition varies with the rigidity requirements in different parts of the plants (Stockton *et al.*, 1980a,b; Craigie *et al.*, 1984).

An exceptional state of the alginate is found in the fruiting bodies of species of the Fucaceae. Here it is found as a viscous solution: it is made up almost entirely of mannuronic acid units and does not gel in the presence of calcium ions. It is considered that it aids dispersion of gametes into the seawater (Haug *et al.*, 1969).

E. Extraction and Purification

1. Principle

The alginate can be extracted from fresh seaweed which has been frozen immediately after collection in liquid nitrogen, or which has been dried. In any case it should be finely divided. A preliminary extraction with cold dilute hydrochloric acid converts the alginate into the acidic form and at the same time removes fucans and other unwanted substances. Addition of formaldehyde before extraction with alkali insolubilises phenolic substances. The alginate is brought into solution as the sodium salt by addition of sodium carbonate and the insoluble residue removed by a combination of settling and filtration. The solution can be dialysed and lyophilised after concentration *in vacuo*, or the alginate can be precipitated as the calcium salt, or as the free acid, or as the sodium salt by addition of an excess of ethanol or 2-propanol. If a pure alginate is required it will be necessary to dissolve the solid obtained and re-precipitate it, preferably using a different method from the first time.

2. Example

A 5 g sample of dried fronds of *Ascophyllum nodosum*, milled to pass through a 500 μm sieve, is mixed with 5 ml 40% formaldehyde solution and allowed to stand overnight. It is then stirred in 50 ml of 0.2 N hydrochloric acid for an hour at room temperature, filtered and the extraction repeated. The residual solids are then stirred in 50 ml 3% sodium carbonate for 3 h, forming a stiff paste which is diluted to 1 litre with water, stirred for a further hour and allowed to settle in a litre measure overnight. Liquid is siphoned off from the settled solids, 5 g Celite added to the supernatant and it is filtered with suction using a coarse filter paper. The volume of filtrate is measured (for calculation purposes) and then poured into 100 ml of 10% calcium chloride solution with stirring. In this way the alginate is precipitated as an adherent gel which is drained on a nylon cloth and pressed by hand. It is transferred to a sintered glass funnel, broken up, and leached with 0.5 N hydrochloric acid until the drainings are free from calcium (ammonium oxalate test), then with deionised water until drainings are neutral to methyl red (pH 5). The alginic acid thus prepared is transferred to a weighed beaker, approximately 50 ml water added with good stirring and titrated to pH 7 with 0.5 N sodium hydroxide. The weight of sodium alginate in solution is calculated using an equivalent weight figure of 216, and water is added to give a 1% solution by weight. The viscosity of a 1% solution of sodium alginate is often included in commercial specifications, and this solution can be used for viscosity measurement, giving an indication of the molecular weight of the alginate (see Section II.G). A viscosity of at least 100 cP is to be expected. The alginate can be purified by precipitation with an equal volume of ethanol.

F. Determination of M/G Ratio and Block Structure

1. Complete hydrolysis

The proportions of M and G residues were first determined by complete acid hydrolysis and separation by paper chromatography (Fischer and Dorfel, 1955). A standard method using column chromatography of the acid hydrolysate on anion-exchange resins, followed by colorimetric determination, was described by Haug and Larsen (1962). Sodium alginate (50 mg) is mixed with 0.5 ml 80% sulphuric acid while it is cooled in an ice bath, and left for 18 h at 20°C. The acid is diluted to 2 N by adding 6.5 ml water while cooling in an ice bath, and the mixture heated in a sealed glass tube in a boiling-water bath for 5 h. After cooling the hydrolysate is neutralised with a slight excess of calcium carbonate. The bulky precipitate is filtered off and washed with water until about twice the original volume of the hydrolysate has passed through the filter.

Mannuronic and guluronic acids are separated on a column using Dowex 1 × 8 anion-exchange resin which has been activated with 2 N NaOH and converted to the acetate form with acetic acid. After using the same method of acid hydrolysis Gacesa et al. (1983) used anion-exchange liquid chromatography to separate the acids and determined them by UV absorbance at 210 nm. There is considerable breakdown of uronic acids during the acid hydrolysis, and as guluronic acid is destroyed more quickly than mannuronic, Haug and Larsen (1962) multiplied the found value of the M/G ratio by 0.66.

2. Partial hydrolysis

Partial hydrolysis using 1 M oxalic acid or 0.3 M hydrochloric acid at 100°C brings part of the alginic acid into solution. The insoluble portion can be fractionated into material soluble and insoluble at pH 2.85. Haug et al. (1966, 1967) examined each of these fractions by complete hydrolysis and found that the material which had been brought into solution by partial hydrolysis contained roughly equal proportions of M and G residues. The fraction that was insoluble at pH 2.85 was made up predominantly of G residues, and the fraction that was soluble at pH 2.85 was made up predominantly of M units. These fractions are therefore referred to as alternating, G and M blocks.

3. Nuclear magnetic resonance methods

To avoid the time-consuming separation of the material resistant to partial acid hydrolysis into M and G blocks Penman and Sanderson (1972) worked out a method of determining the proportions of M and G residues in it by proton nuclear magnetic resonance spectroscopy. Only 40 mg of material in 0.5 ml D_2O was required using a Varian HA-100 spectrometer. It was assumed that in the insoluble material from the partial hydrolysis the M and G residues were in separate blocks and that the soluble portion was made up of equal parts of M and G residues.

It has not been found possible to make nuclear magnetic resonance (NMR) measurements on completely undegraded alginates, but by degradation to a degree of polymerisation (DP) of 20–30 using very mild hydrolysis with acid (30 min, 100°C, pH 3.0) in order to reduce the viscosity of their solutions, Grasdalen et al. (1978) were able to assign peaks in the proton NMR spectrum which differentiated G residues with neighbouring G from those with M residues as neighbours.

Further work using 400 MHz proton NMR (Grasdalen, 1983) gave information on the proportions of the G-centred triplets, designated GGG, MGG, GGM and MGM. Ten milligrams sodium alginate in 0.4 ml D_2O was used; 3 mg EDTA was added to avoid any interaction with calcium ions. Proton NMR can also be used to determine the proportion of M residues, but to obtain information about the proportions of M-centred triplets it is necessary to use carbon-13 NMR. In this case 40 mg sodium alginate in 0.4 ml D_2O is required. Information on the proportions of the different triplets enables average lengths of the M and G blocks to be calculated.

4. Circular dichroism

Morris *et al.* (1980) have shown how circular dichroism (CD) can be used to determine the block structure of < 1 mg sodium alginate. Alginates exhibit CD behaviour in the UV region 190–245 nm, the spectra of D-mannuronate and L-guluronate residues being very different. The CD behaviour is modified by the adjacent residues in the polymer chain so that a mixture of M and G blocks gives a somewhat different spectrum from the same proportion of M and G residues in a mixed polymer. The spectra show a peak at 200 nm and a trough at 215 nm and equations are given to determine the M/G ratio from the observed ratio of peak height to trough depth. In order to find the block structure it is necessary to consider the whole spectrum using an iterative, least-squares, computer technique. Unlike NMR techniques, CD can be used without any degradation of the samples. The solution must be adjusted to a pH of 7 ± 0.3 and divalent cations must be absent. Solutions of 1 mg ml^{-1} are used and for block structure analysis (but not M/G ratio) the uronate concentration must be accurately known. This method was used by Craigie *et al.* (1984) to examine tissues from a number of algae.

5. Enzymic methods

A wide variety of organisms can degrade alginates, and a number of active enzymes have been isolated from bacteria and lower animals (Percival and McDowell, 1967). Some degrading enzymes have also been isolated from seaweeds (Madgwick *et al.*, 1973a; Shiraiwa *et al.*, 1975). All the lyases so far isolated are eliminases, not hydrolases, and the unit on the non-reducing side of the bond is converted into 4,5-unsaturated uronic acid. The same product is therefore formed from both M and G units so that examination of the fragments produced by enzyme action does not give a complete picture of the units originally present. Purified enzymes from different sources may be specific for L-guluronosyl- or D-mannuronosyl-linkages (Fujibayashi *et al.*, 1970).

Caswell *et al.* (1986) separated the alginate lyases obtained from five strains of bacteria by isoelectric focusing. By this means the proteins comprising the enzyme preparations were spaced out on a polyacrylamide gel film according to their isoelectric points. Enzyme activity was assayed by sandwiching agarose gel films containing sodium alginate with the film containing the enzymes, removing it after 1 h at 30°C and staining with ruthenium red. Where the alginate had been degraded a clear zone was observed. The positions of proteins in the polyacrylamide gel film were found by staining with Coomassie Blue. Their results showed that some strains of bacteria produced multiple lyase activities.

The breakdown of whole alginate and alginate fragments by a guluronase obtained from *Klebsiella aerogenes* Type 27 has been studied by Boyd and Turvey (1977, 1978). The organism was grown on a medium containing sodium alginate, and the enzyme precipitated from the centrifuged medium by bringing to 80% saturation with ammonium sulphate. It was redissolved in water and dialysed against 10 mM phosphate buffer at pH 7 and freeze-dried. Further purification was carried out by fractionation on DEAE-Sephadex. The formation of double bonds by the enzymic cleavage of the guluronosyl links was followed by measurement of absorbance at 230 nm. The products were separated on a Sephadex G-10 column which was eluted with water.

Analysis of the fragments obtained indicated that a sequence of at least three G units must be present for the enzyme to bring about degradation. The results showed that in the MG blocks prepared by partial hydrolysis with acid (see Section III.F.2) there were some sequences of at least two M units and two G units. M blocks resulting from the action of the enzyme on whole alginate had a length of about 24 units.

Greene and Madgwick (1986) have studied the preparation and action of alginate lyases and epimerases from a number of algae and suggest that in future ways may be found for algae to provide better sources of these enzymes than the microorganisms or marine animals which have been used until now in the study of alginate structure.

G. Gel Strength

The alginate in algae is present in the form of a gel which provides much of the strength of the plant; gel formation is also of importance in many commercial applications. It is therefore important to be able to determine the gel-forming ability of isolated alginates. Diffusion of calcium chloride into a sodium alginate solution gives a non-homogeneous gel. To overcome this fault Skjåk-Bræk *et al.* (1986) used a solution of sodium alginate containing Ca-EDTA and D-glucono-δ-lactone. The latter hydrolyses to gluconic acid which releases calcium ions from the EDTA complex, thus forming a homogeneous mixed H^+Ca^{2+} alginate gel. This is then converted completely into the calcium form by dialysis against calcium chloride solution. The strengths of gels formed from different sodium alginate preparations were compared using the stress–strain relationship under compression.

IV. GALACTANS

A. General

The galactans, the major polysaccharides of the Rhodophyceae, comprise agars, carrageenans and related hybrid polysaccharides. For comprehensive reviews see Percival (1978a) and Craigie (1990). The agars and carrageenans consist essentially of linear chains of alternating $(1 \rightarrow 3)$-β-galactose and $(1 \rightarrow 4)$-α-galactose, but whereas in agar the $(1 \rightarrow 4)$-linked units are L-galactose, in carrageenan both units are D-galactose. At the same time many of the units are masked by substitution with half ester sulphate, with methoxyl groups and with pyruvic acid groups (as the 4,6-O-carboxyethylidene derivative) and by modification of the 6-sulphated (1,4)-linked units to the 3,6-anhydrosugar. It is believed that this last change in nature is brought about by a sulphohydrolase, whereas in the laboratory it is achieved by the action of alkali (for formulae see Fig. 15.2). It is these differences which determine the overall shape or conformation of the polysaccharides. There are polysaccharides which give stiff gels in dilute solution and others that have no gelling power at all. The presence of the 3,6-anhydro sugar causes gelling, but if this is replaced by the 6-sulphate the gelling power is considerably lessened and the 2,6-disulphate in place of the 3,6-anhydrosugar results in the complete loss of gelling power. Each genus of seaweed metabolises polysaccharides which differ somewhat in their properties. All extracts consist of a family of macromolecules which differ in the fine details of structure, and separation into a single molecular structure has proved to be extremely difficult and in some instances impossible.

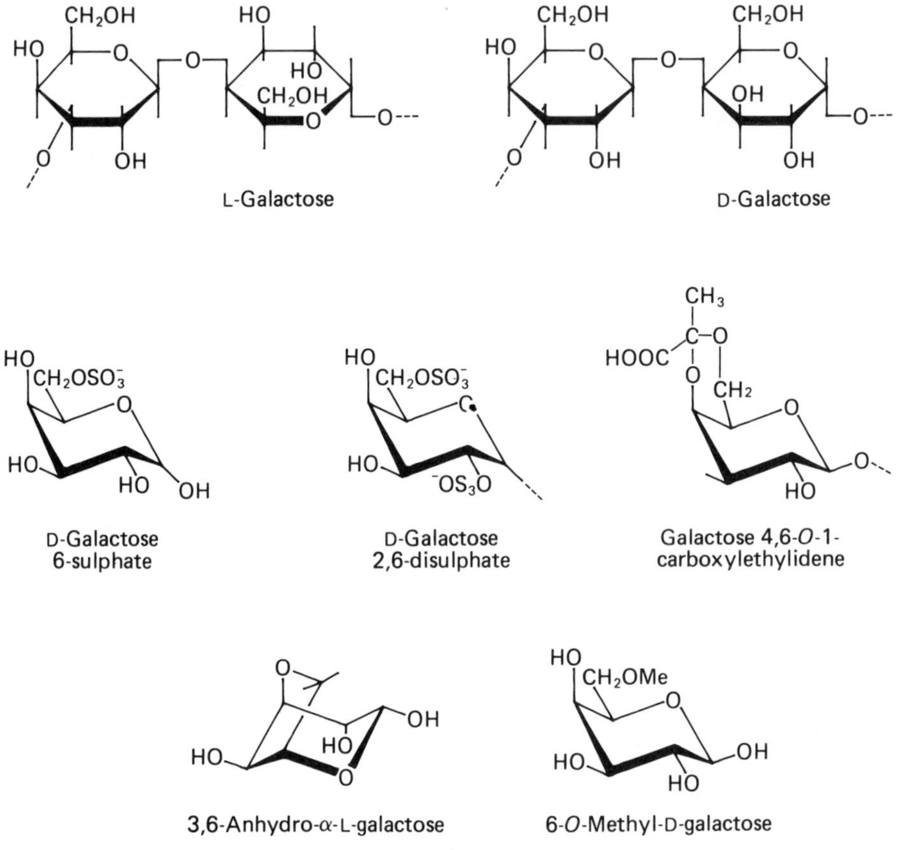

L-Galactose

D-Galactose

D-Galactose
6-sulphate

D-Galactose
2,6-disulphate

Galactose 4,6-O-1-
carboxylethylidene

3,6-Anhydro-α-L-galactose

6-O-Methyl-D-galactose

FIG. 15.2.   Structural units of sulphated galactans

B.   Agar

1.   Composition and structure

Details of the various modifications to the simple alternating 1,3-linked D-galactose and 1,4-linked L-galactose units present in agar can be found in Section IV.A and Fig. 15.2. The first separation of agar into two components—agarose, a non-sulphated fraction and agaropectin, a mixture of various sulphated molecules—was by Araki in 1937 (see Araki, 1966). Various fractionating procedures supported this composition and it was not until 1971 that Duckworth and Yaphe put forward the idea that agar contained a family of polysaccharides which differ in the fine details of structure, as detailed in Section IV.A. Depending on the source of the agar, so the proportions of the different polysaccharides vary. Pyruvate occurs in varying amounts in agarose in most species except those of *Gracilaria* and some of *Pterocladia*. *Gracilaria* agarose contains methyl ethers, the position of which is variable according to species, but occurs most frequently at carbon-6 of the 1,3-linked D-galactose residues and less so as 2-*O*-methyl 3,6-anhydro α-L-galactose (Guiseley, 1987). Work by Guiseley in 1970 indicated that in naturally occurring agars an increase of gelling temperature occurred with increase of methoxyl content. In contrast methylation of a low methoxyl agarose gave a product of lower gelling temperature (Guiseley, 1987).

2. History and uses

Agar is the oldest known phycolloid and was isolated by Japanese peasants centuries ago by boiling the weed in water and freezing and thawing the gelled extract (Guiseley, 1968). It is used as a gelling agent for many forms of food. It is made mainly from species of *Gracilaria* and *Gelidium* and in New Zealand from *Pterocladia* spp. (Furneaux and Miller, 1985). Although a number of other genera have been investigated chemically (Percival and McDowell, 1981) in which 6-*O*- and 2-*O*-methylgalactose and L-galactose 6-sulphate figure predominantly, none of these species, as far as the authors are aware, has found commercial application. The extreme inertness of agar led Koch and Petri in 1882 to use it as a medium in which to grow bacteria. This found wide medical application.

Agar and its more neutral component, agarose, have been and continue to be important in biomedical research and diagnosis of disease states and as a medium for cloning genetically engineered microorganisms (Renn, 1984). Agar finds use in a number of bakery products, confectioneries and canned meats (Glicksman, 1987). Synthetic derivatives such as hydroxyethyl-, allylglyceryl- and glyoxal have broadened the area of applicability of agarose gels (Guiseley, 1987).

3. Extraction

Agar is extracted from the milled seaweed with 50–100 parts of water or 0.1 M phosphate buffer at pH 6.3 in the autoclave at 120°C for 3 h. The extract is filtered under pressure with added Celite, maintaining the temperature above 60°C. On cooling the extract sets to a gel, from which much of the water separates on freezing and thawing. In commercial production water removal by hydraulic press has largely replaced the traditional freezing process. With some species of algae higher gel strengths are obtained by treating the weed with sodium hydroxide and sodium borohydride before extraction. For full details of laboratory preparation, see Craigie and Leigh (1978).

C. Carrageenan

1. Composition and structure

Carrageenan polysaccharides consist of chains of 1,3-linked β-D-galactose and 1,4-linked α-D-galactose units which are variously substituted and modified to the 3,6-anhydro derivative as shown in Fig. 15.2. κ-Carrageenan, a sulphated gelling fraction, is precipitated by potassium ions from whole extracts of *Chondrus crispus* and *Gigartina* spp. (Smith and Cook, 1953). The material left in solution, the non-gelling fraction known as λ-carrageenan, is a mixture of variously substituted molecules which can be further fractionated (Table 15.1). *Chondrus crispus* is the primary source of λ-carrageenan, *Eucheuma cottonii* of κ- and *Eucheuma spinosum* of ι-carrageenan.

TABLE 15.1. Repeating units of carrageenans.

$$—^3A^{1\beta}——^4B^{1\alpha}——^3A^{1\beta}——^4B^{1\alpha}——^3A^{1\beta}——$$

A units	B units	Found in[a]
D-Galactose 4-sulphate	D-Galactose 6-sulphate	μ
	D-Galactose 2,6-sulphate	ν
	3,6-Anhydro-D-galactose	κ
	3,6-Anhydro-D-galactose 2-sulphate	ι
D-Galactose 2-sulphate	D-Galactose 2-sulphate	ξ
	D-Galactose 2,6-disulphate	λ
	3,6-Anhydro-D-galactose 2-sulphate	θ

[a] For additional rare hybrids, see Craigie (1990).

2. History and uses

The seaweed *Chondrus crispus* (Irish moss) was used traditionally to make milk puddings without any preliminary extraction. Until the Second World War, which cut off the supplies of agar to the Western World, very little commercial use was made of carrageenan, but since Japan was the sole supplier of agar and agar-bearing weeds Western chemists were faced with finding an alternative medium. Attention was at once turned to improving the extraction procedure, for example, from *Chondrus crispus* and similar carrageenan-bearing weeds.

κ-Carrageenan (described in Section IV.C.1) gels in the presence of potassium ions to form strong crisp gels, while λ-carrageenan is non-gelling, but forms viscous solutions; ι-carrageenan gels in the presence of Ca^{2+} ions to form elastic gels. Because of their strong anionic nature these carrageenans exhibit a high degree of protein reactivity. The free acids are unstable and rapidly undergo autocatalytic degradation. Commercial products are usually mixtures of sodium, potassium or calcium salts. As a result of these properties they find many uses in foods. More than three-quarters of the carrageenan Marine Colloids makes for hundreds of purposes goes into the food industry. λ-Carrageenan is used as a thickener or viscosifier and potassium κ-carrageenate and calcium ι-carrageenate are used as gelling agents.

Carrageenan is invaluable in keeping the cocoa particles suspended in chocolate milk products. It also finds use in milk puddings, ice cream, infant formulas, evaporated milk, processed cheeses, water dessert gels, low calorie jellies and pet foods. Other uses include binders for toothpastes, bodying agents for shampoos, ingredients in skin creams and lotions, to name only a few of its applications.

3. Extraction and fractionation

For details of extraction, see Craigie and Leigh (1978). In this method the frozen powdered seaweed is extracted with acetone, ethanol and ether before being dried. It is then extracted with 0.5 M sodium bicarbonate at 90°C, followed after filtration by precipitation with cetyltrimethylammonium bromide (CTAB), and the precipitate is then washed with sodium acetate in 95% ethanol. In order to get a more complete extraction Bremond et al. (1987) modified the pre-treatment by shaking the alcohol-extracted weed with 0.5 N hydrochloric acid in 80% ethanol for 5 min at 4°C and washing with 80% ethanol also at 4°C before further washing and drying. By this method they were able to extract λ-carrageenan as well as the predominant κ-carrageenan from gametophytes of *Chondrus crispus*.

Carrageenan was originally separated into κ- and λ-fractions by precipitating with 0.3 M potassium chloride (Smith and Cook, 1953), but Stancioff and Stanley (1969) found that better separation into fractions was achieved by leaching the powdered carrageenan with 0.3 M potassium chloride solution. The soluble fraction can be precipitated from the solution with CTAB or propanol. This was the method adopted by Bremond et al. (1987).

4. Gametophyte and sporophyte carrageenans

In 1973 McCandless et al. showed that *Chondrus crispus* sporophyte plants contained λ-carrageenan and the gametophyte plants mainly κ-carrageenan, with as much as 25% of KCl-soluble material. Work has continued on this aspect which indicates that both forms, to a certain extent, may occur in the two phases of the life history. Studies on the gametophyte of *C. crispus* by Bellion et al. (1983) showed the polysaccharide to comprise κ-carrageenan 73%, ι-carrageenan 17% and other sulphated galactans 10%. Greer et al. (1984), working on *Hypnea musciformis*, obtained similar results, and Bremond et al. (1987) found 0.43 g of λ-carrageenan in 2.95 g of polysaccharide extracted from the gametophytic plants of *C. crispus*. Dawes et al. (1977) and Doty and Santos (1978) did not find any differences in carrageenan composition of the different life phases of a number of *Eucheuma* species, and this has been confirmed by Bert et al. (1988) for the tetrasporophytic and gametophytic stages of *Cystoclonium purpureum*.

D. Analytical Techniques

1. Characterisation of polysaccharides of red algae by pyrolysis

Pyrolysis-capillary gas chromatography (Py-GC) has been applied to purified agars and carrageenans and to intact algal material. It is a rapid and useful technique for characterising simple anhydro, O-methyl and sulphated saccharide units which are present in the galactans of marine plants (Helleur et al., 1985b). It can therefore be useful in aiding taxonomic characterisation studies.

Pyrograms have been obtained using a Chemical Data Systems Pyroprobe 120 equipped with a coil filament interfaced to a Hewlett-Packard 5880A gas chromato-graph equipped with flame ionisation detector and using a fused capillary column [J and W,DB-1701 (bonded phase), 1.0 μm thickness, 3.0 m × 0.329 mm]. Accurately weighed samples of 0.15–0.20 mg for agars, and 0.35–0.40 mg for carrageenans and plants, were used. All plants were seawater-rinsed. Samples were milled to 200–400 mesh particle size. Further experimental details for Py-GC are given by Helleur *et al.* (1985a).

2. *Colorimetric determination of 3,6-anhydrogalactose*

The 3,6-anhydrogalactose is determined with the resorcinol reagent (Yaphe and Arsenault, 1965). A stock solution of resorcinol is prepared by dissolving 150 mg (1.36 mmol) in distilled water (100 ml). The reagent is prepared by adding concentrated hydrochloric acid (sp. gr. 1.188–1.192; 100 ml) to 9 ml of resorcinol stock solution and adding to this mixture 1 ml of an aqueous solution of acetal (1,1-diethoxyethane) (2.78 μmol). The reagent develops a colour on standing but is stable for at least 3 h.

A solution (2 ml) containing 0.25 μmol of 3,6-anhydrogalactose in the free state or combined in agar or carrageenan in a boiling tube 25×150 mm covered with a glass marble is placed in an ice bath and resorcinol reagent (10 ml) is added. The contents are mixed in the ice bath and cooled for at least 3 min but not longer than 30 min. The tube is placed in a 20°C water bath for 4 min and then heated for 10 min at 80°C. It is then cooled for 1.5 min in an ice bath and the absorbance is measured within 15 min at 555 nm. Cold acid degrades 3,6-anhydrogalactose and reproducible results are obtained only with careful attention to the details of the procedure. The absorbance with 0.18 μmol of 3,6-anhydrogalactose was 92% of that of an equimolar amount of fructose. Thus with a correction factor, fructose may be used as a reference sugar to determine the concentration of 3,6-anhydrogalactose in agar, carrageenan and other algal polysaccharides.

3. *Pyruvate analysis*

In one method (Duckworth and Yaphe, 1970) the pyruvic acid is released with dilute hydrochloric acid and the 2,4-dinitrophenylhydrazone derivative prepared. The application of a correction factor for agar and carrageenan is necessary owing to the degradation of the 3,6-anhydrogalactose units to keto acids which give a positive pyruvic reaction.

Absolute values for pyruvic acid can only be obtained by using the definitive lactate dehydrogenase method of Duckworth and Yaphe (1970), modified by Hirase and Watanabe (1972).

4. *Infrared analysis*

Evidence for the site of the half ester sulphate groups can be obtained by infrared analysis. A broad band at $1210 \, \text{cm}^{-1}$ is common to all sulphated polysaccharides and increases in size with sulphate content. The peaks at $930–940 \, \text{cm}^{-1}$ have been assigned to 3,6-anhydrogalactose. Peaks at $820 \, \text{cm}^{-1}$ are characteristic of sulphate at a primary alcoholic group, that is at galactose C-6; at $830 \, \text{cm}^{-1}$ of equatorial sulphate, that is at galactose C-2, and at $850 \, \text{cm}^{-1}$ of axial sulphate, that is at galactose C-4 (Stancioff and Stanley, 1969; Rochas *et al.*, 1986). Films of solutions of the polysaccharide are made on AgCl plates or KBr disks prepared. Perkin-Elmer 177 Grating infrared spectrometers are suitable instruments. Thin transparent films can also be prepared by drying seaweed extract at 37°C (5 mg ml^{-1}; 2 ml) in Teflon wells 0.6 cm deep and 2.5 cm in diameter (Knutsen and Grasdalen, 1987). The films are mounted in cardboard with a centre hole before scanning.

5. Enzymic analysis

Araki in 1957 used enzymic hydrolysis in his studies of agar. More recently Yaphe has used β-agarases isolated from *Pseudomonas atlantica* to degrade agar to a homologous series of oligosaccharides based on neoagarobiose (3,6-anhydro-α-L-galactopyranose-(1 → 3)-D-galactose) and extended this work wherein further enzyme activities have been discovered which digested the oligomers to neoagarobiose by exo-scission (Groleau and Yaphe, 1977). Morrice *et al.* (1984) applied the action of these enzymes to the study of the structure of Porphyran, the polysaccharide from *Porphyra umbilicalis*. Similarly it has been found that κ-carrageenase isolated from *Pseudomonas carrageenovora* hydrolyses the β-linkages in carrageenan to produce a series of oligosaccharides homologous to neocarrabiose 4-sulphate (3,6-anhydro-α-D-galacto-pyranose (1→3)-D-galactose 4 sulphate) and the same organism produces a glycosulphatase which desulphates the above neocarrabiose (Weigl and Yaphe, 1966; McLean and Williamson, 1981). κ- and ι-carrageenases have been used for the analysis of carrageenans from a variety of Rhodophyta (Bellion *et al.*, 1981; Greer and Yaphe, 1984).

6. Carbon-13 NMR analysis

Agar from *Porphyra*, *Ahnfeltia* and *Gracilaria* spp., κ-carrageenan from *Eucheuma* and *Chondrus* spp. and ι-carrageenan from *Eucheuma* hybrid carrageenans, *Furcellaria* and *Phyllophora* spp. were found to yield well resolved carbon-13 NMR spectra in the anomeric region in each case: the peak at 103.2–103.6 p.p.m. can be attributed to C-1 of the D-galactose of D-galactose 4-sulphate residue. In agar the signal at 99.2 p.p.m. is due to C-1 of the 3,6-anhydro-α-L-galactose residue. κ-Carrageenan gave a signal at 96.2 p.p.m. and ι-carrageenan at 93.1 p.p.m., assigned to C-1 of 3,6-anhydro-α-D-galactose and 3,6-anhydro-α-D-galactose 2-sulphate residues, respectively. Other signals are considered to indicate disaccharide units (Bhattacharjee *et al.*, 1978). This method has also been used in a study of the hybrid structure of carrageenans from species of *Eucheuma*, *Ahnfeltia* and *Irideae* (Bellion *et al.*, 1981), and for the characterisation of water-extractable polysaccharides from *Furcellaria lumbricalis* (Knutsen and Grasdalen, 1987).

V. FUCANS

A. General

The fucans occur as matrix polysaccharides in all members of the Phaeophyta. They comprise families of polydisperse heteromolecules based on L-fucose, D-xylose, D-glucuronic acid, together in some species with D-galactose and D-mannose. Application of methylation, periodate oxidation, and partial hydrolysis studies have revealed the essential similarity of fucans from different genera of seaweeds. The L-fucose units are

$(1\rightarrow2)$- and $(1\rightarrow3)$-linked with sulphate on C-4. The glucuronic acid and xylose units are not sulphated and appear to be on the periphery of the molecules (Jabbar Mian and Percival, 1973b). For more details of structural studies and tentative evidence that they exist as proteoglycans see Percival and McDowell (1985). The core of *Pelvetia* fucan is considered to be a helicoidal chain of sulphated fucose units linked through $C-1\rightarrow C-2$, six of them coiling spontaneously without any tension. The sulphate groups point outward in a particularly reactive position while the oxygen atoms gather all around the axis making a sort of tube that is able to complex Ca^{2+} (Lestang and Quillet, 1981). Results of extensive studies (Kloareg *et al.*, 1986) on sulphated homofucans suggest that in aqueous solutions they exist as extended flexible coils.

After pulse labelling experiments with $^{14}CO_2$ on *Fucus vesiculosis*, a brown alga, three different fucans were separated (Bedwell *et al.*, 1972). These authors were able to postulate the order in which the different types of fucan were metabolised. Evans *et al.* (1973) and Evans and Callow (1974) carried out biosynthetic studies with $^{35}SO_4$ on *Pelvetia* and *Laminaria* fucans to determine their site of synthesis. Quatrano *et al.* (1978) investigated the fucans at different stages of growth of the zygotes of species of *Fucus*. No commercial applications for fucans have been reported.

B. Isolation

1.   Extraction

A certain proportion of the fucans present in the alga can readily be extracted by cold or hot water, preferably in the presence of Ca^{2+} ions to insolubilise any alginate that might otherwise be extracted. After dialysis, the extract (the fucan) can be isolated as a white powder by freeze-drying. To avoid contamination with brown phenolic substances, preliminary treatment of the weed with formalin is advised. This polymerises the phenolic compounds and renders them insoluble in water. Complete removal of fucan from the weed involves sequential extraction with aqueous calcium chloride, dilute hydrochloric acid, dilute aqueous sodium carbonate, ammonium oxalate, oxalic acid, and treatment of the residue with chlorite. Each of the extracts contains some fucan (Jabbar Mian and Percival, 1973a; Percival *et al.*, 1983).

2.   Purification and fractionation

This is often achieved with DE52 microgranular cellulose. The columns are eluted with water, followed by an increasing gradient of 0.1 M potassium chloride, which yields fractions with high fucose and sulphate and low glucuronic acid contents and at the other extreme with low sulphate and high glucuronic acid content (Jabbar Mian and Percival, 1973a). Fractionation on DEAE-Sephadex A-25 (Mori and Nisizawa, 1982; Mori *et al.*, 1982) and elution of the column with increasing concentrations of sodium chloride in 0.01 N-hydrochloric acid solutions gave rise to similar fractions to those reported above.

C. Analytical Techniques

1. Determination of fucose content

A specific spectrophotometric procedure for the determination of methylpentoses (6-deoxyhexoses) is described by Dische and Shettles (1948). This involves heating with concentrated sulphuric acid and the development of a specific colour with cysteine hydrochloride. This was modified by Gibbons (1955), but recovery of methyl pentose is low. Quantitative periodate oxidation of fucose and determination of the acetaldehyde released was developed by Nicolet and Shinn (1941). A modification by Cameron et al. (1948) involves the formation of a sodium bisulphite complex and titration with iodine. Later a procedure which enables 0.01–0.5 μmol of methyl pentose to be determined was developed. This absorbs the acetaldehyde with semicarbazide hydrochloride and the resulting semicarbazone absorbs strongly in the ultraviolet and can be measured spectrophotometrically (Bhattacharyya and Aminoff, 1966).

2. Other methods

For other analytical methods applied to fucans see those discussed in Sections II.A–E and VI.E and F.

VI. POLYSACCHARIDES OF GREEN SEAWEEDS

A. General

For more than twenty years one of us (E.P.) has been involved in studying the water-soluble polysaccharides of many genera of the Chlorophyceae or green seaweeds. In addition to starch-type materials, aqueous extracts contained polysaccharides comprising a number of different monosaccharides, some or all of which carried half ester sulphate groups. The results of this work can be found in *Carbohydrate Sulphates* by Elizabeth Percival (1978b). No commercial applications for these polysaccharides have emerged. In water they exist as viscous solutions and only gel on the addition of metallic ions. Space does not permit more than a mention of the methods used to elucidate their structure, but references giving experimental details will be included where appropriate.

B. Fractionation

All attempts to fractionate the complex heterosulphated polysaccharides into homo-polysaccharides proved unsuccessful (Hirst et al., 1965). Only by elution from a DEAE-cellulose column was it possible to separate similar fractions but of different molecular size (Percival and Wold, 1963; Bourne et al., 1974). A sulphated mannan was separated from the acid extract of *Codium pusillum* as the copper complex by the addition of Fehling's solution (Carlberg and Percival, 1977).

C. Desulphation

Reduction and alkali treatment of the *Ulva lactuca* polysaccharide, which comprises rhamnose, xylose, glucuronic acid and sulphate, reduced the last from 14.1% to 12.5% (Percival, 1980) and the recovered material contained 1.1% arabinose. Since this could only have been derived from the xylose units it was deduced that 15% of these units are sulphated.

The action of sodium methoxide on the alkali-treated polysaccharide led to the isolation of 2-*O*-methylxylose (Percival, 1980), proof that it is C-2 in xylose that is sulphated.

D. Partial Hydrolysis

Partial hydrolysis of *Ulva lactuca* polysaccharide, followed by the separation of the barium uronates either on cellulose or Amberlite G45 resin columns, led to the isolation of 4-*O*-β-D-glucuronosyl-L-rhamnose (McKinnell and Percival, 1962). Acidic and neutral fractions were separated from the hydrolysate of *Codium fragile* polysaccharide on Amberlite IR400 resin. 3-*O*-β-L-Arabinopyranosyl-L-arabinose and 3-*O*-β-D-galacto-pyranosyl-D-galactose were then separated from the neutral fraction on Whatman 3MM chromatography paper, while the acidic fraction yielded 4-*O*- and 6-*O*-mono-sulphates of glactose after fractionation first on a charcoal column and then on Whatman 3MM chromatography paper (Love and Percival, 1964). In a similar manner L-arabinose 3-sulphate, D-galactose 6-sulphate, 1,3- and 1,6-linked galactobioses and a 1,4- or 1,5-linked L-arabinobiose 3-sulphate were isolated from *Cladophora rupestris* polysaccharide (Hirst *et al.*, 1965).

E. Smith Degradation

The application of successive Smith degradations (periodate oxidation, followed by borohydride reduction and mild hydrolysis to cleave the acetal but not the glycosidic linkages) to the polysaccharide from *Cladophora rupestris* (Johnson and Percival, 1969) confirmed the highly branched nature of this material and the presence of 1,4-linked and end-group xylose. It also provided evidence of 1,6-linked and/or 6-sulphated galactofuranose units and of mutally linked arabinose and galactose residues.

F. Methylation

By a modified Hakomori methylation (Percival and Smestad, 1972) the polysaccharide and desulphated polysaccharide from the green seaweed *Acetabularia crenulata* were methylated and the derived methylated monosaccharides were characterised as their methylglycosides and alditol acetates by gas–liquid chromatography. From these and other results it was possible to deduce the linkages and site of sulphate of the majority of the galactose, rhamnose, xylose and glucuronic acid units present in this polysaccharide.

REFERENCES

Araki, C. (1966). *Proc. Int. Seaweed Symp., 5th, 1965*, pp. 3–17.

Atkins, E. D. T., Mackie, W., Parker, K. D. and Smolko, E. E. (1971). *Polymer Lett.* **9**, 311–316.
Bellion, C., Hamer, G. and Yaphe, W. (1981). *Proc. Int. Seaweed Symp., 10th, 1980*, pp. 379–384.
Bellion, C., Brigand, G., Prome, J. C., Welti, D., and Bociek, S. (1983). *Carbohydr. Res.* **119**, 31–48.
Bert, M., Ben Said, R., Deslandes, E. and Cosson, J. (1988). *Phytochemistry* **28**, 71–72. and refs cited therein.
Bhattacharjee, S. S., Yaphe, W. and Hamer, G. K. (1978). *Proc. Int. Seaweed Symp., 9th, 1977*, pp. 379–385.
Bhattacharyya, A. K. and Aminoff, D. (1966) *Anal. Biochem.* **14**, 278–289.
Bidwell, R. G. S., Percival, E. and Smestad, B. (1972). *Can. J. Bot.* **50**, 191–197.
Bitter, T. and Muir, H. M. (1962). *Anal. Biochem.* **4**, 330–334.
Bjorndal, H., and Lindberg, B. (1969). *Carbohydr. Res.* **10** 79–85.
Blumenkrantz, N. and Asboe-Hansen, G. (1973). *Anal. Biochem.* **54**, 484–489.
Bociek, S. M. and Welti, D. (1975). *Carbohydr. Res.* **42**, 2117–2126.
Bourne, E. J., Megarry, M. L. and Percival, E. (1974). *J. Carbohydr., Nucleosides Nucleotides* **1**, 235–264.
Boyd, J. and Turvey, J. R. (1977). *Carbohydr. Res.* **57**, 163–171.
Boyd, J. and Turvey, J. R. (1978). *Carbohydr. Res.* **66**, 1187–1194.
Bremond, G. L., Quillet, M. and Bremond, M. (1987). *Phytochemistry* **26**, 1705–1707.
Callow, M. E. and Evans, L. V. (1976). *Planta* **131**, 155–157.
Cameron, M. C., Ross, A. G. and Percival, E. G. V. (1948). *Chem. and Ind. (London)* **67**, 161–164.
Carlberg, G. E. and Percival, E. (1977). *Carbohydr. Res.* **57**, 223–234.
Caswell, R. C., Gacesa, P. and Weightman, A. J., (1986). *Int. J. Biol. Macromol.* **8**, 337–341.
Craigie, J. S. (1990). *In* "Biology of the Red Algae" (K. M. Coale and Bob Sheath, eds), in press. Cambridge University Press, Cambridge.
Craigie, J. S. and Leigh, C. (1978). *In* "Handbook of Physiological Methods" (J. A. Hellebust and J. S. Craigie, eds), pp. 109–131. Cambridge University Press, Cambridge.
Craigie, J. S., Morris, E. R., Rees, D. A. and Thom, D. (1984). *Carbohydr. Polymers* **4**, 237–252.
Dawes, C. J., Stanley, N. F. and Stancioff, D. J. (1977). *Bot. Mar.* **20**, 137–147.
Dische, Z. and Shettles, L. B. (1948). *J. Biol. Chem.* **175**, 595–603.
Dodgson, K. S. (1961) *Biochem. J.* **78**, 312–319.
Dodgson, K. S. and Price, R. G. (1962). *Biochem. J.* **84**, 106–110.
Doty, M. S. and Santos, G. A. (1978). *Aquat. Bot.* **4**, 143–149.
Dubois, M., Gilles, K. A., Hamilton, J. K., Rebers, P.A. and Smith, F. (1956). *Anal. Chem.* **28**, 350–356.
Duckworth, N. and Yaphe, W. (1970). *Chem. and Ind.*, pp. 747–748.
Duckworth, N. and Yaphe, W. (1971). *Carbohydr. Res.* **16**, 189–197.
Evans, L. V. and Callow, M. E. (1974). *Planta* **117**, 93–95.
Evans, L. V., Simpson, Margaret, and Callow, M. E. (1973). *Planta (Berl.)* **110**, 237–252.
Fischer, F. G. and Dorfel, Helmut (1955). *Physiol. Chem* **302**, 186–203.
Fujibayashi, S., Habe, H. and Nisizawa, K. (1970). *J. Biochem. (Tokyo)* **67**, 37–45.
Furneaux, R. H. and Miller, I. J. (1985). *Bot. Mar.* **28**, 419–425.
Gacesa, P., Squire, A. and Winterburn, P. J. (1983). *Carbohydr. Res.* **118**, 1–8.
Gibbons, M. N. (1955). *Analyst* **80**, 268–276.
Glicksman, Martin (1987). *Proc. Int. Seaweed Symp., 12th, 1986*, pp. 31–47.
Gordon Mills, E. M. and McCandless, E. L. (1975). *Phycologia* **14**, 275–281.
Grant, T., Morris, E. R., Rees, D. A., Smith, J. C. and Thom, D. (1973). *FEBS Lett.* **32**, 195–198.
Grasdalen, H. (1983). *Carbohydr. Res.* **118**, 255–260.
Grasdalen, H., Larsen, B. and Smidsrød, O. (1978). *Proc. Int. Seaweed Symp., 9th, 1977*, pp. 309–317.
Greene, A. C. and Madgwick, J. C., (1986). *Bot. Mar.* **29**, 329–334.
Greer, C. W. and Yaphe, W. (1984). *Proc. Int. Seaweed Symp., 11th, 1983*, pp. 563–567.
Greer, C. W., Shower, I., Goldstein, M. E. and Yaphe, W. (1984). *Carbohydr. Res.* **129**, 189–196.
Groleau, D. and Yaphe, W. (1977). *Can. J. Microbiol.* **23**, 672–679.
Guiseley, K. B. (1968). *In* "Kirk–Othmer Encyclopedia of Chemical Technology", Vol. 17, pp. 763–784. Interscience, New York.

Guiseley, K. B. (1970). *Carbohydr. Res.* **13**, 247–256.
Guiseley, K. B. (1987). *Progr. Biotechnol.* **3**, 139–147.
Hakomori, S. I. (1964). *J. Biochem. (Tokyo)* **55**, 205–210.
Hamer, H. (1970). *Acta Chem. Scand.* **24**, 1294–1300.
Haug, A. and Larsen, B. (1962). *Acta Chem. Scand.* **16**, 1908–1918.
Haug, A. and Larsen, B. (1971). *Carbohydr. Res.* **17**, 297–308.
Haug, A., Larsen, B. and Smidsrød, O. (1966). *Acta Chem. Scand.* **20**, 183–190.
Haug, A., Larsen, B. and Smidsrød, O. (1967). *Acta Chem. Scand.* **21**, 691–704.
Haug, A., Larsen, B. and Baardseth, E. (1969). *Proc. Int. Seaweed Symp., 6th, 1968*, pp. 443–451.
Helleur, R. J., Hayes, E. R., Jamieson, W. D. and Craigie, J. S. (1985a). *J. Anal. Appl. Pyrol.* **8**, 333–348.
Helleur, R. J., Hayes, E. R., Craigie, J. S. and McLachlan, J. L. (1985b). *J. Anal. Appl. Pyrol.* **8**, 349–357.
Hirase, S. and Watanabe, K. (1972). *Bull. Inst. Chem. Res. Kyoto Univ.* **50**, 332–336.
Hirst, E. L., Mackie, W. and Percival, E. (1965). *J. Chem. Soc.*, pp. 2958–2967.
Ishikawa, M. and Nisizawa, K. (1981). *Bull. Jpn. Soc. Sci. Fish.* **47**, 889–893; 895–899.
Jabbar Mian, A. and Percival, E. (1973a). *Carbohdyr. Res.* **26**, 133–146.
Jabbar Mian, A. and Percival, E. (1973b). *Carbohdyr. Res.* **26**, 147–161.
Johnson, P. G. and Percival, E. (1969). *J. Chem. Soc.*, pp. 906–909.
Jones, A. S. and Letham, D. S. (1954). *Chem. and Ind. (London)*, pp. 662–663.
Kloareg, B., Demaity, M. and Mabeau, S. (1986). *Inst. J. Biol. Macromol.* **8**, 380–386.
Knutsen, S. H. and Grasdalen, H. (1987). *Bot. Mar.* **30**, 497–505.
Knutson, C. A. and Jeanes, A. (1968). *Anal. Biochem.* **24**, 470–481; 482–490.
Larsen, B., Vreeland, V. and Laetsch, W. M. (1985). *Carbohydr. Res.* **143**, 221–227.
Lestang, G. de and Quillet, M. (1981). *Proc. Int. Seaweed Symp., 8th, 1974*, pp. 200–204.
Love, J. and Percival, E. (1964). *J. Chem. Soc.*, pp. 3338–3345.
McCandless, E. L., Craigie, J. S. and Walter, J. A., (1973). *Planta* **112**, 201–212.
McKinnell, J. P. and Percival, E. (1962). *J. Chem. Soc.*, pp. 2082–2083.
McLean, M. W. and Williamson, F. B. (1981). *Proc. Int. Seaweed Symp., 10th, 1980*, pp. 479–484.
Madgwick, J., Haug, A. and Larsen, B. (1973a). *Acta Chem. Scand.* **27**, 711–712.
Madgwick, J., Haug, A. and Larsen, B. (1973b). *Acta Chem. Scand.* **27**, 3592–3594.
Madgwick, J., Haug, A. and Larsen, B. (1978). *Bot. Mar.* **21**, 1–3.
Mori, H. and Nisizawa, K. (1982). *Bull. Jpn. Soc. Sci. Fish.* **48**, 981–986.
Mori, H., Kamei, H., Nishide, E. and Nisizawa, K. (1982). *In* "Marine Algae in Pharmaceutical Science", (H. A. Hoppe and T. Levring, eds,) Vol. 2, pp. 109–121. Walter de Gruyter, Berlin and New York.
Morrice, L. M., McLean, M. W., Long, W. E. and Williamson, F. B. (1984). *Proc. Int. Seaweed Symp., 11th, 1983*, pp. 576–579.
Morris, E. R., Rees, D. A. and Thom, D. (1973). *Chem. Commun.*, pp. 245–246.
Morris, E. R., Rees, D. A., Thom, D. and Boyd, J. (1978). *Carbohydr. Res.* **66**, 145–154.
Morris, E. R., Rees, D. A. and Thom, D. (1980). *Carbohydr. Res.* **81**, 305–314.
Morris, E. R., Cutler, A. N., Ross-Murphy, S. B., Rees, D. A. and Price, J. (1981). *Carbohydr. Polymers* **1**, 5–21.
Nicolet, B. H. and Shinn, L. A. (1941). *J. Am. Chem. Soc.* **63**, 1456–1458.
Painter, T. J. (1983). *In* "The Polysaccharides" (G. O. Aspinall, ed.), pp. 195–285. Academic Press, New York.
Penman A. and Sanderson, G. R. (1972). *Carbohydr. Res.* **25**, 273–282.
Percival, E. (1978a). *ACS Symp. Ser.* **77**, 213–224.
Percival, E. (1978b), *ACS Symp. Ser.* **77**, 203–212.
Percival, E. (1980). *Methods Carbohydr. Chem.* **8**, 281–285.
Percival, E. and McDowell, R. H. (1967). *In* "Chemistry and Enzymology of Marine Algal Polysaccharides". Academic Press, London, New York.
Percival, E. and McDowell, R. H. (1981). *In* "Encyclopedia of Plant Physiology: New Series" (W. Tanner and F. A. Loewus, eds), Vol. 13B, pp. 277–316. Springer, Berlin and Heidelberg.
Percival, E. and McDowell, R. H. (1985). *In* "Biochemistry of Storage Carbohydrates in Green Plants" (P. M. Dey and R. A. Dixon, eds), pp. 305–348. Academic Press, London.

Percival, E. and Smestad, B. (1972). *Carbohydr. Res.* **25**, 299–312.
Percival, E. and Wold, J. K. (1963). *J. Chem. Soc.*, pp. 5459–5468.
Percival, E., Venegas Jara, M. and Weigel, H. (1983). *Phytochemistry* **22**, 1429–1432.
Quatrano, R. S., Hogsett, W. S. and Roberts, M. (1978). *Proc. Int. Seaweed Symp., 9th, 1977*, pp. 113–123.
Quillet, M. and Lestang Bremond, G. (1985). *Phytochemistry* **24**, 43–45.
Renn, D. W. (1984). *Industrial and Engineering Chemistry Product Research and Development* **23**, 17–21.
Rochas, C., Lahaye, M. and Yaphe, W. (1986). *Bot. Mar.* **29**. 335–340.
Sandford, P. A. (1985). *In* "Biotechnology of Marine Polysaccharides" (R. R. Colwell, E. R. Pariser and R. J. Sinskey, eds), pp. 454–516. Hemisphere, Washington.
Scott, J. E. (1960). *In* "Methods of Biochemical Analysis" (D. Glick, ed.), Vol. 8. p. 163. Interscience, New York.
Shiraiwa, Y., Abe, K., Sasaki, S. F., Ikawa, T. and Nisizawa K. (1975). *Bot. Mar.* **18**, 97–104.
Skjåk-Bræk, G., Smidsrød, O. and Larsen, B. (1986). *Int. J. Biol. Macromol.* **8**, 330–336.
Smidsrød, O. and Haug, A. (1968). *Acta Chem. Scand.* **22**, 797–810.
Smidsrød, O. and Haug, A. (1972). *Acta Chem. Scand.* **26**, 2063–2074.
Smith, D. B. and Cook, W. H. (1953). *Arch. Biochem. Biophys.* **45**, 232–233.
South, G. R. (1978). *Proc. Int. Seaweed Symp., 9th, 1977*, pp. 133–142.
Stancioff, D. J. and Stanley, N. F. (1969). *Proc. Int. Seaweed Symp., 6th, 1968*, pp. 595–609.
Stockton, B., Evans, L. V., Morris, E. R. and Rees, D. A. (1980a). *Int. J. Biol. Macromol.* **2**, 176–178.
Stockton, B., Evans, L. V., Morris, E. R., Powell, D. A. and Rees, D. A. (1980b). *Bot. Mar.* **23**, 563–567.
Taylor, R. L. and Conrad, H. E. (1972). *Biochemistry* **11**, 1383–1388.
Vreeland, V. (1972). *J. Histochem. Cytochem.* **20**, 358–367.
Vreeland, V., Slomich, M. and Laetsch, W. M. (1984). *Planta (Berl.)* **162**, 506–517.
Vreeland, V., Zablackis, E., Deboszewski, B. and Laetsch, W. M. (1987). *Proc. Int. Seaweed Symp., 12th, 1986*, pp. 155–160.
Weigl, J. and Yaphe, W. (1966). *Can. J. Microbiol.* **12**, 874–876; 939–947.
Whyte, J. M. C. and Englar, J. R. (1974). *Anal. Biochem.* **59**, 426–435.
Yaphe, W. and Arsenault, G. (1965). *Anal. Biochem.* **13**, 143–148.

16 Isolation and Analysis of Plant Cell Walls

ROBERT R. SELVENDRAN and PETER RYDEN

AFRC Institute of Food Research, Colney Lane, Norwich NR4 7UH, UK

I. INTRODUCTION

The plant cell wall is the principal structural component of the cell surrounding the protoplast and exterior to the plasmalemma of the cell. Each cell wall interacts with its

METHODS IN PLANT BIOCHEMISTRY Vol. 2
ISBN 0-12-461012-9

neighbours, binding the cells to form various tissue types. The walls of growing tissues are hydrated and are constantly changing in response to the growth, stage of differentiation, environment of the cells and their activities. In order to perform the various functions such as growth, morphogenesis, disease-resistance, recognition and digestibility (e.g. cell separation during ripening and abscission) effectively, cell walls with varied compositions and structures are formed and the walls contain a number of specific enzymes. Further, cells in some regions of the plant, for example vascular and supporting tissues, become differentiated into specialised structures such as xylem and phloem bundles, and sclerenchyma. The xylem cells and sclerenchyma become thickened and hardened by lignification as the plant organ matures, and the water and associated pectic substances in the walls are largely replaced by the hydrophilic filler lignin. The lignified cells are usually dead and their walls have mainly cellulose microfibrils dispersed in hemicelluloses and the whole complex is encrusted with lignin. Therefore, different types of matrix polysaccharides are deposited to cope with the different aqueous environments within the walls. The biochemical processes of differentiation that transform primary walls into secondary walls have been studied by various workers and the relevant information has been summarised by Northcote (1963, 1969). For an account of the chemistry of cell-wall polymers see Aspinall (1980), Darvill *et al.* (1980a), Dey and Brinson (1984), Selvendran *et al.* (1987), Bacic *et al.* (1988) and Fry (1988). The main components of cell walls from different tissue types of a range of plants and the structural features of some of the cell wall polymers are summarised in Tables 1 and 2 of Selvendran and O'Neill (1987) and Table 1 of Selvendran (1985).

It is clear that the cell walls of a plant organ are complex composites of polymeric material bathed in various amounts of water. Unravelling the structures of these walls has proved a formidable task in the past. However, in recent years, the following factors have helped to elucidate the structures of the walls and their constituent polymers.

1. The development of improved methods for preparing gram quantities of relatively pure cell walls from various tissue types.
2. Better techniques for resolving cell-wall polymers by ion-exchange, cellulose and affinity chromatography, gel filtration, electrophoresis and ultracentrifugation. Various precipitation methods to effect partial fractionation of the polymers before further separation by chromatographic methods have proved useful in several instances, and have given better recoveries from columns.
3. The availability of several cell-wall degrading enzymes commercially; the enzymes have to be further purified before use.
4. The development of various techniques for identifying carbohydrate derivatives for improved methods of methylation analysis, coupled with gas–liquid chromatography–mass spectrometry (GLC-MS). Characterisation of derivatives of cell-wall fragments, released by a range of methods, by GLC-MS and HPLC-MS, and proton and carbon-13 NMR spectroscopy has made sequencing of complex polysaccharides possible.
5. Location of polymers within the wall by immunocytochemistry.

As the developments in analysing cell walls have recently been reviewed in some depth (Selvendran and O'Neill, 1987; Fry, 1988), the main aim of this chapter will be to focus on practical problems and pitfalls to be avoided, and aspects which will help researchers to select or devise a method for analysing cell walls mainly from immature, parenchymatous or cultured tissue of dicotyledons.

II. ISOLATION AND FRACTIONATION OF CELL-WALL MATERIAL: GENERAL DISCUSSION OF EXPERIMENTAL PROBLEMS AND WAYS OF OVERCOMING THEM

The purity of the cell-wall material (CWM) is dependent on the plant organ (or tissue) and on the overall objectives of the investigation. It is more difficult to prepare relatively pure CWM from free-growing and parenchymatous tissues compared with lignified tissues, which are usually dead and contain very small amounts of intracellular compounds. The problems with parenchymatous tissues depend on the amount of intracellular compounds present, such as polyphenols, pigments, lipids, nucleic acids, proteins and reserve polysaccharides (e.g. starch) and also on the activity of cell-wall degrading enzymes (e.g. pectin methylesterases and polygalacturonases in certain ripening fruits). Every effort must be made to remove the aforementioned intracellular compounds and also to minimise or eliminate the activity of cell-wall (degrading) enzymes.

A. Isolation of Cell-wall Material

In most of the early work on cell-wall analysis, the alcohol-insoluble residue (AIR) was used as the starting material (Jermyn, 1955). AIR is prepared by immersing the tissue in hot 90% alcohol for 5 min and blending the mixture to disrupt the tissues and solubilise most of the low molecular-weight compounds, and washing the residue with absolute alcohol and ether or acetone; acetone is a better solvent for chlorophyll. As the intracellular proteins, nucleic acids, starch and some polyphenols would co-precipitate with AIR, such residues are only suitable for tissues relatively poor in the above compounds, such as grasses, certain fruits (e.g. tomatoes), vegetables (e.g. immature cabbage leaves) and most suspension-cultured tissues. If required, the co-precipitated proteins and starch can be removed from the AIR by digestion with pronase (Stevens and Selvendran, 1980) and pancreatic or salivary α-amylase (Selvendran and DuPont, 1980), and the small amounts of pectic polysaccharides solubilised by the buffers can be isolated from the dialysed extracts.

AIR has the advantage that most of the cell-wall enzymes would have been inactivated and it is unlikely that the hot alcohol treatment would degrade any of the cell-wall polysaccharides. However, dehydration with alcohol is known to influence the solubility characteristics of the polysaccharides (particularly the pectins) and cause the formation of artifacts. In the case of starch-rich products, such as potato, the bulk (>80%) of the co-precipitated starch can be removed from the AIR by stirring overnight at 20°C in 90% (v/v) aqueous dimethyl sulphoxide (DMSO). To effect complete removal of starch, the AIR should be ball-milled in 80% alcohol to ensure disruption of all starch containing cells, prior to aqueous DMSO extraction. In the case of protein-rich products (e.g. soybean cotyledons), the bulk of the co-precipitated proteins can be extracted from the AIR with phenol–acetic acid–water (2:1:1; w/v/v) (PAW) (Selvendran, 1975a; Selvendran and O'Neill, 1987). PAW is an excellent solvent for intracellular proteins and is a poor solvent for polysaccharides.

An alternative method of preparing CWM involves homogenising the tissue in dilute aqueous detergents (or buffers) (e.g. 1.5% SDS containing sodium metabisulphite) and then ensuring that all the cells are ruptured; the cytoplasmic contents are then washed out with PAW and 90% aqueous DMSO. The steps involved in the isolation of CWM from potato tubers are shown in Scheme 16.1. The advantages of this method are that 1.5% SDS is a good solvent for intracellular proteins and the metabisulphite minimises the oxidation and subsequent co-precipitation of condensed polyphenols with the proteins. Complete disruption of tissue structure is achieved by wet ball-milling and the wall material in the slurry settles readily on centrifugation and can be easily washed by centrifugation. The first extraction with SDS solubilises mainly intracellular compounds and only very small amounts of cell wall polymers; 5–10% of cell-wall polymers are solubilised when the cells are teased apart during wet ball-milling (Step 4, Scheme 16.1). This method has proved useful for isolating CWM from parenchymatous tissues of runner beans, potatoes, apples, immature cabbage leaves and onions.

SCHEME 16.1. Isolation and purification of CWM from potato tubers. The following steps are carried out in the sequence shown below:

---

1. Slice tubers transversely (4 mm thick). Place slices in 5 mM sodium metabisulphite ($Na_2S_2O_5$). Dissect tissue required—pith, cortex or parenchyma—blot dry, freeze in liquid nitrogen and weigh.
2. Put (50 g) frozen tissue in 100 ml 1.5% sodium dodecylsulphate (SDS) + 5 mM $Na_2S_2O_5$. Liquidise in a Waring blender for 2 min, then with an ultraturrax for 3 min.
3. Filter through nylon cloth and wash with two bed volumes of 0.5% SDS + 3 mM $Na_2S_2O_5$. Squeeze out fluid.
4. Resuspend residue in 100 ml 0.5% SDS + 3 mM $Na_2S_2O_5$ and ball mill (Pascall 500 ml pot) 60 rev min^{-1} at 2°C for 15 h.
5. Filter through a coarse nylon sieve and wash balls with water.
6. Centrifuge filtrate at 2°C at 26 000 × g for 15 min.
7. Wash the pellet twice with 3 bed volumes of water by centrifugation and weigh wet pellet.
8. Blend pellet in 150 ml aqueous dimethylsulphoxide (DMSO), final concentration 90%, with an ultraturrax for 1 min, sonicate for 10 min and stir at 20°C for 16 h.
9. Centrifuge at 15°C at 26 000 × g for 15 min and decant supernatant.
10. Blend pellet in 75 ml 90% DMSO, sonicate for 30 min at 20°C and stir for 1 h.
11. Centrifuge. Wash pellet with water six times by centrifugation. Check for the absence of starch with I_2/KI.
12. Suspend pellet in water and dialyse against water for 1 day.
13. Freeze dry an aliquot of suspension to measure cell-wall yield.
14. Store wet pellet at −20°C.

The disadvantages of the method are: (1) the CW enzymes are not fully removed, nor fully inactivated, and can therefore cause wall autolysis if suitable precautions are not taken. Extraction with PAW appears to inactivate and solubilise a proportion of the wall-bound enzymes. For tissues containing pectin-degrading enzymes (e.g. ripe tomatoes), it is better to prepare the AIR and sequentially extract it with PAW and 90% aqueous DMSO. (2) A small proportion (5–8%) of the wall polymers, mainly of intercellular origin, is solubilised during wet ball-milling. The loss of wall polysaccharides from cereal endosperm walls is much higher, but the loss from beeswing wheat bran (DuPont and Selvendran, 1987) and commercial wheat bran is <3% of the dry weight of the CWM (Selvendran et al., 1980). These aspects have been discussed in greater detail in an earlier review (Selvendran and O'Neill, 1987).

Vigorous homogenisation in glycerol at 20% has proved to be a useful technique for some tissues, because glycerol does not precipitate intracellular proteins onto the walls and does not solubilise a significant amount of water-soluble polymers (Kivilaan et al., 1959; Huwyler et al., 1979).

1. CWM preparation for recovery of CW enzymes

Cell-wall enzymes can be obtained by blending and ball-milling the tissue at 1°C in 0.1 M phosphate buffer (pH 7) instead of SDS. Apple peroxidase fractions were extracted with 1 M NaCl before and during ball-milling (A. Whitcombe and R. R. Selvendran, unpubl. res.). Nagahashi and Seibles (1986) prepared large fragments of clean cell walls for cell-wall enzyme studies by blending in 0.3 M sucrose, 0.1 M HEPES-MES (pH 7.8), 5 mM mercaptoethanol and 2 mM $Na_2S_2O_5$; holding in a Parr nitrogen pressure bomb at 10 340 kPa for 10 min; then extruding to obtain almost complete cell disruption; and finally sonicating and washing thoroughly. The method of blending was adapted to suit the tissue to obtain small clumps of cells.

B. Solubilisation and Isolation of Wall Components

1. Pectic substances

In the past pectic substances were obtained from CWM of parenchymatous tissues by extraction with hot aqueous solutions of chelating agents, e.g. ammonium oxalate, ethylene diaminetetra-acetate (EDTA) or hexametaphosphate. As these extraction conditions cause significant β-eliminative degradation of pectins (Albersheim et al., 1960; Barrett and Northcote, 1965; Selvendran, 1985), extraction of pectins with 50 mM 1,2-cyclohexane diaminetetra-acetate (CDTA) at room temperature (20°C) for 6–8 h is preferred (Jarvis et al., 1981; Jarvis, 1982; Redgwell and Selvendran, 1986). CDTA at room temperature completely abstracts Ca^{2+} from the walls and most of the pectic polysaccharides held in the walls by ionic cross-links are gradually solubilised. The amount extracted varies greatly between tissue types (Jarvis, 1982). In our hands two extractions with CDTA solubilised ~40% of the total pectic material from CWM of onions (Redgwell and Selvendran, 1986), potatoes and runner beans (P. Ryden and R. R. Selvendran, unpubl. res.). Treatment with CDTA at 20°C is unlikely to cause side effects, except that some wall enzymes may cause partial autolysis; this can be minimised by incorporating a PAW extraction during isolation of walls. Scheme 16.2 shows the steps in the sequential extraction of CWM from soft tissues (non-lignified).

The CDTA-insoluble pectic polysaccharides are usually highly branched and probably ester cross-linked within the wall matrix. A significant proportion of these pectic polysaccharides can be solubilised by extraction of the residue with dilute Na_2CO_3 (0.05 M) at 1°C for 18 h. Initial extraction with Na_2CO_3 in the cold is preferred because it causes preferential hydrolysis of pectin methyl esters and ester cross-links, and this minimises β-eliminative degradation of the pectic polysaccharides. Additional pectic material can usually be extracted from the residue by sequential extraction with 0.05 M Na_2CO_3 at 20°C for 3 h and 0.5 M NaOH at 20°C for 2 h. Further extraction with stronger alkali (1 and 4 M KOH) solubilises small amounts of additional pectic material along with the hemicelluloses (e.g xyloglucans and proteoglycans). The final α-cellulose residue contains variable amounts of highly branched, cross-linked pectic poly-saccharides and hydroxyproline-rich wall glycoproteins associated with cellulose. The α-cellulose residue from CWM of potatoes contained a significant amount of pectic polysaccharides rich in galactose, whereas the corresponding residue from CWM of runner beans contained the bulk of the hydroxyproline-rich wall glycoproteins (extensin).

SCHEME 16.2. Extraction of CWM from soft tissues. The following steps are carried out in the sequence shown below:

---

1. Suspend wet cell wall material (1 g) in 100 ml CDTA (pH 6.5), final concentration 50 mM. Stir at 20°C for 6 h. Centrifuge and wash with water.
2. Re-extract with 100 ml 50 mM CDTA (pH 6.5) at 20°C for 2 h. Centrifuge and wash.
3. Cool pellet to 1°C, extract with 100 ml 50 mM Na_2CO_3 + 20 mM $NaBH_4$ at 1°C for 16 h. Centrifuge and wash with water.
4. Re-extract with 100 ml 50 mM Na_2CO_3 + 20 mM $NaBH_4$ at 20°C for 3 h. Centrifuge and wash.
5. Filter each peptic extract with its respective washings through glass fibre paper (GF/C), adjust carbonate extracts to pH 5 with HOAc prior to dialysis. Complete removal of CDTA requires exhaustive dialysis.
6. Stir depectinated residue under argon or nitrogen with 100 ml O_2-free 0.5 M KOH + 10 mM $NaBH_4$ at 1°C for 2 h. Filter under suction at 1°C on a coarse sinter.
7. Re-extract residue as above successively with 100 ml 1 M KOH at 1°C for 2 h, 1 M KOH at 20°C, 4 M KOH at 20°C and 4 M KOH + 3% Boric acid at 20°C.
8. Adjust hemicellulose extracts to pH 5 with HOAc (separate precipitate by centrifugation). Dialyse.
9. Adjust α-cellulose residue to pH 6 with HOAc and store at −20°C.

In our experience extraction of the CWM of potatoes with hot water and/or hot oxalate solubilised, in addition to the middle lamellae pectins, significant amounts of the highly branched pectic polysaccharides (rich in galactose) associated with the primary cell walls, particularly the α-cellulose residue. This point should be borne in mind when interpreting results on solubility of pectic polysaccharides by the earlier methods.

2. Material solubilised during delignification

To solubilise the hemicellulosic polysaccharides, delignification of the depectinated CWM is essential for organs containing significant amounts of lignified tissues. In the case of CWM from heavily lignified tissues (e.g. parchment layers of mature runner bean pods), depectination with CDTA may be omitted as only very small amounts of wall polymers are solubilised. Delignification is usually carried out by treatment with sodium chlorite–acetic acid at 70°C for 1–4 h (Jermyn, 1955; Jermyn and Isherwood, 1956). CWM of heavily lignified tissues, such as the parchment layers of mature runner bean pods, require the 4 h treatment, whereas grass cell walls appear to require a shorter treatment (1 h) (Morrison, 1974).

During delignification a small proportion (5–10%) of carbohydrate-containing material is solubilised (Buchala *et al.*, 1972; Selvendran *et al.*, 1975). In the case of CWM of parchment layers of mature runner bean pods some of the acidic xylans and pectic polysaccharides are solubilised. With CWM from mature runner bean pods, the treatment solubilised the bulk of the hydroxyproline-rich glycoproteins from the walls of the soft tissues (Selvendran et al., 1975; Selvendran, 1975b). Following this work, hydroxyproline-rich glycoproteins have been isolated from cultured tissue of tobacco (Mort and Lamport, 1977). Using a shorter treatment (2 × 15 min) with the reagent, we have isolated partially modified wall glycoproteins from parenchymatous tissues of runner bean pods, fractionated the polymers on DEAE-Sephadex, and have elucidated some of their structural features (O'Neill and Selvendran, 1980).

3. Hemicellulosic polymers

The bulk of the hemicellulosic polymers can be extracted from the holocellulose by aqueous solutions of alkali containing 20 mM $NaBH_4$ at 20°C; the latter reduces the latent aldehydic ends of the polysaccharides. The extraction of certain hemicelluloses (e.g. glucomannans) is enhanced by the presence of borate (3–4% boric acid) in the 4 M KOH. The borate presumably forms a complex with the 2,3-cis-hydroxyl groups of D-mannose residues, thus rendering the polymers more acidic and hence more soluble in alkali. A proportion of the glucomannans can be extracted from the depectinated cell walls using chaotropic agents such as guanidinium thiocyanate (Monro et al., 1976; Stevens and Selvendran, 1984a). In the case of the holocellulose from parchment layers of mature runner bean pods, the bulk (>90%) of the acidic xylans can be solubilised by 1 M and 4 M KOH. This suggests that the acidic xylans are ester cross-linked to (degraded) lignin, presumably via the glucuronic acid and 4-O-Me-glucuronic acid residues. Firm evidence for such ester cross-links in the native cell walls has only been obtained in very few instances (Das et al., 1981) and warrants further study. Our work (Stevens and Selvendran, 1988) on the degradability of CWM from wheat bran before and after treatment by faecal bacteria in vitro, clearly suggests ester cross-links between glucurono-arabinoxylans and phenolic hydroxyls of lignin.

The bulk of the hemicellulosic polymers (proteoglycans and xyloglucans) can be extracted from the depectinated CWM of parenchymatous tissues by extraction with 1 M and 4 M KOH; treatment with warm chlorite/HOAc is not necessary for CWM of soft tissues. This sequence of extractions has helped us to isolate proteoglycans and xyloglucans from a range of soft tissues (Stevens and Selvendran, 1984a,b; O'Neill and Selvendran, 1985a; Redgwell and Selvendran, 1986).

4. Cellulose and matrix polysaccharides

Cellulose and matrix polysaccharides can be dissolved by cyclic tertiary amine oxides (Chanzy et al., 1982). These compounds form strong H-bonds between their N—O groups and the primary or secondary OH groups of polysaccharides. N-methylmorpholine N-oxide (MMNO) containing a small amount of water (4–8%; w/w) at 100–150°C is a good solvent for most plant cell-wall polysaccharides. The maximum amount of cell-wall polysaccharides that can be dissolved in MMNO–H_2O ranges from 30 to 45% (w/w). For good dissolution of cellulose the water content should be kept as low as possible, and this could be obtained by mixing MMNO monohydrate and anhydrous MMNO. The mixtures have melting temperatures which range from 76°C (m.p. of

MMNO monohydrate) to 172°C (m.p. of anhydrous MMNO). MMNO–H$_2$O mixtures are alkaline and would hydrolyse ester-linked substituents and also cause significant β-eliminative degradation of acidic polysaccharides (e.g. pectins) at the elevated temperatures used for the dissolution of the polysaccharides. MMNO is potentially a useful solvent for most cell-wall polysaccharides but requires careful study.

5. Glycoproteins

Plant cell walls contain a range of glycoproteins and proteoglycans, and some of these are cross-linked within the wall matrix and require degradative conditions such as treatment with chlorite/HOAc, alkali or wall-degrading enzymes to solubilise them. Arabinogalactan-proteins and some enzymes are water- or buffer-extractable and the bulk of these are solubilised when the homogenised tissue (squeezed dry through muslin cloth to remove intracellular compounds) is wet ball-milled (see Scheme 16.1). A large proportion of the enzymes, lectins and newly deposited extensin associated with the purified walls by ionic bonds can be extracted with salt (Nari et al., 1983; Hatfield and Nevins, 1987); the purification of the walls should not include treatment with PAW, because PAW solubilises a proportion of the bound glycoproteins. Some of the inextractable enzymes can be solubilised by treating the salt-washed cell walls with polysaccharide-degrading enzymes (Stephens and Wood, 1974; Mäder and Schloss, 1979).

The bulk of the hydroxyproline-rich glycoproteins (extensin) are cross-linked within the wall matrix and can be solubilised by treatment with chlorite/HOAc at 70°C for 30 min, which presumably acts by breaking isodityrosine cross-links between extensin molecules (Fry, 1982a). It is also possible that the treatment breaks phenolic cross-links between extensin and other wall polymers. As mentioned before, a large proportion of proteoglycan–polyphenol complexes can be solubilised from the depectinated walls with alkali.

6. Solubilisation of wall polymers by enzymic methods

As an alternative to sequential extraction of walls with aqueous inorganic solvents, enzymatic degradation of the walls can be used to release (degraded) polymers (Talmadge et al., 1973; McNeil et al., 1980; DeVries et al., 1982, 1983; Spellman et al., 1983; Fry, 1988); Selvendran and O'Neill (1987) have given a brief description of the various enzymatic methods used. Hydrolysis with enzymes can be very specific and is carried out under mild conditions. The main problem is the lack of availability of pure enzymes commercially; the enzymes have to be further purified before use. This type of methodology requires highly purified cell-wall degrading enzymes such as pectin methyl esterases, endopolygalacturonases and endoglucanases. Albersheim and co-workers have used purified enzymes in their cell wall studies, and as the methods used by them have been outlined in Chapter 12, we shall not discuss their methodology.

Fry (1982b, 1983, 1988) has used 'Driselase', which contains a mixture of polysaccharide-hydrolases and lacks esterase activity; see Fry (1988) for the range of enzymes present in Driselase. Driselase cleaves wall polysaccharides to give monosaccharides (and sometimes a major disaccharide) but if the sugar residue linked to the backbone is substituted with a phenolic acid (e.g. ferulic acid) then its glycosidic linkage is protected.

For an extended discussion of the details and applications of the method see Fry (1988). Albersheim and co-workers have used cellulase digestion to elucidate the structural features of sycamore xyloglucan (Bauer *et al.*, 1973) and have determined the structure of the nonasaccharide that is resistant to cellulase (Valent *et al.*, 1980). We have used cellulase digestion to determine the fine structure of runner bean xyloglucan (O'Neill and Selvendran, 1985b; Selvendran and O'Neill, 1987). For a detailed discussion of the use of enzymes for structural characterisation of polysaccharides, see Matheson and McCleary (1984).

C. Fractionation of Cell-wall Polymers

The cell-wall polymers present in the neutralised, dialysed and concentrated extracts have to be fractionated into more or less homogeneous individual polymers. In the literature several procedures are described for fractionating neutral and acidic poly-saccharides. The methods used include graded precipitation with alcohol, formation of iodine and copper complexes, fractionation using $Ba(OH)_2$ and quaternary ammonium salts, and chromatography on anion-exchange, cellulose and affinity columns. As most of these methods have been discussed briefly in a recent review (Selvendran and O'Neill, 1987), we shall focus on the relative merits of a few methods which we have found useful, and use routinely for fractionating cell wall polysaccharides.

1. Pectic polysaccharides

We have tested a range of anion-exchange resins for fractionating pectic polysacchar-ides and have found that DEAE-Trisacryl gives the best recovery of CDTA and Na_2CO_3-soluble pectic polysaccharides of onions (Redgwell and Selvendran, 1986), runner beans and potatoes (P. Ryden and R. R. Selvendran, unpubl. res.). The recovery of pectic polysaccharides from the columns ranged from 80–90% for both onion and potato. The recovery of some of the pectic polysaccharides of apple and sugar beet was poor (60%). The poor recoveries may be due to interaction between the polymers themselves, caused by the close proximity on the columns. Such interactions may result in the precipitation on the pectins of the column.

2. Hemicellulosic polymers

The proteoglycans of runner beans were fractionated sequentially on DEAE-Sephadex and hydroxyapatite. The polymers that were bound to hydroxyapatite and eluted with phosphate buffer were of relatively low carbohydrate content and had significant amounts of proteins associated with them. Churms and Stephen (1984) have fraction-ated arabinogalactan-proteins on hydroxyapatite. Proteoglycan complexes have also been obtained by fractionating the hemicellulosic polymers on DEAE-Sephacel (acetate form) (Stevens and Selvendran, 1984c).

The bulk of the alkali-soluble xyloglucans can be isolated from the 'neutral' fractions from DEAE-Sephadex or DEAE-Sephacel columns. The small amount of contaminating 'neutral' pectic polysaccharides (e.g. arabinans) can be separated from xyloglucans by further chromatography on a cellulose column (O'Neill and Selvendran, 1983). The xyloglucan is bound to the column and requires alkaline conditions for its elution from the column, whereas the 'neutral' pectic polysaccharides are not bound and can be washed from the column with water or dilute buffers. The xyloglucans can be further fractionated as borate complexes on DEAE-Sephacel or DEAE-Sepharose (Ruperez *et al.*, 1985). For example, the xyloglucans from apple parenchyma were resolved into two major and five minor components on DEAE-Sephacel(borate).

In order to obtain good separation and recoveries of acidic arabinoxylans and acidic xylans from anion-exchange columns, it is essential to effect partial fractionation of the polymers by graded precipitation with alcohol (10–90%; v/v), before anion-exchange chromatography. For details of the application of this procedure for fractionating a range of hemicellulosic polymers from beeswing wheat bran see DuPont and Selvendran (1987). Using similar procedures we have isolated a range of acidic xylans and some pectic polysaccharides associated with degraded lignin from parchment layers of mature runner bean pods (R. R. Selvendran and S. E. King, unpubl. res.).

III. CHEMICAL CHARACTERISATION OF THE POLYMERS

A. Monosaccharide Composition

1. Aqueous acid hydrolysis

The polysaccharides and glycoproteins of plant cell walls contain a range of sugars linked by a variety of glycosidic linkages (Aspinall, 1980; Darvill *et al.*, 1980a; Dey and Brinson, 1984) that exhibit variable acid lability. No acid hydrolysis procedure will cleave every linkage and give quantitative yield of each component. Further, the monosaccharides released vary in their stability to acid and high temperature—for example deoxysugars, such as rhamnose, degrade in acid, as do acidic sugars and ketoses. The following statements generally apply: furanosidic and deoxypyranosidic linkages cleave more readily than pyranosidic linkages; pentapyranosidic bonds cleave more readily than hexapyranosidic bonds; α-glycosidic linkages are less acid-resistant than their β-counterparts and among pyranosidic linkages 1→4 is more labile than 1→3 which is more labile than 1→2. For further comments and relevant references see Aspinall (1973).

Uronic acids linked to neutral sugars yield acid-resistant aldobiouronic acids. Examples are Gal*p*A-(1→2)-Rha*p* from pectins and Glc*p*A-(1→2)-Xyl*p* from acidic xylans and arabinoxylans. Thus rhamnose residues in pectins and xylose residues in acidic xylans will therefore be underestimated. Insoluble polysaccharides such as cellulose and long-chain mannans (and acidic xylans) require pre-treatment with 72% H_2SO_4 before they can be completely hydrolysed by 1 M acid. Pectins and pectic acids tend to precipitate in acid which could result in incomplete hydrolysis of associated neutral sugars. Hydrolysis can be facilitated by prior treatment with polygalacturonase.

Polysaccharides are usually hydrolysed by refluxing in dilute sulphuric acid for 2–3 h, with or without preliminary dispersion in 72% (w/w) H_2SO_4. Trifluoroacetic acid (TFA, 2 M) and 0.5 M and 1 M H_2SO_4 hydrolyse the bulk of the non-cellulosic polysaccharides (NCP) and 5–10% of the cellulose in the walls. The cellulose can be quantitatively hydrolysed after first dispersing the cell walls in 72% H_2SO_4 for 2–3 h at

20°C (Saeman *et al.*, 1963; Selvendran *et al.*, 1979). Losses of xylose have been reported following 3 h dispersion and 2 h has been preferred by Rasper *et al.* (1981). For a discussion of the merits of acid hydrolysis condition in dietary fibre analysis see Selvendran and Dupont (1984) and Selvendran *et al.* (1988).

To quantify both the neutral sugar and uronic acid residues after aqueous acid hydrolysis, TFA is preferred. Walters and Hedges (1988) have made a thorough study of sugar stability and rates of hydrolysis in 0.3 to 2.5 M TFA at temperatures from 120°C to 140°C. The most reproducible results for uronic acid-containing polysaccharides were obtained with 0.5 M TFA at 135°C for 2 h. Significant degradation of Gal*p*A during hydrolysis did however occur.

Very labile sugars such as ketoses would degrade during complete hydrolysis of a polysaccharide. They may be estimated by a mild partial hydrolysis. 3-Deoxy-D-*manno*-octulosonic acid (KDO) and 3-deoxy-D-*lyxo*-2-heptulosaric acid (DHA) have been released by 1 M HOAc at 40°C for 6 h (York *et al.*, 1985) and 0.1 M TFA at 100°C for 1 h (Stevenson *et al.*, 1988), respectively. KDO and DHA are minor components which have been identified but not quantified in a wide range of plant cell walls (Stevenson *et al.*, 1988). Garegg *et al.* (1988) have developed a method for hydrolysing under reducing conditions in order to prevent degradation of labile sugars. The liberated sugars are immediately reduced to the alditol by 4-methylmorpholineborane, which is fairly stable in acid. Complete recovery of the labile 3,6-anhydro-L-galactose, from agarose, as the alditol was achieved using 0.5 M borane in 0.5 M TFA at 100°C for 15 h. This method is also suitable for quantifying ketoses such as KDO.

2. Determination of sugars released by acid hydrolysis

The neutral sugars released on acid hydrolysis are usually reduced with $NaBH_4$, acetylated and estimated as the alditol acetates by GLC (Jones and Albersheim, 1972; Selvendran *et al.*, 1979). More recently Blakeney *et al.* (1983) and Harris *et al.* (1988) have described a rapid procedure in which the H_2SO_4 hydrolysate is basified with NH_3, and a small fraction (5.5%) is reduced in dimethylsulphoxide. This method obviates the need to remove borate and uses methylimidazole as the catalyst for acetylation, thereby avoiding any evaporations. The hydrolysis conditions of Blakeney *et al.* (1983) are inadequate and those of Harris *et al.* (1988) may cause degradation of pentosans (Selvendran *et al.*, 1989). When the amount of material is small (~ 1 mg) it is obviously better to use a method in which the whole sample is derivatised. The alditol acetate chromatogram shows a single peak for each sugar and the derivatives from cell walls are readily separable on commonly available phases such as OV-225 (Selvendran *et al.*, 1979) and SP-2330 (Englyst *et al.*, 1982). The detection limit using the packed column is approximately 0.1–0.5 µg, depending on the retention time of the alditol acetate. Phthalic esters have been reported to interfere with GLC determination of alditol acetates (Dudman and Whittle, 1976). This can be avoided by using high quality reagents or capillary columns with polar phases (Henry *et al.*, 1983). We have found that 'Megabore' capillary columns (0.53 mm, i.d.) are especially useful.

Uronic acids are usually estimated non-specifically by the modified carbazole method of Bitter and Muir (1962) or by reaction with *m*-phenylphenol in H_2SO_4-tetraborate (Blumenkrantz and Asboe-Hansen, 1973). In our earlier studies we determined the uronic acid content of cell-wall samples using an adaptation of the method of Bitter and

Muir (Selvendran *et al.*, 1979). In subsequent work we improved on this method in two ways: (1) the material dispersed in 1 M acid was heated at 100°C for 1–3 h before an aliquot was taken and analysed, and (2) the *m*-phenylphenol (3-phenylphenol) method was used for the determination of uronic acid, because it gives negligible interference from neutral sugars and no correction needs to be made for them. Englyst *et al.* (1982) and Englyst and Cummings (1984) have used a modification of the Scott (1979) procedure to estimate uronic acid in dietary fibre samples. Theander and Aman (1979) have accurately estimated the uronic acid content of cell walls and polysaccharide samples by the decarboxylation method.

Uronic acids can also be estimated as the silyl derivatives of the aldonolactones by GLC. Earlier methods depended on drying from HCl to form the (1,4) lactones and then silylating (Perry and Hulyalkar, 1965). Ford (1982) has proposed a method whereby Gal*p*A, released from water-soluble pectins by treatment with polygalacturonase, is converted to the aldonolactone and silylated. GLC of the trimethylsilyl derivatives has given satisfactory resolution of glucuronic and galacturonic acids. We have found that the conditions of lactonisation given by Ford (vacuum-dried sample treated with 0.8 M methanolic-HCl at 60°C for 4 h, then vacuum-dried) gave multiple peaks on analysis by GLC. This was shown by GLC-MS of the derivatives to be partly due to the formation of the aldono-1,5-lactone (P. Ryden and R. R. Selvendran, unpubl. res.). Lehrfeld (1985) has prepared the acetylated *N*-alkylaldonamides by heating the aldonic acid *in vacuo* at 80°C for 2 h to form the (1,4)- and (1,5)-lactones. Both lactone forms then react with 1-propylamine in pyridine and the product is acetylated. Each uronic acid gives one peak on GLC. Other primary amines can easily be substituted where there are particular problems of resolution. The method is compatible with neutral sugar determination. Walters and Hedges (1988) prepared the acetylated *N*-hexylaldonamides for simultaneous determination with alditol acetates.

HPLC is not widely used for the analysis of monosaccharides because of difficulties with resolution and detection which have only recently been overcome. Slavin and Marlett (1983) achieved a separation of most of the monosaccharides and this was adequate for the analysis of NDF samples. The cellobiose in the hydrolysates was estimated in the same chromatogram. HPLC columns which can resolve all the neutral sugars commonly found in plant cell walls as alditols are now available. Takeuchi *et al.* (1987) have published a list of retention times on a Shodex SP-1010 column eluted with 20% aqueous ethanol. The alditols were detected by their refractive index, or for high sensitivity, by reduction with NaB^3H_4. Forni *et al.* (1986) have used a Partisil SAX HPLC column for uronic acid determination.

Monosaccharides can also be resolved by HPLC without derivatisation by anion-exchange in 22 mM NaOH on a Dionex Carbopac AS-6 column. The monosaccharides are separated in 15 min (Hardy *et al.*, 1988). The same column can also be used for oligosaccharides (Townsend *et al.*, 1988). The sugars can be detected without derivatisation by pulsed amperometric detection. The eluate passes over a gold electrode and a small fraction of the sample is oxidised and the oxidation current is measured (Rocklin and Pohl, 1983). The method is highly sensitive; < 1 nmol is required.

3. Methanolysis

The problem of low yields of uronic acid obtained after aqueous acid hydrolysis (because of the stability of the uronic acid linkage and degradation by oxidation or decarboxylation) can be overcome by methanolysis. Acidic methanol depolymerises polysaccharides to methyl glycosides which are then protected from further degradation. There have been several recent attempts to improve the method of methanolysis. Chaplin (1982) used 0.625 M HCl in methanol, containing methyl acetate to maintain anhydrous conditions, at 70°C for 16 h. The evaporation of HCl was assisted by the addition of *t*-butyl alcohol. Ha and Thomas (1988) used 0.5 M HCl at 80°C for 16 h. The methyl glycosides and methyl ester of methyl glycosides, from uronic acids, are separated after silylation. These derivatives produce up to four peaks for each sugar on GLC. This is not a problem so long as they are all resolved; fused-silica capillary columns make this possible. The total peak area for each monosaccharide is obtained by summing the peak areas of the isomers. Roberts *et al.* (1987) have extended the use of methanolysis to cellulosic samples by solubilising in 72% (w/w) methanolic sulphuric acid, then adding methanol and boiling under reflux for 16 h. Virtually quantitative yields were reported for cellulosic samples and pectic polysaccharides. In the case of pectins, esters and lactones from the uronic acids were hydrolysed with barium hydroxide to simplify the chromatograms.

4. HF solvolysis

At low temperatures, −40 to −23°C, anhydrous liquid hydrogen fluoride cleaves specific glycosidic linkages but not ester linkages. This is a useful reagent for degrading bacterial polysaccharides to their repeat unit (Mort and Bauer, 1982). Mort *et al.* (1988) are now using HF for the structural determination of pectins.

5. Absolute configurations

The most popular GLC methods for determining absolute configuration are those of Gerwig *et al.* (1978) and Leontein *et al.* (1978), in which the acetyl or silyl derivatives of glycosides of (−)-2-butanol or (+)-2-octanol are separated on achiral columns. The chromatograms of such derivatives are complicated by the presence of multiple peaks corresponding to different anomers and ring sizes.

Newer methods in which the sugar enantiomers are converted to acyclic diastereoisomers by reductive amination provide simple chromatograms. Oshima *et al.* (1983) used L-(−)-α-methylbenzylamine in the presence of sodium cyanoborohydride to prepare the 1-(*N*-acetyl-α-methylbenzylamino)-1-deoxyalditol acetates. This method has been applied to determine the configuration of sap gum sugars of *Rhus vernicifera* (Oshima and Kumanotani, 1984). Hara *et al.* (1986, 1987) prepared the methyl 2-(polyhydroxyalkyl)-thiazolidene-(4R)-carboxylate derivative by heating pyridine solutions of the sugars with L-cysteine methyl ester hydrochloride at 60°C for 1 h, which was then silylated. The procedure of Hara *et al.* (1987) is the simpler method and has given better resolution, by GLC, for most sugar enantiomers.

Alternatively enantiomers can be separated on chiral capillary gas chromatography columns. Chiral stationary phases stable at high temperature are becoming available. A recent promising example uses α-cyclodextrin which is used in liquid chromatography. When stabilised by per-*n*-pentylation, it is suitable in gas chromatography for the enantiomeric separation of trifluoroacetylated derivatives of sugars, alditols and 1,5-anhydrohexitol (König *et al.*, 1988).

B. Glycosyl Linkage Analysis

1. Methylation

The chemical determination of linkage positions in polysaccharides is routinely performed by methylation of all free hydroxyl groups, hydrolysis and characterisation of the methylated monosaccharides as partially methylated alditol acetates. The improved method of methylation developed by Hakomori (1964) uses sodium methylsulphinylmethanide (dimsyl anion) in dimethylsulphoxide to catalyse the ionisation of the hydroxyl groups followed by reaction with iodomethane. For details of the procedure see Lindberg (1972), Lindberg and Lönngren (1978) and the experimental guide prepared by Lindberg's group (Jansson *et al.*, 1976). The procedure which we use routinely for neutral and acidic polysaccharides is given in Ring and Selvendran (1978) and Selvendran and Stevens (1986). Precautions to be taken include ensuring that esterified uronic acids are de-esterified and salts are removed, and that the polysaccharides are well dispersed in dimethylsulphoxide; milling in liquid nitrogen may be necessary with some samples (Lomax *et al.*, 1983). The methylation procedure has been adapted by several workers. Harris *et al.* (1984) prefer the potassium carbanion as it is more easily prepared. Blakeney and Stone (1985) have prepared the lithium carbanion from butyllithium which is available commercially in a much purer form than the alkali metal hydrides, so providing a cleaner chromatogram. Other solvents have been tested when samples are insoluble in dimethylsulphoxide. SO_2-Diethylamine-dimethylsulphoxide is particularly useful for solubilising cellulosic and lignified samples (Isogai *et al.*, 1985). 4-Methylmorpholine *N*-oxide has been proposed as a solvent for methylation (Joseleau *et al.*, 1981) but has not proved to be of much use (Harris *et al.*, 1984).

An alternative to Hakomori methylation which is becoming popular is the method of Ciucanu and Kerek (1984) in which powdered sodium hydroxide and iodomethane are added to the carbohydrate in dimethylsulphoxide and stirred for 7 min. The solubility of sodium hydroxide in dimethylsulphoxide is low but sufficient to catalyse the hydroxyl group ionisation. High yields and clean chromatograms have been obtained. No problems were reported in using this technique for the methylation of rye-bran arabinoxylans (Hromádková *et al.*, 1987) or for acidic oligosaccharides derived from the gum of *Althea officinalis* (Capek *et al.*, 1987). Isogai *et al.* (1985) have reported that cellulosic samples are not depolymerised by methylation with sodium hydroxide, unlike the case when methylsulphinylmethanide is used.

2. Acidic polysaccharides

Acidic polysaccharides must be reduced either before or after methylation to analyse their uronic acids as partially methylated alditol acetates. The uronic acids are normally reduced with deuteride so that they can be distinguished by mass spectrometry. Water-soluble polysaccharides activated with carbodiimide can be reduced with sodium borohydride (Hoare and Koshland, 1967; Taylor and Conrad, 1972). These reactions are slow and the pH must be carefully maintained at 4.75 during formation of the carbodiimide complex and at 7 during reduction. This is most easily done using 2-(*N*-morpholine) ethane sulphonate and Tris buffers (Anderson and Stone, 1985). Repeated treatments are sometimes necessary and some degradation usually occurs. Other researchers have esterified the uronic acid residues with methanolic HCl (0.08 M, 20°C, 24 h) and then reduced with 25 mM sodium borohydride at 4°C for 24 h (Moody *et al.*, 1988).

It is more convenient, where amounts of material are small, to carboxyl reduce the methylated polysaccharide by refluxing with lithium aluminium hydride in dichloromethane–ether (Redgwell and Selvendran, 1986). The method gave complete reduction of GlcpA in the partially hydrolysed gum from the stem pith of the kiwi fruit tree (Redgwell *et al.*, 1986a), but with pectins the recovery of reduced GalpA varied from 50–80%. Waeghe *et al.* (1983) used sodium borohydride in ethanol–tetrahydrofuran (THF). Lithium triethylborohydride in THF is also highly effective on methyl esterified methylated polysaccharides (York *et al.*, 1986).

3. Preparation and identification of partially methylated alditol acetates (PMAA)

The standard hydrolysis method is 90% formic acid for 2 h and 0.25 M sulphuric acid for 12 h at 100°C (Lindberg, 1972); 2 M trifluoroacetic acid (TFA) at 121°C for 1 h is also effective (Harris *et al.*, 1984). Methods of reduction with sodium borodeuteride and acetylation are the same as for unmethylated sugars. Procedures developed to eliminate evaporation, such as acetylation in the presence of borate, catalysed by 1-methylimidazole (Blakeney *et al.*, 1983) or by perchloric acid (Harris *et al.*, 1984), are especially useful to minimise the loss of volatile methylated derivatives.

Derivatives are separated on packed or capillary GLC columns. Derivatives which co-chromatograph on one column can be separated on another. For example, derivatives from (1→2)- and (1→4)-linked xylose and terminal galactose residues co-chromatograph on OV-225, but can be separated on ECNSSM. The relative retention times of PMAA from some pectins on OV-225 and the main diagnostic ions are given in Table 1 of Selvendran and Stevens (1986) and the elution sequence of the PMAA from cabbage pectin is shown in Fig. 5 of Selvendran and O'Neill (1987). Lists of relative retention times on capillary columns of CP-Sil 88, SP-1000 and BP-1 have been published for most of the PMAA commonly found in plant cell walls (Lomax *et al.*, 1985). PMAA can be easily identified from the pattern of fragment ions produced by electron-impact mass spectrometry (Jansson *et al.*, 1976). Lönngren and Svensson (1974) and Aspinall (1982) have explained the mechanisms of fragmentation. Primary fragment ions arise from cleavage of carbon–carbon bonds. C-1 is deuterated so ions derived from different ends of the molecule are distinguished. The most abundant ions are derived from cleavage between methoxylated carbons and the next most abundant from cleavage between a methoxylated and acetoxylated carbon. The PMAA, with their primary fragments derived from potato xyloglucan (Ring and Selvendran, 1981), are shown in Fig. 16.1.

4. Ring size and reductive cleavage

After hydrolysis of a methylated polysaccharide oxygens involved in glycosidic linkages or in ring formation are all acetylated so (1→4)-hexp and (1→5)-hexf residues are not distinguished. By partial hydrolysis of the methylated polysaccharide, realkylation and then complete hydrolysis, the ring size can be determined unambiguously (Darvill *et al.*, 1980b).

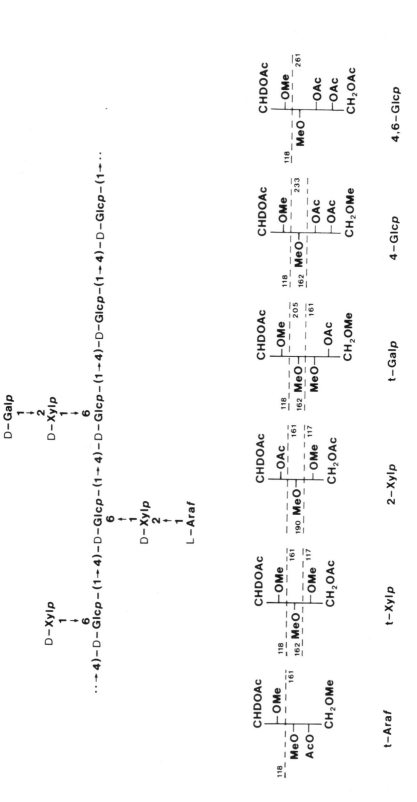

FIG. 16.1. The main structural features of potato xyloglucan and PMAA obtained from the polysaccharide.

A new method of polysaccharide analysis termed reductive cleavage allows the ring size and linkage positions to be determined simultaneously (Rolf and Gray, 1982). By cleaving the glycosidic bonds of a methylated polysaccharide with hydride equivalents pyranose residues are reduced to 1,5-anhydro-alditols and furanose residues to 1,4-anhydro-alditols. The reducing agent is triethylsilane and the catalysts are boron trifluoride etherate or trimethylsilylmethanesulphonate or a mixture of both (Jun and Gray, 1987). The method has been carefully tested on oligo- and polysaccharides of known structure and conditions have been adapted to ensure efficient conversion of all residues to the desired anhydro derivatives without side reactions. This has been achieved for all linkages tested so far, except for (1→4)-GlcpA which rearranges to the 1,4-anhydro derivative (Vodonik and Gray, 1988). Further aspects of the method's versatility are that a specific partial cleavage is possible—only certain linkages are cleaved using boron trifluoride etherate (Rolf et al., 1985) and the reaction does not degrade carboxylic acid esters (Rolf and Gray, 1986). The method's main drawback is that in the EI-MS data, whilst the differences between 1,4- and 1,5-anhydro-alditols are obvious, the positions of methyl and acetyl substitution are not so easily interpreted as they are for PMAA. Their fragmentation pathways are beginning to be understood (Bråthen et al., 1988).

5. Oxidative cleavage of polysaccharides

Periodate cleaves vicinal diol groups to aldehydes. In the Smith degradation the aldehydes are reduced with borohydride and the acyclic linkages are hydrolysed with mild acid. Supporting evidence for sugar linkages, ring sizes and the position of protecting acyl groups can be obtained by analysis of the oxidised fragments which are released and of the periodate-resistant oligo- or polysaccharide remaining. The method is most frequently applied to highly branched plant gums in which repeated Smith degradations can successively remove the non-reducing terminal and less branched sugars of the side chains to leave a polysaccharide derived from the highly branched backbone, allowing the arrangement of sugars in the polysaccharide to be determined (Banerji et al., 1986; Cartier et al., 1987). Among other examples are the following:

1. The acetyl substitutents of the sycamore xyloglucan nonasaccharide were shown to be attached to Gal by comparing the extent of periodate degradation before and after de-esterification (York et al., 1988).
2. The arabinogalactan of *Lolium multiflorum* has blocks of (1→3) Gal$_7$ between periodate-sensitive (1→6) Gal residues (Bacic et al., 1987).
3. The distribution of (1→3)- and (1→4)-linkages in mixed linkage β-glucan has been determined (Parrish et al., 1960).

IV. GLYCOSYL SEQUENCE DETERMINATIONS OF POLYSACCHARIDES

Of the chemical methods which assist in obtaining information on the sequence and branching arrangement of sugar residues in polysaccharides, some are more specific than others. Generally the methods involve partial degradation of the native, acetylated or methylated polysaccharides and characterisation of the oligosaccharide fragments released and of the residual polysaccharides. Fragments containing up to eight residues can be characterised by mass spectrometry. Extracellular bacterial polysaccharides contain repeating units of 3–8 glycosyl residues and a number of methods have been developed to sequence them (Valent *et al.*, 1980; Aman *et al.*, 1981; Kenne and Lindberg, 1983).

Selvendran and Stevens (1986) have collated some of the applications of mass spectrometry for the sequencing of pectic polysaccharides. As we have discussed the applications of MS for the characterisation of oligosaccharides released by cellulase from runner bean xyloglucan (Selvendran and O'Neill, 1987), we shall consider the applications of MS for the structure determination of another group of hemicellulosic polysaccharides, namely, substituted glucuronomannans. These acidic polysaccharides are well known plant mucilages, but have been isolated from cell walls of cultured tissues (Mori and Kato, 1981) and from culture media (Akiyama *et al.*, 1984). The basic principles in glycosyl determination by MS of methylated oligosaccharide derivatives have been discussed at length in the aforementioned reviews and will not be described here.

A. Partial Acid Hydrolysis Studies of the Mucilage from the Stem Pith of the Kiwi Fruit Tree (*Actinidia deliciosa*)

The purified mucilage (KM1) was sequentially hydrolysed with 0.5% $H_2C_2O_4$, 0.5 M and 2 M TFA as shown in Scheme 16.3, and the polymeric materials were separated

SCHEME 16.3. Degradation of stem mucilage from the kiwi fruit tree (see Redgwell *et al.*, 1986a).

Native mucilage (KM1, 412 mg)

| 0.5% oxalic acid, 100°C, 2 h

|—→ Mono- and oligosaccharides

Degraded mucilage (DKM1, 245 mg)

| 0.5 M trifluoroacetic acid, 100°C, 1 h

|—→ Mono- and oligosaccharides

Degraded mucilage (DKM2, 133 mg)

| 2 M trifluoroacetic acid, 100°C, 2 h

Degraded mucilage (DKM3, 40 mg)

from the mono- and oligosaccharides by dialysis against distilled water. The oligo-saccharides in the diffusates were fractionated by gel filtration, reduced with $NaBD_4$, methylated, separated by reverse phase HPLC and examined by EI-MS and FAB-MS (Redgwell *et al.*, 1986a,b). The main fragment ions used to characterise one of the oligosaccharides released from the side chain by oxalic acid hydrolysis are shown in Fig. 16.2. Certain secondary ions provide additional structural information. In particular, the ald J_1 ions (MeO—CH=O^+—alditol, m/z 296) arise from rearrangements involving the retention of the C-1 and the methoxyl group on C-3 of the residue next to the alditol. When the internal residue is 3-linked J_1 ions cannot form but a J_0 ion (HO^+=CH—O—alditol, m/z 282), which is diagnostic for 3-linked residues, is formed (Sharp and Albersheim, 1984). In this example significant J_0 ions (m/z 282 and 486) confirmed that both internal hexosyl residues were substituted at C-3.

The residual polysaccharide after 2 M TFA hydrolysis (DKM3) contained the bulk of the backbone. This was reduced, methylated, carboxyl reduced with $LiAlD_4$, and partially hydrolysed to di-, tri-, tetra- and pentasaccharides with 90% formic acid at 70°C for 50 min. These were reduced with $NaBD_4$, remethylated, separated by HPLC and examined by MS. The EI-MS of the di- and tetrasaccharides shown in Fig. 16.2 helped to establish that the backbone repeat unit was →4)-D-GlcpA-(1→2)-D-Manp-(1→, which confers on it acid stability.

Labile sugar residues of the mucilage were hydrolysed by mild acid (oxalic acid) and their points of attachment were inferred by comparing the methylation analysis of the native and degraded mucilage (DKM1). With hindsight shorter treatment with oxalic acid would have been better. The positions of attachment of labile sugars can be determined more specifically by partially degrading the methylated polysaccharide and labelling the exposed hydroxyls by realkylation. In the potato xyloglucan (Fig. 16.1) the position of arabinose substitution was shown to be to C-2 of xylose. After partial hydrolysis of the methylated xyloglucan with 90% formic acid at 70°C for 40 min and remethylation with CD_3I the 2,3,4-tri-*O*-methylxylose derivative from the residual polysaccharide contained ions at m/z 121 and 165 instead of 118 and 162 (Ring and Selvendran, 1981). The same hydrolysis conditions released the terminal arabinose and terminal fucose residues of cabbage xyloglucan and the positions of attachment were to C-2 of xylose and C-3 of galactose (Stevens and Selvendran, 1984c).

B. Selective Fragmentation of Methylated Glucuronomannans

The carboxylic acid groups in methylated polysaccharides provide sites for specific chemical degradations, and we shall discuss briefly those degradations which have found application in the structural analysis of glucuronic acid containing polysac-charides. Aspinall and co-workers have developed several chemical methods for the fragmentation of plant gums such as gum arabic and leiocarpan A. Leiocarpan A has the same backbone as the stem mucilage of *Actinidia deliciosa*. The same structural feature is also found in an extracellular polysaccharide of tobacco (Akiyama *et al.*, 1984; Aspinall *et al.*, 1989).

The methods involve the decarboxylation of the glucuronic acid residues in the methylated polysaccharide and the polysaccharide is cleaved to xylitol-terminated fragments. An earlier method was the Hofmann reaction in which the uronic acid is converted to the uronamide and this degrades in strong base (Aspinall and Rosell,

1978). This has been improved to obtain the same degradation under neutral non-aqueous conditions (Aspinall *et al.*, 1983). An alternative method uses lead tetra-acetate to decarboxylate the uronic acid directly and this method will be described.

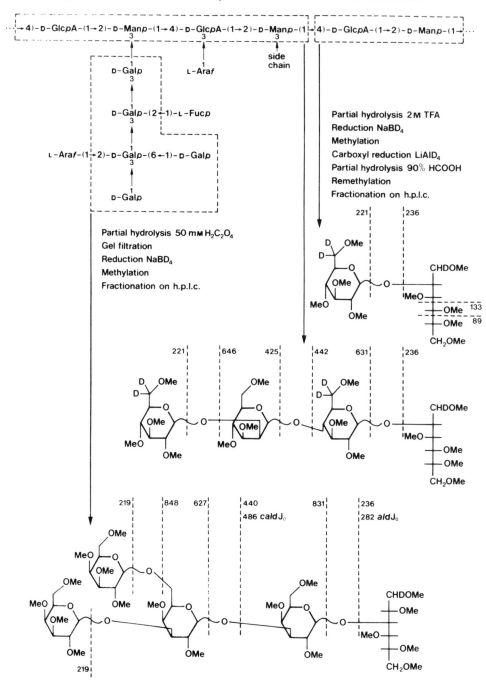

FIG. 16.2. The mass spectral fragmentation patterns of fragments obtained by partial acid hydrolysis of mucilage from the kiwi fruit tree.

Lead tetra-acetate is an oxidising agent. It cleaves vicinal diols in the same way as periodate. In the methylated polysaccharide its action is restricted to the de-esterified carboxylic acid groups. The reaction proceeds by a free radical mechanism (Beckwith *et al.*, 1974). Methylated residues of α- and β-GlcpA and β-GalpA are oxidised, but α-GalpA resists oxidation (Kitagawa *et al.*, 1978). The methylated polysaccharide is de-esterified with NaOH at 0°C to prevent β-eliminative degradation. The pH is maintained at 12 for 2 h with addition of NaOH as required. The de-esterified methylated polysaccharide is heated under reflux with lead tetra-acetate in benzene for up to 20 h. Further additions of lead tetra-acetate are made during the reaction. The uronic acid is decarboxylated and the C-5 is acetoxylated. The product is degraded with sodium borohydride in water–methanol or water–tetrahydrofuran by deacetylation, ring opening and reduction of the xylodialdose (Aspinall *et al.*, 1981). Complete conversion of the GlcpA residues has been achieved. Figure 16.3 illustrates the fragments which are obtained from a glucuronomannan of tobacco after remethylation with CD$_3$I (Aspinall *et al.*, 1989).

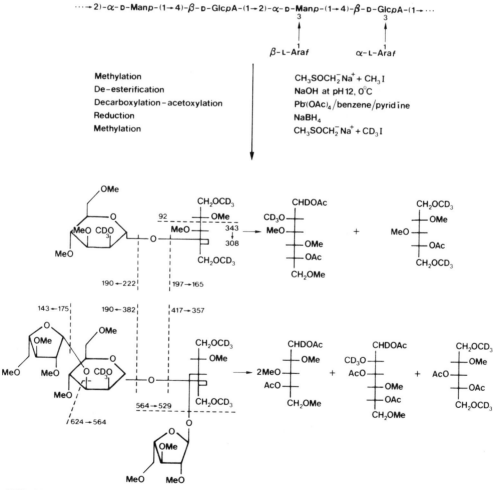

FIG. 16.3. The fragments obtained by degradation of tobacco glucuronomannan with lead tetra-acetate and the PMAA obtained from the oligosaccharide fragments.

V. APPLICATIONS OF NMR

NMR spectroscopy is being developed as a rapid non-destructive alternative to chemical analysis and the dynamic properties of carbohydrates can also be investigated.

The two nuclei studied are ^{1}H and ^{13}C. ^{1}H nuclei have a higher sensitivity than ^{13}C in NMR experiments. The receptivity of ^{13}C relative to ^{1}H is 1.76×10^{-4} (Harris, 1983). Proton spectra can be acquired rapidly with 2 mg of material but resonances fall in a narrow spectral range and are complicated by spin couplings. The more complex the molecule the less useful the ^{1}H spectrum will be. The low sensitivity and low natural abundance of ^{13}C nuclei means that large amounts of material (30–50 mg) are needed and several thousand scans have to be acquired. ^{13}C–^{13}C spin couplings can be neglected. When the protons are decoupled the carbon spectrum gives well-resolved signals over a broad spectral range. NMR spectroscopy has been applied in a variety of ways to purified oligosaccharides, isolated polysaccharides and whole cell walls to obtain structural information on anomeric configurations, linkage, sequence and branching arrangement and to obtain information on the mobilities and crystallinity of carbohydrates.

A. Polysaccharides

Proton spectra are acquired in D_2O to limit the spectra to the non-exchangeable protons. The anomeric proton of an aldose is attached to the hemiacetyl carbon which is bonded to two oxygen atoms, and so is more deshielded and resonates downfield with respect to other protons. This anomeric region is frequently the only usefully resolved region of a polysaccharide ^{1}H spectrum. The chemical shifts of the anomeric protons of polysaccharides are similar to those of the methyl glycosides of the monosaccharides. Lists of these chemical shifts have been published (Bock and Thøgersen, 1982). Protons of α-linked residues have chemical shifts of 4.2–5.5 p.p.m. and protons of β-linked residues have shifts of 4.2–4.9 p.p.m. The anomeric resonances are split by coupling with the vicinal proton. The coupling constant depends on the dihedral angle between the C–H bonds, being 8 Hz for the *trans* and 3.5 for the *gauche* configuration. So the configurations at C-1 and C-2 can be determined from the anomeric signals.

^{1}H spectra of polysaccharides suffer from line-broadening because the molecular motion of the polymer is less than isotropic, so there is incomplete averaging of the localised magnetic fields. Line-broadening is less of a problem with a decoupled ^{13}C spectrum because of the greater dispersion of chemical shifts, though in both cases resolution may be enhanced by increasing the mobility of the polymer. Various methods have been used to increase the mobility of polymers in solution. Spectra are frequently recorded at high temperature, up to 90°C for a galactomannan of *Melilotus officinalis* endosperm (Gupta and Grasdalen, 1988). Many polysaccharides give more mobile solutions in dimethylsulphoxide. Samples are dissolved in DMSO-d$_6$ and spectra are recorded at 33°C (Sierakowski *et al.*, 1987) or 80°C (Joseleau *et al.*, 1983). Polymers can be solubilised by a partial depolymerisation in syrupy orthophosphoric acid (Grasdalen and Painter, 1980). Joseleau and Chambat (1984) have solubilised the whole cell wall of *Rosa glauca* suspension cultured cells in 4-methylmorpholine *N*-oxide (MMNO)-DMSO. This causes a partial degradation and signals of primary alcohol groups in the ^{13}C spectrum are obscured by the MMNO; otherwise the spectrum is well resolved.

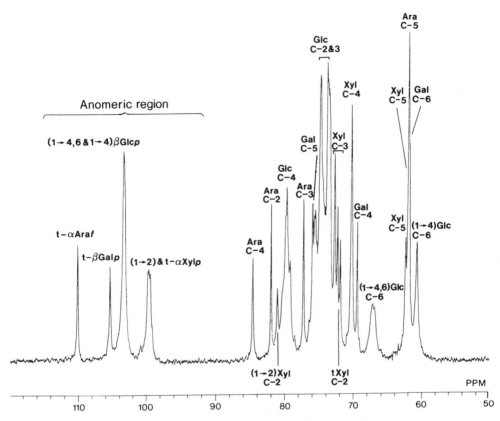

FIG. 16.4. The carbon-13 NMR spectrum of potato xyloglucan.

^{13}C assignments are made by comparison with simpler model structures such as methyl glycosides, oligosaccharides and homopolymers. There is a significant downfield shift of the resonance of a carbon involved in a glycosidic linkage relative to the unsubstituted chemical shift so linkage information is readily obtained. In the 1 M KOH-extracted xyloglucan of potato cell wall (Fig. 16.4) the C-6 signals of glucose at 60.8 and 67.2 p.p.m. are respectively from the $(1\rightarrow4)$-linked and $(1\rightarrow4,6)$-linked residues (for experimental details see Ryden *et al.*, 1989). In this example assignments were made by comparisons with a simpler xyloglucan of tamarind which lacks arabinose, and with other potato xyloglucan fractions with different proportions of arabinose and galactose. The spectrum shown is in D$_2$O (Fig. 16.4). Slightly better resolution was obtained in DMSO. An alternative approach is to simplify the polysaccharide structure by a partial degradation. The assignment of the spectrum of the arabinogalactan of *Rubus fruticosus* was assisted by a mild acid hydrolysis to remove Ara*f*, and by repeated periodate oxidations (Cartier *et al.*, 1987).

Where the polysaccharide has a particularly simple structure the resonance positions in a given sugar unit may be affected by neighbouring sugars. This gives rise to multiplicities which can be interpreted in terms of the probable sequential structure of the polysaccharide. For example, alginates consist of a linear polysaccharide of $(1\rightarrow4)$-linked mannuronic and guluronic acid residues. ^{13}C signals, especially the anomeric signals, are split depending on the type of sugar on either side (Grasdalen *et al.*, 1981). In the ^{1}H spectrum the anomeric and H-5 signals are also split (Grasdalen, 1983). H-5 is deshielded by the carboxylic acid group and resonates in the anomeric region. The extent to which guluronic acid residues are in blocks or are interspersed with mannuronic acid residues has been calculated. In partly esterified polygalacturonic acid the H-1 and H-5 signals are sensitive to the presence or absence of the methyl group on the same or neighbouring sugars (Grasdalen *et al.*, 1988). The endosperm galactomannans of *Gleditsia triacanthos* had a mannan backbone substituted with single galactose residues on the C-6 position. Statistical information on the distribution of galactose side chains could be obtained from the splitting of the mannose C-6 signal (Manzi *et al.*, 1986). In all these examples a partial degradation of the polysaccharide to a degree of polymerisation of 30 was necessary to decrease the viscosity of the solution.

Additional information on the overall structural arrangement of the polysaccharide can be obtained by an examination of the ^{13}C line-widths. Narrow peaks indicate highly mobile sugars of terminal groups or side chains. Broader peaks indicate restricted segmental motion or a greater variety of chemical environments which is often found in the backbone of a polysaccharide. In the potato xyloglucan (Fig. 16.4) the signals for the terminal arabinose and terminal galactose are the narrowest and those of glucose are the broadest. The arabinogalactan mucilage of *Pereskia aculeata* gave narrow arabinose signals and broad galactose signals, showing that the galactan core is much less mobile (Sierakowski *et al.* 1987).

The polygalacturonic acid backbone of pectins is particularly restricted so that spectra will show the side chains but the uronic acid signal may be absent or insignificant, as is the case with tomato pectin (Pressey and Himmelsbach, 1984) and onion pectin (Ryden *et al.*, 1989). The line-broadening of uronic acids can be caused by interaction with traces of divalent cations. A chelating agent, sodium triethylenetetra-mine-hexa-acetate (25 mg ml^{-1}) should be added to prevent this (Grasdalen *et al.*, 1981).

Quantitative analysis of side chain lengths and degree of branching of polysaccharides by NMR depends on resolution of the appropriate signals for the terminal and branched sugars. In onion pectin (Fig. 16.5) minor structural features of the galactan side chains were resolved (for experimental details see Ryden *et al.*, 1989). The side chains are β(1→4)-linked (C-4 78.5 p.p.m.). The unsubstituted C-4 of the non-reducing terminal galactose has a chemical shift of 69.6 p.p.m. and there are minor differences in the shifts of C-1,2,3,5,6 from those of the mid-chain residues. The average side chain length was estimated to be 9 from the integrated intensities of the anomeric peaks of the terminal galactose (104.4 p.p.m.) and mid-chain galactose (105.2 p.p.m.). The conditions required for accurate quantitation of NMR signals are given by Peng and Perlin (1987).

B. Oligosaccharides

In the ^{1}H spectra of oligosaccharides resolution of all ^{1}H resonances is possible. The signals of ring protons are crowded in the range 4.3–3.6 p.p.m. Assignment in this region requires an analysis of the ^{1}H spin couplings by two-dimensional and other techniques to establish which protons are next to each other in the molecule. Complete ^{1}H assignment has been achieved by homonuclear chemical-shift-correlated spectroscopy (COSY) in a trisaccharide glucuronoxylan (Cavagna and Deger, 1984) and tetrasaccharide (Excoffier *et al.*, 1986). Where the sugars of an oligosaccharide are very similar, such as in cello-oligosaccharides, assignment of the ¹H spectrum is a considerable challenge even in a trisaccharide (Ikura and Hikichi, 1987).

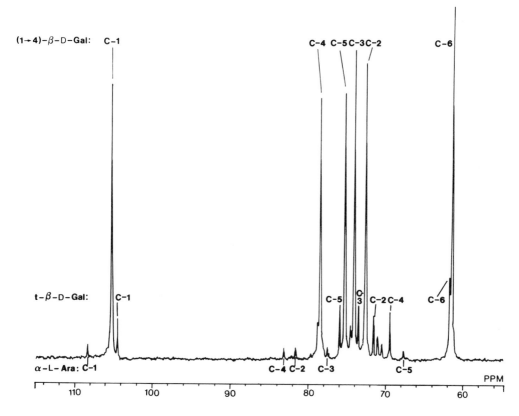

FIG. 16.5. The carbon-13 NMR spectrum of onion pectin.

By heteronuclear chemical-shift correlated experiments the ¹H spectrum can be spread out in the second dimension over the greater range of ¹³C chemical shifts. To observe the ¹³C nucleus would require large amounts of material but newer techniques are now available whereby the effect of the ¹³C nuclei can be observed indirectly via the ¹H spectrum (Bax and Summers, 1986). This allows complete ¹H and ¹³C assignment of 1–2 mg of a trisaccharide (Lerner and Bax, 1987).

Proton NMR has proved useful for the analysis of labile substituents such as acyl groups which are difficult to analyse chemically. The positions of acetylation in the sycamore xyloglucan were determined by comparing ¹H spectra of the nonasaccharide before and after deacetylation. *O*-acylation causes a downfield shift of 0.5–1.5 p.p.m. into the anomeric region of protons attached to the acetylated carbons (York *et al.*, 1988).

In the ¹³C spectra of oligosaccharides complete resolution is easily achieved. Assignments can be made by comparing with spectra of simpler, structurally related compounds, but it is preferable to use a combination of the two-dimensional methods mentioned above for unambiguous assignment (see Kováč and Hirsch, 1982, for the assignment of xylan oligosaccharides). Lists of ^{13}C chemical shifts of several oligosaccharides are available (Bock and Thøgersen, 1982; Bock et al., 1984).

The downfield shift of carbons directly involved in a glycosidic linkage and the minor changes in the shifts of adjacent carbons compared with the unsubstituted monosaccharides are dependent on the type of linkage and the absolute and anomeric configurations of the sugars. These glycosylation effects are being studied systematically in various disaccharides in (1→6)- (Forsgren et al., 1985), (1→3)- (Baumann et al., 1988a) and (1→4)-linkages (Backman et al., 1988). The data bases being constructed will be used for the computer-assisted prediction of oligosaccharide structure from their ^{13}C spectra (Jansson et al., 1987). The data bases are being extended to include branched trisaccharides (Baumann et al., 1988b). Prediction of the linkage structure and sequence of unbranched tetrasaccharides of known sugar composition from bacterial O-specific chains is already possible (Lipkind et al., 1988).

C. Cell Walls

Static samples do not provide resolved ^{1}H spectra, but information on the ^{1}H mobilities within a solid can be obtained from the line shape of the decay of total transverse magnetisation after application of a high power 90° pulse. Specialised NMR equipment, capable of delivering very short (1 μs) 90° pulses and with rapid (< 10 μs) receiver recovery, is required for this type of experiment. Immobile protons have short transverse relaxation times ($T_2 < 100$ μs), whereas mobile protons relax more slowly, so the free induction decay can be resolved into components with different decay constants and the proportions of protons in the different phases of mobility can be calculated. The mobilities of polysaccharides in bean cell walls have been studied after exchange of water with D_2O (Taylor et al., 1983). Sixty per cent of protons are rigid (decay constant < 100 μs) and 40% are mobile ($\gg 100$ μs). Since the cellulose content of the wall accounts for 30% of the wall protons this indicates that as well as the cellulose much of the hemicellulose must also be rigid. More detailed work has shown that the rigid microfibrils have a diameter of 10–40 nm. A model of the wall has been proposed in which the microfibrils have a core of cellulose and are surrounded by a 2–8 nm layer of rigid hemicellulose. The rest of the hemicellulose and the pectic substances are loosely suspended in water (MacKay et al., 1988). The mobile and rigid hemicellulose fractions were assigned to the material solubilised from the depectinated wall with 4% and 24% KOH respectively.

Other possible uses are to determine the amount and mobility of water in the native cell wall. It is important to understand the conditions of water activity in which cell-wall enzymes work. The water activity varies with metabolic activity and with the extent of phenolic cross-linking.

The conditions of rapid isotropic molecular tumbling which provide narrow ^{13}C signals in solution can be simulated in a solid by magic angle spinning (MAS) in which the sample is spun rapidly at an angle of 54.7°. Combination of MAS with cross-polarisation from ^{1}H (to increase ^{13}C sensitivity) and high power ^{1}H decoupling has made high resolution carbon-13 NMR spectroscopy of solids possible. The technique is useful for studying the structure of wood (Taylor, M. G. et al., 1983) and the bonding of lignin in wheat cell walls (Lewis *et al.*, 1987). In the case of wheat, plants were fed specifically labelled [^{13}C]ferulic acid to enhance the lignin signals and these were compared with spectra of model lignins. The polymorphs of cellulose can be distinguished by solid state NMR (Chanzy *et al.*, 1987), complementing information obtainable by diffraction techniques.

VI. CONCLUDING REMARKS

An attempt has been made to discuss the methods available for the isolation and analysis of cell walls, mainly from parenchymatous tissues of dicotyledons. Attention is drawn to the problems associated with co-precipitation of intracellular compounds and the need to inactivate wall-degrading enzymes in certain tissues. The preparation of cell walls from a starch-rich tissue, such as potatoes, is discussed at some length to illustrate the basic principles involved, and also because of the increasing importance of dietary fibre in human nutrition.

The methods which we use routinely to extract and fractionate the cell-wall polymers are described in greater detail, because we are in a position to make specific comments on the applicability of the methods. The use of methylation analysis for linkage analysis of polysaccharides is discussed, and we have illustrated the applications of GLC-MS and HPLC-MS for characterising the oligosaccharides released on partial acid hydrolysis and chemical degradation of (highly) substituted glucuronomannans. These methods obviously have much wider application. The applications of proton and carbon-13 NMR for structural studies are described in slightly greater detail. Although brief, we hope that this chapter will provide a useful guide for cell-wall analysis using modern techniques.

ACKNOWLEDGEMENTS

The authors thank Dr Ian J. Colquhoun for help and advice with NMR spectroscopy, John Eagles and Keith R. Parsley for help with mass spectrometry, and Anne Brown for typing the manuscript. This work was partly funded by the Ministry of Agriculture, Fisheries and Food.

REFERENCES

Akiyama, Y., Eda, E., Mori, M. and Katō, K. (1984). *Agric. Biol. Chem.* **48**, 403–407.
Albersheim, P., Neukom, H. and Deuel H. (1960). *Arch. Biochem. Biophys.* **90**, 46–51.
Aman, P., McNeil, M., Franzén, L.-E., Darvill, A. G. and Albersheim, P. (1981). *Carbohydr. Res.* **95**, 263–282.
Anderson, M. A. and Stone, B. A. (1985). *Carbohydr. Polymers* **5**, 115–129.
Aspinall, G. O. (1973). *In* "Techniques of Chemistry, Vol. 4: Elucidation of Organic Structures by Physical and Chemical Methods" (K. W. Bentley and G. W. Kirby, eds), pp. 379–450. Wiley Interscience, New York and London.
Aspinall, G. O. (1980). *In* "The Biochemistry of Plants, Vol. 3: Carbohydrates—Structure and Function" (J. Preiss, ed.), pp. 473–500. Academic Press, New York.

Aspinall, G. O. (1982). *In* "The Polysaccharides" (G. O. Aspinall, ed.), pp. 52–55. Academic Press, New York.

Aspinall, G. O. and Rosell, K.-G. (1978). *Can. J. Chem.* **56**, 685–690.

Aspinall, G. O., Fanous, H. K., Kumar, N. S. and Puvanesarajah, V. (1981). *Can. J. Chem.* **56**, 935–940.

Aspinall, G. O., Fanous, H. K., Kumar, N. S. and Puvanesarajah, V. (1983). *Can. J. Chem.* **61**, 1858–1863.

Aspinall, G. O., Khondo, L. and Puvanesarajah, V. (1989). *Carbohydr. Res.*, **188**, 113–120.

Bacic, A., Churms, S. C., Stephen, A. M., Cohen, P. B. and Fincher, G. B. (1987). *Carbohydr. Res.* **162**, 85–93.

Bacic, A., Harris, P. J. and Stone, B. A. (1988). *In* "The Biochemistry of Plants" (J. Preiss, ed.), Vol. 14, pp. 297–371. Academic Press, London.

Backman, I., Erbing, B., Jansson, P.-E. and Kenne, L. (1988). *J. Chem. Soc. Perkin Trans. I*, pp. 889–898.

Banerji, N., Sarkar, K. K. and Das, A. K. (1986). *Carbohydr. Res.* **147**, 165–168.

Barrett, A. J. and Northcote, D. H. (1965). *Biochem. J.* **94**, 617–627.

Bauer, W. D., Talmadge, K. W., Keegstra, K. and Albersheim, P. (1973). *Plant Physiol.* **51**, 174–187.

Baumann, H., Jansson, P.-E. and Kenne, L. (1988a). *J. Chem. Soc. Perkin Trans. I*, pp. 209–217.

Baumann, H., Erbing, B., Jansson, P.-E. and Kenne, L. (1988b). *In* "XIV International Carbohydrate Symposium, Stockholm", Abstracts A12.

Bax, A. and Summers, M. F. (1986). *J. Am. Chem. Soc.* **108**, 2093–2096.

Beckwith, A. L. J., Cross, R. T. and Gream, G. E. (1974). *Aust. J. Chem.* **27**, 1673–1962.

Bitter, T. and Muir, H. M. (1962). *Anal. Biochem.* **4**, 330–334.

Blakeney, A. B. and Stone, B. A. (1985). *Carbohydr. Res.* **140**, 319–324.

Blakeney, A. B., Harris, P. J., Henry, R. J. and Stone, B. A. (1983). *Carbohydr. Res.* **113**, 291–299.

Blumenkrantz, N. and Asboe-Hansen, G. (1973). *Anal. Biochem.* **54**, 484–489.

Bock, K. and Thøgersen, H. (1982). *In* "Annual Reports on NMR Spectroscopy" (G. Webb, ed.), Vol. 13. pp. 1–57. Academic Press, London.

Bock, K., Pedersen, C. and Pedersen, H. (1984). *Adv. Carbohydr. Chem. Biochem.* **42**, 193–225.

Bråthen, E., Engebretsen, M. and Kjølberg, O. (1988). *In* "XIV International Carbohydrate Symposium, Stockholm," Abstracts A84.

Buchala, A. J., Fraser, C. G. and Wilkie, K. C. B. (1972). *Phytochemistry* **11**, 1249–1254.

Capek, P., Rosík, J., Kardošová A. and Toman, R. (1987). *Carbohydr. Res.* **164**, 443–452.

Cartier, N., Chambat, G. and Joseleau, J.-P. (1987). *Carbohydr. Res.* **168**, 275–283.

Cavagna, F. and Deger, H. (1984). *Carbohydr. Res.* **129**, 1–8.

Chanzy, H., Chumpitazi, B. and Peguy, A. (1982). *Carbohydr. Polymers* **2**, 35–42.

Chanzy, H., Henrissat, B., Vincendon, M., Tanner, S. F. and Belton P. S. (1987). *Carbohydr. Res.* **160**, 1–11.

Chaplin, M. F. (1982). *Anal. Biochem.* **123**, 336–341.

Churms, S. C. and Stephen, A. M. (1984). *Carbohydr. Res.* **133**, 105–123.

Ciucanu, I. and Kerek, F. (1984). *Carbohydr. Res.* **131**, 209–217.

Darvill, A. G., McNeil, M., Albersheim, P. and Delmer, D. P. (1980a). *In* "The Biochemistry of Plants" (N. E. Tolbert, ed.), Vol. 1, pp. 92–162. Academic Press, New York.

Darvill, A. G., McNeil, M. and Albersheim, P. (1980b). *Carbohydr. Res.* **86**, 309–315.

Das, N. S., Das, S. C., Dutt, A. S. and Roy, A. (1981). *Carbohydr. Res.* **94**, 73–82.

De Vries, J. A., Rombouts, F. M., Voragen, A. G. J. and Pilnik, W. (1982). *Carbohydr. Polymers* **2**, 25–33.

De Vries, J. A., Rombouts, F. M., Voragen, A. G. J. and Pilnik, W. (1983). *Carbohydr. Polymers* **3**, 245–258.

Dey, P. M. and Brinson, K. (1984). *Adv. Carbohydr. Chem. Biochem.* **42**, 265–382.

Dudman, W. F. and Whittle, C. P. (1976). *Carbohydr. Res.* **46**, 267–272.

DuPont, M. S. and Selvendran, R. R. (1987). *Carbohydr. Res.* **163**, 99–113.

Englyst, H. N. and Cummings, J. H. (1984). *Analyst* **109**, 937–942.

Englyst, H. N., Wiggins, H. S. and Cummings, J. H. (1982). *Analyst* **107**, 307–318.

Excoffier, G., Nardin, R. and Vignon, M. R. (1986). *Carbohydr. Res.* **149**, 319–328.

Ford, C. W. (1982). *J. Sci. Food Agric.* **33**, 318–324.

Forni, E., Toreggiani, D., Battiston, P. and Polesello, A. (1986). *Carbohydr. Polymers* **6**, 379–393.

Forsgren, M., Jansson, P.-E. and Kenne, L. (1985). *J. Chem. Soc. Perkin Trans. I*, pp. 2383–2388.

Fry, S. C. (1982a). *Biochem. J.* **204**, 449–455.

Fry, S. C. (1982b). *Biochem. J.* **203**, 493–504.

Fry, S. C. (1983) *Planta* **157**, 111–123.

Fry, S. C. (1988). *In* "The Growing Plant Cell Wall: Chemical and Metabolic Analysis". Longman Scientific and Technical, London.

Garegg, P. J., Lindberg, B., Konradsson, P. and Kvarnströn, I. (1988). *Carbohydr. Res.* **176**, 145–148.

Gerwig, G. J., Kamerling, J. P. and Vliegenthart, J. F. G. (1978). *Carbohydr. Res.* **62**, 349–357.

Grasdalen, H. (1983). *Carbohydr. Res.* **118**, 255–260.

Grasdalen, H. and Painter, J. J. (1980). *Carbohydr. Res.* **81**, 59–66.

Grasdalen, H., Larsson, B. and Smidsrød, O. (1981). *Carbohydr. Res.* **89**, 179–191.

Grasdalen, H., Bakøy, O. E. and Larsen, B. (1988). *Carbohydr. Res.* **184**, 183–191.

Gupta, A. K. and Grasdalen, H. (1988). *Carbohydr. Res.* **173**, 159–168.

Ha, Y. W. and Thomas, R. L. (1988). *J. Food Sci.* **53**, 574–577.

Hakomori, S. (1964). *J. Biochem.* **55**, 205–208.

Hara, S., Okabe, H. and Mihashi, K. (1986). *Chem. Pharm. Bull.* **34**, 1843–1845.

Hara, S., Okabe, H. and Mihashi, K. (1987). *Chem. Pharm. Bull.* **35**, 501–506.

Hardy, M. R., Townsend, R. R. and Lee, Y. C. (1988). *Anal. Biochem.* **170**, 54–62.

Harris, P. J., Henry, R. J., Blakeney, A. B. and Stone, B. A. (1984). *Carbohydr. Res.* **127**, 59–73.

Harris, P. J., Blakeney, A. B., Henry, R. J. and Stone, B. A. (1988). *J. Assoc. Off. Anal. Chem.* **71**, 272–275.

Harris, R. K. (1983). *In* "Nuclear Magnetic Resonance Spectroscopy", p. 73. Pitman, London.

Hatfield, R. D. and Nevins, D. J. (1987). *Plant Physiol.* **83**, 203–207.

Henry, R. J., Blakeney, A. B., Harris, P. J. and Stone, B. A. (1983). *J. Chromatogr.* **256**, 419–427.

Hoare, D. G. and Koshland, D. E. (1967). *J. Biol. Chem.* **242**, 2447–2453.

Hromádková, Z., Ebringerova, A., Petrakova, E. and Schraml, J. (1987). *Carbohydr. Res.* **163**, 73–79.

Huwyler, H. R., Franz, G. and Meir, H. (1979). *Planta* **146**, 635–642.

Ikura, M. and Hikichi, K. (1987). *Carbohydr. Res.* **163**, 1–8.

Isogai, A., Ishizu, A., Nakana, J., Eda, S. and Katō, K. (1985). *Carbohydr. Res.* **138**, 99–108.

Jansson, P.-E., Kenne, L., Liedgren, H., Lindberg, B. and Lönngren, J. (1976). *Chem. Commun. Univ. Stockholm*, No. 8.

Jansson, P.-E., Kenne, L. and Widmalm, G. (1987). *Carbohydr. Res.* **168**, 67–77.

Jarvis, M. C. (1982). *Planta* **154**, 344–346.

Jarvis, M. C., Hall, M. A., Threlfall, D. R. and Friend, J. (1981). *Planta* **152**, 93–100.

Jermyn, M. A. (1955). *In* "Modern Methods of Plant Analysis" (K. Paech and M. V. Tracey, eds), Vol. 2, pp. 197–225, Springer, Berlin.

Jermyn, M. A. and Isherwood, F. A. (1956). *Biochem. J.* **64**, 123–133.

Jones, T. M. and Albersheim, P. (1972). *Plant Physiol.* **49**, 926–936.

Joseleau, J.-P. and Chambat, G. (1984). *Physiol. Veg.* **22**, 461–470.

Joseleau, J.-P., Chambat, G. and Chumpitazi-Hermoza, B. (1981). *Carbohydr. Res.* **90**, 339–344.

Joseleau, J.-P., Chambat, G. and Lanvers, M. (1983). *Carbohydr. Res.* **122**, 107–113.

Jun, J.-G. and Gray, G. R. (1987). *Carbohydr. Res.* **163**, 247–261.

Kenne, L. and Lindberg, B. (1983). *In* "The Polysaccharides" (G. O. Aspinall, ed.), Vol. 2, pp. 287–363. Academic Press, New York.

Kitagawa, I., Yoshikawa, M. and Kadota, A. (1978). *Chem. Pharm. Bull.* **26**, 484–496.

Kivilaan, A., Beaman, T. C. and Bandurski, R. S. (1959). *Nature* **184**, 81–82.

König, W. A., Lutz, S. and Wenz, G. (1988). *Angew. Chem. Int. Ed. Engl.* **27**, 979–980.

Kováč, P. and Hirsch, J. (1982). *Carbohydr. Res.* **100**, 177–193.

Lehrfeld, J. (1985). *Carbohydr. Res.* **135**, 179–185.

Leontein, K., Lindberg, B. and Lönngren, J. (1978). *Carbohydr. Res.* **62**, 359–362.

Lerner, L. and Bax, A. (1987). *Carbohydr. Res.* **166**, 35–46.

Lewis, N. G., Yamamoto, E., Wooten, J. B., Just, G., Ohashi, H. and Towers, G. H. N. (1987). *Science* **237**, 1344–1346.

Lindberg, B. (1972). *Methods Enzymol.* **28**, 178–195.

Lindberg, B. and Lönngren, J. (1978). *Methods Enzymol.* **50**, 3–33.

Lipkind, G. M., Shashkov, A. S., Knirel, Y. A., Vinograd, E. V. and Kochetkov, N. K. (1988). *Carbohydr. Res.* **175**, 59–75.

Lomax, J. A., Gordon, A. H. and Chesson, A. (1983). *Carbohydr. Res.* **122**, 11–22.

Lomax, J. A., Gordon, A. H. and Chesson A. (1985). *Carbohydr. Res.* **138**, 177–188.

Lönngren, J. and Svensson, S. (1974). *Adv. Carbohydr. Chem. Biochem.* **29**, 41–106.

MacKay, A. L., Wallace, J. C., Sasaki, K. and Taylor, I. E. P. (1988). *Biochemistry* **27**, 1467–1473.

Mäder, M. and Schloss, P. (1979). *Plant Sci. Lett.* **17**, 75–80.

Manzi, A. E., Cerezo, A. S. and Shoolery, J. N. (1986). *Carbohydr. Res.* **148**, 189–197.

Matheson, N. K. and McCleary, B. V. (1984). In "The Polysaccharides" (G. O. Aspinall, ed.), Vol. 3, pp. 1–105. Academic Press, New York.

McNeil, M., Darvill, A. G. and Albersheim, P. (1980). *Plant Physiol.* **66**, 1128–1134.

Monro, J. A., Bailey, R. W. and Penny, D. (1976). *Phytochemistry* **15**, 175–181.

Moody, S. F., Clarke, A. E. and Bacic, A. (1988). *Phytochemistry* **27**, 2857–2861.

Mori, M. and Katō, K. (1981). *Carbohydr. Res.* **91**, 49–58.

Morrison, I. M. (1974). *Phytochemistry* **14**, 505–508.

Mort, A. J. and Bauer, W. D. (1982). *J. Biol. Chem.* **257**, 1870–1875.

Mort, A. J. and Lamport, D. T. A. (1977). *Anal. Biochem.* **82**, 289–309.

Mort, A., Komalavilas, P., Maness, N., Ryan, J., Moersbacher, B. and An, J. (1988). In "XIV International Carbohydrate Symposium, Stockholm", Abstracts A6.

Nagahashi, G. and Seibles, T. S. (1986). *Protoplasma* **134**, 102–110.

Nari, J., Noat, G., Ricard, J., Franchini, E. and Moustacas, A.-M. (1983). *Plant Sci. Lett.* **28**, 313–320.

Northcote, D. H. (1963). In "Cell Differentiation" Symp. Soc. Exp. Biol. Vol. 17, pp. 157–174. Cambridge University Press, Cambridge.

Northcote, D. H. (1969). In "Essays in Biochemistry", Vol. 5, pp. 89–137. Academic Press, London.

O'Neill, M. A. and Selvendran, R. R. (1980). *Biochem. J.* **187**, 53–63.

O'Neill, M. A. and Selvendran, R. R. (1983). *Carbohydr. Res.* **111**, 239–255.

O'Neill, M. A. and Selvendran, R. R. (1985a). *Biochem. J.* **227**, 475–481.

O'Neill, M. A. and Selvendran, R. R. (1985b). *Carbohydr. Res.* **145**, 45–58.

Oshima, R. and Kumanotani, J. (1984). *Carbohydr. Res.* **127**, 43–57.

Oshima, R., Kumanotani, J. and Watanabe, C. (1983). *J. Chromatogr.* **259**, 159–163.

Parrish, F. W., Perlin, A. S. and Reese, E. T. (1960). *Can. J. Chem.* **38**, 2094–2104.

Peng, Q.-J. and Perlin, A. S. (1987). *Carbohydr. Res.* **160**, 57–72.

Perry, M. B. and Hulyalkar, R. K. (1965). *Can. J. Biochem.* **43**, 573–584.

Pressey, R. and Himmelsbach, D. S. (1984). *Carbohydr. Res.* **127**, 356–359.

Rasper, V. F., Brillouet, J. M., Bertrand, D. and Mercier, C. (1981). *J. Food Sci.* **46**, 559–563.

Redgwell, R. J. and Selvendran, R. R. (1986). *Carbohydr. Res.* **157**, 183–199.

Redgwell, R. J., O'Neill, M. A., Selvendran, R. R. and Parsley, K. J. (1986a). *Carbohydr. Res.* **153**, 97–106.

Redgwell, R. J., O'Neill, M. A., Selvendran, R. R. and Parsley, K. J. (1986b). *Carbohydr. Res.* **153**, 107–118.

Ring, S. G. and Selvendran, R. R. (1978). *Phytochemistry* **17**, 745–752.

Ring, S. G. and Selvendran, R. R. (1981). *Phytochemistry* **20**, 2511–2519.

Roberts, E. J., Godshall, M. A., Clarke, M. A., Tsang, W. S. C. and Parrish, F. W. (1987). *Carbohydr. Res.* **168**, 103–109.

Rocklin, R. D. and Pohl, C. A. (1983). *J. Liq. Chromatogr.* **6**, 1577–1593.

Rolf, D. and Gray, G. R. (1982). *J. Am. Chem. Soc.* **104**, 3539–3541.

Rolf, D. and Gray, G. R. (1986). *Carbohydr. Res.* **152**, 343–349.

Rolf, D., Bennek, J. A. and Gray, G. R. (1985). *Carbohydr. Res.* **137**, 183–196.

Ruperez, P., Selvendran, R. R. and Stevens, B. J. H. (1985). *Carbohydr. Res.* **142**, 107–113.

Ryden, P. and Selvendran, R. R. (1989). *Carbohydr. Res.*, (in press).

Ryden, P., Colquhoun, I. J. and Selvendran, R. R. (1989). *Carbohydr. Res.* **185**, 233–237.

Saeman, J. F., Moore, W. E. and Millet, M. A. (1963). *In* "Methods in Carbohydrate Chemistry" (R. L. Whistler, ed.), Vol. 3, Ch. 12, pp. 54–69. Academic Press, New York.

Scott, R. W. (1979). *Anal. Chem.* **51**, 936–941.

Selvendran, R. R. (1975a). *Phytochemistry* **14**, 1011–1017.

Selvendran, R. R. (1975b). *Phytochemistry* **14**, 2175–2180.

Selvendran, R. R. (1985). *J. Cell Sci. Suppl.* **2**, 51–88.

Selvendran, R. R. and DuPont, M. S. (1980). *J. Sci. Food Agric.* **31**, 1173–1182.

Selvendran R. R. and DuPont, M. S. (1984). *In* "Developments in Food Analysis Techniques" (R. D. King, ed.), Vol. 3, pp. 1–68. Elsevier Applied Science, New York.

Selvendran, R. R. and O'Neill, M. A. (1987). *Methods Biochem. Anal.* **32**, 25–153.

Selvendran, R. R. and Stevens, B. J. H. (1986). *In* "Modern Methods of Plant Analysis" (H.-F. Linskens and J. F. Jackson, eds), Vol. 3, pp. 23–46. Springer, Berlin and New York.

Selvendran, R. R., Davies, A. M. C. and Tidder, E. (1975). *Phytochemistry* **14**, 2169–2174.

Selvendran, R. R., March, J. F. and Ring, S. G. (1979). *Anal. Biochem.* **96**, 282–292.

Selvendran, R. R., Ring, S. G., O'Neill, M. A. and DuPont, M. S. (1980). *Chem. & Ind.*, pp. 885–888.

Selvendran, R. R., Stevens, B. J. H. and DuPont, M. S. (1987). *Adv. Food Res.* **31**, 117–209.

Selvendran, R. R., Verne, A. V. F. V. and Faulks, R. M. (1989). *In* "Modern Methods of Plant Analysis" (H.-F. Linskens and J. F. Jackson, eds), Vol. 10, pp. 234–259. Springer-Verlag, Berlin.

Sharp, J. K. and Albersheim, P. (1984). *Carbohydr. Res.* **128**, 193–202.

Sierakowski, M.-R., Gorin, P. A. J., Reicher, F. and Corrêa, J. B. C. (1987). *Phytochemistry* **26**, 1709–1713.

Slavin, J. L. and Marlett, J. A. (1983). *J. Agric. Food Chem.* **31**, 467–471.

Spellman, M. W., McNeil, M., Darvill, A. G., Albersheim, P. and Henrick, K. (1983). *Carbohydr. Res.* **122**, 115–129.

Stephens, G. J. and Wood, R. K. S. (1974). *Nature* **251**, 358.

Stevens, B. J. H. and Selvendran, R. R. (1980). *J. Sci. Food Agric.* **31**, 1257–1267.

Stevens, B. J. H. and Selvendran, R. R. (1984a). *Carbohydr. Res.* **135**, 155–166.

Stevens, B. J. H. and Selvendran, R. R. (1984b). *Carbohydr. Res.* **128**, 321–333.

Stevens, B. J. H. and Selvendran, R. R. (1984c). *Phytochemistry* **23**, 339–347.

Stevens, B. J. H. and Selvendran, R. R. (1988). *Carbohydr. Res.* **183**, 311–319.

Stevenson, T. T., Darvill, A. G. and Albersheim, P. (1988). *Carbohydr. Res.* **179**, 269–288.

Takeuchi, M., Takasaki, S., Inoue, N. and Kabata, A. (1987). *J. Chromatogr.* **400**, 207–213.

Talmadge, K. W., Keegstra, K., Bauer, W. D. and Albersheim, P. (1973). *Plant Physiol.* **51**, 158–173.

Taylor, R. L. and Conrad, H. E. (1972). *Biochemistry* **11**, 1383–1388.

Taylor, I. E. P., Tepfer, M., Callaghan, P. T., MacKay, A. L. and Bloom, M. (1983). *J. App. Polymer Sci.* **37**, 377–384.

Taylor, M. G., Deslandes, Y., Bluhm, T., Marchessault, R. H., Vincendon, M. and Saint Germain, J. (1983). *Tappi* **66**, 92–94.

Theander, O. and Aman, P. (1979). *Swed. J. Agric. Res.* **9**, 97–106.

Townsend, R. R., Hardy, M., Olechno, J. D. and Carter, S. R. (1988). *Nature* **335**, 379–380.

Valent, B., Darvill, A. G., McNeil, M., Robertson, B. K. and Albersheim, P. (1980). *Carbohydr. Res.* **79**, 165–192.

Vodonik, S. A. and Gray, G. R. (1988). *Carbohydr. Res.* **175**, 93–102.

Waeghe, T. J., Darvill, A. G., McNeil, M. and Albersheim, P. (1983). *Carbohydr. Res.* **123**, 281–304.

Walters, J. S. and Hedges, J. I. (1988). *Anal. Chem.* **60**, 988–994.

York, W. S., Darvill, A. G., McNeil, M. and Albersheim, P. (1985). *Carbohydr. Res.* **138**, 109–126.

York, W. S., Darvill, A. G., McNeil, M., Stevenson, T. T. and Albersheim, P. (1986). *Methods Enzymol.* **188**, 1–40.

York, W. S., Oates, J. E., Van Halbeek, H., Darvill, A. G., Albersheim, P., Tiller, P. R. and Dell, A. (1988). *Carbohydr. Res.* **173**, 113–132.

17 Anhydrous Hydrogen Fluoride in Polysaccharide Solvolysis and Glycoprotein Deglycosylation

GREGORY L. RORRER[1], MARTIN C. HAWLEY[2],
SUSAN M. SELKE[3], DEREK T. A. LAMPORT[4] and
PRAKASH M. DEY[5]

[1]*Department of Chemical Engineering, Oregon State University, Corvallis, OR 97331, USA*

[2]*Department of Chemical Engineering, Michigan State University, East Lansing, MI 48824, USA*

[3]*Department of Packaging, Michigan State University, East Lansing, MI 48824, USA*

[4]*MSU-DOE Plant Research Laboratory, Michigan State University, East Lansing, MI 48824, USA*

[5]*Department of Biochemistry, Royal Holloway and Bedford New College (University of London), Egham, Surrey TW20 0EX, UK*

METHODS IN PLANT BIOCHEMISTRY Vol. 2
ISBN 0-12-461012-9

I. INTRODUCTION

The use of anhydrous hydrogen fluoride (HF) as a pre-treatment step for the hydrolysis of lignocellulosic biomass can be traced back to the early 1930s. Helferich and co-workers examined the effects of anhydrous HF on various polysaccharide materials (Helferich and Böttger, 1929; Helferich *et al.*, 1930; Helferich and Peters, 1932) and Fredenhagen and Cadenbach (1933) utilised the reagent for wood saccharification, thus opening the process for commercial exploitation. However, after an interruption during the Second World War, but following the oil crisis of the 1970s, the research on this subject was revitalised with a view to utilising ethanol, obtained by fermenting wood-derived glucose, as a fuel source. Research on cellulose saccharification by reacting anhydrous HF with poplar and aspen wood was taken up at Michigan State University (Hardt and Lamport, 1982a,b; Selke *et al.*, 1982, 1983; Rorrer *et al.*, 1987, 1988a,b) who have developed laboratory-scale equipment for studying kinetics and intermediates of the process.

Anhydrous hydrogen fluoride is a potent and selective reagent for the solvolytic cleavage of polysaccharide and glycoprotein *O*-glycosidic linkages. In this chapter, we present two techniques for HF solvolysis of the polysaccharide component of plant cell-wall materials and one technique for the HF solvolysis of primary plant cell-wall glycoproteins. The goal of each HF solvolysis technique is to remove selectively the polysaccharide component of interest from the non-polysaccharide components. HF solvolysis is a unique tool for this purpose because at room temperature anhydrous HF quantitatively cleaves most *O*-glycosidic linkages on neutral, acidic or amino sugar residues (Knirel *et al.*, 1988) but does not cleave peptide linkages, or *N*-glycosidic linkages (Mort and Lamport, 1977). Furthermore, HF does not dehydrate sugars (Hardt and Lamport, 1982a), affect *N*-acyl amino sugar substituents (Mort and Lamport, 1977), or affect *O*-acyl sugar substituents (Knirel *et al.*, 1989). Thus, quantitative recovery of plant cell-wall polysaccharide (in the form of monosaccharides) can be obtained by HF solvolysis with minimal destruction of non-carbohydrate components.

Anhydrous HF, by virtue of its toxicity, corrosiveness, and low boiling point (19.54°C) is often considered difficult to handle. However, in our laboratory, we have developed safe and reliable methods for reacting plant cell-wall materials with anhydrous HF in either the liquid or vapour phase. Our three established techniques for HF solvolysis with application to analysis of plant cell-wall components are: (1) sugar composition of the plant cell wall polysaccharide fractions by liquid-phase HF solvolysis; (2) primary cell-wall glycoprotein deglycosylation by liquid-phase HF solvolysis; and (3) sugar composition and lignin distribution of 'hard-to-hydrolyse' lignocellulosic plant material by vapour-phase HF solvolysis. In this chapter, we describe the HF solvolysis chemistry, the HF solvolysis apparatus and reaction procedure, and the product isolation and analysis procedures relevant to each technique. We also provide sample analysis results and discuss the advantages and limitations of each technique.

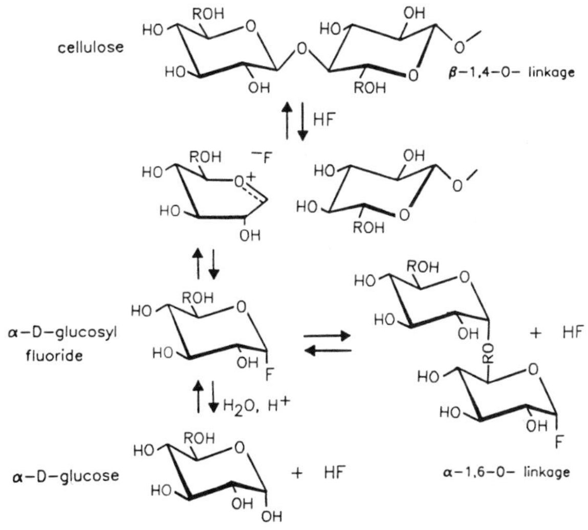

FIG. 17.1. HF solvolysis of cellulose.

II. SUGAR COMPOSITION OF PLANT CELL-WALL POLYSACCHARIDES BY LIQUID-PHASE HF SOLVOLYSIS

A. General Considerations

Anhydrous HF liquid is a potent reagent for quantitative solvolysis of cell-wall polysaccharides. At 0°C and above it dissolves polysaccharides and quantitatively cleaves most neutral sugar O-glycosidic linkages (Knirel et al., 1989). For example, liquid HF at 0°C readily dissolves even the 'hard-to-hydrolyse' polysaccharide, crystalline cellulose, from the cell wall. The HF then solvolytically cleaves the β-1,4-O-glycosidic bonds joining anhydroglucose (glucan) units constituting the cellulose chain to release α-D-glucosyl fluoride in HF solution. The solvolysis proceeds rapidly to completion within 1 h with little apparent glucoside dehydration (Hardt and Lamport,

1982a). However, prolonged exposure of glucose in liquid HF (6% by wt) at 20°C for 72 h can dehydrate about 10% of the glucosyl fluoride to 1,6-anhydroglucose (Kraska and Micheel, 1976). In HF solution, glucosyl fluoride can repolymerise to oligosaccharides of various chain lengths and glycosidic linkage conformations in concentration- and temperature-dependent equilibrium. Glucosyl fluoride can also undergo HF-catalysed Friedel–Crafts alkylation with aromatic compounds in HF solution (Wagner, 1965). The chemistry of liquid-phase HF-solvolysis of cellulose, amylose, and xylan is well documented (Hardt and Lamport, 1982a, 1982b; Defaye et al., 1982, 1983; Franz et al., 1987). The general reaction scheme for liquid-phase HF solvolysis of cellulose is presented in Fig. 17.1.

Anhydrous HF liquid selectively cleaves different O-glycosidic linkages as a function of temperature. In general, as the temperature is lowered, more O-glycosidic linkages are stable. Liquid HF at 23°C cleaves both neutral and amino sugar O-glycosidic linkages within 3 h. However, liquid HF at 0°C cleaves the neutral sugar O-glycosidic linkages but leaves the amino sugar O-glycosidic linkages intact after 1 h (Mort and Lamport, 1977). This selectivity is particularly marked at liquid HF temperatures below 0°C. For example, liquid-phase HF solvolysis of the extracellular polysaccharide *Bradyrhizobium japonicum* (strain 3I1b 138) at −23°C for 15 min resulted in cleavage of α-galactosyl (4-O-methyl) linkage, the α-mannosyl linkage, and the α-glucosyl linkage, but did not result in cleavage of the β-glucosyl linkage or the α-galacturonosyl (4-O-Acetyl) linkage (Mort and Bauer, 1982). Low-temperature, liquid-phase partial HF solvolysis of polysaccharides, when coupled with the appropriate oligosaccharide product analysis techniques, e.g. gel filtration and NMR spectroscopy, is a powerful tool for elucidation of the structure and glycosidic linkage composition of plant cell-wall polysaccharides (Mort et al., 1983). Low-temperature, selective cleavage of polysaccharide O-glycosidic linkages by liquid HF was recently reviewed by Knirel et al. (1989).

Experimentally, liquid-phase HF solvolysis of plant cell-wall material can be carried out on a microgram, milligram, or gram scale. A microapparatus (Sangar and Lamport, 1983) is used for liquid-phase HF solvolysis of 0.1 to 50 mg of cell-wall material, and a Peninsula Laboratories KEL-F vacuum distillation apparatus (Mort, 1983) is used for liquid-phase HF solvolysis of 0.05 to 10 g of cell-wall material. In general, we recommend the micro/milligram scale technique, because the liquid-phase HF solvolysis microapparatus is simple, inexpensive, and easy to use relative to the more complicated gram-scale KEL-F vacuum distillation apparatus. However, liquid-phase HF solvolysis of gram-scale quantities of cell-wall material may be required to provide product volumes sufficient for certain applications. These applications may include chemical analysis of lignin residue (Clark, 1962; Defaye et al., 1983; Smith et al., 1983), complex oligosaccharide product fractionation (especially for low-temperature selective cleavage of polysaccharides in liquid HF), or product utilisation studies (e.g. bioprocessing of sugars, sugar-oligomers or residual lignin to useful chemicals) relevant to development of the HF solvolysis of lignocellulose as a viable biomass-conversion technology (Selke et al., 1983). Experimental methods for liquid-phase HF solvolysis of gram quantities of plant cell wall material in the KEL-F vacuum distillation apparatus are well-documented (Mort and Lamport, 1977; Hardt and Lamport, 1982a, 1982b; Mort, 1983; Selke et al., 1983; Smith et al., 1983).

Our microtechnique for liquid-phase HF solvolysis of 0.1 mg to 20 mg quantities of plant cell-wall material is described in three parts: (1) preparation of plant cell-wall material for HF solvolysis, using *Triticum aestivum* (wheat) as an example; (2) liquid-phase HF-microapparatus and reaction procedure; and (3) sample analysis of the *T. aestivum* cell wall.

B. Cell Wall Preparation

Two-week old wheat leaves (*T. aestivum* var. Chancellor) are ground in liquid nitrogen and then homogenised in 15 mM potassium phosphate buffer (pH 7.0) using a motorised glass homogeniser for 3 min. The homogenate is then centrifuged at $600 \times g$ for 30 s. The pellet is collected and washed twice with distilled water. The solids are resuspended in distilled water and sonicated within a bath sonicator for 1 min or longer if microscopic examination reveals any intact cells or large cellular debris. The final pellet is washed with dry acetone, blown dry with flowing N_2 gas, and then dried in a vacuum desiccator for 24 h.

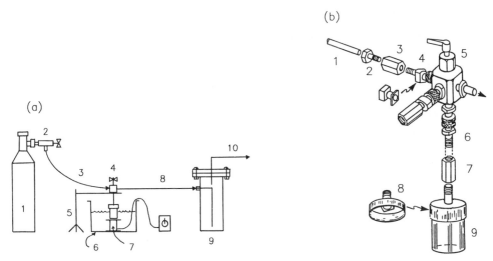

FIG. 17.2. Liquid-phase HF solvolysis microapparatus.
(a) Flow diagram: 1, liquid HF tank (Matheson Gas Products size 4); 2, Monel manual control valve (Matheson); 3, Teflon tubing (3 mm); 4, microapparatus assembly—see Fig. 17.2(b); 5, support stand for microapparatus assembly; 6, ice or controlled-temperature water bath; 7, microsubmersible stir plate (TRI-R Instruments); 8, vacuum line to calcium oxide trap; 9, calcium oxide trap (steel vessel with flanged lid containing 2 kg calcium oxide); 10, vacuum line to vacuum pump.
(b) Microapparatus detail: 1, Teflon tubing (3 mm); 2, Teflon tubing adaptor (3 mm); 3, Teflon coupling; 4, Luer adaptor (female); 5, Hamilton miniature Mininert valve, 6, Luer adaptor (female); 7, Coupling; 8, Teflon tubing adaptor (3 mm) bored into Teflon 'Tuf-Tainer' vial cap (Pierce); 9, 7 ml Teflon 'Tuf-Tainer' vial (Pierce).

C. Liquid-phase HF Solvolysis Microapparatus and Reaction Procedure

1. Microapparatus detail and HF safety precautions

A diagram of the HF solvolysis microapparatus is given in Figs 17.2a,b. Liquid HF is obtained from Matheson Gas Products. The liquid HF is contained within a steel cylinder of 1600 ml capacity equipped with a Monel manual control valve (Matheson model 55-A). Since anhydrous HF liquid is very corrosive, HF-wetted microapparatus parts are constructed of HF-inert Teflon plastic. Also, due to the toxicity and low boiling point of anhydrous HF liquid (19.54°C), the entire apparatus is contained within a fume hood.

Anhydrous HF liquid and its vapour are potentially lethal if ingested or inhaled. HF also burns skin tissue, with the extent of damage dependent on the HF concentration, amount, and time of contact. HF toxicity levels, first-aid procedures, and safe HF handling procedures are well-documented (Braker and Mossman, 1971; Gall, 1980; Allied Chemical Corporation, 1978, 1984). In general, whenever handling anhydrous HF, always work within a fume hood and wear protective clothing, including a full face shield, elbow-length impermeable gloves (viton, neoprene, or polyvinylchloride), and full labcoat.

2. *Reaction procedure*

The liquid-phase HF solvolysis procedure is straightforward and safe provided the microapparatus is constructed with care and the proper HF handling and safety procedures are reviewed. The procedure is divided into three major steps: (1) loading of cell-wall material and HF liquid into the reaction vial; (2) liquid-phase HF solvolysis of cell-wall material for a given reaction time at a given temperature; and (3) evacuation of liquid HF from the reaction products. All steps of the reaction procedure are designed to contain HF within the apparatus and away from the operator.

A pre-weighted quantity of dried cell-wall material (0.1 to 20 mg) and the internal standard myo-inositol (0.05 to 10 mg) are added to the 7 ml Teflon 'Tuf-Tainer' screw-cap vial (Pierce Chemical). For plant cell-wall material loadings from 20 to 50 mg, a 15 ml Teflon 'Tuf-Tainer' screw-cap vial (Pierce Chemical) is needed. For very small quantities of internal standard (< 1 mg), the appropriate volume of standard myo-inositol solution is pipetted into the vial and then blown down to dryness over flowing N_2 gas before the cell-wall material is added. Next, anhydrous methanol (100 μl per 1 ml HF) is pipetted into the vial, and a small Teflon-encased magnetic stirring bar is placed within the vial. The methanol will block sugar fluoride repolymerisation and possible HF-catalysed Friedel–Crafts alkylation of sugar fluorides with phenolic cell-wall components. Immediately after adding the methanol, the vial is screwed into the vial cap mounted on the HF delivery system. The vial is first cooled with liquid nitrogen and then evacuated under cooling. When evacuation is complete, the HF cylinder manual control valve and the Hamilton valve are opened, and HF liquid is distilled from the HF cylinder to the reaction vial. When the loading of HF liquid into the reaction vial is complete, the HF cylinder manual control valve and the Hamilton valve are closed. For 2 ml of HF liquid, the loading time is typically 3–5 min. Good general practice is to add 2 ml HF per 10 mg cell-wall material (0.5% by wt), or 2 ml HF per 20 mg cell wall material (1% by wt). For small sample amounts (less than 5 mg plant cell-wall material) use 1 ml HF. After the HF has thawed (freezing point -83°C), the dissolution and liquid-phase HF solvolysis of plant cell-wall polysaccharides is allowed to proceed under gentle stirring for the desired time at the desired temperature within the sealed Teflon reaction vial.

If the plant cell-wall polysaccharides contain any amino sugars, then liquid-phase HF solvolysis is carried out at room temperature (23°C) for 3 h. If the cell-wall polysaccharides are composed of neutral or acidic sugars, then the HF solvolysis reaction is carried out at 0°C for 1 h. In general, reactions at 0°C are preferred because potential fuming of HF vapour is minimised (the vapour pressure of HF liquid is 364 mm Hg at 0°C and 850 mm Hg at 23°C). For HF solvolysis reactions at 0°C, the microapparatus assembly (which is mounted on a ring stand) is positioned so that the lower half of the reaction vial is immersed in an ice bath and rests on top of a microsubmersible magnetic stir-plate (TRI-R Instruments), also immersed in the ice bath.

When the reaction is complete, the HF liquid is evacuated from the reaction vial. First, the sealed reaction vial and liquid reaction contents are lifted from the ice bath and allowed to reach room temperature. The Hamilton valve is then opened to the vacuum aspirator line, and shortly afterward the vacuum pump is turned on. Under reduced pressure, the liquid HF evaporates and is sucked through a calcium oxide trap which neutralises the HF to calcium fluoride. The HF-evacuation process requires at least 1 h. After HF evacuation, about 50–100 μl of HF is still retained in the reaction product syrup. This residual HF is evaporated from the product syrup by blowing a stream of N_2 gas directly into the open reaction vial within the fume hood.

3. Water-soluble product recovery and post-hydrolysis

After HF removal is complete, 2.0 ml of deionised/distilled water is added to the reaction vial and the contents are thoroughly mixed. The water-soluble fraction (sugar-oligomer product solution) is centrifuged (600 × g for 5 min) from the insoluble residue, decanted and frozen for storage.

During liquid-phase HF solvolysis and HF evacuation, glucosyl fluoride and other sugar fluorides repolymerise to water-soluble reversion oligosaccharides (Defaye et al., 1982; Hardt and Lamport, 1982b). Since the polysaccharide component of the cell-wall material is quantified as its constituent monosaccharide composition, the water-soluble HF solvolysis products are hydrolysed by dilute acid at mild conditions in order to recover monomer sugars from reversion oligosaccharides. For assay of the sugar composition in the water-soluble HF-solvolysis products, an aliquot of the sugar-oligomer product solution containing about 50–100 μg of total sugar is required, e.g. for 10 mg of original cell-wall material containing 6 mg total sugar in 2.0 ml water, 25–50 μl aliquot of the sugar-oligomer product solution is needed. Typically, for 10 mg of original cell-wall material in 2.0 ml water, a 50 μl aliquot of the sugar-oligomer product solution is blown down to dryness over flowing N_2 gas, and then hydrolysed in 200 μl of 2 N trifluoroacetic acid (TFA) at 121°C for 1 h within a septum-sealed 1.0 ml conical glass 'reacta vial' (Pierce Chemical) heated by a dry heating block (half-depth holes, 13 mm diameter). The hydrolysed sample is then blown down to dryness over flowing N_2 gas at 50°C in order to evaporate the volatile TFA (boiling point 72.4°C).

A minor amount of sugar degradation occurs during 2 N TFA post-hydrolysis under the conditions just described. The post-hydrolysis correction factor K_s is defined as

$$K_s = \text{weight of sugar before post-hydrolysis (g)/weight of sugar after post-hydrolysis (g)}$$
$$(1)$$

Typical values of K_s for glucose and xylose are 1.03 and 1.14, respectively (Rorrer et al., 1987). Minor sugar degradation during 2 N TFA hydrolysis at 121°C is quantified in detail by de Ruiter and Burns (1986).

4. Quantification of sugar composition by GLC/alditol-acetates

Gas–liquid chromatography (GLC) is a cumbersome but sensitive method for assaying the sugars released from the vapour-phase HF solvolysis of lignocellulose. Before analysis, the 50–100 μg total sugars in the assay volume (see Section IV.C.8: Post-hydrolysis) are derivatised to their corresponding alditol acetates (Albersheim *et al.*, 1967). The alditol acetates of neutral sugars are separated on a 2 m × 2 mm i.d. glass column packed with PEGS 224 (0.2% polyethylene glycol succinate, 0.2% polyethylene glycol adipate, 0.4% G.E. silicone XF 1150) on Gas Chrom Q. Typically, 1 μl sample (*c*. 1 μg each sugar derivative) and 3 μl ethyl acetate solvent are injected directly onto the column. The column carrier gas is helium at 40 ml min^{-1}, and the column oven temperature is programmed as follows: 130°C initial, 1.0°C min^{-1} ramp to 180°C, 16 min hold at 180°C. The injector temperature is set at 210°C, and the detector temperature is set at 240°C. Column effluent components are detected by a flame ionisation detector (FID). A typical GLC/alditol-acetates chromatogram is presented in Fig. 17.3.

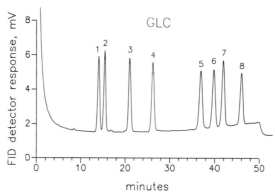

FIG. 17.3. Gas–liquid chomatography (GLC) of the alditol acetates of neutral sugars (standard sugar mix). Column, operating conditions and sample loading are specified in the text. Data were acquired on a Perkin Elmer 910 gas chromatograph (attenuation 8 ×, amplification 10 ×). Sugar identification: 1, rhamnose; 2, fucose; 3, arabinose; 4, xylose; 5, mannose; 6, galactose; 7, glucose; 8, myo-inositol.

The yield (Y_s, μmol sugar per mg plant cell-wall material) of each sugar released by the liquid-phase HF solvolysis of plant cell-wall material is calculated by the 'internal standard' method:

$$Y_s = \frac{A_s R_{F,I} w_I}{A_I R_{F,s} w_0} \tag{2}$$

where A_s is the peak area of a given sugar (μV s per μl injected), A_I is the peak area of the myo-inositol internal standard (μV s per μl injected), $R_{F,s}$ is the detector response factor for a given sugar (μV s μmol^{-1}), $R_{F,I}$ is the detector response factor for the myo-inositol internal standard (μV s μmol^{-1}), w_I is the amount of the myo-inositol internal standard (μmol) in the Teflon reaction vial and w_0 is the *dry* weight of plant cell-wall material in the Teflon reaction vial before reaction (μg or mg).

D. Sample Analysis

The sugar composition of wheat (*Triticum aestivum*) cell walls as determined by the liquid-phase HF solvolysis microtechnique and a traditional 2 N TFA hydrolysis technique (Albersheim *et al.*, 1967; de Ruiter and Burns, 1986) are compared in Table 17.1. In all sugar composition determinations by the liquid-phase HF solvolysis technique, a mild post-hydrolysis (2 N TFA at 121°C for 1 h) on the water-soluble products is required to recover monosaccharides from 'reversion' oligosaccharides. In general, combined liquid-phase HF solvolysis and 2 N TFA post-hydrolysis yields more sugars than 2 N TFA hydrolysis alone, particularly glucose. This is because liquid HF readily dissolves and solvolytically cleaves even 'hard-to-hydrolyse' crystalline cellulose from the plant cell wall under conditions that will not degrade the sugars released.

TABLE 17.1. Sugar composition of *Triticum aestivum* (wheat) cell walls.

Sugar	TFA hydrolysis[a] (mg sugar/100 mg cell wall ±1 SD)	Liquid-phase HF solvolysis/ TFA post-hydrolysis[b] (mg sugar/100 mg cell wall ±1 SD)
Rhamnose	0.3 ± 0.2	0.7 ± 0.1
Arabinose	4.3 ± 0.2	3.1 ± 0.4
Xylose	14.5 ± 0.3	15.4 ± 1.1
Mannose	0.1 ± 0.1	0.8 ± 0.1
Galactose	0.7 ± 0.1	0.6 ± 0.1
Glucose	3.1 ± 0.1	32.4 ± 1.3
Total sugars[c]	23.0	53.0

[a] Hydrolysis of wheat cell wall material in 2 N TFA at 121°C for 1 h.
[b] Liquid-phase HF solvolysis of wheat cell-wall material at 23°C for 3 h; post-hydrolysis of water-soluble reaction products in 2 N TFA at 121°C for 1 h.
[c] Solid residue not quantified.

In this example, the solid residue (e.g. lignin) was not quantified or characterised. In general, the action of HF on phenolic plant cell-wall components such as lignin is not well characterised. However, in preliminary work, Smith *et al.* (1983) and Defaye *et al.* (1983) showed that residual lignin from liquid-phase HF solvolysis of wood is significantly autocondensed but not fluorinated. Defaye *et al.* (1983) further proved that α- and β-aryl-ether linkages in lignin remain intact after exposure to liquid HF. Clark (1962) showed that the Klason content of HF-lignin is at least 95%. Therefore, liquid-phase HF solvolysis can be used to determine the overall lignin composition of lignified plant tissue, but not its native chemical structure.

III. DEGLYCOSYLATION OF PRIMARY CELL-WALL GLYCOPROTEINS BY LIQUID-PHASE HF SOLVOLYSIS

A. General Considerations

Deglycosylation of glycoproteins and mucoproteins in liquid HF is generally more than

90% complete at 1 h at 0°C for *O*-glycosidic linkages of neutral and acidic sugars, while *O*-glycosidic linkages of amino sugars require upwards of 3 h at 23°C. One must emphasise that anisole does not effectively scavenge the released sugar fluorides; anhydrous methanol (10%, v/v, in liquid HF) is better because the products, *O*-methyl glycosides, are water-soluble and dialysable.

The glycoproteins selected for demonstration of the HF-deglycosylation technique are the monomeric extensins, a class of primary cell-wall glycoproteins involved in primary plant cell-wall growth and assembly (Lamport, 1986). Isolation, purification and characterisation of these extensin 'precursors' from tomato cell liquid suspension cultures is described by Smith *et al.* (1984, 1986).

B. Deglycosylation Procedure and Isolation of Deglycosylated Proteins

For deglycosylation of extensin monomers we generally recommend the liquid-phase HF solvolysis procedure described in Section II.C of this chapter, with the following modifications. Instead of HF evacuation, the reaction vial is unscrewed from the microapparatus, and the contents in the vial are immediately and carefully poured in 0.5 ml increments into 10-fold excess of ice water. Extreme caution must be exercised during this step because the heat of solution of HF liquid in water is highly exothermic and will cause fuming of HF and water vapour from the quenched reaction mixture. Be sure that full face protection, gloves and labcoat are worn before attempting this portion of the procedure. Make sure that this procedure is carried within a fume hood using the safety guidelines described in Section II.C.

The HF-deglycosylated proteins are isolated from the ice-water quenched reaction mixture by dialysis and freeze-drying. Specifically, the diluted HF reaction mixture is dialysed (21 mm dialysis tubing, 3500 molecular weight cut-off) against water within a continuously stirred, tap-water-filled 15 litre plastic bucket for 48 h at 4°C (cold room). The tap water is changed in the bucket every 8 h, and the pH of the tap water in the bucket should be maintained above 4.0. When dialysis is complete, the dialysis tubing contents (deglycosylated glycoproteins in aqueous solution) are freeze-dried and weighed. When the sugars need to be measured for analysis, follow the procedure described under Section II.C.

C. Sample Results and Commentary

After HF deglycosylation, extensin monomers gave reproducible tryptic peptide maps and amino acid sequences showing that extensins consist of a few highly repetitive peptides (Smith *et al.*, 1986). Using mild liquid-phase HF solvolysis at 0°C for 1 h there is no evidence of HF-induced peptide cleavage (although we have not tested the very acid-sensitive Asp–Pro linkage) but liquid-phase HF solvolysis at 23°C for 3 h does tend to give (unquantified) peptide cleavage. However this degradation is not necessarily a disadvantage when raising antibodies against the core protein of heavily glycosylated mucoglycoproteins for subsequent screening of cDNA libraries and isolation of cDNA clones corresponding to the core of otherwise intractable proteins, such as heavily glycosylated cancer-related mucins (Gendler *et al.*, 1987). The very high degree of glycosylation prevents cross-reactivity between antibodies raised against the native and deglycosylated mucins. However, we predict that for normal glycoprotein antibodies raised against both the native and the deglycosylated glycoprotein will cross-react; the first demonstration of such cross-reactivity involved extensin monomers (Kieliszewski and Lamport, 1987) and the general method now makes it possible for the first time to quantify the contribution of both glycosylated and non-glycosylated eptitopes to the antigenicity of a glycoprotein.

Although deglycosylation of extensin monomers was shown to lead to a decrease in polyproline II helix content (Van Holst and Varner, 1984), we have demonstrated by electron microscopy that the molecule has a flexous rod shape (Heckman et al., 1988) which no longer behaves as a substrate for the cross-linking enzyme, extensin peroxidase (Everdeen et al., 1988).

Finally, we should mention anhydrous trifluoromethane sulphonic acid (TFMSA) as a possible alternative (Edge et al., 1981; Woodward et al., 1987) to the use of anhydrous HF for glycoprotein deglycosylation. Sojar and Bahl (1987) have described this method.

IV. SECONDARY CELL-WALL COMPOSITION AND LIGNIN DISTRIBUTION IN LIGNOCELLULOSIC PLANT MATERIAL BY VAPOUR-PHASE HF SOLVOLYSIS

A. General Considerations

The secondary cell wall of lignified plant tissue is an intricately constructed, biopolymer composite of cellulose, hemicellulose, and lignin. In general, it is difficult to determine accurately and simultaneously the carbohydrate composition and lignin distribution in lignified plant tissue because Nature designed the lignocellulosic matrix to resist both enzymatic and chemical attack. The micropores in the secondary cell wall severely restrict the mobility of bulky cellulose-hydrolysis and lignin-oxidation enzymes through the lignocellulosic matrix and thus hinder selective enzymatic degradation. The rigid hydrogen-bonding network within the paracrystalline cellulose microfibril severely dampens pyranosyl ring flexure to the cyclic oxonium/carbonium ion intermediate formed upon hydronium-ion catalysed cleavage of the β-1,4-O-glycosidic bond (Harris, 1975). Therefore, hydrolysis of cellulose from the lignocellulosic matrix by aqueous mineral acids (e.g. H_2SO_4 and HCl) occurs only at high temperatures or acid concentrations—conditions which promote acid-catalysed sugar dehydration and lignin condensation.

Anhydrous HF vapour is a potent reagent for the solvolysis of 'hard-to-hydrolyse' secondary cell-wall polysaccharides (Rorrer et al., 1987, 1988a). The solvolysis of lignocellulose by anhydrous HF vapour is briefly described here. Anhydrous HF vapour at ambient conditions readily diffuses into the lignocellulose particle and physically adsorbs onto the lignocellulosic matrix. The physically adsorbed HF disrupts the rigid hydrogen-bonding network of the crystalline cellulose microfibril and 'solvates' individual cellulose chains by displacing OH···OH hydrogen bonds with stronger HO···HF hydrogen bonds. Adsorbed HF then rapidly cleaves the β-1,4-O-glycosidic linkage joining anhydroglucose (glucan) units of the HF-solvated cellulose chain to yield glucosyl fluoride, which resides on the HF-reacting lignocellulosic matrix. The solvolysis is usually complete within 2–12 min, depending on HF vapour conditions, and does not dehydrate the sugars. However, during HF solvolysis, the glucosyl fluoride can repolymerise to oligosaccharides of various chain lengths and glycosidic linkage conformations. The extent of reversion increases with increasing reaction temperature and decreasing HF loading on the lignocellulose particle. Glucosyl fluoride and reversion oligosaccharides are water-soluble and can be readily dissolved from the anatomically intact, residual lignin matrix by washing the HF-reacted lignocellulose particle in water.

Vapour-phase HF solvolysis is a novel and useful technique for composition and lignin-distribution analysis of lignified plant tissue. Specifically, this technique offers: (1) rapid and simple fractionation of polysaccharides from the highly encrusted lignocellulosic matrix; (2) quantitative determination of 'hard-to-hydrolyse' secondary cell-wall polysaccharides; and (3) preparation of a anatomically-intact residual lignin matrix free of polysaccharides. Furthermore, since HF vapour readily penetrates the lignocellulose particle 'whole' lignocellulose samples (not just finely ground material) can be analysed. Therefore, the vapour-phase HF solvolysis technique has obvious advantages over enzymatic degradation, solvent extraction, and acid hydrolysis for selective removal and analysis of the secondary cell-wall components of lignified plant tissue.

Although the vapour-phase HF solvolysis apparatus is much more complicated than the liquid-phase HF solvolysis microapparatus described earlier (Section II.C), the vapour-phase HF solvolysis reaction procedure is simple, rapid, and yields an anatomically intact residual lignin matrix. In contrast, liquid-phase HF solvolysis of lignocellulose is slow and somewhat cumbersome, especially with respect to loading of HF liquid on the plant cell-wall material and removal of HF liquid from the reaction mixture. Furthermore, the process of polysaccharide dissolution in liquid HF disintegrates the architecture of the lignocellulosic matrix.

Our technique for the vapour-phase HF solvolysis of lignocellulose is described in Sections IV.B–E below. Modifications of the technique for specific applications and limitations of the technique are also discussed. Although the hardwood *Populus grandidentata* is the substrate described in the protocol, the general technique is applicable to any lignified plant tisue.

B. Sample Preparation and Anhydrous HF Vapour Properties

1. *Lignocellulose sample preparation*

Roughly-chipped bigtooth aspen wood (*Populus grandidentata*, Michx, 2.5 cm mesh, 6% moisture) is microtomed with the grain of the wood into wafers 0.5 mm thick. The wafers are then cut into squares of side 1.0 cm and dried *in vacuo* over phosphorous pentoxide. Typical weight of the lignocellulose chips (after drying) is 20 mg.

2. *Anhydrous hydrogen fluoride*

Anhydrous hydrogen fluoride (HF) is obtained in liquid form from Matheson Gas Products. The HF liquid is contained in a steel cylinder of 1600 ml capacity equipped with a Monel manual control valve (Matheson model 55-A). The cylinder is stored under its own vapour pressure at room temperature within a fume hood.

Anhydrous HF vapour is potentially lethal if inhaled or ingested. HF also burns skin tissue, with the extent of damage dependent on the HF concentration, amount, and time of contact. HF toxicity levels, first-aid procedures, and safe HF handling procedures are well-documented (Braker and Mossman, 1971; Gall, 1980; Allied Chemical Corporation, 1978, 1984).

The physical properties of HF vapour are also well documented. HF vapour at 1.0 atm. and 20–100°C is non-ideal, because HF molecules in the vapour state associate by hydrogen bonding to form oligomers ranging from 2–6 HF units in size. The extent of monomer/oligomer equilibrium decreases with increasing temperature and decreasing HF partial pressure (Long *et al.*, 1943; Smith, 1958). HF liquid properties (including vapour pressure) are summarised by Simons (1950) and Gall (1980). Detailed thermodynamic properties of HF vapour are provided by Vanderzee and Rodenburg (1970).

C. Vapour-phase HF Solvolysis Apparatus and Reaction Procedure

The vapour-phase HF solvolysis apparatus is illustrated in Figs 17.4–17.6.

1. *Materials of construction*

Anhydrous HF vapour is toxic and corrosive, therefore the entire apparatus must be contained within a fume hood. The apparatus is designed to completely contain the HF vapour even when repeated access to the HF-reacting lignocellulose sample is desired. All HF-wetted tubing (0.64 cm o.d.), fittings (Swagelok, Inc.) and valves (Whitney, Inc.) are fabricated from HF-resistant Monel (alloy 400) metal. The reactor is constructed of plain-carbon steel, which is suitable for HF vapour service provided water vapour is not present. The tubular portion of the reactor is fabricated from ASTM (Americal Society of Testing and Materials) schedule-40 (3.8 cm o.d., 3.2 cm i.d.) steel pipe. The calcium carbonate bed shell is fabricated from polyvinylchloride (PVC) pipe and bushings, and sealed with PVC-cement.

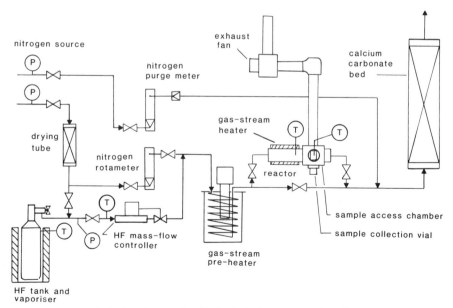

FIG. 17.4. Vapour-phase HF solvolysis apparatus: flow diagram.

2.   HF vapour delivery and control

Although the flow design of the vapour-phase HF solvolysis apparatus is apparent from Fig. 17.4, several features merit special description. Anhydrous HF liquid (normal boiling point 19.54°C) is vaporised directly from the HF cylinder (1600 ml capacity) by a heating mantle under gentle heating. The cylinder skin temperature is monitored by a surface thermocouple and maintained between 30 and 35°C [50°C is the recommended safety limit (Braker and Mossman, 1971)]. This cylinder temperature provides: (1) slow and even HF vapour evolution; and (2) an HF vapour delivery pressure sufficient for driving HF vapour across the mass-flow control valve. The HF vapour flow rate is monitored and controlled by a Matheson Instruments model 8205 ($0-5 \, l \, min^{-1}$) mass flow transducer and control valve. All HF-wetted parts within the transducer and control valve are plated with HF-resistant Monel metal. The transducer is calibrated for mass flow by measuring the weight loss of the HF cylinder as a function of time at a given flow setting. All HF vapour lines (including the HF mass flow transducer and control valve) are wrapped with heating cord and heated at 40°C to prevent HF vapour condensation. The pre-heater bath is used to maintain the mixed HF vapour + N_2 stream at 30°C, because the heat of dilution of HF vapour is endothermic (Briegleb and Strohmeier, 1953). The HF/N_2 gas stream temperature within the tubular portion of the reactor is maintained by a 300 W heating coil (stainless steel sheath of 0.16 cm diameter by 2 m straight length). The heating coil is wrapped around the outside of the reactor so that corrosion of the heating coil by HF vapour is prevented. Power input to the heating coil is provided by a Barber-Colemen model 580A PID temperature controller. All gas stream temperatures are measured by Omega Engineering type-J Inconel-sheathed thermocouples (0.16 cm diameter) interfaced to an Omega Engineering (model 660) digital thermometer.

3.   Reactor detail and sampling mechanism

The solvolysis of a single lignocellulose sample to solid sugars and residual lignin is carried out within the reactor (see Figs 17.5 and 17.6) of the vapour-phase HF solvolysis apparatus. Within the tubular portion of the reactor, a single lignocellulose sample is exposed to an HF vapour stream (in N_2 diluent) of precisely controlled HF vapour flow rate, N_2 diluent flow rate, and HF/N_2 gas stream temperature. Total pressure (P_T) in the reactor is always 1.0 atm. The complete range of operating conditions is summarised in Table 17.2. The HF partial pressure (P_{HF}) in the reactor is calculated by the equation

$$P_{HF} = F_{HF}P_T/(F_{HF} + F_{N_2}) \tag{3}$$

where F_{HF} and F_{N_2} are the molar flow rates of HF vapour and N_2.

A Teflon sampling mechanism (see Fig. 17.6) loads a single lignocellulose wafer into the reactor, and then retrieves the HF-reacted solid for later sample processing and analysis. The sample access chamber contains the Teflon sampling mechanism and sample access port. The lignocellulose wafer (typically 20 mg) rests on the Teflon sample cartridge. The Teflon sample cartridge in turn rests on the tip of the sampling probe. When the Teflon sampling mechanism is inserted into the tubular portion of the reactor (see Figs 17.6a,b), a Teflon plug (with polyethylene gasket not shown) mounted at the

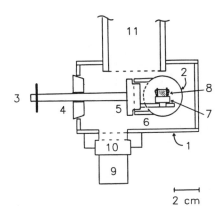

2 cm

FIG. 17.5. Vapour-phase HF solvolysis apparatus: reactor detail (side-view cross-section). Key: 1, sample access chamber; 2, tubular section of reactor running through sample access chamber (conduit for HF vapour of controlled flowrate, HF partial pressure, and temperature); 3, Teflon sampling probe and mechanism; 4, re-sealable Teflon plug for sample access chamber; 5, re-sealable Teflon plug for sample access port; 6, sample access port; 7, Teflon sample cartridge; 8, lignocellulose sample propped upright on Teflon sample cartridge; 9, 15 ml Teflon 'Tuf-Tainer' sample collection vial (Pierce); 10, vial cap for 15 ml vial (mounted onto base of sample access port); 11, conduit to exhaust fan (sucks up HF vapours released during sampling).

base of the sampling probe seals the reactor at the sample access port. The sample cartridge and lignocellulose wafer are now inserted within the tubular portion of the reactor. After a given reaction time, the sampling probe is unplugged and rotated 90° (see Figs 17.6c,d). The HF-reacted solid and Teflon sample cartridge are deposited into a 15 ml Teflon 'Tuf-Tainer' vial (Pierce Chemical) screwed into the base of the sample access chamber. This vial contains 5.00 ml deionised/distilled water (to quench the reaction) and a known inclusion of myo-inositol (typically 6.00 mg), an internal standard used for later chromatographic quantification of the sugar products. Other HF-reacted sample removal options are possible, depending on the specific application desired.

TABLE 17.2. Range of reaction conditions for the vapour-phase HF solvolysis apparatus.

Reaction condition	Range
HF or HF + N_2 gas stream temperature	30–90°C
HF vapour flow rate through reactor[a]	0–5 g min^{-1} 0–0.4 mol min^{-1}
N_2 diluent flow rate through reactor	0–5 l min^{-1} 0–6.25 g min^{-1} 0–0.22 mmol min^{-1}
HF partial pressure[b]	0–1.0 atm
Total pressure	Fixed at 1.0 atm
Typical sample loading	1 wood wafer 0.5 mm thick × 1.0 cm × 1.0 cm, 20 mg

[a] Reactor cross-sectional area = 7.9 cm². Typical HF vapour flow rate = 1.5 to 3.0 g min^{-1}.
[b] HF partial pressure calculated by Equation (3) in text.

(a)

(b)

(c)

(d)

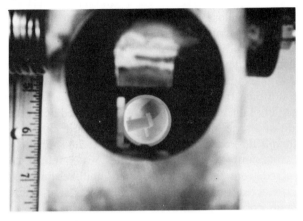

FIG. 17.6. Vapour-phase HF solvolysis apparatus: sampling mechanism operation (top view). (a) Placement of the Teflon sampling mechanism in the sample access chamber. (b) Insertion of the Teflon sample cartridge and lignocellulose sample into the tubular section of reactor seals the sample access port (HF vapour flow in tubular section of reactor only—see Fig. 17.5). (c) Removal of the Teflon sample cartridge and lignocellulose sample from the tubular section of the reactor. (d) Rotation of the Teflon sampling probe deposits the Teflon sample cartridge and lignocellulose sample into the sample collection vial (vial contains 5.0 ml water to quench the rotation).

4. HF disposal

During sampling, the seal between the tubular portion of the reactor and the sampling mechanism is momentarily broken. An exhaust fan plumbed into the sample access chamber sucks up the fugitive HF-vapours and vents them to the fume-hood exhaust. In normal operation, the reactor effluent HF vapour is passed through an 8 kg bed of dry calcium carbonate chips (1–2 cm mesh). The calcium carbonate neutralises the HF vapour to calcium fluoride, water and carbon dioxide. By reaction stoichiometry, 5.0 kg calcium carbonate chips are sufficient to neutralise 2000 ml liquid HF. Usually, the calcium carbonate bed is recharged with fresh calcium carbonate chips whenever the HF cylinder (1600 ml capacity) is changed.

5. Safety precautions

As previously mentioned, anhydrous HF is both toxic and corrosive. The vapour-phase HF solvolysis apparatus must be contained within a fume hood. Protective garments, including full face shield, elbow-length impermeable gloves (Viton, Neoprene or PVC), and labcoat must be worn during all experimental work with HF. All personnel working with HF must be acquainted with safe HF handling procedures, HF toxicity levels, and first-aid in the event of direct exposure to HF via inhalation or skin contact.

6. Standard reaction procedure

Lignocellulose wafers (side 1.0 cm, 0.5 mm thick, 20 mg) are weighed to a precision of

0.1 mg. The HF cylinder is heated to constant skin temperature of 35°C. The HF vapour flow rate, N_2 flow rate, and HF/N_2 gas stream temperature are set and allowed to stabilise. A single lignocellulose sample is then loaded into the reactor via the Teflon sampling mechanism just described. After a specified reaction time (typically within 15 min), the HF-reacted solid is removed from the reactor by the Teflon sampling mechanism and quenched in water in order to stop the reaction. The Teflon sample cartridge is washed with an additional 5 ml deionised/distilled water (washings added to reaction products) and then carefully dried with a cotton swab before reloading. The procedure is repeated until the desired number of HF-reacted lignocellulose samples are obtained. At the end of a series of reactions, the HF vapour flow is turned off and the apparatus flow system is flushed thoroughly with N_2 at $2.0 \, l \, min^{-1}$ for at least 30 min.

When the HF-reacted sample is quenched in water, water-soluble sugar fluorides and oligomers are dissolved from the water-insoluble lignin residue. Any HF physically adsorbed on the HF-reacted sample is also dissolved. The anatomically intact residual lignin shell (brown in colour) is removed from the water-soluble products, rinsed thoroughly and gently with the deionised/distilled water, and then dried *in vacuo* and weighed. The residual lignin shell is flaccid and easily fragmented, and therefore must be handled carefully. Calcium carbonate powder is added incrementally (10–20 mg) to the water-soluble product solution to precipitate dissolved HF to calcium fluoride. The pH-neutral sugar product solution is then centrifuged (4000 rev min^{-1} for 5 min), decanted from the insoluble calcium carbonate and calcium fluoride, and frozen for storage.

7. *Variations to the standard reaction procedure*

Some applications may require that the HF-reacted sample be retrieved as a solid without water quenching. The reaction procedure is easily modified to obtain a solid HF-reacted sample. After a desired reaction time, the HF vapour flow is turned off and an ambient N_2 gas stream at $2.0 \, l \, min^{-1}$ is allowed to pass over the HF-reacted sample for at least 30 min. The N_2 gas stream desorbs the physically adsorbed HF from the sample. The entire sampling mechanism is then removed from the reactor and the HF-reacted solid sample is removed from the Teflon sample cartridge. This HF-reacted sample (solid sugars plus residual lignin matrix) is brittle and easily crumbled, and therefore must be handled carefully.

Delicate lignocellulose samples can be accommodated. The lignocellulose sample is carefully mounted onto a square polyethylene wafer of side 1.0 cm and 1.0 mm thick. The polyethylene backing helps to maintain the overall structural integrity of the sample, particularly if the sample is very thin (less than 50 μm). The polyethylene wafer (plus lignocellulose sample) is then inserted into the reactor via the Teflon sampling mechanism described earlier. HF adsorbed on the HF-reacted solid is desorbed under flowing N_2 gas and then retrieved from the reactor as described in the preceding paragraph. The water-soluble reaction products are then gently washed away from the residual lignin matrix by carefully pippetting deionised/distilled water onto the HF-reacted solid.

8. *Post-hydrolysis*

During both liquid- and vapour-phase HF solvolysis of lignocellulose, glucosyl fluoride and other sugar fluorides repolymerise to short-chain, water-soluble reversion oligosaccharides (Defaye *et al.*, 1982, 1983; Hardt and Lamport, 1982b; Rorrer *et al.*, 1987). In order to recover the parent glucose (and other monomer sugars) from these reversion oligosaccharides, the water-soluble products are hydrolysed in dilute acid under mild conditions. Typically, a 50 μl aliquot of the water-soluble product solution (containing 0–50 μg of each sugar) is blown down to dryness over flowing N_2 gas at 50°C, and then hydrolysed in 200 μl of 2 N trifluoroacetic acid (TFA) for 1 h at 121°C within a septum-sealed 1.0 ml conical glass 'reacti vial' (Pierce Chemical) heated by a dry heating block (half-depth holes, 13 mm diameter). These hydrolysis conditions are sufficient to hydrolyse sugar fluorides and reversion oligomers to monomer sugars (Rorrer *et al.*, 1987). The hydrolysed sample is then blown down to dryness over an N_2 gas stream at 50°C in order to evaporate the volatile TFA (boiling point 72.4°C). The dry sample is then redissolved in 20 μl of degassed HPLC-grade water and vortexed prior to sugar analysis by HPLC or GLC/alditiol acetates.

9. Quantification of sugar composition by HPLC

The sugar composition in the HF-reacted lignocellulose sample is conveniently assayed by high-performance liquid chromatography (HPLC). A Spectra-Physics isocratic HPLC system, consisting of a SP8810 precision isocratic pump, Rheodyne 7125 injector (20 μl sample loop), SP8790 column heater, and SP6040 differential refractive index detector is suitable for carbohydrate analysis by HPLC. Glucose, xylose, mannose, etc. and myo-inositol (the internal standard) are separated on a BIO-RAD Laboratories HPX-87P monosaccharide analysis HPLC column. The 20 μl assay sample (see above) is injected directly onto the column without derivatisation. A BIO-RAD Laboratories de-ashing column is plumbed in front of the analysis column to remove any particulates or residual acid (TFA) present in the 20 μl injection sample. The column is eluted with degassed HPLC-grade water at 0.6 ml min^{-1}, and the column temperature is isothermally maintained at 85°C. Column effluent sugars are detected by differential refractive index with respect to water. A typical HPLC chromatogram is presented in Fig. 17.7. Sugar yields are calculated by the 'internal standard' method via Equation (2) and corrected for loss during 2 N TFA post-hydrolysis via Equation (1).

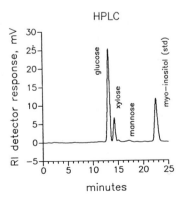

FIG. 17.7. High performance liquid chromatography (HPLC) of sugars released by the vapour-phase HF solvolysis of the hardwood *Populus grandidentata* (aspen) after post-hydrolysis. HPLC hardware, column, column operating conditions, and sample loading are specified in the text. RI detector range: 10 mV = 16×10^{-6} RIU.

D. Sample Analysis

The solvolysis of lignocellulose by anhydrous HF vapour is dependent on several conditions, including: (1) reaction time; (2) HF/N_2 gas stream flow rate; (3) HF/N_2 gas stream temperature; (4) HF partial pressure; and (5) lignocellulose wafer thickness (Rorrer *et al.*, 1988a,b). Detailed kinetic and yield data for vapour-phase HF solvolysis of lignocellulose, along with mathematical modelling of the HF adsorption, sugar product release and energy release processes, are documented by Rorrer *et al.* (1988a,b). Typical reaction conditions and sugar yields for the vapour-phase HF solvolysis of the hardwood *Populus grandidentata* (aspen) are presented in Table 17.3.

TABLE 17.3. Selected reaction conditions and water-soluble glucose and xylose recoveries from the vapour-phase HF solvolysis of the hardwood *Populus grandidentata* (aspen). Detailed data given by Rorrer *et al.* (1987, 1988a,b).

Vapour-phase HF solvolysis conditions[a,b]				Yield[d]	
Reaction time (min)	HF vapour temperature (°C)	HF partial pressure (atm)	Equilibrium HF loading (g HF/g)[c]	Glucose (mg/100 mg wood ±1 SD)	Xylose (mg/100 mg wood ±1 SD)
>2	30	1.00	1.49	51.6 ± 1.1	15.3 ± 0.9
>10	30	0.60	0.70	51.5 ± 1.5	16.3 ± 1.7
>6	50	1.00	0.57	52.5 ± 1.7	14.3 ± 2.6

[a] Total HF + N_2 flow rate fixed at $0.4 \, \text{g cm}^{-2} \, \text{min}^{-1}$ ($3 \, \text{g min}^{-1}$). Lower flow rates can be used, but the reaction time to maximum conversion must be increased because the reaction is subject to convective mass transfer resistances.
[b] Lignocellulose chip thickness fixed at 0.5 mm. Different lignocellulose particle thicknesses or diameters can be used, but the reaction time to maximum conversion must be adjusted accordingly because the reaction is subject to intraparticle mass transfer resistances.
[c] Equilibrium HF loading determined by gravimetric HF adsorption experiments is described by Rorrer *et al.* (1988a).
[d] All yield determinations based on mg of original, *dry* lignocellulose (e.g. mg glucose/100 mg dry lignocellulose). ±1 SD, $n = 7$ determinations. 'Extractives' were not removed prior to HF-solvolysis.

The composition of aspen wood as determined by both the vapour-phase HF solvolysis technique and a well-established two-step sulphuric acid hydrolysis technique (Timell, 1957) are compared in Table 17.4. The lignin recovery obtained from vapour-phase HF solvolysis of wood is high because minor amounts of sugar and the wood's original 'extractive' component are retained in the solid residue, as shown in Table 17.5. The extractive component (e.g. resins, rosins, oils, tannins, terpenes, etc.) is easily removed from the solid residue by extraction with acetone–water (9 : 1, v/v) at room temperature for 24 h.

Methanolysis of glucosyl fluoride to 1-*O*-methyl glucose is known to inhibit glucosyl fluoride reversion during liquid-phase HF solvolysis of cellulose (Franz *et al.*, 1987). *In situ* methanolysis of glucosyl fluoride may likewise inhibit possible HF-catalysed glucosyl fluoride/lignin condensation. In fact, when methanol-absorbed lignocellulose (*c.* 20 µl per 10 mg substrate) is reacted with HF vapour, the sugar yields are increased relative to dry lignocellulose, as shown in Table 17.4. Therefore, the accuracy of carbohydrate composition determination via vapour-phase HF solvolysis is potentially improved if methanol (20 μl per 10 mg) is pipetted onto the lignocellulosic sample just before reaction with HF vapour.

TABLE 17.4. Composition of the hardwood *Populus grandidentata* (aspen): vapour-phase HF solvolysis and two-step sulphuric acid hydrolysis compared.

	Two-step sulphuric acid hydrolysis[a] (mg/100 mg wood ±1 SD)	Vapour-phase HF solvolysis[b]	
Component		Dry (mg/100 mg wood ±1 SD)	+Methanol[c] (mg/100 mg wood ±1 SD)
Glucose	55.6 ± 0.4	51.6 ± 1.1	55.8 ± 1.1
Xylose	19.1 ± 0.6	15.3 ± 0.9	15.7 ± 0.7
Mannose	2.1 ± 0.2	1.9 ± 0.4	—
Galactose	1.1 ± 0.1	1.1 ± 0.2	—
Solid residue (lignin)	17.7 ± 0.1	22.0 ± 1.6	—
Total measured recovery[d]	87.4	84.6	—

[a] The wood sugar determination is carried out by two-step sulphuric acid hydrolysis described by Timell (1957). 200 mg dried wiley-milled wood grindings are hydrolysed in 5.0 ml 72% (wt) sulphuric acid at 30°C for 1 h; products are diluted with water to 3% (wt) sulphuric acid solution and post-hydrolysed in an autoclave at 121°C for 1 h. Solid residue is filtered, washed, dried, and weighed.
[b] Single wood chips (0.5 mm thick × 1.0 cm × 1.0 cm, 20 mg) are reacted with 100% HF vapour at 30°C and 1.0 atm for 3–8 min. The water-insoluble residue is washed, dried, and weighed. The water-soluble products (1% by wt) are hydrolysed in 2 N TFA at 121°C for 1 h.
[c] 10 μl of anhydrous methanol per 10 mg dry wood is pipetted onto the lignocellulose wafer just before reaction with HF vapour. Minor wood sugars and solid residue were not quantified.
[d] Sum of sugar recovery (expressed as their anhydrosugars) and solid residue. Arabinose, uronic acids, wood 'extractives' were not quantified.

TABLE 17.5. Composition of solid residue obtained from the solvolysis of aspen wood by 100% HF vapour at 1.0 atm, 30°C.

Component	% of solid residue	% of original dry wood
Total solid residue	100.0	22.0
Acetone-extractable[a] material in solid residue	14.1	3.1
Glucan residual[b]	7.5	1.7
Xylan residual[b]	4.3	0.9
Glucan + xylan residual	11.8	2.6
Total acetone-extractable and carbohydrate residual	25.9	5.7
Remaining solid (lignin)	74.1	16.3

[a] Solid residue extracted with acetone–water (9:1; v/v) at room temperature for 24 h.
[b] Solid residue hydrolysed in 2 N TFA (100 μl per 1 mg) at 121°C for 1 h. Water-soluble products assayed for glucose and xylose by HPLC.

Rorrer *et al.* (1988a,b) previously showed that HF vapour must first physically adsorb onto the lignocellulosic matrix before HF-solvolysis of cellulose and hemicellulose can occur. The equilibrium HF loading on the lignocellulosic matrix decreases with decreasing HF partial pressure and increasing temperature. The minimum HF loading for complete reaction with cellulose and hemicellulose in aspen wood is about 0.4 g HF per g dry wood. This value represents the HF required to saturate the hydroxyl groups and *O*-glycosidic linkages constituting cellulose and hemicellulose. The effects of HF loading and HF vapour temperature (at 1.0 atm) on the maximum glucose yield attainable from the vapour-phase HF solvolysis of aspen wood are shown in Table 17.6.

TABLE 17.6. Effect of 100% HF vapour temperature at 1.0 atm on: (1) the equilibrium loading of HF physically adsorbed on the lignocellulosic matrix of aspen wood; (2) the maximum average glucose yield from the HF-reacted aspen wood.

HF vapour temperature (°C)	Equilibrium HF loading (g HF/g wood)	Maximum average glucose yield (mg/100 mg wood ± 1 SD)
30	1.49	51.6 ± 1.1
40	0.91	53.5 ± 1.6
50	0.57	52.5 ± 1.7
67	0.31	45.7 ± 2.7
80	0.12	30.1 ± 6.2
90		17.1 ± 1.9

In general, the framework of the whole HF-reacted lignocellulose particle is best preserved if the equilibrium HF loading on the sample is between 0.4 and 1.0 g HF per g dry wood. Higher equilibrium HF loadings tend to disintegrate the architecture of the lignocellulose particle. After solvolysis of wood by 100% HF vapour at 30°C and 1.0 atm (HF loading 1.5 g HF per g wood), the HF-reacted lignocellulose particle is easily broken into several fragments. However, these fragments are easily recovered. In general, for preparation of an anatomically-intact residual lignin framework, HF/N$_2$ gas stream reaction conditions of $P_{HF} = 0.6$ atm (1.5:1 HF/N$_2$ molar flow ratio) at 30°C, or $P_{HF} = 1.0$ atm (undiluted, 100% HF vapour) at 45–50°C are most suitable.

E. Microscopy of the Residual Lignin Framework

The technique for vapour-phase HF solvolysis of lignocellulose generates an anatomically intact residual lignin framework free of polysaccharides. A technique for preparation of the residual lignin framework for visualisation by light microscopy is briefly described here.

A 2 mm × 2 mm specimen is cut from the residual lignin shell. The specimen is dried in 1.0 ml acetone–1% 2,2-DMP for 24 h. The acetone is decanted, and the specimen is impregnated with 1.0 ml (1:1; v/v) dry acetone–'spurrs' resin (22.5% wt % or vol %

FIG. 17.8. Light micrographs (200 ×) of the lignocellulose matrix of the hardwood *Populus grandidentata*. (a) Native lignocellulosic matrix. (b) Partially HF-reacted lignocellulosic matrix. (c) HF-reacted residual lignin shell (water-soluble reaction products are dissolved away from lignocellulosic matrix). Reaction conditions: 100% HF, 50°C, 3 min. Scale bars, 50 μm.

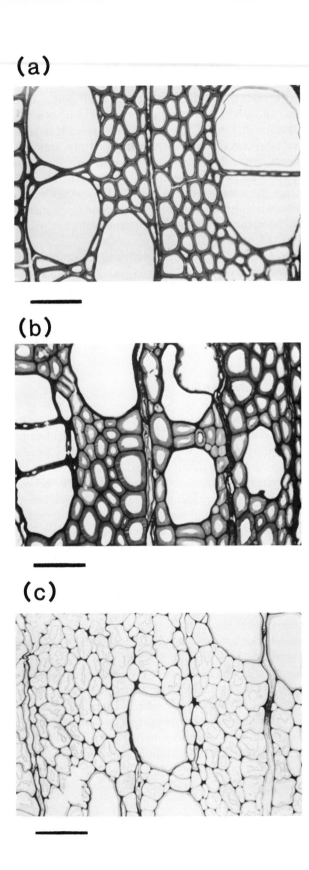

vinyl cyclohexane dioxide, 18.0% diglycidyl ether of propylene glycol, 56.6% nonenyl succinic anhydride, and 0.9% dimethylaminoethanol) for 24 h at room temperature under slow wrist-action shaking. The impregnation procedure is repeated with 1.0 ml (2:1; v/v) dry acetone–'spurrs' resin, 1.0 ml (3:1; v/v) dry acetone–'spurrs' resin, and finally, 1.0 ml 'spurrs' resin. The specimen and resin are then poured into a casting mould, and the 'spurrs' resin is cured in the casting mould within a vacuum oven at 60°C for 24 h. Several 2 μm thick cross-sections are sliced from the embedded specimen using a glass-blade mechanical ultramicrotome. The sections are stained successively with methylene blue and basic fuchsin (lignin, blue/purple; cellulose, red/pink) and visualised directly via light microscopy.

Sample light micrographs (200×) of the native lignocellulosic matrix, partially HF-reacted lignocellulosic matrix, and HF-reacted residual lignin shell of aspen wood are presented in Figs 17.8a,b,c.

In earlier work, Sachs *et al.* (1963) used aqueous hydrogen fluoride (5–80% by wt) to attempt to dissolve polysaccharides away from the lignocellulosic matrix of pine wood. Sachs then observed the lignin distribution in the residual lignin framework by electron microscopy.

V. CONCLUDING REMARKS

The HF solvolysis process is a rapid and low-energy process for desaccharification of polysaccharides and compares favourably with mineral acid-catalysed reaction. By applying this process polysaccharides such as starch (from plant storage organs) and cellulose (from wood and straw) can be readily converted to glucose. The HF is recoverable after the reaction. However, one drawback of the process is the generation of reversion products (oligosaccharides of various DP) which are not readily fermentable as compared to free glucose. Thus, further research is required in order to (1) inhibit the reversion, and (2) re-convert the reversion products into simple monosaccharides. The latter can be achieved either by enzyme-catalysed hydrolysis of the oligosaccharides, probably via immobilised enzyme technology, or by mineral acid-catalysed hydrolysis. In this respect the use of dilute acids, such as sulphuric or hydrochloric acid, may be of much value. On the other hand, reversion oligosaccharides may be isolated from the solvolysed mixture and these can serve as useful substrates for studying the mechanism of action of various glucosyl hydrolases. It would also be interesting to examine whether these 'oligosaccharins' possess biological activities (cf. Albersheim and Darvill, 1985).

It is of much importance that the action of HF on various types of glycosidic linkages can be made selective by altering the reaction temperature and time (Knirel *et al.*, 1988). At low temperatures certain types of glycosidic linkages become resistant to hydrolysis. Thus, HF solvolysis can be used as an important means of determination of polysaccharide structure. In addition, it can also generate interesting oligosaccharide fragments (with biological activity!), for example, from cell-wall fractions.

REFERENCES

Albersheim, P. and Darvill, A. G. (1985). *Scient. Am.* **253**, 58–64.

Albersheim, P., Nevins, D. J., English, P. D. and Karr, A. (1967). *Carbohydr. Res.* **5**, 340–345.

Allied Chemical Corporation (1978). "Hydrofluoric Acid". Industrial Chemicals Division, Morristown, NJ.

Allied Chemical Corporation (1984). "First Aid Treatment for Hydrofluoric Acid Burns". Industrial Chemicals Division, Morristown, NJ.

Braker, W. and Mossman A. L. (1971). *In* "Matheson Gas Data Book", pp. 305–309. Matheson Gas Products, East Rutherford, NJ.

Briegleb, G. and Strohmeier, W. (1953). *Z. Elektrochem.* **57**, 668–674.

Clark, I. T. (1962). *Tappi* **45**, 310–314.

Defaye, J., Gadelle, A. and Pedersen, C. (1982). *Carbohydr. Res.* **110**, 217–227.

Defaye, J., Gadelle, A., Papadopoulos, J. and Pedersen, C. (1983). *J. Appl. Polymer* **Sci. : Appl. Polymer Symp.** **37**, 653–670.

de Ruiter, J. M. and Burns, J. C. (1986). *J. Agric. Food Chem.* **34**, 780–785.

Edge, A. S. B., Faltynek, C. R., Hof, L., Reichert, L. E. and Weber, P. (1981). *Anal. Biochem.* **118**, 131–137.

Everdeen, D. S., Kiefer, S., Willard, J. J., Muldoon, E. P., Dey, P. M., Li, X.-B. and Lamport, D. T. A. (1988). *Plant Physiol.* **87**, 616–621.

Franz, R., Fritsche-Lang, W., Deger, H. M., Erckel, R. and Schlingmann, M. (1987). *J. Appl. Polymer Sci.* **33**, 1291–1306.

Fredenhagen, K. and Cadenbach, G. (1933). *Angew. Chemie* **46**, 113–124.

Gall, J. F. (1980). *In* "Kirk-Othmer Encyclopedia of Chemical Technology", 3rd edn (M. Grayson and D. Eckroth, eds), Vol. 10, pp. 733–753. John Wiley, New York.

Gendler, S. J., Burchell, J. M., Duhig, T., Lamport, D., White, R., Parker, M. and Taylor-Papadimitriou, J. (1987). *Proc. Natl. Acad. Sci. USA* **84**, 6060–6064.

Hardt, H. and Lamport, D. T. A. (1982a). *Phytochemistry* **21**, 2301–2303.

Hardt, H. and Lamport, D. T. A. (1982b). *Biotech. Bioeng.* **24**, 903–918.

Harris, J. F. (1975). *J. Appl. Polymer Sci.: Appl. Polymer Symp.* **28**, 131–144.

Heckman, J. W., Terhune, B. T. and Lamport, D. T. A. (1988). *Plant Physiol.* **86**, 848–856.

Helferich, B. and Böttger, S. (1929). *Justus Liebigs Ann. Chemie* **476**, 150–170.

Helferich, B. and Peters, O. (1932) *Justus Liebigs Ann. Chemie* **494**, 101–106.

Helferich, B., Starker, A. and Peters, O. (1930). *Justus Liebigs Ann. Chemie* **482**, 183–188.

Knirel, Yu. A., Vinogradov, E. V. and Mort, J. A. (1989). *Adv. Carbohydr. Chem. Biochem.* **47**, 167–202.

Kraska, U. and Micheel, F. (1976). *Carbohydr. Res.* **49**, 195–199.

Kieliszewski, M. and Lamport, D. T. A. (1987). *Phytochemistry* **25**, 673–677.

Lamport, D. T. A. (1986). *In* "Cellulose: Structure, Modification and Hydrolysis" (R. M. Rowell, ed.), pp. 77–90. John Wiley and Sons, New York.

Long, W. R., Hildebrand, J. H. and Morrell, W. E. (1943). *J. Am. Chem. Soc.* **65**, 182–187.

Mort, A. J. (1983). *Carbohydr. Res.* **121**, 315–321.

Mort, A. J. and Bauer, W. D. (1982). *J. Biol. Chem.* **257**, 1870–1875.

Mort, A. J. and Lamport, D. T. A. (1977). *Anal. Biochem.* **82**, 289–309.

Mort, A. J., Utille, J. P., Torri, G. and Perlin, A. S. (1983). *Carbohydr. Res.* **121**, 221–232.

Rorrer, G. L., Ashour, S. A., Hawley, M. C. and Lamport, D. T. A. (1987). *Biomass* **12**, 227–246.

Rorrer, G. L., Mohring, W. R., Hawley, M. C. and Lamport, D. T. A. (1988a). *Energy & Fuels* **2**, 556–566.

Rorrer, G. L., Mohring, W. R., Hawley, M. C. and Lamport, D. T. A. (1988b). *Chem. Eng. Sci.* **43**, 1831–1836.

Sachs, I. B., Clark, I. T. and Pew, J. C. (1963). *J. Polymer, Sci. Part C* **2**, 203–212.

Sanger, M. P. and Lamport, D. T. A. (1983). *Anal. Biochem.* **128**, 66–70.

Selke, S. M., Hawley, M. C., Hardt, H., Lamport, D. T. A., Smith, G. and Smith J. (1982). *Ind. Eng. Chem. Prod. Res. Dev.* **21**, 11–16.

Selke, S. M., Hawley, M. C. and Lamport, D. T. A. (1983). "Wood and Agricultural Residues: Research on Use for Feed, Fuels and Chemicals" (J. Soltes, ed.), pp. 329–349. Academic Press, New York.

Simons, J. H. (1950). *In* "Fluorine Chemistry" (J. H. Simons, ed.), pp. 225–256. Academic Press, New York.

Smith, D. F. (1958). *J. Chem. Phys.* **28**, 1040–1056.

Smith, J. J., Lamport, D. T. A., Hawley, M. C. and Selke, S. M. (1983). *J. Appl. Polymer Sci., Appl. Polymer Symp.* **37**, 641–651.

Smith, J. J., Muldoon, E. P. and Lamport, D. T. A. (1984). *Phytochemistry* **23**, 1233–1239.

Smith J. J., Muldoon, E. P., Willard, J. J. and Lamport, D. T. A. (1986). *Phytochemistry* **25**, 1021–1030.

Sojar, H. T. and Bahl, O. P. (1987). *Arch. Biochem. Biophys.* **259**, 52–57.

Timell, T. E. (1957). *Tappi* **40**, 568–572.

Vanderzee, C. E. and Rodenburg, W. W. (1970). *J. Chem. Therm.* **2**, 461–478.

Van Holst, G.-J. and Varner, J. E. (1984). *Plant Physiol.* **74**, 247–251.

Wagner, A. (1965). *In* "Friedel–Crafts and Related Reactions" (G. A. Olah, ed.), p. 235. Interscience, New York.

Woodward, H. D., Ringler, N. J., Selvakumar, R., Simet, I. M., Bhavanandan and Davidson, E. A. (1987). *Biochemistry* **26**, 5315–5322.

18 Techniques for Studying Interactions Between Polysaccharides

M. J. GIDLEY[1] and G. ROBINSON[2]

[1]*Unilever Research Laboratory, Colworth House, Sharnbrook, Bedford, MK44 1LQ, UK*

[2]*Department of Food Research and Technology, Silsoe College, Silsoe, Bedford, MK45 4DT, UK*

I. INTRODUCTION

Polysaccharides are widely distributed in nature and perform a range of biological functions. Particularly important examples of polysaccharide functionality include

METHODS IN PLANT BIOCHEMISTRY Vol. 2
ISBN 0-12-461012-9

maintenance of structural integrity (e.g. cellulose, chitin), energy reserve storage (e.g. starch, glycogen), and biological protection and adhesion (e.g. gum exudates, extracellular microbial polysaccharides). Structural roles are played by polysaccharides in plant, animal and microbial systems. In plants, the majority of the complex and versatile cell-wall matrix is provided by cellulose, pectins, xyloglucans and other polysaccharides (McNeil et al., 1984; Dey and Brinson, 1984; Mackie, 1985). In animals, glycosamino-glycans are important cell-surface and intercellular components (Hook et al., 1984; Fransson, 1985), and in microbial systems polysaccharides are involved in cell-wall architecture (Ward, 1981; Cabib et al., 1982; Nikaido and Vaara, 1985; Shepherd, 1987) as well as often being present as secreted extracellular polymers (Kenne and Lindberg, 1983; Sutherland, 1985).

Interactions between polysaccharides could be either covalent or non-covalent in character. Direct covalent linkages between polysaccharides have been demonstrated in plant (McNeil et al., 1984; Dey and Brinson, 1984) and microbial (Ward, 1981) cell walls. Covalent linkage between polysaccharide chains also occur through bridging bifunctional groups, e.g. in the peptidoglycan matrix of bacterial cell walls (Burchard, 1985). Non-covalent interactions between polysaccharides are much more widespread and in general are more important as they often form the mechanistic basis for natural functionality. For example, the strength of many plant materials can, in part, be traced to the ability of cellulose to form fibrous structures through assembly of individual chains; and the storage of starch in a state which contains comparatively little water (i.e. the starch granule) is accomplished by partial crystallisation of polysaccharides. Less obviously, the ability of gum exudates and many extracellular microbial polysaccharides to form effective viscous microenvironments reflects interactions between polysaccharides based on the overlap and interpenetration of the volumes occupied by hydrated molecules. The above examples represent two general mechanisms involved in non-covalent polysaccharide interactions, i.e. 'specific' chemical interactions (e.g. crystallisation) and 'non-specific' physical interactions (e.g. polymer entanglement).

Non-covalent interactions also form the basis for many commercial applications of polysaccharides. Examples include the ability of low concentrations of polysaccharides (e.g. xanthan, guar gum) to form highly viscous solutions and the tendency of certain polysaccharides to form aqueous gels under appropriate conditions (e.g. agar, carrageenans, alginates, pectins, starches). The commercially valuable gelling properties of polysaccharides in many cases mirror their biological function as network-forming agents (e.g. in cell walls) and utilise specific intermolecular interactions as a basis for the formation of a three-dimensional gel network (Rees et al., 1982).

In principle, there are two types of polysaccharide interactions, those between like components (self-interaction) and those involving unlike components (binary interaction). Examples of specific self-interactions (e.g. double helix formation, crystallisation) and non-specific self-interactions (e.g. polymer entanglement) are well known. The area of binary interactions is less well understood and is the subject of current debate. In general, dissimilar polymers will have a tendency to undergo phase separation when solvated and this has been documented for certain mixed polysaccharide–protein (Clark et al., 1983) and polysaccharide–polysaccharide (Kalichevsky and Ring, 1987) solutions and gels. Specific binary molecular interactions have also been proposed for various polysaccharide mixtures (Dea et al., 1972, 1977; Cairns et al., 1987) but there is no general agreement as yet for any proposed model. One example where Nature might benefit from a specific binary polysaccharide interaction is in plant cell-wall structures where it is thought that complexes of cellulose and xyloglucans are present (Dey and Brinson, 1984).

In this chapter we describe some of the techniques which are most useful for studying non-covalent polysaccharide interactions over various distance scales (i.e. macroscopic, supramolecular and molecular). Due to the complexity of natural systems and the commercial importance of isolated polysaccharides, intact natural systems have received less attention than well-characterised model systems. Such model studies can however provide significant insights into the probable *in vivo* behaviour of polysaccharides and complement direct studies of natural systems.

In the first section, the use of mechanical measurements for characterising the macroscopic consequences of 'specific' and 'non-specific' polysaccharide interactions is discussed. In the second section, techniques applicable to the study of the supramolecular organisation of polysaccharide assemblies are highlighted, and in the final section techniques which can be used to deduce detailed molecular information on the nature of 'specific' polysaccharide interactions are discussed. To illustrate applications of the techniques under discussion, examples will be drawn mainly from plant polysaccharides and their assemblies. The reader is referred to other chapters in this volume and elsewhere (Aspinall, 1985) for detailed chemical descriptions of individual systems.

II. MECHANICAL CHARACTERISATION OF POLYSACCHARIDE SYSTEMS

The consequences of interactions between polysaccharides are often apparent from the mechanical properties of material, e.g. the addition of potassium ions to a mobile solution of κ-carrageenan can result in the formation of a solid-like gel. The mechanical strength of many biological structures is also crucially dependent on polysaccharide interactions in fibrils (e.g. cellulose) and networks (e.g. alginate, carrageenan). In attempts to relate the mechanical properties of polysaccharides to molecular structures and interactions, many insights into the molecular origins of mechanical (rheological) properties have been gained (Morris and Ross-Murphy, 1981; Ross-Murphy, 1984; Clark and Ross-Murphy, 1987). The following sections will describe some of the approaches to characterising the rheological properties of isolated (i.e. non-interacting) polysaccharides, the consequences of both non-specific (i.e. physical) and specific (e.g. chemical) interactions, and the destructive failure properties of polysaccharide networks.

A. Rheological Properties of Isolated Polysaccharides

Studies of dilute solutions provide information on polysaccharide molecular masses and shapes, as well as establishing the properties of non-interacting systems which can then be used to assess whether polysaccharide interactions are occurring in other (more concentrated) systems. The most useful experimental parameter is the intrinsic viscosity, $[\eta]$, which is derived from measurements of relative viscosity, η_{rel} (solution viscosity/ solvent viscosity) and specific viscosity, η_{sp} (relative viscosity $-$ 1) for polysaccharide concentrations (c) having $1.2 < \eta_{rel} < 2.0$ for a range of shear rates using, typically, a coaxial rotational viscosimeter (Morris and Ross-Murphy, 1981). Extrapolation of η_{sp}/c and $\ln\eta_{rel}/c$ to zero concentration and shear rate yields the intrinsic viscosity which has units of reciprocal concentration (i.e. it is not a viscosity).

Intrinsic viscosity is related to relative molecular mass, M_r, through the Mark–Houwink–Sakurada equation:

$$[\eta] = K\,M_r^{\alpha}$$

where α and K are constants for a given polysaccharide/solvent system. The exponential factor α is an indicator of polymer conformation type with typical values of 0.5–0.8 for random coils and 1.5–2.0 for rigid rods, and K is a measure of chain stiffness. A direct relationship between intrinsic viscosity and polymer size is given by the Flory–Fox equation:

$$[\eta] = 6^{3/2}\,\Phi\,R_g/M_r$$

where R_g is the so-called radius of gyration and Φ is a constant for coil-like polymers.- For a given molecular mass, intrinsic viscosity is therefore seen to describe a volume which may be taken as a measure of the hydrodynamic volume of the isolated polymer molecule.

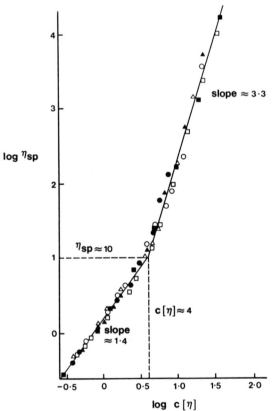

FIG. 18.1. Zero shear viscosity as a function of the coil overlap parameter, $c\,[\eta]$, for a range of random coil polysaccharides. Reprinted with permission from *Carbohydr. Polymers* **1** (1981), 5–21, © 1981 Elsevier Applied Science Publishers.

B. Physical Interactions between Polysaccharides in Solution

Using intrinsic viscosity as a measure of hydrodynamic volume, generalised properties of polysaccharide random coils can be probed as a function of the volume fraction occupied (i.e. concentration, c, $\times$ intrinsic viscosity). A plot of log η_{sp} against log $c[\eta]$ for a range of polysaccharides is shown in Fig. 18.1 (Morris *et al.*, 1981). For concentrations up to $c[\eta] \sim 4$, a slope of ~ 1.4 is obtained: this is termed the 'dilute' solution region and corresponds to isolated polymer molecules. With increasing concentration, a point is reached where polymer molecules are forced to interpenetrate or entangle due to volume constraints: this is known as the coil overlap concentration, although in practice the experimentally observable consequences of entanglement onset appear gradually. For polysaccharide random coils this region occurs at $c[\eta] \sim 4$ (Fig. 18.1). At concentrations significantly above coil overlap, viscosity increases more

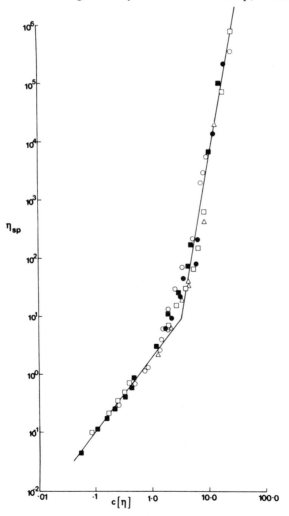

FIG. 18.2. 'Zero shear' viscosity as a function of the coil overlap parameter, $c[\eta]$, for various molecular weight fractions of guar galactomannan. The limiting slope at high overlap is ~ 5.1. Reprinted with permission from *Carbohydr. Res.* **107** (1982), 17–32, © 1982 Elsevier Scientific Publishing Company.

rapidly with concentration as polymer coils are forced to interpenetrate to greater extents: this is termed the 'semi-dilute' region and is an example of physical interaction between polysaccharides in solution. For some polysaccharides, the limiting slope of $\log \eta_{sp}$ against $\log c[\eta]$ in the semi-dilute region is higher than the typical value of ~ 3.3. An example is given in a study of guar galactomannan samples of different molecular weights (Robinson et al., 1982). Figure 18.2 shows a limiting slope of ~ 5.1 in the 'semi-dilute' region. This value is higher than that expected theoretically (3.75) for topological entanglements (DeGennes, 1979) of linear polymers and has been ascribed to more specific polymer–polymer interactions (Robinson et al., 1982).

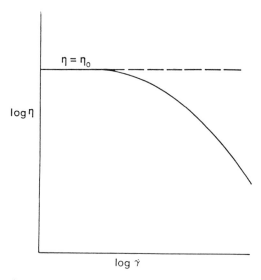

FIG. 18.3. Shear rate ($\dot{\gamma}$) dependence of viscosity for a Newtonian liquid (dashed line) and a typical polysaccharide solution (solid line). Reprinted with permission from Carbohydr. Polymers 1 (1981), 5–21, © 1981 Elsevier Science Publishers.

The shear rate dependence of solution viscosity has also been examined for a number of polysaccharides. For an ideal (Newtonian) solution the ratio of stress (τ) to shear rate ($\dot{\gamma}$) (i.e. the shear viscosity) is linear. For most polymer solutions, the dependence of viscosity on shear rate is as shown in Fig. 18.3. The decrease in viscosity with increasing shear rate is termed shear thinning. To obtain accurate viscosity measurements for intrinsic viscosity determinations etc., it is important that the linear region is accessed for determination of η_0, the zero shear viscosity (Fig. 18.3). Shear thinning can be understood in terms of the timescale of relaxation for intra- or intermolecular effects such as segmental motion or entanglement which lead to viscosity effects. For a polymer of known molecular weight and intrinsic viscosity an average relaxation time t' can be defined as

$$t' = 6\eta_s [\eta] \, M_r/\pi^2 RT$$

where η_s is solvent viscosity, R is the gas constant and T is absolute temperature. This refers to zero concentration and is modified for finite concentrations by replacing $[\eta]$ with $(\eta_0 - \eta_s)/\eta_s c$:

$$t' = 6(\eta_0 - \eta_s)M_r/\pi^2 cRT$$

The effect of increasing shear rate is to decrease the time available for relaxation processes and hence generalised behaviour can be probed through a dimensionless parameter (β) containing the product of shear rate × relaxation time. The parameter β is defined as $t'\gamma\,\pi^2/6$ and therefore

$$\beta = \gamma(\eta_0 - \eta_s)\,M_r/cRT$$

From measurements of η_0 as a function of γ for given concentrations and molecular weights of guar galactomannan, the plot shown in Fig. 18.4 was obtained (Robinson *et al.*, 1982). Such a plot demonstrates that the shear thinning behaviour of guar galactomannan is a predictable property if M_r (or $[\eta]$) is known. Similarly predictable shear thinning has also been shown for a range of random coil polysaccharides (Morris *et al.*, 1981).

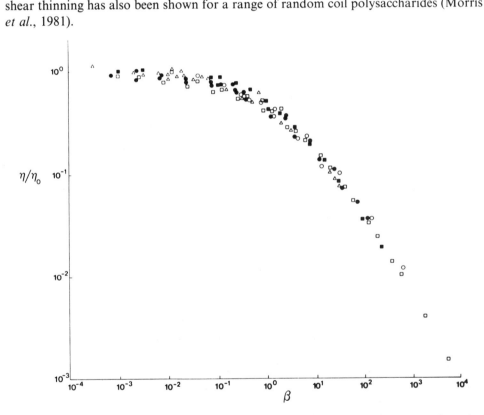

FIG. 18.4. Fractional decrease in viscosity with increases in generalised shear rate, β (see text), for various molecular weight fractions of guar galactomannan. Reprinted with permission from *Carbohydr. Res.* **107** (1982), 17–32, © 1982 Elsevier Scientific Publishing Company.

C. Rheological Probes of Specific Polysaccharide Interactions

The physical interactions such as chain entanglement described above can be detected by rotational viscometers. However, if more permanent structures are present within the system, then purely rotatory experiments are no longer applicable as any structure present will be broken down during the experiment. Clearly, small deformation measurements are required which preserve the structure. It is in this area that considerable progress has been made since the introduction of reliable strain-controlled rheometers and the cheaper stress-controlled instruments. Care must be exercised with the latter type of instrument to ensure that measurements are made within the linear viscoelastic region as strain is not controlled directly. In a typical small deformation experiment a small oscillatory drive is applied to the system and parameters such as G' (the storage modulus), G'' (the loss modulus) and η^* (the dynamic viscosity) are measured. When a sinusoidal strain wave is applied to a perfectly elastic material, the resulting stress is in phase with the strain. For a purely viscous liquid it will be 90° out of phase with the strain as the maximum response occurs when the rate of change of strain with time is at its maximum value. G' is the ratio of the in-phase stress to strain and G'' is the 90° out-of-phase stress to strain. The total or dynamic shear modulus (stress) of the system $G^* = (G'^2 + G''^2)^{\frac{1}{2}}$ and the ratio G'/G'' defines the phase angle (tan δ). We can also define a dynamic viscosity η^* as the ratio of total stress to frequency of oscillation (ω)

i.e.
$$\eta^* = \frac{G^*}{\omega}$$

The variation of such parameters as G', G'' and η^* with frequency of oscillation is called the mechanical spectrum and provides a rheological 'fingerprint' for viscoelastic materials. Figure 18.5 shows typical mechanical spectra for dilute and concentrated solutions and a strong gel. For solutions at low frequency, $G'' > G'$ as might be expected: with increasing frequency and concentration, the timescale of polymer flow decreases and the solution assumes more solid-like character, i.e. G' increases and G'' decreases at high frequencies. For gels $G' \gg G''$ and is nearly independent of frequency.

As the dynamic viscosity η^* is the small deformation equivalent of conventional solution viscosity, values for non-interacting polymer systems should be superimposable at equivalent values of $\dot{\gamma}$ and ω. However, if specific polymer interactions occur then η will be less than η^* as the rotational experiment would destroy some of the structure which contributes to η^*. A typical example of a random coil polymer with equivalence of η and η^* is guar galactomannan (Richardson and Ross-Murphy, 1987a): an example of non-equivalence of η and η^* is that of xanthan gum solutions (Richardson and Ross-Murphy, 1987b) for which specific polymer interactions are thought to occur.

A notable property of many polysaccharides is the ability to form strong gels at low concentration (typically 1–5%). The minimum concentration at which gelation occurs is typically below the onset of entanglement coupling (Fig. 18.1) and is ascribed to specific inter-chain binding of ordered ribbon or helical structures (see Section IV). Techniques for characterising the small deformation mechanical properties of aqueous polysaccharide gels have been reviewed recently (Clark and Ross-Murphy, 1987). Rheological parameters for gels should contain information relating to the strength and number of cross-links within the gel. Theoretical models have been developed recently in an attempt to relate modulus measurements to molecular details (Clark and Ross-Murphy, 1987). The most useful approach seems to be through analysis of the dependence of modulus on polymer concentration. Using one of these approaches, it has been shown recently that modulus/concentration data for amylose gels formed from a range of mono-disperse fractions show a generality of behaviour which can be ascribed to a mechanism in which the distance between cross-links is ~ 100 glucose residues (Clark *et al.*, 1989).

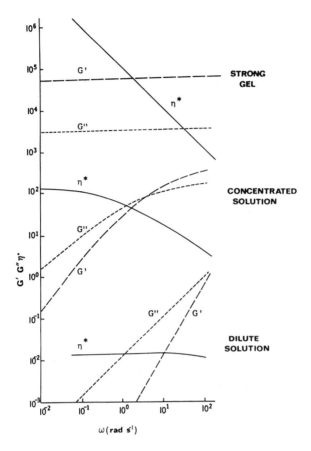

FIG. 18.5. Characteristic features of polysaccharide mechanical spectra for a strong gel (2% agar), and concentrated (5% λ-carrageenan) and dilute (5% dextran) solutions. Reprinted with permission from 'Gums and Stabilisers for the Food Industry—2' (edited by G. O. Phillips, D. J. Wedlock and P. A. William), p. 60, © 1984 Pergamon Press plc.

D. Large Deformation Measurements

Although small deformation measurements can provide non-destructive data relevant to an understanding of molecular interactions, such measurements may not relate directly to mechanical properties important in the natural function or commercial utilisation of polysaccharides. For example measurement of G' for gels at small deformation does not lead directly to an estimate of the breakstrength. Many comparative techniques have been used to assess the large deformation and failure properties of polysaccharide gels: examples include compressive and extensive tensile testing, and penetrometry (Christianson *et al.*, 1986; Clark and Ross-Murphy, 1987). The networks present in polysaccharide gels often mimic those which occur naturally, and the mechanical properties of gels may have counterparts in the maintenance of natural structures. For example, it has been shown that the chemical composition of alginates extracted from various algal tissues differs such that a high proportion of guluronate

residues occurs in tissues where strength is required and a high proportion of mann-
uronnate occurs in tissues where mobility is considered to be important (Stockton *et al.*,
1980). This correlates with the properties of alginate gels, which increase in strength
with guluronate content. Recently, tensile extension experiments on fresh samples of
aglal tissues have shown this effect to be detectable by mechanical measurements
(Dawes, 1988).

The large deformation and failure properties of fibrous polysaccharide structures are
also of great interest. In particular the mechanical properties of cellulosic fibres (as used
in textiles) and of wood and paper have been well studied (Hearle, 1985; Salmen, 1985).

III. SUPRAMOLECULAR ORGANISATION OF INTERACTING POLYSACCHARIDES

Information concerning the nature of polysaccharide macromolecular assemblies can be
gained from a number of techniques. The most direct method involves visualisation by
microscopy, although the interpretation of observed structures has to take into account
observation conditions and any necessary pre-treatments. Scattering techniques (light,
X-ray, neutron, etc.) provide information on the dimensions of polysaccharide aggregates and assemblies, the distance scales over which information can be obtained being
dependent on the wavelength of the scattering radiation. Information on molecular
mobilities in polysaccharides and their assemblies can be obtained from relaxation
measurements (e.g. nuclear magnetic resonance) and thermodynamic data are provided
by calorimetric and other techniques. Each of these types of experimental technique will
be discussed below.

A. Microscopy

Microscopy can be used to study polysaccharide assemblies covering a range of distance
scales depending on the technique employed. Conventional light microscopy can
visualise natural assemblies such as plant cell walls but little direct information
concerning polysaccharide interactions is obtained due to the relatively low magnification employed. Light microscopy can, however, usefully monitor gross changes in
polysaccharide interactions such as those which accompany starch gelatinisation.
Fluorescence-detected microscopy is a useful method for localising specific (fluorescent-
labelled) polysaccharides (Fulcher, 1982). A new methodology which promises to reveal
much detail about the molecular organisation of polymer assemblies involves the
determination of spectral features associated with localised areas of samples as identified
by microscopy. Raman spectra can, for example, be obtained from domains as small as
1 μm: furthermore by rotating the sample relative to the laser radiation beam, spectral
features associated with defined orientations of microscopic structures can be observed.
In this way, evidence was obtained for the specific orientation of lignin residues relative
to the cell-wall surface in samples of black spruce (Atalla and Agarwal, 1985).

The techniques of scanning and transmission electron microscopy (SEM and TEM,
respectively) are suitable direct probes of polysaccharide interactions and have been
used to study isolated molecules, *in vitro* assemblies (e.g. gels and fibrils) and natural

systems such as cell walls and starch granules. SEM visualises surfaces via the secondary radiation produced following electron bombardment and can be used to study samples of variable thickness. TEM utilises ultrathin (typically 1 μm) sections and the sample image is obtained after passage of the primary beam through the sample. Both techniques are performed under high vacuum (i.e. samples are dehydrated). Some samples require relatively mild pre-treatments prior to electron microscopy, e.g. starch granules studied by SEM (Gallant and Sterling, 1976), but many materials require extensive treatments such as fixing, embedding, staining and dehydration before they can be observed by electron microscopy. In order to provide sufficient contrast between areas of a sample, specific labels or stains are often employed which descriminate between molecules or structures. These may be dyes specific for individual components, e.g. protein in the presence of polysaccharide (Clark et al., 1983), chemical treatments such as periodate oxidation followed by derivatisation (Gallant and Sterling, 1976), or specific complexation with for example lectins (Gallant and Bouchet, 1986). Any analysis of the structures observed by electron microscopy should take into account the preparation conditions used, and consider the possibility of artifact structures unrelated to the native sample. For this reason, conclusions reached on the basis of electron microscopic observations should, wherever possible, be corroborated by evidence from other techniques.

Isolated polysaccharides have recently been studied by electron microscopy following vacuum drying of dilute glycerol–water (1 : 1; v/v) solutions sprayed onto mica surfaces (Stokke et al., 1987). The observed structures were used to measure parameters such as polymer persistence lengths and the number of molecules in individual strands. For example, micrographs of xanthan polysaccharide suggested a mixture of single- and double-stranded structures (Stokke et al., 1987). The sample preparation conditions used are likely to preserve relatively stiff native structures such as xanthan rather better than flexible polymers such as amylose.

Networks of interacting polysaccharides such as in agarose beads (Amsterdam et al., 1975; Attwood et al., 1988) and various gelled systems such as carrageenan and konjac mannan (Chanzy and Vuong, 1985) have also been studied by electron microscopy. Such studies can provide information on the extent of network cross-linking and the size distribution of individual strand elements. However, as polysaccharide gels contain typically >90% water, the dehydration required prior to observation leads to the possibility of structural rearrangements.

Polysaccharides which form fibrous structures have been studied extensively by electron microscopy and provide information on fibre dimensions (Chanzy and Vuong, 1985). The most studied system is cellulose, but other fibrillar polysaccharides such as chitin and $(1 \rightarrow 4)$-β-D-mannan have also been examined. When combined with diffraction data, information on both the molecular structure and macromolecular organisation of fibrils can be obtained (Chanzy and Vuong, 1985). Figure 18.6 shows an example of SEM and electron diffraction data for a cellulose specimen.

Aggregation processes in polysaccharide solutions (e.g. amylose retrogradation) can be probed by SEM. Structures of fibrils produced following retrogradation of various amylose chain lengths (Pfannemüller and Bauer-Carnap, 1977) and the particle size distribution of retrograded polydisperse amylose (Kitamura et al., 1984) have been described.

FIG. 18.6. (A) Ultrastructural features of a thin layer of *Valonia ventricosa* cellulose microfibrils shadowed with W/Ta (scale bar 0.5 μm). (B) Corresponding electron diffraction pattern. Reprinted with permission from 'Polysaccharides—Topics in Molecular and Structural Biology' (edited by E. D. T. Atkins), p. 48, © 1985 Macmillan Press.

FIG. 18.7. Transmission electron micrograph (× 44 000) of a corn starch granule cross-section following partial oxidation (periodic acid) and derivatisation (thiosemicarbazide, silver nitrate). Reprinted with permission from 'Examination and Analysis of Starch and Starch Products' (edited by J. A. Radley), p. 42, © 1976 Elsevier Applied Science Publishers.

Starch granules have been the subject of many SEM and TEM studies. The size distribution and surface structure of native and treated (e.g. α-amylase-digested) granules have been characterised by SEM (Gallant and Sterling, 1976; Gallant and Bouchet, 1986) and comparisons have been made between microscopic structure (by SEM) and rheology of swollen starch granules (Christianson *et al.*, 1982). Interesting details of the molecular architecture within starch granules have been deduced from SEM studies of granules partially digested by α-amylase which suggest a layered 'growth-ring' type of structure (Gallant and Sterling, 1976). TEM studies of thin sections of granules (Gallant and Sterling, 1976; Yamaguchi *et al.*, 1979) confirmed the layered structure (Figure 18.7). In a combined TEM and optical diffraction study (Oostergetel and van Bruggen, 1989) a repeated layer thickness of ~ 100 Å was observed, consistent with the proposed 'cluster' model for amylopectin molecular structure (Manners and Matheson, 1981). Thus parallel studies of microscopic and molecular structure have led to a reasonable model for an important feature of the internal architecture of starch granules.

B. Scattering Techniques

The phenomenon of scattering arises from inhomogeneities in a material (i.e. contrast between one region and another) and reflects inhomogeneities in electron density in the case of X-ray scattering, scattering length in the case of neutron scattering, and polarisability for light scattering. Typical wavelengths employed in X-ray, neutron, and light scattering are 0.1–0.2 nm, 0.2–1.0 nm and 400–650 nm, respectively.

In a scattering experiment, a monochromatic radiation source is directed at the sample and the intensity of elastically scattered radiation is measured at various angles (usually either 0–5° or 30–150°). The scattering due to solvent may be subtracted, and from the resulting scattering function due to the macromolecule under investigation, assessment of mass, size and shape can be made (Huglin, 1972; Morris, 1984). As the intrinsic nature of scattering is the same for X-rays, neutrons and light, similar equations apply and the type of information which can be obtained from each technique is determined primarily by the ratio of molecular size to scattering wavelength.

The reason for this is that for scattering particles comparable in size to the irradiation wavelength, radiation scattered from different points within the same particle ('multiple scattering') reaches the detector with different phases. This is illustrated in Fig. 18.8 which shows the effect of scattering from different point sources within a particle. For particular distances between scattering loci the effect of phase angle differences will lead to varying degrees of interference: when the phase difference is 90°, scattering from these points will be completely extinguished. The total observed intensity scattered at a given angle will be the sum of intensities from all scattering centres. In order to compare results from different types of scattering experiments, a 'scattering function' $P(\theta)$ is defined as the scattering intensity for the particle divided by the scattering intensity from an infinitely small particle having the same number of scattering centres (i.e. no interference). Figure 18.9 shows comparisons of predicted scattering functions from particles having various sizes in comparison to the radiation wavelength. As can be seen (Fig. 18.9), for particles which are small compared to the radiation wavelength $P(\theta) \sim 1$. This is a typical case for many isolated polysaccharides studied by light scattering. From the angular dependence of scattering, estimates can be obtained of

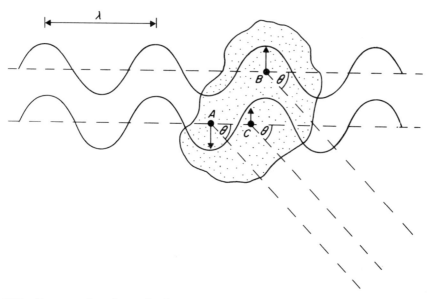

FIG. 18.8. Representation of scattering from a macromolecule with dimensions comparable to the irradiation wavelength. The phase of radiation scattered at A, B and C will differ and further phase shifts will occur as the detector for a given angle will be at different distances from A, B and C. Reprinted with permission from V. J. Morris in 'Biophysical Methods in Food Research' (edited by H. W.-S. Chan), p. 52, 1984 Blackwell.

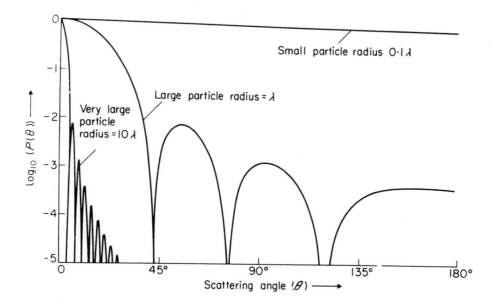

FIG. 18.9. Particle scattering functions $(P(\theta))$ for spherical scatterers of various sizes. Reprinted with permission from V. J. Morris in 'Biophysical Methods in Food Research' (edited by H. W.-S. Chan), p. 53, 1984 Blackwell.

particle size, and from the zero angle extrapolated scattering value the molecular weight is obtained. This data is usually extracted from a Zimm plot (Huglin, 1972). For particles with sizes comparable to the wavelength, $P(\theta)$ becomes markedly dependent on scattering angle (Fig. 18.9) and many methods have been developed for determining size and shape from this information (e.g. Kratochvil, 1972). When the particle becomes very large compared to the wavelength it is possible to probe the internal structures of particle assemblies. This situation is often exploited in small-angle X-ray scattering studies of polysaccharides.

The methods described above measure time-averaged scattering and are referred to as static methods: more recently dynamic methods have become available which measure scattering intensity over short periods of time (ns). As described above, scattering amplitudes at individual angles will differ in phase and the angular dependence of scattered intensity will form a complex interference pattern. Under the influence of particle motion (e.g. Brownian motion), relative phases observed at a given angle are altered which lead to fluctuations in interference patterns. The rate of change in these fluctuations depends on the rate of particle motion and therefore on particle size and diffusion coefficient. Information related to size, shape and dynamics of scattering particles can therefore be obtained.

All of the techniques described above have been used in studies of polysaccharide interactions (Morris, 1984; Clark and Ross-Murphy, 1987), although neutron scattering is only available at a few research centres. To illustrate the use of scattering techniques in defining the structures involved in polysaccharide interactions, two examples will be cited. In the first, a study by Morris and co-workers (Morris et al., 1980) based partly on light scattering studies led to the proposal of a new mechanism for the gelation of the algal polysaccharides ι- and κ-carrageenan. From X-ray diffraction studies, a double-helical structure was proposed to exist in the condensed phase (Arnott et al., 1974) and there was evidence from optical rotation and NMR studies that such structures persist under more hydrated conditions (Bryce et al., 1974). Together with the observation that the thermally reversible sol–gel transition is accompanied by a cooperative coil–helix transition, a model for gel formation via double helix formation was proposed (Rees, 1972). However there are clear topological problems in forming an infinite network solely through double helix formation. This conceptual problem was resolved through light scattering studies of various cation forms of ι-carrageenan segments. For the sodium form, a simple doubling of molecular weight was observed upon double helix formation (as monitored by optical rotation), whereas a 4–10 fold increase in molecular weight was found for the potassium form following double helix formation suggesting significant lateral aggregation of helices. Under the same conditions of temperature and ionic strength, gelation was only observed for those salt forms which showed evidence of aggregation in the segmented form. These results formed the basis for the 'domain' model of carrageenan gelation in which cation-mediated aggregation of double-helical domains is required in order to develop a cohesive three-dimensional network structure (Morris et al., 1980).

In a second example of scattering studies of interacting polysaccharide systems, information concerning the structures present in agarose gels was obtained from X-ray scattering data (Djabourov et al., 1989). Figure 18.10 shows small angle X-ray scattering profiles for typical agarose solutions and gels. Modelling studies showed that the scattering observed from aqueous solutions was consistent with single polysaccharide chains (Fig. 18.10a), and that scattering from gelled systems was different from that computed for isolated double helix formation (Fig. 18.10b). Detailed analysis of the gel scattering results in terms of cross-sectional radii of gyration, suggested a population of aggregates of ~3 nm diameter together with some larger aggregates (~9 nm diameter). The stable 3 nm aggregate was interpreted in terms of a structure containing six double helices, packed laterally (Djabourov et al., 1989).

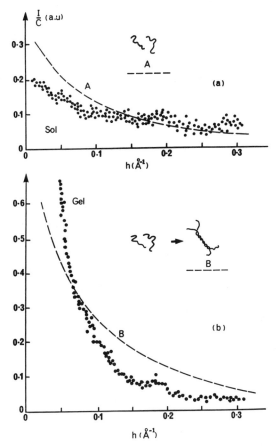

FIG. 18.10. Comparison of calculated small-angle X-ray scattering curves (dashed lines) for an agarose single chain and an agarose double helix with experimental data (I/c, intensity normalised with respect to concentration) for an agarose sol and an agarose gel. Reprinted with permission from *Macromolecules* **22** (1989), 180–188, © 1989 American Chemical Society.

C. Molecular Mobilities from Relaxation Measurements

Information concerning molecular mobilities of polysaccharide assemblies and their components can be obtained by measuring relaxation behaviour following a defined perturbation of the system. Examples of relaxation measurements have already been given in sections on mechanical characterisation and scattering techniques (e.g. dynamic light scattering). Other probes of mobility in polysaccharide assemblies include NMR and ultrasonic relaxation.

Measurements of NMR relaxation behaviour of polysaccharides and their assemblies have focused both directly on the polysaccharide component(s) and also on the water component. The versatility of the NMR technique is reflected in the large number of relaxation parameters which can be measured. A particularly useful probe of polysaccharide assemblies is the polysaccharide proton spin–spin (T_2) relaxation time which is thought to be determined by segmental motion. In a study of polysaccharide solutions, it was found that a range of T_2 relaxation times covering two orders of magnitude (1–100 ms) was required to adequately describe the experimental results (Ablett et al., 1982). In more motionally-constrained environments such as gels, polysaccharides' proton T_2 values are very much shorter—typically 10 μs (Ablett et al., 1982; Gidley, 1988) and this leads to the 'disappearance' of signal in the high resolution NMR spectrum as T_2 is inversely related to NMR signal linewidth. The width at half height of a high resolution proton NMR signal corresponding to a T_2 of 10 μs is ∼30 kHz compared with a typical spectral width of 1–5 kHz. The loss of high resolution NMR signal has been used extensively in characterising thermal or salt-induced conformational transitions in polysaccharides (Rees et al., 1982). Due to recent advances in NMR techniques, the 'lost' signal can now be spectroscopically characterised (Section IV.B).

The NMR relaxation properties of the complex polysaccharide assemblies present in plant cell walls have also been studied recently (Irwin et al., 1984, 1985; Mackay et al., 1988). These studies have provided details of differential mobility of polysaccharide components in the primary cell wall of bean (*Phaseolus vulgaris*) seedlings (MacKay et al., 1988) and mobility changes associated with the ripening of apples (Irwin et al., 1984). Once the relaxation properties of a system have been established, appropriate high resolution NMR experiments can be designed to provide chemical information on components possessing different relaxation times. For example, in partially crystalline cellulose samples, characteristic spectra of crystalline and amorphous components can be obtained based on differences in relaxation parameters (Horii et al., 1984).

NMR relaxation studies of the water component in polysaccharide systems have provided insights into polysaccharide–water interactions and, indirectly, polysaccharide–polysaccharide interactions. For example, a recent study of water proton T_2 values as a function of temperature and ionic strength in various gelling and non-gelling carrageenan systems showed that such measurements reflect polysaccharide aggregation and gelation properties (Lewis et al., 1987). Information on starch–water systems has also been derived from water T_2 measurements in granular starches, amylose and amylopectin gels, and bread (Lechert and Schwier, 1982; Wynne-Jones and Blanshard, 1986).

Ultrasonic relaxation in polysaccharide solutions and gels has also been studied, and has been assigned to solvent motion (Gormally et al., 1982). This technique is sensitive to polysaccharide aggregation and network structure as shown by studies on carrageenan and agarose (Gormally et al., 1982), and can also be used to probe entanglement effects in disordered polymers and define experimental coil overlap concentrations (Pereira et al., 1982).

D. Thermodynamic Measurements

The thermodynamic properties of aggregated polysaccharide systems have been probed by a variety of techniques including calorimetry, swelling and de-swelling experiments, and ion-binding studies. Differential scanning calorimetry (DSC) has proved to be a particularly useful technique for characterising interacting polysaccharide systems. In a DSC experiment, a sample and reference are heated or cooled at a defined rate of temperature change, and the differential heat input required to maintain the defined rate in sample and reference cells is measured. In this way, the difference in heat capacity between sample and reference is measured as a function of temperature. Thermal events such as crystallite melting are characterised by increased uptake of heat by the sample and lead to an endothermic transition (peak). Conversely, crystallisation or ordering events lead to an exothermic transition. Changes in heat capacity without uptake or evolution of heat are characteristic of glass transitions.

DSC has been used in many studies of helix–coil transformation in polysaccharides (Rees *et al.*, 1982; Goodall and Norton, 1987) to define the temperature course of conformational change and to quantify the enthalpy associated with such changes. In combination with optical rotation measurements of the extent of conformational ordering and the application of Zimm–Bragg helix–coil transition theory, detailed insights into the thermodynamic parameters which control polysaccharide conformational ordering can be obtained (Norton *et al.*, 1984b; Goodall and Norton, 1987).

Starch gelatinisation has been the subject of many DSC studies. In the presence of excess water, a relatively sharp endothermic transition is observed typically at 60–80°C; further transitions are observed at higher temperatures in limited water conditions, and a high temperature (typically 100–120°C) melting of lipid inclusion complexes is observed when they are present (Kugimuya *et al.*, 1980). Despite good reproducibility of results, there is no general agreement on the nature of the molecular processes which control the observed melting transitions. In particular, the nature of the 'trigger' which controls the onset of gelatinisation is the subject of current debate. Donovan has suggested that swelling of amorphous regions within granules leads to the onset of crystallite melting (Donovan, 1979; Donovan and Mapes, 1980), whilst Evans and Haisman (1982) have postulated that gelatinisation is initiated by melting of the least stable crystallite which leads to granule swelling and further crystallite melting. More recently, Slade and Levine (1988) have proposed that a glass transition occurs on heating starch granules in water and that this transition immediately precedes (i.e. triggers) gelatinisation. Although heat capacity changes have been demonstrated for frozen solutions of gelatinised starch and starch hydrolysates (Levine and Slade, 1986), there is as yet no convincing experimental evidence for a heat capacity change immediately preceding gelatinisation. The observation (Slade and Levine, 1988) of a heat capacity change associated with (but not separate from) gelatinisation does not prove that a glass transition has occurred, as granular and gelatinised starches would be expected to exhibit different intrinsic heat capacities.

Thermodynamic information on the interaction of ions with charged polysaccharides has been obtained through conductimetry (which measures transport parameters) and ion-selective electrode measurements which record ion activities (Kohn, 1987; Thibault and Rinaudo, 1985). The specific binding of ions (e.g. calcium with pectin) can be deduced from such measurements: the stoichiometry of specific ion binding has also been determined by equilibrium dialysis measurements (Morris *et al.*, 1982).

The volume changes associated with polysaccharide aggregation processes have also been characterised. An apparatus (dilatometer) for measuring volume changes accompanying gelation has been described (Miles *et al.*, 1985) and used to characterise amylose and amylopectin gelation (Ring *et al.*, 1987). Equipment for monitoring the compressive de-swelling of biopolymer gels has recently been described (Lips *et al.*, 1988) and used to study agar and gelatin gels. At 5% w/v agar gels expressed water under applied pressure (2–39 p.s.i) but took *c.* 10^6 s to reach equilibrium. Release of pressure and immersion in excess water resulted in re-swelling of the gel to its original shape and size, thus demonstrating the effective permanence of cross-links within the gel.

IV. MOLECULAR DETAILS OF POLYSACCHARIDE INTERACTIONS

Previous sections have described the mechanical consequences of polysaccharide inter- actions and techniques for characterising the size, mobility, and organisation of polysaccharide assemblies. None of the techniques discussed, however, directly gives information on detailed specific interactions at the molecular level. At the present time, the only general approach to detailed three-dimensional structures of polysaccharides is from diffraction measurements on crystallites or aligned fibres. The most common technique is that of fibre X-ray diffraction, although useful information can also be obtained from electron diffraction. Other probes of molecular structure include nuclear magnetic resonance (NMR), circular dichroism (CD) and other chiroptical techniques, infrared (IR) and Raman spectroscopy. These techniques cannot lead directly to the molecular structures involved in polysaccharide interactions, but are invaluable in corroborating or refining structures deduced from diffraction measurements. In particu- lar they can be used to assess whether diffraction-derived structures persist under various environmental conditions (e.g. in the hydrated or gel state). Spectroscopic techniques can often monitor specific functional groups within molecules and have been used to follow the effect of various treatments on the environment of such functional groups in systems based on polysaccharide interactions. In the following subsections, the individual techniques mentioned above will be discussed and examples given of their application to polysaccharide interactions.

A. Diffraction Methods

Any material which has a regular, repeating, three-dimensional molecular structure will diffract radiation according to the Bragg formula, $n\lambda = 2d \sin \theta$, where λ is the radiation wavelength, d is a repeating distance within the structure, θ is the diffracting angle, and n is a positive integer. X-ray diffraction is the most common technique and has been used to solve numerous three-dimensional structures of small crystalline molecules. Many mono-, di- and trisaccharide crystal structures have been solved and in some cases have been used as models for polysaccharide structures. Higher molecular weight carbohydrates are often difficult to crystallise although recent reports have described the crystallisation of short chain amyloses (Buleon *et al.*, 1984; Pfannemüller, 1987; Gidley and Bulpin, 1987). These materials have not yet, however, yielded perfect crystals of sufficient size (~ 1 mm) for direct X-ray analysis.

For polysaccharides, crystallisation into single crystals has not been achieved, and the method of choice is fibre diffraction in which a polysaccharide sample is drawn into a fibre and annealed to promote the formation of ordered structures typically under high humidity for extended times. The practical aspects of sample preparation, data collection and structure derivation have been reviewed recently (Winter, 1982). Basically, structures are derived through testing conformationally-reasonable models against observed diffraction data and selecting the structure whose predicted diffraction behaviour most closely matches the observed pattern. A measure of the certainty of any proposed structure can be obtained (*R*-factor) but sometimes two or more structures can assume nearly equal probability. In these cases other spectroscopic and scattering studies may be used to select the most likely structure. Another type of uncertainty exists when X-ray diffraction data are not comprehensive enough to lead to a reliable structure; once again spectroscopic probes can be used to limit the range of possible structures. Examples of the use of spectroscopy in these situations will be given later.

A recent example of a structure derived from X-ray fibre diffraction is the double helix formed by the microbial polysaccharide gellan (Chandrasekharan *et al.*, 1988). Figure 18.11a shows the observed diffraction pattern for the lithium salt of ordered gellan. Figure 18.11b depicts the most probable molecular structure, a three-fold double helix, and Fig. 18.11c shows how the helices are packed into the crystallographic unit cell. A large number of other polysaccharides have been similarly analysed by X-ray diffraction and have revealed a wide range of single, double and triple helices and ribbon-type structures (French and Gardner, 1980; Rees *et al.*, 1982). In many cases, the propensity for formation of ribbon or helical structures follows directly from consideration of glycosidic conformation (Rees *et al.*, 1982).

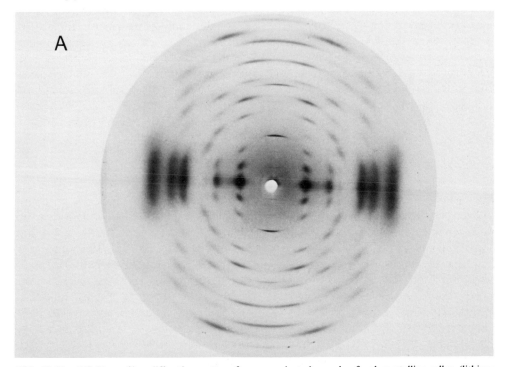

FIG. 18.11. (A) X-ray fibre diffraction pattern from an oriented sample of polycrystalline gellan (lithium salt). (B) Two mutually perpendicular views of the most probable three-fold double helix. (C) Antiparallel packing arrangement of double helices within unit cells. Reprinted with permission from *Carbohydr. Res.* **175** (1988), 1–15, © 1988 Elsevier Science Publishers.

B

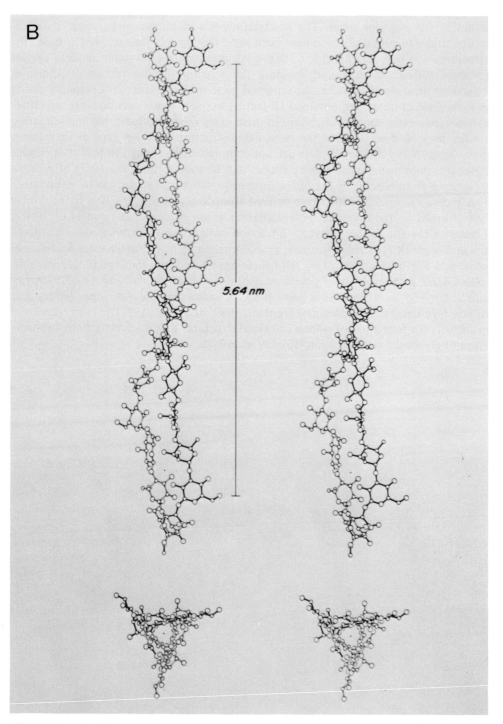

5.64 nm

FIG. 18.11. *continued*

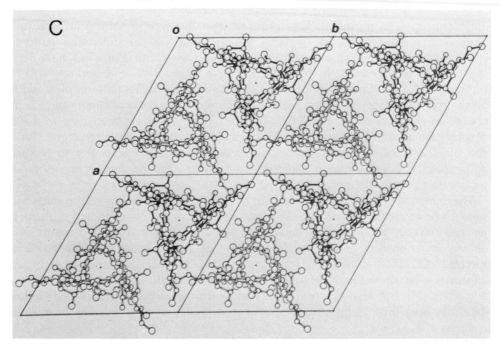

FIG. 18.11. *continued*

X-Ray diffraction studies can also be made of fibres or films prepared from aqueous gels, although the quality of diffraction patterns is markedly inferior to those of conventional fibre diffraction (e.g. Fig. 18.11a). Generally, only very intense reflections can be observed but these may be sufficient to measure unit cell parameters and hence compare gel structures with the more detailed fibre structures. Recently Cairns *et al.* (1987) studied single and binary polysaccharide gels in an attempt to determine whether co-crystallisation occurs between the two components in mixed polysaccharide gels. For a variety of mixtures of galactomannans with algal polysaccharides, diffraction patterns of fibres drawn from mixed gels were composites of the individual component patterns. Although this result provides no evidence for repeated co-crystalline structures, the possibility of non-repeated (in a three-dimensional sense) intermolecular binding cannot be excluded. For mixtures of xanthan with carob or tara galactomannan, however, diffraction patterns from binary gels were not simple composites of the components. This result was taken as evidence of co-crystallisation of the two components (Cairns *et al.*, 1987), although the possibility that the crystallisation behaviour of one component is altered by the presence of the other but without specific co-crystallisation, cannot be ruled out.

X-Ray analysis of collections of small, non-oriented crystallites leads to so-called powder diffraction patterns in which the information present in two-dimensional diffraction patterns of oriented samples (e.g. Fig. 18.11a) is reduced to a single dimension with consequent difficulties in interpretation. The main use of powder diffractometry is to provide a 'fingerprint' of crystallite-type in natural or processed samples. For example, starch granules usually show powder diffraction patterns characteristic of one of two crystalline polymorphs, the structures of which have been shown by fibre diffraction to be six-fold double helices (Wu and Sarko, 1978a,b; Imberty et al., 1988; Imberty and Perez, 1988). Powder diffraction patterns of A and B polymorphs of starches have been analysed in terms of degree of crystallinity and size of crystallites (Komiya et al., 1987; Gernat et al., 1987) and have been used to make detailed comparisons between fibre diffraction data and native starch structures (Wild and Blanshard, 1986). Starch-based samples seem particularly amenable to powder diffractometry, possibly due to a propensity to form large ordered aggregates more readily than many other polysaccharides.

The technique of electron diffraction can be used to study crystals and crystallites too small to be amenable to direct X-ray methods. Samples are observed and diffraction patterns recorded in an electron microscope (see Fig. 18.6); crystalline domains with dimensions as small as 10–100 nm give useful information (Chanzy and Vuong, 1985). Electron diffraction data provide accurate unit cell dimensions and space group information but do not have the wealth of detailed information obtainable from X-ray diffraction. Nevertheless, electron diffraction, when combined with X-ray diffraction data, has been used to propose structural models of, for example, A-type amylose (Imberty et al., 1988).

B.   High Resolution NMR Spectroscopy

High resolution NMR is a powerful spectroscopic method for the determination of molecular structure and conformation. In principle an NMR spectrum contains separate signals for each atomic site within a molecule or repeating structure. The position (chemical shift) of each signal is determined by the local electronic environment and is a sensitive probe of local molecular structure. As the chemical shift effect of for instance a particular glycosyl substitution is similar for a range of carbohydrates, NMR (particularly ^{13}C) chemical shifts are much used in polysaccharide primary structure determination (Gorin, 1981; Casu, 1985). In addition, NMR experiments can detect couplings between atoms through 1, 2 or 3 bonds, and through-space connections (whose strength varies with interatomic distance) using the nuclear Overhauser effect (Casu, 1985). This detailed coupling and connection information is, unfortunately, largely limited to low molecular-weight carbohydrates. Due to their more restricted motion, polysaccharides exhibit shorter relaxation times (Section III.C) compared with oligosaccharides. Such fast relaxation leads to: (1) broad lines which can obscure coupling information; (2) loss of the distance dependence of the steady state nuclear Overhauser effect; and (3) limitations in the type of non-standard NMR experiments which can be performed before rapid relaxation leads to loss of signal.

An example of conformational information which relates to polysaccharide inter-action is the geometry of glycosidic linkages in solution. Relevant information can in principle be obtained through measurement of inter-residue proton distances via the nuclear Overhauser effect (Casu, 1985), e.g. H-1 to H-4' in α-(1→4) glucans (Fig. 18.12) or through measurements of inter-residue coupling constants, e.g. between C-1 and H-4' or between C-4' and H-1 in α-(1→4) glucans (Gidley and Bociek, 1985b; Parfondry et al., 1988). In practice such data are only obtainable for small oligosaccharides; however, a second conformationally sensitive probe such as optical rotation (Section IV.C) can be used to determine whether oligosaccharides whose glycosidic conformation can be probed by NMR are appropriate models for polysaccharide solution conformation (Gidley, 1988).

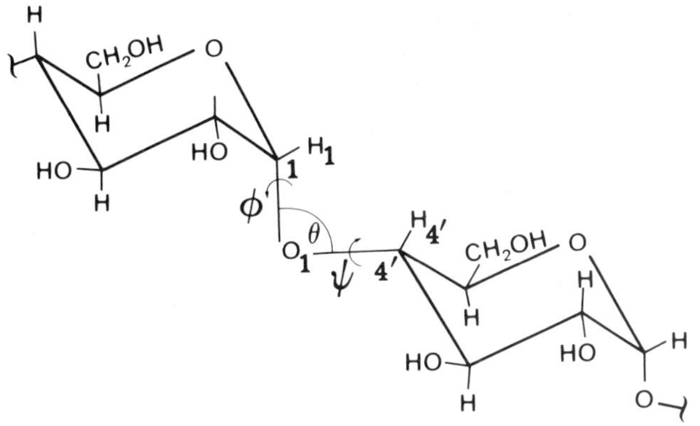

FIG. 18.12. Conformational features associated with the glycosidic linkage in α-(1 → 4) glucans. Torsion angle ϕ is described by (H-1)–(C-1)–(O-1)–(C-4'), and angle ψ is described by (H-4')–(C-4')–(O-1)–(C-1). Reprinted with permission from *J. Am. Chem. Soc.* **110** (1988), 3820–3829, © 1988 American Chemical Society.

Conventional solution-state NMR spectroscopy finds only limited application to the direct characterisation of polysaccharide interactions: a notable exception being the extensive studies by Saito and co-workers on β-(1→3)-D-glucans which have revealed details of the effect of molecular structure on gelation ability (Saito, 1986). This is due to the general restriction of mobility experienced by interacting polysaccharide chains which leads to increases in spin–spin relaxation rates and consequent 'disappearance' of high resolution NMR signal through line-broadening. Useful information can, how-ever, be obtained from interacting systems of low molecular weight, e.g. soluble complexes of short chain amylose with iodine (Jane *et al.*, 1985) and of mycobacterial polysaccharide with fatty acids (Yabusaki *et al.*, 1979; Maggio, 1980). Other aspects of polysaccharide interactions which have been probed by solution-state NMR include inter-residue hydrogen bond formation of amylose in dimethyl sulphoxide (St-Jaques *et al.*, 1976) and interactions of ions with charged polysaccharides such as carrageenans (Norton *et al.*, 1984a; Belton *et al.*, 1986a).

Recent advances in NMR technology now makes it possible to observe high resolution NMR spectra from solid-like materials. The most useful technique observes ^{13}C spectra using the methodology of cross-polarisation, high power dipolar decoupling, and magic angle spinning (^{13}C CP/MAS NMR). This approach is described in detail elsewhere (Yannoni, 1982) but in essence it gives a spectroscopic characterisation of materials whose solution-state spectrum is too broad to observe. ^{13}C CP/MAS NMR is potentially a very powerful probe of polysaccharide interactions and has already been used to characterise a variety of crystalline and amorphous solids and gels. NMR spectroscopy is sensitive to electronic effects on a submolecular scale and therefore spectra of solids provide information at shorter distance scales than those probed by X-ray diffraction (Section IV.A). Several 'crystalline' (i.e. ordered) polysaccharide systems have been analysed by ^{13}C CP/MAS NMR and data obtained which complement those available from diffraction techniques.

Celluloses have been the subject of many studies in an effort to resolve details of molecular organisation in various polymorphic forms. A useful approach to characterising polymeric ordered structures is to prepare highly crystalline materials from low molecular-weight model compounds having the same powder X-ray diffraction pattern. In a study of solid cellulose oligomers related to the type II polymorph, it was found that the characteristic features of the solid state ^{13}C spectrum of polymeric cellulose II were present in a spectrum of cellotetraose (Dudley et al., 1983). This suggested that a single crystal X-ray diffraction study of cellotetraose could yield the full molecular details of the cellulose structure (Dudley et al., 1983). The analysis of ^{13}C CP/MAS spectra of cellulose I and native celluloses is more complex and is the subject of current debate. On the basis of ^{13}C CP/MAS NMR measurements, Atalla and VanderHart suggested that the cellulose I structure should be considered as a composite of two crystalline forms (Atalla and VanderHart 1984). Cael et al., (1985) also suggested a composite structure but proposed different spectral features for the two (non-isolable) limiting structures. On the basis of detailed spectral intensity measurements, however, Horii et al. (1987a) found that neither of these analyses of cellulose I spectra were consistent with observed intensities.

Analysis of starches and other α-(1→4) glucans by ^{13}C CP/MAS NMR has provided much information on solid-state structures. Spectra of highly crystalline samples of A, B and V polymorphs (Fig. 18.13) have been reported (Gidley and Bociek, 1985a; Veregin et al., 1986; Horii et al., 1987b; Gidley and Bociek, 1988a) as well as spectra characteristic of amorphous materials (Fig. 18.13; Gidley and Bociek, 1985a; Horii et al., 1987b). It has been shown that spectra of starch granules are completely accounted for by a combination of features due to crystalline and amorphous materials (Gidley and Bociek, 1985a). Simulation of granule spectra thus allows the degree of molecular ordering within granules to be determined; the values obtained can then be compared with estimates of long-range ordering from X-ray diffraction.

Two features of ^{13}C CP/MAS NMR spectra which differ from those of solution NMR are the occasional occurrence of multiple resonances for chemically equivalent sites (e.g. the C-1 resonance of A and B type starches at ∼100 ppm, Fig. 18.13) and the wide range of chemical shifts which can be observed for a single carbon site (e.g. the C-1 resonance of amorphous starch at 93–105 ppm, Fig. 18.13). The multiplicity effect is thought to be due to crystallographically inequivalent sites within an ordered structure (Veregin et al., 1986) and has been ascribed to helix packing symmetries for the case of A and B type starch (Gidley and Bociek, 1985a, 1988a). Analysis of a range of α-(1→4) glucans (Gidley and Bociek, 1988a) suggested that the major determinant of C-1 chemical shifts is the glycosidic linkage conformation (Fig. 18.12) and that either torsion angle ψ or the sum of φ and ψ (Fig. 18.12) correlated with chemical shifts. Both correlations led to a reasonable simulation of the observed C-1 signal for amorphous starch if the assumption was made that all energetically allowed conformation are present (Gidley and Bociek, 1988a). In a study of cyclodextrin complexes, Veregin et al. (1987) suggested that C-1 and C-4 chemical shifts correlated with torsion angles ψ and φ, respectively. Saito (1986) has also suggested that glycosidic conformation is reflected in solid state ^{13}C chemical shifts. When the relationship(s) between glycosidic site chemical shifts and linkage conformation become established more firmly, ^{13}C CP/MAS NMR may be used to analyse directly solid state glycosidic conformations.

The techniques of ^{13}C CP/MAS NMR can also be used to characterise molecular

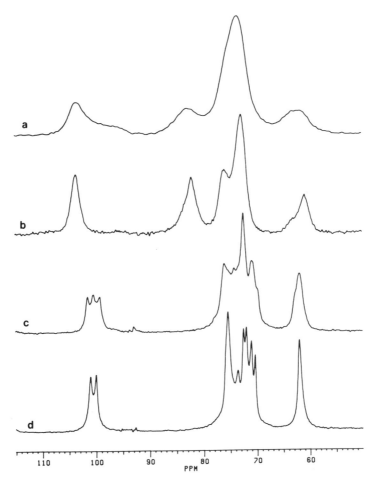

FIG. 18.13. ^{13}C CP/MAS NMR spectra of (a) amorphous starch prepared by ethanol precipitation of gelatinised maize starch, (b) amylose/sodium palmitate V-type complex, (c) crystalline A-type α-(1 → 4) glucan, and (d) crystalline B-type α-(1 → 4) glucan. Reprinted with permission from *J. Am. Chem. Soc.* **110** (1988), 3820–3829, © 1988 American Chemical Society.

structures and conformation in (frozen) solutions and gels. Freezing a polysaccharide solution would be expected to lead to the trapping of a range of rapidly interconverting conformations adopted in free solution. Solution-state NMR yields only time-averaged spectra and information concerning the range of conformations adopted cannot be obtained. Spectra of frozen solutions could be compared directly with those of solids having defined crystallographic structures. In this way, it was shown that α-cyclodextrin adopts an expanded aqueous solution conformation which differs significantly from that determined crystallographically in the solid state (Gidley and Bociek, 1988b).

The application of ^{13}C CP/MAS NMR to studies of polysaccharide gels has been demonstrated for β-(1→6)-D-glucan (pustulan) gels (Stipanovic *et al.*, 1985) and has recently been employed in a study of amylose gels (Gidley, 1989). Figure 18.14 shows NMR spectra of an amylose gel obtained under various experimental conditions. Figure 18.14a is the conventional 'solution-state' spectrum showing broad but observable

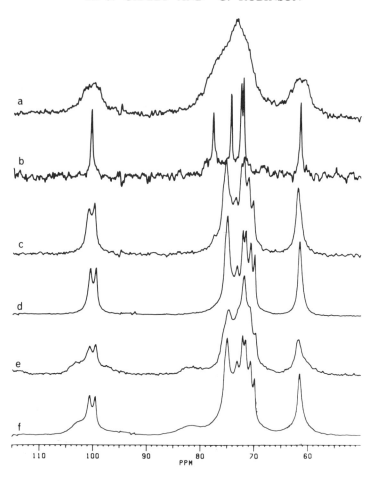

FIG. 18.14. Carbon-13 NMR investigation of molecular structures in amylose gels. (a) Single pulse spectrum of a 10% potato amylose gel (303 K). (b) As (a) with magic angle spinning at 500 Hz. (c) CP/MAS spectrum of the same sample (303 K). (d) CP/MAS spectrum of highly crystalline B-type amylose. (e) CP/MAS spectrum at 233 K of the same gel as (a)–(c). (f) Simulation of spectrum (e) by addition of crystalline (67%) and amorphous (33%) model spectral features. Reprinted with permission from 'Gums and Stabilisers for the Food Industry—4' edited by G. O. Phillips, D. J. Wedlock and P. A. Williams, p. 77, © 1988 IRL Press.

resonances. The fact that these resonances sharpen and exhibit chemical shifts identical to those of amylose solutions following magic angle spinning shows that some of the amylose chains are motionally restricted but access a range of conformations. The ^{13}C CP/MAS NMR spectrum of the same gel (Fig. 18.4c) shows that some of the amylose chains are effectively immobilised and present as double helices characteristic of the B-type crystalline polymorph (Fig. 18.14d). The ^{13}C CP/MAS NMR spectrum of a frozen gel (Fig. 18.14e) shows features due to both B-type crystalline amylose and amorphous material (Fig. 18.13a). This amorphous component represents the immobilised form of the more mobile chain segments characterised in Fig. 18.14a,b. The spectrum of the frozen gel can be simulated by a combination of amorphous and B-type crystalline spectral features (Fig. 18.14f). These spectra (Fig. 18.14) provide a direct characterisation of the molecular structures present in amylose gels, and are consistent with a model involving double-helical junction zones and conformationally mobile interconnecting chain segments (Gidley, 1989). The type of analysis exemplified in Fig. 18.14 promises to provide the first direct experimental approach to assessing whether molecular structures determined by X-ray diffraction persist in hydrated/gel states.

^{13}C CP/MAS NMR analyses of other solid polysaccharide systems such as chitosan and derivatives (Saito *et al.*, 1987) and β-(1 → 3) glucans (Saito, 1986), as well as natural systems such as wood (Haw *et al.*, 1984), seeds (O'Donnell *et al.*, 1981) and plants (Himmelsbach *et al.*, 1983; Maciel *et al.*, 1985) have all been reported recently. It is to be expected that with recent (and future) advances in NMR technology, much useful information concerning the molecular details of polysaccharide interactions both *in vitro* and *in vivo* will be obtained from high resolution solid state NMR.

C. Chiroptical Techniques

Carbohydrates (including polysaccharides) have molecular structures which are invariably chiral, i.e. they cannot be superimposed on their own mirror image. Such asymmetry leads to differential interactions with components of circularly polarised light (i.e. left or right circularly polarised light), and results in an observable optical activity. The nature of the interaction observed depends upon whether light is directly absorbed by the sample. Absorption occurs when the irradiation energy matches the energy requirements for promotion of a particular electron within a molecule from its ground state to an excited state. The structural element which determines this promotion energy is the chromophore. Conventional UV spectroscopy measures the absorption of unpolarised light and circular dichroism (CD) measures the difference in absorption of left and right circularly polarised light by chromophores within a chiral molecule. The observed CD may be positive or negative depending upon whether left or right circularly polarised light is absorbed preferentially. At wavelengths which do not satisfy the energy requirements for absorption, exchange of energy between electronic orbitals within the sample and the electric field of the light beam may still occur; the effect of this exchange is to slow down the rate of propagation (i.e. refraction). The differential refraction of left and right circularly polarised light by chiral molecules is known as optical rotation (OR) at a single wavelength or optical rotatory dispersion (ORD) when measured as a function of wavelength.

Since CD and ORD have a common origin in the interaction of electrons with radiation in a chiral molecular environment they contain the same molecular information in principle, and are in fact related in a well understood way. Figure 18.15 shows typical bandforms for CD and ORD in the vicinity of an isolated electronic absorption. The three parameters of position, height and width completely characterise the simple Gaussian bandform observed for an isolated CD transition. ORD behaviour is characterised by the same three parameters and is mathematically related to CD by the Kronig–Kramers relationship (Morris and Frangou, 1981).

In practice CD and ORD (or OR) have their own advantages and disadvantages. CD bandforms are simpler (Fig. 18.15) and this facilitates resolution of spectral features into component transitions but restricts the use of the technique to those molecules and chromophores which absorb light in an accessible spectral region. For current commercial instruments this involves wavelengths longer than 190 nm (see below). ORD has finite intensity at wavelengths distant from the transition band centre (Fig. 18.15) and could therefore provide information about transitions which are not directly accessible by CD. However ORD at any wavelength will contain contributions from every electronic transition within the sample and therefore resolution may be difficult and assignment is much more complex.

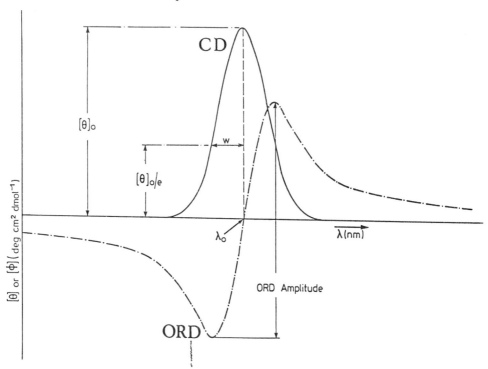

FIG. 18.15.   Fundamental CD and ORD spectral band forms for a single optically active transition. Molar ellipticity [θ] and ORD amplitude at any wavelength are determined by the parameters of position (λ_0), intensity $[\theta]_0$ and width w (half-width at $1/e$ of maximum intensity). Reproduced with permission from *Techniques in the Life Sciences—Biochemistry* **B-308** (1981), 1–51, © 1981 Elsevier Scientific Publishers, Ireland.

Optical rotation (i.e. ORD at a single wavelength) is an established characterisation technique for carbohydrates and is sensitive to details of steric interrelationships between constituent groups. Various empirical rules have been formulated which relate optical rotation (usually at 589 nm, i.e. the sodium D-line) to carbohydrate ring stereochemistry (Shallenberger, 1982). An extension of these rules to glycosidically-linked carbohydrates was proposed by Rees (1970) in which formulae relating observed optical rotation to glycosidic torsion angles ϕ and ψ (Fig. 18.12) were obtained semi-empirically. Despite the lack of strict theoretical justification, Rees' formulae have successfully predicted the observed optical rotation for many oligosaccharides as well as polysaccharides in ordered conformations characterised by X-ray diffraction (Rees *et al.*, 1982). A better understanding of the relationship between polysaccharide conformation and optical rotation is beginning to emerge from studies using vacuum ultraviolet circular dichroism which can detect transitions in the range 140–190 nm. Two main polysaccharide CD transitions are commonly found at ~150 and ~170 nm and are assigned to carbohydrate backbone features (as opposed to pendant chromophores).

These bands have been shown to determine OR at 589 nm for polysaccharides such as agarose (in both sol and gel forms) and a range of galactomannans (Buffington *et al.*, 1980; Morris *et al.*, 1986). Recent progress has also been made in the development of theoretical models (Stevens and Sathyanarayana, 1988) and understanding of the relationships between carbohydrate stereochemistry and vacuum ultraviolet CD transitions (Cziner *et al.*, 1986).

Despite the limitations in detailed understanding, optical rotation studies of polysaccharides are very useful. In particular they provide convenient and sensitive probes of order–disorder transitions in solutions and gels (Rees *et al.*, 1982) which can be correlated with other physico-chemical probes. The kinetics of polysaccharide disorder–order transitions (typically under the influence of added ions) have been studied for a number of systems (Goodall and Norton, 1987) using optical rotation detection of changes in a stopped-flow system. These studies have for example determined the number of polysaccharide strands in the transition state for the disorder to order transition to be two for ɩ- and κ-carrageenan and one for xanthan (Goodall and Norton, 1987).

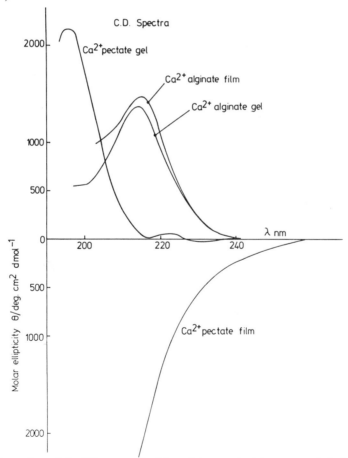

FIG. 18.16. Comparison of the CD spectra of calcium alginate and calcium pectate in both gel and solid film states. Reprinted with permission from *Food Chemistry* **6** (1980), 15–39, © 1980 Elsevier Applied Science Publishers.

Conventional CD studies of polysaccharides have largely focused on $n \rightarrow \pi^*$ transitions of carboxylates and derivatives (e.g. esters, amides) which generally occur in the range 190–240 nm (Morris and Frangou, 1981). Studies of the interaction of the polyuronates alginate and pectin with cations by CD have proved particularly informative. Divalent cations (e.g. Ca^{2+}) induce gelation of both alginate and pectin; spectroscopic and physico-chemical characterisation of calcium alginate under hydrated conditions was consistent with intermolecular interactions based on two-fold chain conformations as previously characterised in the solid state by X-ray diffraction (Morris et al., 1978). Similar characterisation of hydrated calcium pectate systems was also consistent with interactions between two-fold chains (Morris et al., 1982). However, pectin chains had been shown to adopt three-fold conformations in the solid state (Walkinshaw and Arnott, 1981). CD studies of gels and films of both calcium pectate and calcium alginate suggested a rationalisation for these observations. As shown in Fig. 18.16, CD spectra in the carboxyl $n \rightarrow \pi^*$ transition region are similar for films and gels of calcium alginate consistent with the same chain conformation. CD spectra of films and gels of calcium pectate are however dramatically different (Fig. 18.16), thus leading to the suggestion (Morris et al., 1982) that calcium pectate exhibits polymorphism between gel and solid states (two-fold and three-fold chain conformations, respectively).

D. Infrared and Raman Spectroscopy

Infrared and Raman spectroscopy are both probes of molecular vibrational states, spectral features arising from individual bonds stretching or bending as well as longer range segmental vibrations. The energies involved lie between the low energy NMR transitions and the higher energy electronic transitions of UV/CD spectroscopy. The information content and sensitivity of vibrational spectra are also intermediate between those of UV (high sensitivity, low information) and NMR spectroscopy (low sensitivity, high information). Infrared spectroscopy detects vibrations which are associated with a change in electric dipole moment, whereas Raman spectroscopy detects vibrations which involve a change in molecular polarisability. This difference in selection rules means that the two techniques can provide complementary information. Both techniques have been applied to simple carbohydrates with some success (Mathlouthi and Koenig, 1986), but studies of polysaccharides and their interactions are less common.

Observed features in a carbohydrate infrared or Raman spectrum can often be identified with a class of molecular vibration (e.g. C–H bond stretching) or a functional group (e.g. carboxylic acid) but direct interpretation in terms of detailed molecular structures is very difficult and rarely practicable (Mathlouthi and Koenig, 1986). Nevertheless, infrared and Raman spectroscopy have been used to study some aspects of polysaccharide interactions, particularly specific ionic interactions and starch retrogradation. The interaction of cations with both ι- and κ-carrageenans has been investigated by infrared spectroscopy (Norton et al., 1983; Belton et al., 1986b). Significant changes were observed in the sulphate bond-stretching region of carrageenan spectra upon addition of potassium ions but not tetramethylammonium ions; this was interpreted in terms of site binding for the former but not the latter (Norton et al., 1983). This interpretation has been questioned, however, following a study of cation effects (sodium, potassium, caesium, but not tetramethylammonium) on carrageenan infrared spectra (Belton *et al.*, 1986b). The interaction of a range of cations with various heparin samples has also recently been studied by infrared spectroscopy, and evidence obtained for direct cation binding to carboxylate groups and possible changes in hydration patterns around sulphate groups (Grant *et al.*, 1987).

The interactions between starch polysaccharides have been the subject of recent infrared and Raman spectroscopic studies. Bulkin *et al.* (1987) showed that features of the Raman spectra of waxy maize and potato starches changed with time following gelatinisation. The spectral changes were taken as an index of retrogradation and a multi-stage process was proposed on the basis of a kinetic analysis (Bulkin *et al.*, 1987). Wilson *et al.* (1987) studied the retrogradation of waxy maize starch using infrared spectroscopy and found similar time-dependent changes in spectral features. By taking the ratio of two bands as an empirical index of retrogradation, a correlation with shear modulus was demonstrated (Wilson *et al.*, 1987).

With recent improvements in instrumentation (e.g. Fourier transform infrared and laser Raman spectroscopy), studies of polysaccharide interactions are likely to become more frequent. Two other recent instrumental advances are also expected to lead to greater application of vibrational spectroscopy to the study of polysaccharide interactions. The first involves Raman or infrared spectroscopic characterisation of a defined microscopic domain with a sample and has been discussed in Section III.A. The second is the advent of attenuated total reflectance (ATR) in combination with Fourier transformation as an alternative to transmission techniques. ATR measures spectroscopic features of surfaces in contact with (i.e. within a few μm of) a crystal (usually zinc selenide) and is suitable for applications to solid-like materials which cannot be analysed easily by transmission techniques. Although detailed Raman and infrared spectral characterisation (i.e. 'finger-printing') of polysaccharides interactions is now possible in many cases, the interpretation of spectral features in terms of molecular structural details is still likely to be problematic.

REFERENCES

Ablett, S., Clark, A. H. and Rees, D. A. (1982). *Macromolecules* **15**, 597–602.
Amsterdam, A., Er-el, Z. and Shaltiel, S. (1975). *Arch. Biochem. Biophys.* **171**, 673–677.
Arnott, S., Scott, W. E., Rees, D. A. and McNab, C. G. A. (1974). *J. Mol. Biol.* **90**, 253–267.
Aspinall, G. O. (ed.) (1985). "The Polysaccharides", Academic Press, Orlando, FL.
Atalla, R. H. and Agarwal, U. P. (1985). *Science* **225**, 636–638.
Atalla, R. H. and VanderHart, D. L. (1984). *Science* **223**, 283–285.
Attwood, T. K., Nelmes, B. J. and Sellen, D. B. (1988). *Biopolymers* **27**, 201–212.
Belton, P. S., Morris, V. J. and Tanner, S. F. (1986a). *Macromolecules* **19**, 1618–1621.
Belton, P. S., Wilson, R. H. and Chenery, D. H. (1986b). *Int. J. Biol. Macromol.* **8**, 247–251.
Bryce, T. A., McKinnon, A. A., Morris, E. R., Rees, D. A. and Thom, D. (1974). *Faraday Discuss. Chem. Soc.* **57**, 221–229.
Buffington, L. A., Stevens, E. S., Morris, E. R. and Rees, D. A. (1980). *Int. J. Biol. Macromol.* **2**, 199–203.
Buleon, A., Duprat, F., Booy, F. P. and Chanzy, H. (1984). *Carbohydr. Polymers* **4**, 161–173.
Bulkin, B. J., Kwak, Y. and Dea, I. C. M. (1987). *Carbohydr. Res.* **160**, 95–112.
Burchard, W. (1985). *Br. Polymer J.* **17**, 154–163.
Cabib, E., Roberts, R. and Bowers, B. (1982). *Ann. Rev. Biochem.* **51**, 763–793.
Cael, J. J., Kwoh, D. L. W., Bhattacharjee, S. S. and Patt, S. L. (1985). *Macromolecules* **18**, 821–823.

Cairns, P., Miles, M. J., Morris, V. J. and Brownsey, G. J. (1987). *Carbohydr. Res.* **160**, 411–423.

Casu, B. (1985). *In* "Polysaccharides—Topics in Structure and Morphology" (E. D. T. Atkins, ed.), pp. 1–40. Macmillan, Basingstoke.

Chandrasekaran, R., Millane, R. P., Arnott, S. and Atkins, E. D. T. (1988), *Carbohydr. Res.* **175**, 1–15.

Chanzy, H. and Vuong, R. (1985). *In* "Polysaccharides—Topics in Structure and Morphology" (E. D. T. Atkins, ed.), pp. 41–71. Macmillan, Basingstoke.

Christianson, D. D., Baker, F. L., Loffredo, A. R. and Bagley, E. B. (1982). *Food Microstructure* **1**, 13–24.

Christianson, D. D., Casiraghi, E. M. and Bagley, E. B. (1986). *Carbohydr. Polymers* **6**, 335–348.

Clark, A. H. and Ross-Murphy, S. B. (1987). *Adv. Polymer Sci.* **83**, 57–192.

Clark, A. H., Richardson, R. K., Ross-Murphy, S. B. and Stubbs, J. M. (1983). *Macromolecules* **16**, 1367–1374.

Clark, A. H., Gidley, M. J., Richardson, R. K. and Ross-Murphy, S. B. (1989). *Macromolecules* **22**, 346–351.

Cziner, D. G., Stevens, E. S., Morris, E. R. and Rees, D. A. (1986). *J. Am. Chem. Soc.* **108**, 3790–3795.

Dawes, C. P. (1987). PhD Thesis, University of Liverpool.

DeGennes, P. G. (1979). *Nature (London)* **282**, 367–370.

Dea, I. C. M., McKinnon, A. A. and Rees, D. A. (1972). *J. Mol. Biol.* **68**, 153–172.

Dea, I. C. M., Morris, E. R., Rees, D. A., Welsh, E. J., Barnes, H. A. and Price, J. (1977). *Carbohydr. Res.* **57**, 249–272.

Dey, P. M. and Brinson, K. (1984). *Adv. Carbohydr. Chem. Biochem.* **42**, 265–382.

Djabourov, M., Clark, A. H., Rowlands, D. W. and Ross-Murphy, S. B. (1989). *Macromolecules* **22**, 180–188.

Donovan, J. W. (1979). *Biopolymers* **18**, 263–275.

Donovan, J. W. and Mapes, C. J. (1980). *Staerke* **32**, 190–193.

Dudley, R. L., Fyfe, C. A., Stephenson, P. J., Deslandes, Y., Hamer, G. K. and Marchessault, R. H. (1983). *J. Am. Chem. Soc.* **105**, 2469–2472.

Evans, I. D. and Haisman, D. R. (1982). *Staerke* **34**, 224–231.

Fransson, L.-A. (1985). *In* "The Polysaccharides" (G. O. Aspinall, ed.), Vol. 2, pp. 337–415. Academic Press, Orlando. FL.

French, A. D. and Gardner, K. C. H. (1980). "Fibre Diffraction Methods", Am. Chem. Soc. Symp. Ser. 141.

Fulcher, R. G. (1982). *Food Microstructure* **1**, 167–175.

Gallant, D. J. and Bouchet, B. (1986). *Food Microstructure* **5**, 141–155.

Gallant, D. J. and Sterling, C. (1976). *In* "Examination and Analysis of Starch and Starch Products" (J. A. Radley, ed.), pp. 33–59. Applied Science, London.

Gernat, Ch., Reuther, F., Darmaschun, G. and Schierbaum, F. (1987). *Acta Polymerica* **38**, 603–607.

Gidley, M. J. (1988). *In* "Gums and Stabilisers for the Food Industry" (G. O. Phillips, D. J. Wedlock and P. A. Williams, eds), Vol. 4, pp. 70–80. IRL Press, Oxford.

Gidley, M. J. (1989). *Macromolecules* **22**, 351–358.

Gidley, M. J. and Bociek, S. M. (1985a). *J. Am. Chem. Soc.* **107**, 7040–7044.

Gidley, M. J. and Bociek, S. M. (1985b). *J. Am. Chem. Soc. Commun.*, pp. 220–222.

Gidley, M. J. and Bociek, S. M. (1988a). *J. Am. Chem. Soc.* **110**, 3820–3829.

Gidley, M. J. and Bociek, S. M. (1988b). *Carbohydr. Res.* **183**, 126–130.

Gidley, M. J. and Bulpin, P. V. (1987). *Carbohydr. Res.* **161**, 291–300.

Goodall, D. M. and Norton, I. T. (1987). *Acc. Chem. Res.* **20**, 59–65.

Gorin, P. A. J. (1981). *Adv. Carbohydr. Chem. Biochem.* **38**, 13–104.

Gormally, J., Pereira, M. C., Wyn-Jones, E. and Morris, E. R. (1982). *J. Chem. Soc. Faraday Trans. 2* **78**, 1661–1673.

Grant, D., Long, W. F. and Williamson, F. B. (1987). *Biochem. J.* **244**, 143–149.

Haw, J. F., Maciel, G. E. and Schroder, H. A. (1984). *Anal Chem.* **56**, 1323–1329.

Hearle, J. W. S. (1985). *In* "Cellulose Chemistry and its Applications" (T. P. Nevell and S. H. Zeronian, eds), pp. 480–504. Ellis Horwood, Chichester.

Himmelsbach, D. S., Barton, F. E. and Windham, W. R. (1983). *J. Agric. Food Chem.* **31**, 401–404.

Hook, M., Kjellen, L., Johansson, S. and Robinson, J. (1984). *Ann. Rev. Biochem.* **53**, 847–869.

Horii, F., Hirai, A. and Kitamaru, R. (1984). *J. Carbohydr. Chem.* **3**, 641–662.

Horii, F., Hirai, A. and Kitamaru, R. (1987a). *Macromolecules* **20**, 2117–2120.

Horii, F., Yamamoto, H., Hirai, A. and Kitamaru, R. (1987b). *Carbohydr. Res.* **160**, 29–40.

Huglin, M. B. (1972). "Light Scattering from Polymer Solutions". Academic Press, London.

Imberty, A. and Perez, S. (1988). *Biopolymers* **27**, 1205–1221.

Imberty, A., Chanzy, H., Perez, S., Buleon, A. and Tran, V. (1988). *J. Mol. Biol.* **201**, 365–378.

Irwin, P. L., Pfeffer, P. E., Gerasimowicz, W. V., Paressey, R. and Sams, C. E. (1984). *Phytochemistry* **23**, 2239–2242.

Irwin, P. L., Gerasimowicz, W. V., Pfeffer, P. E. and Fishman, M. (1985). *J. Agric. Food Chem.* **33**, 1197–1201.

Jane, J.-L., Robyt, J. F. and Huang, D.-H. (1985). *Carbohydr. Res.* **140**, 21–35.

Kalichevsky, M. T. and Ring, S. G. (1987). *Carbohydr. Res.* **162**, 323–328.

Kenne, L. and Lindberg, B. (1983). *In* "The Polysaccharides" (G. O. Aspinall, ed.), Vol. 2, pp. 287–363. Academic Press, Orlando, FL.

Kitamura, S., Yoneda, S. and Kuge, T. (1984). *Carbohydr. Polymers* **4**, 127–136.

Kohn, R. (1987). *Carbohydr. Res.* **160**, 343–353.

Komiya, T., Yamada, T. and Nara, S. (1987). *Staerke* **39**, 308–311.

Kratochvil, P. (1972). *In* "Light Scattering from Polymer Solutions" (M. B. Huglin, ed.), pp. 333–384. Academic Press, London.

Kugimiya, M., Donovan, J. W. and Wong, R. Y. (1980). *Staerke* **32**, 265–270.

Lechert, H. and Schwier, I. (1982). *Staerke* **34**, 6–11.

Levine, H. and Slade, L. (1986). *Carbohydr. Polymers* **6**, 213–244.

Lewis, G. P., Derbyshire, W., Ablett, S., Lillford, P. J. and Norton, I. T. (1987). *Carbohydr. Res.* **160**, 397–410.

Lips, A., Hart, P. M. and Clark, A. H. (1988). *Food Hydrocolloids* **2**, 141–150.

Maciel, G. E., Haw, J. F., Smith, D. H., Gabrielsen, B. C. and Hatfield, G. R. (1985). *J. Agric. Food Chem.* **33**, 185–191.

Mackay, A. L., Wallace, J. C., Sasaki, K. and Taylor, I. E. P. (1988). *Biochemistry* **27**, 1467–1473.

Mackie, W. (1985). *In* "Polysaccharides—Topics in Structure and Morphology" (E. D. T. Atkins, ed.), pp. 73–105. Macmillan, Basingstoke.

Maggio, J. E. (1980). *Proc. Natl. Acad. Sci. USA* **77**, 2582–2586.

Manners, D. J. and Matheson, N. K. (1981). *Carbohydr. Res.* **90**, 99–110.

Mathlouthi, M. and Koenig, J. L. (1986). *Adv. Carbohydr. Chem. Biochem.* **44**, 7–89.

McNeil, M., Darvill, A. G., Fry, S. C. and Albersheim, P. (1984). *Ann. Rev. Biochem.* **53**, 625–663.

Miles, M. J., Morris, V. J. and Ring, S. G. (1985). *Carbohydr. Res.* **135**, 257–269.

Morris, E. R. and Frangou, S. A. (1981). *Techniques Carbohydr. Metabolism* **B308**, 109–160.

Morris, E. R. and Ross-Murphy, S. B. (1981). *Techniques Carbohydr. Metabolism* **B310**, 1–46.

Morris, E. R., Rees, D. A., Thom, D. and Boyd, J. (1978). *Carbohdyr. Res.* **66**, 145–154.

Morris, E. R., Rees, D. A. and Robinson, G. (1980). *J. Mol. Biol.* **80**, 349–362.

Morris, E. R., Cutler, A. N., Ross-Murphy, S. B., Rees, D. A. and Price, J. (1981). *Carbohydr. Polymers* **1**, 5–21.

Morris, E. R., Powell, D. A., Gidley, M. J. and Rees, D. A. (1982). *J. Mol. Biol.* **155**, 507–516.

Morris, E. R., Stevens, E. S., Frangou, S. A. and Rees, D. A. (1986). *Biopolymers* **25**, 959–973.

Morris, V. J. (1984). *In* "Biophysical Methods in Food Research" (H. W.-S. Chan, ed.), pp. 37–102. Blackwell, Oxford.

Nikaido, H. and Vaara, M. (1985). *Microbiol. Rev.* **49**, 1–32.

Norton, I. T., Goodall, D. M., Morris, E. R. and Rees, D. A. (1983). *J. Chem. Soc., Faraday Trans. I* **79**, 2475–2488.

Norton, I. T., Morris, E. R. and Rees, D. A. (1984a). *Carbohydr. Res.* **134**, 89–101.

Norton, I. T., Goodall, D. M., Frangou, S. A., Morris, E. R. and Rees, D. A. (1984b). *J. Mol. Biol.* **175**, 371–394.

O'Donnell, D. J., Ackerman, J. J. H. and Maciel, G. E. (1981). *J. Agric. Food. Chem.* **29**, 514–518.

Oostergetel, G. T. and van Bruggen, E. F. J. (1989). *Staerke* **41**, 331–335.

Parfondry, A., Cyr, N. and Perlin, A. S. (1977). *Carbohydr. Res.* **59**, 299–309.

Pereira, M. C., Wyn-Jones, E., Morris, E. R. and Ross-Murphy, S. B. (1982). *Carbohydr. Polymers* **2**, 103–113.

Pfannemüller, B. (1987). *Int. J. Biol. Macromol.* **9**, 105–108.

Pfannemüller, B. and Bauer-Carnap, A. (1977). *Coll. Polymer Sci.* **255**, 844–848.

Rees, D. A. (1970). *J. Chem. Soc. B.*, pp. 877–884.

Rees, D. A. (1972). *Chem. Ind. (London)*, pp. 630–636.

Rees, D. A., Morris, E. R., Thom, D. and Madden, J. K. (1982). *In* "The Polysaccharides" (G. O. Aspinall, ed.), Vol. 1, pp. 195–290. Academic Press, Orlando, FL.

Richardson, R. K. and Ross-Murphy, S. B. (1987a). *Int. J. Biol. Macromol.* **9**, 250–256.

Richardson, R. K. and Ross-Murphy, S. B. (1987b). *Int. J. Biol. Macromol.* **9**, 257–264.

Ring, S. G., Colonna, P., L'Anson, K. J., Kalichevsky, M. T., Miles, M. J., Morris, V. J. and Orford, P. D. (1987). *Carbohydr. Res.* **162**, 277–293.

Robinson, G., Ross-Murphy, S. B. and Morris, E. R. (1982). *Carbohydr. Res.* **107**, 17–32.

Ross-Murphy, S. B. (1984). *In* "Biophysical Methods in Food Research" (H. W.-S. Chan, ed.), pp. 138–199. Blackwell, Oxford.

St-Jacques, M., Sundararajan, P. R., Taylor, K. J. and Marchessault, R. H. (1976). *J. Am. Chem. Soc.* **98**, 4386–4391.

Saito, H. (1986). *Magn. Reson. Chem.* **24**, 835–852.

Saito, H., Tabeta, R. and Ogawa, K. (1987). *Macromolecules* **20**, 2424–2430.

Salmen, N. L. (1985). *In* "Cellulose Chemistry and its Applications" (T. P. Nevell and S. H. Zeronian, eds), pp. 505–530. Ellis Horwood, Chichester.

Shallenberger, R. S. (1982). "Advanced Sugar Chemistry". Ellis Horwood, Chichester.

Shepherd, M. G. (1987). *CRC Crit. Rev. Microbiol.* **15**, 7–25.

Slade, L. and Levine, H. (1988). *Carbohydr. Polymers* **8**, 183–208.

Stevens, E. S. and Sathyanarayana, B. K. (1988). *Biopolymers* **27**, 415–421.

Stipanovic, A. J., Giammatteo, P. J. and Robie, S. B. (1985). *Biopolymers* **24**, 2333–2343.

Stockton, B., Evans, L. V., Morris, E. R., Powell, D. A. and Rees, D. A. (1980). *Botanica Marina* **23**, 563–567.

Stokke, B. T., Elgsaeter, A., Skjak-Braek, G. and Smidsrod, O. (1987). *Carbohydr. Res.* **160**, 13–28.

Sutherland, I. W. (1985). *Ann. Rev. Microbiol.* **39**, 243–270.

Thibault, J. F. and Rinaudo, M. (1985). *Biopolymers* **24**, 2131–2143.

Veregin, R. P., Fyfe, C. A., Marchessault, R. H. and Taylor, M. G. (1986). *Macromolecules* **19**, 1030–1034.

Veregin, R. P., Fyfe, C. A., Marchessault, R. H. and Taylor, M. G. (1987). *Carbohydr. Res.* **160**, 41–56.

Walkinshaw, M. D. and Arnott, S. (1981). *J. Mol. Biol.* **153**, 1055–1074.

Ward, J. B. (1981). *Microbiol. Rev.* **45**, 211–243.

Wild, D. L. and Blanshard, J. M. V. (1986). *Carbohydr. Polymers* **6**, 121–143.

Wilson, R. H., Kalichevsky, M. T., Ring, S. G. and Belton, P. S. (1987). *Carbohydr. Res.* **166**, 162–165.

Winter, W. T. (1982). *Methods Enzymol.* **83**, 87–104.

Wynne-Jones, S. and Blanshard, J. M. V. (1986). *Carbohydr. Polymers* **6**, 289–306.

Wu, H-C. H. and Sarko, (1978a). *Carbohydr. Res.* **61**, 7–25.

Wu, H-C. H. and Sarko, (1978b). *Carbohydr. Res.* **61**, 27–40.

Yabusaki, K. K., Cohen, R. E. and Ballou, C. E. (1979). *J. Biol. Chem.* **254**, 7282–7286.

Yamaguchi, M., Kainuma, K. and French, D. (1979). *J. Ultrastruct. Res.* **69**, 249–261.

Yannoni, C. S. (1982). *Acc. Chem. Res.* **15**, 201–208.

Index

The following abbreviations are used in this index:

GC	–	Gas chromatography
GLC	–	Gas–liquid chromatography
GLC–MS	–	Gas–liquid chromatography–mass spectrometry
HPLC	–	High performance liquid chromatography
LPLC	–	Low pressure liquid chromatography
NMR	–	Nuclear magnetic resonance
TLC	–	Thin layer chromatography

Electrophoresis (*cont.*)
  of branched chain monosaccharide, 260, 262
  of disaccharides, planar, 137
  of gums, 494
  of inositol isomers, 224
  of inositol phosphates, 228
  of nucleotide sugars, 51–54
  of planteose, 201
Elson–Morgan assay for amino sugars, 8
  see also Morgan–Elson assay for amino
    sugars
Endoplasmic reticulum, lipid-linked saccharide in, 102–104
Enzymic degradation analysis, 3–4
  see also Acid hydrolysis; Hydrogen fluoride
    solvolysis; Hydrolysis, partial
  of alginates, 534–535
  of cell walls, 556–557
  of cellulose, 306–307
  of chitin, 451, 452
  of disaccharides
    electrochemical biosensors, 166–168
    glycoside hydrolases, 150–154
    hydrolysate determinations 154–160
  enzymes available, 5
  of galactans, 541
  of galactosyl oligosaccharides, 191
  of gums, 499–500
  of *myo*-inositols, 225–226
  of mannose polysaccharides, partial, 390
  of monosaccharides, 11, 12–16
    galactose oxidase, 15–16
    D-glucose oxidase, 13–14
    hexokinase-glucose 6-phosphate dehydrogenase, 14–15
    specificity, 12–13
  of nucleotide sugars, 58
  of pectic polysaccharides, 418–419
  of starch polysaccharides, 340
Erythritols, 277–278
Esterification of gums, 501–502
Esters
  of cellulose, 312–314
  of sucrose, 115–116
  of *myo*-inositols, phosphoric, 228–229
Ethanolamine-boric acid, for HPLC
    detection, 149
Etherification of gums, 500–501
Ethers
  of cellulose, 314–316
  trimethylsilyl derivatives, 22–23
Ethylene diamine for HPLC detection, 149
Exclusion chromatography, *see* Gel
    permeation chromatography
Extensins, hydrogen fluoride solvolysis, 590–591

Exudate gums, *see* Gums

F

Farnesyl-pyrophosphate, 86
Fast atom bombardment (FAB), *see under*
    Mass spectrometry
Fehling methods, 114
Ferricyanide reagent, alkaline, 127
Flory–Fox equation, 610
Fluorescence/fluorimetry
  by amination, reductive, 147
  by dansylation, 147
  glucose hexokinase assay, 156–157
  glucose oxidase assay, 155
  in HPLC detection, 30
  laser-induced, 146
  post-elution derivatisation, 149
  for reducing sugars, 130
Formazan formation for HPLC detection, 149
Fructans, 353–369
  chemical nature, 355–356
  determination, 361–363
  enzyme assays, 363–366
    fructan hydrolases, 366
    fructan-fructan-fructosyl transferase, 364–366
    sucrose-sucrose-fructosyl transferase (SST), 363–364
  history, 354
  industrial applications, 356–357
  inulin, enzyme hydrolysis, 4
  isokestose, 163
  isolation, 357–358
  nomenclature, 354
  occurrence, 354–355
  physiological role, 356
  separation, 358–361
Fructose 2,6-bisphosphate, 119
Fructose
  enzyme assays, D-fructose, 158
  resorcinol assay for, 124
  sensor, 167
β-Fructosyltransferase in sucrose detection, 163–164
Fucans, 541–543
  analysis, 543
  isolation, 542
Fungal chitin synthases, 460–465
  cell-free systems, 461–463
    activation, 462
    in chitosomes, 463
    solubilisation, 462–463
  filamentous, 461
  genetics/regulation, 465